AF182443

# *MASS SPECTROMETRY IN ENVIRONMENTAL SCIENCES*

# *MASS SPECTROMETRY IN ENVIRONMENTAL SCIENCES*

*Edited by*

*F. W. KARASEK*

*University of Waterloo*
*Waterloo, Ontario, Canada*

*O. HUTZINGER*

*University of Bayreuth*
*Bayreuth, Federal Republic of Germany*

*and*

*S. SAFE*

*University of Guelph*
*Guelph, Ontario, Canada*
*and Texas A&M University*
*College Station, Texas, U.S.A.*

*PLENUM PRESS • NEW YORK AND LONDON*

Library of Congress Cataloging in Publication Data

Main entry under title:

Mass spectrometry in environmental sciences.

Bibliography: p.
Includes index.
1. Environmental chemistry—Technique. 2. Mass spectrometry. 3. Pollutants—Analysis. I. Karasek, Francis W., 1919- . II. Hutzinger, O. III. Safe, S.
TD193.M37 1984 628.5 84-13469
ISBN-13: 978-1-4612-9445-0 e-ISBN-13: 978-1-4613-2361-7
DOI: 10.1007/978-1-4613-2361-7

© 1985 Plenum Press, New York
Softcover reprint of the hardcover 1st edition 1985
A Division of Plenum Publishing Corporation
233 Spring Street, New York, N.Y. 10013

All rights reserved

No part of this book may be reproduced, stored in a retrieval system, or transmitted in any form or by any means, electronic, mechanical, photocopying, microfilming, recording, or otherwise, without written permission from the Publisher

# CONTRIBUTORS

*E. ATLAS* Department of Chemistry, Texas A&M University, College Station, Texas, U.S.A.

*H. M. BALBA* Uniroyal Chemical Co., Naugatuck, Connecticut, U.S.A.

*D. A. BIRKHOLZ* Environmental Protection Service, Western and Northern Region, Edmonton, Alberta, Canada

*R. K. BOYD* Department of Chemistry, University of Guelph, Ontario, Canada

*J. A. BUCKLEY* SCIEX, Thornhill, Ontario, Canada

*J. T. BURSEY* Analytical and Chemical Sciences, Research Triangle Institute, Research Triangle Park, North Carolina, U.S.A.

*R. E. CLEMENT* Laboratory Services Branch, Ontario Ministry of the Environment, Ontario, Canada

*W. R. DAVIDSON* SCIEX, Thornhill, Ontario, Canada

*J. M. DESMARCHELIER* CSIRO Division of Entomology, Canberra, Australia

*RALPH C. DOUGHERTY* Department of Chemistry, Florida State University, Tallahassee, Florida, U.S.A.

*J. B. FRENCH* SCIEX, Thornhill, Ontario, Canada

*G. A. EICEMAN* Department of Chemistry, New Mexico State University, Las Cruces, New Mexico, U.S.A.

*E. Ch. Th. GEVERS* Institute of Applied Chemistry TNO, Utrecht, The Netherlands

*C. S. GIAM* Department of Chemistry, Texas A&M University, College Station, Texas, U.S.A.

*C. V. HAMPTON* Chemical Analysis Research Department, Scientific Research Laboratory, Ford Motor Company, Dearborn, Michigan, U.S.A.

*O. HUTZINGER* Departments of Ecological Chemistry and Geochemistry, University of Bayreuth, Federal Republic of Germany

*BO JANSSON* Special Analytical Laboratory, National Swedish Environment Protection Board, Solna, Sweden

*F. W. KARASEK* Department of Chemistry, University of Waterloo, Ontario, Canada

*M. J. LACEY* CSIRO Division of Entomology, Canberra, Australia
*L. M. LIBBEY* Department of Food Science and Technology, Oregon State University, Corvallis, Oregon, U.S.A.
*D. C. G. MUIR* Department of Fisheries and Oceans, Freshwater Institute, Winnipeg, Manitoba, Canada
*K. OLIE* Laboratory of Environmental and Toxicological Chemistry, University of Amsterdam, Amsterdam, The Netherlands
*FRANCIS I. ONUSKA* National Water Research Institute, Canada Centre for Inland Waters, Burlington, Ontario, Canada
*E. D. PELLIZZARI* Analytical and Chemical Sciences, Research Triangle Institute, Research Triangle Park, North Carolina, U.S.A.
*N. M. REID* SCIEX, Thornhill, Ontario, Canada
*B. D. RIPLEY* Provincial Pesticide Residue Testing Laboratory, Ontario Ministry of Agriculture and Food, University of Guelph, Ontario, Canada
*S. SAFE* The Guelph–Waterloo Centre for Graduate Work in Chemistry, Department of Chemistry, University of Guelph, Ontario, Canada
*R. A. SCANLAN* Department of Food Science and Technology, Oregon State University, Corvallis, Oregon, U.S.A.
*D. SCHUETZLE* Chemical Analysis Research Department, Scientific Research Laboratory, Ford Motor Company, Dearborn, Michigan, U.S.A.
*H.-R. SCHULTEN* Fachhochschule Fresenius, Wiesbaden, Federal Republic of Germany
*J. SPARLING* The Guelph–Waterloo Centre for Graduate Work in Chemistry, Department of Chemistry, University of Guelph, Ontario, Canada
*D. L. STRUBLE* Agriculture Canada Research Station, Lethbridge, Alberta, Canada
*SI-EN SUN* Institute of Environmental Chemistry, Academia Sinica, Peking, China
*GÖRAN SUNDSTRÖM* Special Analytical Laboratory, National Swedish Environmental Protection Board, Solna, Sweden
*J. A. SWEETMAN* Statewide Air Pollution Research Center, University of California, Riverside, California
*B. A. THOMSON* SCIEX, Thornhill, Ontario, Canada
*G. R. B. WEBSTER* Pesticide Research Laboratory, Department of Soil Science, University of Manitoba, Winnipeg, Manitoba, Canada

# PREFACE

During the past ten years a variety of methods involving mass spectrometry have been developed for the analysis of environmentally important compounds. Much has been accomplished in that period to solve some of the important problems in the field. Growth of this methodology and its accomplishments has reached the point where an individual scientist can no longer have an in-depth knowledge of all the areas involved. We have attempted to provide this in-depth picture to those scientists concerned by having the important topics treated by experts in the subject matter.

In order to provide all the relevant material in one volume we begin with the general topics which provide the basic background material necessary to understand the techniques discussed in the in-depth topics. These general chapters are kept brief, containing only the essentials needed by the working scientist to deal with the practical applications. References in these chapters are chosen to permit a more complete study of each chapter.

The concept for this book was developed during the activities of two of the editors under a NATO travel grant. These editors gratefully acknowledge this support which made the initial planning of this book possible.

The editors would like to thank the individual authors of each chapter for their cooperation and generously giving of their time for this project.

F. W. KARASEK
O. HUTZINGER
S. SAFE

# CONTENTS

## 5. *USE OF PYROLYSIS/GAS CHROMATOGRAPHY/MASS SPECTROMETRY IN ENVIRONMENTAL ANALYSIS*

*H. M. Balba*



## 6. *ATMOSPHERIC PRESSURE CHEMICAL IONIZATION MASS SPECTROMETRY*

*J. B. French, B. A. Thomson, W. R. Davidson, N. M. Reid, and J. A. Buckley*



## 7. *MASS SPECTRAL RETRIEVAL AND INTERPRETATION SYSTEMS*

*F. W. Karasek and R. E. Clement*

8. *GAS CHROMATOGRAPHY/MASS SPECTROMETRY IN WATER POLLUTION STUDIES*

*E. D. Pellizzari and J. T. Bursey*



9. *GAS CHROMATOGRAPHY/MASS SPECTROMETRY IN AIR POLLUTION STUDIES*

*D. Schuetzle and C. V. Hampton*

## 12. *POLYCHLORODIBENZO-p-DIOXINS AND POLYCHLORODIBENZOFURANS*

*G. R. B. Webster, K. Olie, and O. Hutzinger*



## 13. *DDT AND RELATED COMPOUNDS*

*Göran Sundström and Bo Jansson*

## 24. *MASS SPECTROMETRY OF SOME FEMALE LEPIDOPTERAN SEX PHEROMONES AND RELATED COMPOUNDS*

*D. L. Struble*



## 25. *MASS SPECTROMETRY OF N-NITROSAMINES*

*L. M. Libbey and R.A. Scanlan*



## 26. *OTHER INDUSTRIAL POLLUTANTS AND NATURAL TOXINS*

*S. Safe and J. Sparling*

1

# GENERAL PRINCIPLES OF ANALYTICAL MASS SPECTROMETRY

R. K. BOYD

## 1. INTRODUCTION

The purpose of this introductory chapter is to set the stage for what is to follow. Clearly, it is not possible to give a complete account of all the features of a modern mass spectrometer in a few pages. Accordingly, we shall concentrate on establishing some essential principles, give references to sources of more detailed information, and try to give some indication of more modern trends in instrumentation. Some of these trends will be fully described in later chapters in this book and thus will be treated sketchily here.

Such an approach is feasible only if high-quality monographs and reviews are available as reference materials, and in this respect mass spectrometrists are fortunate; many excellent texts exist. Those mentioned below were chosen as this author's best choice of a small number of broadly based texts which would assist the nonspecialist in acquiring a working expertise as quickly as possible. The first general monograph[1] on techniques of mass spectrometry applied to chemistry is now a classic, over 20 years old, but is still a useful source of information. A more modern text[2] deals in a particularly lucid manner with the various systems which make up a GC/MS/computer. The interpretation of the mass spectrum of a compound in terms of its chemical structure is covered in an elegant exposition[3] written as a self-testing course on the subject. Recent advances in mass spectrometry (MS) are covered in remarkable biennial reviews of a voluminous literature[4–6]; these reviews

*R. K. BOYD* • Department of Chemistry, University of Guelph, Guelph, Ontario N1G 2W1, Canada.

manage to provide complete and critical coverage in a surprisingly condensed form. In addition, the excellent specialist reports on MS published by the Chemical Society of London[7–11] often provide stimulating reviews of selected aspects of the field. A recent monograph[12] includes excellent reviews of modern trends. Finally, the approach to understanding the phenomena observed from a more fundamental point of view is taken in two excellent monographs.[13,14]

It will be useful to break down a mass spectrometer into its major components for purposes of discussion. Such a procedure has its drawbacks, particularly with respect to problems associated with linking the components together. For example, a mass analyzer based upon sector magnetic field normally requires that the ion source be at a few thousand volts with respect to ground. This places some restrictions upon what can be done in such an ion source which are not relevant to systems that use a quadrupole mass filter, for example, for which the ion source need be only a few tens of volts from ground. However, the following discussion will adopt the modular approach. A somewhat terse but highly illuminating discussion patterned on this same general approach has been published.[15]

## 2. *VACUUM SYSTEM*

Accurate control and guidance of ion beams in a manner designed to separate them according to their mass-to-charge ratios requires that the ion beam traverse the analyzer with minimal interference from collisions with neutral gas molecules. Generally this requires maintenance of a vacuum of $10^{-6}$ torr or better in the analyzer region. This is difficult to achieve if a vaporized sample is being supplied continuously to the ion source at rates of up to about 1 ml/min of gas at atmospheric pressure in the case of directly coupled capillary gas chromatography (GC) columns. Differential pumping is the name given to the procedure whereby independent vacuum pumps are attached to each of the ion source housing and the analyzer; the two regions are connected by as small an opening as possible, consistent with the passage of the ion beam.

It is essential for a modern mass spectrometer to have an excellent vacuum system; an inadequate system may severely limit the range of applicability of the instrument. There is no room here to give an extended discussion of this vital but often overlooked component. Fortunately, an excellent discussion has already been given by McFadden[2] and the interested reader is referred there. A rotary mechanical pump used as a forepump for an oil diffusion pump is the most common choice; modern diffusion pump liquids of the polyphenyl ether type can be used without cooling traps or baffles, without giving a background spectrum. The newer turbomolecular pumps do offer some advantages but are considerably more expensive.

## 3. SAMPLE INTRODUCTION

Coupling of a mass spectrometer ion source (pressures as low as $10^{-5}$–$10^{-6}$ torr) to a gas chromatograph (atmospheric pressure, due mainly to carrier gas) is a seemingly impossible task. Dropping the pressure by up to eight orders of magnitude while discarding most of the carrier gas (usually helium) and transmitting most of the sample is achieved by several remarkable devices described in a later chapter. The problem is even more serious in the case of a liquid chromatograph, when large volumes of solvent must be selectively removed; the successes achieved here are also described later in this monograph. The recent development of an interface to a thin-layer chromatography (TLC) plate(16) is noteworthy.

The standard sample inlets, other than interfaces to chromatographs, are the batch inlet and the direct-probe inlet. The former is essentially a heated bulb connected via a molecular leak to the ion source. (The connection must contain a length of quartz or similar insulator in the case of ion sources floating at kilovolts from ground potential.) Volatile liquids are normally introduced to the batch inlet by a syringe through a suitable rubber septum. For less volatile samples the direct insertion probe is used. The sample is placed in a small sample tube (usually of glass or quartz) which is placed in the probe tip and introduced directly into the ion source through a series of vacuum locks. It is important to ensure that the probe design permits the maximum degree of control of sample temperature, including temperatures below that of the ion source.

## 4. IONIZATION OF SAMPLE

Over the last ten years or so there has been a remarkable development in the range of ionization methods available. One convenient classification of ionization techniques is according to their degree of "hardness" or "softness." A soft ionization technique is one which ionizes a molecule without depositing large amounts of internal energy in the newly formed molecular ion; such a technique would produce largely molecular ions with little or no fragmentation. The importance of observing a molecular ion in identifying an unknown compound cannot be overstated. On the other hand, the information contained in the fragmentation pattern often leads to considerable information concerning the unknown substance, and occasionally to its unequivocal identification.[3] Thus, hard ionization techniques, which yield a distribution of internal energies in the nascent molecular ions sufficient to cause appreciable fragmentation, are also essential tools of the trade.

### 4.1. *Electron Impact (EI)*

Historically the first ionization technique in organic MS involved impacting the vaporized sample with a fast electron beam. The total yield of ions is often found to be a maximum if the electron beam is accelerated through a potential drop of about 70 V, and this has become a standard value. Ionization by 70-V electrons is a typically hard ionization technique, usually producing a rich fragment ion spectrum. Reduction of the accelerating voltage applied to the electron beam to much lower values (often 20 or 12 V in some cases) produces a much softer ionization technique at the expense of greatly reduced ionization efficiency. It should be noted that since the electron beam is usually generated by thermal emission from a metal filament heated to about 2000 °C, the electrons possess a statistical spread of energies superimposed upon that given by the accelerating potential applied between the filament and the ion source. Thus, the quoted energy of the ionizing beam is some kind of nominal average value only.

### 4.2. *Chemical Ionization (CI)*

Any process whereby a sample molecule is ionized by transfer of a charged species between it and a component of a preformed plasma is referred to as chemical ionization (CI). The charged species most often transferred is a proton, and the process may then be written as

$$RH^+ + M \rightarrow MH^+ + R$$

where M is the sample molecule and $RH^+$ an appropriate component of the CI reagent gas plasma. The plasma is generally formed by closing up the ion source so that a pressure in the range of (usually) a few tenths to several torr may be maintained within it. This pressure is mostly due to some suitable reagent gas, and the plasma is formed by the action of an intense electron beam, which must be accelerated by several hundred volts in order to penetrate the reagent gas appreciably.

The development and principles of CI, as well as of other ionization techniques, have been ably reviewed.[17–18] The degree of hardness of

CI by proton transfer is controlled by the difference in proton affinity between M and the reagent gas moiety R. For example, $H_2$ has a very low proton affinity, so $H_3^+$ used as a reagent species $RH^+$ provides a fairly hard ionization, while $C_4H_9^+$ (from isobutane) offers a reasonably soft method. Ammonia CI involves a reagent species R with such a high proton affinity that it will effectively protonate only secondary and tertiary amines.

If the charged species transferred is an electron, then the difference in ionization energies controls the degree of hardness of the ionization. However, it is possible to use a wide variety of reagent gases in CI on a chemically much more sophisticated level. For example, use of vinyl methyl ether[19,20] as reagent gas under conditions such that its molecular ion is the charged species transferred can locate the position of a double bond in an olefinic compound.

### 4.3. *Field Ionization (FI)*

When an external electric field is applied to a gaseous molecule such that the external field strength is comparable to that exerted upon the outermost electrons in the molecule by the nuclei and other electrons, there exists a finite probability that the molecule will lose an electron.[21] The electric field necessary for such field ionization (FI) are of the order of $10^8$ V/cm, too large to be applied on a macroscopic basis. However, by taking advantage of sharp points and edges, it is possible to obtain local field strengths of this order through application of some $10^4$ V over gaps of a few millimeters. Field ionization is a soft ionization technique and is desirable for this reason. Despite its potential advantages[17] it is not much used in analytical work due to the irreproducible nature of the phenomenon and the generally low intensity of ions formed.

### 4.4. *Field Desorption (FD)*

A variant of FI is the application of the general technique to involatile molecules of large molecular mass. This technique of field desorption (FD) simply requires that the substance in question be applied to the field emitter, for example, as a solution in a volatile solvent, placed in the ion source, and subjected to the high electric field. This technique is still something of a highly skilled art, particularly with respect to successful preparation and conditioning of the emitter surface.[17,20] Despite the difficulties, some impressive achievements have been reported in obtaining molecular mass information on large molecules, (e.g., polystyrene up to mass 11,000).[21] Until very recently a limiting factor in the usefulness of FD has been the useful mass range of magnetic sector spectrometers (see below), but the development of new high-field magnet technology has removed this obstacle. Field desorption is a soft ionization technique yielding essentially only molecular ions and thus little structural information. However, by use of collision-induced dissociation with linked scans at constant *B/E* (see below), fragmentation spectra of field desorbed ions can be obtained.[22]

### 4.5. *Secondary Ion Mass Spectrometry (SIMS)*

Bombardment of an involatile organic substance on a surface by a fast primary ion beam can lead to emission of secondary ions derived from the involatile organic target.[23]

### 4.6. *Fast Atom Bombardment (FAB)*

An apparently trivial extension of the SIMS technique is the use of fast (2–6 keV) beams of neutral atoms (usually argon) formed by resonant charge exchange from $Ar^+$. The main problem with both techniques, until very recently, was the sporadic, short-term nature of the spectra obtained. However, use of a glycerol matrix appears to offer a self-healing mechanism for the sample, and the FAB spectra of highly polar labile molecules of high molecular weight can now be maintained for minutes or even hours.[24] Together with the new magnet technology (see below), the FAB technique seems likely to revolutionize the analysis of peptides, saccharides, nucleotides, and so on. It is a fairly soft ionization technique, producing molecular ions preferentially with some fragment ions. Preliminary indications are that FAB ionization works best for highly polar compounds and worst for nonpolar materials. This is the precise reverse of the situation for FD, and these two techniques seem likely to complement one another.

### 4.7. *Desorption Chemical Ionization (DCI)*

A relatively cheap and reasonably effective method of obtaining mass spectra of involatile and/or thermally labile organic materials involves placing the sample on a suitable probe tip, which is then inserted directly into the plasma of a CI source. Analytically useful spectra may be thus obtained at least 150 °C below the temperature usually required for volatilization from a conventional probe,[25] though the spectra may depend noticeably upon time, temperature, and the precise position of the sample within the ion source.[26] However, more recent developments appear to have overcome most of these problems.[27]

### 4.8. *Negative Chemical Ionization (NCI)*

In an ion source operated at the high pressures characteristic of CI, many of the electrons in the plasma are thermally equilibrated through collisions. Substances of sufficiently high electronegativity readily attach an electron to form a radical anion. Indeed, if the molecule contains a highly electronegative functional group, or may be derivatized by a moiety containing several fluorine atoms, the sensitivity in NCI mode can be much higher than in the positive-ion CI mode. The NCI technique is generally softer, that is, there is relatively less fragmentation, than positive-ion CI. The ionization process may be simple electron attachment[28] or, by suitable choice of brew as the CI reagent gas, the plasma can be doctored to contain large concentrations of negative ions such as chloride, which can yield attachment spectra.[29] An excellent review covering both fundamentals and applications has recently been published.[30]

### 4.9. *Atmospheric Pressure Ionization (API)*

Atmospheric pressure ionization, which has rather recently come into full prominence, is the subject of an expert review later in this monograph.

### 4.10. *Californium-252 Plasma Desorption*

Californium-252 plasma desorption, a specialized technique for ionization of involatile, labile biological molecules, is unlikely to come into widespread use due to the requirement to combine nuclear physics technology with biological mass spectrometry. The inventor of the technique has recently published an excellent review.[31]

The wide range of ionization techniques described above reflects the difficulties involved in inducing the enormous range of compounds encountered by the analytical chemist to become gaseous ions. It is perhaps worth mentioning here that the renowned sensitivity of mass spectrometry has nothing to do with the ionization process. An ionization efficiency of 1 in $10^4$ is normally considered excellent. The overall sensitivity is gained in the detection chain (see below).

## 5. *ANALYZER SYSTEMS*

In order to obtain a simple one-dimensional mass spectrum (intensity of ions formed in the ion source as a function of their mass-to-charge ratios), the most commonly used analyzers at present are the quadrupole mass filter, the single-focusing magnetic sector, and the double-focusing (magnetic plus electric sectors) mass spectrometer. Time-of-flight (TOF) instruments have some application where very rapid cycle time is necessary, and ion cyclotron resonance spectrometers, particularly those operating in the Fourier transform mode, are currently undergoing rapid development.

### *5.1. Quadrupole Mass Filters*

The popularity of quadrupole mass filters arises from their relatively low cost, compactness, and ease of operation. An appropriate combination of DC and RF electric fields applied to four parallel rods of (generally) circular cross section induces an oscillatory motion in a beam of ions injected approximately axially into the assembly. The oscillating trajectories are mass dependent, and for any one combination of field parameters the range of mass-to-charge ratios transmitted by the device is limited, all other ions being deflected onto the rods. An elementary account of the principles of operation has been given,[2] and a simplified version of the theory has been published.[32] More modern trends in design have been reviewed.[33,34] The quadrupole analyzer is most useful for routine GC/MS analysis at unit mass resolution up to mass 500 or so. Since only electric fields are involved, rapid scanning is routine (e.g., mass 500–25 in a few tenths of a second), and the analyzer may be rapidly switched directly between specified masses without scanning through the intermediate mass ranges to increase the effective sensitivity. The quadrupole analyzer requires no high-quality optical slits and requires that the ion source be at only a few tens of volts from ground potential; it can thus be rapidly switched between positive and negative ion modes, a feature exploited in an ingenious application to simultaneous recording of positive- and negative-ion spectra.[35]

The chief disadvantages of the quadrupole mass analyzer, relative to magnetic sector instruments, lie in its limited resolving power and useful mass range. An upper limit to the useful mass range of a commonly used quadrupole system, with rods 5 in. long by 0.25 in. in diameter, is 500–600 amu; doubling the dimensions increases the usable mass range to well over 1000 amu. This limit is imposed not only by the resolving power of quadrupole mass filters, but also by a fairly dramatic falloff in transmission efficiency with increasing mass. An illuminating comparison of overall sensitivity for a small and large quadrupole analyzer and for a single-focusing magnetic sector analyzer has been published.[15] The sensitivity of the magnetic analyzer was approximately constant over the mass range 0–800. Below about mass 200 the large quadrupole was twice as sensitive, but the sensitivity dropped off steadily with increasing mass and became less than that of the magnetic analyzer for masses above 500; at mass 800 the magnetic analyzer was more sensitive than the large quadrupole by a factor of about 3. The smaller, more common quadrupole analyzer was slightly more sensitive than the magnetic sector analyzer for masses below about 200, but was only half as sensitive at about mass 400 and less than one-tenth as sensitive at mass 800.[15] This behavior is a more complete description of the "high-mass discrimination" feature of quadrupole analyzers, which currently is one of their main limitations.

### 5.2. *Single-Focusing Magnetic Analyzers*

An ion beam subjected to a magnetic field experiences a force in a direction perpendicular to the field and to the beam direction. The radius of curvature of the resulting ion trajectory can be shown[2] to be proportional to the momentum of the ion and inversely proportional to the magnetic field strength. Thus, for a constant radius of curvature fixed by entrance and exit slits, scanning the magnetic field results in a momentum spectrum of the ion beam being successively transmitted through the exit slit to a suitable detector. If the ions are all of the same kinetic energy, as may be accomplished by accelerating the ions out of the ion source by a suitably high applied potential (usually in the kilovolt range), the momentum spectrum is equivalent to a mass spectrum and may be calibrated as such. In contrast the quadrupole analyzer is a true mass spectrometer.

The ion beam emerging from an ion source inevitably has a certain angular spread, which could potentially lead to a loss in the effective resolving power. Fortunately, a sector magnetic field can be shown[36] to act as a thick convergent lens toward ion beams. Thus the entrance and exit slits referred to above are made of high optical quality to exploit the ion optic properties of the magnetic sector. The "single-focusing" property of a magnetic sector analyzer thus refers to its property of refocusing a beam of ions with a directional spread. It is important to note that the focusing action in most conventional magnets acts only in a direction perpendicular to the applied field; there is normally no focusing action in a direction parallel to the field. Recently,[15] commercial magnetic instruments have been produced which also achieve focusing in the direction of the magnetic field by exploiting the fringe fields in the manner proposed some time ago[37]; increases in sensitivity by factors of 3 or 4 have been thus obtained.[15]

The chief disadvantages of magnetic sector instruments when compared with quadrupole mass analyzers are the high-quality ion optics required, the high potential (kilovolt range) relative to ground at which the ion source must be floated, and the relatively slow scan speeds obtainable consistent with hysteresis lag in the magnetic core. Very recently[15] development of laminated cores for magnets has reduced hysteresis effects to the point where magnet scan speeds comparable with those obtainable with quadrupole mass filters can be achieved. In addition, development of new materials for the magnet polefaces has permitted much higher magnetic field strengths to be attained, and correspondingly higher mass ranges. Even more recently the use of nonuniform magnetic fields using laminated magnets has extended the effective mass range of even small magnets to well over 2000 amu at full sensitivity. This latest development, which at the time of writing has barely been announced by two commercial instrument manufacturers, seems likely to revolutionize mass spectrometry over the next few years, particularly in view of the new FAB ion source described above. Thus the useful mass range, which always represented a major advantage of magnetic sector analyzers over quadrupole mass filters, has now become even more significant. It always was possible, with magnets of traditional design, to extend the mass range at the price of sensitivity by lowering the accelerating voltage. Now this sacrifice is no longer necessary.

The other major advantage of magnetic sector analyzers, which is partly related to the mass range issue, lies in the attainable resolving power. As in the case of instrument sensitivity, resolving power is a quantity which must be carefully defined[2] before making comparisons between different instruments. In the case of resolving power, the loose jargon "percent valley" is used as a convenient definition. If two ions of similar mass, centered at mass $M$, are separated by mass $\Delta M$, and are distinguished by an analyzer as two peaks of roughly equal height separated by a valley whose height is the percent of the peak height, the resolving power on the percent valley definition is then $M/\Delta M$. For magnetic sector analyzers $M/\Delta M$ is roughly independent of mass; that is, $\Delta M$ increases with $M$, and is determined mainly by $R_m$, the radius of curvature of the magnetic sector, and by the widths $S_1$ and $S_2$ of the two slits, measured in a direction perpendicular to the directions of both the field and the beam. Then, for an "ideal" magnetic sector analyzer

$$M/\Delta M = KR_m/(S_1 + S_2)$$

where $K$ is a constant. Other factors limiting the resolving power include nonideal fringe fields, second-order aberrations in the ion optic properties for ion beams of large angular divergence, and so on. The most important limitation, however, is the kinetic energy spread in the ion beam, and this is discussed further below. However, even for a single-focusing magnetic sector analyzer, the attainable resolving power can be as high as 5000 on a 10% valley definition for a magnet of reasonable size, and can be traded off for sensitivity by opening the slits. The same trade-off is possible for quadrupole mass filters, of course, by altering the stability range for transmission, but the attainable resolving power is usually at least an order of magnitude less.

### 5.3. *Double-Focusing Magnetic Sector Instruments*

The small spread in kinetic energy, and thus in momentum for ions of a given mass, of an ion beam extracted from even the best ion sources imposes a limitation upon the attainable resolving power of a magnetic sector instrument. A radial electric sector field disperses ion according to their kinetic energy-to-charge ratios, while again acting as a thick convergent lens. By using an electric sector and magnetic sector in series, it is possible to arrange that the combination is direction focusing, in the sense used above, and also velocity focusing. This implies that ions of a given mass with small differences in both directions and velocity can be brought to a common focus and are thus said to be doubly focused. The basic idea is to arrange that the velocity dispersions produced in such an ion beam by electric and magnetic sectors shall be in opposite directions and exactly equal. The requirements for such a double-focusing condition have been described in some detail.(36) There are two basic types of double-focusing design in common use. The first type represents variations on the Nier–Johnson geometry, in which the double-focusing conditions are satisfied at only a single point after the two sectors; this point is, of course, occupied by the final slit of the ion optics, and such instruments must be scanned in order to obtain a mass spectrum. The other type represents variations on the Mattauch–Herzog design, in which the double-focusing condition is achieved in a focal plane; thus photographic plates can be used without scanning the instrument, since the entire mass spectrum is simultaneously in focus. The Mattauch–Herzog type is essential for cases in which it is not possible to scan quickly enough, but is less common than the Nier–Johnson type due to the limitations of photoplate detection.

The scale (radii of sector fields) of such an analyzer is the primary determinant of attainable resolving power. In order to separate certain mass doublets, resolving powers of up to 70,000 are required.(1) In addition, the most precise measurements of molecular mass (to within $10^{-4}$ amu) require high resolving powers; this precision is often necessary to unambiguously specify the atomic composition of an ion.(15) However, more modest (and less expensive) instruments with maximum resolving power in the range 20,000–30,000 can yield excellent results and are a common choice.

### 5.4. *Ion Cyclotron Resonance (ICR) Spectrometers*

Many of the limitations of ICR spectrometers in analytical applications were overcome by the development of Fourier transform techniques.[38] An informative introduction to the subject has been published[39] and recent developments have led to claims that Fourier transform ICR is on the verge of assuming a major role in high-performance, analytical mass spectrometry.[40] Indeed, a commercially produced instrument is presently available. One problem currently remaining concerns the maintenance of the pressure within the ICR cell below $10^{-6}$ torr, in order that the cyclotron motions of the ions are not perturbed by collisions; this is difficult to achieve when the instrument is coupled to a gas chromatograph, for example. Nonetheless, the technique does appear to hold considerable promise.

### 5.5. *Time-of-Flight (TOF) Instruments*

Instruments which disperse ions according to their velocities are collectively known as time-of-flight instruments. The most common type uses a linear flight path, and the basic principles of operation have been described.[2] The mass range and resolving power of such instruments has traditionally been limited, and their use has been restricted to circumstances where very fast cycle times are required. (TOF instruments commonly can give 20,000 mass spectra/sec.) There has been a recent revival of interest in improving the performance of TOF analyzers, based on a need for rapid acquisition of mass spectra of eluent peaks from capillary GC columns, which in some cases can be as little as 1 sec in duration.[41] Even quadrupole analyzers would have difficulty in properly characterizing such a GC peak. Some of the parameters investigated have included electron beam collimation[42] and shaping of the impulse field.[43] One of the most important developments, however, appears to be that of the "mass-reflectron,"[44] in which energy focusing of the ion packets is achieved by use of an electrostatic mirror. Recent applications of this device[31] have led to claims of resolving power (percent valley unspecified) of 10,000.

At present, the most commonly used analyzers are the quadrupole mass filter and the double-focusing magnetic instrument, a combination whose features complement one another very well. However, current developments are progressing so rapidly that it is difficult to predict the instruments of choice 10 years from now.

## 6. ION DETECTORS

The renowned sensitivity of mass spectrometry is entirely due to the detection system. For all but ICR spectrometers, this sensitivity is due to the efficiency with which ion beams can be converted into electron currents.

### 6.1. *Discrete-Dynode Electron Multiplier*

An ion beam with energy in the range 1–10 keV, if impacted on a metal plate, yields about one secondary electron per incident ion. If the secondary electrons are accelerated by about 100 V onto a second plate (or dynode), more electrons are formed. If the process is repeated 10–20 times, there may result as many as $10^8$ electrons per ion incident on the first dynode. This remarkable device thus produces very high amplification with almost zero noise and a very fast rise time. The dynamic range of such a detector is also excellent, from single ions per second (i.e., about $10^{-18}$ A) to about $10^{-9}$ A input (ion beam) current.(15)

### 6.2. *Continuous-Electrode Electron Multiplier (Channeltron)*

A continuous-electrode electron multiplier is fabricated from a specially formulated lead-doped glass.(45) The gain is similar to that obtainable using discrete-dynode multipliers, but on currently available devices the maximum input current before saturation is about $10^{-11}$ A. On the other hand, Channeltrons are considerably more stable toward exposure to air or to organic-laden atmospheres in mass spectrometers.

### 6.3. *Photographic Plates*

Use of a photographic plate as detector for a Mattauch–Herzog double-focusing spectrometer, in nonscanning mode, permits exploitation of the multiplex advantage; that is, a complete spectrum is integrated over the exposure time. This is often the only procedure available for use with ion sources of rapidly varying intensity. For example, a high-resolution mass spectrum over any appreciable mass range of an eluent from a capillary GC would have to be detected in such a fashion since a scanned spectrum would be limited by available ion statistics.(15) The main limitations of photoplate detection are nonlinearity and a very narrow dynamic range (about 50 : 1).

One possible alternative to the photoplate is the microchannel plate detector.[46] This is an array of $10^4$–$10^7$ miniature Channeltron electron multipliers arranged parallel to one another; 12-micron diameter channels with 15-micron center-to-center spacings are typical. Such a device has been used[47,48] as a photoplate replacement, but there are considerable problems in adapting this technique to routine analytical mass spectrometry.

### 6.4. *Faraday Cup*

An electrode shaped as a deep cup, directly connected to a sensitive amplifier, is known as a Faraday cup. Such detectors are used for accurate quantitative work (e.g., in precision isotope ratio measurements).

### 6.5. *Amplification and Recording of Mass Spectra*

The demands placed upon the system which amplifies and records raw data from a mass spectrometer (voltage as a function of time after start of scan) are quite high. These have been ably discussed.[2] The amount of data generated in even a relatively simple GC/MS run is so large that some kind of computer-based data system is essential.[2] A monograph on the applications of digital computers to mass spectrometry has been published.[49] Since the mass spectrometer/computer combination is described in a later chapter in this volume, no further account will be given here.

## 7. *SPECIAL TECHNIQUES*

In this section a brief introduction will be given to some more specialized aspects of mass spectrometric techniques.

### 7.1. *Specific-Ion Detection*

Signal-to-noise can be improved by up to three orders of magnitude by techniques whereby the mass spectrometer is arranged to transmit and detect only a few specified ion masses, instead of repetitive scans over the entire mass range of interest. Dwell-time per peak of interest can thus be markedly increased by discarding spectral regions containing no valuable information. The simplest form of this technique involves low mass-resolution, and quadrupole analyzers have traditionally been superior in this regard due to their ease of switching. The recent development of laminated magnets, together with that of temperature-controlled Hall effect probes, has made feasible magnetic peak switching at rates comparable with those achievable by quadrupoles.[15]

High resolving power specific-ion detection is possible only on double-focusing magnetic instruments, using switching of the accelerating voltage. An interesting comparison of specific-ion detection, at low and high resolution (500 and 5000, 10% valley definition) on the same instrument, has shown a considerable increase in signal-to-noise ratio for the high-resolution mode over the low; this was due to elimination of "chemical noise" present in the low-resolution experiment, which more than made up for the drop by a factor of about 10 in instrumental transmission on closing the slits.[15]

A similar increase in signal-to-noise, through elimination of "chemical noise," can be achieved via specific reaction detection using metastable peak monitoring (linked scan at constant *B/E*, see below). In this technique a mass-specified fragment ion, which arises from fragmentation of a specified precursor ion, is monitored at low resolution.[50]

### 7.2. *Mass Spectrometry/Mass Spectrometry (MS/MS)*

The term *mass spectrometry/mass spectrometry (MS/MS),* coined by analogy to "gas chromatography/mass spectrometry (GC/MS)," refers to techniques in which ions in a mass-selected beam are fragmented, usually via collision with neutral gas, and the fragments formed subsequently mass analyzed. The initial selection of the mass of the ions in the primary beam is a separatory stage, analogous to the GC stage in GC/MS. Persuasive reviews of the method have appeared.[51–53] The use of soft ionization techniques, particularly CI, simplifies the content of the ion beam and provides a degree of selectivity. One advantage claimed for MS/MS over techniques involving chromatographic preseparation is speed of analysis.

Instrumentation used for MS/MS falls into two broad categories, involving low and high ion beam energies. The former category is based upon two quadrupole mass filters, one to mass-select the "reactant" ion and one to analyze its fragments. Often the two quadrupole analyzers are separated by a small quadrupole operated in RF-only mode (zero mass discrimination), used as an ion-trap collision cell; such an arrangement is often referred to as a "triple-quad" instrument. The most commonly used high-energy technique is the so-called "MIKES (mass-analyzed ion kinetic energy spectroscopy) technique,"[54] using a double-focusing instrument in which the magnet precedes the electric sector.

Impressive results have been obtained, for example, by directly pyrolyzing biological samples into the ion source from a direct-insertion probe. However, some cautionary thoughts have been expressed[6] concerning artifacts arising from conversion of the sample into ions (including the pyrolysis). Moreover, the origins of artifact peaks in the MIKES technique itself[55] are such that it is dangerous to attempt an identification of a low-molecular-weight component of a complex mixture by MIKES without a knowledge of the behavior of components of higher mass, since the latter can give rise to interference artifacts in the MIKES spectrum. Linked-scan methods[56,57] suffer from a similar disadvantage. Quadrupole analyzers, being true mass filters, are not subject to this problem, but are, of course, limited in mass range and resolution.

The present author feels that the difficulties faced in analyzing for trace constituents in highly complex mixtures are sufficiently great that the analyst would be foolish to forego any technique which can help. If MS/MS techniques can reduce the amount of sample cleanup required in a particular case, thus reducing the probability of false positives and/or negatives, then they should be used. Moreover, in cases where a large number of samples have been submitted for analysis, the speed of MS/MS methods can offer a fast screening procedure which can pinpoint those samples that should be given priority. However, the limitations of all techniques, mass spectrometric or otherwise, should be understood and kept in mind. The value of GC retention index as an identifying label has recently been reemphasized in a publication from a well-known laboratory,[58] where the interlaboratory reproducibility of 0.2% in this parameter is emphasized. Despite the many attractive features of MS/MS, this degree of reproducibility is not available at present.

### 7.3. *Quantitation*

Space does not permit a detailed discussion of quantitation, an important aspect of mass spectrometry. A book on the subject has been published.[59] More recently[60] an excellent review, containing an extensive bibliography, has appeared.

### 7.4. *Mass Measurement*

If the mass of an ion can be determined to sufficiently high precision, its atomic composition may be thereby established.[1] Traditional methods of mass measurement to within $10^{-3}$ amu or less (e.g., peak matching)[1] require resolving powers of $10^4$ or greater and are time consuming.

More recently, computer-based methods of so-called "accurate mass measurement" (really "precise mass measurement") have been developed, which are applicable to components eluting from a GC, for example. All of these methods depend upon first establishing a time-mass calibration file for the repetitively scanned instrument; such calibrations are normally suitable, without further adjustment, for mass measurement to within the nearest integral value. Due to small drifts in the scan characteristics, traces of suitable substance, providing a number of lock masses, must be introduced to provide a real-time correction to the calibration file, if higher precision is required. The problem is to ensure that the computer can recognize these lock masses as arising from the reference substance and not from the sample of interest.

Historically the first such method exploited the fact that fluorine is a mass-deficient element (true mass of $^{19}F = 18.9984$, slightly lower than the nearest integer), whereas most organic ions are mass sufficient. Thus, use of medium resolving powers (7000 or higher) suffices to distinguish ions arising from commonly used fluorinated reference compounds such as perfluorokerosene (PFK) from the sample ions. Such methods have worked well and give mass measurements to a precision of a few parts per million (ppm), but are of low sensitivity due to the high resolving powers required. More recently, use of polyiodinated compounds as reference standards has permitted the use of much lower resolving powers (1000 or even less) and thus full sensitivity.[61] This is possible because of the very large mass defect of iodine (mass of $^{127}I = 126.9045$).

An ingenious method applicable only to quadrupole instruments because of the low operating voltage of the ion source pulses the polarity of the ion source alternatively positive and negative with a trace amount of PFK bled continuously into the source.[62] In the positive half-cycle the ions detected are predominantly sample ions, the PFK ions being present to the extent of a fraction of a percent. In the negative half-cycle the negative ions detected are predominantly PFK ions due to the enormous efficiency of fluorinated compounds in forming negative ions. Thus, sample and reference ions are readily distinguished, and the calibration file continuously updated.

Such low-resolution methods of mass measurement can give a precision of 10 ppm or so in favorable cases. The results are only meaningful, of course, if the ions thus mass-measured are clearly not multiplets. For samples introduced as eluents from a GC, for example, molecular ions and unambiguous fragment ions (e.g., $M-15$ ions) are likely to fulfill this condition.

## *REFERENCES*

1. J. H. Beynon, *Mass Spectrometry and its Applications to Organic Chemistry*, Elsevier, Amsterdam, 1960.
2. W. H. McFadden, *Techniques of Combined GC/MS: Applications in Organic Analysis*, Wiley, New York, 1973.

3. F. W. McLafferty, *Interpretation of Mass Spectra,* 3rd ed., University Science Books, Mill Valley, CA, 1980.
4. A. L. Burlingame, B. J. Kimble, and P. J. Derrick, *Anal. Chem.* **48,** 368R–403R (1976).
5. A. L. Burlingame, C. H. L. Shackleton, I. Howe, and O. S. Chizhov, *Anal. Chem.* **50,** 346R–384R (1978).
6. A. L. Burlingame, T. A. Baillie, P. J. Derrick, and O. S. Chizhov, *Anal. Chem.* **52,** 214R–258R (1980).
7. D. H. Williams, *Specialist Periodical Report: Mass Spectrometry,* Vol. 1, The Chemical Society, London, 1971.
8. D. H. Williams, *Specialist Periodical Report: Mass Spectrometry,* Vol. 2, The Chemical Society, London, 1973.
9. R. A. W. Johnstone, *Specialist Periodical Report: Mass Spectrometry,* Vol. 3, The Chemical Society, London, 1975.
10. R. A. W. Johnstone, *Specialist Periodical Report: Mass Spectrometry,* Vol. 4, The Chemical Society, London, 1977.
11. R. A. W. Johnstone, *Specialist Periodical Report: Mass Spectrometry,* Vol. 5, The Chemical Society, London, 1979.
12. M. L. Gross (ed.), *High Performance Mass Spectrometry: Chemical Applications,* ACS Symposium Series, 70, American Chemical Society, Washington, D. C., 1978.
13. D. H. Williams and I. Howe, *Principles of Organic Mass Spectrometry,* McGraw-Hill, London, 1972.
14. K. Levsen, *Fundamental Aspects of Organic Mass Spectrometry,* Verlag Chemie, Weinheim, 1978.
15. R. D. Craig, R. H. Bateman, B. N. Green and D. S. Millington, *Phil. Trans. Roy. Soc.* **A293,** 135–155 (1979).
16. L. Ramaley, M. E. Nearing, W. D. Jamieson, and R. G. Ackman, Proceedings of the 29th Annual Conference on Mass Spectrometry and Allied Topics, Paper MPB8, American Society of Mass Spectrometry, 1981.
17. G. W. A. Milne and M. J. Lacey, in *CRC Critical Reviews in Analytical Chemistry,* Vol. 4, CRC Press, Cleveland, 1974, pp. 45–104.
18. W. J. Richter and H. Schwarz, *Angew. Chem. Int. Ed. Engl.* **17,** 424–439 (1978).
19. K. R. Jennings, in *High Performance Mass Spectrometry: Chemical Application,* (M. L. Gross, ed.), ACS Symposium Series, American Chemical Society, Washington, D.C., 1978, pp. 179–187.
20. H.-R. Schulten, *Int. J. Mass Spectrom. Ion Phys.* **32,** 97–283 1979.
21. T. Matsuo, H. Matsuda, and I. Katakuse, *Anal. Chem.* **51,** 1329–1331 (1979).
22. K. M. Straub and A. L. Burlingame, in *Advances in Mass Spectrometry,* Vol. 8, (A. Quayle, ed.), Institute of Petroleum, London, 1980, pp. 1127–1134.
23. K. F. I. Heinrich and D. E. Newbury, Secondary Ion Mass Spectrometry, NBS Special publication No. 427, U.S. Dept. of Commerce, Washington, D.C., 1975.
24. M. Barber, R. S. Bardoli, R. D. Sedgwick, and A. N. Tyler, *J. Chem. Soc. Chem. Commun.,* pp. 325–327 (1981).
25. M. A. Baldwin and F. W. McLafferty, *Org. Mass Spectrom.* **7,** 1353–1356 (1973).
26. G. Hansen and B. Munson, *Anal. Chem.* **50,** 1130–1134 (1978).
27. D. I. Carroll, I. Dzidic, M. G. Horning, F. E. Montgomer, J. G. Nowlin, R. N. Stillwell, J-P. Thenot, and E. C. Horning, *Anal. Chem.* **51,** 1858–1860 (1979).
28. R. C. Dougherty and C. R. Weisenberger, *J. Am. Chem. Soc.* **90,** 6570–6571 (1968).
29. R. C. Dougherty, J. D. Roberts, and F. J. Biros, *Anal. Chem.* **47,** 49–54 (1975).
30. K. R. Jennings, *Phil. Trans. Roy. Soc.* **A293,** 125–133 (1979).
31. R. D. MacFarlane, in *Biochemical Applications of Mass Spectrometry* (G. R. Waller and O. C. Dermer, eds.), Wiley, New York, 1980, pp. 1209–1218.
32. J. E. Campana, *Int. J. Mass Spectrom. Ion Phys.* **33,** 101–117 (1980).

33. J. F. J. Todd, in *Dynamic Mass Spectrometry,* Vol. 4, (D. Price and J. F. J. Todd, eds.), Heyden, London, 1976, pp. 11–24.
34. D. Dixon, in *Dynamic Mass Spectrometry,* Vol. 4, (D. Price and J. F. J. Todd, eds.), Heyden, London, 1976, pp. 83–98.
35. D. F. Hunt, G. C. Stafford, Jr., F. W. Crow, and J. W. Russell, *Anal. Chem.* **48,** 2098–2105 (1976).
36. H. E. Duckworth and S. N. Ghoshal, in *Mass Spectrometry,* (C. A. McDowell, ed.), McGraw-Hill, New York, 1968, pp. 201–274.
37. W. G. Cross, *Rev. Sci. Instrum.* **22,** 717–722 (1951).
38. M. B. Comisarow and A. G. Marshall, *Canad. J. Chem.* **52,** 1997–1999 (1974).
39. C. L. Wilkins, *Anal. Chem.* **50,** 493A–500A (1978).
40. R. L. Hunter and R. T. McIver, *Anal. Chem.* **51,** 699–704 (1979).
41. P. A. Leclercq, G. J. Scherpenzeel, and C. A. Cramers, Proceedings of the 29th Annual Conference Mass Spectrometry and Allied Topics, Paper MPMOA3. American Society Mass Spectrometry, 1981.
42. J. P. Lehman and E. J. Younginger, *Int. J. Mass Spectrom. Ion Phys.* **33,** 95–98 (1980).
43. J. A. Browder, R. L. Miller, W. A. Thomas, and G. Sanzone, *Int. J. Mass Spectrom. Ion Phys.* **37,** 99–108 (1981).
44. B. A. Mamyrin, V. I. Karataev, D. V. Shmikk, and V. A. Zagulin, *Soviet Physics J.E.T.P.* **37,** 45–48 (1973).
45. E. A. Kurz, *Am. Lab.* (March, 1979).
46. J. L. Wiza, *Nucl. Instrum. Methods* **162,** 587–601 (1979).
47. C. E. Giffin, H. G. Boettger, and D. D. Norris, *Int. J. Mass Spectrom. Ion Phys.* **15,** 437–449 (1974).
48. J. H. Beynon, D. O. Jones, and R. G. Cooks, *Anal. Chem.* **47,** 1734–1738 (1975).
49. J. R. Chapman, *Computers in Mass Spectrometry,* Academic, London, 1978.
50. S. J. Gaskell and D. S. Millington, *Biomed. Mass Spectrom.* **5,** 557–558 (1978).
51. R. W. Kondrat and R. G. Cooks, *Anal. Chem.* **50,** 81A–92A (1978).
52. F. W. McLafferty, in (C. R. Waller and O. C. Dermer, eds.), *Biochemical Applications of Mass Spectrometry,* Wiley, New York, 1980, pp. 1159–1168.
53. F. W. McLafferty, *Acc. Chem. Res.* **13,** 33–38 (1980).
54. J. H. Beynon, R. G. Cooks, J. W. Amy, W. E. Baitinger, and T. Y. Ridley, *Anal. Chem.* **45,** 1023A–1031A (1973).
55. T. Ast, M. H. Bozorgzadeh, J. L. Wiebers, J. H. Beynon, and A. G. Brenton, *Org. Mass Spectrom.* **14,** 313–318 (1979).
56. R. K. Boyd and J. H. Beynon, *Org. Mass Spectrom.* **12,** 163–165 (1977).
57. R. K. Boyd and B. Shushan, *Anal. Chem.* **53,** 421–427 (1980).
58. S. C. Gates, N. Dendramis, and C. C. Sweeley, *Clin. Chem.* **24,** 1674–1679 (1978).
59. B. J. Millard, *Quantitative Mass Spectrometry,* Heyden, Lond, 1978.
60. A. P. DeLeenheer and A. Cruyl, in *Biochemical Applications of Mass Spectrometry* (G. R. Waller and O. C. Dermer, eds.), Wiley, New York, 1980, pp. 1169–1207.
61. W. F. Haddon and H. C. Lukens, Proceedings of the 22nd Conference on Mass Spectrometry and Allied Topics, American Society Mass Spectrometry, 1974, pp. 436–439.
62. D. F. Hunt, G. C. Stafford, J. Shabanowitz, and F. W. Crow, *Anal. Chem.* **49,** 1884 (1977).

2

# GAS CHROMATOGRAPHY/ MASS SPECTROMETRY/ COMPUTER INSTRUMENTATION

*R. E. CLEMENT and F. W. KARASEK*

## 1. INTRODUCTION

Gas chromatography/mass spectrometry/computer instrumentation (GC/MS/COMP) is the only technique that can provide qualitative and quantitative information on nanogram quantities of individual components in mixtures containing hundreds of compounds. It is an especially powerful technique in the environmental sciences where many toxic, carcinogenic, and mutagenic organics at trace levels must be determined in complex samples. The applicability and use of GC/MS/COMP techniques has greatly increased in recent years because of this need and the parallel development of all aspects of the instrumentation.

The important features of GC/MS/COMP instrumentation will be discussed in this chapter. Those aspects of the equipment important to environmental chemical analysis are emphasized. Although basic principles of operation are explained, there is no attempt to give a comprehensive treatment of the instrumentation. However, many examples from commercially available instruments are presented to illustrate the design considerations and capabilities of modern GC/MS/COMP systems. More exhaustive treatments of the instrumentation are available in certain books.[1,2]

---

*R. E. CLEMENT* • Laboratory Services Branch, Ontario Ministry of the Environment, P. O. Box 213, Rexdale, Ontario M9W 5L1, Canada. *F. W. KARASEK* • Department of Chemistry, University of Waterloo, Ontario N2L 3G1, Canada.

## 2. BASIC COMPONENTS

Figure 1 is a simplified diagram of a typical GC/MS/COMP instrument. The gas chromatograph consists of a column, injection system, and oven with temperature-programming capability. Column selection and chromatographic methods development are important aspects of GC/MS/COMP analysis, especially for the types of complex samples derived from environmental sources. Recent development of improved injection systems and the introduction of fused-silica capillary columns has improved the ability of GC/MS/COMP instruments to analyze complex mixtures.

Since a gas chromatograph is operated at atmospheric pressure and the mass spectrometer at low pressure ($10^{-5}$–$10^{-6}$ torr) some means of removing most of the GC carrier gas is required. This is usually accomplished by a separator located between the GC column outlet and MS ion source. Poor design of this component will greatly reduce the quality of information which can be obtained. In some cases, it is possible to directly couple the GC column to the ion source, and direct-coupled systems are available commercially.

Sample molecules must be ionized prior to entry into the mass analyzer, which separates ions according to their mass-to-charge (*m/z*) ratios. A recording of ion intensity versus *m/z* for a compound is its mass spectrum. The available ionization methods can be categorized as either "hard" or "soft," depending upon the amount of energy imparted to

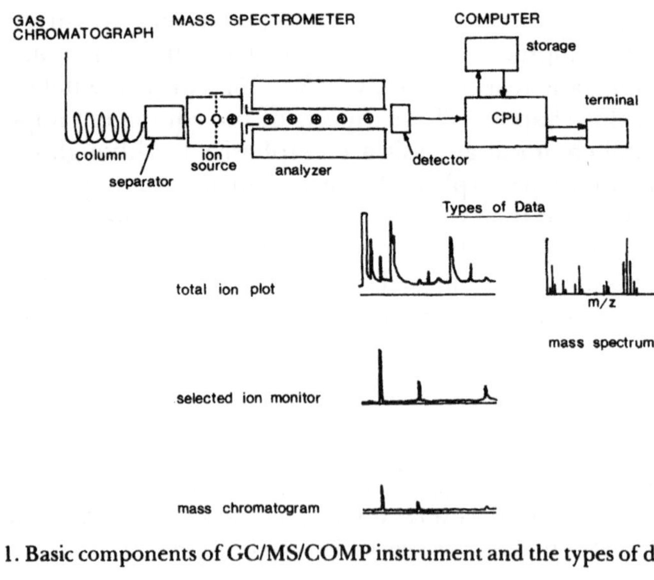


FIGURE 1. Basic components of GC/MS/COMP instrument and the types of data obtained.

molecules during the ionization process. Hard ionization adds excess energy to molecules, which can result in decomposition of the initially formed molecular ion to give many fragment ions. Although these fragment ions are useful in determining molecular structure, the molecular ion is most important, since it gives the compound's molecular weight. Consequently, soft ionization techniques have been developed in order to produce increased yields of the molecular ion. For GC/MS applications to environmental analysis the two ionization methods most frequently employed are electron impact (hard) and chemical ionization (soft). Some of the commercially available GC/MS/COMP instruments allow both electron impact (EI) and chemical ionization (CI) to be performed.

Of the different mass analyzers available, quadrupoles and magnetic sectors are used almost exclusively for environmental analysis. This component represents the heart of the GC/MS/COMP system. Although quadrupole units cannot approach the resolving power of magnetic sector analyzers, they possess other advantages which make them the preferred technique in many applications. Either analyzer can be operated by scanning the ions emerging from the ion source between operator-defined limits to produce a mass spectrum, or by only allowing ions of a chosen specific $m/z$ value to be detected. The latter technique, termed *selected-ion monitoring (SIM),* is useful for quantifying compounds. The different types of data obtained are illustrated in Fig. 1. Optimum results from the mass analyzer are only possible by employing efficient vacuum technology. Although not illustrated in Fig. 1, the vacuum system is an important aspect of ion source and mass analyzer design.

For ion detection, the electron multiplier is used almost universally in qualitative organic analysis by GC/MS/COMP systems. It produces large current amplification with very low background. Further, the fast response time of this device is necessary to achieve good sensitivity at the high scan rates used for capillary column GC/MS. The detector, mass analyzer, ion source, and associated vacuum system together comprise the mass spectrometer.

Control of the MS and the recording and analysis of mass spectral data are best accomplished by computer techniques. Figure 1 illustrates the typical devices associated with GC/MS/COMP system: a central processing unit (CPU), terminal for communication with the operator, and data storage device. An electronic interface allows bidirectional communication between the CPU and the mass spectrometer; scan parameters are sent to the MS and mass spectral data returned to the CPU to be processed. Discussion of the electronics involved is beyond the scope of this chapter. The software used for data acquisition and analysis should also be considered in a discussion of the role of the computer system in GC/MS instrumentation. A large difference exists between programs in terms of ease of use, speed, and ability to process large amounts of data. In some systems the computer completely controls the operation of both the gas chromatograph and mass spectrometer, collects, stores, and analyzes data, and checks the operation of the system by running diagnostic programs.

### 2.1. *Gas Chromatograph Column*

Many analyses of complex samples are inadequate due to inefficient GC techniques. It is imperative that the gas chromatograph column has low-bleed characteristics at elevated temperatures to achieve a high peak: background ratio for optimum results and to keep the ion source clean for extended periods. A rapidly rising background with increasing GC oven temperature is especially detrimental, since the effectiveness of computerized background correction procedures will be reduced.

The use of packed columns is declining for trace organic analysis. However, low-bleed, high-efficiency packed columns are available for these applications.[3–5] If the greater resolving power of capillary columns is not needed, packed columns may be more beneficial to use for their greater capacity and simpler operation. The declining use of packed columns in trace organic analysis is largely a result of the development of fused-silica capillary columns and simple-to-operate, commercially available injection systems.

Fused-silica columns have been recently developed.[6–8] However, their use has gained rapid acceptance and they are now commercially available from several suppliers. The inertness of fused silica is augmented by other desirable properties when the fused-silica capillary is coated with a polymeric material such as polyimide. Such columns are flexible and can be easily installed in GC/MS/COMP instruments without the lengthy end-straightening procedures that are required for glass capillary columns. After installation, they are physically more rugged than regular glass columns. A single fused-silica column can be used in different instruments having various injection configurations without restraightening the ends. A limitation is that columns with polar stationary phases are not available. However, the columns available commercially are applicable to a wide variety of samples. Further improvements to the performance of fused-silica columns can be obtained by using cross-linked phases. The effect of cross-linking the stationary phase is to achieve increased film stabilization and nonextractability of the phase.[9,10] Gas chromatograph operation at higher temperatures with low bleed is possible. On-column injections can be performed with no damage to the stationary phase.

## 2.2. *GC Injection Modes*

A majority of reported applications of GC/MS until recently have utilized flash vaporization injection techniques in which the sample is injected into a heated zone and vaporized before being swept onto the GC column by the carrier gas. Since injection is generally performed by a syringe with a metal needle, sample discrimination effects are possible, including decomposition of thermally labile species on the hot surface of the needle. Selective elution of components out of the needle has been reported.[11] Such problems may be magnified when capillary columns are employed.

The preferred injection technique to achieve maximum precision and minimum discrimination effects is on-column injection with no prior sample vaporization. A recent design for a cool on-column injector is illustrated in Fig. 2. In this design the syringe needle is accurately guided through an isolation valve into the GC column. The duck-bill isolation valve, constructed with a soft elastomeric material, provides a gas-tight seal by the inlet pressure after the syringe is withdrawn. Since this is a cool-injection technique, the valve will not contaminate the system. If necessary, a frit located between the valve and column can be used for purging. The column is sealed at the lower end of the injector by a graphite ferrule and column nut. Relative to many previous systems,

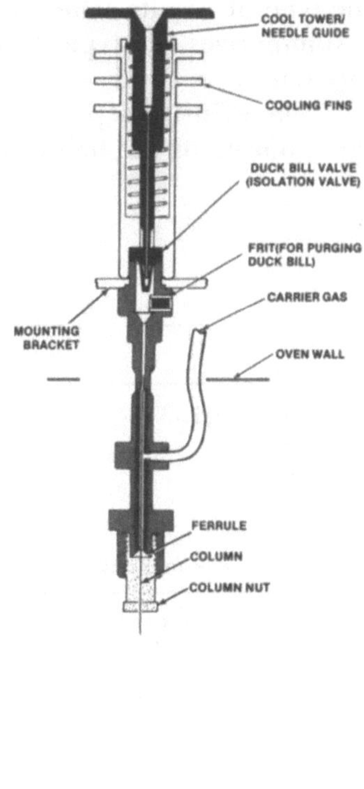


FIGURE 2. Cool on-column injection system for capillary column. (Courtesy of Hewlett-Packard.)

column installation is simple and leaks are easy to track down. A special syringe with a needle constructed of fused silica provides a more inert surface than syringes with metal needles. Since this fused-silica needle has an outside diameter of 0.14 mm, on-column injections can be made with capillary columns having an inner diameter as low as 0.21 mm. The development of commercially available injection systems such as illustrated in Fig. 2 and fused-silica capillary columns has enabled many laboratories to routinely use capillary column GC/MS. Benefits of capillary column technology in GC/MS are greater chromatographic resolution, increased applicability, and better interface techniques.

### 2.3. GC/MS Interface

A critical and difficult aspect of GC/MS instrumentation is the interface between the chromatographic column and MS ion source. Operating principles and relative merits of various interfaces have been described in detail by others.[11,12] Although they vary greatly in design, GC/MS interfaces can be classified into four general groups: effusive (Watson–Biemann); jet orifice (Ryhage); permeable membrane (Llewellyn); and direct coupled. A pressure drop from 760 torr (GC) to about $10^{-4}$ torr (ion source–electron impact) or 1 torr (chemical ionization) must be achieved to ensure proper operation of the mass spectrometer. The performance of the GC/MS interface determines to a large extent the type of results achieved by the entire GC/MS/COMP system. Table 1 summarizes the typical performances of the four principal types of separators.

The effusive separator utilizes the principle that under molecular flow conditions the lighter carrier gas molecules move through orifices

*TABLE 1*
*Comparison of Common Interfaces*[a]

| | Efficiency | Yield | |
|---|---|---|---|
| Classification | *N* | *Y* | *H* |
| Perfect | ∞ | 100 | 1 |
| Jet | $10^2$ | 40 | 1 to 2 |
| Direct | 1 | 1–100[b] | 1 |
| Effusive | $10^2$ | 50 | 1 to 2 |
| Permeable membrane | $10^3$ | 80–95 | 3 |

[a] From Karasek and McFadden (1971).
[b] Equals split ratio:

$$N = \frac{\text{conc. in MS source}}{\text{conc. in GC peak}} \qquad Y = \frac{\text{quantity in MS source}}{\text{quantity in GC peak}} \qquad H = \frac{\text{peak width in MS}}{\text{peak width from GC}}$$

faster than the heavier sample molecules. The jet orifice is available as a standard option on most commercial GC/MS/COMP instruments. GC effluent flows through a nozzle into a lower pressure region as an expanding jet. Directly in line with this nozzle is the exit orifice to the mass spectrometer. Heavier components travel through this orifice while the light carrier gas molecules are selectively removed by the vacuum system. One disadvantage of the jet separator is that the small nozzle or exit orifice may become clogged. In the permeable membrane the organic component of a GC peak passes through a thin elastomer membrane that exhibits a 1000-fold greater conductance to organic compounds than to He or $H_2$ carrier gases. Most of the carrier gas flows over the membrane to the atmospheric vent, while the organic molecules pass through the membrane to the ion source. Although high efficiencies are achieved with this simple design, the membrane limits the upper operating temperature of the GC (250–260 °C for silicone rubber membrane).

The type of interface preferred for use with capillary column GC/MS systems is by direct connection of the column to the MS ion source.[13] By this technique all of an eluting peak will enter the MS ion source, resulting in greater sensitivity compared to other interfaces. This advantage can be realized because of the low flow rates of capillary columns. With the higher flows of packed columns, it is not possible for most pumping systems to keep the ion source at low enough pressure without first splitting off most of the carrier gas (and therefore some of the sample) by the interface. Also, GC peak shape will not be appreciably degraded, ensuring a high concentration of sample in the ion source. The direct-connection interface is especially useful with flexible, low-bleed fused-silica columns.

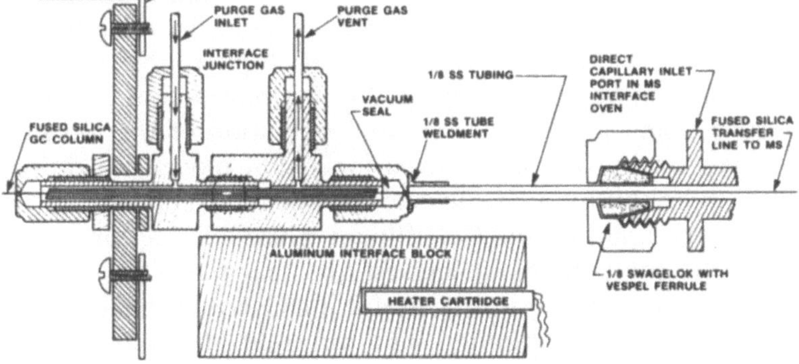


FIGURE 3. Open-split GC/MS interface. (Courtesy of Hewlett-Packard.)

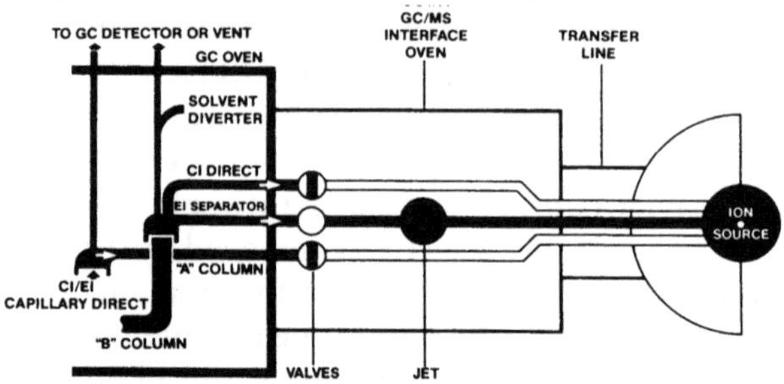


FIGURE 4. Multiple-function GC/MS interface design. (Courtesy of Hewlett-Packard.)

With the direct-connection interface, the system still suffers from some drawbacks. Operation of the GC column exit in a vacuum can produce vacuum effects. Also, the GC/MS system has to be vented each time a column is changed, which results in increased downtime. Samples which have a large dynamic range could produce problems, since operation of the system at high sensitivity for small peaks will cause overloading of the detector for large eluting peaks. Figure 3 shows a recent design for a GC/MS interface which substantially reduces these problems. In this "open-split" design, a fused-silica capillary column is inserted up to the interface junction, at which point a fused-silica transfer line conducts sample peaks directly into the MS ion source. Vacuum effects do not affect GC column performance, since the interface junction is at atmospheric pressure. A purge gas can be used to maintain a constant flow rate into the MS, regardless of the GC column flow rate. In addition, large or unwanted peaks can be purged before entering the MS. Finally, changing GC columns does not require venting of the MS.

The complexity of samples and sample types from environmental sources requires flexibility in instrumental design. No single instrumental configuration will be optimal for every type of sample. For this reason, advanced GC/MS instruments usually have more than one transfer line to the ion source. Figure 4 illustrates a multiple-configuration GC/MS interface allowing capillary column direct connection to the ion source, connection using a jet separator, and operation of the ion source in either electron impact or chemical ionization mode. In addition, the sample can be split to a GC detector to obtain a simultaneous trace by two detectors. Splitting the GC effluent is undesirable for applications in which the components of interest are near the MS detection limit. However, dual detection is very useful if consecutive analyses are to be performed on a single sample using different MS analysis conditions, such as different chemical ionization reagent gases.[14]

## *2.4. Mass Spectrometry—Sample Ionization*

Much work is being performed to develop new methods of ionization in order to attain better limits of detection and to improve the usefulness of mass spectra for compound identification. However, the "hard" technique of electron impact (EI) and "soft" method of chemical ionization are by far the two most often used methods of ionization for environmental analysis by GC/MS/COMP. Application of these and other ionization techniques to mass spectrometry has been reviewed.[15–17]

EI ionization has the advantage of being the first widely accepted technique for use in conventional mass spectrometry. Consequently, large libraries of mass spectra generated by EI source are available. This facilitates the identification of unknown substances by matching the pattern of the unknown spectrum with previously recorded patterns in these reference collections. Ionization is accomplished by collisions of molecules in the gas phase with energetic (70-eV) electrons. This produces a molecular ion usually in a high state of electronic and vibrational excitation. Further decomposition of the molecular ion into fragment ions depends upon the degree of excitation and molecular structure. Because of the simplicity of operation, long-term stability, and precise beam intensity control of the EI source, mass spectral patterns are reproducible and structure specific. The efficiency of ionization is good, although only about 1 molecule in $10^6$ is ionized. Electrons are produced by a filament of tungsten or rhenium wire and pass through the ionizing region of the source to an anode or electron trap in the source. Ions formed in the source are accelerated through ion slits to the analyzer by a repeller electrode. Other electrodes and collimating slits focus the emerging ion beam for optimum resolution. Negative as well as positive ions can be generated and detected.

Despite its many advantages, the EI source is a very drastic means of ionization and often results in excessive fragmentation. This leads to a very small, or in some cases, unobservable molecular ion. Since this ion gives a compound's molecular weight directly, and all fragment ions are derived from the molecular ion, it is the most important single piece of information which can be obtained from a mass spectrum. Consequently, "soft" methods of ionization have been investigated to ensure formation of an observable molecular ion. The technique of chemical ionization (CI) is the foremost of the soft ionization techniques. Ionization proceeds in an ionization chamber that is very similar to that used for EI ionization. However, for CI a reaction gas such as methane is introduced into the ion chamber at pressures of about 1 torr. Primary ionization of this reactant gas proceeds in the same manner as described previously for EI. Sample molecules then react with these reactant gas

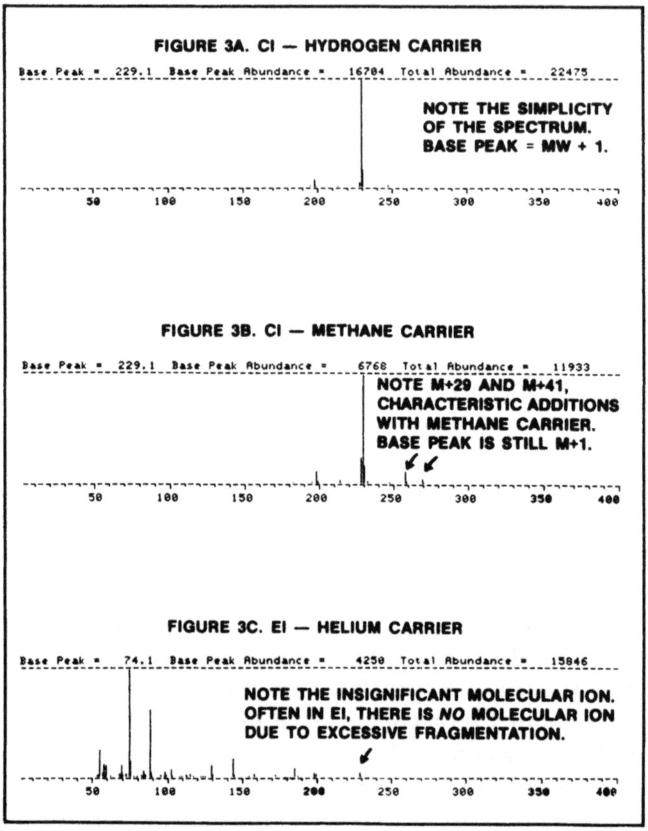


FIGURE 5. Comparison of mass spectra of tridecanoic acid, methyl ester obtained with CI source and $H_2$ reactant gas (top), methane reactant gas (middle), and mass spectrum of same compound with an EI source (bottom). (Courtesy of Hewlett-Packard.)

ions to form products which are detected in the mass spectrum of the sample. The most common type of reactions for methane reactant gas are hydride transfers. For compounds that do not produce an observable molecular ion under EI ionization, a quasimolecular ion is usually produced which is the molecular weight $+1$ of the sample molecule.

Figure 5 shows the CI mass spectra of tridecanoic acid methyl ester using two different reactant gases ($H_2$ and $CH_4$) and the EI spectrum of the same compound. The molecular ion of the compound under EI conditions is very small and will be very difficult to identify during the analysis of a real sample in which ions due to background contaminants or coeluting GC components are also present in the mass spectrum.

The simpler patterns of the CI spectra are not always advantageous. In a survey analysis of all methyl ester compounds in a sample using EI

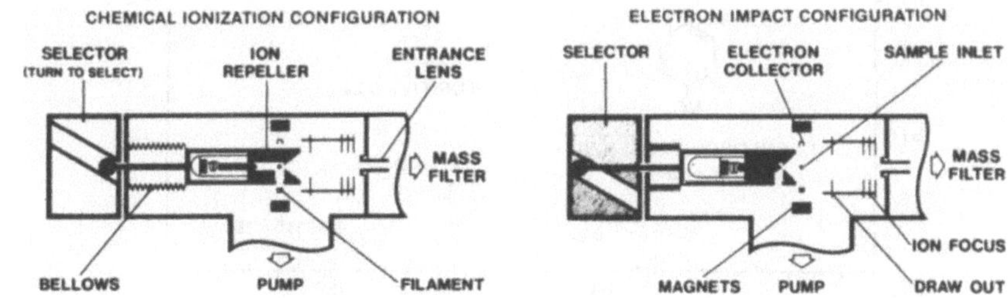


FIGURE 6. Dual EI/CI source design. (Courtesy of Hewlett-Packard.)

ionization, it is only necessary to monitor $m/z = 74$, since the EI spectra of members of this compound class all have the same base peak. Also, determination of the structure of compounds from such simple CI mass spectra as shown in Fig. 5 is difficult because there are so few structural fragments. However, for the optimum use of GC/MS/COMP techniques it is desirable to perform both EI and CI analysis.

Figure 6 is a schematic drawing that shows a dual EI/CI ion source design. By turning the selector, a sliding bellows positions either an EI source or CI source at the sample inlet. All necessary electrical contacts are made when this is done. In a short time interval (1 min) the GC/MS can be switched from EI and CI operation using ion sources optimized for either EI or CI operation without breaking the vacuum. It is practical to obtain consecutive EI and CI analyses to aid the identification of components where one type of mass spectrum alone is insufficient.

Although negative-ion EI has not been reported as an important technique of GC/MS/COMP in environmental analysis, negative-ion CI

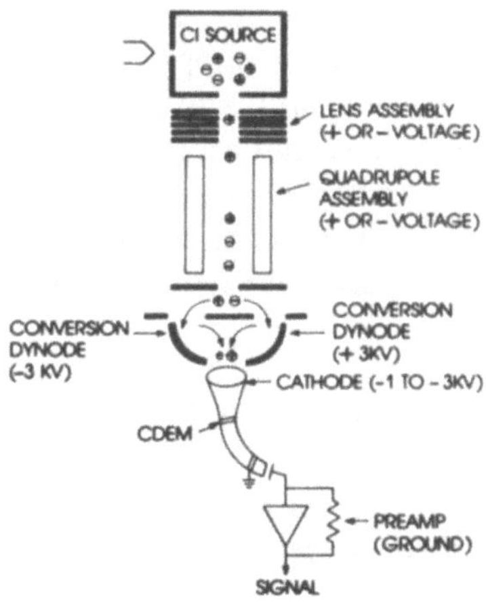


FIGURE 7. Design of dual positive–negative CI source. (Courtesy of Finnigan Corp., by permission of Technical Marketing Associates.)

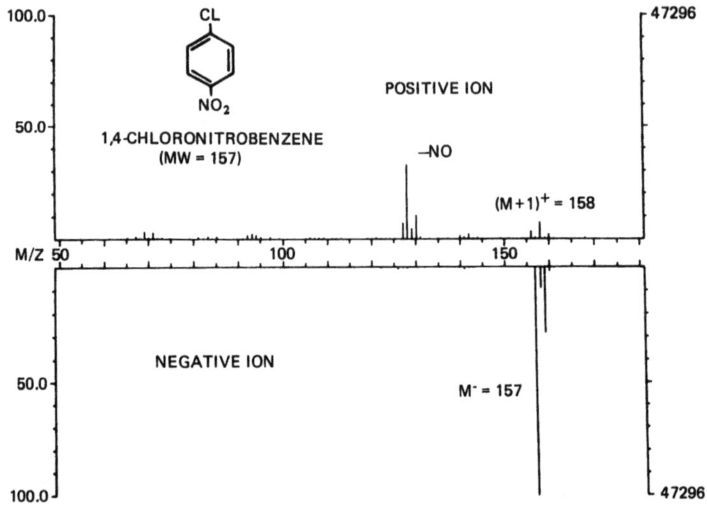


FIGURE 8. Positive and negative CI mass spectra of 1,4-chloronitrobenzene obtained from a single analysis. (Courtesy of Finnigan Corp., by permission of Technical Marketing Associates.)

has been found to be superior to other modes of GC/MS operation for some applications. Many drugs, pesticides, and other toxic compounds contain halogens or other heteroatoms amenable to negative-ion CI operation.[18,19] By choosing the proper reagent gas, it is possible to obtain selective detection of a specific compound in a mixture of isomers.[20]

Figure 7 shows an instrumental design that allows both positive and negative CI spectra to be recorded during a single analysis. Because of the complementary nature of these spectra, this feature allows a large amount of information to be obtained at high sample throughput.

Positive and negative CI spectra are obtained alternately by controlling the voltages to the ion source and lens assembly, quadrupole analyzer, and conversion dynode. Since the quadrupole analyzer can be operated at fast scan rates, positive and negative CI spectra may be obtained for even rapidly eluting capillary column GC peaks. Figure 8 shows the positive and negative CI mass spectra of 1,4-chloronitrobenzene obtained from a single analysis by a GC/MS/COMP system.

## 3. *MASS SPECTROMETER ANALYZERS*

The purpose of the mass analyzer is to distinguish between ions of different *m/z*. The ability with which this can be accomplished is termed the resolution cf the analyzer. Resolution ($R$) is defined as

$$R = M/\Delta M$$

for the two consecutive peaks of highest mass and equal height exhibiting a chosen valley between the peaks (usually 10%). For example, a resolution of 1000 (10% valley) means that ions of mass 1000 and 1001 will have overlapping peaks exhibiting a valley which is 10% of the peak height for equal-sized peaks. Resolution is therefore an important parameter in assessing the performance of a mass spectrometer. Analyzers are generally classified as low (<1000), medium (1000–10,000), or high (>10,000) resolution. For GC/MS analysis other factors such as scan speed, limit of detection, and computer compatibility are also important.

Medium-resolution GC/MS instruments accomplish ion separation by acceleration of the ions by a potential ($V$) through a magnetic field ($H$) with a radius of curvature of the ion path ($R$). An ion having the correct mass-to-charge ratio ($m/z$) will pass through the analyzer to the detector if it satisfies the following relationship:

$$m/z = \frac{H^2R^2}{2V}$$

By scanning $V$ or $H$, ions of varying $m/z$ can be detected to produce a mass spectrum. In practice, $H$ is varied while $V$ is held constant because the efficiency of transmitting ions of low mass is much greater than that for ions of high mass when the accelerating voltage is varied. For example, an ion of $m/z$ 50 will only require one-tenth the accelerating voltage of an ion of $m/z$ 500. Various geometries of this analyzer have been developed.[21] This magnetic sector instrument is called a single-focusing analyzer, since the diverging ion beam from the source converges as it passes through the magnetic field and is focused (in direction) at the exit slit (detector). However, the magnetic sector instrument is a momentum analyzer, not a true mass analyzer. Therefore only limited resolution can be attained, since the ions emerging from the source possess an energy divergence in addition to the angular divergence.

High-resolution instruments use an electric field to counteract the effect of energy dispersion. An ion of mass $m$ and charge $e$ traveling through an electrostatic field of strength $E$ with velocity $v$ will travel with a radius of curvature $r$ such that

$$Ee = \frac{mv^2}{r}$$

For a given electrostatic field, ions of the same mass but different velocity (kinetic energy) will be separated. When combined with a magnetic sector analyzer a double-focusing mass spectrometer is obtained. Very high resolution is obtainable. However, as resolution is increased, the detection limit is increased. Very high resolution also requires slower scan speeds that may not be compatible with GC/MS applications. For environmental analytical problems, the double-focusing instrument is generally used only in special applications.

In many cases low resolution is sufficient to provide a spectrum which will give a fingerprint of the unknown compound. By comparing this fingerprint with those of known compounds, many substances can be identified without performing any additional work. For such low-resolution work, the quadrupole mass spectrometer is the preferred technique. The principle of operation of the quadrupole analyzer is based on alternating electric fields applied to four rods held in a square array. Opposite rods are electrically connected so that the applied voltage consists of a constant DC component ($U$) and a radio frequency (rf) component. Ions entering the quadrupole region will oscillate between adjacent electrodes of opposite polarity because the potential difference between the rods with time ($t$) will be $U \pm V_0 \cos wt$ where $w = 2f$ and $f$ is the radio frequency. Depending on the values of $U$, $V_0$, and $w$, ions of a specified mass will undergo stable oscillations and will be detected. Ions of other masses will undergo oscillations of increasing amplitude until they are neutralized on the quadrupole rods. In practice, $U$ and $V_0$ are usually both varied during a scan to maintain a fixed $U/V_0$ ratio. Depending on the ratio chosen the resolution can theoretically be as high or as low as desired. In practice, the best resolution achievable on current instruments is about 1000. An important advantage of the quadrupole analyzer is that by operating at fixed $U/V_0$ a linear relationship exists between $V_0$ and ion mass. According to the principle of operation, the quadrupole analyzer is more correctly termed a mass filter rather than a mass analyzer.

For GC/MS/COMP applications the most important decision concerning the specific system to employ may be the choice between the quadrupole mass filter and the magnetic sector analyzer. In terms of overall performance, both techniques are capable of achieving good results for the types of samples commonly encountered in environmental applications. Some of the advantages and disadvantages of each are presented in Table 2.

Feser and Kogler[22] have recently presented a discussion of the quadrupole mass filter in GC/MS applications. They state that there is no clear preference for the quadrupole or magnetic sector technique

*TABLE 2*
*Comparison of Quadrupole and Magnetic Sector Analyzers*[a]

| Feature | Quadrupole | Magnetic sector |
|---|---|---|
| Resolution > 1000 | − | + |
| Mass range > 1000 | − | + |
| Peak matching | − | + |
| Fast peak switching | + | − |
| Linear mass scale | + | − |
| Digital control of instrument | + | − |

[a] (+) represents an advantage for the indicated technique.

unless the problem has been clearly defined previously. High resolution and detection of ions with mass greater than 1000 are not of primary importance for many GC/MS applications. The faster scan rates attainable by quadrupole units has been presented as a major advantage, but for most GC/MS applications magnetic sector instruments can scan at rates fast enough for even fast-eluting capillary column GC peaks. Hedfjall and Ryhage, for example, have reported a GC/MS/COMP instrument with a magnetic sector analyzer capable of scanning the mass range *m/z* 50–500 in 0.7 sec.[(23)] These fast scan rates are achieved at the expense of resolution. Since operation of the quadrupole mass filter is readily adapted to digital control, this technique is preferred to routine analysis instruments with high sample throughput and unattended operation. For selected ion monitoring (SIM) applications, the quadrupole instrument is also generally preferred because magnetic sector analyzers are limited in their ability to rapidly switch from one ion to monitoring another of very different mass. However, specialized applications may require high-resolution SIM to distinguish between a specific ion and ions of compounds which have the same nominal mass but a different exact mass. An example is the analysis of 2,3,7,8-tetrachlorodibenzo-*p*-dioxin (2,3,7,8-TCDD) which has a nominal mass of 320. Ions from other compounds such as some polychlorinated biphenyls and DDE also have this same nominal mass. By employing a medium-to-high resolution system, the 2,3,7,8-TCDD exact mass of 319.8965 can be distinguished from these interfering compounds.[(24)] For such applications, it is still preferable to use a quadrupole GC/MS/COMP as a prescreening technique.

A major disadvantage of magnetic sector GC/MS/COMP instruments is their high initial cost. A significant advance in GC/MS/COMP instrumentation was the introduction of the Hewlett-Packard 5992 quad-

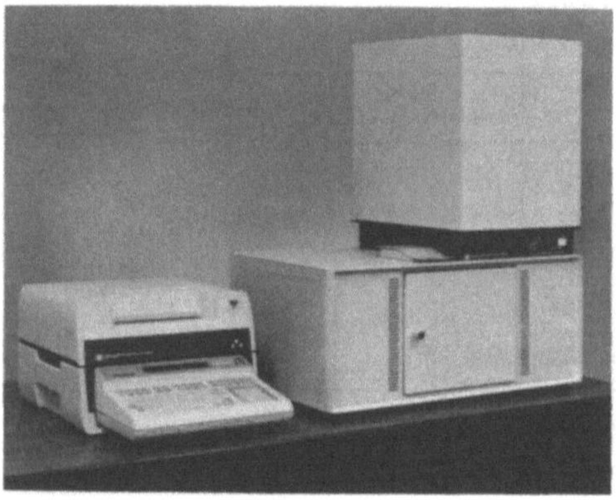

FIGURE 9. Hewlett-Packard 5992 GC/MS.

rupole GC/MS shown in Fig. 9. This instrument incorporated several design features that resulted in its compact size and low cost compared to previous GC/MS/COMP systems. Instrument operation is controlled by a powerful high-speed calculator instead of a computer. The MS analyzer is a quadrupole filter, mounted on top of the GC oven. A rough-vacuum pump is housed in the main frame beside the GC oven, and the diffusion (high-vacuum) pump has been designed around the quadrupole rod assembly. Figure 10 shows a schematic drawing of the quadrupole filter and high-vacuum-pump assembly which can be modified to use a turbomolecular pump.

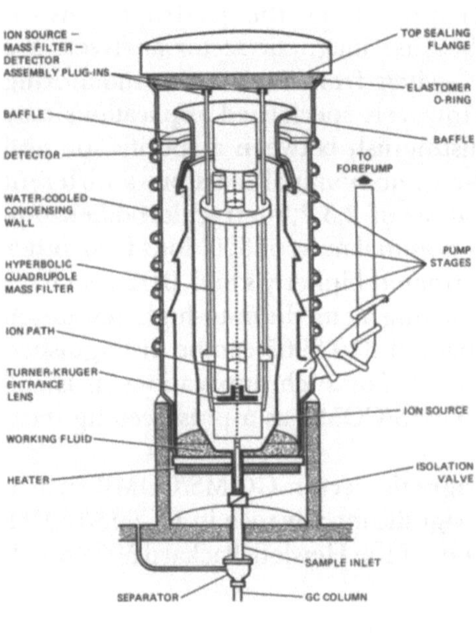


FIGURE 10. Analyzer assembly of bench-top GC/MS/Calculator instrument. (Courtesy of Hewlett-Packard.)

The most important result of the introduction of the Hewlett-Packard 5992 instrument is that it allowed GC/MS/COMP to become available to many laboratories that could not previously consider using this technique on a routine basis. Results obtained from the system are comparable to those from some instruments costing much more. This is a result of the manufacture of true hyperbolic-shaped quadrupole rods instead of the round rods used in some other quadrupole systems. The electric fields around hyperbolic rods allow more efficient passage of higher mass ions than they do with round rods.[25] A comparison of the important features of this system and other commercially available GC/MS/COMP instruments has been recently presented.[26]

## 4. GC/MS APPLICATIONS

Choosing the best instrumentation for a specific application should be done by considering the combinations of different components (type of GC column, GC/MS interface, type of mass analyzer), not the various components separately. The degree of GC/MS/COMP sophistication needed depends upon the sample complexity, possible interferences, and required limit of detection. Low- or high-resolution gas chromatography (LRGC, HRGC) can be combined with low- or high-resolution MS (LRMS, HRMS) to give four principal GC/MS configurations. Relative merits of these various instrumental combinations can be examined by observing their application to a specific problem.

The analysis of polychlorinated dibenzo-*p*-dioxins (PCDDs) has been extensively studied because of the high toxicity reported for some of these compounds, especially 2,3,7,8-TCDD. LRGC/LRMS has been used to provide quantification of total isomers for each group of congeners.[27] Packed-column GC was sufficient to resolve each group of congeners with the aid of LRMS selected ion monitoring. Since the various isomers were present in the ppb to ppm concentration range and no significant interferences were present, analyses could be performed on concentrated sample extracts with no cleanup. Individual isomer determinations could not be performed using this method. By using HRGC, the analytical capability of LRMS can be greatly increased. The HRGC/LRMS combination has been used to determine 2,3,7,8-TCDD in the presence of other tetrachlorinated dibenzo-*p*-dioxin isomers in environmental samples.[28] Lower detection limits can be obtained for HRGC than LRGC because of the narrow capillary column peak widths and better separation of the 2,3,7,8-TCDD from possible interferents. LRGC has been combined with both LRMS and HRMS.[29] LRGC/LRMS was used as a prescreening method and LRGC/HRMS only applied to those samples which could not be positively identified. The lowest detection limits for 2,3,7,8-TCDD can be obtained by HRGC/HRMS. This technique has been used to determine 2,3,7,8-TCDD in fish, water, and sediment samples.[30] Direct connection of the capillary column to the MS ion source was employed. Detection limits for 2,3,7,8-TCDD were lower than 1 ppt for some samples. These examples illustrate that even for the analysis of similar compounds, choice of GC/MS/COMP instrumentation depends largely upon the type of sample and specific objectives of the analysis.

## 5. COMPUTER TECHNIQUES

The role of the computer in the GC/MS/COMP system is important. All GC/MS operating conditions, collection of raw data, and data manipulation can be computer controlled. Tuning of the mass spectrometer and diagnostic checking can also be automated. Modern GC/MS/COMP instruments have reached a high level of computerization, especially systems with a quadrupole mass filter. So much data can be generated by capillary column GC/MS analysis of environmental samples that much information may be lost unless advanced computer techniques are utilized to process this information. The magnitude of the problem can be appreciated by considering a typical analysis situation.

A sample of diesel engine particulate matter exhaust is collected and concentrated solvent extract obtained for analysis. It is desired to obtain a qualitative characterization of this sample, plus a quantitative analysis of any carcinogenic or mutagenic compounds that are identified. HRGC analysis of this sample shows it is a complex mixture of over 200 compounds at widely varying concentrations. Optimum separation of the components requires a 90-min capillary column GC analysis. Typical GC/MS/COMP systems with large data systems can make repetitive scans and store each mass spectrum on a data storage medium such as a disk drive. If a mass spectrum is obtained and stored every 2 sec, the entire analysis will require the storage of 2700 mass spectra. Each peak in a single mass spectrum consists of a mass and abundance value, so a typical mass spectrum of about 150 detected peaks requires the storage of 300 numbers. The entire analysis would require storage space for about $300 \times 2700 = 810{,}000$ numbers. The actual storage needed would probably be greater than this, since other information such as retention time may be stored with each spectrum. Although much of this information can be discarded—for example, scans taken during times in the analysis where no GC peaks were detected—the task of processing such a vast amount of data is too labor intensive to perform manually for even a single analysis.

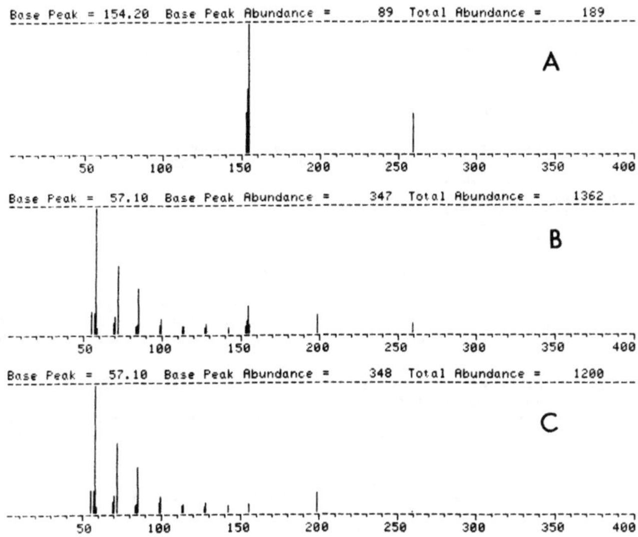


FIGURE 11. Computerized background correction. A—background spectrum, B—original spectrum, C—corrected spectrum (B – A).

Typical data reduction procedures include background subtraction, spectra normalization and display, and library searching. To identify specific components, specialized techniques such as generation of mass chromatograms can be performed.[31] Special algorithms have been developed to improve the quality of the stored spectra by using mass chromatograms to subtract background ions or remove ions of an overlapping component from each stored spectrum.[32] Figure 11 illustrates the value of background subtraction procedures in improving the quality of spectra obtained. Spectrum *A* is a background spectrum taken at a valley between two GC peaks. The mass spectrum of the next peak is shown as spectrum *B*. Spectrum *C* is the difference spectrum, $B - A$, which has been corrected for the overlapping previous GC peak. The mass peak at *m/z* 259, which is background resulting from the diffusion pump oil used, has also been removed.

Computer programs are available on some commercial quadrupole GC/MS instruments that automatically tune the instrument. This enables users who are not GC/MS specialists to obtain results consistent with results of more experienced operators. The instrument illustrated in Fig. 9 uses a perfluorotributylamine (PFTBA) internal calibration to adjust mass spectrometer operating parameters to give optimum peak shape and resolution using the computer program AUTOTUNE. Different compounds have been used for this purpose.[33] Figure 12 shows the final report of a dual EI/CI AUTOTUNE program for an instrument similar to the one shown in Fig. 9. The operator is informed in the report

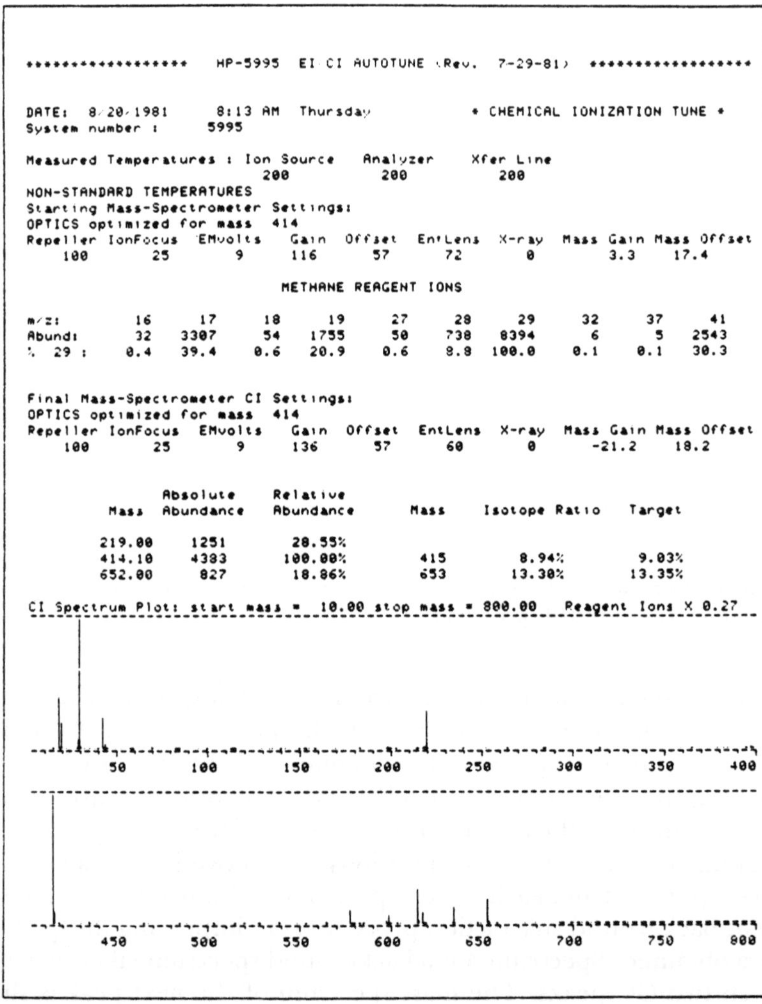

```
******************   HP-5995  EI CI AUTOTUNE (Rev.  7-29-81)  *****************

DATE:  8/20/1981      8:13 AM  Thursday            * CHEMICAL IONIZATION TUNE *
System number :       5995

Measured Temperatures : Ion Source   Analyzer    Xfer Line
                            200         200         200
NON-STANDARD TEMPERATURES
Starting Mass-Spectrometer Settings:
OPTICS optimized for mass  414
Repeller IonFocus  EMvolts   Gain  Offset  EntLens  X-ray  Mass Gain Mass Offset
   100       25        9      116     57      72       0        3.3     17.4

                            METHANE REAGENT IONS

m/z:         16     17     18     19     27     28     29     32     37     41
Abund:       32   3307     54   1755     50    738   8394      6      5   2543
% 29 :      0.4   39.4    0.6   20.9    0.6    8.8  100.0    0.1    0.1   30.3


Final Mass-Spectrometer CI Settings:
OPTICS optimized for mass  414
Repeller IonFocus  EMvolts   Gain  Offset  EntLens  X-ray  Mass Gain Mass Offset
   100       25        9      136     57      60       0      -21.2     18.2


           Absolute   Relative
    Mass  Abundance   Abundance      Mass    Isotope Ratio    Target

  219.00     1251       28.55%
  414.10     4383      100.00%        415        8.94%          9.03%
  652.00      827       18.86%        653       13.30%         13.35%

CI Spectrum Plot: start mass =  10.00 stop mass = 800.00   Reagent Ions X 0.27
```



FIGURE 12. CI AUTOTUNE report from Hewlett-Packard 5995 GC/MS.

if any of the abundances of five peaks chosen for the tuning procedure are out of specified ranges.

The capabilities of a GC/MS/COMP system may be significantly improved through software development. A commercial system in general will allow the user access to its processing power through high-level languages such as BASIC or FORTRAN. An example of this use of program has been reported by Dickson and co-workers.[34] They improved the capabilities of the calculator-based GC/MS system illustrated in Fig. 9 by modifying the software supplied by the manufacturer. Since this instrument is a low-cost calculator-based system, the supplied operating pro-

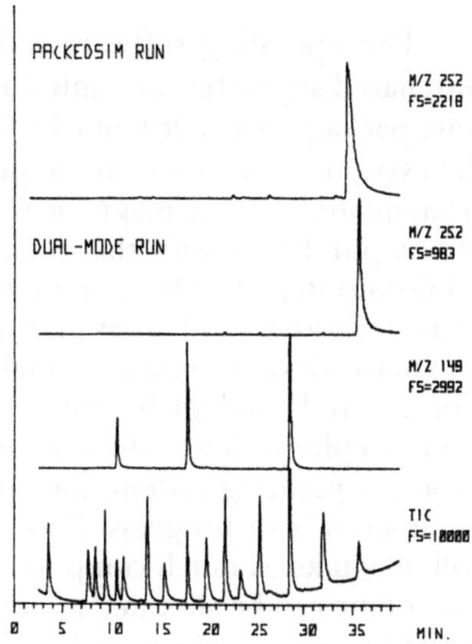


**FIGURE 13.** Comparison of total ion plot (TIC) and mass chromatograms from a single GC/MS analysis by the user-developed program DUAL-MODE with selected-ion monitoring analysis (PACKEDSIM) of the same sample.

grams which were based on a single floppy disk drive did not store every spectrum in a GC/MS analysis due to the limited storage space of a floppy disk. Therefore, mass chromatograms could not be generated and the total ion abundances were not stored. The user-developed programs allowed storage of the total ion abundance data in addition to as many as six preselected mass chromatograms on the floppy disk. A versatile plotting program allowed the user to compare total ion data (TIC), mass chromatograms (MC), and selected ion monitor (SIM) traces in the same plot. Figure 13 shows the comparison of the TIC and two MCs from one GC/MS analysis with SIM results from analysis of the same sample. Through program development, the capabilities of a GC/MS system were increased beyond those originally supplied by the manufacturer.

Large GC/MS/COMP systems are generally equipped with a minicomputer-based data system and large-capacity disk data storage device. The system can be efficiently operated through a video display terminal. A printer–plotter is used to provide hard copy of the results. Alternatively, the printer and plotter can be separate devices. The storage capacity necessary for a system which stores every mass spectrum during an analysis is enormous. Using the previous example of a 90-min capillary column analysis in which a mass spectrum is stored every 2 sec, the minimum total storage required for this analysis would be approximately 1.6 million bytes (1.6 Mbyte). A byte is a standard computer term and represents eight bits of storage, where one bit is a binary 1 or 0. The standard floppy disk which is used for data storage on the calculator-based GC/MS system (Fig. 9) can store a maximum of about 0.5 Mbyte of data.

The operating software with associated computer devices can be purchased separately and interfaced to a GC/MS system. One such software package available from the Finnigan Corporation is called the INCOS data system. Versions of these programs are capable of processing data from quadrupole or magnetic sector mass spectrometers. The data system is purchased with the necessary peripheral devices that include a minicomputer, 10-Mbyte disk drive, graphic video display terminal, printer–plotter, and mass spectrometer electronic interface. Optional hardware devices include a higher capacity disk drive, magnetic tape storage, and modem for direct communication to external computers over telephone lines. Because the system is a true foreground–background operating system, the operator can perform data processing as an analysis is in progress. Mass spectrometer control and data storage will only take as much computer time as necessary; the remaining computer time can be allocated to operator-directed tasks. Table 3 is a partial list of features available with this data system. The quality and quantity of data from GC/MS systems can be increased by upgrading the software instead of the hardware.

Development of the GC/MS/COMP system has increased rapidly over the past decade. Based on a powerful, flexible computer system the Hewlett-Packard 5987 instrumentation is capable of executing all of the usual and special techniques developed for the GC/MS/MS using capillary columns. The system shown schematically in Fig. 14 will provide for six graphic terminals all interfaced to the system and capable of parallel operation to do multitasks. One terminal can be monitoring a sample run in progress, another can be processing data from a previous analysis, and another can be monitoring a second GC/MS unit in which another sample is being run. The huge memory capacity permits storage of an 80,000 reference spectra library serving as a data base for an

*TABLE 3*
*INCOS Features for GC/MS Analysis*

1. Fully automated data acquisition and processing
2. Foreground/background operation
3. Quadrupole or magnetic sector operation
4. Programmable by user using FORTRAN
5. Full scan or selected-ion monitoring operation
6. Generation of mass chromatograms
7. Background subtraction of spectra
8. Variety of data display options
9. Biller–Biemann peak stripping
10. Library search
11. Programs for quantitative analysis
12. Diagnostic programs

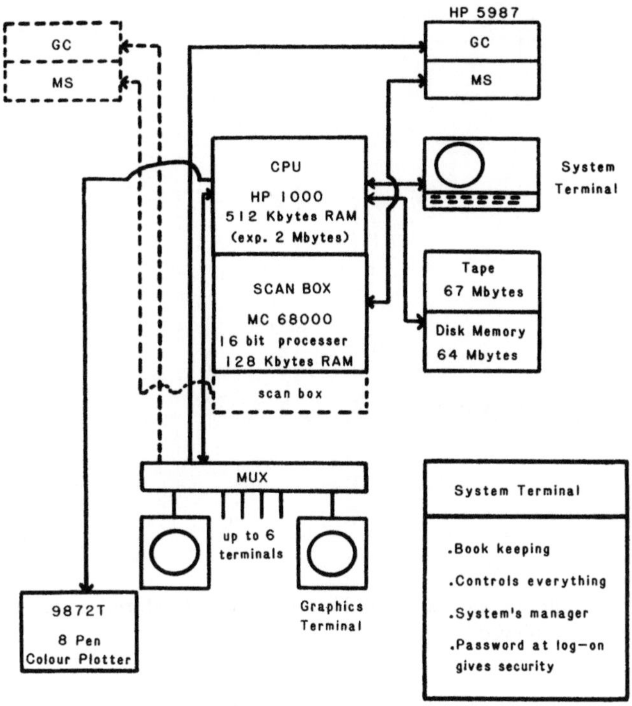


FIGURE 14. Schematic of the large HP 5987 GC/MS/DS system.

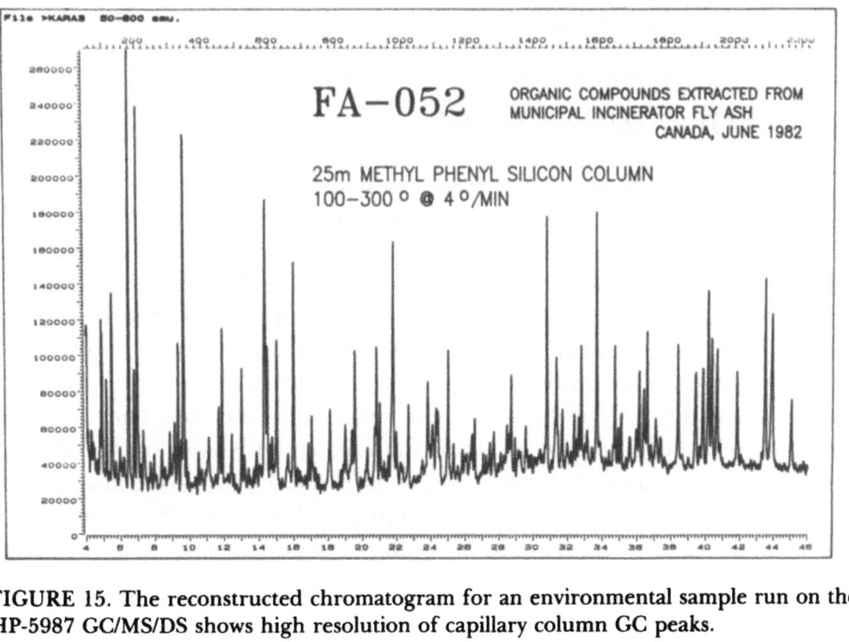


FIGURE 15. The reconstructed chromatogram for an environmental sample run on the HP-5987 GC/MS/DS shows high resolution of capillary column GC peaks.

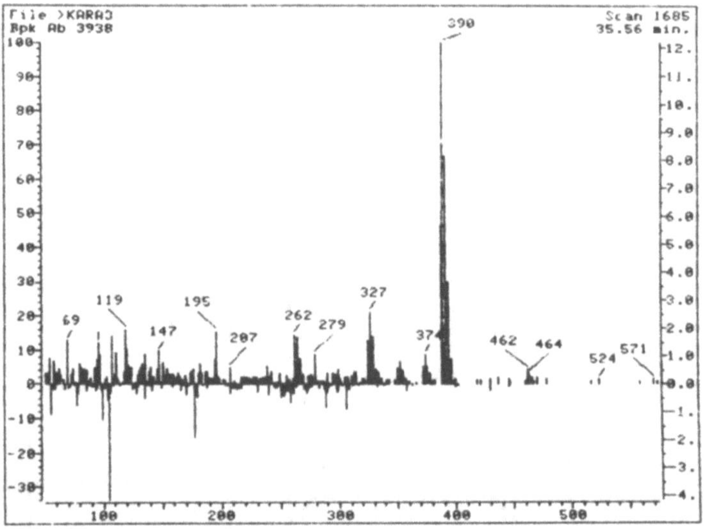


FIGURE 16. Results of an automatic library search of reference spectra in which no identification was achieved.

automatic search routine to identify each GC peak in sequence from samples as complex as the one seen in Fig. 15. Should the GC peak compound not be identified from the library, the interpretive STIRS program will provide the main structural features of the molecule to aid in identification (Fig. 16).

## 6. *SPECIAL INSTRUMENTATION*

Improvements are being made in all areas of GC/MS/COMP instrumentation. Further capillary column developments, especially the ability to coat highly polar phases onto fused-silica columns, will increase the applicability of GC/MS/COMP. Although high-efficiency GC/MS interfaces are available, refinements can be made to increase the concentration of compounds in the ion source. Some current systems can routinely analyze components at the picogram level. Development of a more efficient ionization method could lower detection limits by orders of magnitude. As new methods of data analysis and display are developed, software will be updated, resulting in improved qualitative and quantitative analysis and faster sample throughput. Increased computerization will be a feature of the instrumentation; a separate microprocessor can be used for routine instrument control which will free the main processor for real-time data analysis.

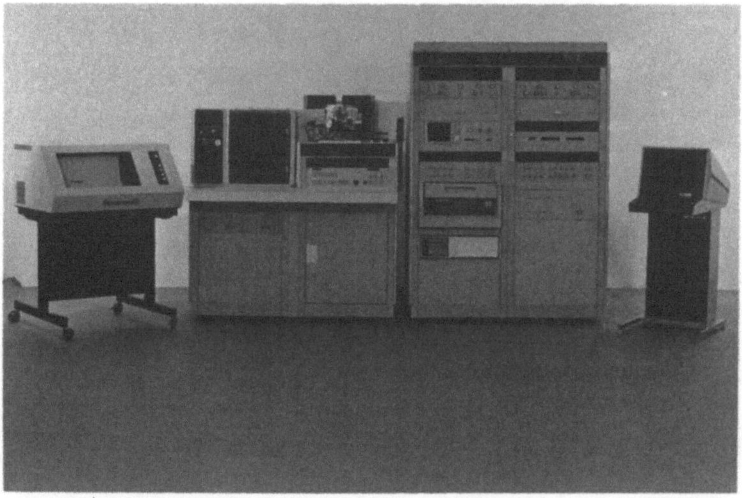

FIGURE 17. GC/MS/MS/COM system with associated peripheral devices. (Courtesy of Finnigan Corp., by permission of Technical Marketing Associates.)

A recent technique which has great potential for environmental analysis is mass spectrometry/mass spectrometery (MS/MS).[35–38] In this technique a sample is ionized and the ions separated by mass dispersion in the first analyzer. Specific ions from the initial spectrum are allowed to enter a reaction region where they are fragmented by a target gas. A subsequent analyzer effects mass dispersion of the ion fragments to produce an MS/MS spectrum. The analyzers can be almost any combination of magnetic, electrostatic, or quadrupole. Since two mass selections are performed, a variety of different types of spectra can be obtained. These include parent spectra, which consist of all parent ions (formed in the source and allowed to pass through the first analyzer) that fragment to a specified daughter ion in the reaction region; daughter spectra, where all daughter ions resulting from a specific parent ion are recorded; and neutral loss spectra, where all parent ions that lose a specified neutral fragment are recorded. The MS/MS can also be operated to obtain conventional mass spectra. With an MS/MS system it is possible to analyze many complex samples for specific compounds without prior chromatographic separation. Figure 17 shows a commercially available GC/MS/MS/COMP system that combines the GC chromatographic separation ability with high selectivity in ion detection by a triple quadrupole MS/MS analyzer. The second quadrupole actually is the reaction region where ions passing the first quadrupole can be made to collide with a target gas. It is operated only in the radio frequency mode

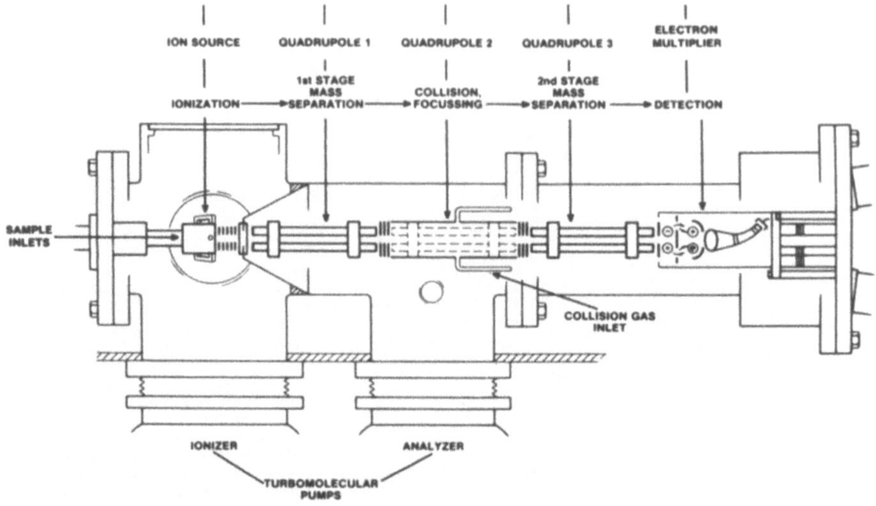


FIGURE 18. Cutaway view of triple quadrupole MS/MS analyzer. (Courtesy of Finnigan Corp., by permission of Technical Marketing Associates.)

and as such it is not used as a mass filter. Figure 18 shows a cutaway view of the triple quadrupole analyzer section. Although currently used primarily as a research tool because of its high initial cost, the GC/MS/MS/COMP technique will enjoy increasing use in the environmental sciences for the rapid specific analysis of ultratrace quantities of compounds in complex mixtures not easily analyzed by conventional GC/MS/COMP.

Although GC/MS/COMP instrumentation has reached a high level of development, potential still exists for considerable improvement. As shown by the applications described in this book, this technique has been extensively applied to environmental analytical problems with a large degree of success. More extensive application of GC/MS/COMP and GC/MS/MS/COMP to difficult environmental problems will closely parallel this rapidly growing need.

## *REFERENCES*

1. W. McFadden, *Techniques of Combined Gas Chromatography/Mass Spectrometry: Applications in Organic Analysis*, Wiley, New York, 1973.
2. B. S. Middleditch, *Practical Mass Spectrometry*, Plenum, New York, 1979.
3. W. A. Aue, C. R. Hastings, and S. Kapila, *J. Chromatogr.* **77,** 299–307 (1973).
4. M. M. Daniewski and W. A. Aue, *J. Chromatogr.* **147,** 119–125 (1978).
5. T. J. Nestrick, L. L. Lamparski, and R. H. Stehl, *Anal. Chem.* **51,** 2273–2281 (1979).
6. R. D. Dandeneau and E. H. Zerenner, *J. High Res. Chromatogr./Chromatogr.* **2,** 351 (1979).

7. G. Schomburg, R. Dielmann, H. Bowvitzky, and H. Husmann, *J. Chromatogr.* **167,** 337 (1978).
8. W. J. Lipsky, W. J. McMurray, M. Hernandez, J. E. Purcell, and K. A. Billeb, *J. Chromatogr. Sci.* **18,** 1–9 (1980).
9. K. Grob and G. Grob, *J. Chromatogr.* **213,** 211–221 (1981).
10. K. Grob, G. Grob, and K. Grob, Jr., *J. Chromatogr.* **211,** 243–246 (1981).
11. K. Grob, Jr., and H. P. Neukom, *J. High Res. Chromatogr./Chromatogr. Commun.* **2,** 15–21 (1979).
12. M. C. Ten Noever De Brauw, *J. Chromatogr. Chromatogr. Rev.* **165,** 207–233 (1979).
13. C. A. Cramers, G. J. Scherpenzeel, and P. A. LeClercq, *J. Chromatogr.* **203,** 207–216 (1981).
14. W. Blum and W. J. Richter, *J. Chromatogr.* **132,** 249–259 (1977).
15. E. M. Chait, *Anal. Chem.* **44,** 77A–91A (1972).
16. H.-D. Beckey and H.-R. Schulten, *Angew. Chem. Int. Ed. Engl.* **14,** 403–415 (1975).
17. R. D. Craig, R. H. Bateman, B. N. Green, and D. S. Millington, *Phil. Trans. Roy. Soc. Lond.* A **293,** 135–155 (1979).
18. H. Budzikiewicz, *Angew. Chem. Int. Ed. Engl.* **20,** 624–637 (1981).
19. R. C. Dougherty, *Anal. Chem.* **53,** 625A–636A (1981).
20. D. F. Hunt, T. M. Harvey, and J. W. Russell, *J.C.S. Chem. Comm.,* 151–152 (1975).
21. J. Roboz, *Introduction to Mass Spectrometry,* Wiley, New York, 1968.
22. K. Feser and W. Kogler, *J. Chromatog. Sci.* **17,** 57–63 (1979).
23. B. Hedfjall and R. Ryhage, *Anal. Chem.* **47,** 666–670 (1975).
24. F. W. Karasek and F. I. Onuska, *Anal. Chem.* **54,** 309A–315A (1982).
25. F. W. Karasek, *Res./Dev.,* **27**(11), 42–46 and 48 (1976).
26. A. M. Greenway and C. F. Simpson, *J. Phys.* **E13,** 1131–1147 (1980).
27. G. A. Eiceman, R. E. Clement, and F. W. Karasek, *Anal. Chem.* **51,** 2343–2350 (1979).
28. H. R. Buser, *Anal. Chem.* **49,** 918–922 (1977).
29. L. A. Shadoff and R. A. Hummel, *Biomed. Mass Spectrom.* **5,** 7 (1978).
30. R. L. Harless, E. O. Oswald, M. K. Wilkinson, A. E. Dupuy, Jr., D. D. McDaniel, and H. Tai, *Anal. Chem.* **52,** 1239–1245 (1980).
31. R. A. Hites and K. Biemann, *Anal. Chem.* **42,** 855–860 (1970).
32. J. E. Biller and K. Bieman, *Anal. Lett.* **7,** 515–528 (1974).
33. J. W. Eichelberger, L. E. Harris, and W. L. Buddle, *Anal. Chem.* **47,** 995–1000 (1975).
34. L. C. Dickson, R. E. Clement, K. R. Betty, and F. W. Karasek, *J. Chromatogr.* **190,** 311–319 (1980).
35. F. W. McLafferty, *Acc. Chem. Res.* **13,** 33–39 (1980).
36. R. G. Cooks and G. L. Glish, *Chem. Engr. News* **59,** 40–52 (1981).
37. J. R. B. Slayback and M. S. Story, *Ind. Res. Dev.* Feb., 128–134 (1981).
38. R. W. Kondrat and R. T. Cooks, *Anal. Chem.* **50** 81a–92a (1978).

3

# FIELD DESORPTION AND FIELD IONIZATION

## H.-R. SCHULTEN and SI-EN SUN

### 1. INTRODUCTION

One important trend in the development of mass spectrometry (MS) during the last ten years has been the search for suitable ionization methods for polar, nonvolatile substances. The need for "soft ionization" has arisen from the difficulties in obtaining clear and complete (molecular ion plus structural details) mass spectra for many biochemically, pharmacologically, and toxicologically relevant molecules.

The conventional analysis procedure by electron impact (EI) ionization transfers a relatively large amount of thermal and electronic excitation energy to the molecules being analyzed. Hence, the electron impact mass spectra of such compounds are characterized by many and varied fragment ions and few (or even no) molecular ions. For the majority of analytical purposes, however, just the opposite is preferred, namely the formation of intense molecular ion species to the exclusion or reduction of fragment ions. Techniques which produce a much higher proportion of molecular ions from samples of biological origin would facilitate the identification and quantification of many environmental substances.

Since chemical ionization (CI) and atmospheric pressure ionization (API) are treated from this viewpoint in other chapters of this book, the principle and techniques of field ionization (FI) and field desorption (FD) are briefly described here. The main emphasis, however, is on real applications of FDMS in environmental research describing the meth-

*H.-R. SCHULTEN* • Fachhochschule Fresenius, Department of Trace Analysis, Dambachtal 20, D-6200 Wiesbaden, Federal Republic of Germany. *SI-EN SUN* • Institute of Environmental Chemistry, Academia Sinica, P.O. Box 934, Peking, China.

odological aspects of organic and inorganic trace analysis of air, food, and water.

## 2. PRINCIPLE AND TECHNIQUES

The theory, physicochemical problems, and some analytical applications of FI and FD have been treated previously in a fundamental monograph by Beckey.[1] FI involves considerably less electronic excitation than conventional EI ionization since the energy transferred is only of the order of a few tenths of an electron volt. If evaporation of the sample is possible (Fig. 1), and at least a fraction of the sample molecules reach the gas phase undecomposed, the FI mass spectra are characterized by reduced fragmentation and high abundance of molecular ions. However, the thermal stress required for evaporation cannot be avoided with this method and may cause complete decomposition of the sample prior to ionization.

Because both FI and CI require evaporation but achieve soft ionization once the neutral molecules are in the gaseous state, these techniques have, in principle, similar potential. In real analytical application, however, CI has clear advantages because the technique is simpler, more sensitive, and of particular utility in combination with chromatographic methods.

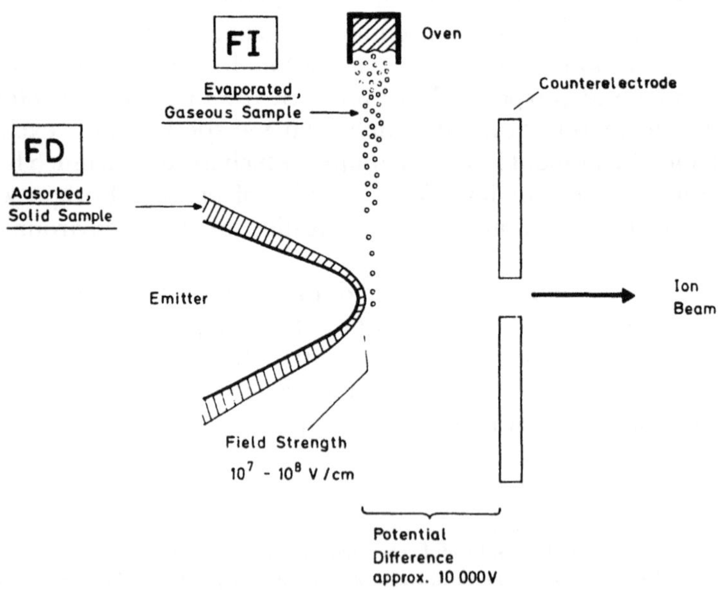


FIGURE 1. The different sample supply in the FI and FD modes, illustrating the definition of field ionization and field desorption.[3]

Field desorption is regarded as the method of choice for the analysis of nonvolatile and thermally labile substances in biochemistry, medicine, and environmental research because FDMS does not require a separate evaporation.[2] FI and FD techniques thus can be distinguished by the manner in which the sample is supplied to the emitter.

Field ionization is the situation where the sample molecules approach the emitter from the gas phase, as shown in Fig. 1. If the sample is adsorbed on the emitter in a solid or liquid layer from solution, suspension, aerosol, or electrolytic deposit and is then ionized in the adsorbed state, the process is termed field desorption.

The special features of FDMS, such as emitter activation, sample handling, and the registration of low and high mass resolution spectra, have been outlined in a comprehensive review article.[2] With respect to the transfer of the sample to the FD emitters mainly two different techniques are employed.

For qualitative analysis, and if enough sample material is available to prepare a 2–3 μl solution of relatively high concentration (≧ μg/μl), use of the dipping technique is a simple and rapid way to coat the FD emitter. The resulting FD spectra, however, give only a rough estimate of the relative abundances and types of ion which can be expected.

For quantitative investigations with FDMS the syringe technique is preferred. This method allows a known amount of sample to be deposited on the emitter surface. A convenient way to transfer the sample droplets to the emitter is to control the deposition of the solution by means of a stereomicroscope at an approximately 100-fold magnification. Figure 2 shows application of the sample from a 10-μl syringe mounted on a small manipulator for careful and accurately adjustable movement in all three dimensions. Thus, possible damage to the FD

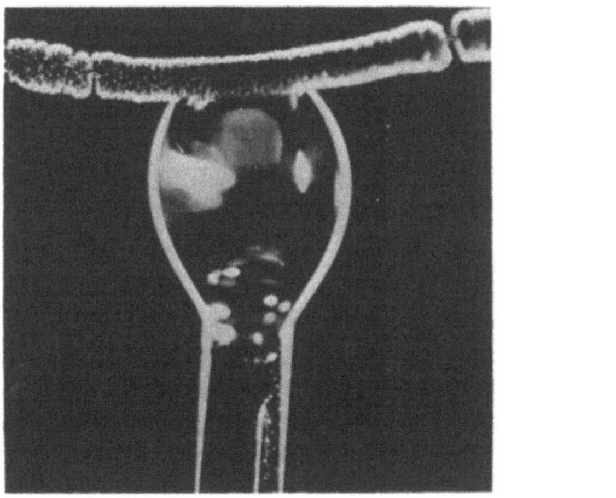

FIGURE 2. The dissolved sample is supplied to the FD emitter with a microliter syringe. After evaporation of the solvent, which can be accelerated by a warm-air blower, the sample molecules are adsorbed in the center of the emitter wire in a reliable, fast, and reproducible manner.[4]

emitter is avoided and the substance can be directed to the central part of the activated emitter portion, preferably to those tips of the dendritic microneedles, which in the mass spectrometer point in the direction of the entrance slit. Further, control of the emitters before and after FD measurements can, with some experience, give a judgment on the quality of the emitter. This is particularly valid if the FD emitter is used repetitively.

Two essential informations are obtained using the syringe technique:

(i) The sensitivity of the instrument can be determined for a standard substance in C/μg. This figure is absolutely critical as the adjustment and performance of the FD mass spectrometer (newly installed or used for a longer period of time in the FD mode) can be checked. Values between $10^{-11}$ and $10^{-12}$ C/μg should be obtained using, for instance, a freshly prepared solution of adenosine. Users of commercial instruments are strongly advised to have the supplier produce these data "on the spot."

(ii) The detection limit for a substance of interest can be evaluated by loading decreasing amounts of sample onto the emitter in

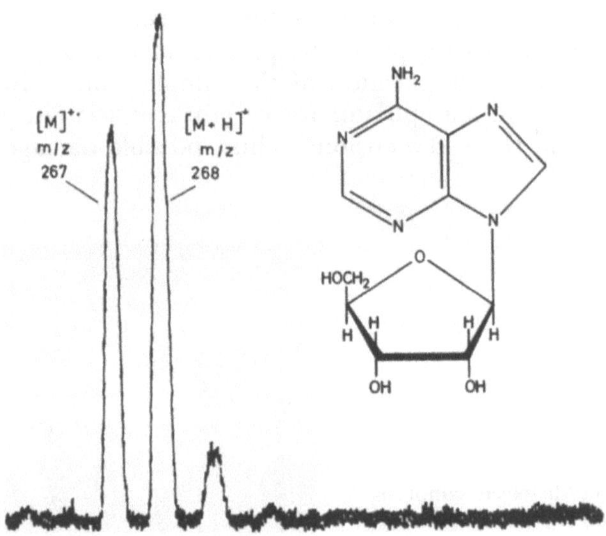


FIGURE 3. Evaluation of the detection limit on a Varian MAT 731 following installation of the instrument (November 1977). Adenosine (3 ng) was applied by microsyringe; multiplier voltage, −3 kV; ion source potentials, +8/−1 kV; 85 repeated magnetic scans were accumulated in a multichannel analyzer; the plotter range is 25 mV/cm.

consecutive measurements. In Fig. 3 the original plot which is obtained by coating 3 ng of adenosine from a standard solution to the FD emitter is displayed.

Essentially the entrance fee one has to pay for joining the "FD club" is the patience and skill to prepare the first satisfying FD emitter. As described above, the sensitivity test will be the decisive measure to indicate whether all experimental and instrumental parameters are optimized. Once this control is achieved, the main pitfalls for failure of FD experiments may be caused by the following two reasons.

As shown schematically in Fig. 4, vapor bubbles, mostly from bound solvent or water generated in the FD experiment by raising emitter temperature, may cause an eruptive and sudden loss of sample. Often a discharge between the emitter and the counterelectrode is observed simultaneously. Reducing the thickness of the sample layer, longer pump-down before the high voltage is applied, and, if possible, exchange of the solvent water against acetone, chloroform, methanol, or acetonitrile are alternatives that eliminate this effect.

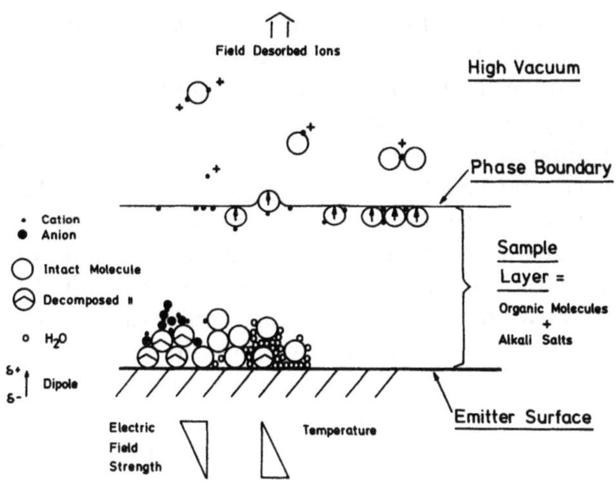


FIGURE 4. Schematic drawing of a sample layer composed of an organic, thermally labile substance and alkali salts. The metal cations migrate in the molten layer alone or attached to organic molecules (solvent or substance) according to the strong field gradient. Since the highest temperature in the sample layer is produced close to the emitter surface, at this site gas elimination may lead to sputtering of the sample material. This is then often accompanied by discharges and the result is loss of the whole sample. The second so-called "salt effect" in FDMS is visualized in the left part of the scheme. The signals of inorganic cations are observed with high intensities; the organic molecules, however, decompose before desorption. Sometimes only stable, pyrolytic fragments are detected, such as water, methylfuran, and clusters of phosphoric acids in the case of oligonucleotides.

Alkali salts may be very beneficial in FDMS if cationization is performed, and thus energetically favored ion formation that leads to $(M + Na)^+$, $(M + 2Na)^{2+}$, and $(2M + Na)^+$ ions is achieved for the cation $Na^+$. In routine FD analysis of environmental samples it will be an exception that inorganic salts have to be added to the genuine samples. If too much salt is present, the organic molecules are encircled by the salt lattice, and when the inorganic part of the sample is desorbed, the required emitter temperatures are so high that they lead to decomposition of the organic molecules. One solution to this problem is the use of high-pressure liquid chromatography prior to the FD analysis as will be discussed later.

## 3. *APPLICATION TO ENVIRONMENTAL ANALYSIS*

In general, the problems encountered in environmental analytical research are particularly difficult since almost any substance may be present in the samples, and concentrations of individual compounds are usually very small. Generally, the substance to be investigated forms a constituent of a very complex mixture. Therefore, identification and quantification requires a sensitive, selective, and reliable instrumental analytical procedure. In recent years mass spectrometry has proved to be an outstanding, successful tool in qualitative identification and quantitative determination of environmental chemicals because of its sensitivity, specificity, and reliable, rapid analytical capability. However, the most severe limitation to the broad application of electron impact MS in environmental studies is its requirement that compounds to be analyzed be convertible into free gaseous ions. As a result, EI analysis is impossible for substances of low volatility which cannot be transferred into the gas phase without thermal decomposition.

The field ionization and field desorption as soft ionization methods in MS are the methods of choice for the analysis of low-volatile, highly polar, and/or thermolabile compounds, and therefore have found a wide area of application in environmental research. In particular, FDMS appears well suited for environmental samples such as aerosols in air, mycotoxins and toxic metals in food, and body fluids and ultratraces of metals in environmental samples such as surface and drinking water.

### 3.1. *Air*

Schulten investigated artificial aerosols from the ozonolysis of 1-butene by high-resolution FDMS using a new direct sampling method for aerosols.[5] Ozonolysis of olefins in the gas phase produces an aerosol

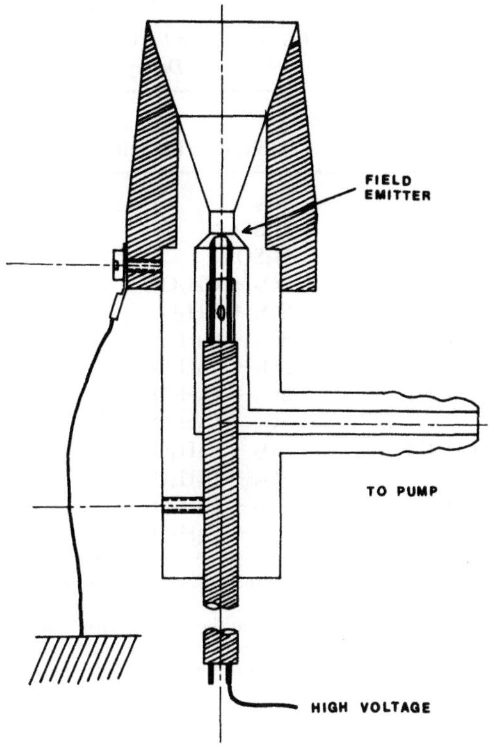


FIGURE 5. Device for the direct impaction of aerosols on the emitter wire of an FD ion source.[5]

when the reactant concentrations are high enough to cause supersaturation of the products (Fig. 5). The condensable products and aerosols generated by ozonolysis of 1-butene at low reactant concentrations in air and in the presence of sulfur dioxide have been identified. A great number of molecular ions up to *m/z* 392 were recorded. The elemental composition of these ions was established up to *m/z* 200 and included several sulfur-containing species (see Table 1).

Aldehydes, organic acids, sulfuric acid, and some probable precursors have been identified. The observed mixtures of sulfur compounds support the assumption that $SO_2$ is oxidized by zwitter-ions in this system. The formation of hydrogen peroxide reported by Atkinson has been confirmed, but only at low pressure.[6] The results show that FDMS offers considerable promise for the analysis of particulate matter, especially for compounds of high polarity suspended in air.

### *3.2. Natural Aerosols*

The chemical composition of aerosols varies over a wide range depending upon the origin of the particulates. Practically any element and a vast number of inorganic compounds as well as many organic com-

*TABLE 1*
*Analysis of Aerosols from the Ozonolysis of 1-Butene by High-Resolution Field Desorption Mass Spectometry*[c]

| Precision mass | Molecular formula | Densitometer reading[a] | Densitometer reading[b] |
|---|---|---|---|
| 63.962 | $SO_2$ | Weak | |
| 97.967 | $H_2SO_4$ | 40 | |
| 98.975 | $H_3SO_4$ | 32 | |
| 110.975 | $SO_2 + CH_3O_2$ | 12 | 60 |
| | $SO_3 + CH_3O$ | | 62 |
| 124.991 | $SO_2 + C_2H_5O_2$ | 8 | 35 |
| | $SO_3 + C_2H_5O$ | | 55 |
| 155.002 | $SO_2 + C_3H_7O_3$ | 10 | Not observed |
| | $SO_3 + C_3H_7O_2$ | | 40 |
| 155.973 | $SO_2 + C_2H_4O_4$ | 15 | 22 |
| | $SO_3 + C_2H_4O_3$ | | 50 |
| 156.981 | $SO_2 + C_2H_5O_4$ | 20 | 8 |
| | $SO_3 + C_2H_5O_3$ | | 35 |

[a] Sulfur-containing ions in the mass spectrum of products generated in the reaction of 1-butene, ozone, and sulfur dioxide.
[b] The organic ions detected in the mass spectrum of the same sample are given for comparison.
[c] Experimental conditions: reaction time, 20 min; flow rates 1-butene 7.6, ozone 12, and sulfur dioxide 0.7 (ml/min). The reaction products were precipitated directly on the emitter surface; dry synthetic air was used as diluent gas.

pounds can be found in aerosols. Analysis of such aerosols by the standard wet chemical method requires large amounts of material which must be collected with high-volume samplers. Modern techniques like neutron activation analysis,[7] x-ray fluorescence,[8] and photoelectron spectroscopy[9] are important tools for the detection of elements in aerosols and provide information on the chemical state of these elements. High-resolution MS has also been adapted to the problem of aerosol analysis.[10–12] The collected aerosol is transferred into the ion source region of the mass spectrometer and heat evaporated. Evaporation and ionization of polar compounds, however, cause extensive thermal and mass spectrometric fragmentation making computer techniques indispensable to the interpretation of the complex spectra.

Analyses of inorganic and organic natural aerosols by high-resolution FDMS have been reported.[13] The most direct way of analyzing aerosol particles for FD analysis is by direct impaction from the gas phase onto an activated emitter. This can be achieved by exposing the emitter to an air blast focused by a jet collector similar to that of a conventional impactor (Fig. 5).

The diameter of the oblong air stream focused on the emitter wire corresponded to roughly 10 times its projected area. Therefore, less than 10% of the aerosol drawn through the jet is impacted on and can

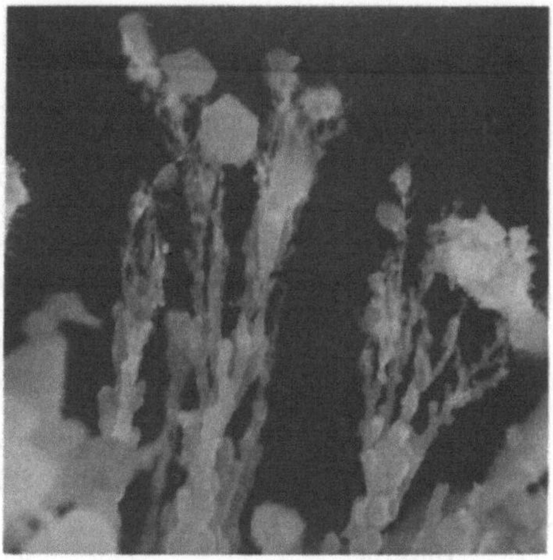

FIGURE 6. Scanning electron micrograph of an FD emitter after impaction of a natural aerosol from topside. Single, crystalline inorganic (left part) and organic (right part) aerosol particles are seen to be caught in the branched structure of the carbonaceous microneedles.(13)

be at best retained by dendritic needles of the emitter. Flow rates of 3–6 l/min have been found to be adequate for sampling a natural aerosol within a few hours. The outcome of this sampling process is shown in Fig. 6 for one characteristic measurement.

The results of the FD mass spectrum of an aerosol which consisted mainly of inorganic substances are shown in Table 2. Inspection of this table shows that $(Na)^+$, $(K)^+$, $(NH_4)^+$, and a trace of $(Rb)^+$ were present as cations in the aerosol. Rubidium is detected as a natural trace element. Since the registration of the FD ions was performed by a photoplate, the densitometer readings give only approximate information on the relative abundances. However, the ammonium compounds may be underestimated because of some evaporation of their neutral compounds, especially at higher temperature.

More quantitative results can be obtained with an appropriate calibration technique. The only anions detected in this aerosol are $(NO_3)^-$, $(SO_4)^-$, $(HSO_4)^-$, and $(Cl)^-$. The predominance of the nitrate anion is noteworthy, stressing the importance of aerosols as a sink for $NO_2$ in the atmosphere. $NO_2$ has been shown to react rapidly with NaCl to form $NaNO_3$ and NOCl as an intermediate of presumably short lifetime in the atmosphere. The product $NaNO_3$ is the most abundant observed in the aerosol except for some unreacted NaCl. For $NO(H_3O)^+$; a signal is observed, which for $(H_2O)^{+\cdot}$ is quite weak due to the fact that the aerosol is dried out by pumping to $10^{-5}$ torr in the ion source. The ion $(H_3SO_4)^+$ for free sulfuric acid is absent in the spectrum. However, this ion has been previously observed in the study of an artificial aerosol (see Table 1) and was also observed by Schuetzle using computer-assisted high-resolution MS.[14] This excludes substantial amounts of free sulfuric acid in the precipitated aerosol, probably as a result of rapid reaction

*TABLE 2*
*Field Desorption Mass Spectrum of a Natural Aerosol Collected December 1974*[a]

| Formula of ion | Densitometer reading | Accurate mass |
|---|---|---|
| $H_2O^+$ | 2.8 | 18.001 |
| $(NH_4)_2SO_4H^+$ | 3.5 | 133.028 |
| i | vw | 135.024 |
| $NH_4HSO_4H^+$ | 2.8 | 116.002 |
| $NH_4^+$ | 2.0 | 18.034 |
| $Na^+$ | S | 22.990 |
| $NaClNa^+$ | 5.5 | 80.948 |
| i | 3.8 | 82.945 |
| $(NaCl)_2Na^+$ | w | 138.907 |
| $NaNO_3Na^+$ | S | 107.967 |
| i | 3 | 108.946 |
| i | 3 | 109.972 |
| $Na_2(NO_3)Na^+$ | 5 | 192.972 |
| i | vw | 193.942 |
| i | vw | 194.949 |
| $Na_3(NO_3)Na^+$ | 3 | 277.923 |
| $Na_4(NO_3)_4Na^+$ | vw | 362.900 |
| $Na_2ClNO_3Na^+$ | 3.5 | 165.926 |
| i | 2 | 167.923 |
| $Na_3Cl(NO_3)Na^+$ | 1 | 250.904 |
| $Na_2SO_4Na^+$ | 2.2 | 164.921 |
| $Na_2SO_4(Na^+)_2$ | 1.5 | 2 × 93.955 |
| $Na_3SO_4NO_3Na^+$ | 2 | 249.899 |
| $Na_3SO_4NO_3(Na^+)_2$ | 1.5 | 2 × 136.444 |
| $K^+$ | S | 38.963 |
| i | S | 40.962 |
| $KNO_3K^+$ | 2.5 | 139.915 |
| i | vw | 141.913 |
| $NaNO_3K^+$ | 5 | 123.941 |
| i | 3 | 125.939 |
| $Na_2(NO_3)K^+$ | 3 | 208.919 |
| i | vw | 210.917 |
| $Na_3(NO_3)K^+$ | vw | 293.897 |
| $NaK(NO_3)_2K^+$ | vw | 224.893 |
| $NaClK^+$ | w | 84.912 |
| $Rb^+$ | w | 86.992 |

[a] The identified ions are arranged in groups of similar chemical nature, in the order of decreasing densitometer readings (w = weak; vw = very weak; S = saturation; i = isotopic ion).

with chlorides and nitrates during the process of sample collection. The ion $(NH_4HSO_4 + H)^+$ indicates some residual acidity in the aerosol. Moreover, the FD spectrum of a natural aerosol which was collected in acetone and transferred in suspension to the emitter shows a mixture of $C_{26}$–$C_{34}$ aliphatic hydrocarbons.

In a first pilot study FDMS has been applied to quantitative determination of polycyclic aromatic hydrocarbons from model mixtures which were constituted of two to three standards.[15] These substances are present mainly in the aerosol and some of these are strongly carcinogenic. The molecular ions of the constituents of a polycyclic hydrocarbon mixture were recorded on the photographic plate at high-resolution FDMS. The blackening intensities obtained were compared with relative concentrations of the employed compounds of the mixture. Samples in the microgram range yielded deviations of about $\pm 5\%$.

### 3.3. *Food and Body Fluids*

*3.3.1. Mycotoxins.* Toxic metabolites from microorganisms in molding food, the mycotoxins, represent a broad spectrum of biologically active substances. Especially aflatoxins, which consist of eight compounds of related molecular configuration—of $B_1$, $B_2$, $B_{2a}$, $G_1$, $G_2$, $G_{2a}$, $M_1$, and $M_2$—are now well recognized for their high carcinogenic activity.

Although a number of analytical methods for the determination of mycotoxins has been reported, still no completely satisfactory method or technique is presently available.[16–20]

Sphon *et al.* have demonstrated that model mixtures of mycotoxins in the range of 10–100 ng/compound can be analyzed directly by FDMS because this technique is very suitable for the analysis of mixtures.[21] Also, the FD mass spectra of a variety of single mycotoxins have been studied. These include the four aflatoxins $B_1$, $B_2$, $C_1$, and $C_2$, rubratoxin B, T-2 toxin, and zearalenone, as well as mixtures of these mycotoxins. The spectra of all of the mycotoxins exhibited intense molecular ions $(M)^{+\cdot}$ under the conditions employed, only rubratoxin B exhibited major fragmentation. In contrast, the EI mass spectrum of T-2 toxin shows strong fragmentation with a very small peak at *m/z* 466 corresponding to the molecular weight of the compound. The FD spectrum which shows no fragmentation and the molecular ion as the base peak shows again the capacity of FDMS for mixture analysis. For rubratoxin B, as previously reported, the highest *m/z* value observed in the EI spectrum is an ion of small abundance at *m/z* 474, corresponding to the loss of carbon dioxide from the molecular ion.[22] However, the *m/z* 474 ion is the base peak in the FD spectrum, and a cluster of ions of small and varying abundance is observed in the region of the molecular weight. Because of the high abundance of the *m/z* 474 ion, FD is the more reliable and sensitive method for the characterization of rubratoxin.

Mixtures of the standards were prepared and examined using FD, showing no loss in sensitivity as compared to that of the single compounds. A mass chromatogram of a mixture of all seven mycotoxins shows the intensities of characteristic ions of the mycotoxins as a function of emitter heating. Further, extracts of foodstuffs were examined by using the dipping technique because of the high viscosity of the extract solutions.

The FD spectrum of a sample naturally contaminated with 15 ppb aflatoxin $B_1$ and 2.2 ppb aflatoxin $B_2$ was obtained after extraction using Method 1 of *Official Methods of Analysis*.[23] The spectrum shows positive responses at *m/z* 312 for the molecular ion of $B_1$ and *m/z* 314 for the molecular ion of $B_2$. In addition, the lack of response at *m/z* 318, 328, and 330 indicates that the sample did not contain a detectable amount of zearalenone, aflatoxin $G_1$, or aflatoxin $G_2$.

An extract of tomatoes naturally contaminated with an unknown quantity of zearalenone was examined by FD. A clear response was noted at *m/z* 318 corresponding to the molecular ion of zearalenone. The lack of response at *m/z* 328 again indicated an undetectable level of aflatoxin $G_1$. The positive responses observed at *m/z* 312, 314, and 330, however, indicated that the sample could contain aflatoxin $B_1$, $B_2$, and $G_2$ and should therefore receive further examination. The potential of EIMS is much lower than for FDMS because extensive fragmentation of any mycotoxins present in extracts is observed. However, it should be noted that FDMS as a screening technique for mycotoxins is not routine at present and some questions remain to be answered before the feasibility of screening can be accurately assessed.

It is clear, however, that the FD method is a suitable tool for a wide variety of chemically different mycotoxins and that the high molecular ion abundances and small fragments particularly favor screening techniques. In Fig. 7 the electrically recorded FD mass spectrum of satratoxin H is shown. This substance gives an extremely weak molecular ion in EIMS which is below 3% relative abundance.[24] The base peak in spectra using this ionization mode is obtained at *m/z* 44 and, as all other signals are below 10%, this technique gives no chance for identification. In contrast, the FD mass spectrum gives the $(M)^{+\cdot}$ ion as base peak and this signal contributes more than 16% to the total ion current. Since this spectrum was recorded at a mass resolution of more than 10,000, the accurate mass determination revealed the ion at *m/z* 510 as due to loss of water and at *m/z* 482 for elimination of $CO_2$. Thus, it appears that the FD method should be very useful for mycotoxin screening, if, and this may be a crucial point, optimal instrumentation and an appropriate direct cleanup step for the food material is worked out.

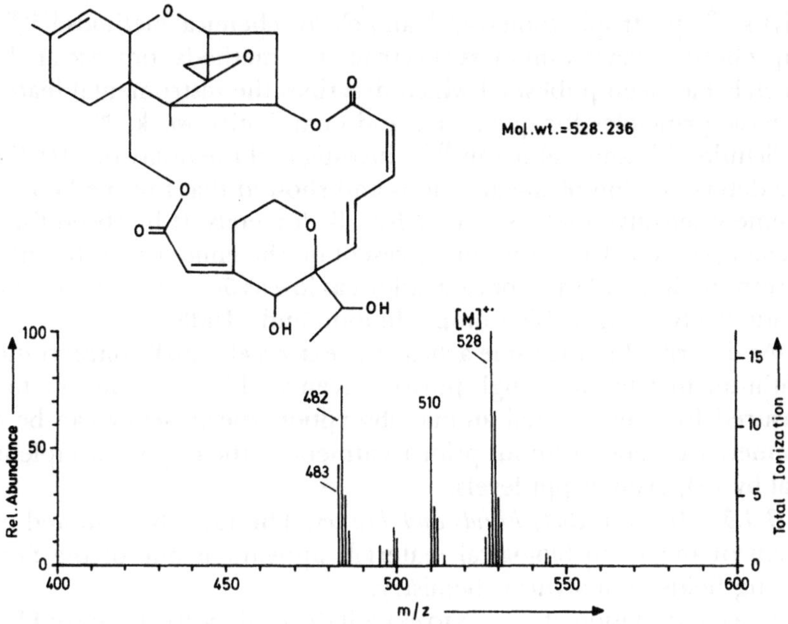


FIGURE 7. FD mass spectrum of satratoxin H, electrical registration on a Varian MAT 731 instrument; emitter heating current, 28 mA.

Recent applications of FD in food analysis include the mass spectral detection of scirpene mycotoxins.[25] Further, FD mass spectrometry has been applied to the identification of amines and amino acids as their fluorescamine derivatives.[26] Reaction with fluorescamine produces single derivatives of a series of mono- and diamines as well as amino acids. Under FD conditions these derivatives exhibit strong molecular ions and little or no fragmentation. The increase in molecular weight of the product largely avoids interferences from chemical noise. Therefore, FDMS is recommended as a specific, sensitive, rapid, and versatile method for the detection and monitoring of amines and amino acids in crude biological extracts. This holds for applications before or following chromatographic separation.

Cyanogenic glycosides can be the source of hydrogen cyanide in foods and when consumed by humans can result in acute or chronic poisoning. The FD mass spectra of several underivatized cyanogenic glycosides recently have been studied.[27] In this work the ability to successfully determine the presence of cyanogenic glycosides by FD has been demonstrated in crude extracts isolated from *Vicia sativa*, a food-contaminating plant material.

*3.3.2. Metals in Food.* A wide range of instrumental analytical techniques has been applied to the analysis of metals including atomic absorption spectroscopy,[28] emission spectroscopy,[29] neutron activation analysis,[30] spectrophotometry,[31] and electrochemical methods.[32–35] A comprehensive review on mass spectrometric methods for trace analysis of metals has been published which describes the instrumental features and basic principles for qualitative and quantitative work.[69]

Schulten[36] and Lehmann[37,38] investigated the utility of FDMS for trace determination of metal cations and showed that this method is an extremely sensitive tool, especially for alkali metals. It has been found, for example, that 1 pg of lithium present on the emitter is sufficient for a determination and that concentrations as low as 50 ng/l can be analyzed quantitatively using stable isotope dilution and FDMS.

As described in the same article, the extremely small concentrations of lithium in wine and high-purity solvents which are too low to be measured by conventional atomic absorption spectroscopy can be determined accurately without prior treatment of the original sample material by FD, even at ppt levels.

*3.3.3. Metals in Body Fluids and Tissues.* The identification and estimation of metals in biological materials appears as one of the fastest growing fields in analytical chemistry.

In a continuation of efforts to exploit the analytical capacity of FDMS as an additional, independent tool for metal cation analysis, this novel method was applied to a number of elements from group IIa, IIb, and IIIb in the periodic system, such as calcium, barium, cadmium, and thallium.[36]

Isotope dilution in combination with spectra accumulation and averaging in a multichannel analyzer is the best means for reliable quantification in FDMS. The first application in the determination of a trace metal in human body fluids has been reported for lithium. The use of FDMS for the determination of lithium in these samples at therapeutic levels—ppm region—as well as at the normal level—ppb region—has been developed.[39] The use of a stable-isotope-enriched standard, together with the outstanding sensitivity of FD for alkali cations and the high specificity of mass spectrometry, allows a quantitative determination of lithium in microliter amounts of plasma, saliva, and urine without pretreatment.

A newly developed technique using electrolytic deposition of trace metals such as chromium, manganese, iron, nickel, copper, gallium, silver, gold, thallium, and uranium on the emitter surface from a complex, biological matrix showed improvements in sensitivity by a factor of approximately 20.[40] As an example the detection and estimation of the nickel isotopes from a biological sample is shown in Fig. 8.

For the determination of thallium cations FDMS appears to be a reliable, sensitive, and fast method. A direct estimation of thallium in homogenized tissue samples is feasible without any pretreatment. This has been reported for the determination of thallium in brain tissue of

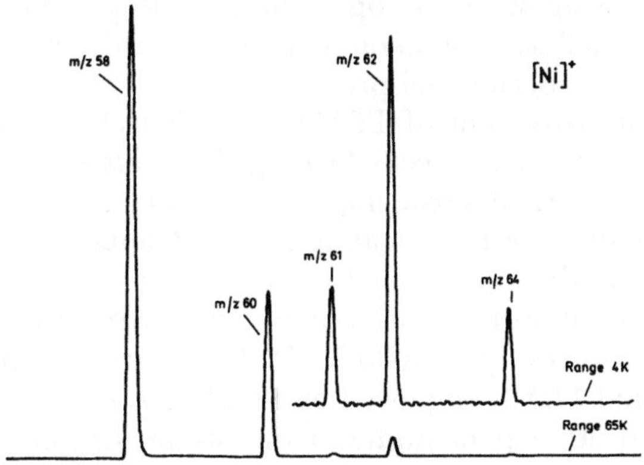


FIGURE 8. Detection and estimation of the nickel isotopes from a tissue sample of a test animal after electrolytic enrichment on the emitter directly from the tissue homogenizate. The deviation of the found isotopic abundances from the natural abundances is less than a few tenths of a percent.

an acutely poisoned test animal[41] and the pharmacokinetics of thallium in mice.[42,43] In a recent quantitative investigation FDMS was used in the studies on placental transfer of thallium to fetal tissue.[44,73]

In summarizing the results of FDMS in trace and ultratrace analyses of metals in biological samples the main advantages are the reliable

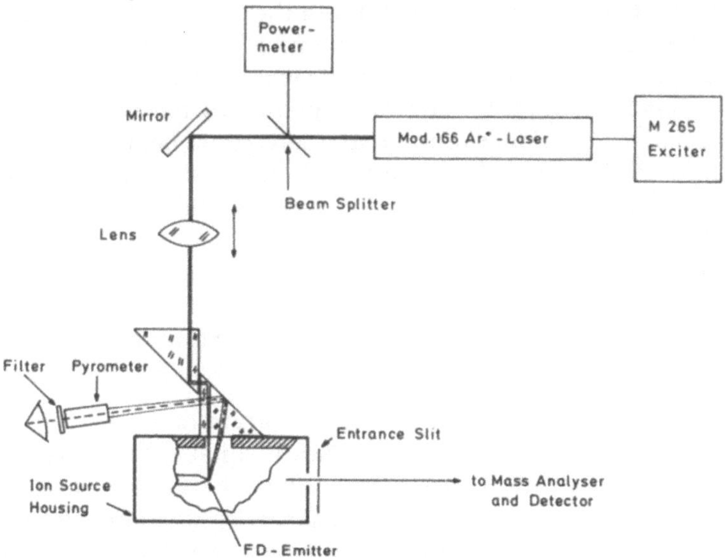


FIGURE 9. Instrumental arrangement for laser-assisted FDMS. Desorption of the sample and observation of the sample temperature between 850 °C and approximately 3000 °C is achieved simultaneously.[46]

quantification using stable isotope dilution, the possibility to estimate levels in very small sample amounts (mg and less), and the speed and simplicity of the experimental procedure.

For the improvement of FDMS as analytical procedure of trace metals, the method of indirectly heating the emitter with a laser introduced into FDMS is of great importance, since from theoretical and practical considerations laser heating offers advantages over direct heating by increasing the ionization efficiency and thus the sensitivity; facilitating high-resolution FD measurement; and giving access to very high emitter temperatures up to 3000 °C.[45] The instrumental arrangement for laser-assisted FDMS employed by Schulten is shown in Fig. 9.[46–48]

These and other pilot studies demonstrate convincingly the utility and versatility of laser-assisted FDMS for the detection and identification of metals, accompanying trace metals and organic/inorganic trace impurities. The total sample amount required for analysis is some micrograms. but only between $10^{-12}$ and $10^{-9}$ g are actually desorbed depending mainly upon the laser adjustment and the kind of metal which is under investigation.

### 3.4. *Water*

*3.4.1. Pesticides.* In recent years MS as such, or in combination with chromatographic techniques, has successfully been used in the qualitative and quantitative analysis of pesticides in the environment. Damico was able to demonstrate in a first comparison of EI and FI mass spectra of a number of pesticides the characteristic features of both techniques for environmental analysis.[49] The main point was to show that a much higher relative abundance of molecular ions could be detected in the FI spectra. With the rapid development of chemical ionization, however, and in particular its high potential in GC/MS studies of pesticides, the use of FIMS for this purpose has declined.

The utility of the FDMS for the analysis of pesticides of low volatility has been demonstrated for a wide variety of chemical classes of compounds. The basic advantages of the FD technique in this field are as follows:

(i) detection sensitivity is in the nanogram range;
(ii) selectivity of the method makes it suitable to analysis of prepurified, crude extracts, and relatively complex mixtures;
(iii) technique reduces strongly thermal degradation and can therefore be used to identify and detect pesticide metabolites and other polar compounds; and
(iv) it is applicable to qualitative and quantitative investigations.

The high molecular ion intensities and the simple, characteristic fragmentation pattern are of particular importance in the identification of unknown substances.

*3.4.1.1. Organochlorine Pesticides.* The first application of FDMS to organochlorine pesticides showed that FD spectra of these substances exhibited high molecular ion intensities and little or no fragmentation when compared with results from other ionization methods.[51,52] In general, they showed that FDMS can be more valuable in the positive identification of pesticidal chemicals than the conventionally applied EIMS.

Figure 10a shows the FI mass spectrum of endrin recorded on a singly focusing instrument. It should be noted that in comparison the EI spectra show the molecular ion at *m/z* 378 with about 1% relative abundance. In the FI mass spectra of a series of chlorinated polycyclic hydrocarbons it was seen that the molecular ions produced by these

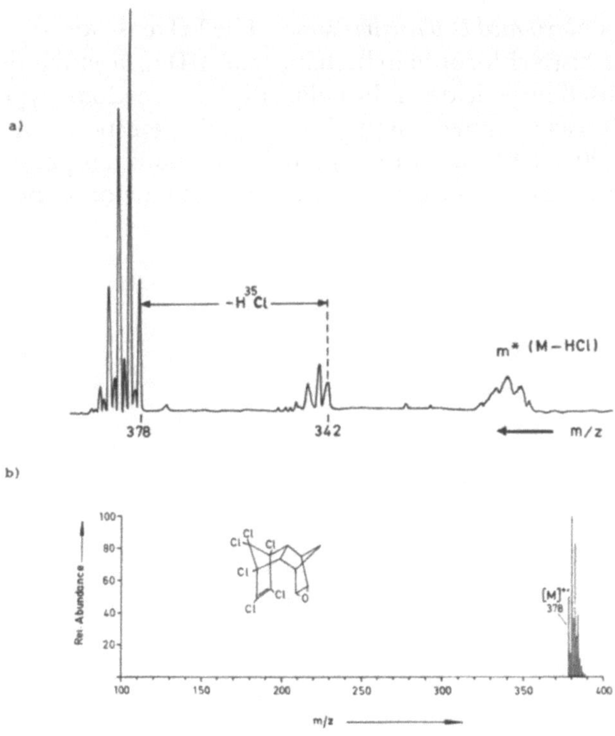


FIGURE 10. (a) FI mass spectrum of endrin, section of the mass range between *m/z* 300 and *m/z* 390, sample temperature 30 °C[49]; (b) FD mass spectrum of endrin, 0 mA emitter heating current, ~30 °C.[50] Both spectra were recorded electrically under low mass resolution conditions.

substances make up the prevailing share of the ion current.[53] Endrin is sufficiently volatile for a FI mass spectrum to be obtained at approximately 30 °C, the signal at *m/z* 342 is observed with about 10% relative abundance. The corresponding mass lines for the metastable ions provide evidence for the loss of hydrochloric acid from the molecule. Figure 10b shows an FD mass spectrum recorded at room temperature with electrical detection and low resolution obtained with a modified Varian MAT-CH4 mass spectrometer.[50] It consists exclusively of the molecular ion group, no fragments being detectable. Under the above-specified conditions, FD spectra can be obtained for other pesticides of the dimethanonaphthalene type such as dieldrin, aldrin, and isodrin in which only the molecular ions appear.[54] Similar FD mass spectra of insecticides of the dimethanoindene type such as chlordene, heptachlor, and endosulfan have also been investigated. With the same experimental conditions as the FD measurements of the dimethanonaphthalene type, the typical picture of no fragments in the FD spectra and of intensive fragmentation with EI has been obtained.

*3.4.1.2. Chlorinated Diphenylmethanes.* The FD mass spectrum of 1,1,1-trichloro-2,2-bis(*p*-chlorophenyl)ethane (*p,p'*-DDT), a widely distributed, extensively used insecticide of the chlorodiphenylmethane type, is shown in Fig. 11. It was obtained using photographic detection at a resolution of about 15,000, 10% valley definition.[55] At room temperature an FD mass spectrum can be recorded displaying virtually only the molecular

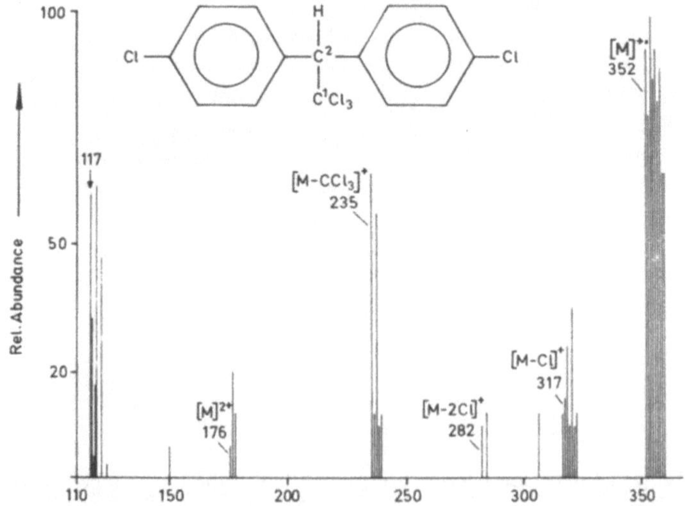


**FIGURE 11.** FD mass spectrum of *pp'*-DDT; emitter heating current, 5 mA; solvent acetone; photographic detection.[55]

ion. In order to induce thermally some fragmentation, the spectrum shown in Fig. 11 was recorded at a raised emitter temperature corresponding to 5 mA emitter heating current.

In the EI spectrum the $M^{+\cdot}$ ion has a relative abundance below 5% in regard to the signals at *m/z* 317 and *m/z* 282 $[M - Cl]^+$ $[M - 2Cl]^+$.[56] The loss of one or two chlorine atoms from the DDT molecule is shown clearly in FD, but the direct bond fission between C(1) and C(2) is still more marked at *m/z* 235 (base peak in the EI spectrum). Since a relatively high peak is also observed at *m/z* 117 for $CCl_3^+$, a typical feature of many FD spectra becomes recognizable, namely the frequent fission of the investigated molecule into two complementary charged particles. This direct bond cleavage appreciably simplifies the interpretation of spectra. In addition, doubly charged ions such as $M^{2+}$ at *m/z* 117.5 for the $(M - CCl_3)^{2+}$ ion have been observed in the spectrum. The occurrence of these multiply charged ions appears to be a characteristic feature of FD and is of special interest with respect to their formation in the high electric field on the emitter surface.

*3.4.1.3. Organophosphorus Pesticides.* The potentialities of the FD method become apparent when the EI, CI, FI, and FD spectra of some pesticides such as delnav, phosphamidon, and temik are compared.[57] These spectra of delnav are shown in Fig. 12. It will be seen that with all ionization methods a signal is recorded in the high mass region at *m/z* 270 (271 for CI) for the process $[M\text{-}(EtO)PS_2H]^+$. The molecular ion is shown only in the FD mass spectrum and as the base peak.

The characteristic behavior of four groups of commonly used organophosphorus pesticides such as phosphates, phosphorothionates, phosphorothiolates, and phosphorodithioates recently have been investigated by FDMS.[74] Their spectra show the high molecular ion abundances and characteristic fragmentation pattern when the emitter heating current is raised. A characteristic feature of many FD spectra produced by these class of compounds shows again the splitting of the molecule into two complementary, charged particles. The one-step processes can be more easily recognized than the complicated rearrangement and consecutive decomposition mechanisms of other ionization methods.[58]

In order to demonstrate the characteristic patterns in the fragmentation of the standards, the FD spectrum of ronnel obtained at an emitter heating current mA was taken as example. As compared with its EI mass spectrum, where the molecular ion was not observed, an intensive molecular ion group (base peak) was recorded with FD.[59]

Organophosphorus pesticides such as dimethoate, trichlorphon, as well as dichlorvos are widely used in China for the protection of agricultural products. In a first application of FDMS to these pesticides the extracts of river water were investigated in collaboration with the Insti-

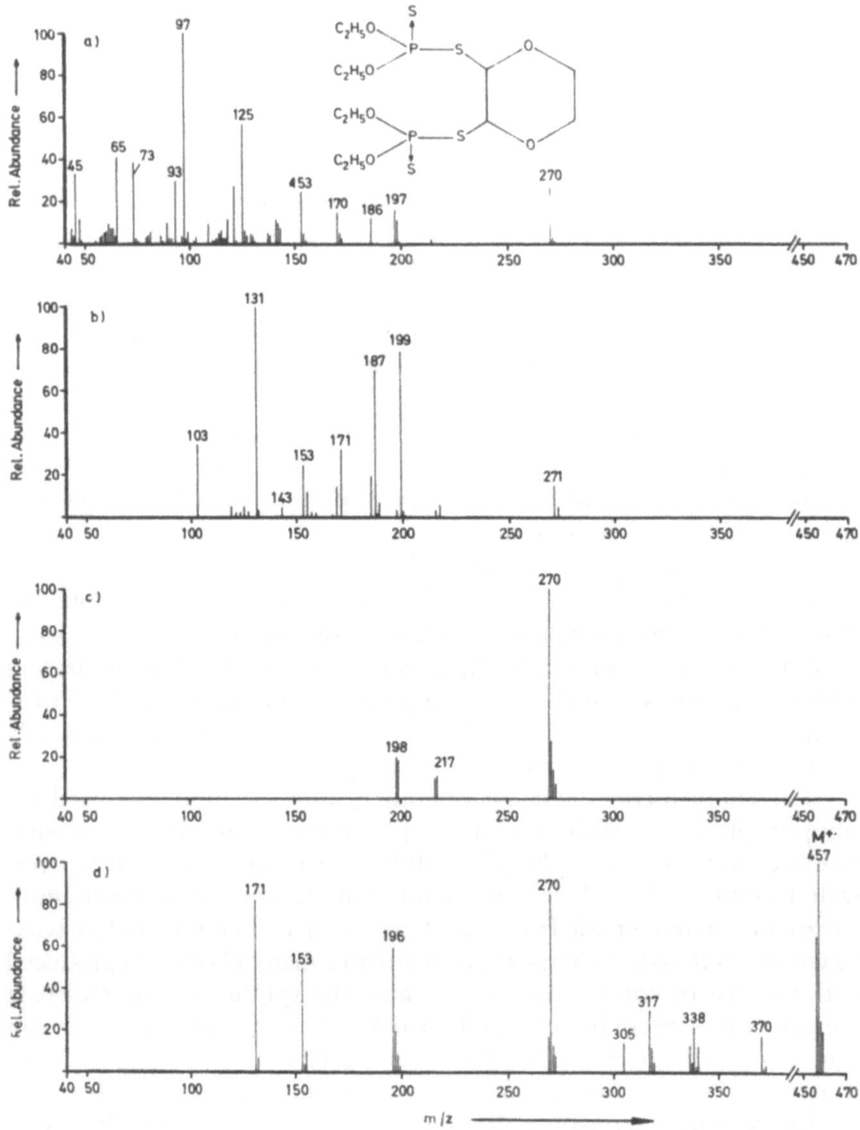


FIGURE 12. (a) EI; (b) CI isobutane, 200 °C; (c) FI; and (d) FD mass spectra of delnav.[57]

tute of Environmental Chemistry, Peking. The identification of dimethoate in the river water was confirmed by high resolution and direct isotope determination.[74] In addition, a $C_{20}$–$C_{35}$ alkane impurity was found.

*3.4.2. Phenylureas, Carbamates, and Thiocarbamates.* The FD mass spectra of herbicides, insecticides, and fungicides of the phenylurea-, carbamate-, and thiocarbamate-type have been reported.[60] The spectra obtained show high molecular ion intensities and characteristic fragmentation. Since these biocides occur in environmental samples, the limits of detection for standard compounds and test mixtures were evaluated. The optimal methodological parameters for trace analysis of these compounds in low- and high-resolution FDMS were described and applied to actual samples.

*3.4.3. Mixtures of Pesticides and Their Metabolites.* The application of FDMS in the investigation of mixtures of environmental chemicals is important because environmental samples are often obtained from complex mixtures. The outstanding capability of the technique in this field has been demonstrated for commercial products such as pesticides[3,61] and biochemicals.[62] When a mixture of 50% 1-naphthyl *N*-methylcarbamate (carbaryl) and 13% 3,4,6,4′-tetrachlorodiphenylsulfone (tetradifon) was examined, the FD mass spectrum showed only the molecular ions of the compounds and one major fragment. The FD mass spectra appeared to be more complex when mixtures of salts (e.g., chlorinated aromatic pesticides) were studied. In this case the mixture consisted of 2,4-D sodium salt and MCPA sodium salt, a technical product used as a herbicide. As expected, the molecular ions of salts were not detected with FDMS, but the spectra showed very abundant cluster ions which allowed the calculation of the molecular weight.

FD spectra of inorganic and many organic salts are characterized by high ion intensities of the cations.[63] In addition, cluster ions with $m/z$ values above the molecular weight ($M = C + A$; M = molecular weight, C = cation, A = anion) of the salt are observed. These cluster ions are produced by field-induced reactions on the emitter surface and normally enable the determination of the anion of the salt.

When the commercial mixture of 2,4-D sodium salt and MCPA sodium salt was investigated by FDMS over a wider range of sample temperatures, the spectra appeared to be more complex. At the beginning the FD spectrum was obtained with photographic detection by varying the emitter heating current between 0 and 17 mA. Only the major cluster ions $(n\text{M} + \text{C})^+$ of the salts were found and these are easily interpretable. When the emitter heating current was raised to 20 mA, more fragmentation and more complex cluster formation occurred. The masses of the cluster ions can be evaluated by high-resolution FDMS. Thus it becomes clear that FDMS can be used for the analysis of mixtures of large organic salts. However, careful control of the emitter quality, sample temperature, and hence the sample desorption rate are prerequisites for reproducible and interpretable results.[64]

More recently FDMS has been applied by other groups to the positive identification of organochlorine insecticides and their related compounds in soil[65] and to the screening of aquatic environment for herbicides.[66] The molecular ion of standards was observed as the base peak. The spectra of the extracts of river water showed positive responses at *m/z* 257 and *m/z* 344 corresponding to molecular ions of benthiocarb and oxadiazon. The authors indicated that these herbicides were simultaneously confirmed by other analytical techniques when present in the environment samples.

*Pesticidal metabolites.* The advantages of using the FD technique for the analysis of polar pesticidal metabolites have been demonstrated in the early stages of FDMS for DDA, a known metabolite of *p,p'*-DDT. The base peak in the FD spectrum of DDA is the molecular ion. The only fragment ions in the mass range above *m/z* 100 are formed by the direct cleavage between C(1) and C(2) and loss of the carboxyl group. Both these signals are also observed in the EI spectrum of DDA[55] but the $(M - COOH)^+$ peak in this case is the base peak and molecular ion is displayed with about 20% relative abundance. The assignment of the molecular and fragment ions was in both methods, EI and FD, confirmed by accurate mass measurements.

High-resolution FDMS is a very useful tool not only for the determination of the precise molecular weight and hence the elemental composition, but also for the acquisition of interesting structure information. Utilizing the high ionization efficiency and the extreme thermal stability of the high-temperature-activated emitters, the FD spectra can be obtained by inducing a controlled thermal fragmentation of the compounds examined.

*3.4.4. Identification of Biocides in Surface Water.* The combined application of high-pressure liquid chromatography (HPLC) and FDMS for the determination and identification of the polar, low-vapor pressure class of substances in environmental samples (e.g., biocides in Rhine water) has been described.[67,68] Specific preseparation and prepurification steps for the HPLC determination of polar substances such as phenylurea-, carbamate-, and thiocarbamate herbicides in polluted surface water have been developed. These steps are essential for reliable identification of biocides in complex environmental samples. The qualitative determination of a single substance in a sample is performed by calibration against corresponding test substances. In view of its specificity, sensitivity, and versatility, FDMS was chosen for the confirmatory step in the analysis of these biocides. It was shown that sample amounts of some nanograms corresponding to 10–20 ppt concentration in the river water sample were sufficient for reliable identification.

The FDMS analysis for the biocides in water extract can be performed in four consecutive steps. First, the complete mass range for the biocides is screened by FD to obtain a first survey of the substances present. Second, complete FD low-resolution mass spectra are recorded at emitter heating currents which are typical for the best anode tem-

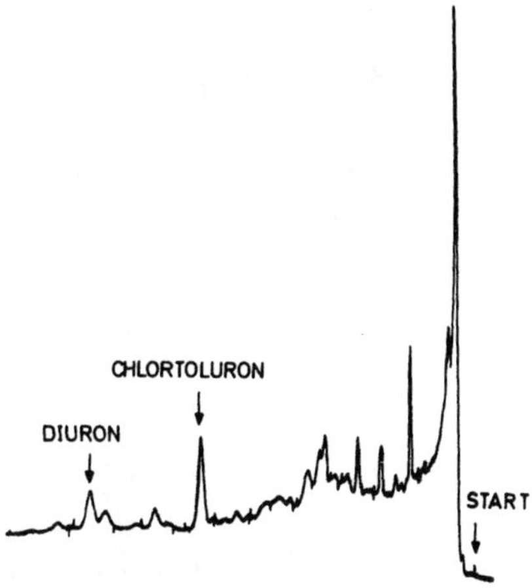


FIGURE 13. HPLC chromatogram of a sample extract from 1 liter of Rhine water (8-30-77). Experimental conditions: instrument, Waters M 6000A; stationary phase, Nucleosil 5—C18; mobile phase, methanol–water 47/53; isocratic operation at 30 °C; column length, 200 mm; inner diameter 4 mm; detector, UV at 254 nm; sample amount found: 8.2 ng chlorotoluron and 3.7 ng diuron.[68]

perature of the compounds to be analyzed from the mixtures. Third, concentrating on the assumed molecular ion region, the mass range is drastically reduced and the ions in this range only recorded by repetitive magnetic scans and signal accumulation with a multichannel analyzer. In most cases abundant molecular ions are observed and confirmed the results from the chromatographic procedure. A direct isotopic analysis often provides further evidence for the correct assignment of the biocide. Fourth, high-resolution mass measurements are made in cases where further information was desirable and the amount of sample available was sufficient.

The HPLC chromatogram of a sample from 1 liter of Rhine water is shown in Fig. 13, and the corresponding FD mass spectra obtained in the third stage described above is shown in Fig. 14. This figure shows the averaged (eight runs) FD mass spectrum of this fraction. These examples demonstrate clearly that combined application of HPLC and FDMS is a useful and reliable method in water analysis. In the low-resolution mode the short analysis time (10 min), the small sample consumption (~nanograms), the high selectivity obtained in multicomponent mixtures, and the unmatched reliability using two independent methods are the essential advantages.

If enough sample material is available and more instrument time can be invested, the ultimate proof of the correct identification of a biocide can be performed using high mass resolution. When only few accurate masses have to be determined, dynamic high resolution is sufficient; however, if a large number of peaks have to be estimated, the integrating, photographic detection is clearly preferred. As an example the photographically recorded lines of an environmental compound are given in Fig. 15.

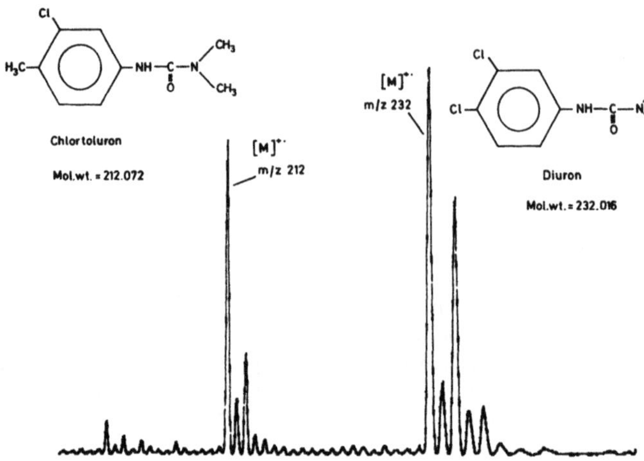


FIGURE 14. FD mass spectrum of the Rhine water extract (8-30-77) obtained by 74 repetitive magnetic scans and accumulation of the FD signals in a multichannel analyzer (range 213). The corresponding HPLC separation is shown in Fig. 13.[68]

## 3.5. *Ultratrace Analysis for Metals in Environmental Samples*

Because of extraordinary sensitivity of FDMS for metal cations, especially, for alkali metal cations, this method has had numerous applications in medical, toxicological, and environmental research.[69,75] The sensitivity of FDMS for alkali metal has long been recognized.[70] Only recently was the capacity of the method demonstrated for the detection

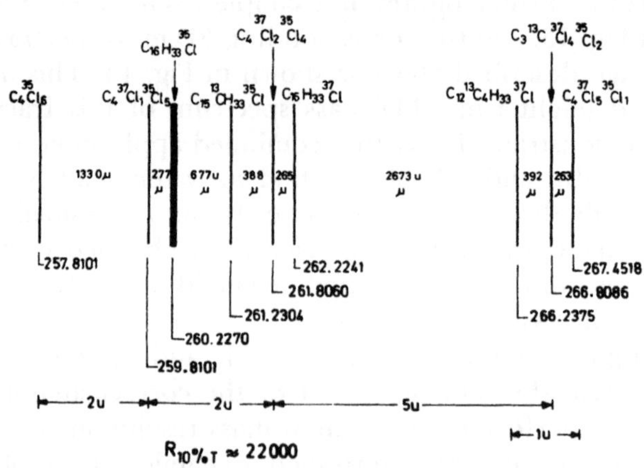


FIGURE 15. Section of a photographically recorded FD mass spectrum of an environmental compound. The achieved mass resolution was 22,000. The distances between the lines are given in μm and the mass differences in nominal masses are indicated at the bottom of the figure in u (mass units). Emission-controlled FDMS was performed.[58]

of metal cations such as Li, Na, K, Rb, Cs, Mg, Ca, Sr, Ba, Cu, Hg, Zn, Cd, Ti, and Y. The detection limit for FDMS determination of cesium lies in the lower femtogram region ($10^{-15}$ g). In this work the FD technique has been applied successfully to quantitative determination of trace cesium in a number of samples: solvents, body fluids, and environmental samples.(37)

The precision for repeated measurements of a standard solution was ±10%; the accuracy of the technique for the determination of unknown concentration was ±20%. A sensitivity between 1.4 and 2.5 $10^{-9}$ C/μg was obtained for cesium.

Field desorption MS using an internal standard was used to determine lithium in Rhine water, mineral and tap water, wine, and high-purity solvents.(71,72) The quantitative assay of trace metals in environmental and biological samples has the following advantages with FDMS:

(i) the ion currents are powerful;
(ii) the quality of the emitter plays only a subordinate role;
(iii) the coupling of the mass spectrometer and the multichannel analyzer can be quickly modified to accommodate a variety of analytical problems; and
(iv) the averaging of several hundred scans requires only 10–30 min—an experimental accuracy of several tenths of a percent was achieved by accumulation of more than 50 scans.

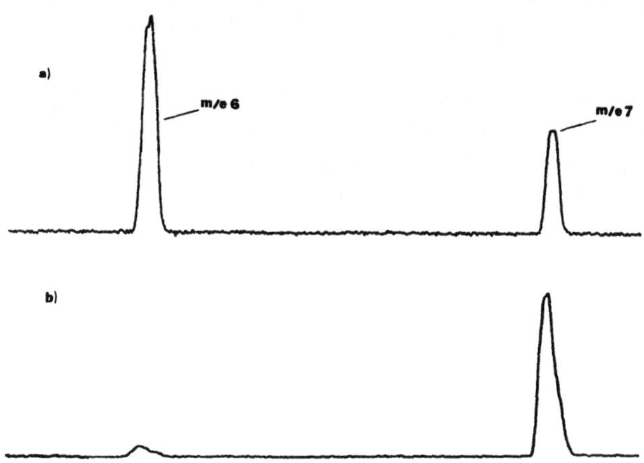


FIGURE 16. Determination of lithium in Rhine water by stable isotope dilution and FDMS. Sample size was 2 μl. (a) Water sample plus internal standard; 36.0 pg lithium standard with an isotopic distribution of $^{6}Li$ (67.5%) and $^{7}Li$ (32.5%) were added; (b) original Rhine water sample; approximately the natural isotopic abundances were found: $^{6}Li$ (7.4%) and $^{7}Li$ (92.6%). This sample was taken at the bank of the river on August 25, 1977. Accumulation of the FD signals in a multichannel analyzer (100 scans) gave a mean experimental error of 0.3%.(71)

Figure 16 demonstrates the lithium isotope distribution in Rhine water obtained by FDMS in conjunction with an integrating analyzer C-1024, Varian (analysis time about 30 min).

Evaluation of the experimental data from this figure gives a concentration of 8.3 μg $(Li)^{1-l}$ in the water sample. The total error of this determination is less than 5%.

### 3.6. *Evaluation and Prospects*

FD–MS allows detection, identification, and quantification of metals *and* organic compounds of low volatility. So far, investigations have concentrated on aerosols, food, body fluids, tissues, and water (river, tap, mineral, etc.). Field desorption is of unique value where only small samples (in the order of milligrams or less) are available, and the concentration of the metallic element or organic compound in these specimens are low (ppm or less). The capacity for multielement and biochemical (pesticidal) mixture analysis is one of the outstanding features of the method.

Further developments are expected in improvements of ion production and transmission, accessible mass range and mass resolution, and a better control of the desorption process, e.g., by emission control, programmable heating devices, and laser-assisted desorption. To fully realize the potential of FD–MS in environmental analyses an interdisciplinary approach is recommended. Joint studies with disciplines such as physiology and toxicology are particularly useful.

ACKNOWLEDGMENTS. This work was financially supported by the Deutsche Forschungsgemeinschaft (Schu 416/1-81) and the Umweltbundesamt. One of the authors (S. Sun) is grateful for the grant and continuous support by the Heinrich-Hertz-Stiftung.

## *REFERENCES*

1. H. D. Beckey, *Principles of Field Ionization and Field Desorption Mass Spectrometry,* Pergamon, Oxford, 1977.
2. H.-R. Schulten, *Int. J. Mass Spectrom. Ion Phys.* **32,** 97–284 (1979) and literature cited therein.
3. H.-R. Schulten and W. D. Lehmann, *Mikrochim. Acta (Wien)* **II,** 113–129 (1978).
4. H.-R. Schulten and H. M. Schiebel, *Naturwissenschaften* **65,** 223–230 (1978).
5. H.-R. Schulten and U. Schurath, *J. Phys. Chem.* **79,** 51–57 (1975).
6. R. Atkinson, B. J. Finlayson, and J. N. Pitts, Jr., *J. Am. Chem. Soc.* **95,** 7592–7599 (1973).
7. J. Bogen, *Atmos. Environ.* **7,** 1117–1125 (1973).
8. R. D. Giaugue, L. Y. Goda, and N. E. Brown, *Environ. Sci. Technol.* **8,** 436–441 (1974).
9. N. L. Craig, A. B. Harker, and J. Novakov, *Atmos. Environ.* **8,** 15–21 (1974).

10. D. Schuetzle, *Biomed. Mass Spectrom.* **2,** 288–296 (1975).
11. D. Schuetzle, A. L. Crittenden, and R. J. Charlson, *J. Air Pollut. Contr. Assoc.* **23,** 704–709 (1973).
12. J. L. Schultz, A. G. Sharkey, Jr., and R. A. Friedel, *Biomed. Mass Spectrom.* **1,** 137–141 (1974).
13. H.-R. Schulten and U. Schurath, *Atmos. Environ.* **9,** 1107–1112 (1976).
14. D. Schuetzle, A. L. Crittenden, and R. L. Charlson, *Air Pollut. Contr. Assoc. Ann.* **72,** 15–21 (1972).
15. S. Pfeifer, H. D. Beckey, and H.-R. Schulten, *Z. Anal. Chem.* **284,** 193–195 (1977).
16. L. Yin, A. D. Campbell, and L. Stoloff, *J. Assoc. Off. Anal. Chem.* **54,** 102–105 (1971).
17. T. V. Reddy, L. Viswanathan, and T. A. Venkitasubramanian, *Anal. Biochem.* **38,** 568–571 (1970).
18. D. Miskovic, *Chromatographia* **13,** 342–344 (1980).
19. J. C. Ayres, H. S. Lillard, and D. A. Lillard, *Food Technol. (Chicago)* **24,** 55–60 (1970).
20. T. Romer, *Feedstuffs* **48,** 18–46 (1976).
21. J. A. Sphon, P. A. Dreifuss, and H.-R. Schulten *J. Assoc. Off. Anal. Chem.* **60,** 73–81 (1977).
22. M. O. Moss, F. V. Robinson, A. G. Wood, H. M. Paisley, and J. Feeney, *Nature (London)* **220,** 767–770 (1968).
23. Official Methods of Analysis, 12th ed., AOAC, Washington, D. C., Secs. 26.014–26.019, 1975.
24. L. R. Dusold, A. E. Pohland, P. A. Dreifuss, and J. A. Sphon, in *Mycotoxins: Mass Spectral Data Bank,* U.S. Food and Drug Administration, Bureau of Foods, Division of Chemistry and Physics, Washington, D.C., 1978.
25. H. M. Stahr, P. F. Ross, R. W. Hyde, and W. Obioha, *App. Spectrosc.* **60,** 73–76 (1977).
26. K. E. Murray and D. L. Ingles, *Chem. Ind. (London)* **476,** 811–812 (1976).
27. P. A. Dreifuss, G. E. Wood, J. A. G. Roach, W. C. Brumley, D. Andrzejewski, and J. A. Sphon, *Biomed. Mass Spectrom.* **7,** 201–204 (1980).
28. G. Lehnert, K. H. Schaller, and T. Haas, *Z. Klin. Chem.* **6,** 174–176 (1968).
29. P. A. Leighton, *Photochemistry of Air Pollution,* Academic, New York, 1961, pp. 152–183.
30. K. W. Lieberman and H. H. Kramer, *Anal. Chem.* **42,** 256–267 (1970).
31. G. Ghersini and S. Mariottins, *Talanta* **18,** 442–446 (1971).
32. L. Sipos, J. Golimowski, P. Valenta, and H. W. Nuernberg, *Z. Anal. Chem.* **298,** 1–8 (1979).
33. L. Sipos, H. W. Nuernberg, P. Valenta, and M. Branica, *Anal. Chim. Acta* **115,** 2–42 (1980).
34. L. Mart, H. W. Nuernberg, and P. Valenta, *Z. Anal. Chem.* **300,** 350–362 (1980).
35. H. W. Nuernberg, *Chem. Ing. Tech.* **51**(7), 718–728 (1979).
36. H.-R. Schulten, R. Ziskoven, and W. D. Lehmann, *Z. Naturforsch.* **33c,** 178–183 (1978).
37. W. D. Lehmann and H.-R. Schulten, *Anal. Chem.* **49,** 1744–1746 (1977).
38. H.-R. Schulten and W. D. Lehmann, in *Quantitative Mass Spectrometry in Life Sciences II,* (A. P. De Leenheer, R. R. Roncucci, and C. van Peteghem, eds.), Elsevier, Amsterdam, 1978, pp. 63–82.
39. W. D. Lehmann, U. Bahr, and H.-R. Schulten, *Biomed. Mass Spectrom.* **5,** 536–539 (1978).
40. H.-R. Schulten, B. Bohl, R. Mueller, and R. Palavinskas, *Int. J. Mass Spectrom. Ion Phys.* **38,** 281–296 (1981).
41. H.-R. Schulten, W. D. Lehmann, and R. Ziskoven, *Z. Naturforsch.* **33c,** 484–487 (1978).
42. C. Achenbach, U. Bahr, F. Koehler, H.-R. Schulten, and R. Ziskoven, *Angew. Chem.* **91,** 944–945 (1979); *Angew. Chem. Int. Ed. Engl.* **18,** 882 (1979).
43. C. Achenbach, O. Hauswirth, C. Heindrichs, R. Ziskoven, F. Koehler, U. Bahr, A. Heindrichs, and H.-R. Schulten, *J. Toxicol. Environ. Health* **6,** 519–528 (1980).

44. R. Ziskoven, C. Achenbach, U. Bahr, and H.-R. Schulten, *Z. Naturforsch.* **35c,** 902–906 (1980).
45. H.-R. Schulten, W. D. Lehmann, and D. Haaks, *Org. Mass Spectrom.* **13,** 361–365 (1978).
46. H.-R. Schulten, R. Mueller, and D. Haaks, *Z. Anal. Chem.* **304,** 15–22 (1980).
47. H.-R. Schulten and W. D. Lehmann, *Trends Biochem. Sci.* **5,** 142–146 (1980).
48. H.-R. Schulten, *GIT Fachz. Lab.* **24,** 916–917 (1980).
49. J. N. Damico, R. P. Barron, and J. A. Sphon, *Int. J. Mass Spectrom. Ion Phys.* **2,** 161–182 (1969).
50. H.-R. Schulten, Research Conference on Pesticide Metabolites—Enzymology Isolation and Characterization, Theory and Application of Field Desorption Mass Spectrometry to Problems in Pesticide Analysis. Fargo, North Dakota, June, 1972.
51. H.-R. Schulten and H. D. Beckey, *J. Agric. Food Chem.* **21,** 372–375 (1973).
52. H.-R. Schulten, H. Prinz, H. D. Beckey, W. Tomberg, W. Klein, and F. Korte, *Chemosphere* **I,** 23–30 (1973).
53. J. N. Damico and R. P. Barron, *Anal. Chem.* **43,** 17–21 (1971).
54. H.-R. Schulten, Conference Proceedings No. 27, Fourth Annual Symposium on Recent Advances in the Analytical Chemistry of Pollutants, Basle, June 17–19, 1974.
55. H.-R. Schulten, in *New Methods in Environmental Chemistry and Toxicology,* (F. Coulston, F. Korte, and M. Goto, eds.), International Academic Printing, Tokyo, 1973, pp. 31–42.
56. J. A. Sphon and J. N. Damico, *Org. Mass. Spectrom.* **3,** 51–62 (1970).
57. J. M. Fales, G. W. A. Milne, H. U. Winkler, H. D. Beckey, J. N. Damico, and R. Barron, *Anal. Chem.* **47,** 207–219 (1975).
58. H.-R. Schulten and N. M. M. Nibbering, *Biomed. Mass Spectrom.* **4,** 55–61 (1977).
59. E. Stenhagen, S. Abrahamsson, and G. W. McLafferty (eds.), *Registry of Mass Spectral Data,* Vol. 3, Wiley-Interscience, New York, 1974, p. 2011.
60. H.-R. Schulten, *Z. Anal. Chem.* **293,** 273–281 (1978).
61. H.-R. Schulten, *J. Agric. Food Chem.* **24,** 743–749 (1976).
62. H. D. Beckey and H.-R. Schulten, in *Practical Spectroscopy Series,* Vol. 3, *Mass Spectrometry,* Part A, (C. N. McEwen and C. Merrit eds.), Marcel Dekker, New York, 1979, pp. 145–264.
63. H.-R. Schulten and F. W. Roellgen, *Angew. Chem.* **87,** 544–545 (1975); *Angew. Chem. Int. Ed. Engl.* **14,** 561 (1975).
64. H.-R. Schulten, in *Advances Mass Spectrometry,* (N. R. Daly ed.), 7A, Heyden, London, 1978.
65. M. Suzuki, Y. Yamato, and M. Koga, *Biomed. Mass Spectrom* **6,** 205–207 (1978).
66. Y. Yamato, M. Suzuki, and T. Watanabe, *Biomed. Mass Spectrom.* **6,** 205–207 (1979).
67. I. Stoeber and H.-R. Schulten, *Sci. Total Environ.* **16,** 249–262 (1980).
68. H.-R. Schulten and I. Stoeber, *Z. Anal. Chem.* **293,** 370–376 (1978).
69. U. Bahr and H.-R. Schulten, in *Topics in Current Chemistry, Mass Spectrometric Methods for Trace Analysis of Metals,* Vol. 95, (F. L. Boschke ed.), Springer, Heidelberg, 1980, pp. 1–48.
70. H.-R. Schulten and H. D. Beckey, *Org. Mass Spectrom.* **6,** 885–895 (1972).
71. W. D. Lehmann and H.-R. Schulten, *Angew. Chem.* **89,** 890–892 (1977).
72. H.-R. Schulten, U. Bahr, and W. D. Lehmann, *Mikrochim. Acta (Wien)* **I,** 191–198, London, pp. 83–96 (1979).
73. R. Ziskoven, C. Achenbach, H.-R. Schulten, and R. Roll, *Toxicol. Lett.* **19,** 225–231 (1983).
74. H.-R. Schulten and Si-en Sun, *Environ. Anal. Chem.* **10,** 247–263 (1981).
75. H.-R. Schulten, U. Bahr, and R. Palavinskas, *Z. Anal. Chem.* **317,** 497–511 (1984), and literature cited therein.

4

# POSITIVE AND NEGATIVE CHEMICAL IONIZATION MASS SPECTROMETRY

## RALPH C. DOUGHERTY

### 1. INTRODUCTION

#### 1.1. Chemical Ionization Mass Spectrometry

Chemical ionization (CI) mass spectrometry developed from studies of ion–molecule reactions in simple hydrocarbon gases.[1] CI mass spectra depend upon the fact that the cross sections for ion–molecule reactions involving electron or ion transfer are considerably larger than the corresponding cross sections for interaction of molecules with free electrons.[2] If a substrate is present in an ion source at concentrations ranging from $10^{-3}$ to $10^{-5}$ torr, and the same source contains a reagent gas at a pressure of approximately 1 torr, the initial ionization caused by interaction of high-energy electrons with the gases in the source will occur primarily in the reagent gas. Because the gas pressure in the ion source is maintained at approximately 1 torr, ions formed by interaction with the primary electron beam will experience up to thousands of collisions with neutral molecules prior to exiting the source. Since the cross sections for ion–molecule reactions are considerably larger than the cross sections for interactions of electrons and molecules, many of these collisions will be reactive, with the result that the spectrum of ions emerging from the source will be substantially altered from that obtained at low source pressures. The ionization process in CI mass spectrometry is best reviewed by examining the reactions that occur in the high-pressure mass spectrum of methane, one of the most common CI reagent gases.

---

*RALPH C. DOUGHERTY* • Department of Chemistry, Florida State University, Tallahassee, Florida 32306.

The electron impact (EI) mass spectrum of methane obtained at low pressures with 70-eV electrons is dominated by the molecule ion and the methyl cation, $CH_3^+$. These two ions have roughly equal abundance and account for approximately 90% of the total ionization. The remainder of the ionization occurs at lower masses including $CH_2^+$, $CH^+$, and $C^+$. When the positive-ion mass spectrum of methane is obtained with source pressures of the order of 1 torr, collisions between methane ions and neutral molecules suppress the formation of high-energy fragment ions such as $CH^+$ and result in the formation of ions with masses greater than that of methane, through ion-molecule reactions. Figure 1 illustrates a typical high-pressure mass spectrum of methane. The spectrum is dominated by two intense ions $CH_5^+$ and $C_2H_5^+$, both of which result from ion-molecule collisions within the ion source. The reactions which result in the formation of these ions are illustrated in Table 1(b).

The even-electron ions $CH_5^+$ and $C_2H_5^+$ are exceptionally strong proton acids in the gas phase. The proton affinity of methane and ethylene are 127 and 159 kcal/mol, respectively.[3,4] These proton affinities are lower than virtually all hydrocarbons and all neutral molecules which contain oxygen, nitrogen, or other electron-rich heteroatoms.[5] A substrate which is in the chemical ionization source at a pressure between $10^{-3}$ and $10^{-5}$ torr will be readily ionized by collision with either $CH_5^+$ or $C_2H_5^+$ by one of the processes illustrated in Table 1(b).

The two dominant reagent ions in methane CI can also form adducts or displace groups in the ionization of a molecule. The atom X in the sixth equation in Table 1(b) can be a hydrogen atom, a halogen atom, or in rare cases a heteroatomic functional group. One more process occurs in the formation of the primary CI spectrum, namely Lewis acid adduct formation [last equation, Table 1(b)]. Reactions of this type are more prominent when isobutane is used as the reagent gas where the

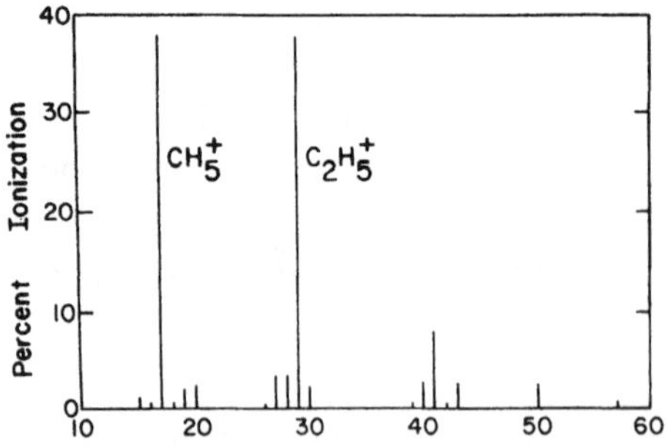


FIGURE 1. Methane CI mass spectrum (1 torr).

*TABLE 1*
*Ion-Forming Reactions in Chemical Ionization Mass Spectrometry*[a]

(a) Reactions common to all systems; and dominant in charge exchange CI (e.g., $N_2$, CO, $CF_4$, or He reagent gases)

$$G + e^- \rightarrow G^{+\cdot} + e^- + e_s^-$$
$$G^{+\cdot} + S \rightarrow G + S^{+\cdot}$$
$$G^{+\cdot} + AH \rightarrow GH^+ + A^\cdot$$

(b) Methane CI

$$CH_4^{+\cdot} \rightarrow CH_3^+ + H^\cdot$$
$$CH_4^{+\cdot} + CH_4 \rightarrow CH_5^+ + CH_3^\cdot$$
$$CH_3^+ + CH_4 \rightarrow C_2H_5^+ + H_2$$
$$CH_5^+ + S \rightarrow SH^+ + CH_4$$
$$C_2H_5^+ + S \rightarrow SH^+ + C_2H_4$$
$$C_2H_5^+ + A\text{-}X \rightarrow A^+ + C_2H_5X$$
$$C_2H_5^+ + S \rightarrow SC_2H_5^+$$

(c) Isobutane CI

$$i\text{-}C_4H_{10}^{+\cdot} \rightarrow i\text{-}C_4H_9^+ + H^\cdot$$
$$i\text{-}C_4H_9^+ + S \rightarrow SH^+ + C_4H_8$$
$$i\text{-}C_4H_9^+ + S \rightarrow S\text{-}C_4H_9^+$$
$$i\text{-}C_4H_9^+ + RCl \rightarrow R^+ + i\text{-}C_4H_9Cl$$

The major difference between isobutane CI and methane CI is the fact that protonated methane can transfer 68 kcal/mol more energy in the ionization of a substrate than the *t*-butyl cation; as a result, there is usually less fragmentation with isobutane than with methane.

(d) Ammonia–hydrocarbon (1 : 10) CI

$$RH^{+\cdot} + NH_3 \rightarrow NH_4^+ + R^\cdot$$
$$NH_4^+ + S \rightarrow SH^+ + NH_3$$
$$NH_4^+ + S \rightarrow S\cdot\cdot NH_4^+$$

The ammonium ion can transfer a proton only to compounds with a higher proton affinity than $NH_3$, namely, stronger gas phase bases. Ion–molecule association between $NH_4^+$ and substrates occurs very readily with very little excitation of the substrate so that intact adducts of ammonium with pentasaccharides have been seen. This plasma is extremely insensitive to molecules with low proton affinities such as the Freons or PFK. Water is very similar to ammonia except for the higher proton affinity of the latter gas.

[a] G, reagent gas; S, substrate.

dominant ion in the spectrum is the *t*-butyl cation $C_4H_9^+$; its adducts with substrates commonly occur as shown in Table 1(c).

The amount of energy transferred to a substrate molecule in the ionization process is considerably smaller under CI reaction conditions than with electron ionization. Excited states of ionized molecules also can be relaxed by collision with neutrals in a high-pressure ion source. These two factors significantly restrict the fragmentation that occurs under CI conditions as compared to a corresponding EI spectrum.

The major differences between CI mass spectra and EI mass spectra are the following:

1. Odd-electron ions are virtually absent from CI mass spectra because odd-electron species generally react by ion–molecule reactions to form their more stable even-electron counterparts.
2. Fragmentation of complex organic structures occurs to a smaller extent in CI mass spectra than in EI mass spectra because the amount of energy deposited in a molecule on ionization is smaller using the former technique.
3. The amount of fragmentation observed in CI mass spectra does depend upon the nature of the reagent gas and to a lesser extent on temperature and substrate concentrations. Fragmentation decreases as the reagent gas is changed from methane to isobutane to isobutane-ammonia. The proton affinity of the reagent gas neutral increases in the same series.
4. Although CI mass spectra contain significantly fewer fragment ions than their EI counterparts, the spectrum of the reagent gas is always superimposed on that of the substrate.

### 1.2. *Negative Chemical Ionization Mass Spectrometry*

The low-density ionic plasma which is created when gas at 1 torr is irradiated with 500-eV electrons contains a cascade of both positive and negative ions. The ions that are observed by the mass spectrometer depend upon the polarity of the accelerating fields in the ion source.

The concentration of positive and negative ions in a chemical ionization source must be approximately equal if a mass spectrum is to be observed. If the concentration of electrons and anions considerably exceeds the concentration of positive ions, the positive ions will be trapped in the ion source by the "space-charge." The same is true for positive ions trapping negative ions if their numbers vastly exceed the negative space charge. Although the numbers of positive and negative ions must be of comparable magnitude, the species of positive and negative ions are vastly different. The large differences in the positive- and negative-ion beams from a given ionic plasma are due to the fact that all molecules contain relatively high-energy electrons while only certain molecules have low-energy vacant orbitals or low-energy virtual vacant orbitals. Thus, virtually all molecules will give intense positive-ion beams while only certain molecules will have high sensitivity in negative chemical ionization (NCI) mass spectrometry. The ionization potentials for most molecules vary between 8 and 15 eV. If 70-eV electrons are used to initiate the ionization, to a first approximation all molecules will give equivalent positive-ion sensitivities. The electron affinities of molecules vary from negative values to roughly 4 eV. The numbers of molecules with positive-electron affinities is relatively small. Molecules with positive-electron affinities and other molecules that give intense negative-ion mass spectra can be generally classed as oxidizing agents or alkylating agents or both. These classes of molecules span a large portion of the spectrum of the toxic substances that appear in the environment and, as a result, one of the major applications of NCI mass spectrometry has been in the detection and analysis of toxic substances.

The major ion-forming processes in NCI mass spectrometry are illustrated in Fig. 2 and Table 2. These processes are (a) resonance capture of an electron to give a vibrationally excited molecule anion, which must relax by internal redistribution of the vibrational energy or collision with a neutral if the molecule anion is to avoid autodeionization (ejection of an electron); (b) dissociative capture of an electron to give a fragment ion and a neutral; and (c) ion molecule association to give an anion molecule adduct. There are two additional negative-ion-forming processes that can be important under specific conditions: (d) ion–molecule exchange reactions such as the reaction of superoxide anion with aromatic polyhalides which replaces halogen with $O^-$ and (e) decomposition of highly excited neutral molecules to give positive- and negative-ion pairs—ion pair formation is very common in low-pressure negative-ion mass spectrometry and virtually unknown in negative chemical ionization mass spectrometry because the abundance of highly excited neutrals under NCI conditions is vanishingly small due to collisional relaxation of vibronic excitations.

The versatility of CI and NCI mass spectrometry is illustrated in Tables 1 and 2. The selectivity of the positive-ion or negative-ion plasmas for specific classes of molecules can be directly manipulated by variations

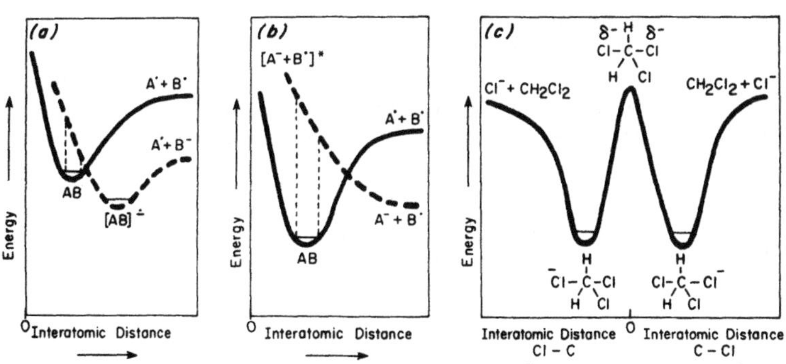


FIGURE 2. Negative CI ion-forming processes.

TABLE 2

*Ion-Forming Reactions in Negative Chemical Ionization Mass Spectrometry*

(a) Reactions common to all systems.

$$M + e_s^- \rightarrow [M]^{*\cdot -} \xrightarrow{N} M^{\cdot -}$$
$$AB + e_s^- \rightarrow A^- + B^\cdot$$

(b) Hydrocarbon (93%), methylene chloride or methylchloride (5%), oxygen (2%). Chloride/oxygen reagent.

$$RCl + e_s^- \rightarrow R^\cdot + Cl^-;\ ZH + Cl^- \rightarrow Z^- + HCl$$
$$ZH + Cl^- \rightarrow [ZH \cdots Cl]^-;\ ACl^{\cdot -} + O_2 \rightarrow AO^- + ClO^\cdot$$
$$RCl + Cl^- \rightarrow RCl_2^-$$

This gas mixture has the same high sensitivity for electron-capturing molecules as pure hydrocarbon or hydrocarbon–oxygen mixtures, and it also has high sensitivity for alkylating agents such as aliphatic polyhalides, phosphate esters, and carbamates. The selectivity of the gas mixture is very high—neutral lipids are virtually transparent.

(c) Hydrocarbon (95%), water (5%). Hydroxide reagent.

$$H_2O + e_s^- \rightarrow OH^- + H^\cdot;\ M + OH^- \rightarrow MOH^-$$
$$AH + OH^- \rightarrow A^- + H_2O$$

Hydroxide can be generated by the following reactions:

$$N_2O + e_s^- \rightarrow N_2 + O^{\cdot -};\ O^{\cdot -} + RH \rightarrow OH^- + R^\cdot$$

Reactions of hydroxide are mimicked by methoxide and amide reagents:

$$CH_3ONO + e_s^- \rightarrow CH_3O^- + NO;\ NH_3 + e_s^- \rightarrow NH_2^- + H^\cdot$$

All these reagents give high sensitivity for many structural classes of compounds. The selectivity for toxic substances or electron-capturing molecules is low, because these reagents all react with neutral lipid esters and alcohols and can even produce intense spectra with saturated hydrocarbons.

(d) Fluorocarbon, $CHF_3$, $C_2F_3Cl_3$, etc. Fluoride or chloride reagent.

$$CHF_3 + e_s^- \rightarrow F^- + 2F^\cdot + CH;\ M + F^- \rightarrow MF^-$$
$$AH + F^- \rightarrow A^- + HF;\ M^{\cdot -} + F^\cdot \rightarrow M + F^-$$

Fluorocarbons will produce either $F^-$ or $Cl^-$ as the reactive reagent, depending on their structure. $CHF_3$ gives $F^-$, $CF_2Cl_2$ and $CFCl_2CF_2Cl$ give $Cl^-$ with some $Cl_2^{\cdot -}$. Fluoride is an exceptionally strong gas phase base, and mimics $OH^-$ in reactivity and low selectivity. All of the fluorocarbon reagents appear to have decreased sensitivity for electron-capturing substances because of reactions like these:

$$M^{\cdot -} + F^\cdot \rightarrow M + F^-;\ M^{\cdot -} + Cl^\cdot \rightarrow M + Cl^-$$
$$Cl_2^{\cdot -} \rightarrow Cl^- + Cl^\cdot;\ Z^- + Cl^\cdot \rightarrow Z^\cdot + Cl^-$$

Similar reactions can occur with other radicals with high electron affinities in the source.

in the reagent gas which is used to develop the plasma. The variation in selectivity can change the sensitivity for a given compound by as much as four orders of magnitude. Scientists who use these techniques should be aware of this wide variation in sensitivity, and if quantitative data are required, calibration of the sensitivity in the instrument to be used with the reagent gases is a necessity for reliability.

Not all reagent gases can be successfully used in CI and NCI mass spectrometry. In our experience, when ethylene is used as the reagent gas, the conditions in the ionic plasma catalyze polymerization of the ethylene with the ultimate result that very fine fibers of carbon are produced within the ion source. These carbon fibers destroy the sensitivity of the source within minutes.

When we attempted to use hydrogen sulfide as a reagent gas for NCI, the gas oxidatively polymerized and gave abundant quantities of $S_8$ in the ion source. Once again, these reactions very rapidly destroyed the sensitivity of the source.

Strongly oxidizing reagent gases such as pure oxygen or nitric oxide very dramatically decreased the source life for conventional CI and NCI sources. These problems can be alleviated by using mixtures of hydrocarbons with oxidizing gases in ratios of approximately 10 : 1. It is also possible to use Townsend discharge as a source of the ionic plasma; however, fluctuations in the stability of the Townsend discharge may substantially reduce the quantitative reliability of the spectra obtained.

The major applications of CI and NCI mass spectrometry take advantage of the relative simplicity of the mass spectra generated under these conditions for studies of chemical reactivity in the gas phase, the analysis of complex biological molecules, and the analysis of complex mixtures. All of these types of investigations will be briefly reviewed in the following sections.

## 2. *APPLICATIONS*

### 2.1. *Studies of Chemical Reactivity*

Since CI mass spectra result from ion–molecule reactions between ions and neutrals in the ion source, it is possible to use CI mass spectrometers to study the rates and thermochemistry of ion–molecule reactions that occur in the absence of solvation. The availability of CI mass spectrometry along with the complimentary techniques of ion cyclotron resonance spectrometry and mass spectra of flowing afterglows has resulted in a dramatic revision of our understanding of the inherent reactivity of ions, that is, reactivity in the absence of solvation. This data is particularly valuable because until recently chemical theory treated molecules and transition states as if chemical reactions occurred in the absence of solvation. The background for understanding the development of the ion equilibrium method for investigating the thermodynamics of proton transfer reactions in the gas phase has recently been extensively reviewed.[6] This review also discusses the similarities and differences between data obtained by CI mass spectrometry using a pulsed electron beam and similar data obtained by ion cyclotron resonance or flowing afterglow techniques.

Studies of gas phase ion thermochemistry have been of signal importance in testing the validity of theoretical predictions of chemical reactivity in the gas phase and in experimentally bridging the gap between gas phase and solution chemical reactivity. From the point of view of an analytical chemist, the major interests in these studies comes from their development of our understanding of relative reactivity of anions and cations in the gas phase. The inversion in the acidity of alcohols on going from solution to the gas phase was first demonstrated using ion cyclotron resonance techniques[7] and has been confirmed by CI techniques on a number of occasions.[6] In the series of alcohols where the alkyl group is changed from methyl to ethyl, isopropyl, and finally *t*-butyl, the acidity of the alcohols decreases monotonically in solution. In the gas phase, however, *t*-butyl alcohol is the most acidic, and acidity decreases monotonically to methyl alcohol. The reason for this reactivity inversion can be qualitatively described in terms of the influence of solvation on the energies of the highest energy electrons in the alcoholate anion.[8]

Studies of the proton affinities of molecules, that is, the reactivity of cationic species in the gas phase as compared to the solution, have shown a number of quantitative differences in reactivity but have failed to show qualitative reversals like those of anions, with the exception of the relative proton affinities of the alkyl amines. In solution it has been well known for many years that both primary and tertiary alkyl amines are less basic than their secondary amine counterparts. Measurements of the proton affinities of alkyl amines using ion resonance or pulsed CI techniques have very clearly shown that amine basicity increases monotonically from primary through tertiary amines.[6] In this case, the change in the order of basicity must be due to steric inhibition of solvation in the tertiary amine cations.

The collection of data on gas phase ionic equilibria is proceeding at a rapid pace, and we can reliably anticipate that when this database matures, it will provide one of the more solid guides for the development of our theoretical understanding of chemical reactivity.

### 2.2. *Studies of Complex Biological Molecules*

The applications of CI mass spectrometry to structural elucidation for complex natural products are so numerous that a comprehensive review of this subject is beyond the scope of this chapter. The situation is somewhat analogous to the application of high-resolution mass measurement to the determination of the elemental composition of unknown compounds. The basis for the application of CI mass spectrometry and NCI mass spectrometry to the determination of the structure of unknowns stems from the relative simplicity of these spectra as compared to EI mass spectra of the same compounds. When one is confronted with an unknown in microquantities, it is very often the case that a preliminary examination by EI mass spectrometry will not give a clear indication of the molecular weight of the unknown component. In current practice it is more or less routine in these situations to obtain a CI or NCI mass spectrum which generally provides direct information concerning the mass of the substance under investigation. The uses of CI and NCI mass spectrometry in structure determination for complex compounds will be illustrated very briefly with examples from the chemistry of lipids, carbohydrates, and peptides.

The isobutane CI mass spectrum of 4-D-hydroxysphinganine *bis*-methylboronate(9) was dominated by its protonated molecule ion at *m/z* 366 (Fig. 3). In contrast, the EI mass spectra of related lipids are generally dominated by ions which are the result of fragmentation of the molecular structure. The abundance of molecule ions for corresponding derivatives of complex lipids generally have relative intensities of less than 10%.(10) In typical situations when one is dealing with microquantities of an unknown substance, the purity of the material is often less than 90% and ions with 10% relative abundance can easily be overlooked in attempts to establish molecular weight.

Bile acids and related steroidal derivatives are known to undergo complex fragmentation reactions in EI mass spectrometry.(11) In contrast, isobutane CI mass spectra of more than 50 bile acid derivatives contained only three or rarely four ions of any significant intensity.(12) Of these ions the dominant ions were typically either the protonated

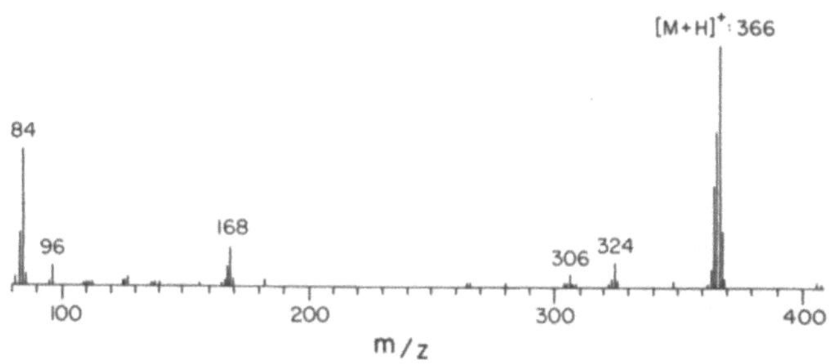


FIGURE 3. Isobutane CI mass spectrum of 4-D-hydroxysphinganine-bis-methyneboronate.(9)

molecule or the protonated molecule less the elements of acidic acid. Both of these ions are highly characteristic of the molecular weight of the substance involved and give substantial aid to the investigation of the structure of unknowns.

In investigations of the structure of oligosaccharides a variety of CI reagents have been used. Mixtures of ammonia and hydrocarbons give spectra that are dominated by adducts of the ammonium ion. These spectra are most valuable where determination of molecular weights or formulas is an overriding consideration. Methane and isobutane give spectra that contain protonated molecules and substantially more fragmentation than when the ammonium ion is the reagent. The hydrocarbon CI spectra are often more useful in structural investigations; however, structural information is also easily obtained from ammonium ion mass spectra. Chemical ionization mass spectra of D-glucose have been obtained using ammonium ion, methane, and isobutane CI techniques.(13) The ammonium ion adduct of glucose was the base peak in the ammonium ion spectra. Protonated molecules did not appear in the spectra generated with methane or isobutane plasmas. For higher polysaccharides derivatization is generally necessary if the molecules are to be desorbed without more or less complete thermolysis. Intense ammonium ion adducts have been obtained for a series of 20 oligosaccharide acetates including four tetrasaccharides.(14) In all of these spectra intense ammonium adducts were observed. Thermal fragmentation of the oligosaccharides yielded ammonium ion adduct ions which contained substantial sequence information. In particular, it was always possible to determine the nature of the reducing and nonreducing sugars in polysaccharide chains.

In contrast to EI mass spectra of derivatives of amino acids which often have molecule ions of only 1% or 2% relative intensity,(15) the isobutane CI mass spectra of the trifluoroacetamide butyl esters of a series of -amino acids were generally dominated by protonated molecules.(16)

Electron impact mass spectrometry has been most widely used for

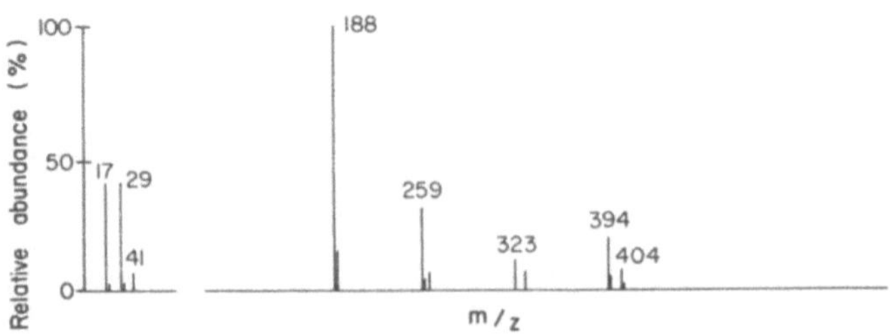


**FIGURE 4.** CI (methane) mass spectrum of per methylated *N*-acetyl-Met-Gly-Met-Met methyl ester.(17)

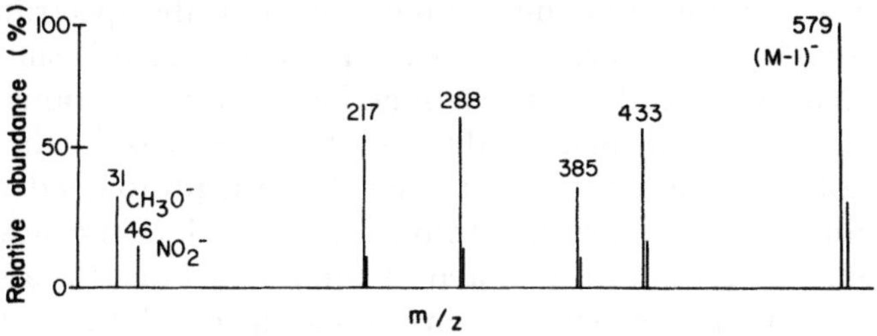


FIGURE 5. NCI (methoxide) mass spectrum of per *N*-methylated *N*-acetyl-Met -Gly-Met-Met methyl ester.[17]

sequencing oligopeptides. Both CI and NCI mass spectrometry appear to be equally well suited to this task. Figures 4 and 5 show the positive and negative CI mass spectra obtained from the tetra *N*-methyl derivative of a tetrapeptide acetate methylester.[17] The CI and NCI plasmas were generated with methane containing a trace of methylnitrite which resulted in formation of abundant quantities of methoxide anion in the gas phase. By use of milder CI conditions, it is possible to obtain similar sequence information in the positive-ion spectra and also to obtain abundant quantities of protonated molecules.

## 2.3. *Analysis of Complex Mixtures*

Fales and his co-workers were the first to use CI screening for rapid analysis of complex biological mixtures.[18] Their original analysis scheme was designed to determine the molecular nature of abused drugs in overdose cases. To this end, they obtained samples of gastric contents in emergency room situations, subjected the gastric fluid to extraction with chloroform, and directly analyzed the evaporated chloroform extract. Figure 6 illustrates a spectrum obtained from an overdose case which clearly indicated the presence of aspirin, phenacetin, and codeine

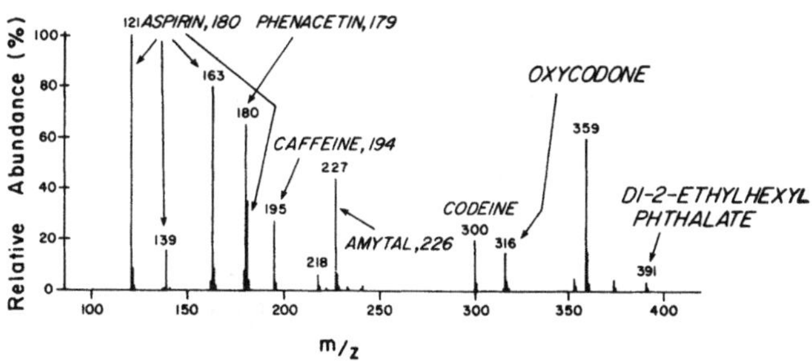


FIGURE 6. Isobutane CI mass spectrum of an extract of gastric contents in an overdose case.[18]

in the gastric contents. One interesting feature of the spectrum in Fig. 6 is that the CI mass spectrum of codeine in this case is substantially different than when codeine is examined alone using isobutane as the reagent gas. Under normal conditions, the base peak in the codeine spectrum corresponds to the loss of water from the protonated molecule; the spectrum in Fig. 6 shows the protonated molecule as the most intense codeine ion. The reason for this stems from the fact that the abundance of basic molecules in the extract completely suppressed the *t*-butyl cation in the plasma so that the ionizing reagent instead of being the *t*-butyl cation was a protonated phenacetin for similar molecule. The excess energy of proton transfer to codeine from weak acids like protonated phenacetin is insufficient to drive the water elimination reaction.

CI mass spectra are very often used to screen complex mixtures for the presence of specific molecules. Murata et al. reported the use of isobutane CI mass spectra to examine the mixture of glucocerebrosides obtained from the spleen of a patient with Gaucher's disease.[19] By using CI mass spectra, it was possible to obtain evidence for the presence of a series of glucocerebrosides in the complex mixture. It was also possible to identify the major components of the mixture.[19]

Negative chemical ionization mass spectrometry shows very high selectivity for particular classes of molecules and high selectivity for these classes. The classes of molecules that have intrinsically high NCI sensitivities are substances that are oxidizing agents, alkylating agents, or both. Because the molecules of biology are very highly reduced, materials that are oxidizing agents or alkylating agents span a large portion of the spectrum of toxic substances. As a result of these features, one of the major applications of NCI mass spectrometry has been the detection and analysis of toxic substances in complex mixtures.

Negative chemical ionization (NCI) techniques have been successfully used to detect the presence of polynuclear aromatic hydrocarbons,[20] trace metals in the atmosphere,[21] and phosphate and carbamate pesticides.[22,23] The largest application of NCI screening of complex mixtures has been in the detection of polychlorinated organic molecules in biological matrices.[24–26]

Figure 7 illustrates the direct-probe NCI mass spectrum of an extract of Lake Ontario trout. This extract was prepared by extracting the lipids from the fish and separating the neutral lipids from the toxic substances using gel permeation chromatography.[27] The NCI plasma was generated with isobutane containing approximately 10% methylene chloride and 2% oxygen. Under these conditions, aromatic polyhalides appear as the $(M - 19)^-$ ions. Aliphatic polyhalides appear at molecule anions and chloride adducts. The concentration of the polychlorinated organic compounds in Fig. 7 are all in the ppb range. One problem with using

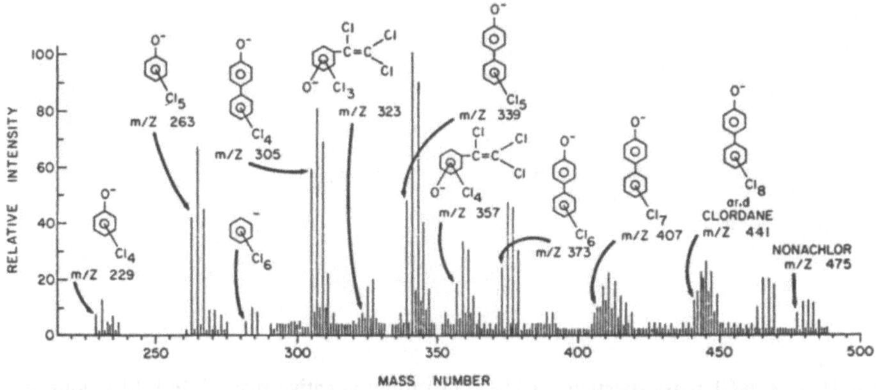


FIGURE 7. NCI mass spectrum ($CH_2Cl_2,O_2$) of a gel permeation cleaned-up extract of Lake Ontario trout.[27]

direct-probe screening as compared to GC/MS analysis is the relatively lower dynamic range in the former technique. If the samples contain ppb concentrations of polychlorinated organic compounds, it is virtually impossible to see contaminants in the ppt range because the ionization is captured by the higher concentration species.

It is possible to examine specific classes of toxic substances in the high ppt range if they are isolated from the other compounds in the matrix. Figure 8 illustrates the direct-probe NCI mass spectrum of the

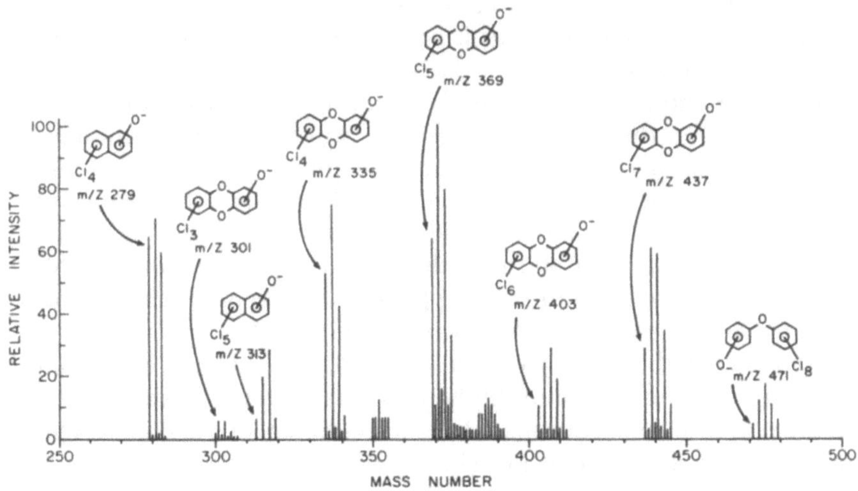


FIGURE 8. NCI mass spectrum ($CH_2Cl_2,O_2$) of planar aromatics isolated from a Tittabawassee River carp.[28]

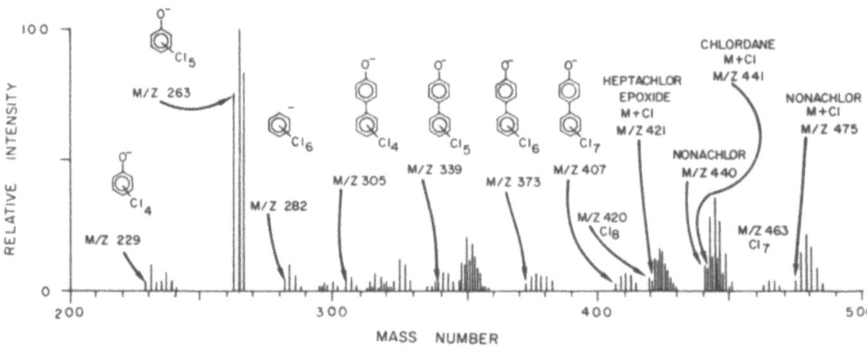


**FIGURE 9.** NCI mass spectrum ($CH_2Cl_2,O_2$) (only negative mass defect 12 C ions are shown) of chlorinated compounds isolated from human adipose tissue.[27]

planar polychlorinated organic molecules isolated from a sample of carp obtained from the Tittabawassee River in Michigan. The finding of ions corresponding to the polychloronaphthalenes was consistent with findings for a series of fish obtained from Midwestern and Northeastern sources. The finding of ions corresponding to polychlorodioxins was unique to the Tittabawassee River fish.[28] The mass spectrum in Fig. 8 is not an accurate record of the relative abundance of specific dioxins in the mixture. The NCI sensitivity for the 2,3,7,8-tetrachlorodibenzo-*p*-dioxin is lower than that for any of the other tetrachloroisomers and is also lower than the sensitivity for the higher chlorinated dioxins. Capillary GC/MS analysis has shown that the only tetra-isomer in this extract was the 2,3,7,8-isomer and that it was more abundant than the other dioxins in the mixture.[29]

Figure 9 shows a final example of an application of direct-probe NCI screening to the examination of toxic substances in environmental substrates. This spectrum was generated from an extract of 250 mg of human adipose tissue that had been freed from lipids by steam distillation with continuous liquid–liquid extraction.[27] When higher than 250 mg of lipids are used in steam distillation, recoveries for polychlorinated organics are substantially suppressed by the oil layer that forms on the top of the steam distillation reflux system. The spectrum in Fig. 9 clearly shows the presence of a series of polychlorophenols, hexachlorobenzene, the polychlorobiphenyl series as oxygen exchange ions, and the polycyclic polychlorinated pesticides heptachloroepoxide, chlordane, and nonachlor. The ion cluster at *m/z* 463 corresponds to the chloride adduct of *tris*-dichloropropylphosphate, a mutagenic compound that is commercially used as a flame retardant.

## *REFERENCES*


1. M. S. B. Munson and F. H. Field, *J. Am. Chem. Soc.* **88,** 1621 (1966).
2. F. H. Field and M. S. B. Munson, *J. Am. Chem. Soc.* **87,** 3289 (1965).
3. M. A. Haney and J. L. Franklin, *J. Chem. Phys.* **73,** 4328 (1969).
4. J. L. Franklin, J. G. Dillard, H. M. Rosenstock, J. T. Herron, K. Draxl, and F. H. Field, "Ionization Potentials, Appearance Potentials and Heats of Formation of Gas Phase Positive Ions," NSRDS-NBS 26, U.S. Dept. of Commerce, Washington, D.C., 1969.
5. G. W. A. Milne and M. J. Lacey, in *CRC Critical Reviews in Analytical Chemistry,* Vol. 4, Chemical Rubber Company, Cleveland, 1974, p. 45.
6. P. Kebarle, *Ann. Rev. Phys. Chem.* **28,** 445–476 (1977).
7. J. I. Brauman and L. K. Blair, *J. Am. Chem. Soc.* **91,** 2126–2127 (1969).
8. M. J. S. Dewar and R. C. Dougherty, *The PMO Theory of Organic Chemistry,* Plenum, New York, 1975, pp. 512–514.
9. S. J. Gaskell and C. J. W. Brooks, *Org. Mass Spectrom.* **12,** 651 (1977).
10. S. J. Gaskell, C. G. Edmonds, and C. J. W. Brooks, *J. Chromatogr.* **126,** 591 (1976).
11. W. H. Elliott, *Biochemical Applications of Mass Spectrometry,* (G. R. Waller, ed.), Wiley-Interscience, New York, 1972, p. 291.
12. P. A. Szczepanik, D. L. Hochey, and P. D. Klein, *J. Lipid Res.* **17,** 314 (1976).
13. A. M. Hogg and T. L. Nagabhushan, *Tetrahedron Lett.* 4827–4831 (1972).
14. R. C. Dougherty, J. D. Roberts, W. W. Binkley, O. S. Chizhow, V. I. Kadentsev, and A. A. Solovyov, *J. Org. Chem.* **39,** 451–455 (1974).
15. W. Vetter, *Biochemical Applications of Mass Spectrometry* (G. R. Waller, ed.), Wiley-Interscience, New York, 1972, pp. 387–404.
16. W. Vetter, *Biochemical Applications of Mass Spectrometry* (G. R. Waller and O. C. Dermer, eds.), Wiley-Interscience, New York, 1980, pp. 439–467.
17. D. F. Hunt, G. C. Stafford, Jr., F. W. Crow, and J. W. Russell, *Anal. Chem.* **48,** 2098 (1976).
18. G. W. A. Milne, H. M. Fales, and T. Axenrod, *Anal Chem.* **43,** 1815–1820 (1971).
19. T. Murata, T. Ariga, M. Oshima, and T. Miyatake, *J. Lipid Res.* **19,** 370–375 (1978).
20. R. C. Dougherty, S. Howard, and J. D. Wander, in *Polynuclear Aromatic Hydrocarbons in the Environment* (N. L. Richards, ed.), Dekker, New York, 1982.
21. T. H. Risby, *Envir. Health Pers.* **36,** 39–46 (1980).
22. R. C. Dougherty and J. D. Wander, *Biomed. Mass Spectrom.* **17** 401–404 (1980).
23. R. C. Dougherty and A. Mitch, unpublished.
24. R. C. Dougherty and K. Piotrowska, *Proc. Natl. Acad. Sci.,* **73,** 1777–1781 (1976).
25. R. C. Dougherty, M. J. Whitaker, L. Smith, D. L. Stalling, and D. W. Kuehl, *Enriron. Health Persp.* **36,** 103–117 (1980).
26. R. C. Dougherty, M. J. Whitaker, S.-Y. Tang, R. Bottcher, M. Keller, and D. W. Kuehl, in *Environmental Health Chemistry* (J. D. McKinney, ed.), Ann Arbor Science, Ann Arbor, 1980, pp. 263–277.
27. D. W. Kuehl, M. J. Whitaker, and R. C. Dougherty, *Anal. Chem.* **52,** 935–940 (1980).
28. D. L. Stalling, L. M. Smith, and C. Rappe, in *Environmental Health Chemistry* (J. D. McKinney, ed.), Ann Arbor Science, Ann Arbor, 1980, pp. 245–261.
29. H. R. Buser, personal communication.

5

# *USE OF PYROLYSIS/ GAS CHROMATOGRAPHY/ MASS SPECTROMETRY IN ENVIRONMENTAL ANALYSIS*

*H. M. BALBA*

## *1. INTRODUCTION*

In general, chemists resort to pyrolysis/gas chromatography (PY/GC) and pyrolysis/gas chromatography/mass spectrometry (PY/GC/MS) when they exhaust all other conventional methods of gas chromatography or mass spectrometry. It is well known that polar and nonvolatile high-molecular-weight chemicals are not amenable to conventional gas chromatography and/or mass spectrometry. However, using controlled thermal degradation, these chemicals often yield volatile and gas chromatographable products. Although this technique has been used in several analytical fields, such as the analysis of synthetic and biological polymers, oil shales, petroleum, and others, as described by May, Pearson, and Scothern,[1] its use for environmental analysis is still limited. However, recently this technique was employed for the analysis of some environmental samples. This chapter describes the technique of this analysis and outlines some of the recent applications for the analysis of (1) bound residues of pesticides in plant and soil samples and (2) humic substances from soil samples.

## *2. COMMON METHODS OF PYROLYSIS*

The most common techniques used for pyrolytic analysis are the use of pyroprobes and furnace-heated tubes. The use of pyroprobes is ideal for small samples (μg quantities). Two types of pyroprobes are

*H. M. BALBA* • Uniroyal Chemical Co., Naugatuck, Connecticut 06770.

available commercially: (a) hot filament or ribbon probes and (b) Curie-point probes. These two types of pyroprobes, although different in the theoretical basis of heating the samples, are alike in several aspects of the technique. Both are good for sample size close to 100 μg. Both could be mounted in the injector of any of several gas chromatographs. When used, the sample undergoes flash pyrolysis on the order of fractions of a second.

### 2.1. *Hot Ribbon or Filament Pyroprobes*

Hot ribbon or filament pyroprobes consist of an electrically heated resistor made of platinum or nichrome coil or foil mounted on a probe-like rod. The probe is designed to fit into an interface connected to the inlet port of the gas chromatograph. The carrier gas of the gas chromatograph is introduced to the interface which flushes all the pyrolysis products into the column. The temperature of the coil or foil is electrically heated to the desired pyrolysis temperature in seconds or fractions of seconds. The heating is carried out by applying a variable voltage at the two ends of the coil or foil. To overcome the low heating rate, the voltage applied is raised by a voltage sweep unit which boosts up the applied voltage to obtain the final temperature within the desired time.

In most cases the sample is placed as a solution on the wall of a quartz tube that fits in the coil. Solid sample is placed either in the coil or inside the quartz tube. When ribbon pyroprobe is used, the sample solution is placed on the ribbon with a microsyringe to form a thin layer on the surface of the ribbon. The solvent is allowed to evaporate before applying the pyrolysis temperature. Usually the heat of the interface and the stream of the carrier gas is sufficient to evaporate the solvent.

Chemical Data System (CDS), Oxford, Pennsylvania, has developed a satisfactory unit that can be fitted on several gas chromatographs with minimum modification. The same unit is now being sold by several other companies.

### 2.2. *Curie-Point Pyroprobes*

Curie-point probes are equipped with a ferromagnetic coil made of alloys containing iron, cobalt, or nickel. The coil is placed in a quartz tube. A high-frequency induction coil is placed around this tube which heats the ferromagnetic wire by induction. The wire heats very quickly until it becomes paramagnetic—that is, up to its Curie point. Accordingly the probe will not heat by induction beyond its Curie point. Selecting wire of different Curie points allows the use of this pyroprobe at specific pyrolysis temperatures. This pyroprobe has the advantage of having a well-defined upper temperature.

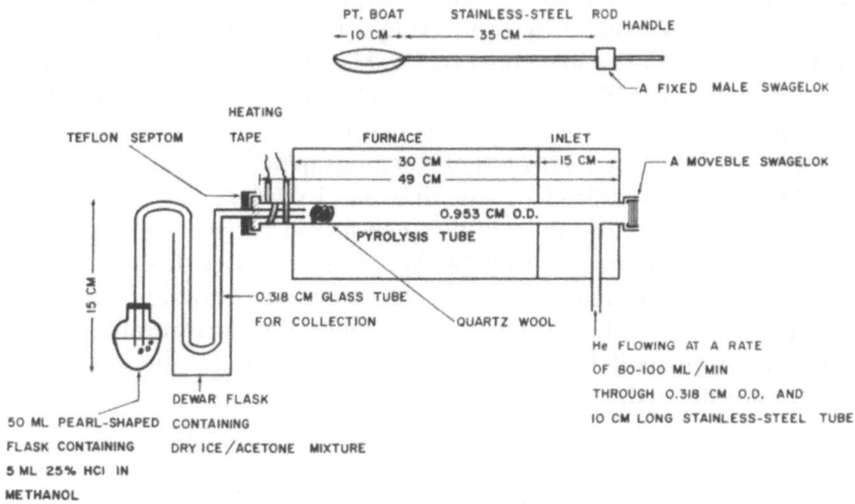


**FIGURE 1.** Diagram of a furnace-heated pyrolyzer.[4]

## 2.3. *Furnace-Heated Tube*

A furnace-heated tube is usually a homemade device, consisting of a quartz or stainless steel tube heated by an outside furnace. The sample is delivered by different techniques usually in a platinum boat attached to a pushing rod to place the sample boat in the hot zone. An inert gas is employed to sweep the pyrolysis products, as they form, out of the tube (Fig. 1). In most cases pyrolyzate is collected in a cold receiver and later chromatographed.

This technique has the advantage of using a larger sample (about 100 mg) and obtaining greater quantities of the pyrolysis products. However, there are sources of potential errors such as (1) occurrence of secondary reaction of the pyrolysis products because the time spent under pyrolysis temperature is far greater than that of coil or ribbon pyroprobes and (2) the possibility of contamination by the products of the previous samples.

# 3. *GENERAL TECHNIQUES*

Pyrolysis/gas chromatography/mass spectrometry (PY/GC/MS) could be carried out simultaneously in one single unit by interfacing all these elements in series. Interfacing gas chromatographs with mass spectrometers (GC/MS) is described in other chapters of this book. PY/GC/MS set-up can be assembled by connecting a pyroprobe, described earlier in this chapter, to the inlet port of GC/MS unit. Gallegos described an interfaced PY/GC/MS used to analyze coal samples.[2] Uden, Henderson, and Lloyd have published a comprehensive description of several pyrolysis units interfaced with gas chromatographs.[3]

An alternate method of carrying out PY/GC/MS is to collect the pyrolyzate from a hot-tube pyrolyzer in a cold receiver. The pyrolysis products are then analyzed by GC/MS. In some cases this technique is preferred over the single-unit assembly. In addition to the large size of samples that could be pyrolyzed, it permits the derivatization of the pyrolysis products if far better chromatographable products and mass spectra are needed.

### 3.1. *Use of Pyrolysis in the Analysis of Bound Residues of Pesticides*

Bound residue of pesticides refers to the fraction retained by the sample after extraction. This fraction is often named unextractable, insoluble, or bound residue. The magnitude of this fraction depends on, among others, the solvent used for extraction and the degree of the extraction. This fraction has often been disregarded in spite of the fact that in some cases it accounts for a significant percentage of the total residue.

Until recently there was no successful method to analyze for this fraction. Initial attempts were targeted to solubilize this fraction by hydrolytic means. However, this approach had limited success and was not used as a regular analytical method.

Balba, Still, and Mansager[4] and Still, Balba, and Mansager[5] were able to determine the bound residue of chloroaniline chemicals in plant tissues by PY/GC. Also, they reported that this technique was used to study the molecular binding of chloroaniline with lignin.

Using a furnace-heated tube pyrolyzer these authors were able to pyrolyze, among other samples, the bound residue fractions of [$^{14}$C] 3-chloroaniline and [$^{14}$C] 3,4-dichloroaniline-treated rice plants. Upon pyrolysis, samples released more than 80% of the radioactivity as volatile products. GC/MS of the pyrolyzates indicated that the major radioactive products were the unchanged chloroaniline compounds. The pyrolyzates were collected in cooled 25% HCl methanol, purified, and derivatized to the corresponding trifluoroacetanilide. The concentration of these derivatives was determined by gas chromatography and electron capture detection. The sensitivity of the procedure as described was reported to be about 0.3 ppm of bound chloroaniline.

Still, Balba, and Mansager[5] also utilized this technique to elucidate the nature of binding of chloroaniline compounds with synthetic lignin-chloroaniline copolymers. These copolymers were prepared as model chemicals to simulate the chloroaniline-lignin-bound residue. While furnace-heated tube pyrolysis yielded mainly the corresponding aniline as mentioned earlier, the hot ribbon pyroprobe PY/GC of [$^{14}$C] 3-chloroaniline-synthetic lignin copolymer produced 1,2-bis(3-chloroaniline) ethylene.[4] Several other studies on the same copolymer gave evidence that the nitrogen of 3-chloroaniline was bound to the benzylic carbon (the -carbon) of coniferyl alcohol (4-hydroxy-3-methoxycinnamyl alcohol) used as a monomer in the copolymerization reaction. Accordingly, if this theory is right, it is likely that the two carbon atoms in the ethylene moiety of the pyrolysis product were the benzylic carbon atoms. Therefore, in order to pinpoint the carbon atom in the side chain that was binding with the aniline nitrogen, a systematic labeling of each carbon in the chain was carried out. [$^{14}$C − B] and [$^{14}$C − γ] coniferyl alcohol were prepared and copolymerized with nonradioactive 3-chloroaniline.[6] The copolymer was analyzed by PY/GC using hot ribbon pyroprobe. The same pyrogram was obtained; however, the isolated bis-chloroanilino-ethylene was not radioactive. Naturally, results would have been more informative and conclusive if [$^{14}$C] bis-chloroanilino-ethylene had been obtained from the pyrolysis of a [$^{14}$C] lignin-3-chloroaniline copolymer, as indicated in Fig. 2. Unfortunately the authors did not complete this investigation. However, the authors presented other results that gave satisfactory evidence of the postulated binding of aniline nitrogen with the benzylic carbon.

Khan and Hamilton reported that the pyrolysis of a bound residue fraction of soil treated with the herbicide Prometryn [2-(methylthio)-4,6-bis(isopropylamino)-*s*-triazine] yielded volatile products.[7] Treated soil samples were pyrolyzed at 800 °C under helium in a furnace-heated tube and pyrolysis products were collected in four traps. The pyrolyzate was analyzed by GC/MS and proven to contain Prometryn and its metabolite hydroxypropazine. The two products were quantitated by gas chromatography. Soil fortified with Prometryn and hydroxypropazine gave upon pyrolysis followed by gas chromatography 80.5 and 52.5% recoveries of these compounds, respectively.

In a related study, Worobey tested several pyrolytic techniques to the bound residue of 4-chloroaniline with soil humic acid.[8] His study concluded, among other things, that pyrolysis of humic-acid-bound 4-chloroaniline at 550 °C in a furnace-heated tube yielded about 50% 4-chloroaniline in the complex. However, if the pyrolysis of the same complex was preceded by base hydrolysis, the recovery of 4-chloroaniline was considered diminished.

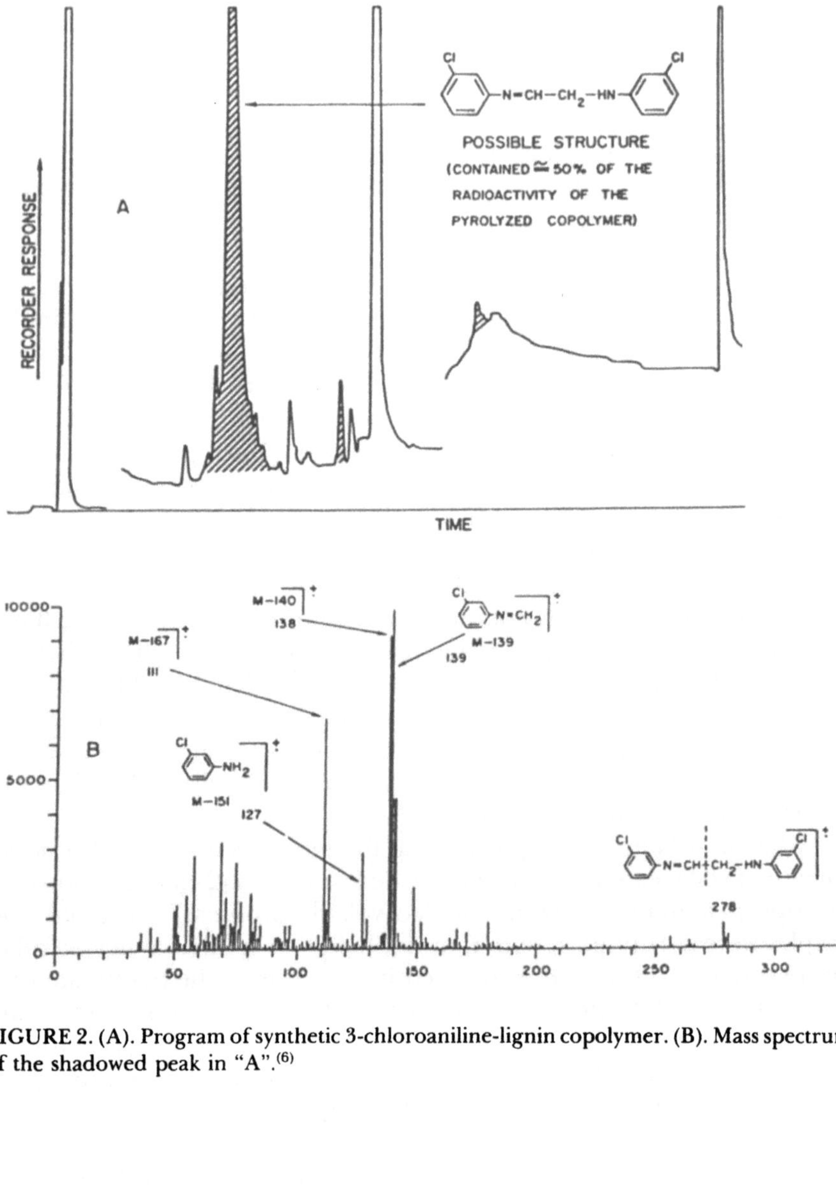


FIGURE 2. (A). Program of synthetic 3-chloroaniline-lignin copolymer. (B). Mass spectrum of the shadowed peak in "A".[6]

### 3.2. *Characterization of Soil Organic Matters*

One of the potential applications of pyrolysis in environmental analysis is the use of this technique in studying the soil organic matters. Nagar,[9,10] Simmonds, Shulman, and Stembridge,[11] Kimber and Searle,[12] Martin,[13] and more recently Meuzelaar[14] all have used PY/GC, PY/MS, or PY/GC/MS to analyze soil organic matters.

Organic matter in soil is a wide spectrum of organic materials. Soil organic matter could be classified into nonhumic and humic substances. The former consists of recognizable remains of soil flora and fauna and is the precursor of the humic materials in soil. Humic substances are formed by biological and nonbiological degradative and synthetic chemical reactions. This class plays an important role in determining soil productivity. It has a profound effect on soil structure, drainage, aeration, water, and buffering capacity and serves as a storehouse of chemical elements essential for plant. In addition to that it has a significant environmental importance because it binds with organic pesticides and pollutants. Because of these important characteristics, efforts were directed to classify soil organic matters in different soil types.

Humic substances are classically fractionated into four distinct portions based upon their solubility: (1) humic acid, soluble in alkali; (2) fulvic acid, soluble in both alkali and acid; (3) humic, insoluble both in alkali and acid; and (4) hematomelanin, soluble in ethanol. All of these fractions represent a complex mixture of polymeric substances having molecular weights from about 1000 to 300,000. Since humic acid is the major component in this mixture, numerous attempts have been made to devise a structural formula to represent it, but none were entirely satisfactory. Nevertheless, several pyrolytic studies on this fraction were conducted to recognize the chemical moieties included in its structure.

Among the most recent reported studies on the pyrolysis of humic acid is that published by Meuzelaar.[14] Many studies on the pyrolysis of humic acid were performed using PY/MS without GC. In these cases low-voltage electron impact ionization (15 eV) was employed to minimize fragmentation to identify the parent ions of each pyrolysis product. Using this technique, Meuzelaar analyzed five humic acid fractions isolated from five different soils. It was found that on the whole, the soil humic acids yield ions characteristic of sulfides, pyrroles, benzene, alkyl benzenes, phenols, and indoles. Although there were no differences in the pyrolysis products of the five humic acids analyzed, there were observable differences in the relative abundance of the pyrolysis products.

PY/GC/MS can play an important role in detecting, identifying, and quantitating the bound residues of pesticide chemicals in soil, plant, and animal samples. Also it is possible to apply this technique to analyze environmental samples of polymeric nature such as humic substances in water or soil and particulate matter in the atmosphere.

## REFERENCES

1. R. W. May, E. F. Pearson, and D. Scothern, *Pyrolysis–Gas Chromatography,* The Chemical Society, London, 1977, pp. 53–71.

2. E. J. Gallegos, in *Analytical Chemistry of Liquid Fuel Sources* (P. C. Uden, S. Siggia, and H. B. Jensen, eds.), Advances in Chemistry Series, No. 170, American Chemical Society, Washington, D.C., 1978, pp. 13–36.
3. P. C. Uden, D. E. Henderson, and R. J. Lloyd, *J. Chromatogr.* **126,** 225–237 (1976).
4. H. M. Balba, G. G. Still, and E. R. Mansager, *J. Assoc. Off. Anal. Chem.* **62,** 237–240 (1979).
5. G. G. Still, H. M. Balba, and E. R. Mansager, *J. Agric. Food Chem.* **29,** 739–746 (1981).
6. H. M. Balba and G. G. Still, *J. Lab. Compds. Radiopharm* **15,** 309–319 (1978).
7. S. U. Khan and H. A. Hamilton, *J. Agric. Food Chem.* **28,** 126–132 (1980).
8. B. L. Worobey, Ph.D. thesis, University of Manitoba, Winnipeg, Manitoba, Canada, 1980.
9. B. R. Nagar, *Nature* **199,** 1213–1214 (1963).
10. B. R. Nagar, E. S. Wright, H. L. C. Meuzelaar, and P. G. Kistemaker, *Plant Soil* **43,** 681–685 (1975).
11. P. G. Simmonds, G. P. Shulman, and C. H. Stembridge, *J. Chromatogr. Sci.* **7,** 36–41 (1969).
12. R. W. L. Kimber and P. L. Searle, *Geoderma* **4,** 47–55 (1970).
13. F. Martin, *Z. Pflanzenern. Bodenk.* 407–415 (1975).
14. H. L. C. Meuzelaar, K. Haider, B. R. Nagar, and J. P. Martin, *Geoderma* **17,** 239–252 (1979).

6

# *ATMOSPHERIC PRESSURE CHEMICAL IONIZATION MASS SPECTROMETRY*

*J. B. FRENCH, B. A. THOMSON, W. R. DAVIDSON, N. M. REID, and J. A. BUCKLEY*

## *1. INTRODUCTION*

Atmospheric pressure chemical ionization (APCI) is one of the growing repertoire of soft ionization techniques available to the mass spectroscopist. APCI offers many advantages particularly in environmental applications, some of which are only partially developed at present.

Many environmental analyses require determination of only one or at most a few compounds or classes of compounds in a complex mixture, or "targeted compound determination" rather than complete sample analysis.[1] Further, as the attention of environmentalists and regulatory agencies becomes more focused on the movement in the biosphere of carcinogens and mutagens which are of concern at trace levels, the requirements are for both selectivity and sensitivity. Chemical ionization carried out at high pressures, including atmospheric pressure, can provide high sensitivity along with selectivity through control of the ion–molecule reactions via reagent addition or deletion. APCI sources provide mass spectra which are predominantly composed of the ions equal or close to the mass of the unfragmented parent ion. While this information does not provide as positive an identification of the compound as the fragmentation pattern resulting from electron impact, it has been shown in many important environmental monitoring applications that the combination of APCI response to chemical reagent ad-

*J. B. FRENCH, B. A. THOMSON, W. R. DAVIDSON, N. M. REID,* and *J. A. BUCKLEY* • SCIEX, 55 Glen Cameron Road, Unit 202, Thornhill, Ontario L3T 1P2, Canada.

dition plus the molecular weight identification constitutes a very powerful air-monitoring tool. It is thus often possible to analyze complex gaseous, liquid, and solid samples directly, eliminating the time and sample work-up constraints imposed by more conventional gas chromatograph/mass spectrometry (GC/MS) techniques. Such direct sampling permits a real-time monitoring system. A number of examples in this chapter illustrate this point.

An APCI mass spectrometer system can be used for continuous multiple ion monitoring or to determine specific traces rapidly in a large number of samples whether these samples are precollected air, water, or solid extracts. In all cases sample introduction and ionization at atmospheric pressure permit a speed, flexibility, convenience, and freedom from memory effects that are very difficult to achieve at lower pressures. Sample introduction and ionization methods which are particularly useful for environmental applications are described.

The recent advent of multiple mass spectrometer systems has dramatically increased the applicability of APCI techniques to environmental analysis.[2] These systems constitute a double filter in which molecular ions selected in the first mass spectrometer are fragmented by collision-induced dissociation (CID) to produce ion fragment spectra in the second mass spectrometer, providing positive identification of the molecular ion. They inherently require an intense source of molecular ions such as a chemical ionization source. It is clear that CI/MS/MS systems are going to revolutionize environmental mass spectroscopy in the coming decade in the same way that GC/MS did in the previous decade because of the tremendous increase in sample throughput they permit and because they can provide continuous monitoring capability with extreme sensitivity, positive identification, and freedom from cross chemical interference (chemical noise).

## 2. *THE DEVELOPMENT OF APCI MASS SPECTROSCOPY*

It is appropriate to review in rather summary fashion the evolution of the atmospheric pressure chemical ionization/mass spectrometry (APCI/MS) combination, not only because this evolution provides some insights into the gradual process by which this technique has come to be applied to environmental problems, but also because the process has involved many interesting contributions from various research groups. It is hoped that the publications cited will serve as a bibliography for those interested in pursuing the subject further.

Papers on ions sampled from atmospheric pressure flames provide some of the early examples of situations in which APCI processes were coupled to mass spectroscopy. In Knewstubb and Sugden[3] and Hayhurst and Sugden[4] a differentially pumped flame sampling system is described. In both papers the ion population sampled is discussed in terms of reactions among the ions and neutral species in the flame; in the latter paper alkali metals are added, providing one of the earliest examples of deliberate modification of the CI processes and a demonstration of sensitivity to traces, in this case the alkali metals.

In the early 1960s Kebarle and co-workers began a series of studies which involved high pressure (up to one-half atmospheric pressure). Their ionization source involved α particle irradiation, described in Kebarle and Godbole and Hogg and Kebarle.[5,6] These papers, aimed at establishing gas phase parameters such as ion proton affinities, clearly describe a number of the ion–molecule reactions involved and point to "very great sensitivity to trace amounts of impurities."

Shahin introduces for the first time a corona discharge as the ionization source, and contributes to the discussion of the ion-clustering phenomena, both in the atmospheric pressure reaction region and during the rapid adiabatic expansion into vacuum (discussed later in this chapter).[7,8] His work also clearly demonstrates trace sensitivity, for example to deuterated water.

During the late 1960s and early 1970s development by Cohen, Carroll, and co-workers of the combined technique of gas phase ion mobility separation in tandem with mass spectroscopy proceeded at the Franklin GNO Corporation Laboratories. Inherently involving APCI/MS, this work is conveniently summarized by Karasek and co-workers.[9–11] These papers not only describe the plasma chromatography/mass spectrometry (PC/MS) combination, but also give examples of analytic trace detection using both the positive-ion and negative-ion populations, for narcotics and explosive detection into the picogram range. Carr also contributed to the application of the PC/MS technique, particularly in the negative-ion mode.[12]

During much the same period, from 1973 to 1978, Horning and co-workers at the Baylor College of Medicine began to publish an important series of papers on APCI/MS, utilizing both a $^{63}$Ni β emitter ion source and a needle-to-plane corona discharge source. Griffin (at Baylor) and co-workers provide insight into the ion–molecule chemistry related to atmospheric nitrogen, of fundamental importance in the application of this technique to environmental air monitoring.[13] Here a plasma chromatograph was interposed between the ion source and the mass spectrometer. However, Horning and co-workers deleted the ion mobility preseparator and describe results demonstrating picogram sensitivity levels achieved with a $^{63}Ni$ source.[14,15] This was improved to provide a 150-fg detectability and to show the combination of APCI/MS with liquid chromatography for the first time.[16,17]

Incidentally, this was the first time the API acronym was introduced. Although use has begun to be more widespread, it is now suggested that APCI is a better term, in order to properly distinguish the use of chemical ionization techniques extended to very high pressures from other high-pressure ion sources such as plasma discharges, and so on.

This series of publications continued with work by Dzidic and co-workers on the chemistry of $O_2^-$ acting as a strong base and in the production of the phenoxide ion, both topics of importance in the utilization of APCI in the negative-ion mode in air environmental monitoring.[18–20] Further descriptions of the corona discharge source and comparisons of the corona discharge and $^{63}Ni$ source operation are given in Dzidic and co-workers and Carroll and co-workers.[21,22] Horning reported the coupling of a gas chromatograph to the APCI/MS system.[23]

During this period several other laboratories were involved in APCI/MS studies. One important class of reactions of atmospheric ions is the clustering or addition of molecules such as water to form a series of ions. Imeson and Harden studied the propensity of protonated trace ions to form clusters, termed ion cluster mass spectroscopy, and examined whether there were related analytical possibilities.[24].

Siegel and Fite used a combination of elevated temperature plus careful cleaning and drying of the carrier and sample gas to reduce clustering in a terminal plasma (in which no significant fields are applied in the APCI source and ion–molecule reactions are allowed to go to completion).[25] This approach is not generally applicable to direct air sampling or to other complex mixtures without chromatographic pre-separation.

Kambara and Kanomata studied APCI corona discharge sources and attacked the clustering problem by making use of CID in the intermediate-pressure (differentially pumped) stage between the ion source and the mass spectrometer.[26,27] Applying accelerating voltages of the order of 10 V/cm at pressures of 1 torr, they demonstrated stripping of water molecules from various molecular and quasimolecular ions.

Over this same time period work in French's laboratories had evolved the combined technique of interposing a gas curtain or membrane between the ion source and the expansion into vacuum, and applying electric fields to produce CID during the expansion in the vacuum.[28,29] Transferring the clustered product ions into the dry curtain gas strongly shifts the cluster equilibrium distribution toward low-cluster members, and elimination of significant water concentration reduces clustering occurrence due to the rapid adiabatic cooling during expansion.[30] The electric field CID completes the declustering. The gas curtain enhances the cleanliness of the vacuum since only ions, and not sample molecules, are admitted to the vacuum chamber. A cryogenic pumping system was used for the first time, and sensitivities of pg/cm$^3$ for organics in air were demonstrated as early as 1974. This was later improved to fg/cm$^3$.[31,32]

## 3. *ION SOURCE DESIGN AND CHARACTERISTICS*

The atmospheric pressure ion source is notable for its simplicity of design and ease of sample introduction, both because of the absence of vacuum pump requirements and because it can be interfaced directly to the real world without vacuum locks or variable leaks. Samples are introduced either by sampling vapor sources such as ambient air or GC effluents directly, or by volatilization of liquid and solid samples into a carrier gas which flushes the source. Because the source operates at high pressure, hot filaments cannot be used to produce electrons. Suitable electron sources which have been used are coronas and β foils, both of which require a minimum amount of maintenance, and neither of which are particularly sensitive to the presence of corrosive or oxidative gas mixtures.[7,14,29,33,34] In addition to these, atmospheric pressure ionization based on field emission from liquid surfaces and on a high-temperature microwave discharge has been successfully developed for specialized analytical applications.[35,36]

The development of APCI sources has followed two divergent paths. One design, pioneered by the Baylor College of Medicine Group and Extranuclear Laboratories is similar to an electron capture detector (ECD). It consists of a low-volume (about 1 cm$^3$) closed source coupled to the vacuum through a small (approximately 25 μm) orifice. This type of source is flushed with a pure carrier gas at 10–100 ml/min and must be heated to minimize absorption of involatile compounds on the walls. Quantitative analysis is best performed by introducing only one compound at a time into the source, so that applications have focused primarily on coupling this source to a gas chromatograph or liquid chromatograph to act as an ultrasensitive MS detector or on introducing single-component samples in a solvent.[17,23]

The second type of APCI source, developed at the University of Toronto by Buckley, French, and Reid, is now used in the TAGA® APCI/MS system manufactured by SCIEX.[30] This source is coupled to the vacuum system through a larger orifice (approximately 100 μm), which is separated from the reaction region by a stream of high-purity dry nitrogen. This gas membrane helps to decluster the ions before they reach the mass spectrometer and allows moist air to be sampled and ionized directly, while still presenting the bare unclustered ions to the mass filter. The ion source itself is designed to allow the use of carrier or sample gases ranging from ambient air (for direct analysis) to various pure gas and reagent gas mixtures, at flow rates of up to 9 l/sec. The ionization source (generally a corona) is suspended in the large source chamber in the flowing gas stream, with the reaction volume being defined by the region between the corona point and the gas curtain (which quenches the reactions). The ions and sample molecules react in this wall-less region, which is continuously swept by the carrier gas to exclude molecules that desorb from the distant walls. This technique allows the source to be operated at ambient temperature with little or no memory effect even with very involatile or "sticky" compounds. Samples can be introduced directly (as ambient air, for example), in the effluents of GCs or LCs without splitting, or from direct probes, by entraining the vaporized sample in the center of the carrier gas flow.

The response of any CI source is directly proportional to concentration (molecules of sample per unit volume). An APCI source is therefore potentially more sensitive than a conventional CI source in which the sample is at lower density. A consideration of the kinetics of the reaction reveals the fundamental differences between the β foil source and the corona, which influence the overall response of the sources. The β foil produces a bipolar ion distribution in the low-volume source, with equal positive and negative ion (or electron) densities governed by the balance between the ionization and recombination process. Siegel and Fite, and Carroll and co-workers, have shown that for a typical ion density of the order of $3 \times 10^8$ ions/cm$^3$, the concentration of sample ions ($n_s^+$) is directly proportional to the concentration of the sample molecules ($n_s$), but is independent of the source strength and of the source geometry:

$$n_s^+/n_s = k_1/k_r$$

where $k_1$ is the reaction rate constant and $k_r$ is the recombination rate constant.[25,16] As pointed out by Siegel and Fite, this response is only true for the case of a single reaction; the presence of two competing sample species in the source changes the response to each individual species by competing for the available charge.[25] The ion lifetime in this source is approximately 5 msec.[16] The eventual sensitivity in terms of ion signal at the detector per unit sample concentration is largely determined by the ion sampling and focusing efficiency. Carroll and co-workers have demonstrated a sensitivity to methyl-γ-pyrone of approximately $10^3$ ions/sec for 0.2 ppb in the carrier gas, or 5 ions/sec per 1 ppt.[16]

The corona employed in the open ion source (and which has also been operated in the low-volume source) generates a current of unipolar ions through the reaction region in which recombination plays no role. A simple calculation shows that the corona produces an average ion density of $10^{10}$ ions/cm$^3$, larger by a factor of 10–100 than the ion density in the β foil source. Furthermore, the average lifetime of an ion is governed by the drift time of the ions between the needle and orifice; for the high electric field between the corona point and the orifice, and for a reaction gap of 1 cm, this lifetime has been calculated to be approximately 0.5 msec. It can be shown that under reasonable assumptions of source conditions the sample ion concentration depends on sample molecule concentration, reagent ion concentration ($n_R^+$), reaction rate constant ($k_1$), and drift time ($t$).

$$n_S^+ = n_R^+[1 - \exp(-k_1 n_s t)]$$

Because of the shorter reaction time, the response to one species is not influenced by the presence of others at trace levels, making the source particularly useful for direct complex mixture analysis. The sensitivity can be optimized by changing the drift time through adjustment of the position of the corona relative to the orifice. This was investigated by Dzidic and co-workers.[21] By substituting appropriate values for drift time (0.5 msec), ion densities ($10^{10}$/cm$^3$) and reaction rate constant ($10^{-9}$ cm$^3$/sec), one can calculate that 1 ppb of sample in the carrier gas should produce a sample ion concentration which is approximately 2% of the reagent ion concentration. This corresponds to a sensitivity of about 40 ions/sec for 1 ppt, a sensitivity which has been demonstrated with the TAGA® system for a variety of compounds.

Samples are introduced into the APCI source in a flow of carrier gas, in which gas dynamic effects play a major role. Diffusion of the sample molecules to surrounding surfaces is much slower at atmospheric pressure than at low pressure or in a vacuum. This is an effect which, when properly used, greatly minimizes problems of source contamination. By introducing the sample so that it is surrounded by a flow of carrier gas, and ensuring that it reaches the ionization region before the molecules can diffuse to the walls, a true wall-less injection, sample transfer, and ionization system can be obtained. In the low-volume, closed source, this is not possible. The dimensions of the source and the flow rate are such that the sample becomes immediately well mixed in the carrier gas, and is in intimate contact with the walls of the tubing and the source, resulting in the previously mentioned need to keep the walls hot. In the open design source, in which the inlet tube can be up to 2 cm in diameter, samples can be introduced into the carrier gas just upstream of the ionization region, and if the flow is laminar, the sample will remain confined within a thin streamtube as it passes through the source to be operated at ambient temperature. Direct-insertion probes, GC interfaces, and other discrete-sample inlet systems have been designed to take advantage of this effect.

Since the sensitivity of an APCI source depends directly on the concentration of the sample in the carrier gas which passes through the reaction volume, the flow rate must be optimized for minimum sample dilution. The low flow rates used in the closed source (a few cc/sec) are well suited for satisfying this condition; a pulse injection of 100 fg of a compound of molecular weight 200 into a flow of 3 cc/sec gives a relative (v/v) concentration of about 35 ppt in the carrier gas. Detection limits in the range of a few tens of femtograms have been demonstrated by both the Baylor group and the Extranuclear Laboratories group. In the open-source design, in which higher total flow rates (generally 50–100 cc/sec) are used, the same effect is achieved by ensuring that the sample is only contained in the center part of this flow. Using this technique, detection limits in the low-picogram to sub-picogram range have been achieved.[37] For sampling "infinite sources" (such as ambient air), where the sample gas is also the carrier, high flow rates are required in order to minimize absorption on the walls of the sample line. The high flow rate both minimizes the contact time with the walls of the glass tubing and acts to dilute any molecules which do desorb from the walls. The ambient temperature open source can therefore be used both to measure parts per trillion levels of compounds in ambient air, as well as picogram to femtogram levels of samples which are volatilized into the center of a lower flow of carrier gas.

## 4. CHEMICAL IONIZATION AT ATMOSPHERIC PRESSURE

The analysis of trace constituents in complex matrices is achieved in an APCI source by the "soft" ionization of the traces by a series of reactant ions prior to mass separation. The specificity of the analysis to particular classes of compounds and the sensitivity of an APCI system to these classes is highly dependent on the ion–molecule chemical characteristics of both the reactant ions and the trace components of the matrix. The ion–molecule chemistry described in this section relates to the direct sampling of ambient air using the TAGA® system rather than the use of ultrapure, ultradry carrier gases in a closed APCI source. The reader is referred to Siegel and Fite for descriptions of the ion–molecule chemistry in such systems.[25]

Both positive and negative reactive ions can preferentially be formed in the APCI source using a corona discharge, with the polarity of point-to-plane field determining the polarity of the ions analyzed. In ambient air the reaction sequence for the creation of the reactant positive ions—the proton hydrates ($H_3O^+[H_2O]_n$)—is shown in Fig. 1. This sequence is derived from experimental results summarized by Huertas and Fontan.[38] The major primary ions ($N_2^+$, $O_2^+$, $H_2O^+$, $NO^+$) are formed by electron impact of corona-created electrons on the major neutral components in air. Although NO is normally not a major constituent of clean ambient air, the concentration of this species in the source is enhanced due to neutral reactions initiated by the corona discharge. The primary ions react extremely rapidly (within $10^{-6}$ sec) through the reactions described in Fig. 1 to form a terminal equilibrium set of proton hydrates. These hydrates dominate the total ion spectrum within $10^{-6}$ sec after the initial ionization and are the ions most likely to react with trace compounds (the total reaction time in the source is approximately $5 \times 10^{-4}$ sec).

In the negative-ion mode, electrons created by the corona are rapidly thermalized (i.e., loss of excess energy) through collisions with neutrals and are readily captured by electronegative species such as $O_2$ to form $O_2^-$. The superoxide anion ($O_2^-$), its hydrates ($O_2^-[H_2O]_n$), and its clusters ($O_2^-[O_2]_n$) are the major reactant negative ions in an APCI source

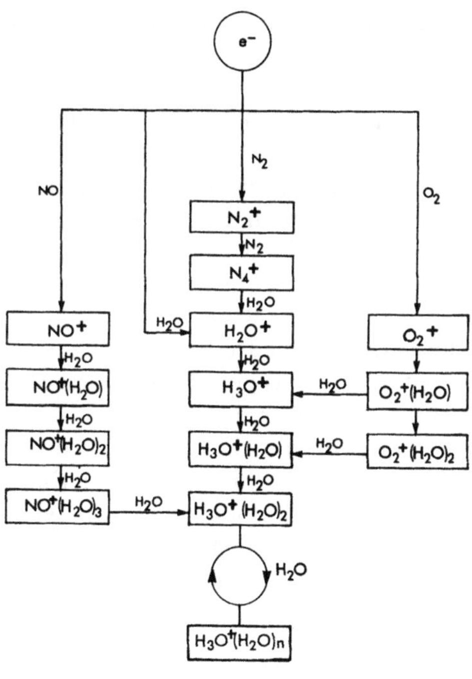


FIGURE 1. Ion–molecule reaction pathways for the atmospheric pressure ionization of air.

directly sampling ambient air. The neutral species formed in the corona (NO, $NO_2$, $O_3$, CO, etc.) also have significant electronegativity and may react with $O_2^-$ or $O^-$ (also formed upon electron capture by $O_2$) to form a series of minor reactant ions ($NO_2^-$, $NO_3^-$, $O_3^-$, $CO_3^-$, etc.). Techniques have been developed by Kotasek to inhibit production of those neutral species and thus simplify the ion–molecule chemistry.[39]

The reactant ions may react with trace components in the ambient air in several manners: charge transfer, where the positive or negative charge on the reactant ion is transferred to the trace, and proton transfer, where a proton is transferred from a positive ion to the trace, or a proton is transferred from the trace to a negative reactant ion. These are summarized in the following equations.

Charge transfer:

$$R^+ + T \rightarrow T^+ + R \tag{1}$$
$$R^- + T \rightarrow T^- + R \tag{2}$$

Proton transfer:

$$RH^+ + T \rightarrow TH^+ + R \tag{3}$$
$$R^- + TH \rightarrow T^- + RH \tag{4}$$

where R represents the reactant and T the trace.

Positive-ion proton transfer reactions use the proton hydrates as reactant ions. A proton or a proton–water cluster is transferred [reaction (5)]:

$$H_3O^+(H_2O)_n + T \rightarrow TH^+(H_2O)_m + (n - m + 1)H_2O \tag{5}$$

The water molecules clustered to the trace are stripped off in the gas curtain and in the declustering lens region before mass analysis. Thus, only $TH^+$ is observed at the detector even though its hydrates are formed in the source. Reaction (5) will only proceed if the gas phase acidity of $H_3O^+(H_2O)_n$ is greater than $TH^+(H_2O)_m$. This is generally true if the gas phase basicity (proton affinity) of T is greater than that of water, which is generally the case with most organic compounds containing oxygen or nitrogen.

The proton hydrates have recombination energies (ability to transfer positive charge) which are lower than most organic compounds and as a result they will not charge-transfer to these compounds. The reactant ions for charge transfer in ambient air are mainly $O_2^+$ and its hydrates (plus $N_2^+$, $H_2O^+$, and $NO^+$ to a lesser degree). Since these ions react rapidly with water vapor to form the proton hydrates, the charge transfer reactions do not occur over the complete reaction time ($5 \times 10^{-4}$ sec) but over the 1-μsec period when the $O_2^+(H_2O)_n$ ions are present in significant concentrations. Thus, the sensitivity of the APCI system is approximately three orders of magnitude less for charge transfer than for proton transfer. For example, benzene is detected as the $M^+$ ion with a detection limit in the 1 ppb range, whereas aniline is detected as the $MH^+$ in the 1 ppt range. To overcome this sensitivity problem in charge transfer reactions, a suitable chemical ionization reagent may be added to the ambient air flow entering the source region. For example, in the analysis of polychlorinated biphenyls (PCBs), benzene is added to the air in ppm levels. The benzene then competes with water vapor in reacting with $O_2^+[H_2O]_n$ (and other primary ions) to form a stable $C_6H_6^+$ molecular ion. The recombination energy of benzene is high enough that it readily charge-transfers to the PCBs, resulting in detection limits in the ppt range. Other charge transfer reagent ions with various recombination energies may be added to highlight other classes of compounds (e.g., $C_6F_6^+$ is used in the detection of chlorinated benzenes).

The superoxide anion has a low electron affinity and will charge-transfer to compounds of higher electron affinities (generally containing electronegative groups). In addition, $O_2^-$ readily abstracts protons from strong gas phase acids such as carboxylic acids, phenols, amides, and inorganic acids (HCl, $NHO_3$, etc.). For many chemicals the charge transfer reaction is accompanied by a dissociation reaction:

$$O_2^- + CHCl_3 \rightarrow Cl^- + O_2 + CHCl_2 \tag{6}$$

In general, the negative-ion reactions lead to detection limits for traces in the ppb to ppt range. A summary of detection limits for various compound classes of environmental interest is presented in Table 1.

As stated earlier, chemical ionization reagents may be added to the ambient air to alter the ion chemistry. In the example mentioned previously, benzene was added to enhance the sensitivity of the source to compounds which undergo charge transfer reactions. Similarly, a source of fluoride ions may be added to the ambient air to enhance the sensitivity of the system to compounds which are more acidic than $HO_2$, the conjugate acid of $O_2^-$, since $F^-$ more readily extracts protons than does $O_2^-$. On the other hand, the selectivity for ionization may also be altered by adding reagents which will ionize fewer classes of compounds than would the normal ambient air reagent ions. For example, if ammonia is added to ambient air in ppm concentrations or greater, the reactant ions become ammonium hydrates, $NH_4^+[H_2O]_n$, rather than proton hydrates. Since ammonia is a much stronger gas phase base than water, the ammonium hydrates will only protonate compounds of higher proton affinity (greater gas phase basicity) than ammonia. Compounds whose basicities fall above

*TABLE 1*
*Inherent Detection Limits Using the TAGA APCI/MS System*

| Compound class | Example | Inherent[a] detection limit |
|---|---|---|
| Alcohols | Methanol | 0.5 ppb |
| Aldehydes | Benzaldehyde | 10 ppt |
| Amides | Demethylformamide | 20 ppt |
| Amines | Pyridine | 2 ppt |
| Aromatic hydrocarbons | Toluene | 5 ppb |
| Aromatic halohydrocarbons | Dichlorobenzene | 1 ppb |
| Carboxylic acids | Acetic acid | 20 ppt |
| Dioxins, dibenzofurans | 2,3,7,8-TCDD | 1 pg |
| Esters | Ethyl butyrate | 0.3 ppb |
| Ethers | Diethyl ether | 50 ppt |
| Halogens | Bromine | 70 ppt |
| Inorganic acids | Sulfuric acid | 10 ppt |
| Ketones | Acetone | 40 ppt |
| Mycotoxins | Aflatoxin B1 | 10 pg |
| Nitriles | Benzonitrile | 4 ppt |
| Nitrosamines | Dimethylnitrosamine | 10 ppt |
| Organometallics | Trimethylarsine | 1 ppt |
| Pesticides | Sulfotep | 0.1 ppb |
| Polycyclic aromatic hydrocarbons | Anthracene | 50 ppt |
| Sulfur compounds (organic) | Dimethyl sulfoxide | 10 ppt |
| Sulfur compounds (inorganic) | Sulfur dioxide | 10 ppt |
| Terpenes (terpenelike compounds) | Linalool | 0.5 ppb |

[a] Based on a signal-to-noise ratio of 3 under optimum conditions.

water but below ammonia are thus not protonated when the $NH_3$ is added. For example, both dimethyl nitrosamine and diethyl ether are protonated by the proton hydrates to form an ion ($MH^+$) at mass 75. The addition of ammonia to the ambient air completely removes the peak due to diethyl ether, since the proton affinity of ammonia is greater than that of the ether, yet does not affect the nitrosamine which has a higher proton affinity. The same technique could be used in the negative-ion mode by adding reagents which produce weak basic anions such as $Cl^-$ or $Br^-$. Under such conditions, ionization would only occur for acids stronger than HCl or HBr.

There are several other chemical reagents which may be added to the ambient air carrier gas to provide high compound (or compound class) specificity. These reagents are used to produce ion–molecule reactions such as halide addition (in the detection of inorganic acids), nucleophilic substitution (in the detection of hexachlorobenzene),[(40)] and the use of photosensitization in the detection of mercaptans.[(41)] Since chemical ionization at atmospheric pressure is highly dependent on both the chemical characteristics of the reactant ions and the trace neutral, not all compounds may be ionized at high sensitivity in an ambient air APCI source. Aliphatic saturated hydrocarbons, for example, have proton affinities lower than water and thus cannot be protonated. Similarly, charge transfer reactions will not produce molecular ions from these hydrocarbons under ambient air APCI conditions. Although this may be undesirable if one wishes to detect hydrocarbons in an air matrix, it offers the advantages that the APCI mass spectrum is not congested with these ubiquitous compounds, which could potentially interfere with the analysis of other organic compounds. Similarly, aliphatic halohydrocarbons do not form positive ions under ambient air APCI conditions, and they tend to undergo dissociate charge transfer to halide ions when the analysis is done in the negative-ion mode. Although the system is sensitive to these compounds, the specificity is poor since several halohydrocarbons form the same halide ion. For applications in which air monitoring of such compounds is important, lower pressure CI sources should be used.[42]

A second problem which may be encountered in atmospheric pressure chemical ionization is a matrix effect caused by the sample matrix of compounds at high concentrations which alter the ion–molecule chemistry by reacting with the normal reactant ions to form a new series of reactant ions with different chemical characteristics. Thus, it is necessary either to monitor the reagent ion concentration during analysis to ensure that the concentration is not being depleted, or to add an internal standard to the sample. In practice, this problem has been encountered in very few cases and is tractable. To remove the matrix effect in severe cases, the sample may be diluted, or the interfering chemical may be removed by filtering or by specific collection of the trace on some medium which is then analyzed in ultrapure air.[41]

## 5. *ENVIRONMENTAL APPLICATIONS OF APCI/MS*

During its brief history, APCI/MS has been applied to a wide variety of environmental problems. In some cases mass spectrometry was able to be used for the first time because of the ability to sample ambient air or atmospheric pressure gases directly. In other cases the use of APCI/MS resulted in higher sensitivity or specificity than was achievable by other means. This combination of high sensitivity with the ability to sample air directly led in addition to the development of a mobile APCI/MS system which was able to be used for *in situ* air monitoring.[31]

One of the first demonstrations of the potential of APCI/MS in environmental analysis was by Carroll and co-workers.[33] They showed that an APCI ion source could tolerate the direct introduction of the liquid eluted from a liquid chromatograph. As an example, the separation and detection of three polycyclic aromatic hydrocarbons was shown, and a sub-nanogram detection limit demonstrated using selected-ion-monitoring (SIM) techniques.

One of the earliest applications of mobile mass spectrometry was performed with a TAGA 3000 in a van.[43] The system was used to produce an isoplethic map of the concentration of several organic vapors emitted from a chemical plant. With the instrumented van parked downwind of the plant, positive and negative mass spectra of the ambient air were obtained. After identifying the peaks which were associated with the plant (by subtracting the background spectrum obtained upwind), SIM was used to monitor the concentration of six targeted components as the vehicle was driven slowly in front of the buildings. Traverses were performed at increasing distances downwind, with some of the compounds identified above background levels at a distance of more than 2 miles from the site. The cross-plume profiles allowed isopleths of the concentration to be plotted and individual plumes and sources of each chemical identified. Figure 2 shows the recording from one traverse, where ammonia, dimethylamine, aniline, benzothiazole, and diphenylamine were monitored at *m/z* 18, 46, 94, 136, and 170, respectively.

A similar capability was demonstrated with the same system in a situation where the plume emitted from a tall stack was monitored and tracked at ground level. With the analysis being performed and the data displayed in real time, the levels of sulfur dioxide, hydrogen chloride,

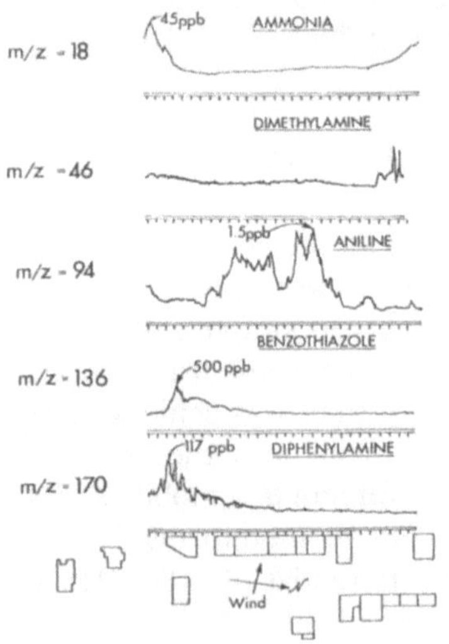


FIGURE 2. Selected-ion-monitoring traces of five ions recorded during a traverse across the front of a chemical manufacturing plant. The buildings are shown schematically at the bottom; horizontal axis represents distance, while the vertical axis represents intensity.

and ammonia were monitored while the vehicle was driven downwind of the stack. The impingement point of the plume could be identified by the levels of the compounds, and the ground-level plume could be identified and tracked even in varying wind situations. The sensitivity allowed the plume to be detected in situations where other dedicated specific compound detectors proved inadequate.

In addition to air monitoring, the use of a mobile mass spectrum system has been proven useful for *in situ* analysis of water and soil. Thomson and Roberts, for example, report on a technique developed for an APCI/MS system which allowed a soil analysis program to be performed at the site at which soil samples were obtained.[44,45] Soil plugs were analyzed for the presence of PCBs by thermal desorption, and more than 700 samples were analyzed in a nine-day period. The on-site monitoring produced immediate measurements of contamination levels in the soil, thus aiding in directing the survey, helping to protect the workers on site, and providing a continuously updated picture of the contamination pattern in the soil.

The same mobile TAGA® APCI/MS system was used by Lane to monitor the levels of sulfotep, an insecticide, in a large greenhouse during and after spraying operations.[46] With the system parked outside the greenhouse, heated glass sampling lines were installed to allow air to be drawn into the ion source from various locations in the greenhouse. Over a 48-hr period, the levels were monitored by cyclically sampling from each location. Figure 3 shows how the levels varied throughout

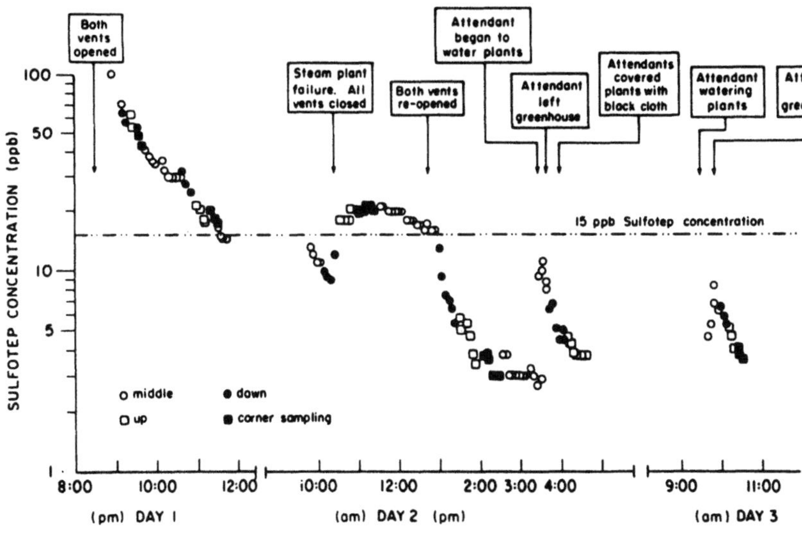


FIGURE 3. Concentration of sulfotep at different locations in a greenhouse after fumigation.

the time period and the correlations between the concentration and activities inside the greenhouse. The results showed that the concentration in the air could significantly exceed the recommended maximum 8-hr average level over short periods of time, and helped to indicate what type of protection should be instituted for the greenhouse workers.

Spills of hazardous chemicals have produced situations requiring immediate monitoring of concentrations in the air around the spills. One example is reported by Lane and Thomson, in which two mobile TAGA systems were used to monitor levels of chlorine in the air around the site of a train derailment and fire in which a tanker of liquid chlorine was leaking.[44] Chlorine was detected in the air and monitored at levels down to a few ppb during the 11-day program. The situation involved the evacuation of more than 250,000 people from their homes as a precaution against the potential release of a large amount of chlorine,

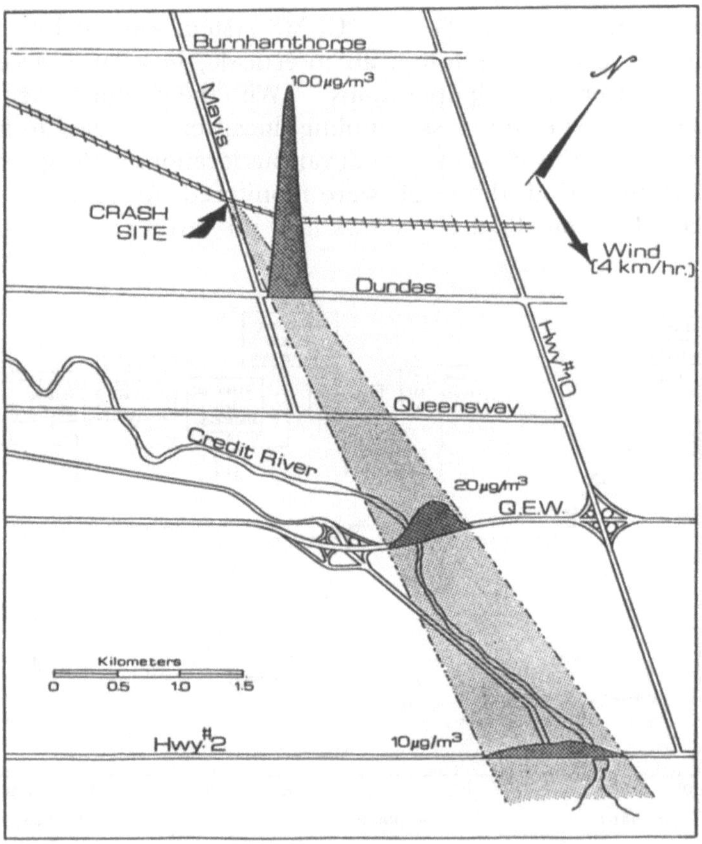


FIGURE 4. Ground-level profiles of the chlorine plume at three distances downwind of the site of a train derailment.

and the use of the APCI/MS systems provided the only sensitive measurements of chlorine levels in the air around the site (and up to 5 miles away). Figure 4 shows the profile of the chlorine plume obtained by traversing the ground-level plume at three different distances downwind.

Several other applications involving the use of APCI/MS for direct environmental analysis have been reported. Thomson and co-workers have reported on a technique for the analysis of stack gas for the presence of PCBs and PAHs, and the use of such a system in a cement kiln stack.[48] A system for the detection of sulfuric acid aerosols in air using an APCI/MS system was reported by Thomson and French.[49] Other air-monitoring applications have been discussed by Karasek, Reid and co-workers, and Thomson, Davidson, and Lovett.[43,31,40]

Mitchum, Moler, and Korfmacher reported the use of gas chromatography coupled to an APCI/MS system to analyze organic extracts of complex environmental samples.[50] The specificity of negative chemical ionization at atmospheric pressure was shown to be advantageous for detecting picogram levels of TCDD, using SIM techniques.

Recently, the technique of tandem mass spectrometry has been developed as a tool for trace organic analysis.[51,52] The specificity and reduced chemical noise which is afforded by this technique greatly enhances the utility of the APCI source for direct mixture analysis,[37] since sufficient information can be generated by the APCI/MS/MS combination to allow unambiguous compound identification in complex matrices. For example, TCDD has been detected at the low picogram level even in the presence of higher levels of other chlorinated aromatics. Figure 5 shows the response of the system to 40 pg of TCDD desorbed from

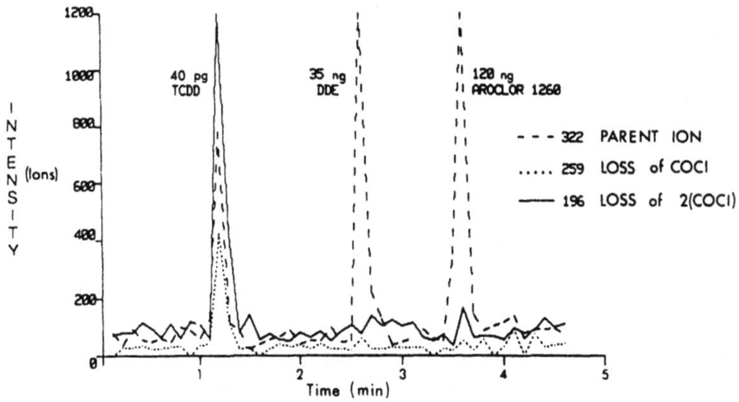


FIGURE 5. Selected-ion monitoring of parent ion (*m/z* 322) and two daughter ions (*m/z* 196 and 259) of TCDD during pulse desorption of TCDD and thousandfold excesses of DDE and PCB.

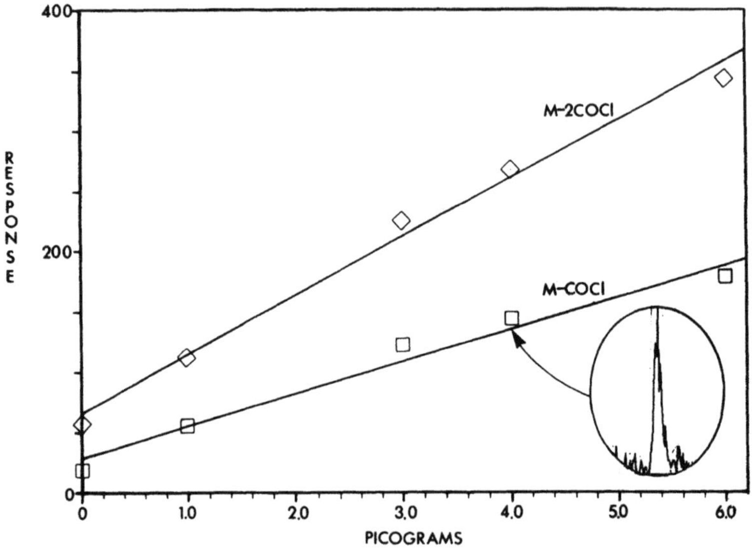


FIGURE 6. Calibration curve for TCDD, measured at two daughter ions.

a direct-insertion probe into the APCI source. Shown beside the TCDD response (detected at the parent *m/z* 322 ion and two daughters at *m/z* 196 and *m/z* 259) are signals due to 35 ng of DDE and 120 ng of PCB, monitored at the same *m/z* values. The response at the two daughters of TCDD, and the absence of a response from DDE and PCB, shows the high discrimination of the system against high levels of other chemicals. Figure 6 shows a calibration curve of the response of the system to 1,2,3,4-TCDD in the range of 1–6 pg. The use of this technique (direct desorption of environmental samples and extracts and analysis by MS/MS) has been proposed for rapidly screening large numbers of samples for the presence of trace levels of targeted compounds.[52] Extracts of fish tissue have been spiked and analyzed using isotope dilution for quantitation.[37] The same technique of direct-probe APCI/MS/MS has been used for the detection of polychlorodibenzofurans and polychlorodibenzodioxins in the matrix of 2,4-D formulation. One of the advantages that the APCI source has for routine direct-probe analysis is that the source can readily tolerate the introduction of high levels of involatile material without serious loss of performance or contamination and is easily cleaned when necessary.[40,32]

The use of APCI/MS in the analysis of combustion products for the presence of nitrated polycyclic aromatic hydrocarbons has also been explored.[53] The APCI source was employed both to sample combustion gases from a diesel engine directly and to analyze the organic extracts from trapped organic particulates. High specificity was demonstrated in the presence of a complex organic matrix by monitoring specific daughter ions corresponding to the loss of $NO_2$ and $(NO_2)_2$ from $M^+$ parents and monitoring $NO_2^-$ from $M^-$ parents. Constant neutral loss scans and parent ion scans for the $NO_2^-$ daughter allowed samples to be rapidly screened for the presence of trace levels of nitropolycyclic aromatic hydrocarbons, even in the presence of much higher levels of polycyclic aromatic hydrocarbons.

## 6. *CONCLUSIONS*

Requirements for the detection of trace contaminants in environmental samples are likely to continue to grow during the next several years. Mass spectrometry currently offers the combination of flexibility, sensitivity, and specificity needed for such analysis, and the use of this technique for routine analysis is certain to increase in the future. The particular combination of atmospheric pressure chemical ionization with mass spectrometry (APCI/MS) is relatively young but already has opened new avenues to the techniques of mass spectroscopy. For example, the ability to do mobile sampling and real-time analysis of air (or other gas streams) for trace levels of organics allows better temporal and spatial resolution than was previously available. In addition, there are many instances where sample collection and work-up time can be eliminated by directly analyzing the raw sample or extract for the components of interest. Although many of the earlier applications focused on air analysis, recent work has demonstrated the suitability of the APCI source for direct solid and liquid analysis, and it is expected that such analysis will dominate the applications in the future.

The recently introduced technique of MS/MS appears to substantially widen the potential applicability of atmospheric pressure ionization by providing structural information on the quasimolecular ions generated in the source. This increased specificity makes the technique attractive for situations in which unknown chemicals are sought, or in which chemical classes are to be identified, rather than just for those situations involving targeted compound analysis. The reports which are now emerging demonstrate that sufficient specificity and sensitivity is often attainable to provide unambiguous compound identification, even in a complex matrix. In many other situations highly specific screening can be performed in a short time by examining relatively raw samples. This allows only the few which indicate a positive response to be subjected to more exhaustive analytical procedures (GC/MS/MS or LC/MS/MS, for example) which provide structural and isomeric information for final identification. Atmospheric pressure chemical ionization appears to have the characteristics of sensitivity and flexibility in sample introduction which will allow MS/MS to reach its potential for rapid and direct trace environmental analysis.

## REFERENCES


1. W. L. Budde and J. W. Eichelberger, *Anal. Chem.* **51,** 567A–574A (1979).
2. F. W. McClafferty, *Science* **214,** 280–287 (1981).
3. P. F. Knewsnubb and T. M. Sugden, *Proc. Roy. Soc. London* **A255,** 520–537 (1960).
4. A. N. Hayhurst and T. M. Sugden, *Proc. Roy. Soc.* **A293,** 36–50 (1966).
5. P. Kebarle and E. W. Godbole, *J. Chem. Phys.* **39,** 1131–1132 (1963).
6. A. M. Hogg and P. Kebarle, *J. Chem. Phys.* **43,** 449–456 (1965).
7. M. M. Shahin, *J. Chem. Phys.* **46,** 2600–2605 (1966).
8. M. M. Shahin, *J. Chem. Phys.* **47,** 4392 (1968).
9. F. W. Karasek, *Res./Dev.,* Dec., 25–27 (1970).
10. F. W. Karasek, *Anal. Chem.* **46,** 710A–720A (1974).
11. F. W. Karasek, H. H. Hill, Jr., and S. H. Kim,, *J. Chromatogr.* **117,** 327–336 (1976).
12. T. W. Carr, *Anal. Chem.* **49,** 828–831 (1977).
13. G. W. Griffin, I. Dzidic, D. I. Carroll, R. N. Stillwell, and E. C. Horning, *Anal. Chem.* **45,** 1204–1209 (1973).
14. E. C. Horning, M. G. Horning, D. I. Carroll, I. Dzidic, and R. N. Stillwell, *Anal. Chem.* **45,** 936–943 (1973).
15. E. C. Horning, M. G. Horning, D. I. Carroll, I. Dzidic, and R. N. Stillwell, *Adv. Biochem. Psychopharmacol.* **7,** 15–31 (1973).
16. D. I. Carroll, I. Dzidic, R. N. Stillwell, M. G. Horning, E. C. Horning, *Anal. Chem.* **46,** 706–710 (1974).
17. E. C. Horning, D. I. Carroll, I. Dzidic, K. D. Haegele, M. G. Horning, and R. N. Stillwell, *J. Chromatogr.* **12,** 725–729 (1974).
18. I. Dzidic, D. I. Carroll, R. N. Stillwell, and E. C. Horning, *J. Amer. Chem. Soc.* **96,** 5258–5259 (1974).
19. I. Dzidic, D. I. Carroll, R. N. Stillwell, and E. C. Horning, *Anal. Chem.* **47,** 1308–1312 (1975).
20. I. Dzidic, D. I. Carroll, R. N. Stillwell, M. G. Horning, and E. C. Horning, in *Advances in Mass Spectrometry,* Vol. 7B, Heyden, London, 1976, pp. 359–365.
21. I. Dzidic, D. I. Carroll, R. N. Stillwell, and E. C. Horning, *Anal. Chem.* **48,** 1763–1768 (1976).
22. D. I. Carroll, I. Dzidic, R. N. Stillwell, and E. C. Horning, *Anal. Chem.* **47,** 1956–1959 (1975).
23. E. C. Horning, D. I. Carroll, I. Dzidic, S.-N. Lin, R. N. Stillwell, and J.-P. Thenot, *J. Chromatogr.* **142,** 481–495 (1977).
24. T. C. Imeson and C. S. Harden, Edgewood Arsenal Technical Report EATR 4642, 1972.
25. M. W. Siegel and W. L. Fite, *J. Phys. Chem.* **80,** 2871–2881 (1976).
26. H. Kambara and I. Kanomata, in *Mass Spectrosc.* **24,** 229–236 (1976).
27. H. Kambara and I. Kanomata, *Mass Spectrosc.* **24,** 271–282 (1976).
28. J. A. Buckley, Ph.D. thesis,, University of Toronto, Toronto, Canada, 1974.
29. J. A. Buckley, J. B. French, and N. M. Reid, *Can. Aeronaut. Space J.* **20,** 231–233 (1974).

30. J. B. French and N. M. Reid, in *Dynamic Mass Spectrometry,* Vol. 6, Heyden, London, 1980, pp. 220–233.
31. N. M. Reid, J. B. French, J. A. Buckley, D. A. Lane, and A. M. Lovett, Proceedings, Fourth Joint Conference on Sensing of Environmental Pollutants, American Chemical Society, 1978.
32. N. M. Reid, B. A. Thomson, W. R. Davidson, J. B. French, and J. A. Buckley, Presented at the 27th Annual Conference on Mass Spectrometry and Allied Topics, Seattle, Washington, June, 1979.
33. D. I. Carroll, I. Dzidic, R. N. Stillwell, K. D. Haegele, and E. C. Horning, *Anal. Chem.* **47,** 2369–2373 (1975).
34. M. McKeown and M. W. Siegel, *Am. Lab.* (Nov., 1975).
35. J. V. Iribarne, P. J. Dziedzic, and B. A. Thomson, Presented at the 29th Annual Conference of Mass Spectrometry and Allied Topics, Minneapolis, May 24–29, 1981.
36. D. J. Douglas and J. B. French, *Anal. Chem.* **53,** 37–41 (1981).
37. G. A. V. Rees, H. Tosine, T. Sakuma, W. R. Davidson, B. A. Thomson, L. M. Danylewych, B. Shushan, J. E. Fulford, and N. M. Reid, Presented at the 29th Annual Conference on Mass Spectrometry and Allied Topics, Minneapolis, May 24–29, 1981.
38. M. L. Huertas and J. Fontan, *Atmos. Environ.* **9,** 1018 (1975).
39. V. Kotasek, M.Sc. thesis, University of Toronto, Toronto, Canada, 1981.
40. B. A. Thomson, W. R. Davidson, and A. M. Lovett, *Environ. Health Pers.* **36,** 77–84 (1980).
41. S. D. Tanner, B. A. Thomson, G. B. Debrou, and N. H. Hijazi, Presented at the 29th Annual Conference on Mass Spectrometry and Allied Topics, Minneapolis, May 24–29, 1981.
42. S. D. Tanner and B. A. Thomson, Presented at the 29th Annual Conference on Mass Spectrometry and Allied Topics, Minneapolis, May 24–29, 1981.
43. F. W. Karasek *Ind. Res./Dev.* (Dec., 1978).
44. B. A. Thomson and J. R. Roberts, Jr., *Int. J. Environ. Anal. Chem.* **11,** 139–151 (1981).
45. NRCC 17586, National Research Council of Canada, Report No. 17586, 1980.
46. D. T. Williams, H. V. Denley, and D. A. Lane, *Amer. Ind. Hygiene Assoc. J.* **41,** 647–651 (1980).
47. D. A. Lane and B. A. Thomson, *J. Amer. Poll. Control Assoc.* **31,** 122–127 (1981).
48. B. A. Thomson, T. Sakuma, J. Fulford, D. A. Lane, N. M. Reid, and J. B. French, in *Advances in Mass Spectrometry,* Vol. 8B, Heyden, London, 1980, pp. 1422–1428.
49. B. A. Thomson and J. B. French, Presented at the 74th Annual Air Pollution Control Association Meeting, Philadelphia, June 21–26, 1981.
50. R. K. Mitchum, G. F. Moler, and W. A. Korfmacher, *Anal. Chem.* **52,** 2278–2281 (1980).
51. T. H. Maugh II, *Science* **209,** 675–677 (1980).
52. W. R. Davidson, J. Fulford, N. M. Reid, T. Sakuma, B. Shushan, and B. A. Thomson, Presented at the 29th Annual Conference on Mass Spectrometry and Allied Topics, Minneapolis, May 24–29, 1981.
53. J. E. Fulford, T. Sakuma, and D. A. Lane, Proceedings of the 6th International Symposium on Polynuclear Aromatic Hydrocarbons, Battelle Columbus Labs, 1981.

7

# MASS SPECTRAL RETRIEVAL AND INTERPRETATION SYSTEMS

F. W. KARASEK and R. E. CLEMENT

## 1. INTRODUCTION

Identification of an organic compound from its mass spectrum is now a successful and well-established practice. Although it may not always be possible to determine the molecular structure of an unknown compound, there is always some useful information to be derived from the mass spectrum. An observable molecular ion will give the molecular weight and under high resolution the elemental composition. Fragment ions, either singularly or in related clusters, reveal some of the building blocks that are comprised in the molecule. Isotopic clusters indicate the presence of certain elements, such as the halogens, sulfur, and carbon.

There is a stepwise procedure which is usually followed in interpretation of a mass spectrum. This procedure follows a sequence of six general steps, each step increasing in difficulty.

1. *Inspection.* Some mass spectra are so simple or familiar that identity is achieved immediately. Simple molecules such as $CH_4$, CO, $CH_3OH$, or $C_6H_6$ are typical examples.

2. *Comparison to reference spectra in an atlas.* A number of atlas-type publications tabulate reference mass spectra by mass and abundance values of eight to ten of the most abundant peaks. The presence or absence of a match between the spectrum of an unknown compound and its counterpart in the tabulation is readily determined.

*F. W. KARASEK* • Department of Chemistry, University of Waterloo, Ontario N2L 3G1, Canada. *R. E. CLEMENT* • Laboratory Services Branch, Ontario Ministry of the Environment, P.O. Box 213, Rexdale, Ontario M9W 5L1, Canada.

3. *Computer search of reference versus unknown spectra via dedicated system.* Most GC/MS/COMP systems have sufficient capacity to automate the search and comparison techniques for reference versus unknown spectra. A rapid, accurate result is obtained in which the quality of the match is given using several different criteria.

4. *Computer search of reference versus unknown spectra via remote terminal to central facility.* A large computer in a centralized facility can carry a large data bank of mass spectra and execute a number of different types of search programs which will enhance the scope and assurance of the information obtainable from the matching technique.

5. *Computerized interpretation of a mass spectrum.* When there is no definitive match between reference spectra and an unknown spectrum, it is possible to apply an interpretive procedure through a computer program and recreate the type of compound and its component parts. STIRS (Self-Training Interpretive and Retrieval System) is available from Cornell University using a remote terminal and telephone coupler.

6. *Interpretation by spectroscopist.* When all of the above steps fail to achieve a satisfactory result, then a lengthy interpretation procedure must be done by a mass spectroscopist. This involves the application of known fragmentation rules, experience and the use of the auxiliary methods of infrared and NMR spectroscopy, chromatographic retention behavior, and synthesis of compounds of suspected or related structures.

Mass spectral identification by comparison of unknown and reference spectra has reached an advanced stage. Although many studies in this area have been reported, the emphasis here will be on the techniques which are generally available and used.

## 2. *USE OF PUBLISHED INDEXES*

Several compilations of mass spectral data have been published, some of which are very specialized. Information on the various data collections can be found in *A Guide to Collections of Mass Spectral Data* by the American Society for Mass Spectrometry Committee VI (second edition, 1978). Two compilations of mass spectral data are especially useful for the identification of unknown spectra. These are the Eight-Peak Index of Mass Spectra (8PI),† compiled by the Mass Spectrometry Data Centre, Aldermaston, U.K., and the Environmental Protection Agency/National Institute of Health (EPA/NIH) Mass Spectral Data Base.

† The 8PI can be obtained from Her Majesty's Stationery Office, P. O. Box 569 (P7B), London, SE1 U. K., or from Pedragon House, 220 University Ave., Palo Alto, Calif. 94031.

The 8PI contains a listing of the eight most abundant ions in 31,101 mass spectra, indexed by molecular weight, elemental composition, and most abundant ions. The total number of different compounds represented is less than 31,101, since some duplication spectra are included. For identification of unknowns, a listing of spectra in order of most abundant ions is used. Figure 1 shows the data format. Once the *m/z* of the base peak has been located, the remaining peaks are matched in order of abundance until the identity of the unknown is reduced to a small number of possibilities. Intensities do not have to match exactly for a good fit. Spectra of the same compound obtained from different instruments can be appreciably different although experience indicates that most spectra of the compounds obtained from a quadrupole GC/MS with electron impact (EI) source have matched well with spectra of these same compounds reported in the index. How closely unknown and reference spectra should match depends largely upon the experience of each user. It is not always necessary to determine a single unambiguous structure for the unknown. For some applications the identification of compound class or a few selected structural features is sufficient. If a close match is achieved between the unknown spectrum and spectra in the index, a further confirmation of identity is required by comparing the mass spectrum of the unknown with the complete spectra of the compounds of closest match. This is necessary because the eight largest peaks are not always the most structurally significant, particularly for isomeric compounds and those in homologous series. The EI mass spectra of normal alkanes, for example, have a small molecular ion that is important for identification of a specific alkane, but it will not be one of the eight most abundant ions.

The EPA/NIH Mass Spectral Data Base is a compilation of verified mass spectra of 25,556 different compounds, with no duplication. The full spectra are presented in bar graph form by increasing molecular weight. Figure 2 shows the data format of these spectra. This compilation can be used to find the mass spectrum of a substance. It is not useful for identifying unknown mass spectra unless the molecular weight or molecular formula is known. Therefore, the EPA/NIH data and the 8PI are complementary publications. Their combined use can provide effective identification of unknowns, if the unknown is included in the compilations.†

An index is included which lists the spectra by substance name, molecular formula, molecular weight, and chemical abstracts registry number. The original volumes were published in 1978. A supplement

† The EPA/NIH Mass Spectral Data Base can be obtained from the U.S. Government Printing Office, Washington, D.C. 20402 (SD Stock No. 003-003-01987-9).

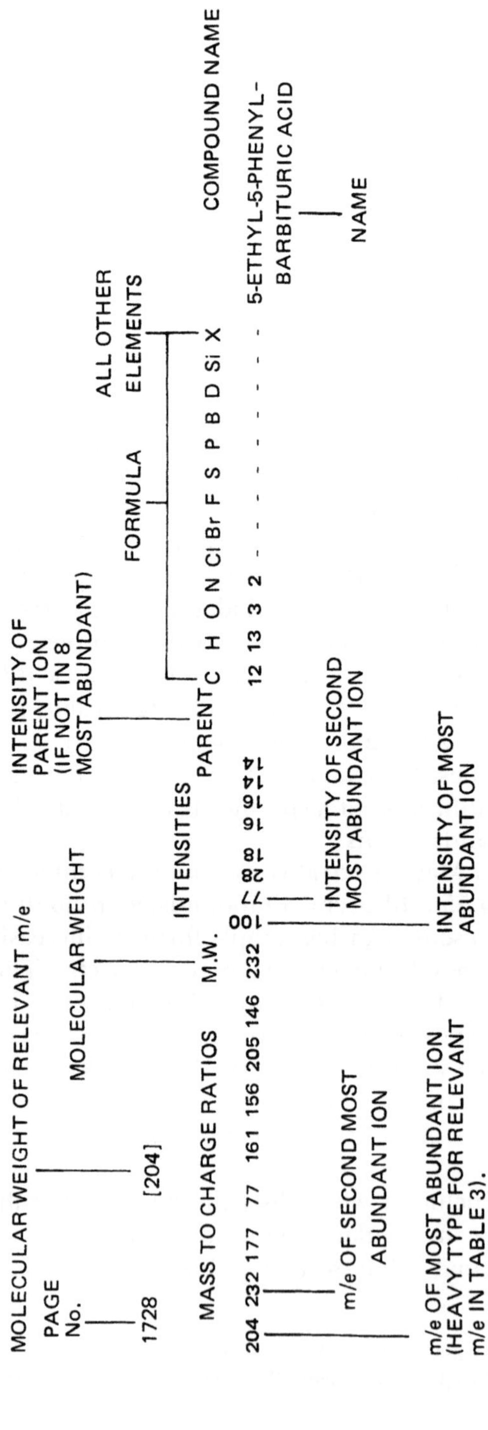


FIGURE 1. Format of spectra in 8 PI.

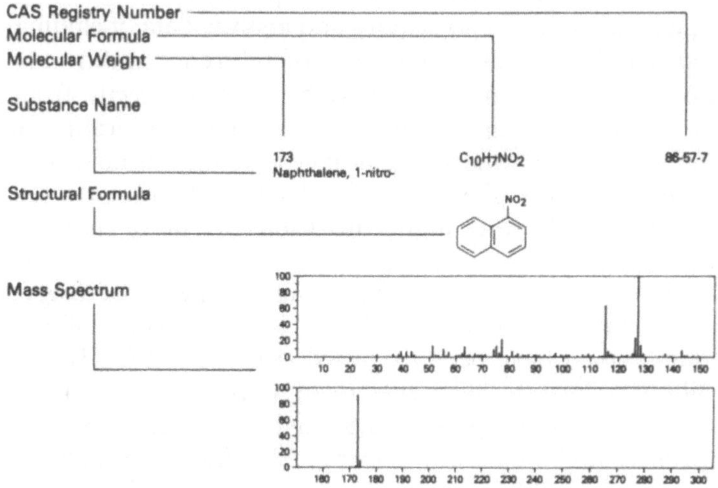


FIGURE 2. Format of spectra in EPA/NIH indexes.

has been issued in 1980 which contains an additional 8807 mass spectra.

A limitation of the reference collections is the small number of available spectra. Although the 8PI and EPA/NIH Mass Spectral Data are periodically updated, obtaining accurate, verified mass spectra of new compounds is a slow process. The usefulness of available data compilations for manual searching is essentially limited to the specific compounds in these published collections. Furthermore, manual spectra matching is laborious and error prone.

## 3. *COMPUTER SPECTRA MATCHING VIA DEDICATED SYSTEM*

Problems associated with manual matching of spectra are reduced or eliminated by computerizing the comparison process. Rapid and accurate comparisons can be made, and the basis for matching can be complex. A useful feature is the calculation of factors used to distinguish good, average, and poor matches. A human operator will generally stop searching at the first good match, but the computer finds all possibilities and can list as many as necessary. If no reasonable match is found some information may still be obtained from the search, since the compounds of closest match usually have some structural features in common.

Instrument manufacturers now include some type of computerized mass spectral search as part of the GC/MS/Computer instrumentation. An example of such a search is the Hewlett-Packard 5992 benchtop GC/

MS/CALC (calculator) system.[1] This completely automated instrument is equipped with software to acquire and analyze data, including a computerized search procedure. Libraries of reference spectra are stored on floppy disks, and unknowns may be chosen directly from stored analyses. In the matching procedure the ten most significant peaks from the unknown are used where significance is defined as mass times abundance. This procedure gives better results than choosing the ten largest peaks for the search, since higher mass ions are more characteristic of specific compounds.[1]

Initially, reference libraries containing 375 spectra of drugs and 517 spectra of pollutants were provided. A series of specialized libraries have been added which have been extracted from the EPA/NIH data base. These libraries include 3948 spectra of sulfur and phosphorus compounds, halogens (4328 spectra), U.S. Environmental Protection Agency (EPA) priority pollutants (227 spectra), EPA industrial pollutants (971 spectra), pesticides (387 spectra), and nitrogen-containing compounds (12,483 spectra). Provisions exist for the creation of new libraries by the analyst, to which spectra of compounds identified during sample analysis can be added. Due to the storage limitations of the floppy disk, these libraries contain only the ten most significant peaks of the reference spectra. The entire EPA/NIH library of complete mass spectra is available for the larger minicomputer-based GC/MS systems as well as a more powerful search procedure.[2]

Figure 3 illustrates the data treatment available on large minicomputer-based GC/MS systems. Raw ion abundance data is plotted as a function of mass spectrum scan number in the upper trace. Repetitive scans were performed on a Ribermag R10-10 GC/MS, stored on disk, and replotted to give the reconstructed gas chromatogram (RGC) seen in the figure. Below the RGC is an ion plot derived from the same mass spectra as the RGC after application of the Biemann stripping technique.[3] This method of reconstructing mass spectra utilizes mass chromatograms to identify all $m/z$ values which maximize at any of the spectra recorded during a GC/MS analysis. Ions which do not fulfill this criterion in a specific scan are stripped from that spectrum. After the stripping process mass spectra are practically free of the contributions of unresolved components, column bleed, and other background contamination.

The GC peaks in the stripped RGC shown in Fig. 3 are narrower and have more apparent resolution than the corresponding peaks in the original RGC. Peak A in the RGC is shown to consist of two components which are resolved in the stripped RGC. The mass spectrum of peak A seen in the figure is free of peaks from the previously overlapping peak. Results of library searching are improved when mass spectra are stripped of interfering peaks. The best matches from the EPA/NIH mass spectral

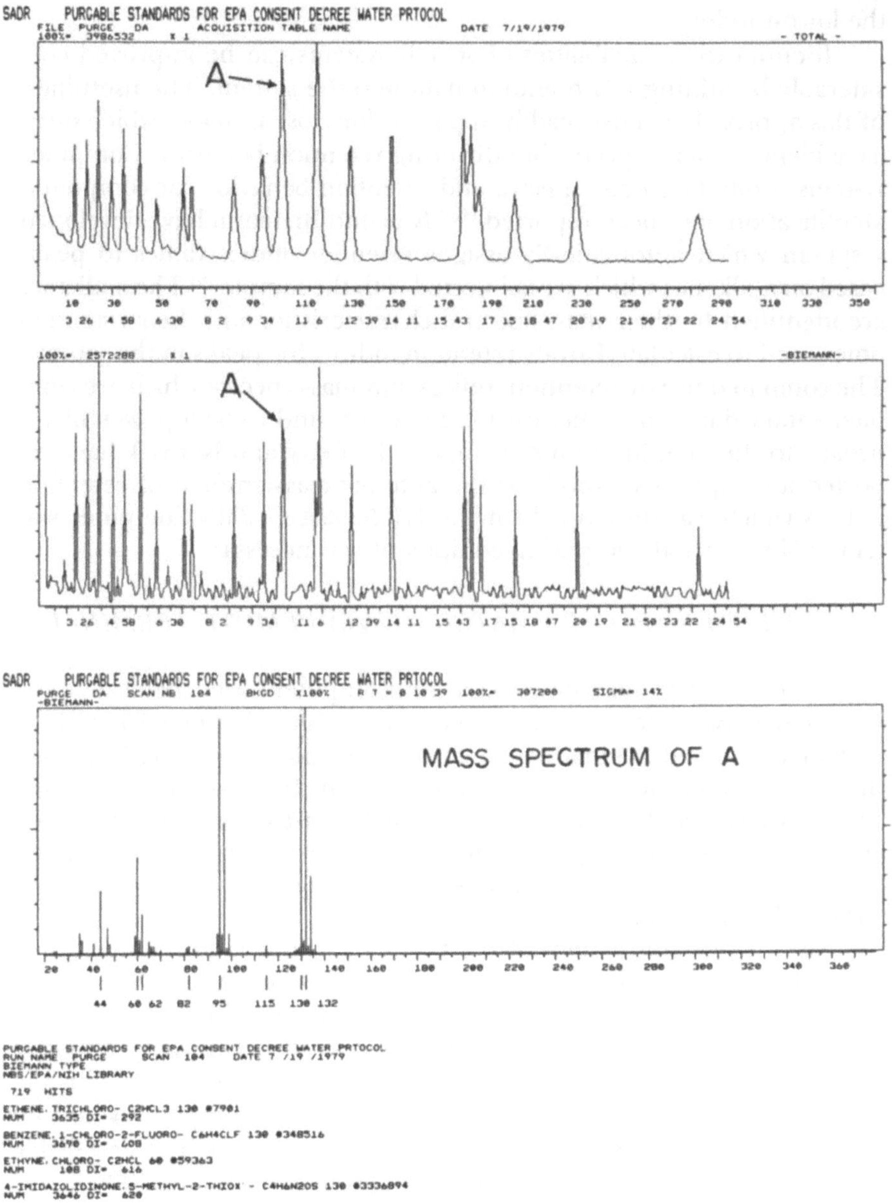


FIGURE 3. Data obtained using the Nermag GC/MS system with its specialized search system for analysis of EPA priority pollutants.

data base for peak A after stripping are shown. A dissimilarity index was calculated for each match, and the closest matches are those with the lowest index.

Identification capabilities of search systems can be improved considerably by adding GC retention data into the system. The usefulness of this approach is most readily apparent for close isomers, which often have identical mass spectra but differing retention behavior. Computer systems combining mass spectra and retention behavior for compound identification have been reported.[4,5] Nau and Biemann have developed a system which automatically assigns retention index values to peaks based on *n*-alkanes which are coinjected with the sample.[6] The *n*-alkanes are identified by their mass spectral characteristics and their retention times used to calculate Kovats retention indices for peaks in the sample. The combined use of retention indices and mass spectra which are complementary data and structure specific was found to be a powerful approach to the identification of compounds. Betty and Karasek have reported a computer method for the automatic assignment of retention indices which was developed for the HP-5992A GC/MS/Computer system.[7] No external computing equipment was necessary.

## 4. COMPUTER SPECTRA MATCHING VIA REMOTE TERMINAL

Computerized searching via a minicomputer-based dedicated system is not available to some users. The results obtained from these systems are sometimes inadequate due to the limited reference files available or the computer search program applied. In these situations computer spectra matching may be performed by use of more sophisticated procedures based on large computers accessible over the telephone lines. The systems available are the mass spectral search systems (MSSS) and probability-based matching (PBM).

MSSS is one component of the Chemical Information System (CIS), which has been developed as a joint project between the U.S. National Institute of Health (NIH) and the U.S. Environmental Protection Agency (EPA). CIS is available in countries served by the TELENET computer communications network. The reference file used by MSSS contains the electron impact mass spectra of 33,000 individual compounds and no isotopically labeled compounds.[8] Different search options are available with MSSS. The two most commonly employed are PEAK and KB.

### 4.1. PEAK Search

PEAK is an interactive search in which peaks of the unknown spectrum are entered one at a time until the number of possible matches is reduced to a few. Each mass is entered with its maximum and minimum abundance. The standard file is searched for all entries having the specified *m/z* value with an intensity in the allowed range. The number of successful matches is reduced as each new peak is entered. Usually less than a dozen spectra remain after only five or six entries. A large intensity range for entered peaks should be used since the library spectra were all obtained from different instruments than the unknown spectrum. If a spectrum of the unknown compound is present in the reference file, it will be retrieved as one of the best matches unless a mistake is made during peak entry, or if the reference or unknown spectrum is distorted.

### 4.2. *KB Search*

KB uses the most abundant ions in each 14-amu division of a mass spectrum beginning at *m/z* 6 for matching.[9] Storage space for spectra is reduced to 25% of the space required to store full spectra without sacrificing the most important structural information. The 14-amu division was chosen because the basic building unit of organic molecules is the $—CH_2—$ structure. Also, since the lowest mass loss from the parent molecule is usually 15 ($CH_3—$), the molecular ion, if present, will be one of the two most abundant in a 14-amu division. KB utilizes more of the mass spectrum than does PEAK, and more reliable results are obtained. The best matches are ranked by a similarity index which is calculated for each comparison of the unknown with a reference spectrum.

### 4.3. *PBM Search*

The Probability-Based Matching (PBM) system is available from Cornell University through the TYMNET computer network. Comparison of unknown and reference spectra is based on the statistical occurrence of mass and abundance values in mass spectra.[9–11] An important aspect of PBM is that it employs a reverse-search procedure. Most search systems attempt to match mass peaks of the unknown with those of reference spectra, but PBM determines whether peaks of the reference spectrum are present in the unknown spectrum. Mass peaks of the unknown not present in a reference spectrum could be from background contamination or an overlapping component. Therefore, PBM is capable of identifying specific compounds in mixtures. After a search the best match is automatically subtracted from the reference spectrum, and the remaining or residual peaks can then be input to PBM.

Since PBM uses reverse searching and a statistically determined basis for matching this system is superior to PEAK and KB for analysis of unknowns that are not represented in the reference file. The closest matching compounds generally have some structural features in common with the unknown. While PEAK and KB use a limited number of mass peaks from the unknown spectrum, users of PBM are encouraged to enter as many significant peaks as practical.

## 5. COMPUTERIZED INTERPRETATION OF A MASS SPECTRUM

Retrieval systems such as PEAK, KB, and PBM give limited results for unknowns not represented in the reference file. The most extensive reference compilations currently available contain mass spectra of 30,000–40,000 compounds. Since this only represents a small fraction of the total number of organic compounds, effective methods for interpretation of mass spectra not in the reference files are needed. Mathematical procedures used to interpret mass spectra have been reviewed.(10) The Self-Training Interpretive and Retrieval System (STIRS) is presently the only generally available interpretive system.(12) Like PBM, it is available over telephone lines from Cornell University. PBM and STIRS are complementary techniques, since STIRS is employed to provide additional information for unknowns not successfully identified by PBM.

STIRS uses a series of match factors to retrieve compounds having common mass spectral features with the unknown spectrum. For a specific match factor, if the retrieved compounds contain similar structural groups, then these substructures are probably also present in the unknown. Each match factor represents a specific mass spectral feature such as low, medium, and high mass characteristic ions, small primary neutral losses, and large primary neutral losses. The system is self-training, since specific structural features are determined on the basis of compounds retrieved by the various match factors. STIRS is used as an interpretive aid for the identification of possible substructures in the unknown.(13,14) Some incorrect substructures will generally be found by STIRS in addition to several correct ones. Therefore, STIRS results should be used to supplement other available information, not replace it. STIRS has been used to predict rings-plus-double-bonds values(15) and molecular weights from mass spectra.(16)

## 6. APPLICATION TO ANALYTICAL PROBLEMS

Few applications of the various mass spectral retrieval methods have been reported. Brenner compared the use of PBM, STIRS, and MSSS for mass spectral identification of compounds in drinking water.(17) They advise others to begin by testing the ability of these commercial systems to interpret known spectra obtained from their own instrumentation. Mathews compared different computer methods for the identification of alkylbenzenes from mass spectra, and concluded that a combination of approaches should be used.(18) Library retrieval methods gave good results if the unknown spectrum was in the reference file. A simplified STIRS procedure predicted structures identical or similar to the unknown even if the unknown was not in the reference file.

## 6.1. *Illustration of Search Procedures*

To illustrate the use of available procedures for the interpretation of mass spectra, the unknown spectrum in Fig. 4 was chosen to be identified. This spectrum will be run stepwise through the available procedures.

The only positive structural information about the unknown was that it is a trimethylsilyl (TMS) derivative. The base peak at *m/z* 73 [—$Si(CH_3)_3$] supports this, and the large peak at *m/z* 147 usually indicates a di-TMS compound. The highest mass ion of the unknown is at *m/z* 288, which could be the molecular ion. There are many neutral losses of 15 amu, which is indicative of the —$CH_3$ structure. Other prominent

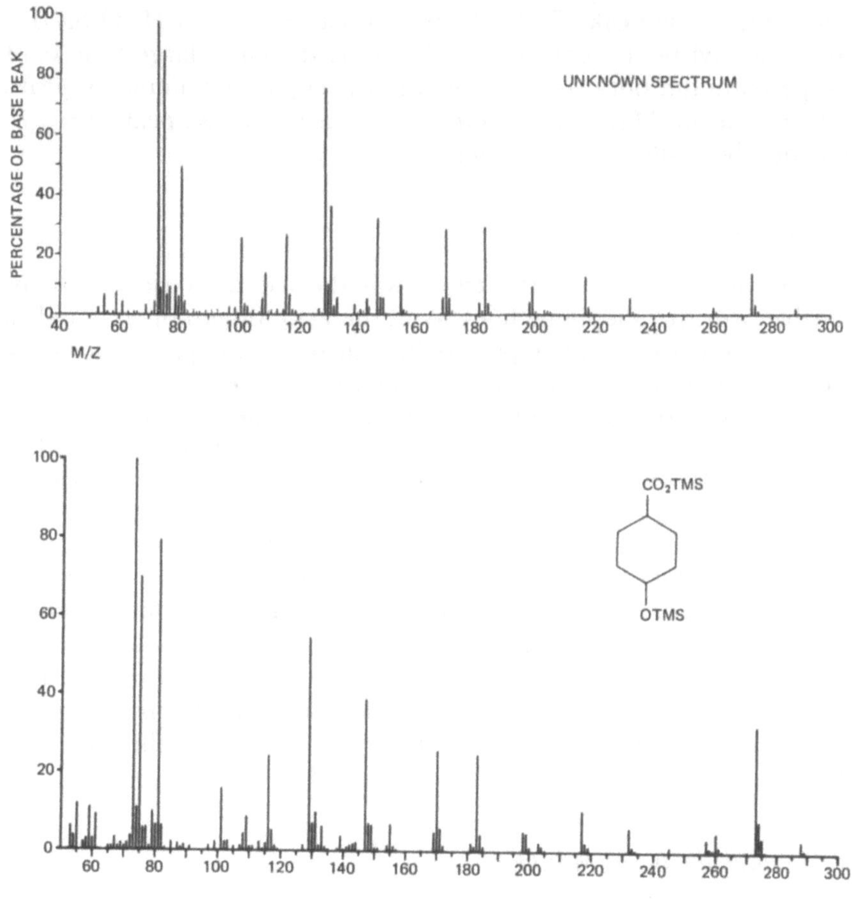


FIGURE 4. Top: Mass spectrum of unknown compound. Bottom: Structure and mass spectrum of compound synthesized by Bindel and co-workers.[19]

mass losses include 13 (CH) and 16 and 18 (indicative of oxygen). The relative abundances of peaks 273, 274, and 275 indicate the presence of two silicon atoms. After correcting for the contribution of these to the *m/z* 274 peak, the remaining abundance indicates that approximately ten carbon atoms are present in the 273 peak.

### 6.2. *Eight-Peak Index*

The unknown spectrum was then manually compared to spectra in the Eight-Peak Index (8PI). Of the five most abundant peaks on the unknown spectrum (*m/z* 73, 75, 129, 81, and 131), no more than two or three could be found among the eight largest peaks of any of the reference spectra. As a further check, the mass 288, which may be the molecular weight of the unknown, was located in the table of molecular weights of the 8PI. Only one spectrum was found with a molecular weight of 288 and a base peak of *m/z* 73 which has a formula $C_{12}H_{24}O_4Si_2$ (Di-TMS 3-methyl pent-2-enedioate). However, the eight largest peaks of this spectrum did not match closely with the eight most abundant peaks of the unknown. Many of the compounds having a base peak of *m/z* 73 contained both silicon and oxygen.

### 6.3. *PEAK Search*

The next step in interpretation was the use of retrieval systems available over the telephone lines. With the PEAK option of MSSS, only a few peaks were necessary to rapidly reduce the number of possible matches. The *m/z* 273 peak was entered first (intensity range 5–20%), as it is a high mass peak of high abundance. High mass peaks of low abundance were avoided, since some of these could be due to background. After entering *m/z* 217 (intensity range 8–18%), the number of possible matches was reduced to 33. When *m/z* 183 (25–35%) was entered, no matches were found. By entering the same peak but extending the intensity range (5–50%), only two compounds were retrieved. One match with a molecular formula of $C_{10}H_6Cl_6$ could be eliminated since no chlorine isotope pattern is observed in the unknown spectrum. The other compound was an epoxy with the formula $C_{15}H_{26}N_2O_6Si_2$ (molecular weight 386). This was not the correct compound, since its spectrum as found in the printed EPA/NIH indexes was considerably different from that of the unknown, although some similarities were also observed.

### 6.4. *Probability-Based Matching*

Figure 5 shows the results obtained from a PBM search of the unknown spectrum. PBM prints the best five matches according to their

```
39338 SPECTRA WERE SEARCHED (  3447 RELEVANT) BY PBM:
 20 SPECTRA WERE RETRIEVED (K >= 20)

SERIAL# NAME                          RELIABILITY   CONTAM-    COMP-
        MW,        FORMULA            CL 4   CL 1   INATION    ONENT

#19454  TRANS-CYCLOHEXANE-1,3-DI       23      9    ( 69%)    ( 32%)
        OL TMS (PART CIS
        MW: 260, C12 H28 02 SI2
# 5910  PROPIONIC ACID-MONOTMS         16      5    ( 64%)    ( 74%)
        MW: 146, C6 H14 02 SI
#35967  MONOTRIMETHYLSILYL PROPI       14      5    ( 64%)    ( 74%)
        ONIC ACID
        MW: 146, C6 H14 02 SI
#35888  MONOTRIMETHYLSILYL ACRYL       14      5    ( 41%)    ( 77%)
        IC ACID
        MW: 144, C6 H12 02 SI
# 5642  ACRYLIC ACID-MONOTMS           14      5    ( 41%)    ( 77%)
        MW: 144, C6 H12 02 SI
SPECTRUM  #19454 BEING SUBTRACTED

LIST OF SIGNIFICANT RESIDUALS
```

| M/E | ABD. | M/E | ABD. | M/E | ABD. | M/E | ABD. | M/E | ABD. |
|---|---|---|---|---|---|---|---|---|---|
| 108 | 600 | 109 | 1400 | 129 | 6457 | 130 | 771 | 131 | 3371 |
| 133 | 600 | 181 | 400 | 183 | 2900 | 184 | 400 | 198 | 400 |
| 199 | 900 | 232 | 600 | 233 | 100 | 234 | 100 | 273 | 1400 |

**FIGURE 5. Results from PBM search of unknown mass spectrum seen in Fig. 4.**

reliability. CL4 is the probability expressed as a percentage that the unknown has a structure closely related to that of the retrieved compound. CL1 is the probability that the retrieved compound (or stereoisomer) has a structure identical to that of the unknown. The best match found here could not be identical to the unknown, since its molecular weight is too low. However, the closest matching structures all contain oxygen and silicon. The significant residuals in Fig. 5 are major peaks in the unknown spectrum not present in the best matching reference spectrum. Low abundance peaks such as those at *m/z* 274, 275, and 288 are not included in this list. PBM data, in addition to earlier information, supports the presence of the —OSiTMS substructures. It is unlikely that a spectrum of the unknown compound is present in the 8PI or either of the reference files used by MSSS or PBM.

## 6.5. *Interpretation Using STIRS*

Following the PBM search, STIRS was used to provide additional information. Figure 6 is a partial listing of the STIRS analysis. The major substructures from prior examination of the unknown [$—CH_3$, $—Si(CH_3)_3$, and $—O—Si(CH_3)_3$] were all reported by STIRS to be present with a 99% reliability. A previously unsuspected substructure, $—OCH_2—$, was also reported by STIRS with a high reliability (87%). The substructure analysis was based on a statistical study of 587 substructures selected on the basis of STIRS to recognize them in unknown mass spectra.[13] The reliability is based on an evaluation of the occurrence of each substructure among 15 compounds used to calculate the most reliable molecular weight. STIRS

```
RELIAB-            S U B S T R U C T U R E S
ILITY          DESCRIPTION                                        ID #
  99%     -O-SI(CH3)3                                                3
  99%     -SI(CH3)3                                                  8
  99%     METHYL                                                    16
  87%     -OCH2-                                                   101
 *** OCCURRENCE OF SUBSTRUCTURES IN THE RETRIEVED COMPOUNDS

 COMPOUND #       SUBSTRUCTURES  -  ID #
                 3    8   16  101
  19454          +    +    +
  26235          +    +    +
  19455          +    +    +
  19453
  29930
  30698          +    +    +    +
  30942
  29277          +    +    +    +
  26234          +    +    +    +
  13699
  30587          +    +    +
  29933
  32130          +    +    +    +
  32584          +    +    +    +
  19458
** THE MOST PROBABLE NUMBER OF CARBON ATOMS FOR THE UNKNOWN, N, IS
**    0 >= N >= 10.
** K-VALUE FOR THE PREDICTION IS   8.

THE MOST RELIABLE ELEMENTAL COMPOSITION AMONG  3 POSSIBILITIES
BASED ON MOLECULAR WEIGHT 288 IS
   C 12  H 24  O  4 SI  2

   K   ELEMENTAL COMPOSITION
  26   C 12  H 24  O  4 SI  2
   8   C 14  H 25  O  4  P  1
   3   C 16  H 32  O  4

COMBINATION 1,2,3,4,7 MATCH : MF11.3
  MF  MF  MF  MF  MF  MF  MF  MF  MF  MF  MF  MF  MF  MF   OVERALL  MF  11
   1  2A  2B  3A  3B  4A  4B  5A  5B  6A  6B  5C  6C   7   .1  .2   .3   .0

 698 602 665 773 551 444 472 605 166 390 196   0   0   0 610 339  458  489
#19454  TRANS-CYCLOHEXANE-1,3-DIOL TMS (PART CIS)
L6TJ AO-SI-1&1&1 CO-SI-1&1&1 -T
MW  260  C12 H28 O2 SI2

 854 732 727 718 567 338 153 210 235 354 190   0   0   0 609 247  457  448
#26235  2-DEOXYRIBOPYRANOSE-1,3,4-TRITMS
T60TJ BO-SI-1&1&1 DO-SI-1&1&1 EO-SI-1&1&1 -B&BDE
MW  350  C14 H34 O4 SI3

 521 804 828 718 393 494 390 307   0 400   0   0   0   0 606 177  455  415
#19455  CIS-CYCLOHEXANE-1,4-DIOL TMS (CALLED TRANS)
L6TJ AO-SI-1&1&1 DO-SI-1&1&1 -C
MW  260  C12 H28 O2 SI2
```

FIGURE 6. STIRS analysis of unknown mass spectrum (Fig. 4).

predicts the most probable molecular weight for the unknown as 288 (this portion of STIRS analysis not shown) and the most reliable elemental composition based on this molecular weight to be $C_{12}H_{24}O_4Si_2$. This is the same formula previously found in the 8PI. The STIRS prediction of the most probable number of carbon atoms in the unknown (N) is incorrect ($0 \geqslant N \geqslant 10$). Since the solution to this expression is undefined, the expression may be the result of an error in the program.

STIRS shows the occurrence of each substructure in the 15 reference compounds selected by their overall match factors (perfect match = 1000). A list of these compounds is given at the conclusion of the STIRS analysis. Only the top three are shown in Fig. 6.

It is interesting to note that the compound of highest overall match is the same one selected as the best match in the PBM analysis. This compound or a close isomer was found by STIRS four times, including three of the top four matching substances. For these four compounds, the substructure —$OCH_2$— is not present. Most of the 15 compounds selected by STIRS have a nonaromatic cyclic structure.

## 7. STRUCTURE DETERMINATION

Since the unknown compound is not present in the reference files, examined by MSSS or PBM/STIRS, a positive identification was not obtained. However, enough information was provided to make a reasonable structural assignment. The formula proposed by STIRS ($C_{12}H_{24}O_4Si_2$), is close to the correct one shown in Fig. 4, which was found by Bindel *et al.*[19] They have documented the extensive analytical work that was required to identify this substance, including synthesis of the three isomers. This work may have been unnecessary if the mass spectrum of the compound had been present in a mass spectral data base that was then examined by a computerized retrieval and interpretation system.

## REFERENCES


1. F. W. Karasek, *Res./Dev.* **27,** 42–46, 48 (1976).
2. J. A. Michnowicz, (1979). Hewlett-Packard Company Application Note AN 176–28.
3. J. E. Biller and K. Biemann, *Anal. Lett.* **7,** 515–528 (1974).
4. Y. Tsuchiya, J. Boulanger, and K. Sumi, *Chromatographia* **10,** 154–156 (1977).
5. S. C. Gates, M. J. Smisko, C. L. Ashendel, N. D. Young, J. F. Holland, and C. C. Sweeley, *Anal. Chem.* **50,** 433–441 (1978).
6. H. Nau and K. Biemann, *Anal. Chem.* **46,** 426–434 (1974).
7. K. R. Betty and F. W. Karasek, *J. Chromatogr.* **166,** 111–122 (1978).
8. G. W. A. Milne, S. R. Heller, R. S. Heller, and D. P. Martinson, in *Advances in Mass Spectrometry,* Vol. 8B (A. Quale, ed.), Heyden, London, 1979, pp. 1578–1581.
9. H. S. Hertz, R. A. Hites, and K. Biemann, *Anal. Chem.* **43,** 681–691 (1971).
10. G. M. Pesyna and F. W. McLafferty, in *Determination of Organic Structures by Physical Methods* (F. C. Nachrod, J. J. Zucherman, and E. W. Randall, eds.), Academic, New York, 1976, pp. 91–155.
11. G. M. Pesyna, R. Venkataraghavan, H. E. Dayringer, and F. W. McLafferty, *Anal. Chem.* **48,** 1362–1368 (1976).
12. K. S. Kwol, R. Venkataraghavan, and F. W. McLafferty, *J. Am. Chem. Soc.* **95,** 4185–4194 (1973).
13. H. E. Dayringer, G. M. Pesyna, R. Venkataraghavan, and F. W. McLafferty, *Org. Mass Spectrom.* **11,** 529–542 (1976).
14. K. S. Haraki, R. Venkataraghavan, and F. W. McLafferty, *Anal. Chem.* **53,** 386–392 (1981).
15. H. E. Dayringer and F. W. McLafferty, *Org. Mass Spectrom.* **12,** 53–54 (1977).
16. I. K. Mun, R. Venkataraghavan, and F. W. McLafferty, *Anal. Chem.* **53,** 179–182 (1981).
17. L. Brenner, B. Silver, and I. Suffet, *J. Environ. Sci. Health* **A13,** 149–166 (1978).
18. R. J. Mathews, *Int. J. Mass Spectrom. Ion Phys.* **17,** 217–231 (1975).
19. T. H. Bindel, P. V. Fennessey, B. S. Miles, and S. I. Goodman, *Clin. Chim. Acta* **66,** 209–217 (1976).

8

# GAS CHROMATOGRAPHY/ MASS SPECTROMETRY IN WATER POLLUTION STUDIES

*E. D. PELLIZZARI and J. T. BURSEY*

## 1. INTRODUCTION

Interest and activity in a field of research or application can be gauged by the frequency with which developments are reviewed. By this criterion, application of gas chromatography/mass spectrometry/computer (GC/MS/COMP) techniques to the analysis of aqueous matrices must be at an all-time high, with no indication that interest in the use of the combined technique is waning. The application of GC/MS/COMP to the qualitative and quantitative analysis of organic water pollutants has been reviewed, and recent reviews have also stressed the value of capillary gas chromatography coupled with automated data-handling techniques and accumulation of spectral reference data for organic trace pollutants in water.[1–6] Special techniques interfaced to GC/MS/COMP systems (e.g., purge and trap methods) have been reviewed. Bishop has discussed recent modifications in the purge and trap procedure.[7]

## 2. COMPUTER-ASSISTED ANALYSIS

### 2.1. *Qualitative Analysis*

In recent years GC/MS/COMP has become a routine tool for the analysis of water for organic compounds. Use of retention indices coupled with MS data for identifying organics was reported by Nau and

*E. D. PELLIZZARI* and *J. T. BURSEY* • Analytical and Chemical Sciences, Research Triangle Institute, Research Triangle Park, North Carolina 27709.

Biemann [8] and others. Subsequently, researchers have reported the use of automated systems which rely on retention data and MS data for identification of organic compounds. The success of this technique requires locating the internal standard in a complex mixture. Sauter further developed the automated data-adaptive background subtraction technique for analysis of data derived from wastewater.[9]

A further development in the use of computers is the automation of GC/MS data reduction. Instrument manufacturers have attempted to use their computerized systems to their full capability to make data reduction fast, efficient, and cost-effective. Most GC/MS/COMP systems now allow positive identification of compounds of interest by comparison with a library of known spectra (using either forward or reverse search, according to user preference) and automatic quantitation by comparing responses for compounds of interest to responses of standards. Systems that will allow comparison of relative retention times to add confidence to the identification have been reported by Beggs.[10–14]

A goal for automated data analysis has been the development of algorithms to evaluate the data as an experienced human would, using computer assistance. Sophisticated MS data analysis systems have been developed to operate on large computers. An extensive program to perform this type of analysis has been developed and its application has been described by Shackelford.[15] It incorporates a sequence of algorithms which consists of four basic parts: (1) software to locate and identify the internal standard in every set of GC/MS data; (2) an analysis program to find and extract the spectra of components observed in the GC/MS analysis; (3) a spectrum-matching system to allow identification of components; and (4) a program to collate and report results. Data obtained by Shackelford from the application of this program to analysis of aqueous industrial effluent extracts show that the automated techniques are at least equal to manual techniques for pointing out components and identifying these components by spectrum matching with a reference library.[15] However, the more convoluted the chromatographic analysis, the more difficulty the automated system encounters because of the overlap of common ions and problems with peak shape and the ability to distinguish one chromatographic peak from another where there is extensive overlap in elution. Probability calculations by Rosenthal on coelution show, for example, that a 200-component mixture will, on the average, have 20% of the components occurring in the form of multiplets, even with the use of capillary columns.[16] In practice, the occurrence of multicomponent multiplets may be substantially worse because of the presence of a large number of closely related (and, therefore, closely spaced in elution) isomers. There would thus be a theoretical limit for the effectiveness of automated methods of GC/MS data analysis when component identifications are made by direct search against a reference library. For these reasons caution in the usage of automated methods is necessary since their statistical accuracy remains ill-defined for a variety of chemical functionalities found in water.

Another rapidly expanding area is the application of computers to the analysis of water pollutant data. A case in point is the continuing development of the WaterDROP program described by Carson.[17] The function of this computer program is to provide a comprehensive list of organic pollutants which have been found in water, so that the database can be searched by country, state, or city, as well as water type. Representive fields for searching would include water type, sampling date, concentrations, identification method, indication of confirmation of identification, chemical information, literature reference, and several others. The database has been designed for ease of searching and updating, and has its foundation in a list published by Shackelford and Keith.[18] The WaterDROP database is publicly available as a part of the U.S. NIH/EPA Chemical Information System, and the database is continually being expanded.

### 2.2. *Quantitative Analysis*

Computerized automated quantitation procedures for GC/MS-derived data can provide a great savings in time for the analyst, and are very cost-effective where an in-house computer can be used to do a reverse library search of a library of standard compounds to provide qualitative analysis. Coleman has reported a library which contains user-generated reference spectra, relative retention times (as an additional reference for accurate identification), response factors (to allow quantitative calculations), and selected quantitation masses for a number of compounds of interest found in water.[19] In Coleman's study featuring Grob Closed Loop Stripping analysis of water samples with capillary GC/MS/COMP, statistical analyses were performed on GC relative retention times, gas chromatographic peak areas, MS detector response factors, amounts, and recoveries observed from water under conditions of replicate analysis.[20] Gas chromatography relative retention times showed excellent reproducibility ($\pm 0.7\%$ of the average value), MS response factors varied from $\pm 3.5$ to $\pm 14.2\%$ of the average values (with some effect on MS response factors due to the abundance of the ion selected for use in quantitation), and gas chromatographic peak areas varied on an average of $\pm 13\%$, with amounts varying an average of $\pm 9\%$. Using an internal standard for quantitation compensated for these changes. Computerized automatic quantitation requires accurate MS library spectra and reproducible GC retention times and MS detector response factors. Close reproducibility of retention times allows the automated procedures to use narrow-search windows which should result in greater success in distinguishing isomers which elute in close proximity. Coleman found that compound recoveries from drinking water vary over a wide range, depending on the nature of the compounds, differences in water solubility, and vapor pressure of the compound.[20] Error in determination of low nanogram MS detector response factors has been found to be analyte dependent: compounds similar to the internal standard used (anthracene-$d_{10}$) show relative standard deviations for single-level response factors of approximately 1.2%, while compounds different from the standard in chemical and physical properties can be determined with a relative standard deviation 20% over the short term.

## 3. STUDIES TO IMPROVE QUANTITATIVE ANALYSES

A number of methods have been used for quantification, among which are those using internal and external standards, regression analysis, and single-point calibration. Ryan has examined a number of parameters to determine the extent of their impact on the precision of quantitation for the internal standard method.[21] These included (1) the chemical class of internal standard; (2) the retention time separation between an internal standard and the compound of interest; (3) chromatographic peak area integration for total ion current and extracted ion current; and (4) short- and long-term stability and reproducibility of MS detector response factors. Ryan concluded:

(1) The precision of quantitation for chemicals in GC/MS data was not significantly improved by using internal standards chemically similar to the unknown compounds.
(2) The precision of measurement for compounds is improved by using an internal standard that elutes close to the target compound, rather than an internal standard that elutes farther away.
(3) The precision of GC/MS data is similar between total ion current peak area and extracted ion current peak area integrations.
(4) The MS detector response factors should be checked daily and corrected, if necessary, before calculating concentrations of analytes in water.

The significant improvements in the quality of chromatography with the introduction of fused-silica capillary columns have also impacted on qualitative and quantitative precision and accuracy of analysis. Excellent interlaboratory accuracy when multiple internal standards are used to bracket the chromatographic elution range can be obtained.[22,23] Also, Sauter observed a relative standard deviation of <10% at the 50-ng level for response factors of most target compounds over a long term.[23] In addition, it has been demonstrated that MS relative molar response factors determined at one concentration are valid over a wide concentration range (from the limits of MS detection up to capillary column capacity).[22]

The determination of priority pollutants in industrial wastewaters using several different quantification methods and GC/MS/COMP has been described by Colby and Rosecrance.[27] A replicate series of wastewater samples was analyzed by GC/MS/COMP and quantification was performed using external standard, internal standard, and isotope dilution techniques. Furthermore, each approach included four methods for calibration curve generation: (1) direct proportionality; (2) two data points to provide an intercept; (3) standard additions to compensate for matrix effects; and (4) linear regression. Colby and Rosecrance observed the greatest error when no internal standard was employed and the least when isotope dilution was used.[24] When no internal standard was incorporated into the sample, matrix effects were not compensated and the accuracy was poor. Thus, the following trends were evident in accuracy: isotope dilution > internal standard > external standard > no internal–external standard and linear regression (weighted) > two data points > standard addition > one data point.

## 4. *WATER ANALYSIS*

Specialized schemes for the GC/MS/COMP analysis of single compounds or single chemical classes have been developed to meet specific needs. Matsunaga has developed a method for the analysis of nitrophenol isomers in fresh water and ocean sediments, and Ingram reported a similar method for pentachlorophenol.[25,26] Frame studied the distribution of alkylnaphthalene isomers in complex groundwater samples from chemical spills using GC/MS/COMP techniques.[27]

GC/MS/COMP has been used to identify fluorocarbons in the lower Main River by Maier.[28] Baehn compared data obtained by use of DuPont and LKB instrumentation in the analysis of organic contaminants.[29] When sampling and enrichment conditions were comparable, results were satisfactory on both systems; differences in performance were seen chiefly in the comparison of capillary and packed-column chromatography, with more unambiguous identifications for components of complex mixtures being observed in the capillary mode. Trace amounts of organic materials in tap water were detected by GC/MS/COMP to assess the health hazard in drinking water in a study by Shinohara.[30] Trace organics in tap water, river water, seawater, rain water, sewage, and sediment were characterized by Koga using GC/MS/COMP.[31] A gas phase stripping technique has been described by Brooks for the concentration and subsequent measurement of trace quantities of volatile organic compounds in water by GC/MS/COMP.[32] Analysis by glass capillary GC/MS/COMP was the final step for identification, confirmation, and quantification of individual organic compounds isolated from drinking water concentrates and advanced waste treatment water concentrates following extraction of the concentrate into acidic, basic, and neutral fractions in the research reported by Lin.[33] Low-boiling organic compounds in tap water have been adsorbed on Amberlite XAD-4, then eluted and identified by GC/MS/COMP in a study by Shinohara.[34] A semiquantitative method for determining trace organic compounds in raw and treated surface water, groundwater, and industrial wastewater which uses purge and trap techniques for isolation of volatile components with subsequent GC/MS/COMP analysis has been described by Pereira and Hughes.[35] Trace organic compounds in Philadelphia drinking water were characterized by GC/MS over a 2-year period by Suffet.[36] In a report by Hartmetz wastewater and water purification plant effluents from chemical and pharmaceutical plants were analyzed by capillary GC/MS and found to contain volatile nitrosamines.[37]

Computerized GC/MS has been applied to other problems in water pollution: a chemical plant wastewater containing organic compounds was discharged into a city sewer, mixed with other industrial effluents, and ultimately treated in a Philadelphia plant using classical secondary methods. Treated effluent was discharged into the Delaware River; during high tides the treated effluent was carried upstream where it was able to enter the city drinking water treatment plant. GC/MS/COMP analysis of the water at various stages by Hites showed that concentrations of chemical pollutants were reduced by dilution but not by the water treatment methods employed.[38] In a report by Lopez-Avila and Hites industrial organic compounds in wastewater, river water, and sediment were identified using MS (electron impact (EI), methane chemical ionization (CI), high-resolution MS), as well as gas and high-pressure liquid chromatography.[39] When GC and GC/MS/COMP results for priority pollutant analysis were compared by Oldham in the characterization of selected surface waters, industrial wastewaters from coal-fired boiler facilities, and wastewaters from chemical processing industries, detection limits by GC were found to be low ppb to sub-ppb levels, while GC/MS/COMP detection limits were at the 1–10-pg level (low ppt range) for many samples.[40]

Trace organic constituents of raw and potable waters were quantified using GC/MS/COMP by James using deuterated internal standards, with the best precision of measurement obtained with peaks appreciably above the detection limit.[41] Shinohara explored the presence and behavior of organic pollutants in an aqueous environment surrounded by an industrialized area using capillary GC/MS/COMP.[42] The same authors studied trace organic components of tap water by extracting with an Amberlite XAD-2 resin column and identifying and quantifying the components using GC/MS/COMP techniques. GC/MS/COMP has been applied to the characterization of organic constituents of municipal wastewater treatment effluents of river water and other water supplies.[7,43,44] Halogenated organic compounds were found by GC/MS/COMP in water and in sediment samples taken from three tributaries of the Niagara River adjacent to hazardous waste disposal areas in Niagara Falls, New York.[44]

New mass spectrometric techniques have been applied to the analysis of organic constituents of aqueous media. In a study by Schulten and Sun the analyses of standard mixtures and some wastewater samples indicated the suitability of field desorption mass spectrometry for identification of organophosphorus pesticides at nanogram levels in environmental samples without preliminary separation and purification.[45] Freudenthal used the mass spectrometer as an element-specific detector for a gas chromatograph, where the GC effluent was atomized in a microwave-induced discharge located in the GC/MS/COMP interface.[46] The elements investigated were fluorine, chlorine, bromine, iodine, sulfur, and nitrogen, but the method is applicable to other elements as well.

## 5. *A NEW HORIZON IN COMPREHENSIVE ANALYSIS*

In contrast to specialized schemes which are tailored for a few chemicals or classes of chemicals in water, the Master Analytical Scheme (MAS) represents the first effort to develop a comprehensive qualitative–quantitative scheme for the analysis of volatile organic compounds by GC/MS/COMP. Although the MAS is designed to span the complexity encountered in drinking and surface waters and municipal, industrial, and energy-related effluents, procedures are included that define the water quality and allow for optimal detection limits for that water sample.

The MAS is a set of analytical protocols that includes organics of any volatility class that will, or can be derivatized to, pass through a gas chromatograph of most functional group types in almost any water sample. The prospective user has the latitude of applying all the protocols or just those that cover organic group types of interest. Thus, each protocol stands alone containing the elements for determining water quality, collecting the sample, adding internal standards, and processing

*TABLE 1*
*Organization of MAS Protocols*

| MAS symbol | Chemical/physical property group | Operational protocol | |
|---|---|---|---|
| VO (volatile organics) | Volatile, low water solubility | Purge and trap | Sampling and sample handling for VO<br>Purge, trap, and analysis of VO |
| NEWS (neutral water soluble) | Volatile, neutral, high water solubility | Elevated temperature purge and trap | Sampling and sample handling for NEWS<br>Elevated temperature purge, trap, and analysis of NEWS |
| WABN (weak acids and bases and neutrals) | pH 8 extractable weak acids, weak bases and neutrals | Sorbent (accumulator) column partitioning (pH 8.0) (WABN-SC) | Sampling and sample handling for extractable organics<br>Extraction and analysis of WABN<br>Sorbent accumulator of WABN GC/MS/COMP |
| | | Batch liquid–liquid partitioning (pH 8.0) (WABN-BL) | Sampling and sample handling for extractable organics<br>Extraction and analysis of WABN<br>Batch liquid–liquid cleanup GC/MS/COMP |
| | | Flow-under liquid–liquid partitioning (pH 8.0) (WABN-FU) | Sampling and sample handling for extractable organics<br>Extraction and analysis of WABN<br>Flow-under GC/MS/COMP |

| | | | |
|---|---|---|---|
| ESSA (extractable semivolatile strong acids) | pH 1 extractable, semivolatile, strong acids | Batch liquid–liquid partitioning (pH 1.0) | Sampling and sample handling for extractable organics |
| | | | Batch liquid–liquid extraction and analysis of semivolatile strong acids |
| VOSA (volatile strong acids) | Volatile carboxylic acids | Anion exchange/distillation | Sampling and sample handling for other ionic compounds |
| | | | Anion exchange, distillation, and analysis of VOSA |
| NOVA (nonvolatile acids) | Very strong, nonextractable, nonvolatile acids | Anion exchange | Sampling and sample handling for other ionic compounds |
| | | | Anion exchange and analysis of NOVA |
| SAM (strong amines) | Primary and tertiary strong amines (SAM-PT) | Cation exchange | Sampling and sample handling for other ionic compounds |
| | Secondary amines (SAM-S) | | Cation exchange and analysis of strong amines |
| | All groups | | GC/MS/COMP analysis procedure—general instructions for all protocols |

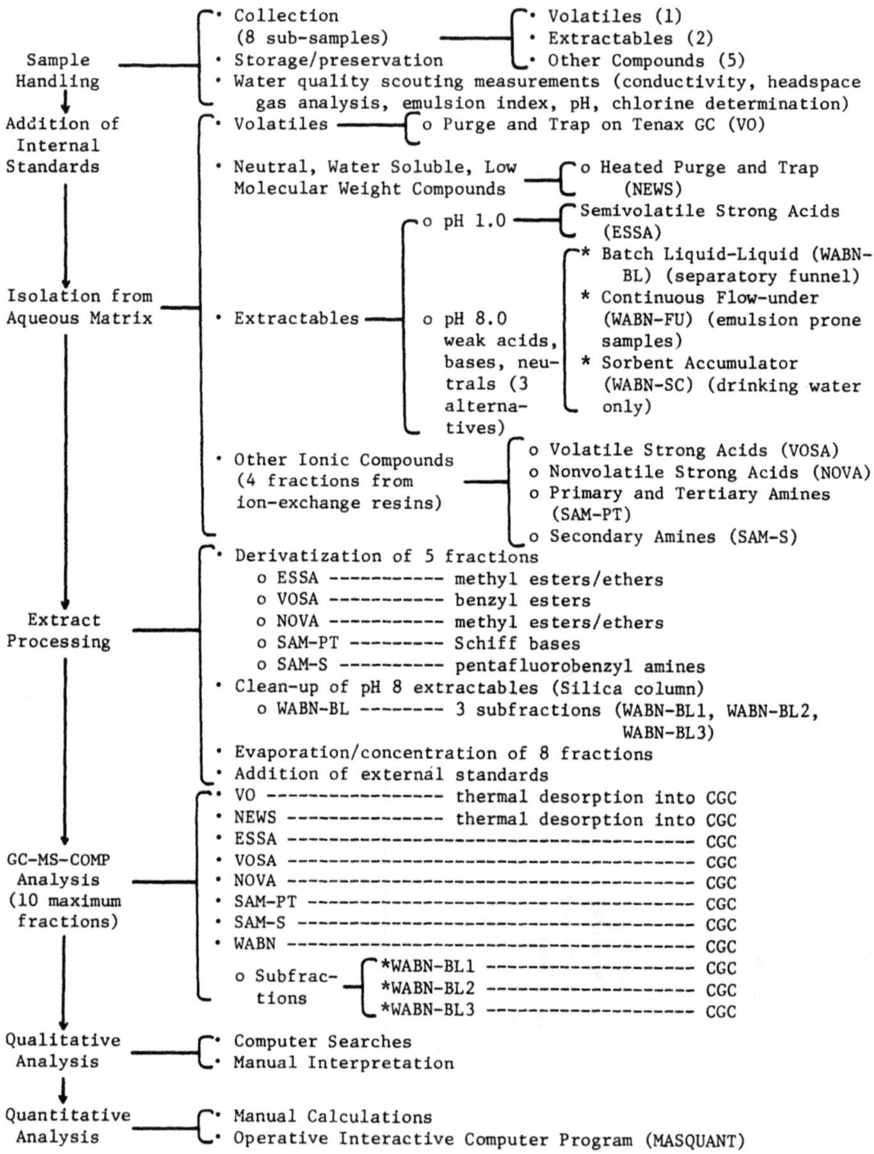


FIGURE 1. Master analytical scheme flow diagram. (From Ref. 47.)

the sample with subsequent analysis according to prescribed GC/MS/COMP conditions.

Table 1 presents some of the MAS terminology, MAS protocol descriptors, and applicable chemical class or group types. An understanding of the physical–chemical classifications of organics and how these relate to the analytical operational protocols is necessary for application of the MAS to a water sample. Symbols given in Table 1 are for convenience in identifying each major MAS fraction corresponding to a chemical–physical property group. Along with the operational protocols Table 1 also gives brief descriptions of the technique for separation of organics from the water sample.

Figure 1 depicts a flow diagram of the procedures for implementation of the MAS. Each step is summarized below.

### *5.1. Sample Handling*

Eight subsamples are required for a comprehensive sample analysis: one for each protocol. Procedures are prescribed for sample collection, storage, and preservation. Volatile organic (VO fraction) samples are collected in septum-capped bottles with no headspace. Methylene chloride is added to extractable and ionic compound samples as a bactericide, while hexane is used as a "keeper" solvent layer for extractable compounds. Chlorine determination (e.g., using a Hach Chlorine Test Kit) indicates the level of sodium thiosulfate necessary to stoichiometrically reduce any residual chlorine left from water treatment. All samples are stored at 4 °C in the dark.

Various water quality scouting measurements help in the selection of appropriate analytical procedures optimized according to water quality rather than sample "type" (e.g., drinking water or municipal effluent). Headspace gas analysis by GC of a separate small sample is employed to determine the dilution necessary for VO purge and trap analysis, and indicates levels at which internal standards are to be added. A trial shakeout with methylene chloride of a small aliquot of the extractable (WABN) sample shows whether emulsion formation is a problem, and thus whether the flow-under extractor must be used. Conductivity measurements indicate maximum sample volume allowable for isolation of ionic compounds by ion exchange resins without exceeding resin capacity.

### *5.2. Internal Standards*

Prescribed deuterated internal standards (Table 2) are added to each subsample before processing or storage, preferably in the field. Selection and packaging assures that from 1 to 9 standards, of the total of 20, will appear in each extract for GC/MS analysis; retention times are such that the standards span the chromatographic window. These standards are used for monitoring recovery during analysis, for quantifying sample components, and for relative retention time calculations. The initial sets of MAS standards were prepared by the National Bureau of Standards (NBS). After assuring chemical and isotopic purity of each

*TABLE 2*
*Internal and External Standards Employed in the Master Analytical Scheme*

| Operational protocol | Chemical group type (symbol) | Internal standards | Low (μg/liter[a]) | High (μg/liter[a]) | External standard |
|---|---|---|---|---|---|
| Purge and trap | Volatile, low water solubiorganics (VO) | Bromoethane-$d_5$ | 0.56 | 6.6 | Perfluorotoluene |
| | | Anisole-2,4,6-$d_3$ | 4.2 | 25.6 | |
| | | Chlorobenzene-$d_5$ | 7.2 | 5.4 | |
| | | Naphthalene-$d_8$ | 5.8 | 48.7 | |
| Elevated temperature purge and trap | Volatile, neutral, highly water soluble organics (NEWS) | *t*-butanol-$d_9$ | 112.9 | 554.0 | Perfluorotoluene |
| | | Nitrobenzene-$d_5$ | 20.1 | 103.5 | |
| Sorbent column, batch L-L, flow-under partitioning (pH 8.0) | Neutrals (WABN) | *o*-xylene-$d_{10}$ | 2.8 | 104.7 | 4-fluoro-2-iodotoluene |
| | | Naphthalene-$d_8$ | 20.2 | 500.7 | |
| | | Nitrobenzene-$d_5$ | 94.2 | 538.8 | |
| | | 1-phenyl-$d_5$-ethanol | — | 520.9 | |
| | | Acetophenone-$d_5$ | — | 104.7 | |
| | | Propiophenone-$d_5$ | — | 98.1 | |
| | | Perylene-$d_{12}$ | 1.1 | 98.1 | |
| | Weak bases | Acridine-$d_9$ | 19.8 | 100.3 | |
| | Weak acids | Phenol-$d_5$ | 112.7 | 400.4 | |
| Batch L-L partitioning (pH 1.0) | Semivolatile strong acids (ESSA) | Heptanoic acid-$d_{13}$ | 13.8 | 111.0 | 4-fluoro-2-iodotoluene |
| | | Benzoic acid-$d_5$ | 1.08 | 100.3 | |
| Anion exchange | Very strong nonvolatile acids (NOVA) | 2-naphthalene-sulfonic acid-$d_7 \cdot H_2O$ | 99.4 | 497.4 | 2-fluorobiphenyl<br>4-fluoro-2-iodotoluene |
| Anion exchange/distillation | Volatile strong acids (VOSA) | Butyric acid-$d_7$ | 15.7 | 535.9 | 2-fluorobibiphenyl |
| Cation exchange | Primary, secondary, and strong amines (SAM-PT) | *n*-butylamine-$d_9$ | 118.3 | 554.3 | 2-fluorobiphenyl |
| | | 2-phenylethyl-1,1,2,2-$d_4$-amine | 121.2 | 530.3 | 4-fluoro-2-iodotoluene |
| | Secondary and strong amines (SAM-S) | *N*-ethyl-2-fluorobenzylamine | 107.2 | 522.1 | 2-fluorobiphenyl<br>4-fluoro-2-iodotoluene |

[a] Concentration after dosing into the appropriate volume of sample.

standard compound, NBS determined its stability and compatibility with the other internal standards. Methanol solutions (water solutions for highly water soluble compounds) were prepared and packaged for sample dosing, as shown in Table 2.

For samples to be analyzed for purgeable components, internal standards are added at the concentration (Table 2) closest to the concentrations of sample components, as indicated by the results of headspace gas-scouting analysis of a separately collected sample aliquot. Volatile organic standards are packaged in glass capillary ampoules that are added to the sample and crushed with a magnetic stirbar. For other sample aliquots internal standards are packaged in vials such that emptying the entire contents of the vial into the prescribed sample volume produces the concentrations of standards shown in Table 2. Low-concentration standards sets are generally used for drinking and surface waters; high-concentration sets are used for effluents.

## 6. *ISOLATION OF ORGANICS*

After addition of internal standards, the eight subsamples are processed as follows. (Fraction/protocol symbols are in parenthesis.)

*Volatile organics (VO).* Highly volatile (purgeable) organics are analyzed by a modification of the Bellar–Lichtenberg (EPA's method 624) method, using a custom built closed system. Sodium sulfate is used to "salt out" the organics in a 200-ml sample, which is purged at 30 °C. The sample is first diluted, if necessary, in accordance with the total concentration of purgeable organics as indicated by GC scouting of the headspace gas. Dilution prevents saturation of the GC/MS/COMP and decreases foaming potential. Purged organic vapors are collected on a Tenax GC sorbent trap, from which they are thermally desorbed into a liquid nitrogen cold trap. An "external" standard, perfluorotoluene, is added to the cold trap through an injection port system installed between the sorbent trap and the cold trap, before desorption of the purged sample components onto the cold trap. The total condensate is then flash-evaporated onto a fused-silica capillary for analysis by GC/MS/COMP. Comparison of MS signals for the external standard with those for the internal standards purged from the sample allows calculation of recoveries of the internal standards, thus monitoring performance of the entire analytical operation.

*Organics extracted at pH 8 (WABN).* Compounds of intermediate volatility, most of which are water insoluble, have usually been analyzed by conventional liquid–liquid extraction. Batch liquid–liquid extraction of 1 liter of sample with methylene chloride in a separatory funnel is also the basic MAS method. Adjustment of the sample pH to 8.0 allows reproducible extraction of weak acids (e.g., most phenols) and weak bases (e.g., most anilines) as well as neutral compounds.

For some samples, however, batch liquid–liquid extraction is not suitable. An initial trial solvent extraction in a stoppered graduated cylinder indicates whether emulsion formation is likely to be a problem. For emulsion-prone samples, continuous liquid–liquid extraction with methylene chloride in a flow-under extractor should be used. For samples in which extractable organic concentration is expected to be low, such as drinking water, XAD-4 resin sorbent columns are designated for sorption/concentration from 10 to 15 liters of water. The organics are desorbed using methanol followed by methylene chloride.

*Other ionic compounds (VOSA, NOVA, SAM).* Compounds that are easily dissociated in water have not been included in analytical schemes because of difficulties with extraction and chromatography. New techniques, however, have been developed to allow inclusion of these compounds in the MAS. Ion exchange resins are used to separate four classes of ionic compounds from the sample matrix using three separate aliquots of the sample. Outlined below are procedures for elution of these compounds from the resin columns, cleanup, concentration, and derivatization.

*"Volatile" carboxylic acids (VOSA).* Butyric acid, methacrylic acid, and other volatile carboxylic acids (through C-8) are separated from the water on Biorad AG 1-X8 anion exchange resin, then eluted with sodium bisulfate in an acetone–water solution. The volatile acids are distilled from the eluate, converted to their nonvolatile salts, then derivatized with benzyl bromide to form the benzyl esters.

*"Nonvolatile" strong acids (NOVA).* These compounds (e.g., naphthalene sulfonic acids) are also separated from the water on Biorad AG 1-X8 resin. They are eluted with HCl in methanol, the solvent is evaporated, and the acids are methylated with diazomethane.

*Strong amines (SAM).* Compounds such as hexylamine, dibutylamine, and piperidine are isolated from the water sample on Biorad AG 50W-X8 cation exchange resin, then eluted with sodium hydroxide in acetonitrile–water solution. The eluent is acidified, the solution is evaporated to dryness, and the amine hydrochloride salts are dissolved in base and extracted with methyl *t*-butyl ether. The extract is split; half is derivatized with pentafluorobenzyl bromide to make pentafluorobenzyl tertiary amines from the secondary amines (SAM-S), and half is derivatized with pentafluorobenzaldehyde to make Schiff bases of the primary amines (SAM-PT).

Tertiary amines are also separated by this protocol, and are quantified (underivatized) in the primary amine fraction. Certain weak bases may be detected in this fraction, but are measured in the pH 8 extractable fraction (WABN), where they are extracted more efficiently.

### 6.1. *Extract Processing*

Extractable and ionic fractions require further processing before GC/MS/COMP analysis. The necessary derivatization steps, for example, are described above and summarized in Fig. 1.

The pH 8 extractable fraction (WABN-BL) of many industrial effluents will require cleanup and subfractionation before effective separation can be achieved, even on capillary columns. First, however, a scouting procedure is implemented to determine whether cleanup is necessary. The crude extract is analyzed by GC using a packed column and flame ionization detection; baseline rise relative to a separately run standard is the evaluation criterion. Cleanup, if necessary, is on a silica gel column from which three fractions (WABN-BL1, -BL2, and -BL3) are eluted using hexane, methylene chloride, methanol, and their mixtures.

Concentration of extracts for GC/MS analysis is by Kuderna–Danish evaporation down to 4 ml, followed by nitrogen blowdown to 0.5 ml using a modified Snyder column. "External" standards (Table 2) are added to each final extract just before GC/MS/COMP analysis to monitor recovery of the deuterated internal standards that were added to the original water samples. The external standard for the purgeable fractions (VO and NEWS) is perfluorotoluene. External standards for all the other extracts are 2-fluorobiphenyl and/or 4-fluoro-2-iodotoluene.

### 6.2. *Gas Chromatography*

As shown in Fig. 1, as many as 10 extracts or fractions may be obtained from one sample if the entire MAS protocol is applied. (This may be reduced to 7 if cleanup of the pH 8 (WABN) extractables is not necessary, and if the primary and secondary amine fractions can be mixed for a single GC/MS analysis.) Glass or fused-silica capillaries are prescribed. The newly available bonded phase (e.g., Durabond, DB-1 or DB-5), wide-bore, thick-film (1 ml), 30- or 60-m fused-silica columns are recommended for inertness, stability, and sample capacity. In all cases performance standards rather than specific columns are specified (see Sec. 7) in the analytical protocol. No more than four different GC columns should be necessary for the entire MAS. The analytical protocol for each fraction prescribes optimum GC conditions.

### 6.3. *Qualitative Analysis*

Sample components are identified by established GC/MS/COMP techniques. Gas chromatography/mass spectrometry data are stored on tape or disk; internal standards in each extract are used as reference points for retention time measurements as well as for quantitation. Compounds are identified by computer searching of mass spectral data banks or by manual interpretation.

### 6.4. Quantitative Analysis

Extensive recovery studies were conducted during development of the MAS. Approximately 340 model compounds from a wide variety of chemical classes (Table 3) were dosed into representative samples of all major types of water (drinking water, surface water, and municipal, industrial, and energy-related effluents); recoveries were determined and recovery factors stored in a computer data bank. Relative molar response (RMR) factors (relative to the deuterated internal standards) were also determined and stored in the data bank. The MAS user can use these data banks and the computer program developed for the MAS (MASQUANT) to calculate the concentration in the original water sample of these model compounds as they are identified. For compounds that are not in the data bank, concentration can be estimated by using RMR and recovery factors for structurally similar compounds in the data bank.

During development of the MAS, it was determined that GC/MS data should be collected in the full scan mode rather than by selected-ion monitoring. It was also shown, however, that quantitation should be based on the peak area (or height) of selected ions of a compound rather than on the total ion current for that compound. Thus, RMR factors involve ratios of ion peak areas and moles of the model compounds (those in the data bank) to ion areas and moles of the internal standard selected for quantitation (usually the closest eluting standard). These ion areas depend strongly upon the GC/MS instrumentation and MS tuning. Thus it is necessary for each MAS user to either develop his own RMR values, or, alternatively, measure a few compound RMRs at the MS tune specified by the MAS and use a linear regression plot of these against the tabulated RMRs of the same compounds to develop a correction factor (slope) for the tabulated RMRs. Tabulated RMR values, the same as those in the computer data bank, will be provided as part of the MAS protocols.

RMR values developed by this linear regression correction procedure will not be as accurate as those measured directly by the user. Even if the user develops his own RMR data bank, it is necessary for him to check the tune of his mass spectrometer daily by chromatographing a performance standard prescribed by the MAS. If the tune is significantly different from that used when his RMR values were measured, the same linear regression procedure should be used to correct his tabulated RMRs for the current tune. (The linear regression and correction can be done with MASQUANT.)

*TABLE 3*
*Types of Model Compounds Used for Recovery and Relative Molar Response Determinations*

| Protocol class | Number of compounds studied |
|---|---|
| Purgeables (VO) | 78 |
| pH 8 extractables (WABN) | 136 |
| Acid extractables (ESSA) | 39 |
| Volatile acids (VOSA) | 23 |
| Nonvolatile acids (NOVA) | 19 |
| Primary and tertiary amines (SAM-PT) | 16 |
| Secondary amines (SAM-S) | 9 |
| Neutral intractables (NEWS) | 20 |
| TOTAL | 340 |

To use MASQUANT for quantitation, the user must provide sample component identities, some information as to the specific MAS analytical steps used, selected-ion peak areas for each sample component and the appropriate (nearest) internal standard, and the concentration of the internal standard in the original water sample. If the identified sample component is one of the model compounds, the computation program will select the appropriate RMR and recovery factor for that component and calculate its concentration in the original sample.

As mentioned above, if the identified sample component is not a model compound, RMR and recovery factors of a structurally similar compound in the data bank may be used for an estimate of concentration. In addition to this obvious source of error, an additional error is involved in using any recovery factors from the data bank. Since sample matrices used for recovery studies were only representative of the various water types, errors will occur in applying these factors to other samples dependent upon the matrix differences between the sample being analyzed and the representative recovery sample.

## 7. *QUALITY ASSURANCE*

Emphasis on quality assurance during development, initial testing, and improvement of the MAS produced an analytical method having a known scope, and within that scope, statistical estimates of accuracy and precision. During development of the interim MAS, statistical analysis techniques were applied to all experiments; this allowed determination of an error propagation factor for the entire scheme.

*System performance standards.* One important assurance step is the use of standard performance solutions to check performance of the GC/MS/COMP each day. The MAS prescribes such performance standard solutions and corresponding criteria of acceptance for each analytical protocol. Solutions include compounds to measure GC peak asymmetry, separation number, resolution, polarity, and column acidity and basicity; inertness of the GC-to-MS transfer line; capillary capacity; limits of detection of the GC/MS/COMP system; and tune of the MS. These solutions also contain the deuterated internal standards and the external standards appropriate to each protocol for RMR verification and periodic determination of the RMR correction factor by linear regression, if necessary.

*Internal and external standards.* Comparison of the recovered quantity of deuterated internal standards to the quantity of external standard added to the extract just before GC/MS analysis reveals recovery deficiencies, thus serving as a check to indicate malfunctions of the MAS analytical procedures. The primary use of deuterated internal standards, however, is for quantitation; reference to internal standards is generally accepted as the most accurate quantitation technique available for the GC/MS analysis of organics in water. These standards are also useful as retention time indices for an aid in compound identification.

*Sample scouting.* As described earlier, several sample scouting measurements are prescribed to characterize water quality and, in turn, allow selection of the appropriate and optimal analytical techniques for a particular water sample. Scouting also indicates the correct level (relative to levels of endogenous sample components) of internal standards to add to the purgeable sample, thus increasing quantitative accuracy.

*Blanks.* Procedures for field and procedural blanks and for quality control of each analytical operation are prescribed in each MAS protocol.

MAS *test samples.* For each MAS protocol, instructions are given for preparing test mixtures for the user to dose into distilled water, thus providing a test (surrogate) sample. A certified standard sample may eventually be available for test analysis by MAS users.

The MAS is a set of analytical protocols for the collection and GC/MS analysis of VO, NEWS, WABN, VOSA, NOVA, and SAM organics in surface and drinking water and in industrial, energy-related, and municipal effluents. Nominal lower qualitative limits are 0.1 g/liter for drinking water, 1 g/liter for surface water, and 10 g/liter for the effluents. One unique feature of the MAS is its comprehensiveness; another is its qualitative–quantitative aspect (an extensive database of response and recovery factors allows computer estimation of concentration without recourse to standards for each analyte). MAS protocols may be used as separate entities for analysis of organic fractions of particular interest (e.g., WABN-type compounds), or the complete set of protocols may be applied for a comprehensive analysis for all volatile compound types.

## *REFERENCES*

1. H. Knoeppel, *Ann. Chim. (Rome)* 67(9–12), 649 (1977).
2. R. A. Hites, *Adv. Chromatogr. (NY)* **15,** 69 (1977).
3. D. Meek. *Water Res. Top.* **1,** 164 (1981).
4. J. M. McGuire and R. G. Webb, *Pollut. Engr. Technol.* **18,** 1 (1981).
5. L. H. Keith. *J. Chromatogr. Sci.* **17,**(1), 48 (1979).
6. A. Cornu, J. C. Leclerc, and R. Masot, *Bull. Inf. Sci. Tech., Commis. Energ. At. (Fr.)* **230,** 181 (1979).
7. D. F. Bishop. U.S. EPA Report, EPA-600/2-80-196, PB81-127813, NTIS, Springville, VA, 1980.
8. H. Nau and K. Biemann. *Anal. Chem.* **46**(3), 426 (1974).
9. A. D. Sauter, L. Faas, G. R. Kurth, W. M. Shackelford, and D. M. Cline, *Advances in the Identification and Analysis of Organic Pollutants in Water,* Vol. 2, (L. H. Keith, ed.), Ann Arbor Science, Ann Arbor, MI, 1981, pp. 571–593.
10. D. Beggs, *Am. Lab.* **10**(7), 81–84, 86–87 (1978).
11. D. Beggs, National Bureau of Standards Spec. Publ. (U.S.), **519,** 169 (1979).
12. D. Beggs, *Oesterr. Chem. Z.* **81**(4), 98 (1980).
13. D. Beggs, *Pergamon Ser. Environ. Sci.* **3,** 317 (1980).
14. D. Beggs, *Environ. Sci. Res.* **16,** 303 (1980).
15. W. M. Shackelford, D. M. Cline, L. Burchfield, L. Faas, G. Deuth, and A. D. Sauter, in *Advances in the Identification and Analysis of Organic Pollutants in Water,* Vol. 2, (L. H. Keith, ed.), Ann Arbor Science, Ann Arbor, MI, 1981, pp. 527–554.
16. D. Rosenthal, *Anal. Chem.* **54,**(1), 63 (1982).
17. B. L. Carson, J. E. Going, V. Lopez-Avilla, J. L. McCann, C. J. Cole, and J. L. Holt, in *Advances in the Identification and Analysis of Organic Pollutants in Water,* Vol. 2, (L. H. Keith, ed.), Ann Arbor Science, 1981, Ann Arbor, MI. p. 497.
18. W. M. Shackelford and L. H. Keith. U.S. EPA Report, EPA-600/4-76-062, PB-265470, NTIS, Springfield, VA, 1976.
19. W. E. Coleman, W. K. Allen, R. W. Slater, J. J. Voto, R. G. Melton, F. C. Kopfler, and T. A. Aurand, in *Advances in the Identification and Analysis of Organic Pollutants in Water,* Vol. 2, (L. H. Keith, ed.), Ann Arbor Science, Ann Arbor, MI, 1981, p. 675.
20. R. G. Melton, W. E. Coleman, R. W. Slater, F. C. Kopfler, W. K. Allen, T. A. Aurand, D. E. Mitchell, S. J. Voto, S. V. Lucas, and S. C. Watson, in *Advances in the Identification and Analysis of Organic Pollutants in Water,* Vol. 2, (L. H. Keith, ed.), Ann Arbor Science, Ann Arbor, MI, 1981, pp. 597–673.
21. J. F. Ryan, J. E. Gebhart, L. C. Rando, D. L. Perry, K. B. Tomer, E. D. Pellizzari, and J. T. Bursey, in *Advances in the Identification and Analysis of Organic Pollutants in Water,* Vol. 1, (L. H. Keith, ed.), Ann Arbor Science, Ann Arbor, MI, 1981, pp. 49–66.
22. K. B. Tomer, J. T. Bursey, L. C. Michael, L. S. Sheldon, E. D. Pellizzari, A. L. Alford, J. D. Pope, and A. W. Garrison, in *Advances in the Identification and Analysis of Organic Pollutants in Water,* Vol. 1, (L. H. Keith, ed.), Ann Arbor Science, Ann Arbor, MI., 1981, pp. 67–85.
23. A. D. Sauter, L. D. Betowski, T. R. Smith, V. A. Strickler, R. G. Beimer, B. N. Colby, and J. E. Wilkinson, *J. High Resolut. Chromatogr./Chromatogr. Commun.* **4**(8), 366 (1981).
24. B. N. Colby and A. E. Resecrance, in *Advances in the Identification and Analysis of Organic Pollutants in Water,* Vol. 2, (L. H. Keith, ed.), Ann Arbor Science, Ann Arbor, MI., 1981, p. 221.
25. K. Matsunaga, F. Shibata, N. Saito, N. Torigashi, K. Kenmochi, and Y. Ogino, *Okayama-ken Kankyo Hoken Senta Nempo* **3,** 192 (1979).
26. L. L. Ingram Jr., G. D. McGinnis, and S. V. Parikh, *Anal. Chem.* **51,**(7), 1077 (1979).

27. G. M. Frame, Control Hazard. Water. Spills, Proc. Natl. Conf., 180, 1980.
28. E. J. Maier, G. Fritschi, and H. Kussmaul, *Vom. Wasser* **51,** 227 (1978).
29. U. Baehn, B. Bohn, G. Fritschi, and E. Maier, *Hydrochem. Hydrogeol. Mitt.* **3,** 75 (1978).
30. R. Shinohara, M. Koga, A. Kodo, S. Eto, T. Hori, and T. Akiyama, *Koenshu-Iyo Masu Kenkyukai* **3,** 69 (1978).
31. M. Koga, R. Shinohara, A. Kido, and S. Eto, *Koenshu-Iyo MASU Kenyikai* **4,** 153 (1979).
32. B. I. Brookes, S. M. Jickells, and R. S. Nicolson, *J. Assoc. Public Anal.* **16**(4), 117 (1979).
33. D. C. K. Lin, R. L. Foltz, S. V. Lucas, B. A. Petersen, L. E. Slivon, and R. G. Melton, ASTM Spec. Tech. Publ., STP 686, 68, 1979.
34. R. Shinohara, A. Kido, S. Eto, and T. Hori, Kankyo Osen. Busshitsu to Sono Tokishikorajii Shinpojumu, (Koen. Yoshishu), 6th, 34, 1979.
35. W. E. Pereina and B. A. Hughes, *J. Am. Water Works Assoc.* **72**(4), 220 (1980).
36. I. H. Suffet, L. Brenner, and P. R. Cairo, *Water Res.* **14**(7), 853 (1980).
37. G. Hartmetz and J. Slemrova, *Bull. Environ. Contam. Toxicol.* **25**(1), 106 (1980).
38. R. A. Hites, *Pergamon Ser. Environ. Sci.* **3,** 281 (1980).
39. V. Lopez-Avila and R. A. Hites, *Pergamon Ser. Environ. Sci.* **3,** 293 (1980).
40. R. G. Oldham, R. L. Spraggins, K. W. Lee, and L. H. Keith, *Prep. Pap. Natl. Meet. Am. Chem. Soc.,* Div. Environ. Chem., **18**(2), 626 (1978).
41. H. A. James, M. Fielding, T. M. Gibson, and C. P. Steel, *Adv. Mass Spectrom.* **8B,** 1429 (1980).
42. R. Shinohara, M. Koga, A. Kido, S. Eto, T. Tabata, T. Hori, and T. Akiyama, *Environ. Int.* **4**(2), 163 (1980).
43. M. T. Dmitriev, E. G. Rastyannikov, and S. A. Volkov, *Vodosnabzh. Sanit. Tekh.* **4,** 5 (1981).
44. V. A. Elder, B. L. Proctor, and R. A. Hites, *Biomed. Mass Spectrom.* **8**(9), 409 (1981).
45. H.-R. Schulten and S. E. Sun, *Int. J. Environ. Anal. Chem.* **10**(3–4), 247 (1981).
46. J. Freudenthal, *Int. J. Environ. Anal. Chem.* **5**(4), 311 (1978).
47. G. M. Pesyna, R. Venkataraghavan, H. E. Dayringer, and F. W. McLafferty, *Anal. Chem.* **48**(9), 1362 (1976).

AUTHOR'S NOTE: Although the research described in this article has been funded wholly or in part by the United States Environmental Protection Agency under Contract No. 68-03-2704 to the Research Triangle Institute, it has not been subjected to the Agency's required peer and administrative review and, therefore, does not necessarily reflect the view of the Agency and no official endorsement should be inferred.

9

# GAS CHROMATOGRAPHY/ MASS SPECTROMETRY IN AIR POLLUTION STUDIES

D. SCHUETZLE and C. V. HAMPTON

## 1. INTRODUCTION

Air pollutant samples are very complex in nature, containing chemical species which range in concentration from low ppt to high ppm levels.[1] Chemical compounds present in the atmosphere include both gaseous components and those present in dispersions of liquid or solid particles. Samples may be collected from ambient and interior (work space) air environments and mobile and stationary source emissions.

In most cases gas chromatography (GC) alone is not suitable for resolving complex air pollutant mixtures. The utilization of computer-controlled mass spectrometry in conjunction with GC or other preseparation techniques is necessary for the definitive analysis of such mixtures.

Figure 1 depicts the protocol which will be described for the GC/MS analysis of air pollutant samples in this chapter. Sampling, extraction, fractionation, and derivatization techniques are typically used before GC/MS analysis. The use of short-term microbiological assays for prioritizing chemical fractions for subsequent GC/MS analysis is also discussed.

The introduction of chemical contaminants may occur at any point in the analytical procedures. Contamination from filters,[2] solvents,[2–4] adsorbents,[5–7] sample blow-down gases, and glassware has been reported. Common contaminants include siloxanes, phthalate esters, saturated hydrocarbons, and fatty acids. Introduction of these contami-

D. SCHUETZLE and C. V. HAMPTON • Chemical Analysis Research Department, Research Staff, Ford Motor Company, Dearborn, Michigan 48121.

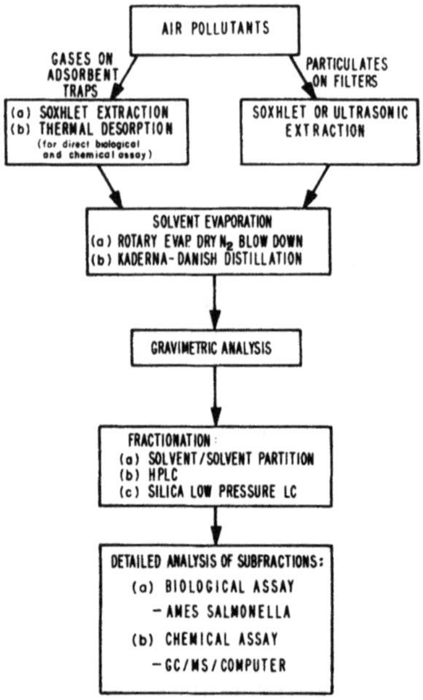


FIGURE 1. Analytical protocol for the GC/MS analysis of air pollutants.

nants does not usually represent more than a few percent of the total extractable mass. However, some of these contaminants may concentrate in certain chemical fractions, representing 50% or more of the fraction mass. The presence of these contaminants often makes GC/MS spectral interpretation of trace components difficult, if not impossible.

## 2. SAMPLING TECHNIQUES

Since many chemical compounds present in the atmosphere may be present as both gases and particulates, it is important that collection techniques be used to sample these two phases simultaneously.[8–10] Since the concentrations of the individual components in the air are usually low, sample concentration procedures are required.

Collection of particulates is undertaken using filters or impactors. Sampling of gaseous components in the $C_2$–$C_{10}$ volatility range may be undertaken using cryogenic (freeze-out) collection.[11] Polymer adsorbents are typically used for the collection of air pollutants in the $C_6$–$C_{30}$ votality range.[5,12–18] These include Tenax, XAD resins, and chromosorb adsorbents. Activated charcoal has been preferred for personnel exposure studies in work place environments.

A number of precautions must be taken when collecting samples for GC/MS analysis. A number of studies have shown that chemical transformations (artifacts) may occur during filter[19,20] and adsorbent[5–7] sampling of organics. Care must be taken to reduce or eliminate such artifacts. Some filter media[2] and adsorbents contain substantial concentrations of background contaminants which must be removed before sampling. Enough blank samples must be taken to determine the level of contaminants throughout the entire analytical procedure.

## 3. EXTRACTION TECHNIQUES

The soluble organic fraction (SOF) is generally separated from the carbonaceous or inorganic particulates by solvent extraction procedures. Extraction may be carried out using Soxhlet apparatus or by ultrasonic agitation. The qualitative and quantitative nature of the extracted organics is influenced by the types of solvent used and the temperature and the time used for extraction. Jennings[21] has shown that supercritical fluid extraction with $CO_2$ is an excellent technique for the isolation of specific PAH from particulates. However, comparative data for this extraction and Soxhlet techniques are limited. Gaseous PAH collected on porous polymer traps may be separated by thermal desorption[5] or solvent extraction procedures.[22]

### 3.1. *Filter Samples*

The Soxhlet technique is most widely used in particulate extractions. The size of the apparatus should be scaled according to the size of the filter used for particulate collections in order to minimize the solvent used. Generally Soxhlet thimbles are not used in order to minimize possible contamination. Instead particulate-laden filters are folded with the particulates sheathed inside the wrapped filter and then placed directly inside the extraction chamber. This procedure can be used for glass fiber or Teflon filters for which the material is strong enough to remain unshredded during extraction. More fragile filters such as Tissuquartz can be wrapped in another blank Teflon filter (e.g., Ghia Teflon filter) to avoid the introduction of shredded filter into the extraction solution.

Ultrasonication has been used for the solvent extraction of air particulates.[23,24] Golden[24] reported that by using sonication he was able to extract 15% more PAH from air particulates compared to using the Soxhlet method with cyclohexane as an extraction solvent. Schuetzle[25] found that Soxhlet extraction gave slightly higher mass recoveries than sonication for diesel particulates when using a number of solvents.

Benzene, cyclohexane, or dichloromethane (DCM) are the most common organic solvents used for the extraction of organics from particulates. Hill[26] found that methanol was more effective than cyclohexane for this extraction; it extracted more components in larger quantities and in less time. Schuetzle[25,27] found that a binary azeotropic mixture of an aromatic solvent and alcohol such as toluene-propanol was the best extractant for diesel particulates.

The mass of extracted material (SOF) after Soxhlet extraction may be determined by both a weight difference of the filter before and after extraction and by weighing the extract after evaporation of the solvent. Only the latter method can be used in the case of sonication extractions because it is difficult to weigh the shredded filter pieces.

### 3.2. *Adsorbent Samples*

The removal of gas phase material from adsorbents such as Tenax is undertaken using thermal desorption techniques and is described in Sec. 6.3. Solvent extraction is preferred in cases where several different analyses are desired on a single sample or when sample derivatization is used. However, thermal desorption is inherently more sensitive than solvent extraction.

## 4. *FRACTIONATION*

The composition of the SOF is extremely complex, consisting of hydrocarbons, oxygenated organics, PAH, and N or S heterocyclics and a variety of other organic as well as a few inorganic compounds such as sulfuric acid. These samples are typically too complex to allow direct GC/MS analysis. Fractionation procedures employing solvent–solvent partitioning, solvent–aqueous partitioning, liquid chromatography (LC), and high-performance liquid chromatography (HPLC) techniques are used to separate the sample into a number of nonpolar, moderately polar, and highly polar fractions before GC/MS analysis. A typical scheme is given in Fig. 2.[28] A number of other liquid–liquid partition schemes have been published for air particulates.[15,29,30] The procedure involves the use of open-column liquid chromotographic separation of the neutral SOF after the removal of acidic and basic fractions. Typically, a glass column of 10–20 cm in length packed with 2–10% water-deactivated silica gel or alumina has been employed. The sample is then separated on the column into aliphatic, aromatic, PAH, and polar fractions through the use of solvents with increasing polarities. The fraction designation in parentheses (see Fig. 2) is given for the purpose of comparing this technique to the normal-phase HPLC separation technique, as discussed later in this section.

The main disadvantages of open-column LC are its slow speed and lack of reproducibility. The separation is controlled manually and the column-to-column variation makes clean-cut separation of compound

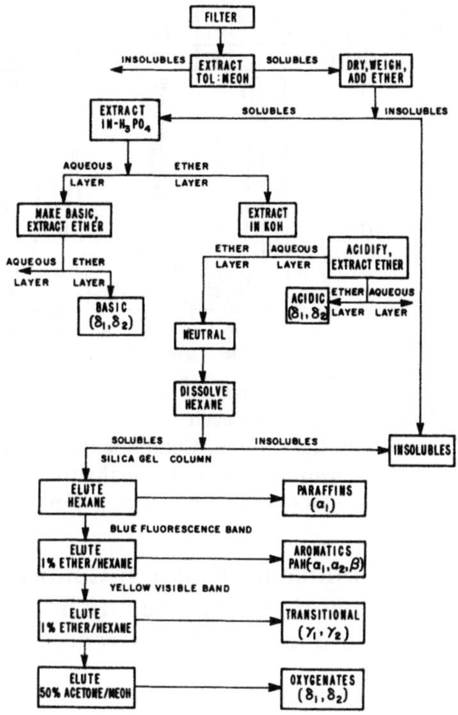


FIGURE 2. Fractionation scheme for particulate extracts from diesel exhaust emissions using liquid–liquid extraction and column–liquid chromatography.

classes an art difficult to reproduce. Since a new column is normally packed for each sample, column activity varies and background contamination is a problem.

A normal-phase (NP) HPLC procedure for the fractionation of particulate samples has been developed recently.[27] Normal-phase HPLC was chosen because of its ability to provide adequate resolution in both analytical and semipreparative scale separations needed in particulate analysis. This procedure has several advantages over that of the open-column LC and liquid–liquid partition procedures described earlier. The use of fluorescence and UV detectors in HPLC allows one to monitor the elution of component mixtures and check individual component retention times. This procedure has been applied in this laboratory for the separation of particulates collected from diesel engines and from atmospheric samples.

The mobile-phase solvents used for the procedure are hexane, dichloromethane (DCM), acetonitrile, and methanol. Figure 3 gives a profile for diesel engine particulate extract. The various chemical fractions are designated as $\alpha_1$, $\alpha_2$, $\gamma_1$, $\gamma_2$, $\delta_1$, and $\delta_2$, which are directly comparable to the fractions designated in Fig. 2.

This procedure has been used in a round-robin experiment by 11 laboratories organized by the Diesel Chemical Characterization Panel of

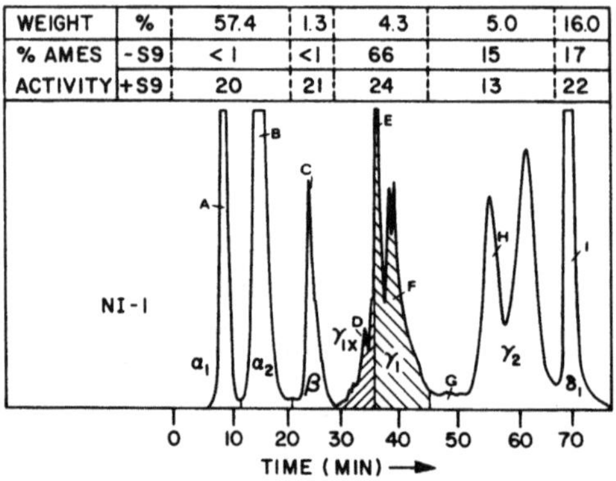


FIGURE 3. HPLC fractionation of a diesel particulate extract on a 3 mm × 30 cm Biosil A(20–44 μ) column with fluorescence detection at 313 nm excitation and 418 nm emission. The solvent flow is 1 ml/min. Solvent program conditions are 5% methylene chloride in hexane for 17 min, 5%/min $MeCl_2$ gradient to 100% $MeCl_2$, 100% $MeCl_2$ for 17 min, 2.0 ml of acetonitrile to elute the $\delta_1$ peak and 2.0 ml of methanol injected to elute the $\delta_2$ peak (not shown). The percentage of weight and Ames activity represented by each fraction is calculated by dividing the values determined for each fraction by the weight and Ames activity of the total extract injected into the column. Components eluted at the designated retention times are: A. pyrene; B. benzo[a]pyrene; C. methylcoronene; D. 1-nitropyrene; E. 9-fluoreneone; F. anthracene quinone; G. 1,6-dinitro pyrene; H. naphthalic acid anhydride; I. acridine.

the Coordinating Research Council.[27] The results obtained from this study show that the procedure has good reproducibility in terms of both retention times and fluorescence peak area measurements.

## 5. *BIOASSAY SCREENING*

Since 1977, short-term bioassay methods have been increasingly used for determining mutagenicity of air pollutants.[31] Epler[32] has reviewed methods for chemical separation and mutagenicity testing, and Chrisp[33] has reviewed mutagenicity data for some air pollutant samples.

These bioassay tests are very useful for prioritization of chemical fractions for GC/MS analysis. In the last section the normal-phase HPLC fractionation of a diesel particulate extract was illustrated in Fig. 3. The *Salmonella typhimurium* Ames activity (TA98), with activation (+S9) and without activation (−S9), are given in percent with respect to the unfractionated extract. The $\gamma_1$ fraction represents a major portion of the direct-acting (66%) and indirect-acting (24%) mutagenicity. GC/MS analysis was used to determine the composition of this fraction in which approximately 60 PAH derivatives were identified.[34] Further subfractionation, GC/MS, and bioassay tests led to the conclusion that 1-nitropyrene was a major mutagen present in the particulate extract.[35,36]

## 6. *GC/MS/COMPUTER ANALYSIS*

Computer-controlled GC/MS (GC/MS/COMP) is the most comprehensive analytical technique available for the analysis of complex mul-

*TABLE 1*
*Summary of Some GC/MS/COMP Techniques Used for Multicomponent Air Pollutant Analysis*

| Reference | Sample source | Sampling media | Pretreatment[a] | GC[b] | MS[c] | Compounds analyzed[d] |
|---|---|---|---|---|---|---|
| 39 | Ambient air | Carbopack B adsorbent | Thermal desorb. | 3m-SP1000 glass packed | EI; HRMS; SIM | Fluorocarbons |
| 36 | Diesel particulates | Teflon filter | DCM extraction; NPHPLC | 30m-SE54 fused silica | EI; LRMS; HRMS | Nitrated PAH |
| 34 | Diesel particulates | Teflon filter | DCM extraction; NPHPLC | 2M-SP2250 packed 25m-Dexsil 300 | EI; LRMS; HRMS | PAH; PAH derivatives (≈100 species) |
| 7 | Vehicle tunnel | Tenax adsorbent | Thermal desorb. | 50m-SF96 | EI; LRMS | $C_6$–$C_{20}$ oxy-HC; hal-HC (≈300 species) |
| 40 | Ambient air | Tenax; polymer plug adsorbents | Thermal desorb.; hexane extraction | 30–50m-SF96, UCON 50 HB 5100 | EI; LRMS | $C_9$–$C_{17}$ HC |
| 41 | Industrial dust; fly ash; urban particulate matter | Not specified | Benzene Sox.; RPHPLC; NPHPLC | 2m-OV-17 packed glass | EI; LRMS; SIM | $C_{14}$–$C_{35}$ HC; PAH; PAH derivatives |
| 20 | Diesel particulates | Teflon filter | DCM extraction; NPHPLC | 2m-SP2250, Dexsil 300 packed | EI; LRMS | PAH derivatives |
| 42 | Ambient air (Love Canal) | Glass fiber filter; Tenax adsorbent | DCM Sox.; thermal desorb. | 100m-SE30 SCOT | EI; LRMS | Hal-HC |
| 43 | Ambient air vehicle garage | Porapak Q adsorbent | Thermal desorb. | 2m-poly-*m*-phenyl ether packed | EI; LRMS; SIM | Hal-HC |

(*continued overleaf*)

*TABLE 1 (continued)*

| Reference | Sample source | Sampling media | Pretreatment[a] | GC[b] | MS[c] | Compounds analyzed[d] |
|---|---|---|---|---|---|---|
| 44 | Municipal incinerator | Electrostatic precipitator | Benzene Sox. | 2.6m-Carbowax 20m-Aue packing | EI; LRMS; SIM | PCB; PAH; dioxins; phenols; phthalate esters; polychlorinated dibenzofurans (≈100 species) |
| 29 | Ambient air—7 sites | Tenax adsorbent | Thermal desorb. | 100m-SE30 SCOT | EI; LRMS | Hal-HC |
| 45 | Coal gasifier particulates | Tenax adsorbent; glass filter | Pentane, DCM extraction Sephadex LH-20; RPHPLC | 30m-Dexsil 300 | EI; LRMS | PAH |
| 46 | Ambient air particulates | Glass filter | Cyclohexane Sox.; DMSO L–L partition | 50m-Dexsil 300 | EI; LRMS | PAH (≈140 species) |
| 47 | Tobacco pyrolyzate | Glass filter | | 35m-SE54 | | PAH |
| 48 | Ambient air particulates | Glass filter | Cyclohexane Sox.; L–L partition | 50m-SE54 | EI; LRMS | PAH |
| 49 | Ambient air particulates | Glass filter | | 20m-OV101 | | PAH |
| 50 | Ambient air | Cryogenic | | 7m-Durapak | EI; LRMS; SIM | $C_4$–$C_8$ HC |
| 51 | Polymer pyrolysis | Glass filter | Cyclohexane Sox.; NPHPLC | 3m-Ultrabond packed | EI; LRMS | PAH |
| 10 | Ambient air | Glass filter; Tenax adsorbent | Benzene + methanol Sox. | 3m-Dexsil 300 packed | EI; LRMS | HC; PAH; acids; heterocyclics; phthalate esters |

| | | | | | | |
|---|---|---|---|---|---|---|
| 52 | Coke oven particulates | Anderson impactor | Benzene Sox. | 3m-Dexsil 300 packed | EI; LRMS; SIM | PAH (14 species) |
| 53 | Ambient air | Glass filter | DCM Sox.; Sephadex LH-20 NPLC | 20m-SE52 | EI; LRMS | PAH |
| 54 | Aluminum plant particulates | | | 50m-SE54 | | PAH |
| 55 | Ambient air | Glass filter; polyurethane foam adsorbent | Toluene extraction | 1.8m-OV101 packed | EI; LRMS; SIM | Polychlorinated naphthalenes |
| 54 | Ambient air | | | 20m-OV101 | | PAH |
| 56 | Coal, wood, kerosene combustion particulates; ambient air particulates | Glass filter | Sox. | 19m-SE52 | EI; LRMS; HRMS | PAH |
| 14 | Ambient air | Tenax adsorbent | Thermal desorb. | | EI; LRMS | HC; aromatics; hal-HC; oxy-HC |
| 57 | Ambient air | Tenax, Carbopack B-HT, Ambersorb XE-340 adsorbents | Thermal desorb. | 40m-OV101 | EI; LRMS | HC; aromatics; chlor-HC; oxy-HC |
| 57–59 | Ambient air particulates | Glass filter | L–L partition | 5m-Dexsil 300 | EI; LRMS | $C_{14}C_{35}$ HC; PAH; PAH derivatives |
| 12, 13 | Ambient air | Tenax adsorbent | Thermal desorb. | 60–120m-OV17, OV101 | EI; LRMS | HC; hal-HC; oxy-HC; nitro-HC, sulfur-HC (≈60 species) |
| 60 | Cigarette particulates | Glass filter | DCM Sox. | 11m-SE52 | EI; LRMS | |

(*continued overleaf*)

*TABLE 1 (continued)*

| Reference | Sample source | Sampling media | Pretreatment[a] | GC[b] | MS[c] | Compounds analyzed[d] |
|---|---|---|---|---|---|---|
| 61 | Ambient air particulates | Glass filter | DCM Sox.; Sephadex LH-20 NPLC | 11m-SE52 | EI; LRMS | PAH; alkyl-PAH (≈100 species) |
| 62 | Tobacco smoke | Condensation | L–L Partition; NPLC; GF | Dexsil 300 | EI; LRMS | PAH |
| 63 | Ambient air | Tenax adsorbent | Thermal desorp. | 60m-OV101, OV225, Silar 10C | EI; LRMS | $C_6$–$C_{25}$ HC |
| 64 | Ambient air particulates | Glass filter | Sox.; Silica CC | 7m-*n*-octanol | EI; SIM | $C_4$–$C_8$ HC |
| 65 | Ambient air particulates | Glass filter | L–L partition; NPTLC | Not specified | EI; LRMS | $C_{11}$–$C_{28}$ HC, PAH, fatty acids, heterocyclics (≈120 species) |
| 66 | Ambient air particulates | Glass filter | Cyclohexane Sox. | 4m-Dexsil 300, 400 packed | EI; LRMS | PAH |
| 18 | Indoor air | Tenax and Carbopack B adsorbent | Thermal desorp. | 1.8m-SP1000 packed | EI; LRMS | HC; aromatics; chlor HC |
| 67 | Ambient air | Stainless-steel cannister | None | | EI; SIM | Chlor HC |
| 66 | Coal tar particulates | Glass filter | | Dexsil 400,410 | EI; LRMS | PAH (49 species) |
| 68 | Automobile particulates | Glass filter | Diethyl ether Sox.; NPCC | 4m-Dexsil 300 packed | EI; LRMS | PAH |

| | | | | | | |
|---|---|---|---|---|---|---|
| 69 | Auto exhaust | | | 50m-OV101 | | PAH |
| 70 | Diesel particulates | Glass wool pad; condensor (0 °C) | Chloroform Sox.; alumina CC | 160m-OV101 | EI; LRMS | PAH; oxy-HC; HC (64 species) |
| 71 | Ambient air | Tenax adsorbent | Thermal desorp. | 100–300m Emulphor on 870 nickel | EI; LRMS | $C_5$–$C_{16}$ species (≈100 species) |
| 30 | Ambient air | Glass filter | L–L partition; NPLC (Sephadex) | 70m-SE52 | | PAH |
| 72 | Ambient air | Carbon black | Thermal desorp. | 100m-OV101 | EI; LRMS | $C_8$–$C_{18}$ HC |
| 73 | Ambient air particulates | Glass filter | Cyclohexane Sox.; NPLC | 100m-Dexsil 300 | EI; LRMS | PAH (≈70 species) |
| 74 | Ambient air particulates | Glass filter | Benzene Sox. | 100m-Dexsil 300 | 22 eV EI; LRMS | PAH; HC; heterocyclics |
| 75 | Automobile exhaust | Glass filter | Sox. | OV-17 packed SS | EI; LRMS | PAH |
| 76 | Ambient air | Glass filter; impactor | DCM Sox.; silica gel CC | 3m-OV17 packed | EI; LRMS | HC and oxy-HC (50 species) |
| 77 | Ambient air | Charcoal adsorbent | Carbon disulfide extraction | 120m-UCON 5100; 80m-UCON HB LB550 | EI; LRMS | $C_6$–$C_{20}$ HC and oxy-HC (108 species) |

[a] Abbreviations are: DCM, dichloromethane; Sox., Soxhlet extraction; CC, column chromatography; NP, normal phase; RP, reverse phase; HPLC, high-performance liquid chromatography; L–L, liquid–liquid; thermal desorb., thermal desorption; GF, gel filtration; TLC, thin-layer chromatography.
[b] GC columns are glass capillary unless otherwise noted. Abbreviations are: SS, stainless steel.
[c] Abbreviations are: EI, electron impact; LRMS, low-resolution mass spectrometry; HRMS, high-resolution mass spectrometry; SIM, selected-ion monitoring. EI is at 70 eV unless otherwise noted.
[d] Abbreviations are: HC, hydrocarbons; PAH, polynuclear aromatic hydrocarbons; oxy, oxygenated; hal, halogenated; chlor, chlorinated.

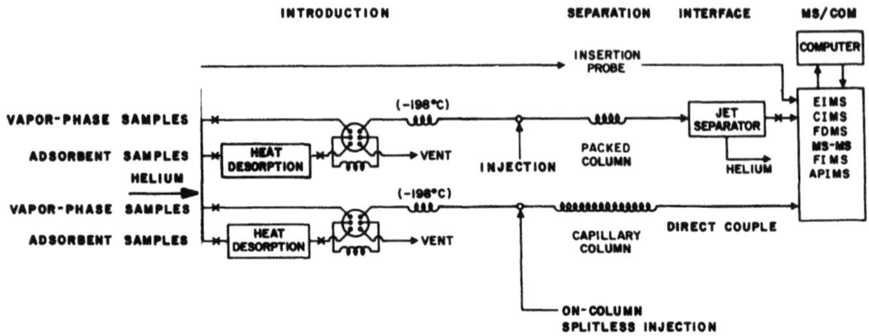


FIGURE 4. Schematic representation of techniques used for the GC/MS analysis of air pollution samples.

ticomponent air pollutant samples.[37] Since 1971 a number of investigators have used GC/MS/COMP for the multicomponent analysis of air pollutants. Table 1 summarizes details for a number of representative air pollution studies undertaken during the last 10 years.

### 6.1. *Gas Chromatography*

Figure 4 gives a schematic for a GC/MS/COMP system used in this laboratory for air pollutant analysis. The choice of sample introduction, columns, and GC/MS interface techniques is dependent upon the complexity, concentration, lability, and polarity of the sample as described in the following sections.

### 6.2. *Sample Derivatization*

Particulate extracts typically contain significant quantities of moderately polar and polar organic substances. Gas chromatography columns containing polar liquid phases can be used for moderately polar organics. However, fused-silica columns are only now becoming available with highly polar phases which have good temperature stability up to 320 °C.

Derivatizing agents can be used to form derivatives with amines, acids, and alcohols in order to facilitate GC/MS analysis. Solvent extracts of Tenax adsorbent tube and particulate filter samples may be subjected to trimethylsilylation (TMS), trifluoroacetylation (TFA), or methylation. The reagents used for these reactions are bis(trimethylsilyl)trifluoroacetamide (BSTFA) for the TMS derivatization, trifluoroacetic anhydride (TFA) for the TFA derivatization, or diazo methane for the

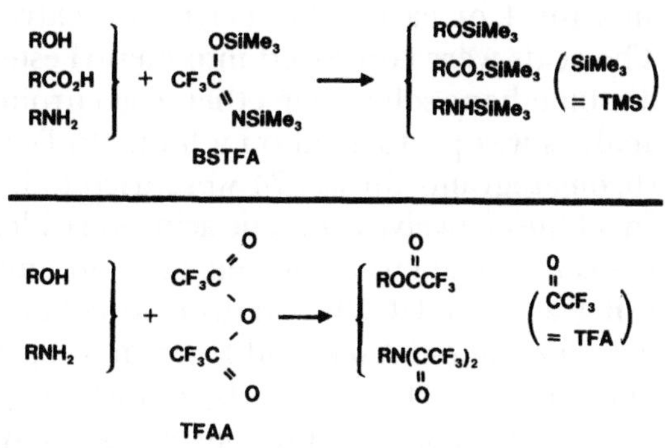


FIGURE 5. Reaction for the derivatization of air pollutant polar species by trimethylsilylation and trifluoroacetylation prior to GC/MS analysis.

methylation.[38] These reactions are given in Fig. 5. Figure 6 gives the GC/MS profile for a hexane/acetone extract of a Tenax adsorbent tube sample treated with BSTFA and some of the TFA derivatives identified in this sample.

Several investigators have used derivatization with diazomethane to determine acids in air pollutant samples. Cautreels[10] collected gas and particulate phase acids from the atmosphere using Tenax and glass fiber filters, respectively. Organic compounds were removed from the particulate samples by successive Soxhlet extractions with benzene and meth-

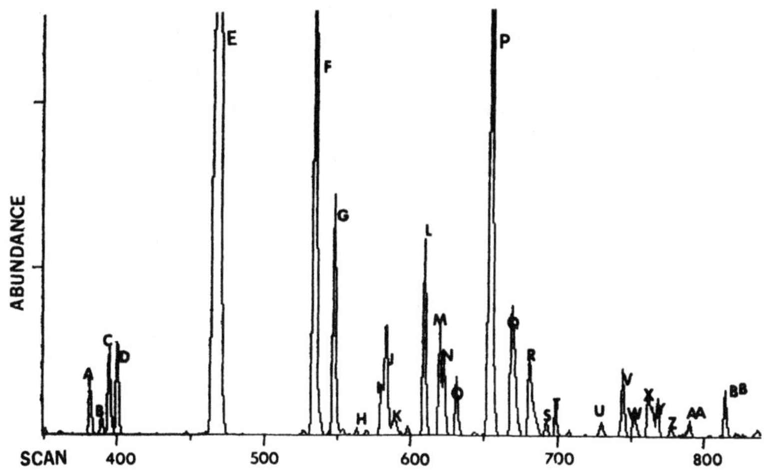


FIGURE 6. Reconstructed GC trace from GC/MS analysis of hexane/acetone extract from a Tenax adsorbent tube treated with TFA. Sample collected from a pilot metal casting operation. GC/MS analysis performed on a 30-m OV-101 capillary column. Some peak identifications are E. creosol-TFA; G. xylenol-TFA; H. benzoic acid-TFA; M. aniline-TFA and P. toluidine-TFA.

anol for 4 hr each. The Tenax was extracted with methanol for 8 hr. The acids were converted into methyl esters by addition of an excess of diazomethane solution in ether. Gas chromatography/mass spectrometry analysis was performed on a 3-m, 4% Dexsil 300 packed column. Mass chromatograms for *m/z* 74 were used to help identify 16 aliphatic acids. In addition, twelve aromatic acids were identified. Grosjean[78] identified $C_3$–$C_{10}$ aliphatic dicarboxylic acids in ambient air in a similar manner using a 30-m UCON50-HB-5100 column for GC/MS analysis of the diazomethane-treated acid fraction of particulate extracts. Van Vaeck[9] determined carboxylic acids in ambient particulate extracts using diazomethane. For quantitation, two standards, $^2H_3$-methyloctadecanoate and $^2H_9$-pyrene, were added. Gas chromatography/mass spectrometry analysis was performed on a 3-m, 3% Dexsil 300 on Chromosorb Q packed column. The samples were run in the selected ion monitoring mode using *m/z* 87 to detect the carboxylic acids.

### 6.3. *Sample Introduction*

A comparison of various sample introduction techniques with regard to discrimination and decomposition effects has been described by Schomburg.[79] Splitters have several characteristics which may be undesirable or even unacceptable. We have found that the splitter shows discrimination for air particulate extracts having compounds with a large range of volatilities. These discrimination effects make quantitation for the entire range of compounds virtually impossible. Labile compounds such as nitrated PAH[36] are decomposed in the hot injection port. It is recommended that the splitter is left out and only on-column injection is used.

On-column injection has been described by Grob[80,81] and is rapidly becoming the preferred technique since discrimination caused by differences in volatility, polarity, or concentration is virtually eliminated. On-column injection consists of inserting the syringe needle directly inside the capillary column so that the sample components are deposited directly on the column head at low temperature, as opposed to insertion into a heated injection port. Most capillary column GC systems can be modified to accommodate these types of injection systems (J. and W., Rancho Cordova, California).

On-column injection of air particulate extracts contain some nonvolatile material which can build up on the head of the column, severely reducing component resolution. However, on-column injection of Tenax adsorbent extracts contain very little nonvolatile material and typically little column degradation is noted. Grob[82] has reported that charcoal adsorbent extract samples of up to 5 μl have been injected daily without splitting for more than a year without causing damage to a capillary column. A standard mixture may be injected on a regular basis to check for this problem. When significant degradation is noted, approximately 0.3 m of the front end of the column is broken off, which usually restores the column to its original operating conditions.

Samples collected on adsorbents may be thermally desorbed into the gas chromatograph.[5,7] Figure 4 shows a schematic of a thermal desorption system interfaced with a GC/MS system. In a typical thermal desorption cycle the sampling cartridge is placed in a preheated (~270 °C) desorption zone, and helium gas is channeled through the cartridge (~20 ml/min) to purge the vapors into a liquid-nitrogen-cooled nickel capillary loop of a sampling valve. After the desorption, the sample is injected and the temperature of the capillary loop is rapidly raised using a flash heater, so that carrier gas introduces the vapors as a slug onto the GC column.

### 6.4. *Column Selection*

Both packed and capillary columns have been used for the GC/MS analysis of air pollutants. However, glass or fused-silica capillary columns are preferred for the analysis of complex air pollutant samples because of their high resolution, chemical inertness, and thermal stability. Grob[77,82] was one of the first investigators to use these columns on a routine basis for the trace analysis of air pollutants. In subsequent publications, Grob[77,80–86] has applied these techniques in a number of air pollution studies.

Dandeneau[87] and Schomburg[79] have demonstrated that fused-silica glass provided a degree of chemical inertness unmatched by other glasses. In addition, when this column is fabricated with a very thin wall and externally coated with a suitable polymer to protect its outer surface, the chemically pure fused-silica glass exhibits a high degree of flexibility which facilitates handling and interfacing with the mass spectrometer.

Lipsky[88] has shown that trace compounds that are either unresolved or obscured under baseline noise in conventional packed-column gas chromatography can be adequately resolved with glass capillaries. This has been demonstrated by work from this laboratory for the analysis of moderately polar organic species in diesel particulate extracts. Figure 7 gives mass chromatograms (see Sec. 7) for *m/z* 247 ($C_{16}H_9NO_2$) used for the GC/MS identification of 1-nitropyrene in a moderately polar HPLC fraction of a diesel particulate extract utilizing both a packed and fused-silica column. The packed column is unable to resolve the 8-nitrofluoranthene (8-NF) from the 1-nitropyrene (1-NP). In addition, there is noise in the mass chromatogram for the *m/z* 247 ion. The fused-silica

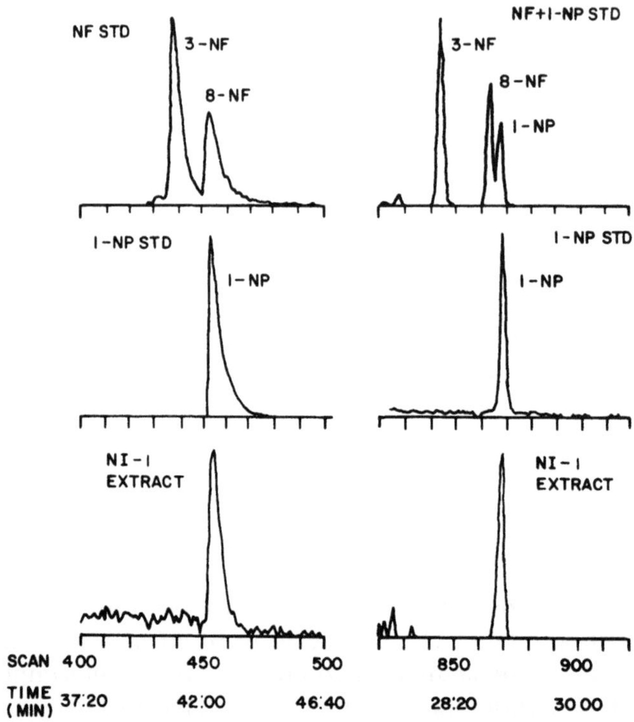


FIGURE 7. GC/MS mass chromatograms of the *m/z* 247 ion generated from 3-nitrofluoranthene (3-NF), 8-nitrofluoranthene (8-NF), and 1-nitropyrene (1-NP) standards and a HPLC-fractionated diesel particulate extract (NI-1).

capillary column improved the resolution by approximately threefold allowing separation of 8-NF from 1-NP, increased the signal-to-noise level, and eliminated tailing of peaks.

Although there are over 100 stationary phases available for GC/MS, only a few need to be considered for the separation of samples containing nonpolar, moderately polar, and polar compounds. Table 1 lists capillary columns which have been used for the GC/MS analysis of air pollutant samples. Stationary phases recommended for the GC/MS analysis of air pollutants are also given in Table 2. Compound types and suitable columns for their separation are designated into three main categories of polarity.

SE-30, OV-101, SP-2100, SE-52, and SE-54 phases are suitable for the analysis of the chlorinated organics, aliphatic and aromatic HC and PAH species. The two most widely used phases for the PAH analysis are SE-52 and SE-54. These two columns are chemically very similar and the McReynolds constants for these phases are nearly the same. The

TABLE 2
*Stationary Phases Recommended for the Capillary Column GC/MS Analysis of Air Pollutants*

| Stationary phase | Temp. limit | Compound types |
|---|---|---|
| *Nonpolar* | | |
| SE-30[a] | 350 | Chlorinated organics; aliphatic HC; aromatic HC; PAH |
| OV-101[a] | 260 | |
| SP-2100 | 350 | |
| SE-52, SE-54[a] | 300 | |
| SF-96 | 250 | |
| *Moderately polar* | | |
| Carbowax 20M[a] | 200 | Aldehyde, ketone, nitro derivatives of HC and PAH; PAH quinones; acid anhydrides; TMS, methylated, TFA derivatives of highly polar species |
| OV-17[a] | 240 | |
| Dexsil 300, 400[a] | 325 | |
| DB5[a] | 275 | |
| SP1000[a] | 275 | |
| SP2250 | 375 | |
| *Highly polar* | | |
| SP-2340 | 275+ | Acid, alcohol, amine derivatives of HC and PAH; nitrogen heterocyclics |
| XE-60 | 250 | |
| Versamid-900 | 275 | |
| Poly S-179 | 390 | |

[a] Available in fused-silica capillary columns.

temperature limits on these columns are approximately 300 °C. Grob[85] extended the temperature range of capillary columns in order to chromatograph high-molecular-weight PAH. In that study compounds from coronene ($C_{20}H_{12}$) to rubrene ($C_{42}H_{28}$) were chromatographed on a short (5.5-m) OV-101 glass capillary. However, we have found that the GC/MS interface limits PAH above MW 350 from being eluted. SE-30 and OV-101 phases are widely used for halogenated organics. Any of the above columns are suitable for the separation of aliphatic and alkyl-substituted benzenes.

Several liquid phases are available for the GC/MS analysis of moderately polar air pollutants. However, very little work has been undertaken to identify this class of compounds. Schuetzle[36] used SP-2250 and Dexsil 300 packed columns for the analysis of oxygenated and nitrated PAH in diesel particulate extracts. A large number of compounds were identified but column resolution was poor, making specific isomer identification difficult. Cautreels[89] also used a 5-m packed Dexsil 300 column but identified only a few compounds of moderate polarity.

Four liquid phases appear to be well suited for the GC/MS analysis of polar organics in particulates (Table 2). However, only limited work has been undertaken on the application of these and similar phases to the analysis of polar organic compounds in air pollutant samples. Wakeham[90] has used a 20-m SP-2340 column for the characterization of azaarenes in roadside particulates. Liquid–liquid and column chromatography separation procedures were used to isolate the basic fraction containing the azaarenes before analysis. The SP-2340 column showed good resolution with no tailing. Compounds were identified up to a molecular weight of 279 (1,2,7,8-dibenzacridine). Compounds of higher molecular weight were eluted; however, standards were not available for identification.

### 6.5. *GC/MS Interfaces*

A variety of techniques have been developed to interface capillary columns with the source of a mass spectrometer.[91] Packed columns typically require separators whereas capillary columns may be directly interfaced.

### 6.6. *Packed Columns*

Although fused-silica capillary columns are now preferred for the analysis of air pollutant samples, there are some cases where packed columns will provide suitable results. Much of the GC/MS work performed today uses packed columns at a flow rate of 15–25 cc/min. Some CI mass spectrometers are able to tolerate up to 20 cc/min of gas; however, an enrichment device is needed in most packed-column applications. The two most commonly employed devices include the jet separator[92] and Llewellyn[93] separator. We have found that the transfer efficiency for these two interfaces is poor for compounds above a molecular weight of approximately 275.

### 6.7. *Capillary Columns*

The lower flow employed in most capillary GC columns (1–2 cc/min) allows direct introduction of the entire sample. Chemical ionization mass spectrometers will allow direct column flow of up to 10–15 cc/min. Platinum capillaries have been used to interface capillary columns. This serves as a restriction which maintains the pressure in the column at a specified level in order to reduce band spreading. These interfaces have high mechanical stabilities but tend to degrade components and cause some band spreading. We have found that gold-plated nickel capillaries or deactivated glass-lined stainless steel greatly reduces that sample degradation problem.

Fused-silica capillary columns are flexible enough to allow their use as an interface. The end of the capillary column is fed directly into the source. The chromatogram shown in Fig. 7 was obtained from a fused-silica column interfaced in this manner.

### 6.8. *Mass Spectrometry*

Gas chromatography coupled with either magnetic sector or quadrupole mass spectrometers have been used for the analysis of complex air pollutant samples. Most magnetic sector and quadrupole mass spectrometers produced since 1976 can achieve scan rates of less than 1 sec/decade needed for capillary GC/MS analysis.

### 6.9. *Ionization Modes*

Most of the GC/MS analyses reported to date on the analysis of air pollutants have utilized 70-eV electron impact ionization techniques. There are only a few studies which have demonstrated the great potential of using alternative forms of ionization for these analysis.

Negative chemical ionization mass spectrometry (NCIMS) can be used to determine molecules that have high NCI sensitivities such as oxidizing and alkylating agents.[94] Harvey[95] has demonstrated the selectivity of NCI for PAH species in the presence of aliphatic hydrocarbons which are present at several orders of magnitude higher concentration than the PAH. Schuetzle[36] has used NCI to reduce interference and increase sensitivity for analysis of nitrated PAH in diesel particulate extracts.

Positive chemical ionization mass spectrometry (PCIMS) has also been used to obtain molecular weight information from the GC/MS analysis of air pollutants.

### 6.10. *Instrument Operation Modes*

Low-accuracy ($\pm 0.1$ amu) GC/MS data are obtained for most air pollutant analysis. However, some data systems may be used to routinely acquire high accuracy ($\pm 0.003$ amu) GC/MS data with the mass spectrometer in the low- or high-resolution modes of operation.[70,96–101] This technique has been used to identify PAH derivatives in diesel particulates[25,27,34,36] and for the analysis of volatile nitrosamines.[102]

High-accuracy–high-resolution mass chromatograms are basically "elemental composition" chromatograms and can be generated for any characteristic ion with a given elemental composition. The identification of complex air pollutant samples using GC/MS is very difficult without this elemental composition data. Mass chromatograms may be used to determine a family of ions which belong to the unknown. Figure 8 shows the elemental composition mass chromatogram for *m/z* 194.073 and several fragments of this ion. The best elemental composition fit for *m/z* 194.073 is $C_{14}H_{10}O$. There are a number of species with this elemental composition. However, there are a great number of isobaric possibilities if only the nominal mass 194 is known, as illustrated in Fig. 9. A mass spectrometer resolution of 4000 is needed to separate the $C_{14}H_{10}O$ isomers from the $C_{15}H_{14}$ isomers and 18,000 to separate $C_{14}H_{10}O$ isomers from $C_{13}H_{10}N_2$ isomers. A lower resolution operating condition can be used if the sample is prefractionated by HPLC, since the $C_{13}H_{10}N_2$ and $C_{15}H_{14}$ species would be eluted in different polarity fractions than the $C_{14}H_{10}O$ species.

Selected-ion-monitoring (SIM) techniques have been developed by several investigators during the past few years to improve the sensitivity of repetitive-scanning MS. Sensitivity may be improved by one to three orders of magnitude, depending upon the number of peaks to be monitored. SIM may be undertaken to obtain low- or high-resolution spectra.[103]

Peterson[104] has used GC/MS in a SIM mode to determine PAH in diesel particulate extracts. In this mode the mass spectrometer concurrently monitors one or more ions characteristic of a specified compound

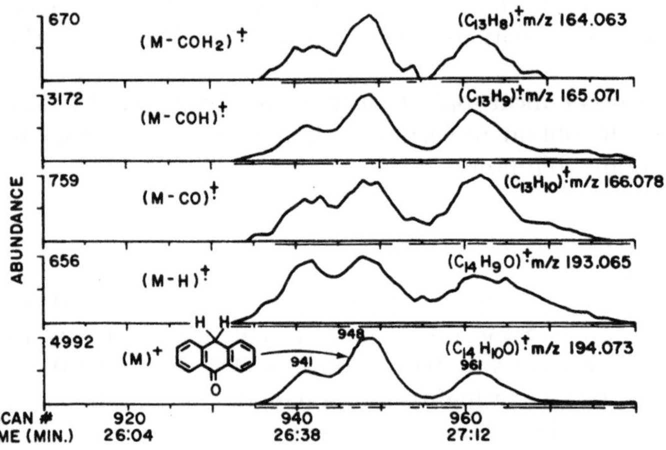


FIGURE 8. High-accuracy mass chromatograms (±30 ppm; $M/\Delta M$; 2000) used for the identification of 9-anthrone and isomers in the $\gamma_1$ HPLC fraction of a diesel particulate extract utilizing a 6′ SP2250 column.

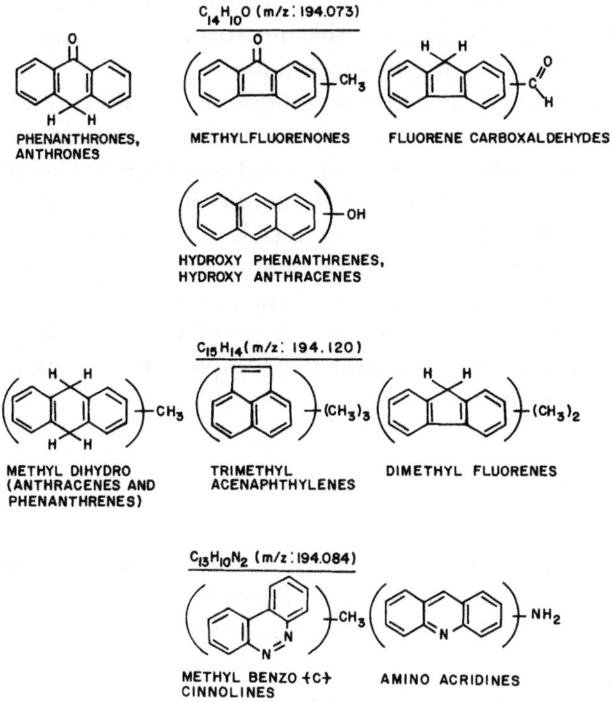


FIGURE 9. Possible PAH and PAH derivatives that have a nominal mass of 194.

during its expected elution time from the gas chromatographic column. The sensitivity of this technique is two to three orders of magnitude more sensitive than the continuous scanning technique. Their method for the analysis of PAH in exhaust is based on the use of stable isotopically labeled PAH as internal standards. The stable isotope used is the perdeutero derivative of these PAH where each proton ($^{1}$H) on the PAH nucleus is replaced with a deuteron ($^{2}$H). Deuterated benzo[*a*]pyrene, benzo[*e*]pyrene, benzo[*a*]anthracene, and pyrene were added to the diesel extracts. Since the chromatographic properties of the natural abundance and deuterated PAH are similar, they coelute into the mass spectrometer. By monitoring these ions at their retention time, the natural abundance PAH can be unequivocally identified. Quantitation is accomplished by comparison of the integrated ion current response of the native PAH to that of the deuterated PAH and relating to a standard curve.

Erickson[(35)] used SIM in conjunction with GC/MS to monitor eight polychloronaphthalenes at levels as low as 50 pg (0.3 ng/m$^{3}$) in particulate and adsorbent extracts. Lamparski[(41)] used SIM at unit mass resolution to determine dioxins in particulate samples. $^{13}C$-labeled dioxins were added to the sample for quantitative measurement. Sensitivity for dioxin analysis was 100 ppt using this technique.

The sensitivity of the SIM technique in GC/MS is often limited by the background contribution of other unresolved components which give ions at the same nominal mass as those of the compound being measured. This situation occurs in air pollution samples in which it is often necessary to detect very small quantities of materials in the presence of large quantities of other components. The use of high-resolution selective-ion-monitoring techniques (HRSIM) in conjunction with GC has been used to overcome some of these problems. A sensitivity of 1–10 pg for single components in complex environmental samples can be achieved at 10,000 resolution in this matter.

## 7. *DATA ANALYSIS*

The data generated in a low-resolution repetitive-scanning GC/MS analysis of an air pollutant sample may exceed 5000 spectra, each spectrum containing 100–400 ion/abundance values. Therefore, computer techniques are necessary to reduce and interpret data files each containing up to $1 \times 10^6$ words of data. Traditionally there have been two problems associated with complete GC/MS profile analysis of such complex matrices. Even though high-resolution capillary columns may be used, unambiguous identifications are not obtained for all components because many are incompletely resolved. Only the major peaks in an air pollution sample are usually quantitated because of the tedious nature involved with quantitation of the hundreds of minor components present. Recent advancements in automated data reduction techniques have made these analyses feasible. Computer techniques used for the qualitative and quantitative interpretation of these large data sets are presented in this section.

### 7.1. *Qualitative Analysis*

McLafferty[105] describes computer algorithms used for compound identification in terms of mass spectral interpretation systems and mass spectral retrieval systems. Interpretive systems can be applied to obtain structural information about the unknown compound. Factor analysis, a statistical technique to remove interfering mass peaks, can be used to separate compounds into structural classes. A number of learning machine systems such as the K-NEAREST NEIGHBOR pattern recognition program necessitates pretraining of the algorithm for each functionality group sought. Artificial intelligence methods have also been applied to interpret mass spectra. Stanford University's heuristic DENDRAL program uses empirical mass spectrometry rules to recognize mass spectra. McLafferty's Self-Training Interpretive and Retrieval System (STIRS) trains itself to identify an unknown based on its match with reference spectra in 15 classes of mass spectral data.

Mass spectral retrieval systems match a mass spectrum unknown against a library of reference spectra rating their similarity based on the number of corresponding peak masses and, in some cases, ion abundance values. Either forward or reverse searches may be utilized for the GC/MS/COM identification of compounds in a complex air pollutant sample. Recent applications of these techniques to automatic processing of GC/MS data are described by Rosenthal(106) and Slayback.(107) The forward search(108) is used for identification of compounds in an unknown mixture with little or no knowledge of its constituents. It involves matching various peaks of the unknown mass spectrum against a reference spectral library. The reverse search is used for identification of specific compounds in an unknown mixture for which mass spectra of standards and GC retention indices are available. A comparison of the various peaks in a reference spectrum is made against all the mass peaks in a segment of an unknown mixture. McLafferty(105) describes three retrieval systems available through MSDC/CYPHERNET and Cornell/TYMNET computer networking systems. PBM, a Probability-Based Matching system, uses weighting of mass and ion abundance data in conjunction with reverse spectral searching. PEAK is a forward-search routine involving user interaction to set mass and abundance value limits for the search. KB is a forward-search routine that forces the match of peaks in all parts of the mass spectrum by using the two most abundant peaks in each 14 amu range of the unknown mass spectrum.

Rosenthal(106) compared forward and reverse searching procedures for the analysis of an ambient air sample. They concluded that forward search is the better method for unknown mixtures but the reverse search can detect low levels of components, even if their presence is obscured by unresolved elements. However, other information is needed for compound confirmation.

Mass spectrometric data is not always sufficient to distinguish between closely related compounds in an air pollutant sample. Gas chromatography retention times are used to supplement the spectral data. The Kovats-type retention index (RI) provides useful, relatively invariant retention time data. This technique assigns relative retention times (RRT) to compounds, referenced to the elution times of neighboring *n*-alkanes or other homologous series of interest. Several computer programs have been developed to use RI values for identification.(8,37,98,99,109–111)

Spivakovskii[112] and Randic[113] have calculated retention indices (connectivity index) based not on a reference compound but on the molecular structure of the compounds. The method works well for aliphatic alcohols and saturated hydrocarbons. Eggertsen[114] describes a chromatographic distillation method used to resolve hydrocarbons according to their boiling points. Stuckey[115] has devised a computer program to search for compounds based on a retention index calculated from the boiling point of the compound. Schuetzle[8,98,99] used relative elution times for $C_2$–$C_8$, $C_9$–$C_{15}$, and $C_{15}$–$C_{35}$ species using hexafluorobenzene, dodecane, and hexahydropyrene standards to help establish compound identification. Betty[111] used RI values calculated from a standard sample containing all the desired *n*-hydrocarbons identified in a GC/MS run by the presence of the $C_6H_{13}$ (*m/z* 85.1) and molecular ions. These standard retention data were used to automatically calculate RI values for samples extracted from airborne particulate matter run under identical GC conditions.

Retention indices are used in several ways. They can be used to help speed library search by providing a search window to limit the range of library compounds which need to be considered for search purposes. In this case the retention data need only be accurate to $\pm 5\%$. More precise retention data ($\pm 0.1\%$) is needed for determining isomers and for highly complex samples. Relative retention times depend on a great number of variables, including column efficiency, column polarity, film thickness, mode of injection, solvent volume, and reproducibility of the GC temperature programming. Because of the high variance risk of not controlling these variables, caution must be used when comparing laboratory values with published retention indices.

### 7.2. *Quantitative Analysis*

Computer algorithms for quantifying a large number of compounds in a complex environmental matrix were first described by Hites,[116] Burlingame,[97] and Schuetzle.[8] Multicomponent analyses became more of a routine technique when these algorithms were combined with procedures for automated interpretation and quantitation of these large data sets. Techniques have been recently presented by Smith,[117] Gates,[118] Ryan,[119,120] and Ouchi[121] that describe automatic compound location and quantitation. The task of analyzing the complete GC/MS profile is simplified by using a library of reference spectra acquired on the same instrument and under the same conditions as the sample.

Reverse library search techniques using the acquired reference spectra are used to locate the target compound.[121] The relative retention time of the compound determines a window within which the search is performed. Compound location is confirmed when fragment ions in the RIC maximize at a preselected match probability with the reference spectra.

The fragment ion peak height or relative abundance (RA) values should be predetermined using standards. This assures little variance in the compound areas determined for quantitation. The accuracy of the concentration measurement depends on the reproducibility of the designated ion intensity relative to the total ionization.[118]

Each compound is quantitated by comparing the ratio of an ion peak area in the unknown to its relative abundance with the ratio of the ion peak area and relative abundance of a quantitative internal standard. Because of the complex nature of the sample matrix, many compounds coelute with other species that generate interfering fragment ions. Therefore, to obtain a statistically valid concentration free from contaminant peaks, it is recommended that more than one fragment ion be used to generate the unknown peak area.

A quantitation factor known as a response factor ($R_f$) or a molar response factor (MRF) is used to relate the standardized area to a reference compound. Molar response factors should be determined experimentally using a pure reference compounds.

Literature values of response factors may be used for semiquantitative determinations. Cornu and Massot[122] list MRF for hundreds of compounds determined on various instruments. These values are referenced to butane, hexadecane, and toluene. The relative standard deviation of these MRFs may be as high as 50%. Ryan and others[119,120] have calculated response ratios and surveyed values from other instruments for a number of phthalates, polynuclear aromatics, and TMS-phenols referenced to $d_{10}$-anthracene. They found a coefficient of variation of 11–46%. Determinations to better than 10% relative standard deviation require that MRF be calculated using the same instrument with standardized tuning and constant source temperature.[123] Tomer[124] has evaluated the deviation in MRF at various instrument conditions and has found that linear regression analysis of MRF at one condition versus response at another condition yields a correction function that would bring a set of data obtained under the two different conditions into relatively good agreement.

Budde[125] has reviewed a number of techniques used for quantitating components in environmental samples. The technique to be employed depends on the level of components present, the number of components to be quantitated, and the accuracy level needed. Techniques used for quantitative GC/MS analysis include external standardization, internal standardization, standard addition, and isotope-labeled internal standardization.

The external calibration method for individual components is based on the direct responses for known amounts of a pure compound analyzed without the sample matrix. It is assumed that there is no significant change in mass spectrometer response between sample and standard analysis. The ratio of the response for the sample and standard is used for the quantitation. A number of factors may influence quantitation including variations in injection volume[(126)] and the effect of sample matrix on the response of individual components. In the multicomponent analysis often found in air pollution studies, it is not feasible to obtain multipoint calibration curves for the hundreds of components that may be present. Instead a single-point calibration is often used. Budde[(125)] reports that quantitative analysis using repetitive scanning GC/MS and a single external standard gave an average bias of −30%.

The internal standard method (ISM) is considered more advantageous[(127)] than the previous method because it compensates for fluctuations in the instrumental operating conditions. The internal standard is a known amount of a pure compound that is added to an environmental sample but is chemically different than any of the components in the sample. Fluorinated analogs of some of the species under quantitation have proved to be reliable standards.[(7,124)] A calibration curve is prepared with a series of mixtures each containing a known quantity of the analyte and a known quantity of the standard. The ratio of the amount of analyte to amount of standard is plotted versus the ratio of the area of analyte to area of standard. This is equivalent to independently measuring response factors ($R_f$) for each compound using the following equation:

$$\text{Amount analyte} = \frac{(\text{area analyte}) \times (\text{amount standard})}{(\text{area standard}) \times R_f}$$

The standard addition technique (spike) has been used also for quantitation in air pollution analysis. A known quantity of a pure compound is added to a sample matrix. The compound used as the standard is also contained in the sample. This gives a measured analyte response which can be used as a measure of the accuracy of the total analytical method. If it is assumed that the addition of the spike causes no significant change in volume, the percent recovery of the spike can be calculated as follows:

$$\text{Percent recovery} = 100 \times (A' - \text{A})/C_s$$

where $A'$ = response for analyte with spike, $A$ = response for analyte without spike, and $C_s$ = amount of spike added.

If a series of spike samples containing a range of spike concentrations are analyzed, then a calibration curve can be constructed. Extrapolation of a line represented by $A'$ versus $C_s$ to a point where $C_s = 0$ yields the response of the analyte in the sample. This technique is used when the sample cannot be obtained without the analyte present, and "matrix effects" (such as differences in source pressure and therefore ionization) occur when a series of analyte concentrations plus internal standard is run.

A special case of the internal standard method which does not require external calibration is the isotope-labeled internal standard or isotope dilution method. This technique involves a perturbation of the normal isotopic composition of the sample through the addition of an isotopically enriched analog of the analyte.(128) If the labeled compound is used to quantitate compounds that are chemically different, response factors are required for calibration. In environmental analyses this stable or radioactive standard is added at some point in the collection, isolation, or derivatization process(104) to determine analytical recovery. Colby and McCaman(128) and Schoeller(129) have discussed the significant differences which exist in data-handling procedures used to convert the measured isotope ratios into analyte mole ratio and standard mole ratio. Because isotope ratios are very sensitive to operating conditions such as mass spectrometer resolution, temperature changes, and electrometer offset, differences in mole ratios may vary depending on the choice of data reduction technique. Interference due to the presence of ions of common mass values in both the natural abundance compound and its labeled analog causes reduction in linearity between the isotope ratio and mole ratio. When a labeled standard is contaminated with the natural abundance compound, a decrease in the linear range between mole ratio and isotope ratio is effected. Falkner(130) cautions that possible exchange or rearrangement of the label can occur with deuterium-labeled standards during fragmentation or under CI conditions. With these precautions in mind, Claeps(131) affirms that stable-isotope-labeled internal standards appear to produce the lowest variance factors due to long-term, instrumental, and sample manipulation errors.

The identification and quantitation of unknown compounds in a complex air pollutant mixture is approached using a number of computer algorithms. The use of some of these algorithms is demonstrated with an example.

Figure 10 present reconstructed ion chromatograms (RICs) of vehicle exhaust gases from the Allegheny Mountain Tunnel in Pennsylvania. The sample was collected using two Tenax adsorbent cartridges connected in series. The experiment(7) generated over 200 RICs of this nature. Samples were analyzed by a thermal desorption GC/MS tech-

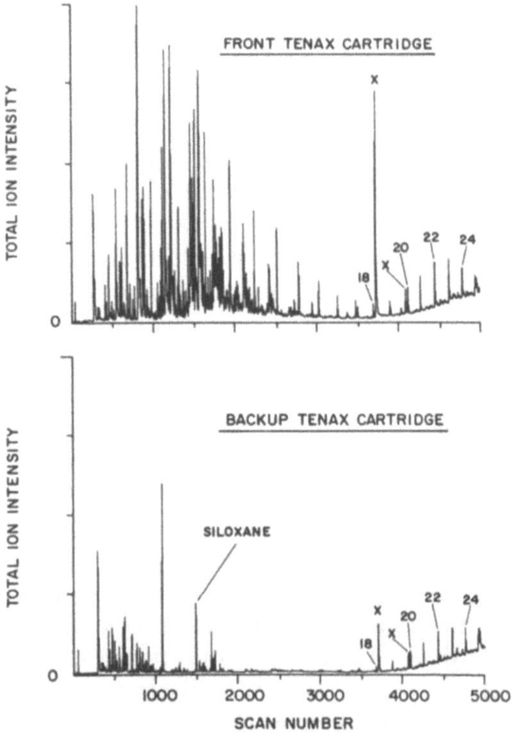


FIGURE 10. Reconstructed ion chromatograms (RIC) of exhaust gases from the Pennsylvania Allegheny vehicle tunnel collected on two Tenax tubes in series, GC/MS conditions: 50-m SF-96 glass capillary, temperature programmed from −20 °C to 230 °C at 3 °C/min. The 70-eV EI mass spectra were taken every second. Peaks labeled × are contaminants. Carbon numbers are indicated on the even-carbon number *n*-alkanes $C_{13}$ through $C_{24}$.

nique using a 50-m SF-96 glass capillary column subambiently programmed from −20 °C (1 min) to 230 °C at 3 °C/min.

Despite the chromatographic complexity of the samples, there is little qualitative variation from sample to sample. However, there are considerable quantitative differences among the samples, since they were collected from the same source at different times.

The analysis task involved intensive compound identification using software techniques available through the Finnigan INCOS mass spectral data system. A description of these methods to facilitate compound identification in such a matrix has been described by Kan[132] and Weiss.[133]

Even though a high-resolution capillary column was used to resolve the components, significant peak overlap occurred. Over 400 peaks in the chromatogram are recognizable using an isometric ion-mapping technique.

Mass spectra of coeluting compounds were resolved using background subtraction techniques in combination with a Biller–Biemann[134] peak maximizing routine. Base peak mass chromatograms were helpful in locating homologous series such as *n*-alkanes (*m/z* 57). The mapping

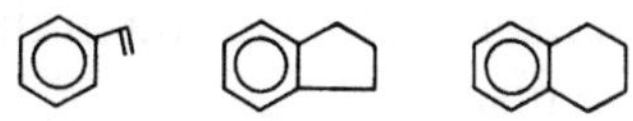

| MW | STYRENES | INDANS | TETRALINS |
|---|---|---|---|
| 104 | $C_0$ | — | — |
| 118 | $C_1$ | $C_0$ | — |
| 132 | $C_2$ | $C_1$ | $C_0$ |
| 146 | $C_3$ | $C_2$ | $C_1$ |
| 160 | $C_4$ | $C_3$ | $C_2$ |

FIGURE 11. Homologous series of coeluting isobaric compounds which exhibit nonunique mass spectra. $C_0$ represents the unsubstituted compound; $C_1$–$C_4$ represent methyl- through tetramethyl-substituted compounds.

of molecular ions was a more definitive compound location technique for compounds with high-abundance molecular ions.

Even though these sophisticated spectral subtraction techniques were used, it was found that mass spectral analysis in itself was not sufficient for positive identification of all the components. There are many problems inherent in such a complex matrix which obscure mass spectral assignments. Components that elute in the same chromatographic region and are chemically similar often exhibit the same fragmentation patterns. Figure 11 lists homologous series of isobaric compounds, the mass spectra of which contain similar ions with slight differences in abundance values. For example, tetralin and isomers of $C_2$ alkyl styrene and methyl indan (MW 132) all have major *m/z* values at 132, 117, 115, and 91. Coeluting isomers of methyl indene and divinyl benzene contribute interferences at *m/z* 115 and 117. The more concentrated $C_4$ and $C_5$ alkyl

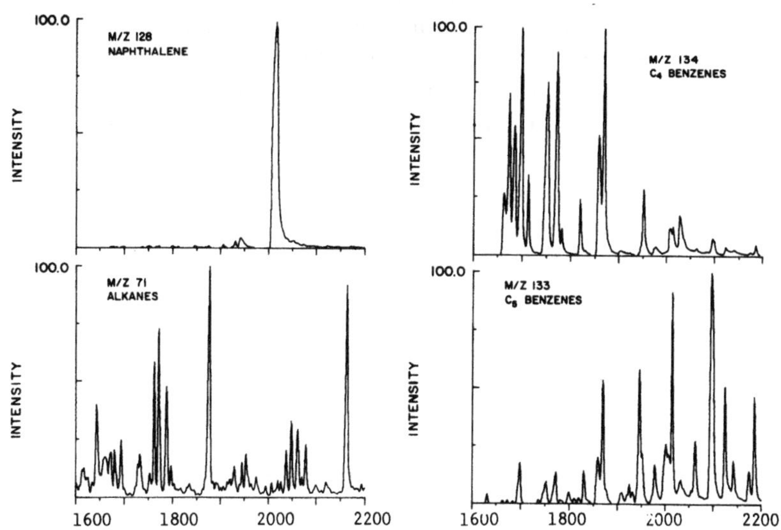


FIGURE 12. Mass chromatograms representing some of the compound classes coeluting in a 10-min segment of the 1.5 hour chromatogram shown in Fig. 10.

benzenes also eluting in this region also introduce interfering mass fragments.

This situation is described in Fig. 12; a display of mass chromatograms representing coeluting compound classes in a 10-min segment of the tunnel sample RIC is shown in Fig. 10. To resolve this problem Eggertsens's[(114)] chromatography distillation technique was used. Compounds elute according to their boiling points on a SF-96 column. A plot of retention temperature versus retention time for each compound is linear for compounds eluting in the higher than ambient temperature region of the chromatogram. In this manner, a boiling point range can be assigned to each compound. By using this technique to aid in mass spectral assignments, over 300 compounds were able to be identified in the tunnel sample.

For quantitation purposes each cartridge was spiked with an internal standard, hexafluorobenzene, to which the response of each compound was referenced. There were three methods (see Table 3) used to obtain quantitative peak values:

(i) the area obtained from one fragment mass divided by the relative abundance of that mass. This method (A) can be used successfully in situations where coeluting peaks cause ion interferences or excessive data rates during acquisition results in saturation for specific *m/z* values. This is described in Table 3. The area obtained from *m/z* 43 for heptane is low due to peak saturation. If this peak was solely relied on for quantitation, the concentration obtained would be 5% less than it would be if the average area of *m/z* 43, 57, and 71 were used.

(ii) the sum of the area under all fragment masses divided by the total relative abundances for all ions. This is method B in Table 3. It was used successfully in chromatographic regions containing low background levels and no interfering ions from coeluting peaks.

*TABLE 3*
*Comparison of Methods for Determining Peak Areas Using Mass Chromatograms*[a]

| | Method A | | | Method B |
|---|---|---|---|---|
| | 43 *m/z* | 57 *m/z* | 71 *m/z* | All ions in spectrum |
| RA[b] | 1.00 | 0.44 | 0.45 | 3.54[c] |
| Area | 96,989[d] | 50,882 | 6,090 | 373,748 |
| Area/RA | 96,989 | 115,641 | 102,422 | 105,578 |

[a] Data shown are for a saturated peak of heptane.
[b] All relative ion abundances (RA) were determined from standard compounds.
[c] RA for method B represents the sum of the RA for all ions in the spectrum.
[d] Low number for area is due to saturation of peak.

(iii) the area obtained from a $C^{13}$ peak divided by the corresponding isotopic relative abundance. This natural-abundance isotope ratio measurement has been used by others for identification[107] and quantitation purposes. However, as described in an earlier section, it may be sensitive to instrument tune, source temperature, and electrometer offset. If instrument operation is stable, it can be successfully applied to situations where data was acquired during detector saturation.

## 8. RECENT AND FUTURE DEVELOPMENTS

We have attempted to describe the latest GC/MS methodologies used for the qualitative and quantitative multicomponent analysis of air pollutants. There are a number of new advancements in mass spectrometry which have shown that moderately polar and polar chemical fractions of environmental particulate samples contain a major portion of the direct-acting mutagenicity for the total extract.[33,34] The survey of GC/MS analysis given in Table 2 has shown that only limited work has been undertaken to identify moderately polar and polar species. Based upon these results, it is projected that analytical emphasis will shift away from the traditional measurement of PAH and hydrocarbons to the oxygenated derivatives of these species. A number of available techniques could be used to identify such compounds, some of which have been described in this chapter.

Since this review was written, a number of developments have occurred. Lee[135] has reviewed the applications of mass spectrometry to PAH analysis. A good deal of work has been undertaken on the GC/MS analysis of nitrated-polycyclic aromatic hydrocarbons in environmental samples, and this subject has been reviewed recently by Schuetzle and Jensen.[136] A number of investigators are now actively pursuing the identification of moderately polar and polar PAH derivatives in environmental samples. For instance, Ramdahl[137] has published recently a study on the GC/MS identification of PAH-ketones, up to a molecular weight of 330., in a variety of environmental samples. Work in these areas is expected to continue at an active pace.

## REFERENCES

1. D. Schuetzle, in *Biochemical Applications of Mass Spectrometry* (G. R. Waller and O. C. Dermer, eds.), Wiley-Interscience, New York, 1980, pp. 970–1001.
2. D. J. Robertson, R. H. Groth, D. G. Gardner, and E. G. Glastris, *J. Air Pollut. Contr. Assoc.* **29,** 143 (1979).
3. W. D. Bowers, M. L. Parsons, R. E. Clement, G. A. Eiceman, and F. W. Karasek, *J. Chromatogr.* **206,** 279 (1981).
4. A. Bjorseth and G. Eklund, *High Resoln. Chromatogr./Chromatogr. Commun.* **2,** 22 (1979).

5. E. D. Pellizzari, J. E. Bunch, B. H. Carpenter, and E. Sawicki, *Environ. Sci. Technol.* **9,** 553 (1975).
6. M. B. Neher and P. W. Jones, *Anal. Chem.* **49,** 512 (1977).
7. C. V. Hampton, W. R. Pierson, T. M. Harvey, W. S. Updegrove, and R. S. Marano, *Environ. Sci. Technol.* **16,** 287 (1981).
8. D. Schuetzle, *Biomed. Mass Spectrom.* **2,** 288 (1975).
9. L. van Vaeck, G. Broddin, and K. van Cauwenberghe, *Biomed. Mass Spectrom.* **7,** 473 (1980).
10. W. Cautreels and K. van Cauwenberghe, *Atmos. Environ.* **12,** 1133 (1978).
11. R. A. Rasmussen and R. S. Hutton, *Bioscience* **22,** 294 (1972).
12. E. D. Pellizzari, J. E. Bunch, R. E. Berkley, and J. T. Bursey, *Biomed. Mass Spectrom.* **3,** 196 (1976).
13. E. D. Pellizzari, J. E. Bunch, R. E. Berkley, and J. McRae, *Anal. Chem.* **48,** 803 (1976).
14. J. T. Bursey, E. D. Pellizzari, D. Smith, J. Bunch, R. Williams, and R. Berkley, *Am. Lab.* **35,** 35 (1977).
15. E. D. Pellizzari, *Measurement of Carcinogenic Vapors in Ambient Atmospheres,* EPA 600/7-78-062, Environmental Protection Agency, Washington, D.C. (1978).
16. J. A. Parsons and S. Mitzner, *Environ. Sci. Technol.* **9,** 1053 (1975).
17. P. W. Jones, J. E. Wilkinson, and P. E. Strup, *Measurement of Polycyclic Organic Materials and other Hazardous Organic Compounds on Stack Gases,* NTIS–U.S. Department of Commerce, PB-274013 Washington, D.C. (1977).
18. P. Ciccioli, G. Bertoni, E. Brancaleoni, R. Fratarcangeli, and F. Bruner, *J. Chromatogr.* **126,** 757 (1976).
19. J. N. Pitts, Jr., K. A. van Cauwenberghe, D. Grosjean, J. P. Schmid, D. R. Fitz, W. L. Belser, Jr., G. B. Knudson, and P. M. Hynds, *Science* **202,** 515 (1978).
20. F. S.-C. Lee, T. M. Harvey, T. J. Prater, M. C. Paputa, and D. Schuetzle, *Chemical Analysis of Diesel Particulate Matter and an Evaluation of Artifact Formation in Sampling and Analysis of Toxic Organics in the Atmosphere,* ASTM STP 721, American Society for Testing and Materials, 1980, pp. 91–110.
21. W. G. Jennings, *J. High Res. Chromatogr.* **2,** 221 (1979).
22. D. Schuetzle, T. J. Prater, T. M. Harvey, and D. E. Hagge, *Adv. Mass Spectrom.* **7,** 231 (1977).
23. G. Chatot, *Anal. Chem. Acta* **53,** 259 (1971).
24. C. Golden and E. Sawicki, *Int. J. Environ. Anal. Chem.* **4,** 9 (1975).
25. D. Schuetzle, and F. S.-C. Lee, in *Informational Report of the Measurement and Characterization of Diesel Exhaust Emission* (J. M. Perez, F. J. Hills, D. Schuetzle, R. L. Williams, eds.), Coordinating Research Council, Atlanta, Georgia, 1981, pp. 65–92.
26. H. H. Hill, Jr., K. W. Chan, and F. W. Karasek, *J. Chromatogr.* **131,** 245 (1977).
27. D. Schuetzle and J. Perez, Proceedings of the 74th Annual APCA Meeting, paper #81-56.4, 1981.
28. F. S.-C. Lee and D. Schuetzle, in *Handbook of Polycyclic Aromatic Hydrocarbons* (A. Bjorseth, ed.), Marcel Dekker, New York, 1983, pp. 27–94.
29. E. D. Pellizzari, *Analysis of Organic Air Pollutants by Gas Chromatography and Mass Spectroscopy,* NTIS–U.S. Department of Commerce, PB-296 835 Washington, D.C., 1979.
30. M. Novotny, M. L. Lee, and K. D. Bartle, *J. Chromatogr. Sci.* **12,** 606 (1974).
31. T. J. Hughes, E. Pellizzari, E. Little, C. Sparacino, and A. Kolber, *Mutat. Res.* **76**(1), 51 (1980).
32. J. L. Epler, in *Chemical Mutagens; Principles and Methods for Their Detection,* Vol. 6 (F. J. de Serres, ed.), Plenum, New York, 1980, pp. 239–270.
33. C. E. Chrisp and G. L. Fisher, *Mutation Research* **76,** 143 (1980).
34. D. Schuetzle, F. S.-C. Lee, T. J. Prater, and S. B. Tejada, *Int. J. Environ. Anal. Chem.* **9,** 93 (1981).

35. I. Salmeen, A. M. Durison, T. J. Prater, T. Riley, and D. Schuetzle, *Mut. Res.* **104,** 17 (1982).
36. D. Schuetzle, T. J. Prater, T. Riley, T. M. Harvey, and D. Hunt, *Anal. Chem.* **54,** 265 (1982).
37. F. W. Karasek, D. W. Denney, K. W. Chan, and R. E. Clement, *Anal. Chem.* **50,** 82 (1978).
38. J. Drozd, *Chemical Derivatization in Gas Chromatography,* Elsevier, New York, 1981.
39. F. Bruner, G. Grescentini, and F. Mangani, *Anal. Chem.* **53,** 798 (1981).
40. R. Eichmann, G. Ketseridis, G. Schebeske, R. Jaenicke, J. Hahn, P. Warneck, and C. Junge, *Atmos. Environ.* **14,** 695 (1980).
41. L. L. Lamparski and T. J. Nestrick *Anal. Chem.* **52,** 2045 (1980).
42. J. Barkley, J. Bunch, J. T. Bursey, N. Castillo, S. D. Cooper, J. M. Davis, M. D. Erickson, B. S. H. Harris III, M. Kilpatrick, L. C. Michael, S. P. Parks, E. D. Pellizzari, M. Roy, D. Smith, K. T. Comer, R. Wagner, and R. H. Zweidinger, *Biomed. Mass Spectrom.* **7,** 139 (1980).
43. A. Jonsson and S. Berg, *J. Chromatogr.,* **180,** 133 (1980).
44. G. A. Eiceman, R. E. Clement, and F. W. Karasek, *Anal. Chem.* **51,** 2343 (1979).
45. R. L. Hanson, R. E. Royer, R. L. Carpenter, and G. J. Newton, in *Polynuclear Aromatic Hydrocarbons* (P. W. Jones and P. Leber, eds.), Ann Arbor Science, Ann Arbor, MI, 1979, p. 3.
46. R. C. Lao and R. S. Thomas, in *Polynuclear Aromatic Hydrocarbons* (P. W. Jones and P. Leber, eds.), Ann Arbor Science, Ann Arbor, MI, 1979, p. 429.
47. R. F. Severson, W. S. Schlotzhauer, O. T. Chortyk, R. F. Arrendale, and M. E. Snook, in *Polynuclear Aromatic Hydrocarbons* (P. W. Jones and P. Leber, eds.), Ann Arbor Science, Ann Arbor, MI, 1979, p. 277.
48. A. Bjorseth and G. Eklund, *Anal. Chim. Acta* **105,** 119 (1979).
49. A. Bjorseth and G. Lunde, *Atmos. Environ.* **13,** 45 (1979).
50. T. Fujii, T. Y. Yokouchi, and Y. Ambe *J. Chromatogr.* **176,** 165 (1979).
51. J. C. Liao and R. F. Browner *Anal. Chem.* **50,** 163 (1978).
52. G. Broddin, L. van Vaeck, and K. van Cauwenberghe, *Atmos. Environ.* **11,** 1061 (1978).
53. W. Giger and C. Schaffner, *Anal. Chem.* **50,** 245 (1978).
54. A. Bjorseth, in *Carcinogenesis,* Vol. 3 (P. W. Jones and R. I. Freudenthal, eds.), Raven Press, New York, 1978, p. 75.
55. M. D. Erickson, L. C. Michael, R. A. Zweidinger, and E. D. Pellizzari, *Environ. Sci. Technol.* **12,** 927 (1978).
56. M. L. Lee, G. P. Prado, J. B. Howard, and R. A. Hites, *Biomed. Mass Spectrom.* **4,** 182 (1977).
57. Holzer *et al, J. Chromatogr.* **142,** 755 (1977).
58. W. Cautreels and K. van Cauwenberghe, *Water, Air Soil Poll.* **6,** 103 (1976).
59. W. Cautreels and K. van Cauwenberghe, *Atmos. Environ.* **10,** 447 (1976).
60. M. Lee, M. Novotny, and K. D. Bartle, *Anal. Chem.* **48,** 405 (1976).
61. M. L. Lee and R. A. Hites, *Anal. Chem.* **48,** 1890 (1976).
62. R. F. Severson, M. E. Snook, H. C. Higman, O. T. Chortyk, and F. J. Akin, in *Carcinogenesis,* Vol. 1 (R. I. Freudenthal and P. W. Jones, eds.), Raven Press, New York, 1976, p. 253.
63. B. Versino, H. Knoeppel, M. DeGroot, A. Peil, J. Poelman, H. Schavenberg, H. Vissers, and F. Geiss, *J. Chromatogr.* **122,** 373 (1976).
64. R. C. Pierce and M. Katz, *Anal. Chem.* **47,** 1743 (1976).
65. G. Ketseridis, J. Hahn, R. Jaenicke, and C. Junge, *Atmos. Environ.* **10,** 603 (1976).
66. R. C. Lao, R. S. Thomas, and J. L. Monkman, *J. Chromatogr.* **112,** 681 (1975).
67. E. P. Grimsrud and R. A. Rasmussen, *Atmos. Environ.* **9,** 1010, 1014 (1975).
68. K. W. Boyer and H. A. Laitinen, *Environ. Sci. Technol.* **9,** 457 (1975).
69. T. Doran and N. G. McTaggart, *J. Chromatogr. Sci.* **12,** 715 (1974).

70. F. W. Karasek, R. J. Smythe, and R. J. Laub, *J. Chromatogr.* **101,** 125 (1974).
71. W. Bertsch, R. C. Chang, and A. Zlatkis, *J. Chromatogr. Sci.* **12,** 175 (1974).
72. A. Raymond and G. Guiochon, *Environ. Sci. Technol.* **8,** 143 (1974).
73. R. C. Lao, R. S. Thomas, H. Oja, and L. Dubois, *Anal. Chem.* **45,** 908 (1973).
74. A. Herlan, *Erdgas. Petrochem. Brennst.-Chem.* **27,** 138 (1974).
75. G. Grimmer and H. Bohnki, *Z. Anal. Chem.* **261,** 310 (1972).
76. S. P. Pavlou and D. Schuetzle, *Finnigan Spectra* **2,** 1 (1972).
77. K. Grob and G. Grob, *J. Chromatogr.* **62,** 1 (1971).
78. D. Grosjean, K. van Cauwenberghe, J. P. Schmid, P. E. Kelley, and J. N. Pitts, Jr., *Environ. Sci. Technol.* **12,** 313 (1978).
79. G. Schomburg, R. Dielmann, H. Bowvitzky, and H. Husmann *J. Chromatogr.,* **167,** 337 (1978).
80. K. Grob and K. Grob, Jr., *J. High Resoln. Chromatogr./Chromatogr. Commun.* **1,** 57 (1978).
81. K. Grob and K. Grob, Jr., *J. Chromatogr.* **151,** 311 (1978).
82. K. Grob and G. Grob, *Chromatographia* **4,** 422 (1971).
83. K. Grob and G. Grob, *Chromatographia* **5,** 3 (1972).
84. K. Grob and H. Jaeggi, *Chromatographia* **5,** 382 (1972).
85. K. Grob, *Chromatographia* **7,** 94 (1974).
86. K. Grob and G. Grob, *J. High Resoln. Chromatogr.* **2,** 109 (1970).
87. R. D. Dandeneau and E. H. Zerenner, *J. High Resoln. Chromatogr./Chromatogr. Commun.* **2,** 351 (1979).
88. S. R. Lipsky, W. J. McMurray, M. Hernandez, J. E. Purcell, and K. A. Billeb, *J. Chromatogr. Sci.* **18,** 1 (1980).
89. W. Cautreels and K. van Cauwenberghe, *J. Chromatogr.,* **131,** 253 (1977).
90. S. G. Wakeham, *Environ. Sci. Technol.* **13,** 1118 (1979).
91. W. H. McFadden, *J. Chromatogr. Sci.* **17,** 2 (1979).
92. R. Ryhage, *Anal. Chem.* **36,** 759 (1964).
93. P. M. Llewellyn and D. P. Littlejohn, Proceedings of the Pittsburgh Conference on Analytical Chemistry and Applied Spectroscopy, 1966.
94. R. C. Doughtery, *Anal. Chem.* **53,** 625A (1981).
95. T. M. Harvey, D. Schuetzle, and S. P. Levine, in *Monitoring of Toxic Substances,* ACS Symposium Series, No. 94, (D. Schuetzle, ed.), American Chem. Society, Washington, D.C., 1979, p. 195.
96. F. W. Karasek, *Res./Dev.* **25,** 30 (1974).
97. A. L. Burlingame, R. W. Olsen, and R. McPherron, *Adv. Mass Spectrom* **6,** 1053 (1974).
98. D. Schuetzle, D. G. Cronn, A. L. Crittenden, and R. J. Charlson, *Environ. Sci. Technol.* **9,** 838 (1975).
99. D. Schuetzle, T. J. Prater, D. E. Hagge, and R. L. Knights, Proceedings of the 23rd Annual Conference on Mass Spectrometry, American Society of Mass Spectrometry, 1975, p. 208.
100. T. J. Prater, T. M. Harvey, and D. Schuetzle, Proceedings of the 24th Annual Conference on Mass Spectrometry, American Society of Mass Spectrometry, 1975, p. 625.
101. J. Meili, F. C. Walls, R. McPherron, and A. L. Burlingame, *J. Chromatogr. Sci.* **17,** 29 (1979).
102. B. Crathorne, M. W. Edwards, N. R. Jones, C. L. Walters, and G. Woolford, *J. Chromatogr.* **115,** 213 (1975).
103. D. Schuetzle and D. E. Hagge, Proceedings of the 24th Annual Conference on Mass Spectrometry, American Society of Mass Spectrometry, 1976, p. 115.
104. B. A. Petersen, C. C. Chuang, G. W. Kinzer, P. W. Meeham, R. M. Riggin, and D. A. Trayser, "Polynuclear Aromatic Hydrocarbon Emissions from Diesel Engines,"

Coordinating Research Council Project CAPE-24-72 Progress Report, Atlanta, Georgia, 1980.
105. F. W. McLafferty and R. Venkataraghavan, *J. Chromatogr. Sci.* **17,** 24 (1979).
106. D. Rosenthal and W. F. Hargrove, Jr., Proceedings of the 28th Annual Conference on Mass Spectrometry, American Society of Mass Spectrometry, 1979, p. 266.
107. J. Slayback, J. Jugovich, and M. N. Kan, Proceedings of the 27th Annual Conference on Mass Spectrometry, American Society of Mass Spectrometry, 1979, p. 269.
108. H. S. Hertz, R. A. Hites, and K. Biemann, *Anal. Chem.* **43,** 681 (1971).
109. F. W. Karasek, *Res./Dev.* **27,** 42 (1976).
110. M. L. Lee, D. L. Vassilaros, C. M. White, and M. Navotny, *Anal. Chem.* **51,** 768 (1979).
111. K. R. Betty and F. W. Karasek, *J. Chromatogr* **166,** 111 (1978).
112. G. I. Spivakovskii, A. I. Tishcanko, I. I. Zaslavskii, and N. S. Wilfson, *J. Chromatogr.* **144,** 1 (1977).
113. M. Randic, *J. Chromatogr.* **161,** 1 (1978).
114. F. T. Eggertsen, S. Groernings, and J. J. Holst, *Anal. Chem.* **32,** 904 (1960).
115. C. L. Stuckey, *J. Chromatogr. Sci.* **16,** 482 (1978).
116. R. A. Hites and K. Biemann, *Anal. Chem.* **42,** 855 (1970).
117. D. H. Smith, M. Achenbach, W. J. Yeager, P. J. Anderson, W. L. Fitch, and T. C. Rindfleisch, *Anal. Chem.* **49,** 1623 (1977).
118. S. C. Gates, M. J. Smisko, C. L. Ashendel, N. D. Young, J. F. Holland, and C. C. Sweeley, *Anal. Chem.* **350,** 433 (1978).
119. J. E. Ryan, R. A. Novak, B. J. Dowty, and R. M. Segesta, Proceedings of the 28th Annual Conference on Mass Spectrometry, American Society of Mass Spectrometry, 1979, p. 343.
120. J. F. Ryan, R. A. Novak, and W. A. Farr, Proceedings of the 28th Annual Conference on Mass Spectrometry, American Society of Mass Spectrometry, 1979, p. 212.
121. G. I. Ouchi, Proceedings of the 28th Annual Conference on Mass Spectrometry, American Society of Mass Spectrometry, 1980, p. 129.
122. A. Cornu and R. Massot, Compilation of Mass Spectral Data, 2nd ed., Heydon, London, 1975.
123. D. Souter, C. Kieda, R. Devine, and H. Nowicki, ASTM Spec. Tech. Publ. STP 686, Meas. Org. Pollut. Water Wastewater, 221–233 (1978).
124. K. B. Tomer, J. T. Bursey, and E. D. Pellizzari, Proceedings of the 28th Annual Conference on Mass Spectrometry, American Society of Mass Spectrometry, 1980, p. 165.
125. W. L. Budde and J. W. Eichelberger, An EPA Manual for Organics Analysis Using Gas Chromatography–Mass Spectrometry, EPA-600/8-79-006, 1979.
126. A. Shatkay and S. Flavian, *Anal. Chem.* **49,** 2222 (1977).
127. A. Shatkay, *Anal. Chem.* **50,** 1423 (1978).
128. B. N. Colby and M. W. McCaman, *Biomed. Mass Spectrom.* **6,** 225 (1979).
129. D. A. Schoeller, *Biomed. Mass. Spectrom.* **7,** 457 (1980).
130. F. C. Falkner, *Biomed. Mass Spectrom.* **8,** 43 (1981).
131. M. Claeps, S. P. Markey, and W. Maenkaut, *Biomed. Mass Spectrom.* **4,** 122 (1977).
132. M. N. Kan, J. Jogovich, and J. Slayback, Proceedings of the 27th Annual Conference on Mass Spectrometry, American Society of Mass Spectrometry, 1979, p. 269.
133. M. Weiss and S. Sokolow, Proceedings of the 28th Annual Conference on Mass Spectrometry, American Society of Mass Spectrometry, 1980, p. 183.
134. J. E. Biller and K. Biemann, *Anal. Lett.* **7,** 515 (1974).
135. M. L. Lee, M. V. Novotny, and K. D. Battle, *Mass Spectrometry in Analytical Chemistry of Polycyclic Aromatic Compounds,* Academic Press, New York, 1981, pp. 242–289.
136. D. Schuetzle and T. E. Jensen, in *Nitrated Polycyclic Aromatic Hydrocarbons* (C. M. White, ed.), Huthig Verlag, Heidelberg, 1984.
137. T. Ramdahl, *Environ. Sci. Technol.* **17,** 666(1983).

# POLYCYCLIC AROMATIC HYDROCARBONS

*J. A. SWEETMAN and F. W. KARASEK*

## 1. INTRODUCTION

Polycyclic aromatic hydrocarbons (PAH) are ubiquitous in the environment. Since many of these compounds are known to be carcinogens or cocarcinogens, the identification and quantification of PAH in the environment are continuing concerns.[1–5] The mechanisms of PAH formation during high-temperature pyrolysis of organic matter have been described.[6] Natural sources of PAH include incomplete combustion of organic matter—for example, forest fires, diagenesis of organic material, and biosynthesis. Anthropogenic sources of PAH include coal refuse burning, coke production, residential fireplaces and coal-fired furnaces, commercial incinerators, vehicle exhaust, and tobacco smoke.[7–10]

The toxicological properties of individual PAH are structure specific. For example, benzo[*a*]pyrene is a potent animal carcinogen and the structural isomer benzo[*e*]pyrene is not. 5-methylchrysene is a strong carcinogen while chrysene and other methylchrysenes studied are weak carcinogens. A summary of PAH structures and carcinogenic activity has been compiled.[11] The importance of isomer-specific identification of PAH has required the use of mass spectrometry and in particular gas chromatography/mass spectrometry (GC/MS) for the analysis of complex PAH mixtures such as those found on urban air particulates,[5,12] diesel exhaust particulates,[13–16] tobacco smoke,[11] and sediments.[7,17]

---

*J. A. SWEETMAN* and *F. W. KARASEK* • Department of Chemistry, University of Waterloo, Waterloo, Ontario N2L 3G1, Canada.

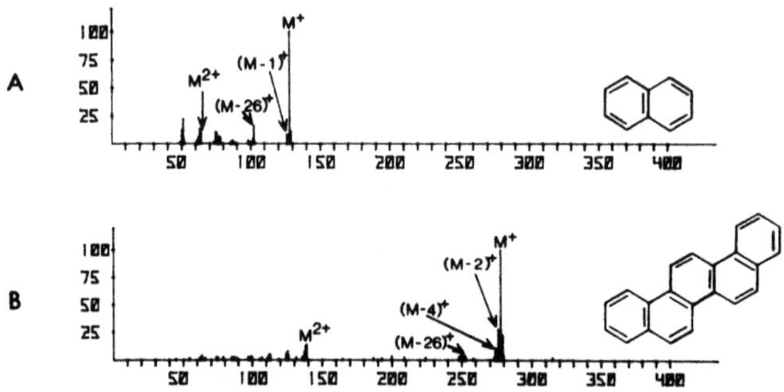


FIGURE 1. EI mass spectra of naphthalene (A) and picene (B).

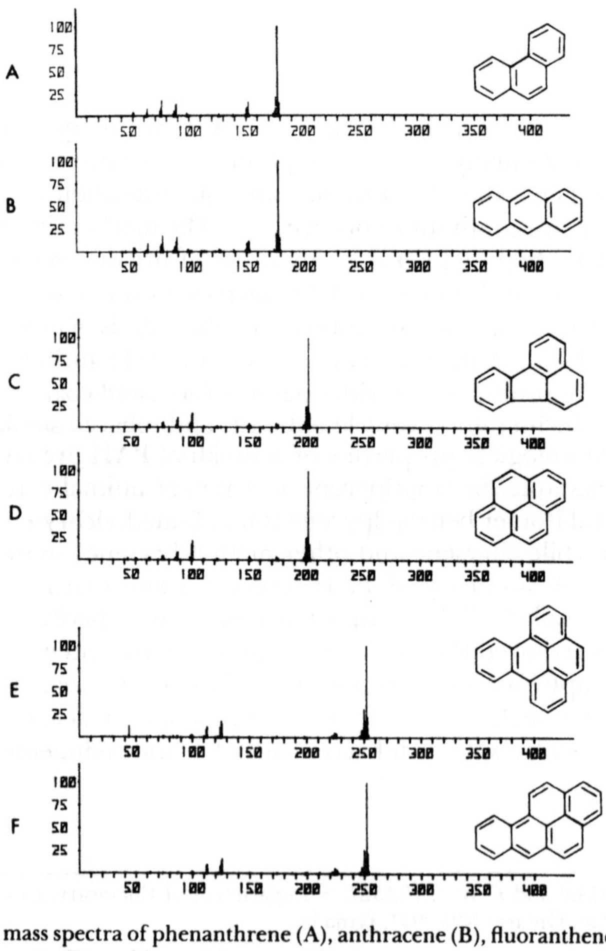


FIGURE 2. EI mass spectra of phenanthrene (A), anthracene (B), fluoranthene (C), pyrene (D), benzo[*e*]pyrene (E), and benzo[*a*]pyrene (F).

## 2. ELECTRON IMPACT MASS SPECTROMETRY

### 2.1. *Polycyclic Aromatic Hydrocarbons*

The electron impact (EI) mass spectra of PAH are characteristically simple, featuring a high molecular ion peak, $M^+$, ion peaks due to the loss of one to four hydrogen ions, and small ion peaks resulting from the expulsion of $C_2H_2$. This common PAH fragmentation pattern is depicted below.

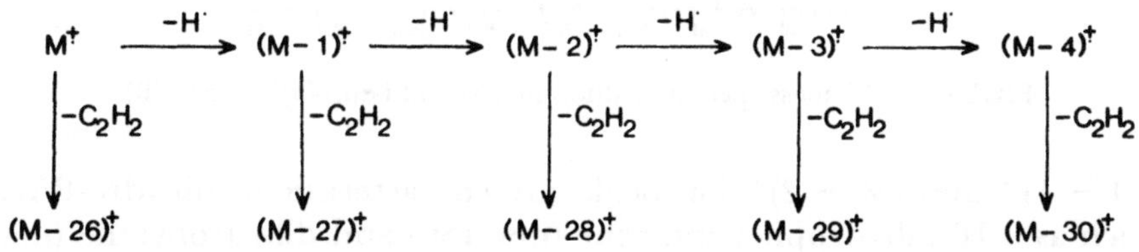


A relatively large percentage (up to 20%) of the molecular ion or the fragments depicted in the above structure may be doubly charged. A $(M + 1)^+$ ion peak will always be present reflecting the $^{13}C$ natural isotopic abundance.[18]

The spectra shown in this chapter were produced by scanning from 400 to 40 amu at 300 amu/sec with a Hewlett-Packard 5992 quadrupole GC/MS system in the authors' laboratory. The *x* axis of each spectrum is mass/charge (*m*/*z*) and the *y* axis is relative abundance. Figure 1 shows the spectra of the simplest PAH, naphthalene, and a high-molecular-weight PAH, picene (mw 278).

The increased abundance of the $(M - 1)^+$ through $(M - 4)^+$ ion peaks for the higher molecular weight PAH can be seen from this figure. The low-mass fragment ion series characteristic of aromatic hydrocarbons ($C_nH_{0.5n}$ to $C_nH_n$) at *m*/*z* 38–39, 50–52, 63–65, and 75–78 is evident in the naphthalene spectrum.

The EI mass spectra of isomeric PAH are often indistinguishable from one another. Figure 2 gives the spectra of the isomer pairs: phenanthrene/anthracene (mw 178), fluoranthene/pyrene (mw 202), and benzo[*e*]pyrene/benzo[*a*]pyrene (mw 252). Perylene, benzo[*e*]acephenanthrylene, benzo[*j*]fluoranthene, benzo[*k*]fluoranthene and $C_{20}H_{12}$ compounds give spectra virtually identical to those shown for the benzopyrenes.[19] As will be discussed in more detail later, the combination of high-resolution gas chromatography with mass spectrometry has allowed the resolution and specific identification of many isomeric PAH from complex environmental samples.

The facile loss of a proton from the saturated carbons of hydroaromatic compounds can be seen from the high $(M - 1)^+$ abundances in the mass spectra of fluorene and benzo[*a*]fluorene given in Fig. 3. Intense

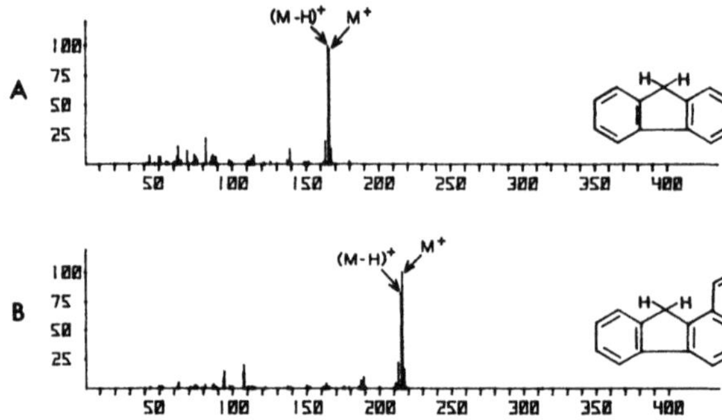


FIGURE 3. EI mass spectra of fluorene (A) and benzo[*a*]fluorene (B).

$(M - 1)^+$ and $(M - 2)^+$ ion peaks are characteristic of dihydro-PAH such as 9,10-dihydrophenanthrene (mw 180) and dihydropyrene (mw 204). Another example of distinctive loss of hydrogens giving structural information is illustrated by the higher $(M - 1)^+$ and $(M - 2)^+$ ion abundances of PAH where steric interaction may facilitate loss of two protons as shown for benzo[*c*]phenanthrene:

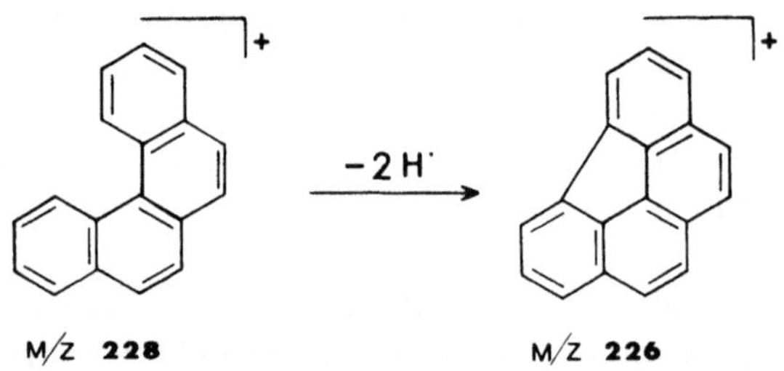


Thus the spectra of benzo[*c*]phenanthrene can be distinguished from the nearly identical spectra of chrysene, benzo[*a*]anthracene, triphenylene, and naphthacene.

### 2.2. *Alkyl-Polycyclic Aromatic Hydrocarbons*

PAH formed by high-temperature pyrolysis of organic matter are usually characterized by a predominance of unalkylated parent PAH. PAH in fossil fuels including petroleum, peat, and coal produced by diagenesis of organic material at low temperatures are characterized by abundant alkyl-PAH homologs with unsubstituted species nearly absent. Recent findings have shown that diesel exhaust, unlike other combustion emissions,, may contain alkyl-PAH as major components.

The lack of standard compounds often makes isomer-specific iden-

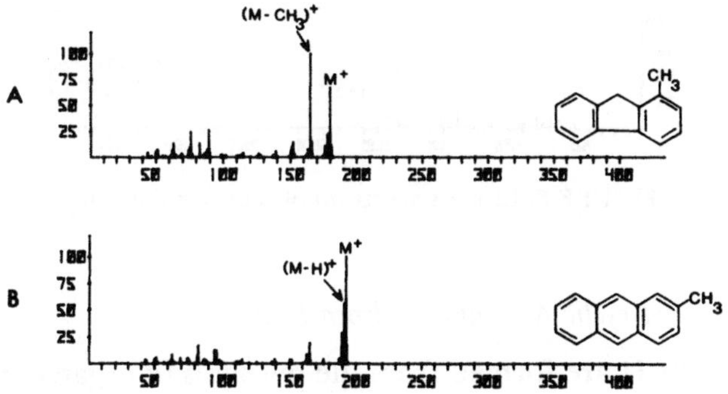


FIGURE 4. EI mass spectra of 1-methylfluorene (A) and 2-methylanthracene (B).

tification impossible even with the use of high-resolution GC/MS. The possible sample complexity is evident if one considers that adding one methyl group to the twelve isomeric five-ring structures of mw 278 results in 117 possible isomers.

Alkyl-PAH follow the fragmentation pattern of alkyl chains with $(M - 15)^+$, $(M - 29)^+$ fragment ions. For methyl-PAH the $(M - 15)^+$ abundance may be very low or, in the case of methyl-fluorenes, this fragment may be the base peak (Fig. 4). The favorable loss of a proton from 2-methylanthracene followed by ring expansion forms a tropylium ion. As can be seen from Fig. 4, the $(M - H)^+$ ion peak abundance is high and a fragment from the further loss of $C_2H_2$ is also present.

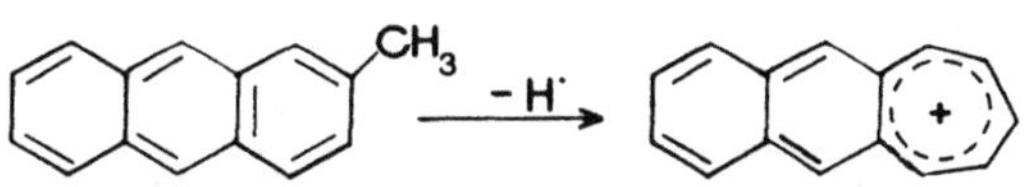


M/Z 192 M/Z 191

For methylfluorenes a tropylium ion can be formed by the loss of $CH_3$ as is shown below for 1-methylfluorene. Mass 165 is then the base peak in these spectra. Abundant *m/z* 165 ions are also seen in the spectra of some dimethylfluorenes. For dimethylfluorenes, an ion at *m/z* 179 results from the loss of a methyl group forming a tropylium ion, and a fragment ion at *m/z* 165 results from the further loss of $CH_2$, also forming a tropylium ion.

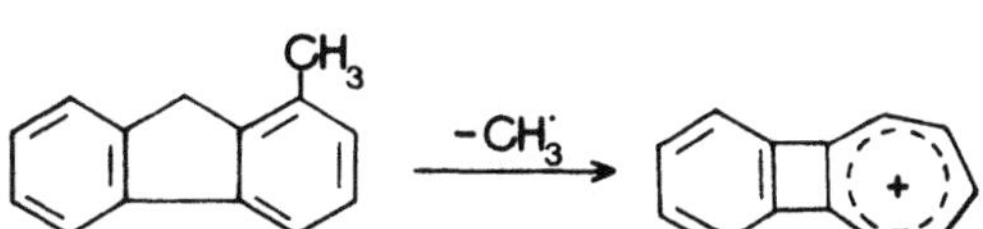


M/Z 180 M/Z 165

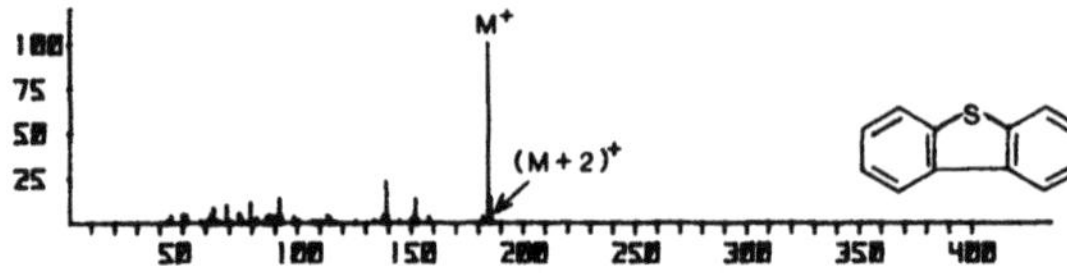


FIGURE 5. EI mass spectrum of dibenzothiophene.

## 2.3. *Hetero-Polycyclic Aromatic Hydrocarbons*

Hetero-PAH are formed from the pyrolysis of organic starting material which contains heteroatoms such as nitrogen, sulfur, and oxygen. The substitution of a nitrogen or sulfur atom into a PAH ring structure does not change the typical PAH spectral pattern with its intense molecular ion. A single nitrogen in a ring will result in a compound with an odd molecular weight according to the nitrogen rule. Sulfur heterocyclics will have an $(M + 2)^+$ ion peak reflecting the 4.4% natural abundance of $^{34}S$ (Fig. 5).

## 2.4. *Derivatives of Polycyclic Aromatic Hydrocarbons*

Findings that the carcinogenicity of air particulates[(20,21)] and the mutagenicity of diesel exhaust particulates[(22)] could not be accounted

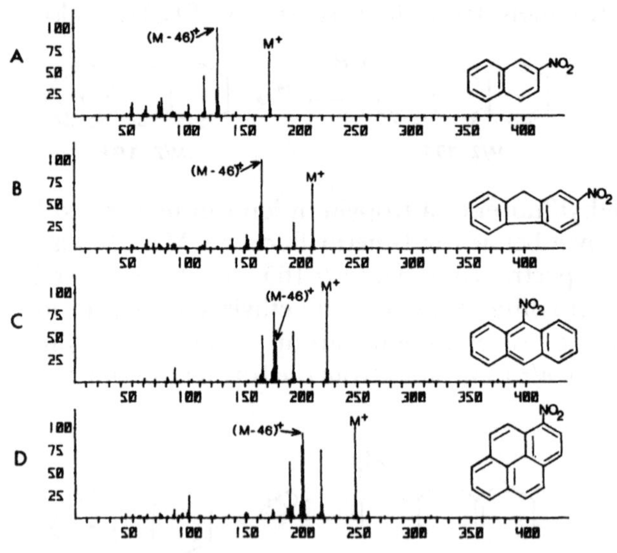


FIGURE 6. EI spectra of 2-nitronaphthalene (A), 2-nitrofluorene (B), 9-nitroanthracene (C), and 1-nitropyrene (D).

for solely by the PAH present has led to the mass spectral identification of many PAH derivatives in these sample types.[15,23,24] Metabolites of PAH, such as phenols, quinones, and dihydroxy-benzo[*a*]pyrenes have also been investigated utilizing mass spectrometry.[25,26]

The exposure of PAH to gaseous pollutants found in photochemical smog produced nitro-PAH species. Several nitro-PAH have been identified in diesel exhaust particulates using a variety of mass spectral techniques.[27] The EI mass spectra of four nitro-PAH are shown in Fig. 6. The molecular weights are 173, 211, 223, and 247 for nitronaphthalene, nitrofluorene, nitroanthracene, and nitropyrene, respectively. Nitro-PAH spectra are characterized by abundant molecular ion peaks and strong fragment ions from the loss of 46 mass units, $(M - NO_2)^+$. Other important ions are those arising from $(M - NO)^+$, $(M - HNO_2)^+$, and $(M - NO - CO)^+$ fragmentation processes.

The spectra of Figs. 7 and 8 are of compounds identified by GC/MS in diesel particulate extracts.[28] The spectra of anthraquinone and phenanthrenequinone (mw 208) are given in Fig. 7. PAH-quinones can be formed from PAH by photochemical oxidation.[29] In addition to their presence in diesel exhaust, PAH-quinones have been found in cigarette smoke condensate[30] and air particulates.[24] PAH-quinones show characteristic losses of one and two molecules of CO from the molecular ion and these two fragments may vary greatly, aiding in isomer identification.

Figure 7 shows the mass spectrum of 9-phenanthrenecarboxaldehyde (mw 206). The loss of a hydrogen atom from the aldehyde group gives an abundant $(M - H)^+$ peak that distinguishes PAH-carboxalde-

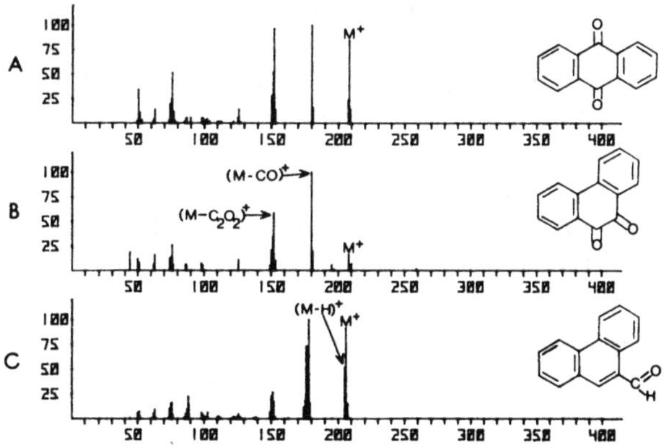


FIGURE 7. EI mass spectra of 9,10-anthracenedione (A), 9,10-phenanthrenedione (B), and 9-phenanthrenecarboxaldehyde (C).

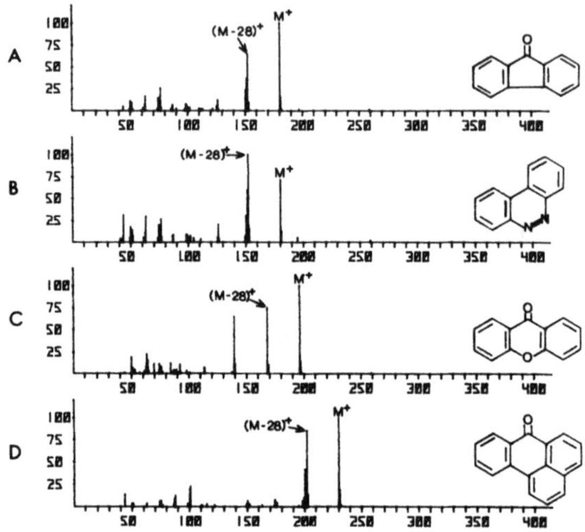


FIGURE 8. EI mass spectra of 9*H*-fluoren-9-one (A), benzo[*c*]cinnoline (B), 9*H*-xanthen-9-one (C), and 7*H*-benzo[*de*]anthracene-7-one (D).

hydes. Loss of carbon monoxide produces $(M - CO)^+$ and $(M - COH)^+$ fragments.

Erikson confirmed the presence of alkyl-9-fluorenones in a mutagenic fraction of diesel exhaust particulate extract using several complementary techniques including GC/MS, high-resolution MS, and GC/Fourier transform infrared spectroscopy.[23] Figure 8 shows the mass spectra of 9-fluorenone and the hetero-PAH, benzo[*c*]cinnoline. Both spectra show a molecular ion peak at *m/z* 180 and an abundant fragment ion peak at $(M - 28)^+$. The alkyl derivatives of the two species also give spectra which are difficult to distinguish from one another. Normal-phase high-performance liquid chromatography (HPLC) has been utilized to isolate 9-fluorenone from the more polar benzo[*c*]cinnoline species.[15,28]

The spectrum of the oxygen-heterocyclic, xanthone (mw 196) shown in Fig. 8, like the other semiquinones shown in this figure, has a strong molecular ion and an abundant $(M - CO)^+$ ion peak. The third strong ion peak in this spectrum is from $(M - C_2O_2H)^+$ fragmentation. Several compounds with molecular weight 230 and $(M - 28)^+$ ion fragment peaks were found in an HPLC fraction of diesel exhaust.[28] One of these compounds was identified as 7*H*-benz[*de*]anthracen-7-one (see Fig. 8) through matching of the mass spectrum and GC retention time with those of an authentic standard.

## 3. GAS CHROMATOGRAPHY/MASS SPECTROMETRY

To separate environmental PAH mixtures consisting of hundreds of components with widely varying concentrations requires the use of capillary column chromatography.(31,32) Gas chromatography/mass spectrometry using high-resolution capillary columns has been very successful in the identification of PAH in complex environmental samples. Fused-silica capillary columns, particularly new cross-linked phases which allow high-temperature operation, provide easy installation and good results for PAH up to at least mw 300. Figure 9 shows the gas chromatogram of the PAH fraction of a diesel particulate sample on a 30-m fused-silica DB-5 column. The labeled peaks include the compounds whose spectra are given in Fig. 2. In addition to resolving benzo[*e*]pyrene and benzo[*a*]pyrene in this diesel sample (peaks 8 and 9, respectively), three additional peaks of molecular weight 252 were resolved and identified as benzo[*j*]fluoranthene (peak 5), benzo[*e*]acephenanthrylene (peak 6), and benzo[*k*]fluoranthene (peak 7) based on their elution order.

Lee developed a retention index system for PAH on temperature-programmed SE-52 capillary columns based on the series of internal standards: naphthalene, phenanthrene, chrysene, and picene.(33) Reproducible retention index values were reported even under conditions of different temperature programming rates and stationary-phase film thicknesses. Their compilation of over 200 retention index values can

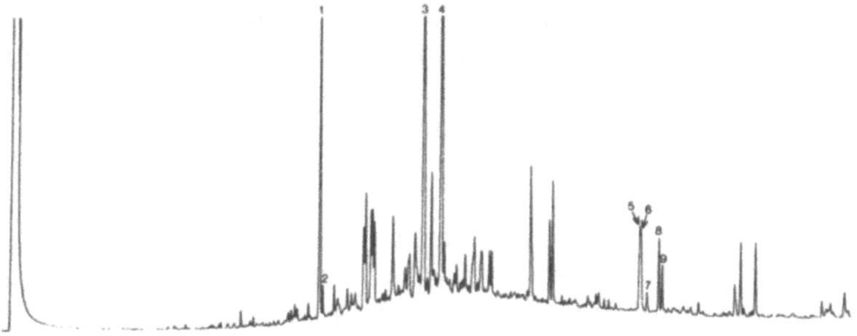

FIGURE 9. Capillary column chromatogram of the PAH-containing HPLC fraction of a diesel particulate extract. Column: fused-silica Durabond DB-5, 30 m × 0.32 mm (J&W Scientific). Conditions: on-column injection, initial column temperature 80 °C for 1 min, then 3 °C/min to 300 °C, helium carrier gas. Peak identifications and retention times (RT) in minutes: (1) phenanthrene, RT 29.7; (2) anthracene, RT 29.9; (3) fluoranthene, RT 39.1; (4) pyrene, RT 40.7; (5) benzo[*j*]fluoranthene, RT 58.6; (6) benzo[*e*]acephenanthrylene, RT 58.7; (7) benzo[*k*]fluoranthene, RT 59.2; (8) benzo[*e*]pyrene, RT 60.3; (9) benzo[*a*]pyrene, RT 60.6.

provide complementary information where GC/MS results show similar spectra and authentic standards are not available for retention time matching.

## 4. SOFT IONIZATION TECHNIQUES

The usual advantage of soft ionization techniques, such as chemical ionization (CI) and field ionization (FI), over EI ionization is that the "soft" techniques produce mainly molecular ion or "pseudomolecular" ion peaks with little fragmentation. This allows molecular weight determination of compounds which in EI spectra may give no molecular ion peak. As the spectra in this chapter demonstrate, PAH show strong molecular ion peaks even with EI ionization. However, in specific applications soft ionization techniques may have additional advantages for the analysis of PAH and PAH derivatives.

Charge exchange chemical ionization shows promise for distinguishing between isomers which give identical EI spectra and identical conventional CI spectra.[34] Conventional CI with methane reagent gas where $[CH_5]^+$, $[C_2H_5]^+$, and $[C_3H_5]^+$ ions predominate produce PAH spectra consisting of protonated molecular ions, $(M + 1)^+$, and less abundant $(M + 29)^+$ and $(M + 41)^+$ ions.[8,35] With 5–10% methane in argon as the CI reagent gas, the most prominent reagent ions are $C_2H_5^+$ and $Ar^+$. Charge exchange with $Ar^+$ produces molecular ions, $M^+$, and reaction with $C_2H_5^+$ gives the protonated molecular ion. The rate of ionization from $Ar^+$ ions is affected by the ionization potential of the PAH. Therefore, the ratio of the protonated molecular ion to the molecular ion will be characteristic of the PAH and may be used to distinguish between isomers. For example, anthracene and phenanthyrene have $(M + 1)^+/M^+$ ratios of 0.82 and 1.59, respectively.

In negative chemical ionization mass spectrometry (NCIMS) the reagent gas may have three functions: slowing down the primary electrons to give thermal electrons, charge transfer, and true chemical ionization where a chemical reaction occurs between an ion of the reactant gas and a molecule of the substance being examined.[36] NCIMS proved advantageous over EIMS for the analysis of nitro-PAH in diesel exhaust particulates. Negative-ion formation is by electron capture with methane reagent gas producing the thermal electrons necessary for capture. In NCI the hydrocarbon background which had made interpretation of the EI spectra difficult is eliminated. The NCI spectra of the nitro-PAH consisted of only two major peaks, the molecular ion peak and the $NO_2$ fragment ion at $m/z = 46$.[27]

Sulfur-containing heterocyclic-PAH can be distinguished from PAH by pulsed, positive-ion–negative-ion chemical ionization (PPINICI) using hydrogen and oxygen reagent gases.[37] Electron bombardment of oxygen at 1 torr produces $O_2^+$ and $O^+$ in the positive-ion mode, and $O_2^-$ and $O^-$ in the negative-ion mode. Addition of hydrogen eliminates $O^-$ from the reagent spectrum

In the positive-ion mode $O_2^+$ reacts to give $M^+$ ions. In the negative-ion mode $O_2^-$ reacts with non-sulfur-containing aromatics to give abundant $(M + O_2 - OH^{\bullet})^-$, or $(M + 15)^-$ ions, either by the aromatic capturing a thermal electron and reacting with neutral oxygen, or by direct reaction between $O_2^-$ and the neutral aromatic. In contrast, sulfur-containing PAH, like dibenzothiophene, exhibit ions corresponding to $M^-$ and $(M + O_2)^-$.

Field ionization mass spectrometry produces virtually fragment-ion-free mass spectra. The FI sensitivities of over 60 PAH and hetero-PAH have been measured and were found to vary less than EI sensitivities.[38] Hogg and Payzant, using FI mass spectrometry to analyze the PAH fraction of Athabasca Tar Sands, found the sensitivities of PAH were relatively independent of instrument parameters.[39] Accordingly, FI mass spectrometry is advantageous for determining molecular weight distributions in mixtures and giving information on the general composition rather than on individual components.

The recently developed technique of mass spectrometry/mass spectrometry (MS/MS) will certainly find applications in solving problems of PAH analysis in complex environmental samples.[40] The ability of MS/MS to give unique identification of isomeric PAH has been demonstrated. The applicability of MS/MS for the screening of complex samples for specific PAH derivatives has been shown with the analysis of nitro-PAH in diesel particulates.[41] Simplified sample preparation and rapid analysis are attractive features of MS/MS.

The lack of authentic standards for the high number of structural isomers in complex PAH mixtures remain a problem both for MS/MS analysis and GC/MS methods. The problem becomes more acute as mutagenicity testing suggests the importance of PAH derivatives as well as PAH.

## *REFERENCES*

1. A. Bjorseth and A. J. Dennis (eds.), *Polynuclear Aromatic Hydrocarbons: Chemistry and Biological Effects,* Battelle, Columbus, Ohio, 1980.
2. R. Freudenthal and P. W. Jones (eds.), *Carcinogenesis—A Comprehensive Survey: Polynuclear Aromatic Hydrocarbons,* Vol. 1. Raven, New York, 1976.
3. P. W. Jones and R. Freundenthal (eds.), *Carcinogenesis—A Comprehensive Survey: Polynuclear Aromatic Hydrocarbons,* Vol. 3. Raven, New York, 1978.

4. P. W. Jones and P. Leber (eds.), *Polynuclear Aromatic Hydrocarbons,* Ann Arbor Science, Ann Arbor, 1979.
5. M. L. Lee, M. Novotny, and K. D. Bartle, *Anal. Chem.* **48,** 1566–1572 (1976).
6. M. Zander, *The Handbook of Environmental Chemistry,* Vol. 3, Part A (O. Hutzinger ed.), Springer-Verlag, Berlin, 1980, pp. 109–131.
7. R. E. Laflamme and R. A. Hites, *Geochim. Cosmochim. Acta* **42,** 289–303 (1978).
8. M. L. Lee, M. V. Novotny, and K. D. Bartle, *Analytical Chemistry of Polycyclic Aromatic Hydrocarbons,* Academic, New York, 1981.
9. J. M. Neff, *Polycyclic Aromatic Hydrocarbons in the Aquatic Environment,* Applied Science Publishers, London, 1979.
10. M. J. Suess, *Sci. Total Environ.* **6,** 239–250 (1976).
11. M. L. Lee, M. Novotny, and K. D. Bartle, *Anal. Chem.* **48,** 405–416 (1976).
12. R. C. Lao, R. S. Thomas, H. Oja, and L. Dubois, *Anal. Chem.* **45,** 908–915 (1973).
13. D. R. Choudhury and B. Bush, in *Chemical Hazards in the Workplace—Measurement and Control* (G. Choudhury, ed.), ACS Symposium Series, No. 149, 1981, pp. 357–368.
14. D. R. Choudhury, *Environ. Sci. Technol.* **16,** 102–106 (1982).
15. D. Schuetzle, F. S.-C. Lee, and T. J. Prater, *Int. J. Environ. Anal. Chem.* **9,** *93–144 (1981).*
16. M.-L. Yu and R. A. Hites, *Anal. Chem.* **53,** 951–954 (1981).
17. R. A. Hites, R. E. Laflamme, and J. W. Farrington, *Science* **198,** 829–831 (1977).
18. S. Safe and O. Hutzinger, *Mass Spectrometry of Pesticides and Pollutants,* CRC Press, Cleveland, Ohio, 1973, pp. 77–86.
19. S. R. Heller and G. W. A. Milne (eds.), EPA/NIH Mass Spectral Data Base, U.S. Government Printing Office, Washington, D.C., 1978, (Supplement 1, 1980).
20. A. E. Freeman, P. J. Price, R. J. Bryan, R. J. Gordon, R. V. Gilden, G. J. Kelloff, and R. J. Huebner, *Proc. Natl. Acad. Sci.* **68,** 445–449 (1971).
21. J. N. Pitts, K. A. van Cauwenberghe, D. Grosjean, J. P. Schmid, D. R. Fitz, W. L. Belser, Jr., G. B. Knudson, and P. M. Hynds, *Science* **202,** 515–519 (1978).
22. J. Lewtas, *Toxicological Effects of Emission from Diesel Engines,* Elsevier Biomedical, New York, 1982.
23. M. D. Erikson, D. L. Newton, E. D. Pellizzari, K. B. Tomer, and D. Dropkin, *J. Chromatogr. Sci.* **17,** 449–453 (1979).
24. R. C. Pierce and M. Katz, *Environ. Sci. Technol.* **10,** 45–51 (1976).
25. D. McCaustland, D. L. Fischer, K. C. Kolwyck, W. P. Duncan, J. C. Wiley, Jr., C. S. Menon, J. F. Engel, J. K. Selkirk, and P. P. Roller, in *Carcinogenesis—A Comprehensive Survey: Polynuclear Aromatic Hydrocarbons,* Vol. 1, (R. Freudenthal and P. W. Jones, eds.), Raven, New York, 1976.
26. G. Takahashi, K. Kinoshita, K. Hasimoto, and K. Yusuhira, *Cancer Res.* **39,** 1814–1818 (1979).
27. D. L. Newton, M. D. Erickson, K. B. Tomer, E. D. Pellizzari, P. Gentry, and R. B. Zweidinger, *Environ. Sci. Technol.* **16,** 206–213 (1982).
28. F. W. Karasek, J. A. Sweetman, R. E. Clement, F. P. Vance, and H.-Y. Tong, "Chemical characterisation of organic compounds associated with diesel engine particulate emissions," Report prepared for Ontario Ministry of the Environment, Air Resources Branch, 1982.
29. M. Katz, C. Chan, H. Tosine, and T. Sakuma, in *Polynuclear Aromatic Hydrocarbons* (P. W. Jones and P. Leber, eds.), Ann Arbor Science, Ann Arbor, MI, 1979, pp. 171–189.
30. I. Schmeltz, J. Tosk, G. Jacobs, and D. Hoffman, *Anal Chem.* **49,** 1924–1929 (1977).
31. W. Jennings, *Gas Chromatography with Glass Capillary Columns,* 2nd. edition, Academic, New York, 1980.

32. M. L. Lee and B. W. Wright, *J. Chromatogr. Sci.* **18,** 345–358 (1980).
33. M. L. Lee, D. L. Vassilaros, C. M. White, and M. Novotny, *Anal. Chem.* **51,** 768–773 (1979).
34. R. A. Hites and G. R. Dubay, in *Carcinogenesis—A Comprehensive Survey: Polynuclear Aromatic Hydrocarbons,* Vol. 3, (P. W. Jones and R. Freudenthal, eds.), Raven, New York, 1978.
35. D. F. Hunt, C. N. McEwen, and T. M. Harvey, *Anal. Chem.* **47,** 1730–1734 (1975).
36. H. Budzikiewicz, *Angew. Chem. Int. Ed. Engl.* **20,** 624–637 (1981).
37. D. F. Hunt, G. C. Stafford, Jr., F. W. Crow, and J. W. Russell, *Anal. Chem.* **48,** 2098–2105 (1976).
38. S. E. Scheppele, P. L. Grizzle, G. J. Greenwood, T. D. Marriot, and N. B. Perreira, *Anal. Chem.* **48,** 2105–2113 (1976).
39. A. M. Hogg and J. D. Payzant, *Int. J. Mass Spectrom. Ion. Phys.* **27,** 291–303 (1978).
40. R. G. Cooks and G. L. Glish, *Chem. Engr. News* **59,** Nov. 30, 1981, 40–52.
41. B. Shushan, S. H. Safe, and R. K. Boyd, *Anal. Chem.* **51,** 156–158 (1979).

11

# POLYCHLORINATED BIPHENYLS

## G. R. B. WEBSTER and D. A. BIRKHOLZ

### 1. MASS SPECTRA OF PURE COMPOUNDS

#### 1.1. Electron Impact Mass Spectrometry

Electron impact mass spectra of polychlorinated biphenyls (PCBs) are characterized by several features:[1]

1. Most PCBs show relatively intense $[M]^{+\cdot}$ and $[M - 70]^{+\cdot}$ ionic species.
2. The characteristic isotopic distribution of atomic chlorine (i.e., 75.8% $^{35}Cl$ and 24.2% $^{37}Cl$) permits easy recognition of chlorinated species. Molecular and fragment ions exhibit isotopic distribution patterns as shown in Fig. 1.
3. Doubly charged ions are observed.
4. Cl scrambling is a feature of many PCBs; for example, the decomposing molecular ions of all three monochlorobiphenyls are of similar internal energies.
5. Fragmentation of PCBs proceeds usually via successive expulsion of Cl atoms; the odd-electron species are favored (i.e., $[M]^{+\cdot}$, $[M - Cl_2]^{+\cdot}$, $[M - Cl_4]^{+\cdot}$, etc., are more abundant than $[M - Cl]^+$, $[M - Cl_3]^+$, $[M - Cl_5]^+$, etc.).
6. Loss of $Cl_2$ from odd-electron ions is favored. For less chlorinated isomers, the loss of HCl is also seen.
7. Loss of Cl (M − 35) is favored in the case of *ortho*-Cl PCBs in which three of the four ortho positions are chlorinated. This [ortho effect] extends to include other asymmetrically substituted ortho-chloro PCBs.

*G. R. B. WEBSTER* • Pesticide Research Laboratory, Department of Soil Science, University of Manitoba, Winnipeg, Manitoba R3T 2N2, Canada. *D. A. BIRKHOLZ* • Environmental Protection Service, Western and Northern Regions, 5320 122 Street, Edmonton, Alberta T6H 3S5, Canada.

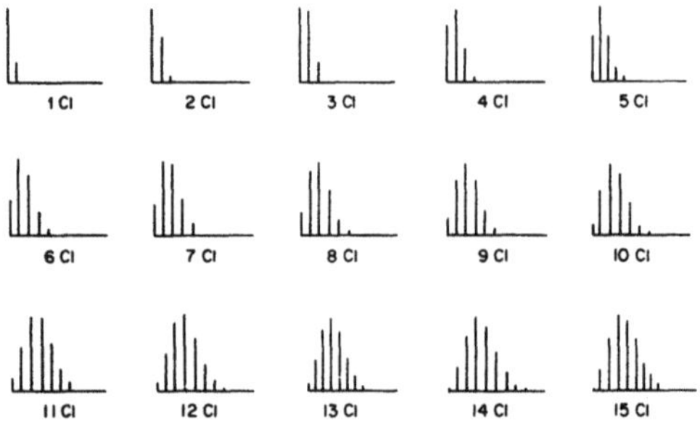


FIGURE 1. Calculated Cl abundance ratios for compounds containing 1 to 15 Cl atoms. (From Ref. 1.)

Many of the mass spectral properties of PCBs and commercial Aroclors are discussed in detail by Hutzinger.[1] Typical EI mass spectra for tetra- and pentachlorobiphenyls are shown in Fig. 2 with the fragmentation mechanisms detailed in Figs. 3 to 6. The fragmentation pathways have been further discussed with evidence drawn from metastable ion and ion kinetic energy data.[1]

The mass spectra of PCB mixtures are distinctive and reflect the data obtained for pure isomers with strong molecular ions and accompanying $[M - Cl_2]^{+\cdot}$ present for each component (Fig. 7). Gas chromatography separation plus mass spectral data allows PCBs to be distinguished from each other and capillary GC coupled with mass spectrometry enables virtually all PCB isomers to be characterized in one analysis. Aroclors have also been examined directly by high-resolution mass spec-

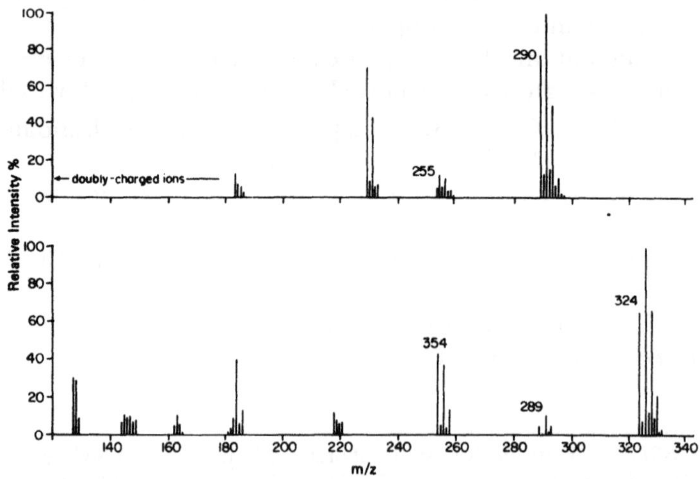


FIGURE 2. Mass spectra of 2,2′,5,5′-tetrachlorobiphenyl and 2,2′,4,5,5′-pentachlorobiphenyl. (From Ref. 1.)

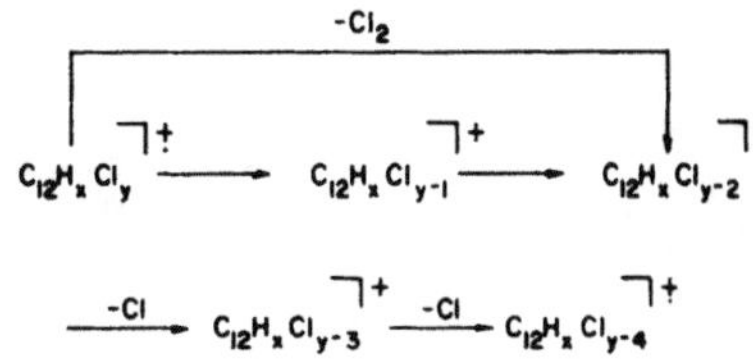


FIGURE 3. General fragmentation pathways for PCBs. (From Ref. 1.)

trometry utilizing the relatively large mass defect of both chlorine isotopes ($^{35}$Cl, −32 mmu; $^{37}$Cl, −34.1 mmu). The exact masses of ions which contain chlorine are usually less than those of the same nominal *m/z* value which contain only isotopes of C, H, N, or O, the elements most commonly present in compounds of biological origin. Differences in GC retention indices are useful in distinguishing between PCBs and other xenobiotics with similar masses and numbers of chlorines.

*1.1.1. The Ortho Effect.* All three monochlorobiphenyls give essentially the same spectrum, and no reliable data on the location of the Cl atom in the ring is available.(2) The dichlorobiphenyls also generate much the same spectra as one another; however, the 2,2′-dichloro isomer forms a very prominent $[M - 35]^+$ ion.(3) Among tetrachlorobiphenyls, the 2,2′,5,5′-isomer can be uniquely identified since the $[M - 35]^+$ ion is approximately 10 times that from other tetrachloro isomers. The 2,2′,6,6′-isomer has a similar $[M - 70]^{+\cdot}$ ion intensity, but much lower $[M - 35]^+$ and $[M - 36]^+$ intensities than the 2,2′,5,5′-isomer.(2) Trichlorobiphenyls exhibit much the same mass spectra as one another except when ortho chlorine atoms are involved; e.g., 2,2′,3- and 2,4′,5-trichlorobiphenyl have markedly differing $[M]^{+\cdot}$, $[M - 35]^+$, and $[M - 70]^{+\cdot}$ intensities (Figs. 8 and 9).

Oswald explained the observed "ortho effect" in terms of an ortho–ortho interaction of chlorine atoms on adjacent rings, rather than between ortho chlorines on the same ring.(4) They further state that *o,o′*-chlorinated biphenyls which have an asymmetric distribution of chlorine atoms are noted for very abundant $[M - Cl]^+$ or $[M - 35]^+$ ions. In

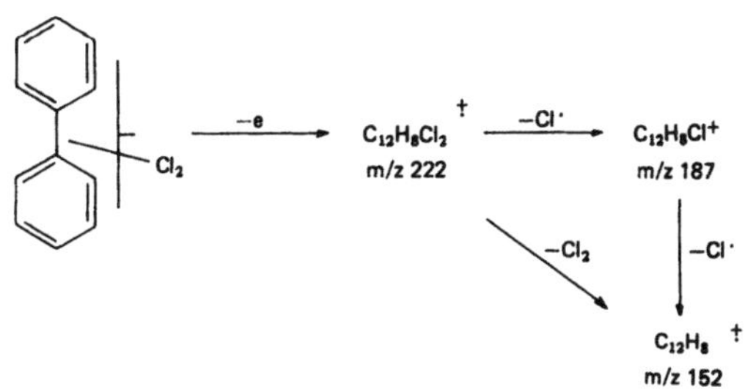


FIGURE 4. Fragmentation pathways for dichlorobiphenyls. (From Ref. 1.)

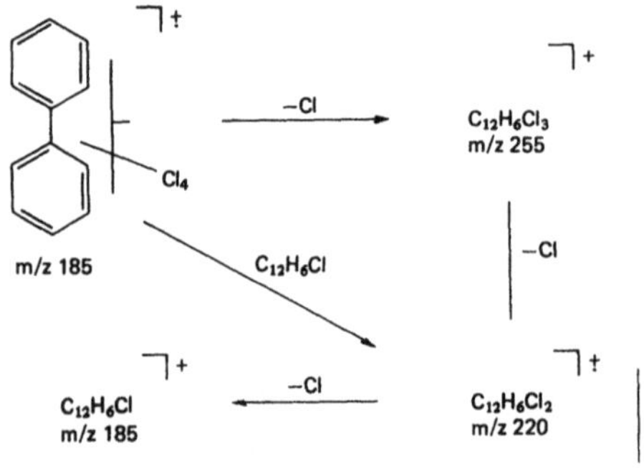


FIGURE 5. Fragmentation pathways for tetrachlorobiphenyls. (From Ref. 1.)

addition, the presence of the $[M - Cl]^+$ ion is a characteristic for numerous PCBs associated not only with *o,o'*-chlorine substitution but also with the number of chlorine atoms present in the ring.

Levy and Oswald studied the "ortho effect" further and in particular examined the mass spectra of biphenyls with three ortho chlorine substituents. In contrast to the virtual absence of $[M - Cl]^+$ fragments from isomers containing zero, one, two, or four ortho chlorines, isomers with three such substituents generated $[M - Cl]^+$ ions up to 60% of the base peak (Table 1).[5]

The unusually high intensity of the $[M - 35]^+$ ion and its almost unique appearance associated with the presence of three ortho chlorines can be explained by considering the steric interaction between the ortho chlorines in the two rings of the molecular ion relative to the daughter ion. In those PCBs with zero or one ortho chlorine, there is no substantial internuclear interaction; thus, the $[M - Cl]^+$ ion is not favored energetically and loss of the ortho chlorine does not generate a more favored species. Where there are two ortho chlorines, the interaction is again low since Cl–Cl interaction of chlorines on adjacent rings prevents this racemate from being preferred. As a result, an insignificant $[M - Cl]^+$ fragment intensity is observed. In the special case of 2,4,5,2',4',5'-hex-

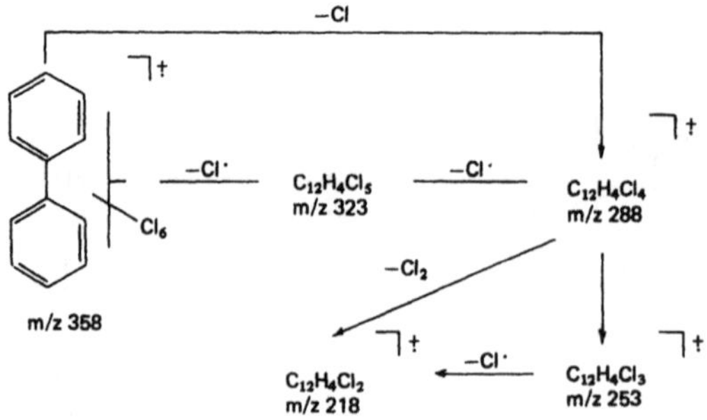


FIGURE 6. Fragmentation pathways for hexachlorobiphenyls. (From Ref. 1.)

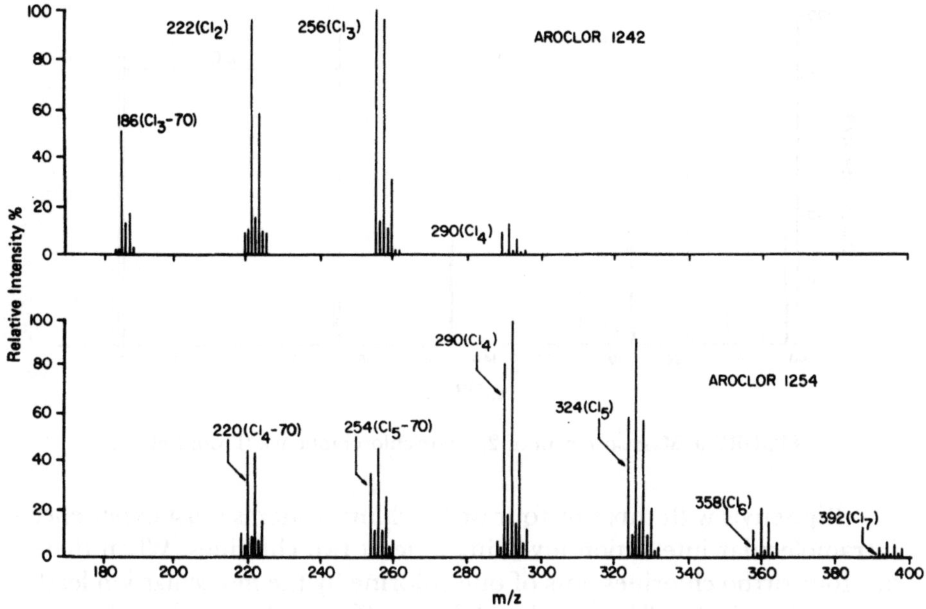


FIGURE 7. Mass spectra of Aroclor 1242 and Aroclor 1254. (From Ref. 1.)

achlorobiphenyl, an isomer with two ortho chlorines, the $[M - 35]^+$ ion has an intensity of 11% of the base peak (Table 1). Levy explains this observation by noting the effect of other chlorine constituents in the ring.[5] He also notes the case of the 2,3,5,2′,3′,5′-hexachloro isomer which shows an $[M - Cl]^+$ peak of 35% of the base peak.

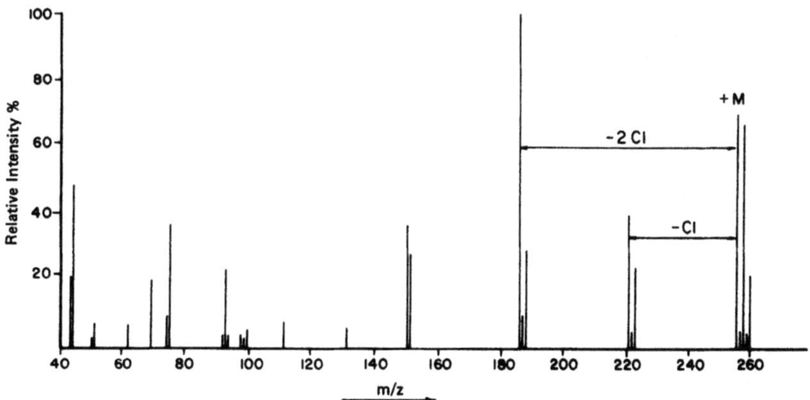


FIGURE 8. Mass spectrum of 2,2′,3-trichlorobiphenyl. (From Ref. 2.)

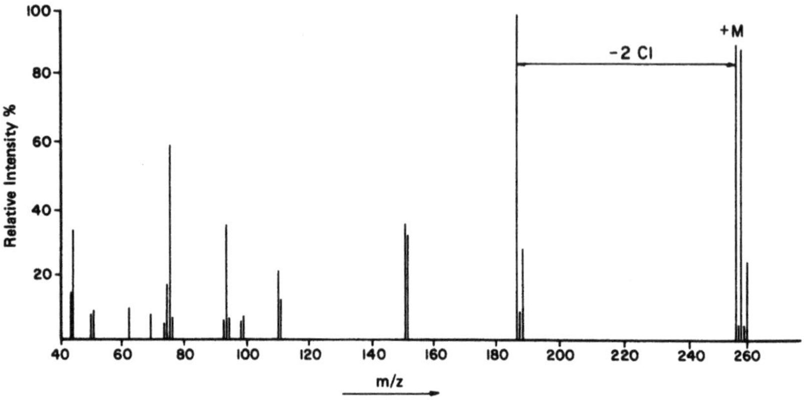


FIGURE 9. Mass spectrum of 2,4′,5-trichlorobiphenyl. (From Ref. 2.)

Biphenyls with three or four ortho chlorine atoms must experience intramolecular interaction involving at least two chlorines. When there are four ortho chlorines, loss of one chlorine by the molecular ion leads to a species which still has the destabilizing effect of three ortho chlorines, that is, a structure no more favored than the parent ion. In contrast, loss of one of three ortho chlorines yields a species with two ortho chlor-

*TABLE 1*
*Mass Spectral Data for Chlorinated Biphenyls*[a]

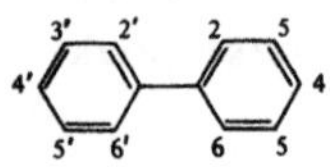


| | Chlorinated biphenyl | Percent relative intensity | | |
|---|---|---|---|---|
| | | $[M]^{+\cdot}$ | $[M-Cl]^{+}$ | $[M-Cl_2]^{\cdot}$ |
| 1 | 2,6,2′- | 60 | 50 | 100 |
| 2 | 2,3,4,5,6,2′- | 80 | 40 | 85 |
| 3 | 2,3,4,5,6,3′- | 80 | 5 | 85 |
| 4 | 2,3,4,5,6,4′- | 100 | 2 | 70 |
| 5 | 2,3,4,5,6,3′,4′- | 60 | 6 | 60 |
| 6 | 2,3,6,2′,3′,6′- | 29 | 7 | 76 |
| 7 | 2,4,5,2′,4′,5′- | 39 | 11 | 88 |
| 8 | 3,4,5,3′,4′,5′- | 35 | 2 | 67 |
| 9 | 2,4,6,2′,4′,6′- | 41 | 2 | 72 |
| 10 | 2,3,5,2′,3′,5′- | 100 | 35 | 40 |
| 11 | 2,3,4,5,6,2′,4′-methyl | 75 | 65 | 25 |
| 12 | 2,3,4,5,6,2′-methyl | 50 | <5 | 80 |
| 13 | 2,3,4,5,6,4′-methyl | 100 | <4 | 60 |

[a] Reference 5.

ines which involves only chlorine–hydrogen interaction, a situation which is greatly favored. Thus, biphenyls with three ortho chlorines, as expected, generate relatively intense $[M - 35]^+$ ions. The simplest biphenyls containing three ortho chlorines, 2,6,2′-trichlorobiphenyl, has an $[M - 35]^+$ ion with 50% of the intensity of the base peak.

Data from metastable peaks and ion kinetic energy spectra support the conclusions above and invoke the participation of a five-membered chloronium ion intermediate.[5,5a]

*1.1.2. Isotopic Abundance Ratios.* Because of the characteristic isotopic distribution of atomic chlorines, chlorinated species are relatively easy to recognize by mass spectrometric techniques. Using these isotopic abundance data for PCBs, Rote and Morris calculated the theoretical probability of the occurrence of ions of different masses in the molecular clusters for PCBs[6] (Table 2). This information, though very useful in itself, does not include the mass spectral data for fragments of higher molecular weight chlorinated species which may interfere with the molecular ion clusters of the lower chlorinated PCBs. Hence, the use of GC/MS is mandatory to enable MS to be used to greatest advantage when analyzing for individual PCBs.

### 1.2. *Positive Chemical Ionization Mass Spectrometry*

Positive chemical ionization (PCI) mass spectrometry of PCBs offers little if any advantage over electron impact ionization (EI) or negative chemical ionization (NCI) mass spectrometry. Very little structural information can be gained using PCI compared to EI. In comparing the sensitivity of PCI with NCI, Pellizzari et al.[7] found PCI to be two orders of magnitude less sensitive. In order to obtain a reasonable spectrum in the pulsed positive CI mode, approximately 10–100 ng of pure chlorobiphenyl isomer is required. This amount of material is sufficient to completely saturate the instrument in the NCI mode.

### 1.3. *Negative Chemical Ionization Mass Spectrometry*

In screening environmental samples for PCBs, it is important that one choose a method which provides not only sensitive detection, but also structural verification of the entity being measured. Mass spectrometry is thus the method of choice; however, conventional mass spectrometry suffers from less than ideal sensitivity for PCB analysis in environmental and biological samples. The limits of detection of EI mass spectrometry and electron capture GC differ by as much as two to three orders of magnitude.[7]

TABLE 2
*Molecular Ion Clusters for PCBs*[a]

| m/z | Intensity | m/z | Intensity | m/z | Intensity |
|---|---|---|---|---|---|
| mono[a] | | hexa | | nona | |
| 188 | 100.00 | 358 | 51.12 | 460 | 25.89 |
| 189 | 13.53 | 359 | 6.83 | 461 | 3.48 |
| 190 | 33.86 | 360 | 100.00 | 462 | 75.55 |
| 191 | 4.34 | 361 | 13.32 | 463 | 10.01 |
| di | | 362 | 80.95 | 464 | 100.00 |
| 222 | 100.00 | 363 | 10.35 | 465 | 13.27 |
| 223 | 13.28 | 364 | 35.61 | 466 | 72.07 |
| 224 | 64.84 | 365 | 4.65 | 467 | 9.69 |
| 225 | 8.59 | 366 | 8.71 | 468 | 35.93 |
| 226 | 10.54 | 367 | 1.21 | 469 | 4.95 |
| 227 | 1.95 | 368 | 1.17 | 470 | 11.59 |
| tri | | 369 | 0.15 | 471 | 1.42 |
| 256 | 100.00 | hepta | | 472 | 2.43 |
| 257 | 13.45 | 392 | 43.78 | 473 | 0.25 |
| 258 | 92.61 | 393 | 5.90 | 474 | 0.33 |
| 259 | 12.13 | 394 | 100.00 | deca | |
| 260 | 29.81 | 395 | 13.02 | 494 | 25.59 |
| 261 | 3.43 | 396 | 96.09 | 495 | 3.34 |
| 262 | 3.16 | 397 | 13.08 | 496 | 78.55 |
| 263 | 0.26 | 398 | 53.02 | 497 | 9.84 |
| tetra | | 399 | 6.51 | 498 | 100.00 |
| 290 | 77.14 | 400 | 16.51 | 499 | 14.28 |
| 291 | 10.44 | 401 | 2.12 | 500 | 92.42 |
| 292 | 100.00 | 402 | 3.18 | 501 | 11.64 |
| 293 | 13.67 | 403 | 0.45 | 502 | 51.69 |
| 294 | 48.84 | 404 | 0.30 | 503 | 6.86 |
| 295 | 6.82 | octa | | 504 | 20.28 |
| 296 | 10.84 | 426 | 34.36 | 505 | 2.71 |
| 297 | 1.50 | 427 | 4.56 | 506 | 5.42 |
| 298 | 0.96 | 428 | 88.10 | 507 | 0.75 |
| penta | | 429 | 11.53 | 508 | 1.04 |
| 324 | 61.84 | 430 | 100.00 | 509 | 0.14 |
| 325 | 7.85 | 431 | 13.43 | | |
| 326 | 100.00 | 432 | 65.33 | | |
| 327 | 12.96 | 433 | 8.73 | | |
| 328 | 66.33 | 434 | 26.11 | | |
| 329 | 8.72 | 435 | 3.39 | | |
| 330 | 21.94 | 436 | 6.99 | | |
| 331 | 2.99 | 437 | 0.87 | | |
| 332 | 3.86 | 438 | 1.16 | | |
| 333 | 0.37 | 439 | 0.11 | | |
| 334 | 0.24 | 440 | 0.11 | | |

[a] Reference 6.
[b] 2-mono; 2,2′di; 2,4,4′-tri; 2,5,2′,5-tetra; 2,4,5,2′,5′-penta; 2,4,5,2′,4′,5′-hexa; 2,3,4,6,2′,3′,5′-hepta; 2,3,4,6,2′,3′,5′,6′-octa; 2,3,4,5,6,2′,3′,4′,5-nona; decachlorobiphenyl.

Negative chemical ionization mass spectrometry is uniquely suited to screening environmental samples for PCBs because of the inherent sensitivity of electron capture detection and the virtual transparency of potentially interfering biomolecules.

Negative chemical ionization mass spectra are obtained by operating a CI mass spectrometer (ion source pressure ~ 1 torr) in the negative-ion mode. Under these conditions, there are five major negative-ion-forming reactions which occur, assuming that the ion source contains oxygen and organochlorine compounds.[8] These reactions are as follows:

| | |
|---|---|
| 1. $M + e_s^- \rightarrow M^{*\cdot -} \rightarrow [M]^{\cdot -}$ | resonance capture |
| 2. $M + Cl^- \rightarrow MCl^-$ | chloride attachment |
| 3. $RH + Cl^- \rightarrow R^- + HCl$ | deprotonation |
| 4. $RCl^{\cdot -} + O_2 \rightarrow RO^- + ClO^\cdot$ | oxygen chloride exchange |
| 5. $MX + e_s^- \rightarrow M^- + X^\cdot$ | dissociative capture |

The major process in NCI is dissociative electron capture.[9,10] Pellizzari studied the effect several moderating and reagent gases had on (i) enhancing the formation of molecular ions of the individual PCBs or (ii) enhancing the fragmentation of these PCBs to yield $^{35}Cl$ and $^{37}Cl$ ions.[7] In addition, a comparison of two ion source designs was made, that is, an LKB 2091 magnetic sector instrument and a Finnigan 4023 quadrupole instrument. The LKB 2091 instrument does not have a pulsed positive–negative capability. Typical negative-ion CI conditions on this instrument utilize an open-source design as opposed to the closed-source design on the Finnigan 4023.

Using a solution containing 2-chlorobiphenyl, 2,3,4-trichlorobiphenyl, 2,2′,3,3′,4,4′,2,2′,3,3′,4,4′-hexachlorobiphenyl, 2,2′,3,3′,4,4′,5,5′-octachlorobiphenyl, and decachlorobiphenyl as a test mixture, Pellizzari investigated a number of moderating gases.[7] These gases include methane; isobutane; air–methane; 40% oxygen, 60% nitrogen–methane; 60% oxygen, 40% nitrogen–methane; 1040 ppm $N_2O$ in nitrogen; tetrafluoromethane; and difluorodichloromethane.

Pellizzarri found that neither difluorodichloromethane nor tetrafluoromethane gave any evidence of reaction products with the polychlorinated biphenyls under either electron capture (open-source) or CI (closed-source) conditions. Furthermore, nitrous oxide in nitrogen yielded NCI spectra qualitatively similar to those obtained with methane.

Analysis of the commercial mixture Aroclor 1260 on an SP 2100 fused-silica capillary column using the pulsed positive–negative-ion

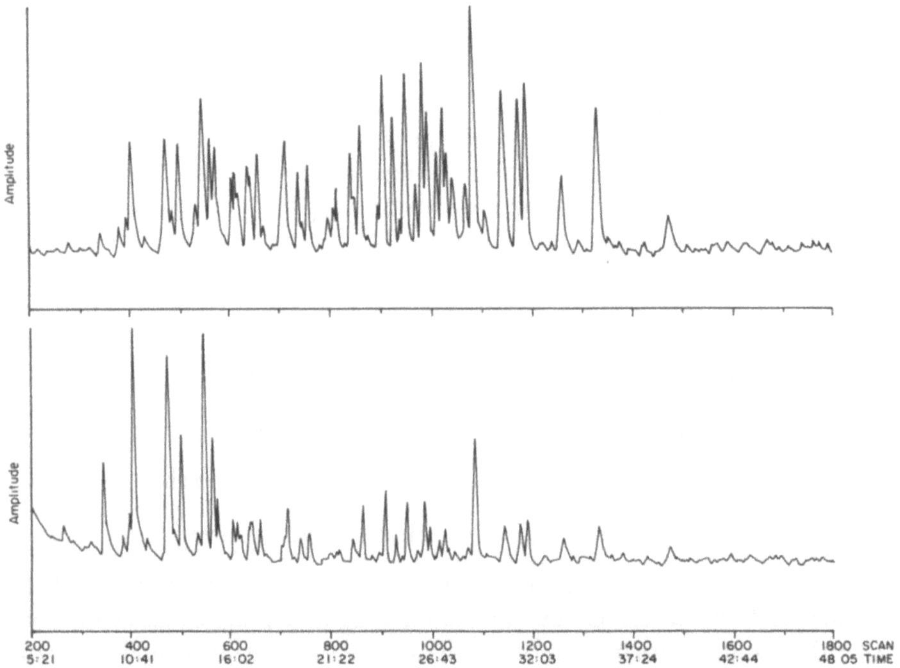


**FIGURE 10.** PPNICI analysis of Aroclor 1260 on an SP 2100 fused-silica capillary column (0.23 mm i.d. × 50 m) using methane reagent gas—upper, pulsed negative; lower, pulsed positive CIMS.[7]

chemical ionization (PPNICI) mode is depicted in Fig. 10; methane was used as the moderating gas. As might be expected, significant differences in sensitivity were observed between pulsed positive (lower tracing) and negative (upper tracing) modes.

Mass spectra obtained for compounds with one to five chlorines showed the predominant ions to be at *m/z* 35 and 37. Molecular ion intensities were low. Beginning with PCB isomers containing six or more chlorines, the molecular ion becomes apparent; for higher degrees of chlorination, the molecular ion becomes the base peak. Oxygen-enhanced methane was used as the reagent gas on both instruments. It was observed that by increasing the oxygen concentration from 40 to 60%, the relative abundance of the molecular ion was substantially increased. The enhancement of the molecular ions for the PCB isomers was more pronounced using the magnetic sector instrument than using the quadrupole instrument. Furthermore, the intensity of the $^{35}Cl^-$ and $^{37}Cl^-$ ions using only methane as the reagent gas was more pronounced in the magnetic sector instrument. Only PCB isomers with 8, 9, and 10 chlorines exhibited molecular ions on the magnetic sector instrument.

In comparing the overall sensitivity of the quadrupole mass spectrometer to the magnetic sector instrument in the NCI mode, it was observed that tetrachlorobiphenyl could be detected in the 1–5-pg range using the magnetic sector instrument; whereas, 100 pg were required for detection using the quadrupole instrument. In both cases the $^{35}Cl^-$ ion was monitored.

In comparing different Aroclors (e.g., 1242 and 1260), 50–100 pg of each Aroclor could be detected using the magnetic sector mass spectrometer and monitoring the $^{35}Cl^-$ ion. Approximately 5–10 ng of each Aroclor was detected using the quadrupole mass spectrometer and monitoring the $^{35}Cl^-$ ion.[7] The more open source design of the magnetic sector mass spectrometer resulted in longer operational time (approximately one month's operation before the source had to be cleaned); whereas, the quadrupole source had to be cleaned every second day.

Tindall and Winiger found that, while interference of hydrocarbons on the PCB parent ion is diminished using the negative-ion mode, the signal for negative parent ions for most PCBs was found to be much weaker than for the positive molecular ions.[11] These workers used methane as the reagent gas and reported that negative PCB ions readily lose $Cl^-$. By monitoring the negative ion at *m/z* 35 or 37, extremely small amounts of PCBs could be detected (less than 1 pg/l) in the presence of hydrocarbons. The authors concluded that, since many other chlorinated compounds also give strong mass 35 and 37 negative ions, and because the mass 35 or 37 ion gives no information as to the identity of the parent molecule, the negative-ion approach is not useful for qualitative work. As indicated earlier, however, for PCBs with higher degrees of chlorination, the molecular ion becomes the base peak, and qualitative work is facilitated.

Pellizzari recognized the limitations inherent in negative-ion mass spectrometry.[7] In developing methodology for the simultaneous detection and measurement of PCBs, PBBs, PCTs, and toxaphene using NCI mass spectrometry, these workers found that utilization of a cleanup procedure using high-pressure liquid chromatography followed by gas chromatography on a 50-m silanized Pyrex glass WCOT capillary column coated with Apiezon L gave good results with very few interferences from extraneous material.

## 2. GAS CHROMATOGRAPHY/MASS SPECTROMETRY/DATA SYSTEM

### 2.1. *Gas Chromatography/Mass Spectrometry/Data System (GC/MS/DS) Analysis*

Analyses based solely on gas chromatographic retention times have an inherent element of ambiguity. The possibility always exists that for any given GC system two compounds will have identical retention times and thus be indistinguishable. In the analysis of chlorinated biphenyls, this possibility becomes highly significant. DDT and its metabolites, various cyclodiene pesticides, chloronaphthalenes, polychlorinated terphenyls, phthalate esters, polychlorodibenzo-*p*-dioxins, and polychlorodibenzofurans have quite similar retention times on the commonly used stationary phases. For example, Shaw found that a pentachlorobiphenyl elutes at the retention time of *p,p'*-DDE and that a hexachlorobiphenyl elutes at the retention time of *p,p'*-DDT on a 1/8-in. × 6-ft aluminum column packed with 8% SE-30 on Gas Chromosorb Q 100/120 mesh.

Rote and Morris noted that the mass spectra of PCBs, polychlorinated terphenyls (PCTs), and chlorinated naphthalenes (CNs), are characterized by a strong parent ion cluster with an isotopic abundance ratio determined by the number of chlorine atom substituents in the rings.(6) Organochlorine insecticides, on the other hand, tend to have weaker parent ions relative to PCBs and can be detected by their characteristic base peaks. From the natural abundance ratios of chlorine and carbon isotopes these workers calculated the theoretical probability of the occurrence of ions of different masses in the molecular cluster for all degrees of chlorination in CNs, PCBs, and PCTs, and compared observed isotopic abundance ratios to theoretical values. Chlorodioxins and chlorofurans can easily be distinguished by their mass spectra, both readily losing a CO + Cl fragment upon ionization. Phthalate esters have a characteristic spectrum with a base peak usually at *m/z* 149 and have no isotopic cluster.

One can take advantage of such information and employ mass monitoring techniques which will not only increase sensitivity but also eliminate the need for elaborate separations. In order to accomplish this, the mass spectrometer must be able to focus rapidly on several ions of interest. This is accomplished by devices which rapidly switch the accelerating voltages between present voltages to focus ions with specific *m/z* values at the detector or by applying voltages corresponding to the *m/z* values of interest.(12) These devices allow the monitoring of many selected ion peaks over the entire mass range of the mass spectrometer. The greater the number of selected ions being monitored, the greater the specificity for the compound being analyzed for, but the lower the sensitivity of the method.

The advent of the computer opened the possibility of program-controlled data acquisition with sampling of specific ions or sequences of ions with varying dwell times on each.

A typical sample analysis on a GC quadrupole mass spectrometer with data system commences with the adjustment of the mass spectrometer current amplifier zero and mass calibration using either perfluorotributylamine (PFTBA or FC-43), perfluoro kerosine (PFK), or decafluorotriphenylphosphine (DFTPP) as a reference compound.[13] The *m/z* values for the major peaks are stored along with their corresponding quadrupole rod voltages in a calibration file in the computer. The *m/z* values in subsequent scans are assigned by means of this calibration file.[14] During a GC/MS/DS run, mass spectral data are acquired continuously, regardless of the emergence of gas chromatographic fractions, and all spectra are stored on a magnetic disc. Over a prespecified duration of data acquisition, a real-time chromatogram based on total ion current or spectral area integration is displayed on a cathode-ray screen. At the termination of a chromatographic run, the mass spectral data may be reduced and printed in the form of tables or recorded graphically by an incremental plotter. Computer data reduction also allows the subtraction of spectral background, thus preventing misinterpretation or confusion of mass spectra with spurious ion interferences contributed by column bleed or other impurities.

### 2.2. *Subset Data Acquisition Techniques*

The application of subset data acquisition techniques to the analyses of PCBs is attractive because of the gain in sensitivity without sacrificing too much of the qualitative information content inherent in a complete mass spectrum. Subset data acquisition techniques should not be confused with the "mass chromatogram" concept which uses conventional data acquisition and displays the change in abundance of selected ions under program control to facilitate interpretation.

In applying subset data acquisition techniques to the analysis of PCBs, the analyst must first choose the ions which are the most valuable for the detection and characterization of the PCBs in question. In deciding upon which ions to monitor, the following criteria should be met; namely, the ions must be relatively unimportant in the mass spectra of other compounds, including chlorinated hydrocarbon pesticides which are likely to be present in environmental extracts, and the ions chosen must provide a means of distinguishing between chlorinated biphenyl compounds with different degrees of chlorination. Eichelberger dem-

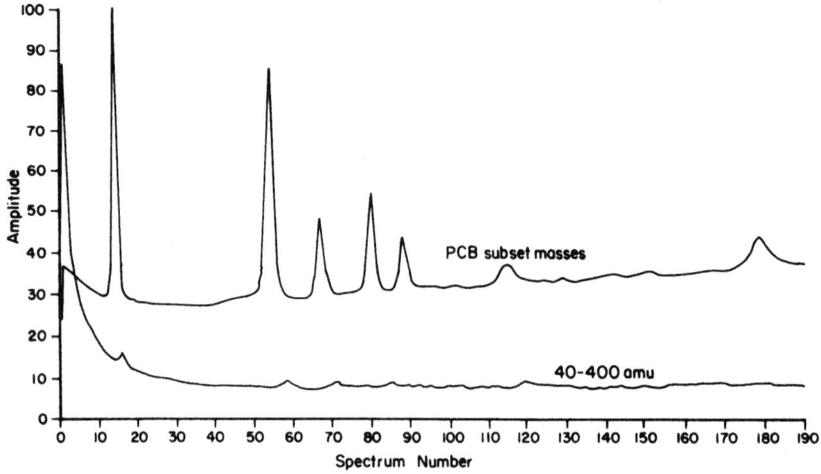


FIGURE 11. Ion abundance chromatograms of a mixture of 5 ng of each of seven pure PCB isomers. Lower chromatogram monitors masses 40–400; upper chromatogram monitors the PCB subset masses.(15)

onstrated that repetitive scans of the subset masses 190, 224, 260, 294, 330, 362, and 394 with an integration time of 540 msec on each was ideally suited for the analysis of PCBs.(15) The gain in signal strength as a result of subset monitoring with time averaging relative to total ion monitoring (40–400 amu) is demonstrated in Fig. 11.

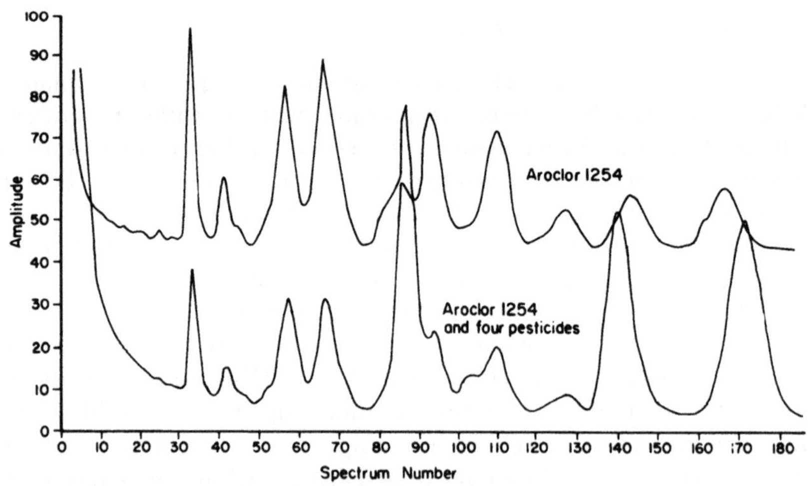


FIGURE 12. Ion abundance chromatograms from repetitive scans of the mass range 40–400 amu. Upper chromatogram monitors a PCB mixture, Aroclor 1254. Lower chromatogram monitors Aroclor 1254 mixed with four chlorinated hydrocarbon pesticides.(16)

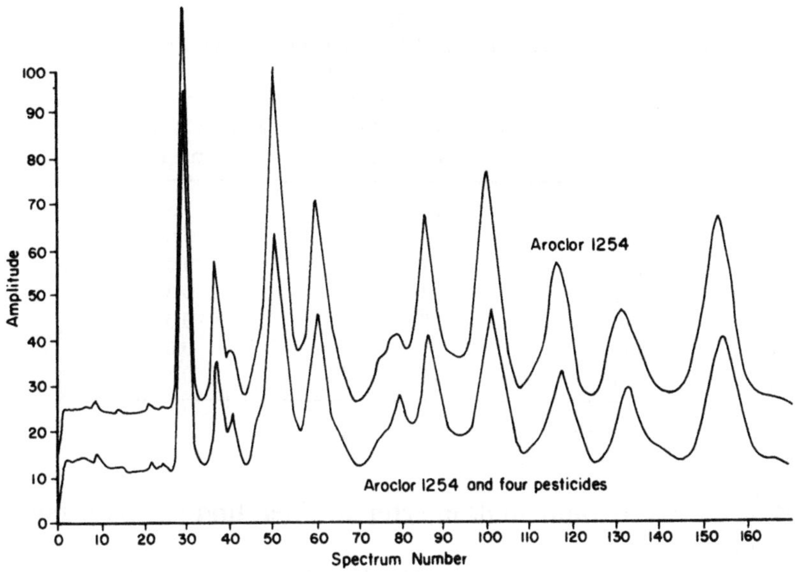


FIGURE 13. Ion abundance chromatograms from repetitive scans of the PCB subset masses. Upper chromatogram monitors Aroclor 1254. Lower chromatogram monitors Aroclor 1254 mixed with four pesticides.[15]

To test the effect structurally similar persistent chlorinated hydrocarbon pesticides would have on the subset data acquisition technique in the analysis for PCBs, four pesticide compounds, namely, DDE, DDD, DDT, and dieldrin, were added to a commercial mixture of PCBs, namely, Aroclor 1254. Figure 12 shows the ion abundance chromatograms from repetitive scans of the mass range 40–400 amu for the commercial PCB mixture and the PCB mixture plus added pesticides. The chlorinated hydrocarbon pesticides coelute with the PCBs, thus creating considerable analytical difficulties for the analyst who must rely solely on characteristic peak patterns to identify PCB mixtures with conventional electron capture GC.

The same two mixtures were rerun using the subset data acquisition technique. The chromatograms are portrayed in Fig. 13. Clearly, there is no significant interference in the PCB chromatograms from the pesticides since the subset masses do not occur to a significant extent in the mass spectra of the latter compounds.

The same authors compared the U.S. National Institute of Health (NIH) file of 8782 mass spectra to the subset ions monitored in order to evaluate the level of interference expected from compounds which are not PCBs. They found that at mass 190, only 1.3% of the compounds in the file interfered with the monochloro PCB monitor mass; at mass

*TABLE 3*
*Mass Spectra in the NIH File Which Contain PCB Subset Ions*[a]

| Mass | Number of non-PCB spectra found | Non-PCB spectra in file (%) |
|---|---|---|
| 190 | 110 | 1.3 |
| 224 | 50 | 0.6 |
| 260 | 37 | 0.4 |
| 294 | 24 | 0.3 |
| 330 | 26 | 0.3 |
| 362 | 3 | 0.03 |
| 394 | 11 | 0.1 |

[a] Reference 15.

394, 0.1% of the compounds interfered. Their findings are summarized in Table 3.

In order to field test this method, a sediment extract was compared to 50 ng of Aroclor 1254 standard. Figure 14 indicates the presence of PCBs in the sediment extract with a profile very similar to the Aroclor standard.

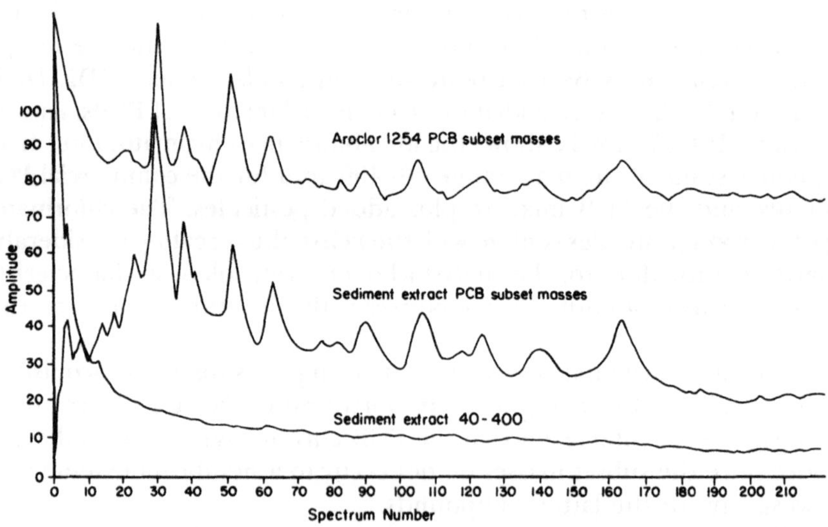


FIGURE 14. Ion abundance chromatograms from repetitive scans of the PCB subset masses and the mass range 4–400 amu. Lower chromatogram monitors 40–400 amu during sediment extract elution. Middle chromatogram monitors PCB subset during sediment extract elution. Top chromatogram monitors PCB subset during elution of a PCB mixture, Aroclor 1254.[15]

Erickson and Pellizzari selected ten ions for detecting and quantitating PCBs in municipal sewage sludge sample extracts.[16] Using the selected ion monitoring mode, ions from the parent cluster for each of the chlorinated biphenyls ($Ph_2H_9Cl$ through $Ph_2Cl_{10}$) were chosen. Ions were chosen from the $[M]^{+\cdot}$, $[M + 2]^{+\cdot}$, or $[M + 4]^{+\cdot}$ values according to an optimum combination of greatest intensity and least interference from other PCBs.

## 3. ANALYSIS OF PCBs BY GLASS CAPILLARY GC/MS

Commercial mixtures of PCBs are complex mixtures of up to 60 or more individual compounds with widely varying toxicological properties. The toxicological evaluations of PCBs have, in recent years, shifted emphasis from assaying commercial mixtures to assaying single compounds and known combinations of individual compounds.

The conventional packed columns used by most laboratories for the analysis of environmental samples for PCBs lack the resolving power necessary to separate the individual components of a complex mixture of commercial PCBs to enable identification. A fish extract might exhibit 25–30 resolved peaks using a conventional packed column, whereas the number of compounds actually present in the extract may well exceed 100.

Glass capillary columns offer greater selectivity (greater resolution) and sensitivity than packed columns. Furthermore, fewer types of column materials are required to effect good separation as opposed to packed columns. Disadvantages such as relative fragility of the columns, greater cost per column, initial cost to modify current instrumentation, and longer analysis times for optimum resolution are more than compensated for by the advantages. The recent introduction of fused-silica columns has all but alleviated the problem of fragility associated with glass capillary columns.

Unlike packed columns, capillary columns are pressure controlled rather than flow controlled. Furthermore, the choice of carrier gas can make a significant difference in resolution. Helium is most often used, but Mullin has reported that hydrogen is the carrier gas of choice.[17]

In comparing stainless steel WCOT capillary columns (45 m) to glass capillary columns (50 and 60 m) coated with Apiezon L and OV-101, Krupcik noted rapid aging with the metal columns resulting in a decreased resolving power and the elution of asymmetric peaks of PCBs.[2] In addition, these workers noted that the metal walls of the capillary column can cause catalytic dechlorination and/or redistribution of the chlorine atoms in the isomeric PCBs.

Krupcik also reported that the characterization of PCB components in a mixture is more precisely done by the comparison of their Kovats' indices with those of standard materials.[2] Kovats' retention indices can be determined for any PCB mixture by adding a mixture of normal paraffins to the PCB mixture in question. Chromatographing this dual mixture then allows one to calculate the retention index of a particular chlorobiphenyl according to the following equation[18]:

$$I = 100z + 100\frac{\log t'_{R(x)} - \log t'_{R(z)}}{\log t'_{R(z+1)} - \log t'_{R(z)}}$$

where $t'_R$ is the adjusted retention time—retention time of component of interest minus retention time of methane; $x$ is the chlorobiphenyl of interest; $z$ is the normal paraffin with $z$ carbon atoms emerging before the substance of interest; and $z + 1$ is the normal paraffin with $z + 1$ carbon atoms emerging after the substance of interest

Krupcik determined the retention indices of the components of Aroclor 1242 using a 50 m × 0.25 mm WCOT glass capillary column coated with Apiezon L.[19] The capillary column was connected to a Finnigan 4000 quadrupole mass spectrometer. Figure 15 depicts the gas chromatogram obtained.

With the aid of the mass spectrometer, the normal paraffins can be readily distinguished from the PCBs because of their unique fragmentation pattern. The retention index associated with peak number 24 was calculated to be 1932.0. The retention index associated with pure 2,3,4'-trichlorobiphenyl was determined to be 1929.1. Since these values are

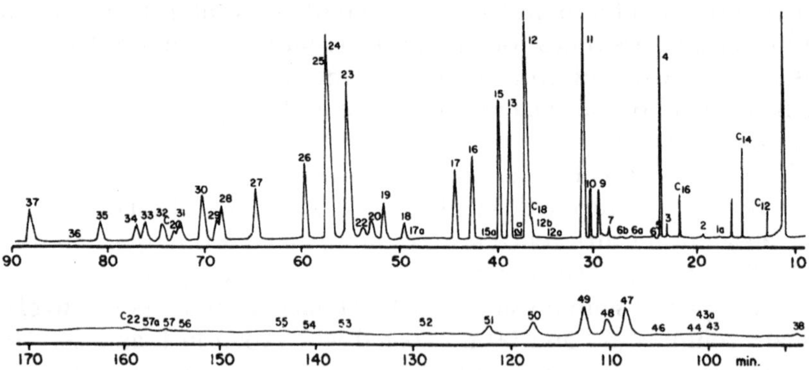


FIGURE 15. Separation of Aroclor 1242 by capillary GC on Apiezon L at 200 °C. $n$-Alkanes as $C_i$, with $i$ = number of carbon atoms. The PCB peaks are numbered from 1 to 57a.[19]

within experimental error, peak number 24 is assigned the structure 2,3,4′-trichlorobiphenyl. For a detailed discussion of the retention index concept and its application, the reader is referred to the papers by Ettre.[18,20,21]

Using a 60-m glass capillary column coated with OV-101, Krupcik ascertained the repeatability of retention index measurements of standard PCBs at three different temperatures.[2] The retention indices measured during a three month period on one column remained virtually unaltered. To be able to take advantage of the published retention indices for the identification of PCBs in mixtures, it was necessary to evaluate the reproducibility of the retention index determination. Kovats' retention indices of some isomeric PCBs were determined on three capillary columns coated with OV-101. The index reproducibility was an order of magnitude worse than their repeatability. These workers speculated that incorrect oven temperature, column material, deactivation method, and chemical composition of the stationary phase are parameters which may be responsible for the errors and may significantly affect the reproducibility of the retention index measurements. By direct coupling of a glass capillary column to a mass spectrometer, these workers observed good mass spectra from nanogram quantities of PCBs. The spectra coupled with the Kovats retention index can be used for the identification of isomeric PCBs.

Pellizzari evaluated wall-coated open tubular columns made of glass (barium carbonate/Carbowax 20M deactivated, silanized Pyrex) and fused silica for optimum resolution of PCBs, PBBs, PCTs, and Halowaxes.[7] Column coatings of C87 Kovats phase, SE-30, SE-52, SE-54, Apiezon L, and Apiezon M were examined with film thicknesses ranging from 0.1 to 0.3 microns.

Capillary columns were evaluated according to the following criteria:

1. Peak asymmetry factor (PAF), using 2,2′,4′,5-tetrachlorobiphenyl:

$$\text{Percent PAF} = B/F \times 100\%$$

where $B$ is the area of the back half of the chromatographic peak at 10% above baseline, and $F$ is the area of the front half of the chromatographic peak at 10% above baseline.

2. Effective height equivalent to a theoretical plate ($HETP_{eff}$), using 2,2′,4′,5-tetrachlorobiphenyl:

$$HETP_{eff} = (L/5.54)(x/y)^2$$

where $x$ is the retention distance of 2,2′,4′,5-tetrachlorobiphenyl minus retention distance of methane, $y$ is the peak width at half height, $L$ is the column length (mm)

3. Separation number (SN), using 2,2′,4′,5-tetrachlorobiphenyl and 2,2′,4,4′,6,6′-hexachlorobiphenyl:

$$\mathrm{SN} = \frac{\mathrm{RT}_B - \mathrm{RT}_A}{W_{1/2_A} + W_{1/2_B}} - 1$$

where $\mathrm{RT}_A$ is the retention of 2,2′,4′,5-tetrachlorobiphenyl; $\mathrm{RT}_B$ is the retention of 2,2′,4,4′,6,6′-hexachlorobiphenyl; $W_{1/2_A}$ is the peak width at half height; and $W_{1/2_B}$ is the peak width at half height.

4. Resolution ($R$), using 2,2′,4′,5-tetrachlorobiphenyl and 2,2′,5,5′-tetrachlorobiphenyl:

$$R = \frac{2(\mathrm{RT}_A - \mathrm{RT}_B)}{1.699(W_{1/2_A} + W_{1/2_B})}$$

where $\mathrm{RT}_A$ is the retention of 2,2′,5,5′-tetrachlorobiphenyl; $\mathrm{RT}_B$ is the retention of 2,2′,4′,5-tetrachlorobiphenyl; $W_{1/2_A}$ is the peak width at half height; and $W_{1/2_B}$ is the peak width at half height.

Film thickness and acid–base characteristics of the columns were assessed using Grob's mixture.[22] These workers found that a 50-m silanized Pyrex glass column coated with Apiezon L gave superior performance relative to the other combinations tried for optimum resolution of PCBs, PBBs, PCTs, and toxaphene. Using this phase, good separations could be achieved using temperatures of 50–75°C lower than other phases, thus allowing for longer column life. Good separation of PCBs could be achieved by temperature programming from 150 to 290 °C at 1 °C/min. The inside diameter of this column was 0.20 mm and the film thickness was 0.1 microns.

Krupcik evaluated glass capillary columns coated with SE-30, OV-101, Carbowax 20M, and Apiezon L for the analysis of mixtures of chlorobiphenyls, alkylbiphenyls, and alkylchlorobiphenyls by GC and GC/MS.[19] Two GC/MS systems were used, a Finnigan 4000 quadrupole mass spectrometer interfaced via a Pt–Ir capillary (60 cm × 0.1 mm i.d.), and an AEI MS-12 single-focusing magnetic sector instrument coupled directly via an all glass interface.

The best separation of an Aroclor 1242 mixture was observed on an Apiezon L glass capillary column. The reproducibility of Kovats' retention indices of PCBs was within only a few units by standardizing the column preparation procedure. This was achieved by etching the glass capillary column with gaseous HCl at precise temperatures using a forced-air circulation oven for fixed periods of time prior to surface coating.[23]

The prediction of Kovats' indices using structural increments was found to be inadequate for all trichlorobiphenyl and tetrachlorobiphenyl derivatives. The largest differences between predicted and measured Kovats' indices were found for compounds having chlorine atoms at para positions with respect to the phenyl–phenyl bond. Differences between predicted and measured Kovats' indices were even larger on Carbowax 20M columns. Retention indices for a number of pure chlorobiphenyls were determined using an Apiezon L glass capillary column (50 m × 0.25 mm i.d.).[19] While interpreting the mass spectra obtained on the Finnigan 4000 quadrupole instrument, it became clear that the Pt–Ir interface was causing catalytic dechlorination and catalytic redistribution of the chlorine substituents. No such effects were observed using the AEI MS-12 with an all glass interface. This effect may be overcome by using fused-silica columns in conjunction with a quadrupole instrument such as the Finnigan 4000. The flexibility of these columns allows the operator to thread these columns right into the ion source, thus eliminating contact with the Pt–Ir interface.

The combination of glass or fused-silica column GC with a mass spectrometer allows the identification of single PCB components in complex mixtures. The mass spectrometer quickly reveals the level of chlorination of the component and a quick determination of its retention index, with appropriate reference to the literature, will reveal the corresponding substitution pattern. This technique provides information to environmental and toxicological scientists in assessing the biohazards associated with specific PCBs.

## 4. *GC/MS OF PCBs IN BIOLOGICAL SAMPLES*

The presence of PCBs in environmental samples was first reported by Jensen.[24] As a consequence of this finding, scientific investigators have identified PCB residues not only in water and sediment samples, but also in birds, fish, animals, human milk, blood, and adipose tissue. In comparing the gas chromatographic patterns of the PCB residues in animals from different trophic levels, Jensen discovered that early eluting components gradually disappeared when going from lower to higher food chain levels, for example, from plankton via fish and birds to birds of prey.[25] Furthermore, biological residue chromatograms have certain changes in their patterns compared to Aroclor standards.

Call compared the residue patterns from extracts of the brains of quail fed Aroclor 1260 to the patterns from Aroclor 1260 itself.[26] Figure

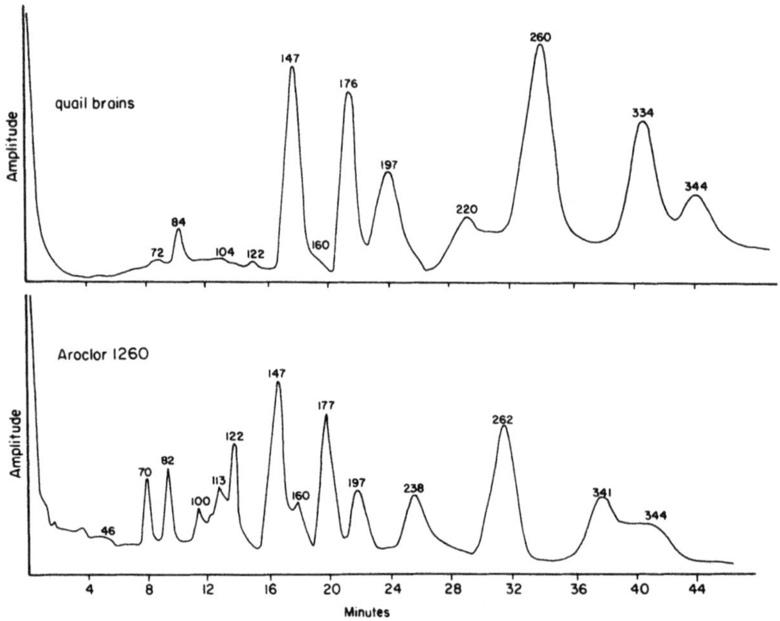


FIGURE 16. Gas chromatograms of extract from brain of Japanese quail fed Aroclor 1260. A. Quail brain extract from quail fed 1000 ppm Aroclor 1260. B. Aroclor 1260 standard. Average retention times (*p,p'*-DDE = 100) are given above the component peaks. Conditions: 1/8 in. o.d. × 5 ft glass column with 1 : 1 mixture of 15% QF-1 and 10% DC-200 and 80–100 mesh Chromosorb W at 200 °C with nitrogen carrier gas flow rate of 40 ml/min.[26]

16 depicts GC/ECD chromatograms of both the standard and the quail brain extract. As is obvious from the chromatogram, some of the Aroclor peaks present in the standard are diminished or nonexistent in the quail brain extract. Furthermore, some peaks are more prominent in the biological extract relative to the standard. In addition, new peaks are present in the biological extract relative to the standard. The relatively larger size of the peaks with longer retention times in the brain tissue indicates that these more highly chlorinated compounds are preferentially stored in the brain lipids over the less chlorinated compounds, or are less readily metabolized.

Greichus compared the residue patterns of double-crested cormorant carcasses (*Phalacrocorax a. auritas*) and eggs to standard Aroclor mixtures 1254 and 1260.[27] Fat obtained from white pelicans (*Pelecanus erythrorhynchos*) was also extracted and the residue patterns obtained were also compared to the commercial PCBs. Comparisons of the PCB patterns of the cormorant carcasses, cormorant eggs, and pelican fat are shown in Fig. 17.

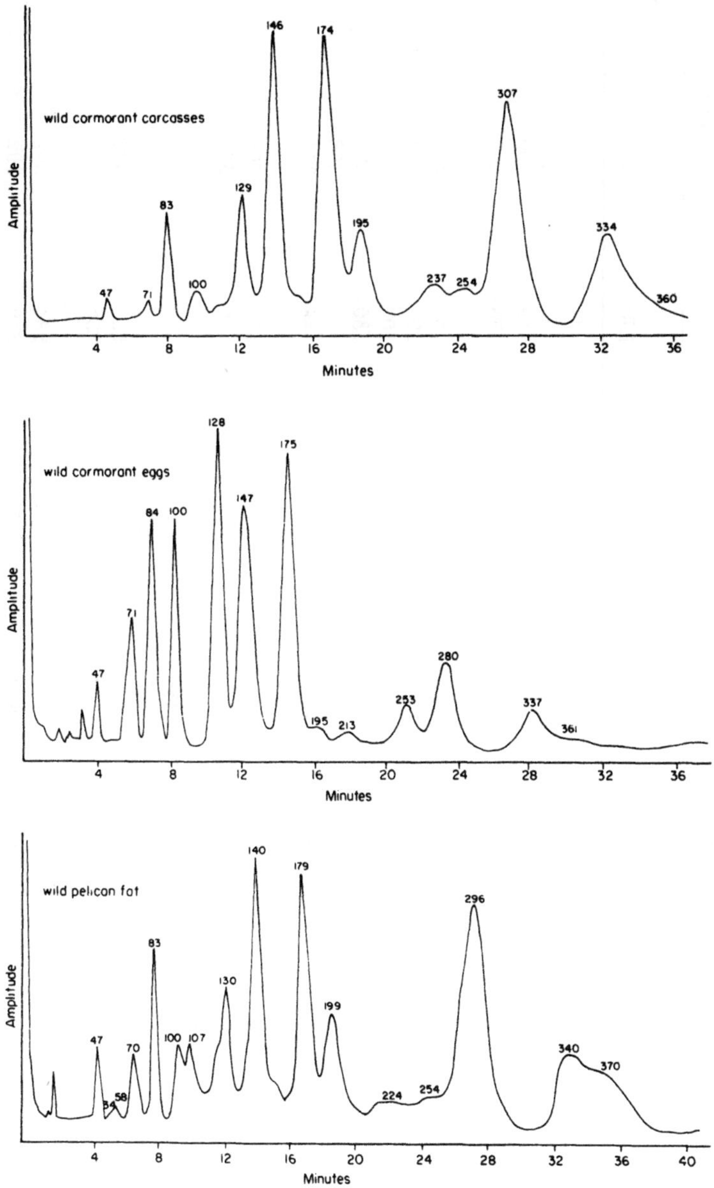


FIGURE 17. Gas chromatograms of extracts from cormorant carcasses, cormorant eggs, and pelican fat. Relative retention times, *p,p'*-DDE = 100. Conditions: 1/8 in. o.d. × 6 ft glass column with 1 : 1 mixture of 15% QF-1 and 10% DC-200 on 80–100 mesh Chromosorb W operated at 200 °C with nitrogen carrier gas flow rate of 40 ml/min.[27]

*TABLE 4*

*Relative Retention Times and Mass Spectra for PCB Residues in Bird Tissues*[a]

| Cormorant carcasses | | | Cormorant eggs | | | Pelican fat | | |
|---|---|---|---|---|---|---|---|---|
| RRt[a] | *m/z* | Number of chlorines | RRt | *m/z* | Number of chlorines | RRt | *m/z* | Number of chlorines |
| 47 | | | 47 | | | 47 | 290 | 4 |
| 53 | | | 53 | | | 53 | 290 | 4 |
| 57 | | | 57 | | | 58 | 290 | 4 |
| 71 | | | 71 | 290 | 4 | 70 | 290 | 4 |
| | | | | 324 | 5 | | 324 | 5 |
| 83 | 324 | 5 | 84 | 324 | 5 | 83 | 324 | 5 |
| 100 | | | 100 | 324 | 5 | 100 | 324 | 5 |
| | | | | | | 107 | 324 | 5 |
| 129 | | | 128 | | | 130 | 324 | 5 |
| | | | | | | | 358 | 6 |
| 146 | 324 | 5 | 147 | 324 | 5 | 148 | 358 | 6 |
| | 358 | 6 | | 358 | 6 | | | |
| 176 | 358 | 6 | 175 | 358 | 6 | 179 | 358 | 6 |
| 195 | 358 | 6 | 195 | | | 199 | 392 | 7 |
| | 392 | 7 | | | | | | |
| | | | 213 | 358 | 6 | | | |
| | | | | 392 | 7 | | | |
| 237 | | | | | | 234 | 392 | 7 |
| 254 | 392 | 7 | 253 | 358 | 6 | 256 | | |
| | | | | 392 | 7 | | | |
| 282 | 392 | 7 | 280 | | | 284 | 392 | 7 |
| 336 | 392 | 7 | 337 | 392 | 7 | 340 | 392 | 7 |
| 368 | 426 | 8 | 361 | | | 370 | 426 | 8 |

[a] Reference 27.

[b] Retention times relative to *p,p'*-DDE = 100 as measured from first appearance of solvent peak.

In comparing the residue pattern obtained from the cormorant carcasses to that obtained from the cormorant eggs, one can see pronounced differences. Referral to the mass spectral data depicted in Table 4 shows that the residue patterns in the eggs suggest preferential storage (or decreased metabolism) of chlorobiphenyls containing five, six, and seven chlorines. The residue pattern of the cormorant carcasses suggests higher levels of chlorobiphenyls containing six and seven chlorines than are observed in the residue patterns from the cormorant eggs.

In comparing the residue pattern obtained from the wild pelican fat to that from the cormorant carcass, one can see a general degree of similarity. One notable difference is the presence of a peak with a relative retention time of 107 (*p,p'*-DDE = 100) in the wild pelican fat residue pattern which is absent in the wild cormorant carcass residue pattern. Referral to the gas chromatogram depicted in Fig. 18 reveals that peak 107 is present in the commercial mixture Aroclor 1254. Greichus obtained a mass spectrum of this peak and found it to be a chlorobiphenyl containing five chlorines.[27] This peak thus represents an isomer (or isomers) of pentachlorobiphenyl which is absent in the residue pattern of the wild cormorant carcass. One other important feature to note is the presence of a peak with a relative retention time of 100 in all three

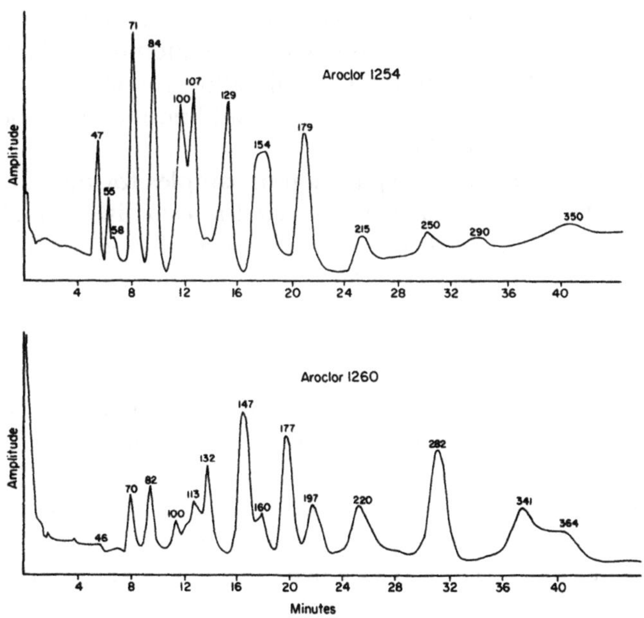


FIGURE 18. Relative retention times in gas chromatograms of Aroclors 1254 and 1260. Conditions as in Fig. 17.[27]

biological extracts, indicating the presence of *p,p'*-DDE This exemplifies the need for mass spectrometric detection of PCBs in order to dispel any ambiguity caused by the coelution of interfering compounds.

Extraction of total-body homogenates from cormorants found dead in the field by ten Noever de Braum indicated the presence of hexachlorobenzene, *p,p'*-DDE, octachlorostyrene, and PCBs.[28] Analysis was carried out using capillary GC/MS and demonstrated the utility of this powerful combination when analyzing complex environmental extracts.

In monitoring field-collected bald eagle specimens (*Haliaetus leucocephalus*) for organochlorine pesticide residues using thin-layer chromatography (TLC) plus GC, Bagley detected a number of chlorine-containing compounds which could not be identified.[29] Analysis by GC/MS demonstrated conclusively that most chemicals in the eagle samples were identical to components of Aroclor 1254. PCBs and the commonly occurring organochlorine pesticides have similar retention times, therefore, prohibiting identification by GC alone. Bagley observed that chlorobiphenyls containing different amounts of chlorine could elute at the same retention time using a conventional GC column.[29] This could easily be discerned with a mass spectrometer. Furthermore, even though some peaks had similar retention times, mass spectral data showed them not to be the same compound, but isomers presumably through differences attributed in subsequent publications to the ortho effect. This work suggests that high-resolution GC coupled with mass spectrometry would be the method of choice for positive confirmation of PCBs in biological extracts. The high-resolution GC would allow the identification of specific isomers by relative retention time in addition to mass spectral characteristics such as the ortho effect.

Methodology for the preparation of samples for multiresidue analysis by GC/MS has been largely restricted to established analytical scale techniques. The unequivocal identification of potentially hazardous chemicals at concentrations below 0.05 μg/g can best be achieved using selected-ion monitoring techniques. Although selected-ion monitoring greatly increases the sensitivity, the technique also increases the uncertainty of identification, particularly in samples containing much coextracted material. Greater specificity can be achieved by working at greater than 10,000 resolution, although this requires the use of a high-resolution mass spectrometer, which is much more expensive than the low-resolution instruments which are more likely to be present in environmental monitoring laboratories. Where the sensitivity of existing GC/MS facilities is optimal, efforts to improve the detection limits have generally centered around scaling-up analytical cleanup techniques. In this connection Veith developed a cleanup technique using Micro Cel-E in conjunction with gel permeation which permits readily interpretable mass spectra to be obtained from chemicals present at ng/g concentrations.[30] Approximately 300 g of fish tissue can be cleaned up by this procedure in a normal work day.

Residues of polychlorinated biphenyls have been reported in the environment by a great number of authors in different parts of the world during the past decade. The findings of these researchers have shown PCBs to be ubiquitous in the environment. This is not surprising since PCBs have seen such varied uses as industrial chemicals. As a result of these findings and as a consequence of the "Yusho" incident in Japan, recent emphasis has been placed on analyzing human blood, adipose tissue, and milk for PCBs in attempts to better understand the impact these chemicals may have on our lives.

Mes used low-temperature precipitation followed by florisil column chromatography to clean up adipose tissue extracts for PCB and organochlorine residues.[31] PCBs were confirmed by pooling samples and applying TLC followed by high-resolution GC/MS. The technique of selected-ion monitoring, using the most intense ion in the mass spectrum of the tetrachlorobiphenyl being monitored, $m/z$ 291.919 ± 0.058, was used.

Yakushiji used a rigorous cleanup technique in analyzing human milk for the presence of PCBs by GC/MS.[32] Fats present in the milk extract were first degraded with 1 *N* alkaline ethanol solution and extracted with *n*-hexane, followed by florisil column cleanup. The *n*-hexane eluent of each sample was mixed and concentrated to about 5 ml and treated with 8% fuming sulfuric acid, followed by silica gel column cleanup. Using a 2% OV-1 column and selected-ion monitoring for $m/z$ 256, 258, 290, and 292, these workers confirmed the presence of trichlorobiphenyls and tetrachlorobiphenyls in human milk. By comparing retention times to available and synthesized chlorobiphenyl isomers, the authors reported the finding of 2,4,4′-trichlorobiphenyl, 2,4,4′,5-tetrachlorobiphenyl, and 2,4,3′,4′-tetrachlorobiphenyl in considerable amounts in human milk. According to these workers, all major residual components have specific structures with two chlorine atoms para to the biphenyl bridge. The 2,4,5-substitution pattern in one ring and the 4-substitution in the other ring are assumed to be one of the most stable structures for chlorinated biphenyls retained by the human body.

Although the production of PCBs has been stopped, there is still a potential for environmental contamination for sediments, transformers, and recirculation of PCBs from commercial products. As part of their overall concern with organic compounds in drinking water, Williams and Benoit analyzed household products which are disposed of via the sewage system.[33] Such materials might contribute PCBs to raw water systems from which potable water suppliers are taken. Following extraction of toilet paper, detergents, fabric softeners, cleansers/disinfectants, and soap with hexane, the hexane solution was concentrated to 2 ml and analyzed by GC. Those samples which showed peaks with retention times similar to PCBs were cleaned up using florisil column chromatography. Confirmation was effected using GC/MS operated in the selected-ion monitoring mode by monitoring for *m/z* 186, 222, 256, and 292. Quantitation was achieved by comparing the total integrated area for six to eight selected peaks from the ion scans to the total area of the corresponding peaks from standard Aroclor 1248. These workers found that most of the cleaning products contained no detectable levels of PCBs except for one soap powder. The toilet paper samples, however, were almost all positive for PCBs and contained levels of PCBs ranging from trace amounts to as high as 21 μg/g.

In conclusion, it can be seen that because of the complexity inherent in biological and environmental samples, they are best analyzed for PCBs by a combination of rigorous cleanup techniques followed by GC/MS using EI or NCI (where the degree of chlorination permits the use of the molecular ion) mass spectrometry and selected-ion monitoring. This technique lends itself readily to the problem inherent with interfering coextractives.

## 5. PCB METABOLITES

The principal metabolic mechanism for PCBs is hydroxylation[34]; in addition, a number of sulfur-containing PCB metabolites have been described.[35] The metabolism of chlorobiphenyls has been reviewed by Sundström; GC/MS analysis using several derivatization methods is discussed.[36]

### 5.1. Hydroxylated PCB Metabolites

Although the presence of a hydroxyl group in a chlorobiphenyl can be demonstrated easily by mass spectrometry, no information is provided on the position of this group in the biphenyl molecule; most isomeric hydroxychlorobiphenyls give similar mass spectra. In contrast, methyl ethers of these compounds give characteristic fragmentations allowing unambiguous identification of these isomers.[37,38] The mass spectra for a complete series of monomethoxy monochlorobiphenyls are shown in Table 5, for monomethoxy dichlorobiphenyls in Table 6, and for dimethoxy dichlorobiphenyls in Table 7.[39,30]

The methyl ethers of hydroxylated chlorobiphenyls provide straightforward structural identification through their highly characteristic fragmentation patterns. The 4-methoxy chlorobiphenyls yield in-

*TABLE 5*

*Mass Spectral Data and GC Retention Times for Monomethoxymonochlorobiphenyls*[a]

| No. | Structure | Mass spectral data[b] (mass fragments $m/z$) $M^{+\cdot}$ | −15 | −35 | −43 | −50 | $R_t$ (I) | $R_t$ (II) |
|---|---|---|---|---|---|---|---|---|
| 1 | 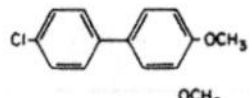 | 100 | 55 | — | 38 | — | 3.6[c] | — |
| 2 | 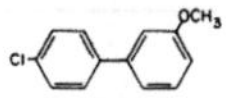 | 100 | — | — | 32 | — | | |
| 3 | 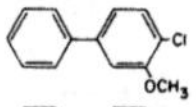 | 100 | — | — | 23 | — | 4.1[d] | 5.4[e] |
| 4 | 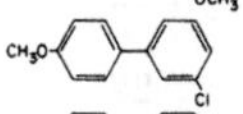 | 100 | 56 | — | 28 | — | 4.5 | 5.6 |
| 5 | 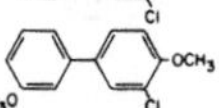 | 100 | 56 | — | 28 | — | 4.5 | 6.4 |
| 6 | 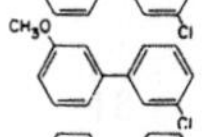 | 100 | — | — | 39 | — | 4.2 | 4.8 |
| 7 | 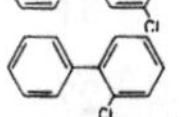 | 100 | — | — | — | 97 | 3.4 | 3.9 |
| 8 | 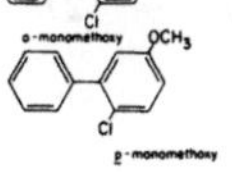 | 100 | 9 | 5 | 49 | 24 | 3.4 | 4.1 |
| 9 | | 100 | 55 | — | 26 | — | 6.4 | 11.8 |
| 10 | 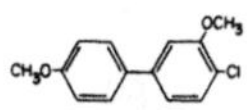 | 100 | 55 | — | 36 | 4 | 8.2 | 14.4 |
| 11 | 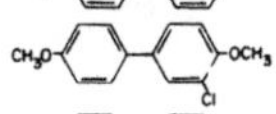 | 100 | 62 | — | 24 | — | 8.7 | 17.4 |
| 12 | 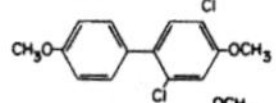 | 100 | 40 | — | — | — | 6.8 | 13.2 |
| 13 | 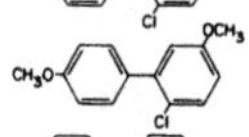 | 100 | 36 | — | 30 | 3 | — | 14.4 |
| 14 | 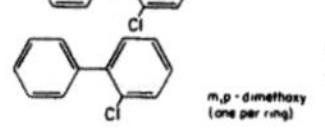 | 100 | 13 | — | — | — | 6.4 | 11.8 |

[a] References 39 and 40.
[b] Figures in percent of base peak; fragments based on $^{35}Cl$ only.
[c] 3 m × 1.5 mm glass column; 3% OV-225 on Gaschrom Q; 200 °C, 2 min, 8 °C/min to 240 °C; 22 ml/min He.
[d] 1.8 × 2mm i.d. glass column; Ultrabonded Carbowax 20m on Gaschrom Q; 160 °C, 8 min, 4 °C/min to 210 °C; 40 ml/min He.
[e] 1.8 m × 4 mm i.d. glass column; 3% OV-1 on Gaschrom Q; 190 °C, 4 °C/min to 210 °C; 40 ml/min He.

*TABLE 6*

*Mass Spectral Data and GC Retention Times for Monomethoxydichlorobiphenyls*[a]

| No. | Structure | $M^{+\cdot}$ | Mass spectral data[b] −15 | −35 | −43 | −50 | $R_t$ |
|---|---|---|---|---|---|---|---|
| 15 | 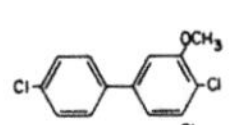 | 100 | — | — | 35 | — | 5.7[c] |
| 16 | 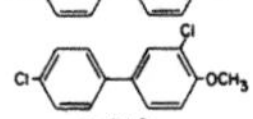 | 100 | 76 | — | 35 | — | 6.1 |
| 17 | 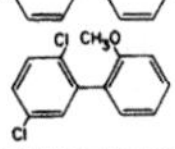 | 73 | — | 24 | — | 100 | 2.7 |

[a] Reference 40.
[b] Figures in percent of base peak; fragments based on $^{35}Cl$ only.
[c] 3 m × 1.5 mm glass column; 3% OV-225 on Gaschrom Q; 200 °C, 2 min, 0 °C/min to 240 °C; 22 ml/min He.

*TABLE 7*
*Mass Spectral Data and GC Retention Times for Dimethoxydichlorobiphenyls*[a]

| No. | Structure | $M^{+\cdot}$ | Mass spectral data[b] −15 | −35 | −43 | −50 | −65 | $R_t$ |
|---|---|---|---|---|---|---|---|---|
| 18 | 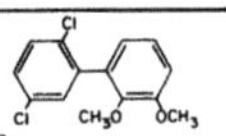 | 78 | 13 | — | — | 100 | 15 | 4.1[c] |
| 19 | 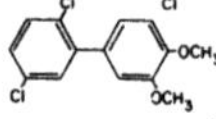 | 100 | 22 | — | 20 | — | — | 6.1 |
| 20 | 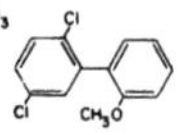 | 100 | — | — | 10 | 51 | 27 | 5.4 |
| 21 | 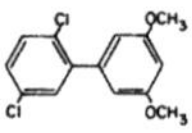 | 100 | — | — | 25 | — | — | 6.0 |
| 22 | 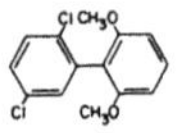 | 58 | — | 13 | — | 100 | 31 | 4.4 |
| 23 | 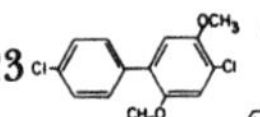 | 75 | 30 | — | — | 100 | 27 | 6.7 |
| 24 | 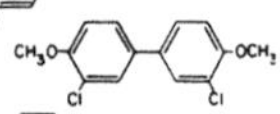 | 78 | 100 | — | 18 | — | — | 9.7 |
| 25 | 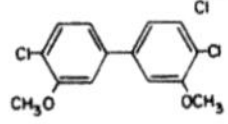 | 100 | — | — | 41 | — | — | 9.4 |
| 26 | 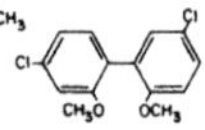 | 100 | — | 33 | — | 89 | 60 | 5.9 |
| 27 | 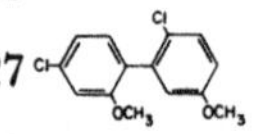 | 100 | 56 | — | 35 | — | — | 8.4 |
| 28 | 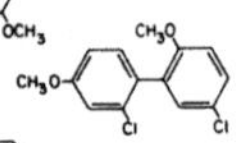 | 100 | 27 | — | 15 | 87 | 35 | 7.6 |
| 29 | 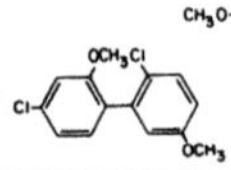 | 100 | — | — | 13 | 62 | 33 | 7.4 |

[a] Reference 40.
[b] Figures in percent of base peak; fragments based on $^{35}$Cl only.
[c] 3 m × 1.5 mm glass column; 3% OV-225 on Gaschrom Q; 200 °C, 2 min, 0 °C/min to 240 °C; 22 ml/min He.

tense $[M - 15]^+$ and $[M - 43]^+$ ions; the 3-methoxy chlorobiphenyls yield only the $[M - 43]^+$ ion; and the 2-methoxy chlorobiphenyls fragment to yield relatively intense $[M - 35]^+$ or $[M - 50]^+$ ions; that is, $[M - Cl]^+$ or $[M - CH_3X]^+$ ions.(40,37) When there are two methoxyl groups in one or both rings, the 2-methoxyl group appears to dominate the 3-methoxyl in determining the fragmentation pattern, the 4-methoxyl dominates the 3-methoxyl, a 2,5-substitution provides an $[M - 15]^+$ and an $[M - 50]^+$, and di-ortho, -meta, or -para substitution reinforces the pattern already described for monomethoxy substitution. This characteristic fragmentation behavior only occurs with methyl ethers of hydroxychlorobiphenyls; ethyl ethers fragment like the parent hydroxy compound as described elsewhere in this section. The same mass spectral behavior has been noted for methoxy bromobiphenyls.(39,40)

The EI mass spectral behavior of methoxyhalobiphenyls has been investigated using a double-focusing VG Organic 7070F mass spectrometer and the effect of the methoxyl group on the fragmentation pathways has been confirmed and extended.(40)

### 5.2. *Hydroxylated Monochlorobiphenyls*

In a study of the avian metabolism of 4-chlorobiphenyl, Jones identified 4′-chloro-4-biphenylol and 4′-chloro-3,4-biphenyldiol as metabolites in chicken excreta.[41] The metabolites separated by TLC were analyzed by direct-probe mass spectrometry. The dihydroxy metabolite (major) had the expected molecular ion at *m/z* 220; the minor metabolite, the monohydroxy compound, exhibited a molecular ion at *m/z* 204. In both cases oxidation occurred preferentially in the unsubstituted ring.

Safe examined the metabolism of 4-chlorobiphenyl in the rat; the major metabolite was 4-hydroxy-4′-chloro biphenyl with the structure being confirmed by GC, mass spectrometry, and NMR.[42] Hass identified a minor monohydroxy metabolite as its trimethylsilyl ether obtaining the expected *m/z* values of 276 and 261 with the appropriate relative abundances for one chlorine.[43] The metabolite was converted to its methyl ether and shown by GC to coelute with an authentic standard of 3-methoxy-4′-chlorobiphenyl on OV-101 at 220°C and OV-225 at 190°C but was clearly resolved from the 2- or 4-methoxy isomers.

The mass spectrum of the trimethylsilyl derivative of the dihydroxy metabolite gave the expected peaks at *m/z* 364 and 366, consistent with one chlorine and two derivatized OH groups. This information, coupled with $^{1}H$ NMR, $^{13}C$ NMR, and IR spectra led to the metabolite being identified as the 3,4-dihydroxy-4′-chlorobiphenyl. Similar work with the monomethoxy hydroxy metabolite as its trimethylsilyl ether yielded the parent ion at *m/z* 306 with intense peaks at 276 and 278 corresponding to loss of $CH_2O$ from the molecular ion. When the metabolite was derivatized with $CH_3I/K_2CO_3$, the resulting compound coeluted with a similar derivative of the dihydroxy metabolite when gas chromatographed on OV-101 (220°C) or OV-225 (190°C) and had the expected molecular ion.

4′-Chloro-4-biphenylol has been shown to give partially methylated metabolites in rats[44] and rabbits.[45] The metabolites in rat urine were prepared for GC/MS in four ways: methylation, ethylation, deuteriomethylation, and the metabolites themselves.[46] Gas chromatography retention times and mass spectral data of the metabolites and their derivatives are given in Table 8.

The mass spectra of the methylated compound showed that Me-1 was methylated starting material; Me-2 was shown to be the dimethyl ether of m-2 by comparison of the alternative structure with the same mass spectral characteristics, I, which had a different retention time and different mass spectral intensities. Me-4 was shown to be methylated m-4 by comparison of the alternate structure with the same mass spectral characteristics, II, which had a different retention time and different mass spectral intensities.

Gas chromatography/mass spectrometry examination of the ethy-

*TABLE 8*

*GC Retention Times and Mass Spectral Data of Metabolites of 4-Chlorobiphenyl*[a]

| Compound[b] | Molecular formula | Retention[c] time (min) | | Mass spectrum[d] | | | | | | | |
|---|---|---|---|---|---|---|---|---|---|---|---|
| | | A | B | *m/z* | $M^{+\cdot}$ | M-15 | M-28 | M-43 | M-56 | M-71 | M-99 |
| Me-1 | $C_{13}H_{11}OCl$ | 7.0 | 3.4 | 218 | 100 | 55 | – | 41 | – | – | – |
| Me-2 | $C_{14}H_{13}O_2Cl$ | 11.0 | 7.2 | 248 | 100 | 31 | – | 24 | – | – | – |
| I | $C_{14}H_{13}O_2Cl$ | – | 8.5 | 248 | 100 | 44 | – | 22 | – | – | – |
| Me-4 | $C_{15}H_{15}O_3Cl$ | 12.8 | 9.7 | 278 | 100 | 78 | – | 14 | – | – | – |
| II | $C_{15}H_{15}O_3Cl$ | – | 12.9 | 278 | 100 | 21 | – | 28 | – | – | – |
| Et-1 | $C_{14}H_{13}OCl$ | 7.9 | 4.0 | 232 | 55 | – | 100 | – | – | – | – |
| Et-2 | $C_{15}H_{15}O_2Cl$ | 10.9 | 7.4 | 262 | 100 | – | 40 | 20 | – | 36 | – |
| Et-3 | $C_{16}H_{17}O_2Cl$ | 11.2 | 7.6 | 276 | 95 | – | 19 | – | 100 | – | – |
| Et-4 | $C_{17}H_{19}O_3Cl$ | 13.0 | 10.2 | 306 | 100 | 16 | – | 22 | 8 | 89 | 31 |
| DM-1 | *e* | 6.8 | – | 221[e] | +[e] | + | – | + | – | – | – |
| DM-2 | *e* | 10.8 | – | 251 | + | + | – | + | – | – | – |
| DM-3 | *e* | 10.8 | – | 254 | + | + | – | + | – | – | – |
| DM-4 | *e* | 11.7 | – | 284 | + | + | – | + | – | – | – |
| m-1 | $C_{12}H_9OCl$ | 12.6 | *f* | 204 | 100 | – | – | – | – | – | – |
| m-3 | $C_{13}H_{11}O_2Cl$ | 10.1 | *f* | 238 | 100 | 39 | – | 37 | – | – | – |

[a] Reference 46.
[b] Me = methylated, Et = ethylated, DM = deuteriomethylated, m = untreated = true metabolite. For structures m-1-m-4 and for structures of compounds I and II, see previous page.
[c] A = Column A: 180 × 0.4 cm 0.2% Carbowax 20M on Chromosorb W 100–120 mesh. B = Column B: 200 × 0.4 cm 3% Apiezon L on Chromosorb W 100–120 mesh. Temperature program: A, 150 °C, 2 min, 8 °C/min to 240 °C; B, 170 °C, 2 min, 4 °C/min to 240 °C.
[d] Calculations based on fragments containing $^{35}$Cl only.
[e] Mixtures of the $-CD_3$, $-CD_2H$, and $-CDH_2$ ethers. "Molecular ions" were recorded at *m/z* values corresponding to each isomer. Intensities could not be calculated.
[f] Could not be eluted.

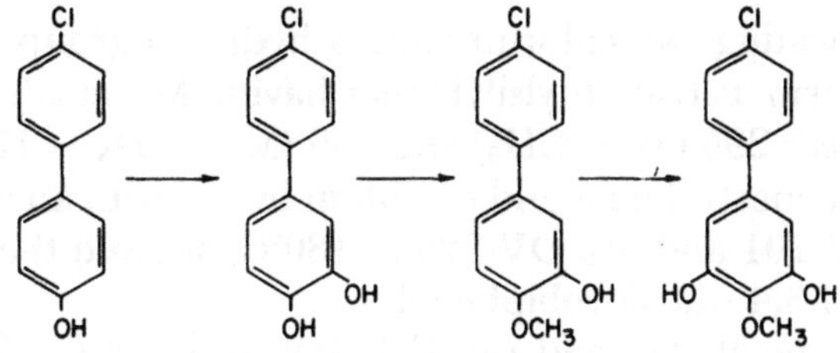


m-1 m-2 m-3 m-4

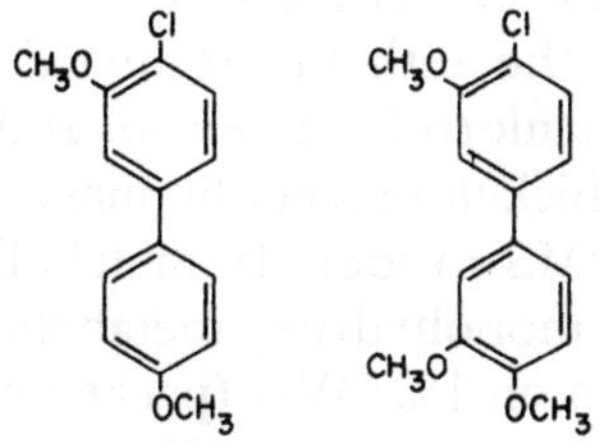


I II

lated products showed two partially methylated metabolites, Et-2 and Et-4. The first, Et-2, had both a methoxy and an ethoxy group and the mass spectrum showed $[M - 43]^+$ and $[M - 71]^+$; that is, $[M - (28 + 15)]^+$ and $[M - (28 + 43)]^+$ indicating that the methoxy group was in the 4-position. Et-4 had two ethoxy groups and one methoxy group. The mass spectrum contained both $[M - 71]^+$ and $[M - 99]^+$; that is, $[M - (56 + 15)]^+$ and $[M - (56 + 43)]^+$ indicating that the methoxy group was again in the 4-position. Et-1 proved to be the ethylated starting material, the ethyl ether of m-1, and Et-3, the diethyl ether of m-2. Compounds with *m/z* corresponding to Me-1, Me-2, and Me-4 were not seen in the GC/MS of the ethylated sample; thus, the methyl ethers of m-1, m-2, and m-3 are not, themselves, metabolites. Triethoxychlorobiphenyl was also not found, showing that m-4 is formed from m-3, since the triol of m-4 would have yielded the triethyl ether.

The deuteromethylated sample indicated similar conclusions. DM-2 with *m/z* 251 had one methoxy and one deuteromethoxy group; DM-4 had one methoxy and two deuteriomethoxy groups.

Attempted GC/MS of the untreated sample showed that derivatization of free OH groups renders these phenolic metabolites much easier to analyze by GC/MS.

## 5.3. *Hydroxylated Dichlorobiphenyls*

Hass examined the metabolites of 4,4′-dichlorobiphenyl in the rat.[43] The major metabolite showed molecular ions at *m/z* 238, 240, and 242

(9 : 6 : 1) indicating two chlorines and a hydroxyl group which was derivatized to form its trimethylsilyl ether having $M^{+\cdot}$ at *m/z* 310 with fragment ions at *m/z* 295 (M − $CH_3$) and 259 (M − $CH_3$ − HCl). This data plus $^1$H NMR spectral data and coelution by GC with an authentic standard (3% OV-101 and 3% OV-225 at 180°C) showed the metabolite to be 4-hydroxyl-3,4′-dichlorobiphenyl.

A minor metabolite had *m/z* 310, 312, and 314 (9 : 6 : 1) as its trimethylsilyl ether; its methyl ether was found to coelute with 3-hydroxy-4,4′-dichlorobiphenyl on 3% OV-101 and 3% OV-225 and to exhibit a mass spectrum like that of the authentic standard.

Jones identified 4,4′-dichloro-3-biphenylol as the only metabolite of 4,4′-dichlorobiphenyl in chicken excreta by mass spectrometry and $^1$H NMR.[41] Kamal used GC/MS to identify metabolites of 2,2′-dichlorobiphenyl in the rat. Three monohydroxy metabolites gave the expected *m/z* 238 after GC separation on 1% OV-1 (packed column) at 150°C and their methyl ethers (diazomethane) at *m/z* 252 (GC on 1% OV-1 packed column at 185°C).[47] Three dihydroxy metabolites could not be gas chromatographed (1% OV-1 packed column at 185°C). Their dimethyl ethers gave *m/z* 282 (2 Cl). Similarly, the single trihydroxy metabolite, when fully methylated, yielded *m/z* 312 (2 Cl).

Tulp and Hutzinger investigated the characterization of hydroxylated chlorobiphenyls by GC/MS of methyl and ethyl ethers.[46] Three monohydroxy isomers of 2,5-dichlorobiphenyl were examined.

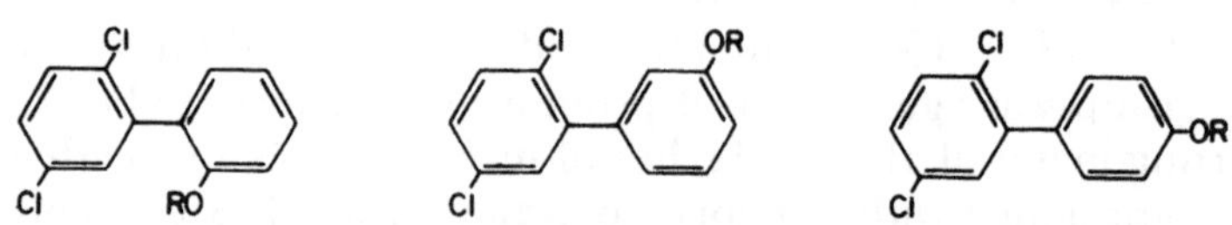


R = H, Me or Et

Ethylated compounds all gave molecular ions 30–80% of the base peak intensities, $[M - 28]^+$, formed by loss of the ethyl group with concomitant hydrogen atom rearrangement to the phenolic analog. The mass spectra of the ethyl ethers are virtually identical with those of the underivatized hydroxy compounds with the addition of the new parent ion in the former 28 mass units higher (Fig. 19). Mass spectra of ethyl ethers of chlorobiphenylols do not allow further structural assignments in contrast to the mass fragmentation of methyl ethers.

In the case of biphenyldiols with one hydroxyl group methylated and one ethylated (e.g., the ethyl ether of a hydroxymethoxybiphenyl) (Fig. 20), the base peak and the spectra are virtually identical except for the $[M]^{+\cdot}$ at 28 mass units higher for the ethyl ether. The ethylation is required to enable GC separation of the metabolite; the mass spectrum is essentially unaltered from that of the methoxy-hydroxy compound, but the position of the methoxyl group is unambiguously shown.

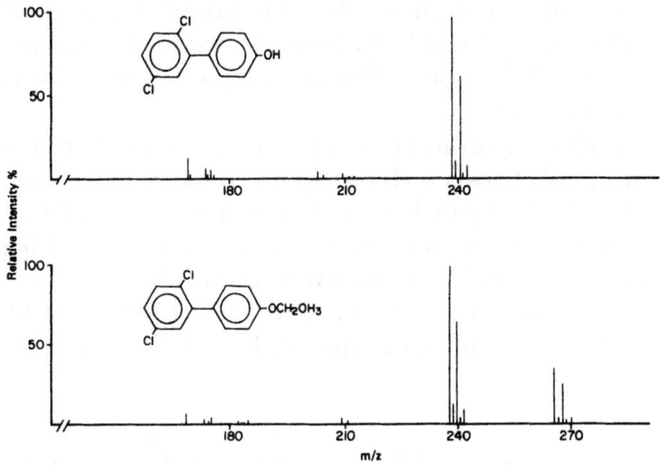


FIGURE 19. Partial mass spectra of 2′,5′-dichloro-4-biphenylol and its ethyl ether.(46)

Deuteromethylation has also been achieved for chlorobiphenylols.(46) However, because the deuterodiazomethane is not isotopically pure, a mixture of parent derivatives is formed (—$OCD_3$, —$OCHD_2$, —$OCH_2D$) giving a mixture of parent ion peaks in the mass spectrum

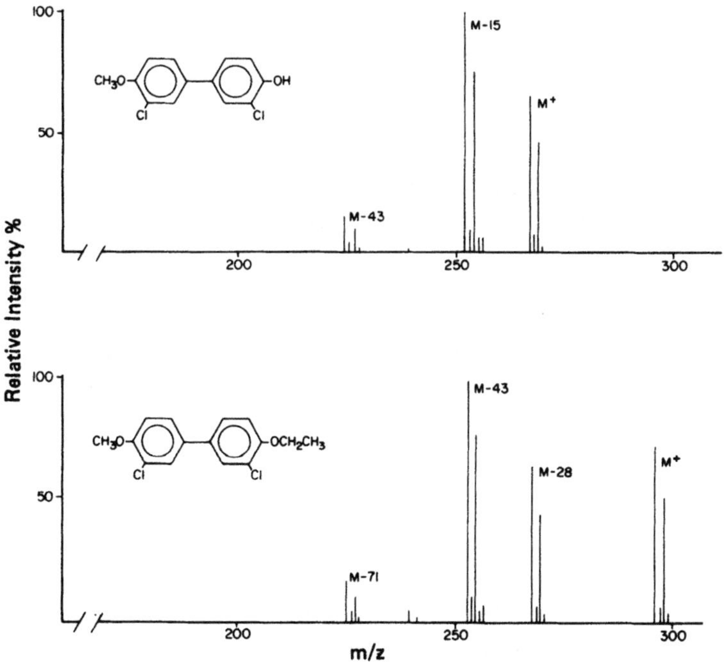


FIGURE 20. Partial mass spectra of 3,3′-dichloro-4′-methoxy-4-biphenylol and its ethyl ether.(46)

which are very difficult or impossible to interpret. Fragmentation yields similar spectra to methyl ethers; however, rapid deuterium exchange quickly alters the $M^{+\cdot}$ cluster after the sample has stood or the solvent has partially evaporated.

Derivatization of ortho diols to form the *n*-butyl boronates can be used to distinguish between chlorobiphenyls which otherwise yield the same spectra.[46] The methyl ethers of compounds A and C, and those of B and D give identical mass spectra; however, only A and B give cyclic *n*-butyl boronates, E and F, respectively, giving sharp total ion monitor peaks on GC/MS with *m/z* 320, 322, and 324.[40] Retention times for E and F are 8.3 and 8.4 min (3% Apiezon L, 170 °C for 2 min, 8 °C/min to 240 °C).

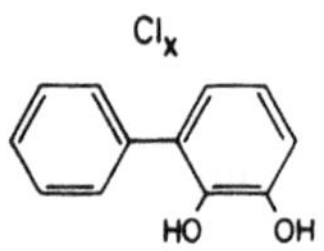


A

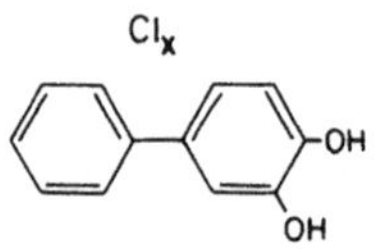


B

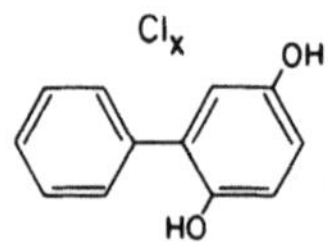


C

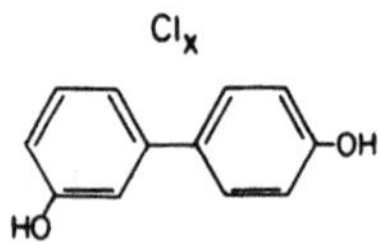


D

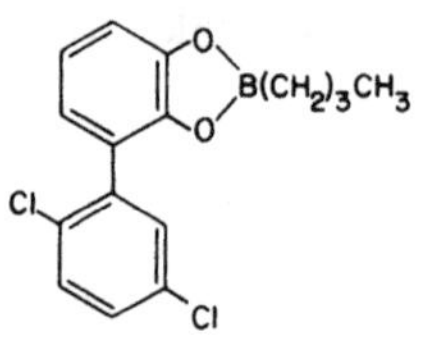


E

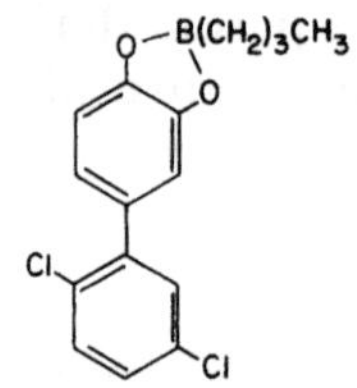


F

In each case the base peak was at M − 56 corresponding to loss of $C_4H_8$ which is characteristic for *n*-butyl boronates.

### 5.4. *Hydroxylated Tetrachlorobiphenyls*

Methyl ethers of tetrachlorobiphenyls were synthesized by Phillipson and Puma.[48] Consistent with earlier work, the 4-methoxy tetrachlorobiphenyl they examined, 3,5,3′,4′-tetrachlorobiphenyl, showed major peaks at M − 15 and M − 43, while the 3-methoxy isomer, 3-methoxy-2,4,3′,4′-tetrachlorobiphenyl, showed only the $[M - 43]^+$ ion with little evidence of the $[M - 15]^+$ ion.

Stalling and Hogan prepared tetramethylsilyl (TMS) derivatives of hydroxylated tetrachlorobiphenyls using *N-O-bis*-trimethylsilyltrifluoroacetamide (BSTFA).[49] Compounds so derivatized had mass spectra with base peaks at $[M - 15]^+$ or 93. The molecular ion was 8–35% of the base peak (Fig. 21). The ratio of $93/[M - 15]^+$ was in-

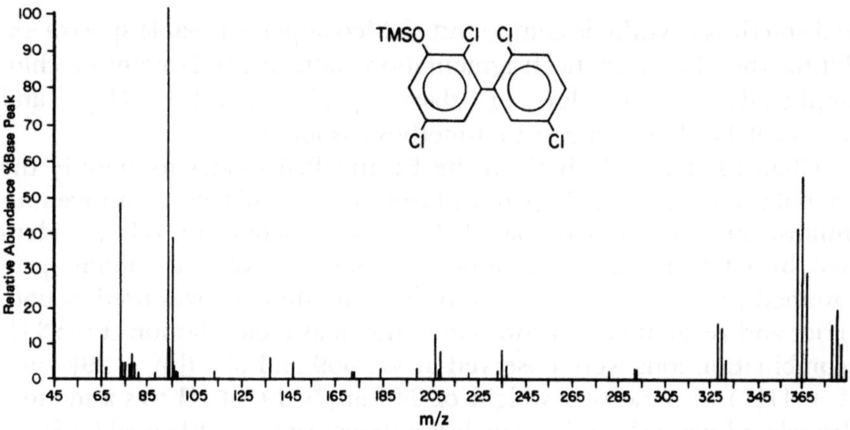


**FIGURE 21.** Mass spectrum of the TMS derivative of 3-hydroxy-2,5,2′,5′-tetrachlorobiphenyl.[49]

dicative of the presence of a chlorine atom ortho to the *O*-TMS group. The ion ratios for *o*-chloro phenols were in the range 0.71–6.3, and for non-ortho–chloro phenols, 0.17–0.29. While all of the non-ortho–chloro phenols and several of the *o*-chlorophenols studied by Stalling and Hogan are not substituted biphenyls, they imply that this observation applies to hydroxylated biphenyls.[49]

### 5.5. *Hydroxylated Pentachlorobiphenyls*

A hydroxylated metabolite 2,4,5,2′,5′-pentachlorobiphenyl in mice was identified by Sundström and Wachtmeister with the assistance of GC/MS of the methyl ether.[38] Three hydroxylated analogs were synthesized and characterized:

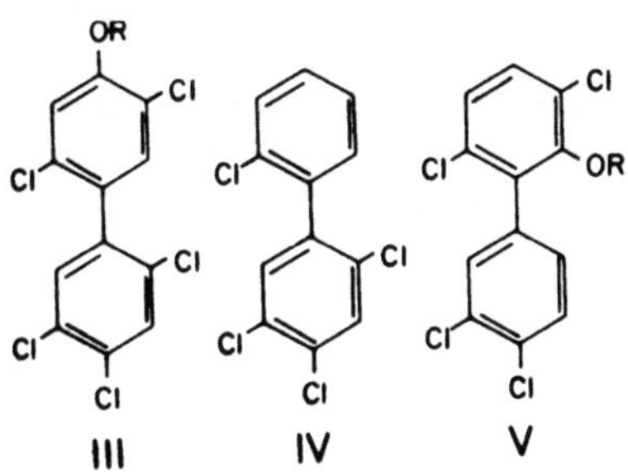


III IV V

R = H or $CH_3$

Comparison of the mass spectra of the methylated metabolite and the 3′-methoxy synthetic compound yielded superimposable spectra exhibiting the characteristic fragmentation pattern for 3-methoxy chlorobiphenyls; namely, low or absent peaks at $[M - CH_3]^+$ and $[M - COCH_3]^+$ in contrast to 4-methoxy isomers.

Chen used GC/MS both in the EI and PCI modes to identify the metabolites of 2,4,5,2′,5′-pentachlorobiphenyl following intravenous administration to the male rat.(50) The major metabolite, III, was identified by GCPCI mass spectrometry using a Dexil 300 column programmed from 160 to 240 °C at 10 °C/min. Methane was used as both carrier and reagent gas; in addition to the quasimolecular ion at *m/z* 341, recombination ions were observed at *m/z* 369 and 381 ($[M + 29]^+$ and $[M + 41]^+$). A molecular weight of 340 implied that III was a monohydroxylated pentachlorobiphenyl; the structure was confirmed by Fourier transform $^1H$ NMR.

The second metabolite, a suspected dihydrodiol, was subjected to acid dehydration, converting 75% of IV to a material which cochromatographed with III. A further metabolite unaffected by acid dehydration was derivatized with diazomethane and examined by direct-insertion EIMS. A molecular ion at *m/z* 384 and a fragment at *m/z* indicating the loss of a methyl group indicated that the metabolite was most likely the dihydroxypentachlorobiphenyl, V.

### 5.6. *Hydroxylated Hexachlorobiphenyls*

Rats fed 2,2′,4,4′,5,5′-hexachlorobiphenyl excreted in their feces, in addition to the unchanged starting material, a hydroxylated compound which was subsequently methylated with diazomethane.(51) Analysis by GC/MS showed a molecular ion at *m/z* 338 (6 Cl) corresponding to a methoxyhexachlorobiphenyl, the structure of which could be either VI or VII, providing no chlorine migration had occurred.

Mass spectra of 2-methoxy chlorobiphenyls show intense $[M - 50]^+$ peaks; those of 3-methoxy chlorobiphenyls, $[M - 43]^+$ only. Although this work of Jensen and Sundström predates the publishing of the above general rules for methoxy chlorobiphenyls, their data is consistent with the observed metabolite being VI, R = H; the methyl ether showed an $M^{+\cdot}$ of 388 (base peak) and an intense fragment at $[M - 43]^+$ (345, 6 Cl.)(51)

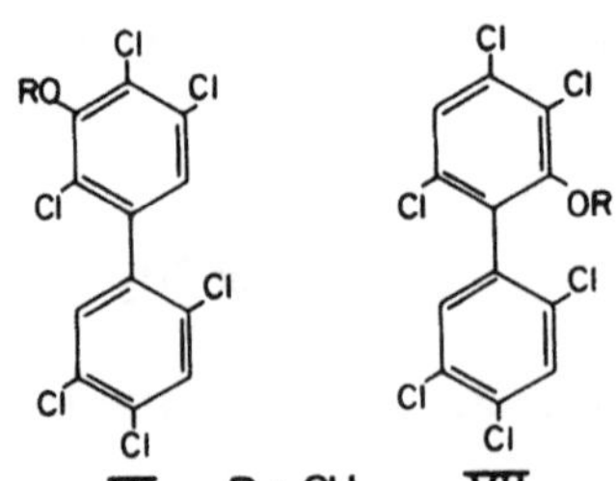


VI R = $CH_3$ VII

### 5.7. Sulfur-Containing Metabolites of PCBs

Methylsulfonyl-PCBs have been identified in environmental samples,[52] and methylthio- and methylsulfonyl-metabolites of PCBs have been reported in laboratory feeding trials.[53] In the mouse, thiol and methylsulfinyl metabolites have been identified in feeding trials with 2,2′,5,5′-tetrachlorobiphenyl.[54] These metabolites are present only in small amounts relative to the major hydroxylated metabolites and GC and GC/MS are required to enable identification.

As in the case of methoxy polychlorobiphenyls, data collected during the identification of sulfur-containing metabolites of PCBs indicate differences in the abundance of individual mass fragments of isomers of both thiol and sulfone metabolites.

### 5.8. Methylthio-PCBs

Bergman has prepared 25 methylthio-PCBs and described their GC properties (capillary column, SE-30, 25 m, 130–190 °C at 1 °C/min) and their EI mass spectral characteristics (Table 9).[55] In general, the molecular ions are the base peaks except for structures 1–7 which give intense peaks at $[M - 50]^+$ (loss of $CH_3Cl$). For compounds 1, 2, 4, and 6 the $[M - 50]^+$ peak is the base peak, whereas, for compounds 3, 5, and 7 the $[M - 35]^+$ (loss of Cl) is the base peak. The intense peak at $[M - 50]^+$ is probably due to a ring closure giving a dibenzothiophene structure in the same manner that dibenzofurans are formed in the mass spectrometer from the corresponding methoxy-PCBs. Compounds 1, 3, 5, and 7 all gave low molecular ion peaks in the same manner as PCBs with 2,2′-substitution. The biphenyls 2, 4, and 6 have no chlorine atom in the 2-position, and more readily lose a methyl group than a chlorine atom to yield $[M - 15]^+$ peaks.

The compounds with the MeS group in the 3-position have the molecular ions as the base peak. A loss of SH is also prominent, probably explained by the formation of a tropylium ion as reported for 3-methylthioanisole.[56] Loss of chlorine plays little importance in the fragmentation of these compounds, and therefore the chlorine substitution pattern does not influence the mass spectrum. Since compounds with the methylthio group in the 3- or 4-positions cannot form a stabilized tricyclic ion, the loss of $CH_3Cl$ $[M - 50]^+$ is less favored and the abundance of $[M - 82]^+$, loss of $CH_3S + Cl$, for compounds 8–15 is of essentially equal magnitude. Loss of HS· $[M - 33]^+$ and of $CH_2S$ $[M - 46]^+$ is common for the 3-methylthio-PCBs.

### 5.9. Methylsulfonyl-PCBs

Bergman also prepared 28 methylsulfonyl-PCBs and described their GC properties (2% OV-17 on Chromosorb W, AW, DMCS, 100–120

*TABLE 9*
*Structures, and Mass Spectral and Gas Chromatographic Data for the Methylthio-PCBs*[a]

| No. | Structure | $M^{+\cdot}$ | −15 | −33 | −35 | −46 | −50 | −70 | −82 | $R_t$ | $R_t$(rel) |
|---|---|---|---|---|---|---|---|---|---|---|---|
| 1 | 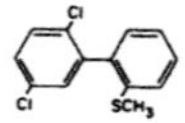 | 3 | — | 35 | 94 | 2 | 100 | 23 | 5 | 12.2[b] | 1 |
| 2 | 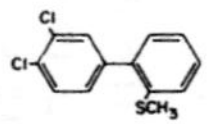 | 59 | 14 | 2 | 4 | 3 | 100 | 5 | 6 | 17.0 | 1 |
| 3 | 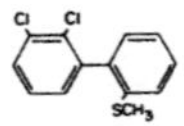 | 2 | — | 34 | 100 | 2 | 90 | 28 | 6 | 13.6 | 1 |
| 4 | 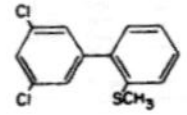 | 55 | 23 | 2 | 4 | 3 | 100 | 6 | 6 | 15.0 | 1 |
| 5 | 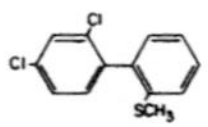 | 5 | — | 38 | 100 | 2 | 94 | 29 | 6 | 12.4 | 1 |
| 6 | 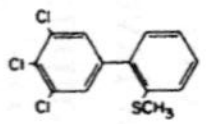 | 48 | 10 | 2 | 5 | 15 | 100 | 5 | 6 | 27.0 | 1 |
| 7 | 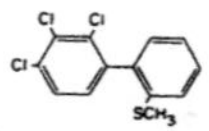 | 2 | — | 74 | 100 | 13 | 86 | 32 | 5 | 23.2 | 1 |
| 8 | 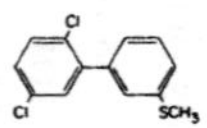 | 100 | 4 | 16 | 2 | 22 | 20 | 3 | 20 | 17.2 | 1.41 |
| 9 | 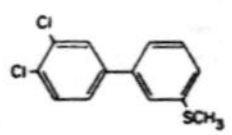 | 100 | 3 | 21 | 2 | 19 | 20 | 4 | 18 | 24.4 | 1.44 |
| 10 | 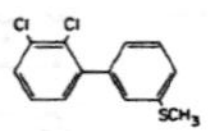 | 100 | 5 | 20 | — | 26 | 19 | 2 | 20 | 19.2 | 1.40 |
| 11 | 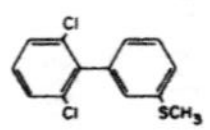 | 100 | 4 | 17 | 2 | 22 | 25 | 3 | 25 | 14.4 | — |
| 12 | 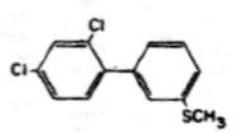 | 100 | 3 | 19 | 2 | 25 | 22 | 3 | 24 | 17.4 | 1.40 |
| 13 | 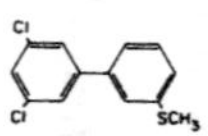 | 100 | 5 | 27 | 2 | 25 | 19 | 3 | 26 | 21.4 | 1.42 |
| 14 | 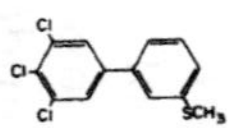 | 100 | 2 | 19 | — | 17 | 17 | 1 | 21 | 35.4 | 1.31 |
| 15 | 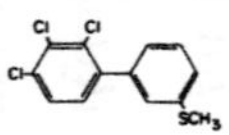 | 100 | 2 | 15 | 2 | 25 | 17 | 3 | 25 | 30.6 | 1.31 |
| 16 | 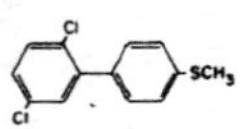 | 100 | 26 | 4 | — | 11 | 23 | 3 | 16 | 19.6 | 1.60 |
| 17 | 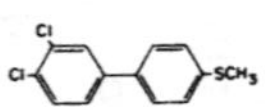 | 100 | 43 | — | 1 | 8 | 22 | 1 | 14 | 26.8 | 1.57 |
| 18 | 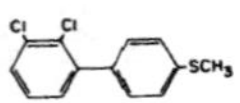 | 100 | 24 | 4 | 1 | 11 | 23 | 3 | 16 | 21.4 | 1.57 |

| | | | | | | | | | | | |
|---|---|---|---|---|---|---|---|---|---|---|---|
| 19 | 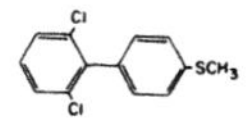 | 100 | 16 | 6 | 1 | 15 | 27 | 3 | 20 | 16.2 | — |
| 20 | 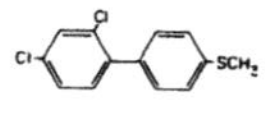 | 100 | 34 | 3 | — | 11 | 21 | 2 | 18 | 19.2 | 1.55 |
| 21 | 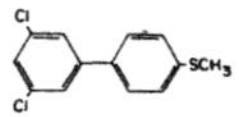 | 100 | 30 | 4 | 3 | 13 | 18 | 3 | 16 | 23.6 | 1.57 |
| 22 | 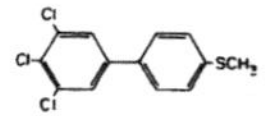 | 100 | 34 | 3 | 1 | 8 | 21 | 2 | 13 | 38.4 | 1.42 |
| 23 | 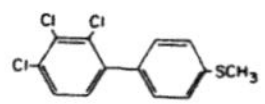 | 95 | 24 | 4 | — | 14 | 19 | 3 | 18 | 32.8 | 1.41 |
| 24 | 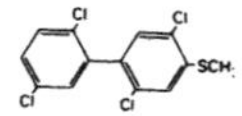 | 62 | 6 | 5 | — | 8 | 17 | 5 | 11 | 36.0 | — |
| 25 | 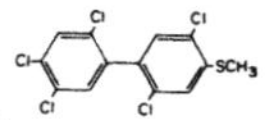 | 51 | 5 | 6 | 2 | 11 | 17 | 5 | 10 | 48.6 | — |
| 26[b] | 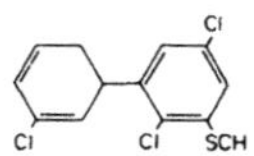 | 96 | 4 | 10 | — | — | 28 | — | — | 6.0[d] | 1.0 |
| 27 | 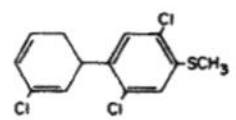 | 100 | 18 | 5 | — | — | 39 | — | — | 6.2 | 1.03 |
| 28 | 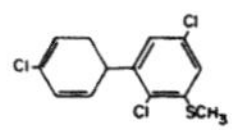 | 97 | 5 | 11 | — | — | 29 | — | — | 6.5 | 1.08 |
| 29 | 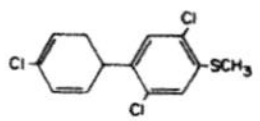 | 92 | 18 | 4 | — | — | 43 | — | — | 6.7 | 1.12 |
| 30 | 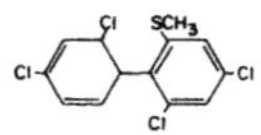 | 51 | 22 | 6 | — | — | 100 | — | — | 7.7 | 1.28 |
| 31 | 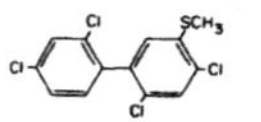 | 73 | 9 | 5 | — | — | 29 | — | — | 9.3 | 1.55 |
| 32 | 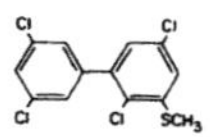 | 84 | 2 | 10 | — | — | 22 | — | — | 9.9 | 1.65 |
| 33 | 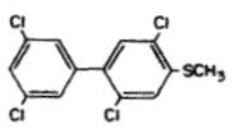 | 65 | 10 | 7 | — | — | 21 | — | — | 10.3 | 1.72 |
| 34 | 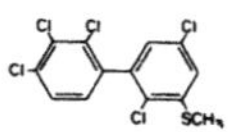 | 60 | 5 | 8 | — | — | 18 | — | — | 20.0 | 3.33 |
| 35 | 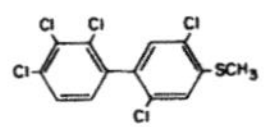 | 58 | 8 | 5 | — | — | 20 | — | — | 22.1 | 3.68 |
| 36 | 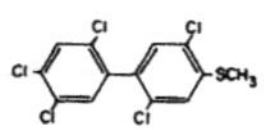 | 61 | 3 | 8 | — | — | 15 | — | — | 17.4 | 2.90 |

[a] References 52 and 53.
[b] 25 glass capillary column, 130–190 °C at 1 °C/min, 1 ml $N_2$/min.
[c] Methylthio-PCB metabolites from mouse feces.
[d] 1 m × 2 mm glass column, 2% OV-1 on Chromosorb W, 200 °C, 1 $kg/cm^2$ He.

*TABLE 10*

*Structures, and Mass Spectral and Gas Chromatographic Data for the Methylsulfonyl-PCBs[a]*

| No. | Structure | $M^{+\cdot}$ | Relative abundance | | | | | | | | $R_t$ | $R_t$ (rel) |
|---|---|---|---|---|---|---|---|---|---|---|---|---|
| | | | −15 | −35 | −62 | −63 | −79 | −91 | −98 | −114 | | |
| 41 | 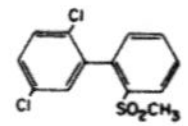 | 5 | — | 70 | — | 2 | 3 | — | 86 | 46 | 5.7[a] | 1.0 |
| 42 | 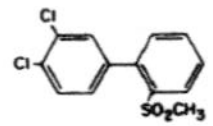 | 59 | 6 | — | 6 | 4 | 13 | — | 41 | 100 | 8.0 | 1.0 |
| 43 | 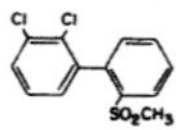 | 2 | — | 100 | — | — | 2 | — | 74 | 50 | 6.7 | 1.0 |
| 44 | 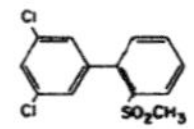 | 47 | 6 | — | 7 | 8 | 8 | — | 45 | 100 | 6.4 | 1.0 |
| 45 | 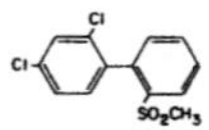 | 4 | — | 96 | — | 3 | 3 | — | 56 | 60 | 5.7 | 1.0 |
| 46 | 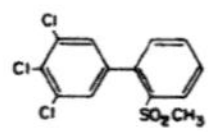 | 15 | 4 | — | 5 | 4 | 5 | — | 24 | 62 | 13.7 | 1.0 |
| 47 | 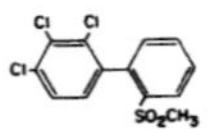 | — | — | 36 | — | — | — | — | 28 | 19 | 12.3 | 1.0 |
| 48 | 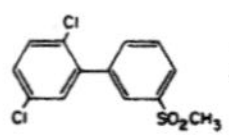 | 32 | 7 | — | — | — | 21 | 18 | 6 | 97 | 10.3 | 1.79 |
| 49 | 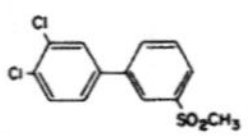 | 20 | 13 | 7 | — | — | 22 | 14 | — | 79 | 12.8 | 1.59 |
| 50 | 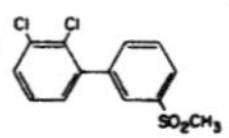 | 16 | 7 | — | 7 | — | 20 | 11 | 6 | 64 | 13.0 | 1.94 |
| 51 | 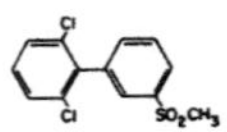 | 6 | — | — | 5 | 3 | 8 | 6 | 2 | 30 | 8.7 | — |
| 52 | 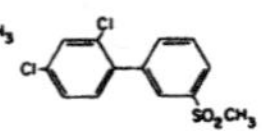 | 21 | 4 | — | — | — | 9 | — | 13 | 69 | 10.9 | 1.92 |
| 53 | 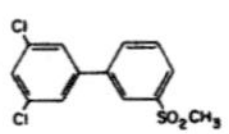 | 27 | — | — | — | 2 | 20 | 4 | 2 | 81 | 10.9 | 1.72 |
| 54 | 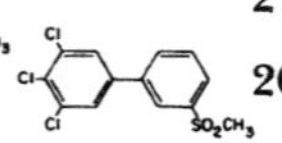 | 20 | — | — | — | — | 18 | 26 | 8 | 50 | 24.6 | 1.79 |
| 55 | 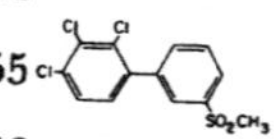 | 26 | 4 | — | 5 | — | — | 17 | 2 | 59 | 24.6 | 2 |
| 56 | 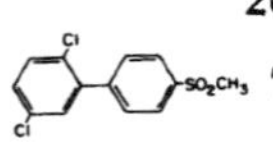 | 7 | — | — | 1 | 8 | 4 | 3 | — | 32 | 12.4 | 2.16 |

| | | | | | | | | | | | | |
|---|---|---|---|---|---|---|---|---|---|---|---|---|
| 57 | 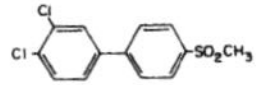 | 2 | 2 | — | — | 4 | 9 | 4 | 2 | 10 | 17.2 | 2.15 |
| 58 | 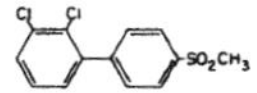 | 15 | 2 | — | — | 10 | 13 | 6 | — | 73 | 15.0 | 2.24 |
| 59 | 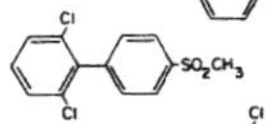 | 10 | 9 | — | 2 | 10 | 12 | 6 | 2 | 49 | 9.9 | — |
| 60 | 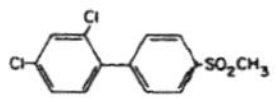 | 14 | 4 | — | — | 5 | 7 | 3 | — | 41 | 12.4 | 2.19 |
| 61 | 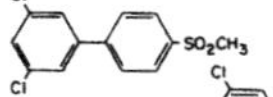 | 5 | 1 | — | 2 | 9 | 4 | 6 | 1 | 27 | 15.4 | 2.42 |
| 62 | 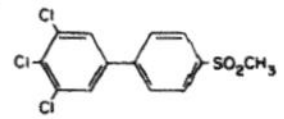 | 13 | 4 | — | 3 | 8 | 2 | 4 | — | 31 | 36.1 | 2.63 |
| 63 | 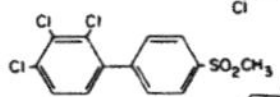 | 3 | 2 | — | — | 4 | 3 | 2 | — | 11 | 29.3 | 2.38 |
| 64 | 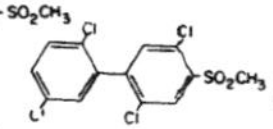 | 57 | — | — | 7 | 29 | 10 | 39 | 1 | 99 | 20.9 | — |
| 65 | 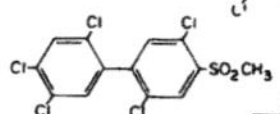 | 23 | 2 | — | — | 27 | 7 | 20 | — | 52 | 37.2 | — |
| 66 | 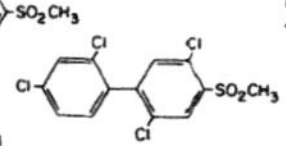 | 5 | — | — | 2 | 6 | 2 | 13 | — | 56 | 22.7 | — |
| 67 | 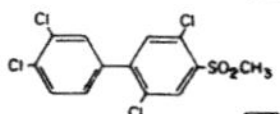 | 7 | — | — | 2 | 10 | 2 | 12 | — | 32 | 35.6 | — |
| 68 | 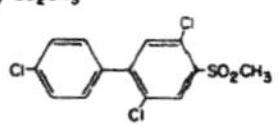 | 42 | 3 | — | 6 | 41 | 8 | 30 | 1 | 95 | 19.0 | — |
| 69[b] | 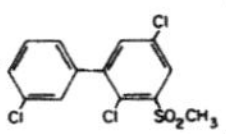 | 79 | — | — | 7 | — | — | 27 | — | 100 | 13.2[c] | 1.0 |
| 70 | 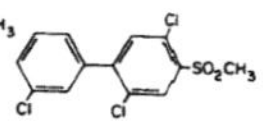 | 83 | 5 | — | — | 39 | — | 27 | — | 100 | 14.6 | 1.11 |
| 71 | 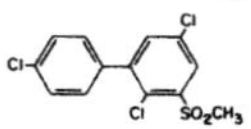 | 79 | — | — | 7 | — | — | 27 | — | 100 | 14.6 | 1.11 |
| 72 | 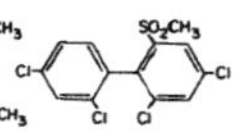 | 26 | — | — | — | 2 | — | — | — | 44 | 11.1 | 0.84 |
| 73 | 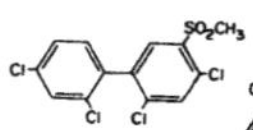 | 77 | 5 | — | — | 17 | — | 17 | — | 69 | 18.9 | 1.43 |
| 74 | 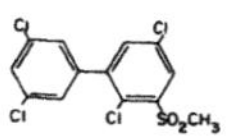 | 78 | — | — | 12 | — | — | — | — | 67 | 20.6 | 1.56 |
| 75 | 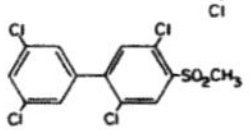 | 71 | 6 | — | — | 39 | — | 21 | — | 97 | 22.5 | 1.70 |

[a] References 52 and 53.

(*continued overleaf*)

*TABLE 10 (continued)*

| No. | Structure | $M^{+\cdot}$ | Relative abundance −15 | −35 | −62 | −63 | −79 | −91 | −98 | −114 | $R_t$ (I) | $R_t$ (II) |
|---|---|---|---|---|---|---|---|---|---|---|---|---|
| 76 | 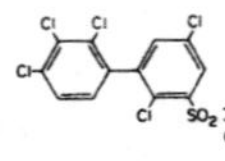 | 57 | — | — | 13 | — | — | 25 | — | 72 | 40.1 | 3.04 |
| 77 | 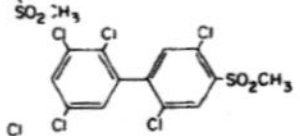 | 63 | 5 | — | — | 26 | — | 22 | — | 74 | 43.1 | 3.27 |
| 78 | 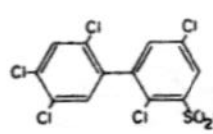 | 69 | — | — | 7 | — | — | 21 | — | 69 | 31.3 | 2.37 |
| 79 | 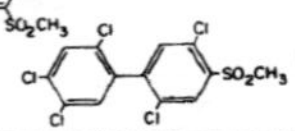 | 63 | 3 | — | — | 18 | — | 18 | — | 60 | 34.0 | 2.58 |

[a] 1.8 m × 1.8 mm i.d. column, 2% OV-17 on Chromosorb W, AW DMCS, 210 °C, 20 ml Me/Ar/min.
[b] Methylsulfonyl-PCBs from mouse feces.
[c] 1 m × 2 mm glass column, 2% OV-1 on Chromosorb W, 200 °C, 1 kg/cm²/He.

mesh, 1.8 m × 1.8 mm i.d., inj. 250 °C, det. 300 °C, carrier at 20 ml/min) and EI mass spectral characteristics (Table 10).[52]

Methylsulfonyl-PCBs 41–68 gave molecular ions from 2 to 59% of the base peak. In all cases loss of $CH_3SO_2Cl$ $[M - 114]^+$ gave more intense peak (11–100%). The spectra for these compounds were dominated by the fragments at *m/z* 63 $[CH_3SO]^+$ and 79 $[CH_3SO_2]^+$ which were not reported for methylaryl sulfones[57,58] nor for methylsulfonyl-PCB metabolites isolated from mice.[53] Bergman suggests that the latter investigation did not see the fragments at *m/z* 63 and 79 because of the low ionization energy (20 eV) used.[52]

In 2-methylsulfonyl-PCBs containing an *o*-chlorine in the second ring, loss of chlorine $[M - 35]^+$ is observed. No ring closure to form a dibenzothiophene on loss of $CH_3Cl$ is observed as for the methylthio-PCBs, but formation of a dibenzofuran structure $[M - SOCH_3Cl]^+$, $[M - 98]^+$, takes place after rearrangement of the sulfone group.[58] Compounds 48–68 with the methylsulfonyl group in the 3- or 4-position give ions corresponding to loss of $CH_3SO + CO$, $[M - 91]^+$, with the formation of the dibenzofuran ion being of much lower intensity or absent altogether allowing these sulfones to be readily distinguished from the 2-methylsulfonyl-PCBs.

It is difficult to distinguish 3- and 4-methylsulfonyl-PCBs from their fragmentation patterns, but Mizutani has noted that 4-methylsulfonyl-PCBs lose a methyl group to yield an $[M - CH_3]^+$ peak (6%) which is not seen in the 3-isomer.[53] In addition, where 4-methylsulfonyl-PCBs show a prominent $[M - CH_3SO]^+$ peak, the 3-isomers in many cases show $[M - CH_2SO]^+$ peaks. Since the GC retention times for 3- and 4-isomers are different, the position of the methylsulfone group can be pinpointed.

## REFERENCES


1. O. Hutzinger, S. Safe, and V. Zitko, *The Chemistry of PCBs,* CRC Press, Cleveland, 1974.
2. J. Krupcik, P. A. Leclercq, A. Simova, P. Suchanek, M. Collak, and J. Hrivnak, *J. Chromatogr.* **119,** 271–283 (1976).
3. S. Safe and O. Hutzinger, *J. Chem. Soc. D* **9,** 446–448 (1971).
4. E. O. Oswald, L. Levy, B. J. Corbett, and M. P. Walker, *J. Chromatogr.* **93,** 63–90 (1974).
5. L. A. Levy and E. O. Oswald, *Biomed. Mass Spectrom.* **3,** 88–90 (1976).

5a. S. Safe, O. Hutzinger, and W. D. Jamieson, *Org. Mass Spectrom.* **7,** 169–176 (1973).

6. J. W. Rote and W. J. Morris, *J. Assoc. Off. Anal. Chem.* **56,** 188–199 (1973).
7. E. D. Pellizzari, K. B. Tomer, and M. A. Mosely, *Advances in the Identification and Analysis of Organic Pollutants in Water* (L. H. Keith, ed.), Ann Arbor Science, Ann Arbor, MI, (1981).

8. R. C. Dougherty and E. A. Hett, *Environm. Sci. Res.* **12,** 339–350 (1978).
9. J. R. Hass, M. D. Frieser, and M. K. Hoffman, *Org. Mass Spectrom.* **14,** 9–16 (1979).
10. K. L. Busch, A. Norstrom, M. M. Bursey, J. R. Hass, and C. A. Nillson, *Biomed. Mass Spectrom.* **6,** 157–161 (1979).
11. G. W. Tindall and P. E. Winniger, *J. Chromatogr.* **196,** 109–119 (1980).
12. E. J. Bonelli, *Anal. Chem.* **44,** 603–606 (1972).
13. J. W. Eichelberger, L. E. Harris, and W. L. Budde, *Anal. Chem.* **47,** 995–1000 (1975).
14. R. C. Lao, R. S. Thomas, and J. L. Monkman, *Dynam. Mass Spectrom.* **4,** 107–131 (1976).
15. J. W. Eichelberger, L. E. Harris, and W. L. Budde, *Anal. Chem.* **46,** 227–232 (1974).
16. M. D. Erickson and E. D. Pellizzari, *Bull. Environ. Contam. Toxicol.* **22,** 688–694 (1979).
17. M. D. Mullin and J. C. Filkins, 180th Meeting American Chemical Soc., Div. Environm. Chem., Las Vegas, August, 1980.
18. L. S. Ettre, *Chromatographia* **6,** 489–495 (1973).
19. J. Krupcik, P. A. Leclercq, J. Garaj, and A. Simova, *J. Chromatogr.* **191,** 207–220 (1980).
20. L. S. Ettre, *Chromatographia* **7,** 39–46 (1974).
21. L. S. Ettre, *Chomatographia* **7,** 261–268 (1974).
22. K. Grob, Jr., G. Grob, and K. Grob, *J. Chromatogr.* **156,** 1–20 (1978).
23. J. Krupcik, M. Kristin, M. Valachovicova, and S. Janiga, *J. Chromatogr.* **126,** 147–160 (1976).
24. S. Jensen, *New Scientist* **32,** 612 (1966).
25. S. Jensen, *Ambio* **1,** 123–131 (1972).
26. D. J. Call, Y. A. Greichus, and J. J. Worman, *Bull. Environm. Contam. Toxicol.* **11,** 333–338 (1974).
27. Y. A. Greichus, J. J. Worman, M. A. Pearson, and D. J. Call, *Bull. Environm. Contam. Toxicol.* **11,** 113–120 (1974).
28. ten Noever de Braun, "Detection of Chlorinated Pollutants in Cormorant Tissue," Varian MAT Product Information Application No. 6, January, 1976.
29. G. E. Bagley, W. L. Reichel, and E. Cromartie, *J. Assoc. Off. Anal. Chem.* **53,** 251–261 (1970).
30. G. D. Veith, D. W. Kuehl, and J. Rosenthal, *J. Assoc. Offic. Anal. Chem.* **58,** 1–5 (1975).
31. J. Mes, D. S. Campbell, R. N. Robinson, and D. J. A. Davies, *Bull. Environm. Contam. Toxicol.* **17,** 196–203 (1977).
32. T. Yakushiji, I. Watanabe, K. Kuwabara, and S. Yoshida, *J. Chromatogr.* **154,** 203–210 (1978).
33. D. T. Williams and F. M. Benoit, *Bull. Environm. Contam. Toxicol.* **21,** 179–184 (1979).
34. O. Hutzinger, D. M. Nash, S. Safe, A. S. W. DeFreitas, R. J. Norstrom, D. J. Wildish, and V. Zitko, *Science* **178,** 312–314 (1972).
35. T. Mio, K. Sumino, and T. Mizutani, *Chem. Pharm. Bull.* **24,** 1958–1960 (1976).
36. G. Sundström, O. Hutzinger, and S. Safe, *Chemosphere* **5,** 267–298 (1976).
37. B. Jansson and G. Sundström, *Biomed. Mass Spectrom.* **1,** 386–392 (1974).
38. G. Sundström and C. A. Wachtmeister, *Chemosphere* **3,** 7–11 (1975).
39. J. Sparling, D. Fung, and S. Safe, *Biomed. Mass Spectrom.* **7,** 13–19 (1980).
40. M. Th. M. Tulp, K. Olie, and O. Hutzinger, *Biomed. Mass. Spectrom.* **4,** 310–316 (1977).
41. D. H. Jones, J. Kohli, and S. Safe, *Xenobiotica* **9,** 733–736 (1979).
42. S. Safe, O. Hutzinger, and D. Ecobichon, *Experientia* **30,** 720–721 (1974).
43. J. R. Hass, L. T. Jao, N. K. Wilson, and H. B. Matthews, *J. Agric. Food Chem.* **25,** 1330–1333 (1977).
44. S. Safe, O. Hutzinger, D. J. Ecobichon, and A. A. Grey, *Can. J. Biochem.* **53,** 415–420 (1975).
45. S. Safe, O. Hutzinger, and D. Jones, *J. Agric. Food Chem.* **23,** 851–853 (1975).
46. M. Th. M. Tulp and O. Hutzinger, *J. Chromatogr.* **139,** 51–59 (1977).

47. M. Kamal, I. Weisgerber, W. Klein, and F. Korte, *J. Environ. Sci. Health* **B11,** 271–287 (1976).
48. D. W. Phillipson and B. J. Puma, *Anal. Chem.* **52,** 2328–2332 (1980).
49. D. L. Stalling and J. W. Hogan, *Bull Environm. Contam. Toxicol.* **20,** 35–43 (1978).
50. P. R. Chen, J. D. McKinney, and H. B. Matthews, *Drug. Metab. Disposit.* **4,** 362–367 (1976).
51. S. Jensen and G. Sundstrom, *Nature* **251,** 219–220 (1974).
52. A. Bergman, B. Jansson, and I. Bamford, *Biomed. Mass Spectrom.* **7,** 20–27 (1980).
53. T. Mizutani, K. Yamamoto, and K. Tajima, *J. Agric. Food Chem.* **26,** 826–866 (1978).
54. K. Sumino and T. Mio, *Proc. 1st Mtg. Japan. Soc. Med. Mass Spectrom.* **1,** 67 (1976).
55. A. Bergman and C. A. Wachtmeister, *Chemosphere* **7,** 949–956 (1978).
56. J. H. Bowie, S.-O. Lawesson, J. O. Madsen, G. Schroll, and D. H. Williams, *J. Chem. Soc. B,* 951–956 (1966).
57. J. H. Bowie, D. H. Williams, S.-O. Lawesson, J. O. Madsen, C. Nolde, and G. Schroll, *Tetrahedron* **22,** 3515–3525 (1966).
58. W. H. Baarschers and B. W. Krupay, *Can. J. Chem.* **51,** 156–161 (1973).

# POLYCHLORODIBENZO-*p*-DIOXINS AND POLYCHLORODIBENZOFURANS

*G. R. B. WEBSTER, K. OLIE, and O. HUTZINGER*

## 1. ELECTRON IMPACT MASS SPECTRA

The electron impact (EI) mass spectra of polychlorodibenzo-*p*-dioxins (PCDDs) and polychlorodibenzofurans (PCDFs) are characterized by several features:

(i) intense parent ions $[M]^{+\cdot}$ (usually the base peak);
(ii) strong doubly charged molecular ions $[M]^{2+}$;
(iii) major fragment ions due to loss of CO and $Cl^{\cdot}$ with formation of $[M - COCl]^{+}$ and $[M - 2COCl]^{+}$ ions;
(iv) minor fragment ions due to loss of $Cl^{\cdot}$, $Cl_2$, and sometimes HCl leading to $[M - Cl]^{+}$, $[M - Cl_2]^{+\cdot}$, $[M - COCl - Cl]^{+\cdot}$, $[M - COCl - Cl_2]^{+}$, $[M - 2COCl - Cl_2]^{+\cdot}$, and so on; and
(v) little structural information regarding position of substituent chlorine atoms.

Mass spectra indicative of individual isomeric series can be seen to exhibit these features.

In the region of the mass spectrum used for the identification of PCDDs and PCDFs (*m/z* 200), positional isomers are essentially indistinguishable. Differences have been reported, however, in the lower mass range of the mass spectra of many mono- to hepta-chlorodibenzo-*p*-

*G. R. B. WEBSTER* • Pesticide Research Laboratory, Department of Soil Science, University of Manitoba, Winnipeg, Manitoba R3T 2N2, Canada. *K. OLIE* and *O. HUTZINGER* • Laboratory of Environmental and Toxicological Chemistry, University of Amsterdam, Nieuwe Achtergracht 166, 1018 Amsterdam, The Netherlands.

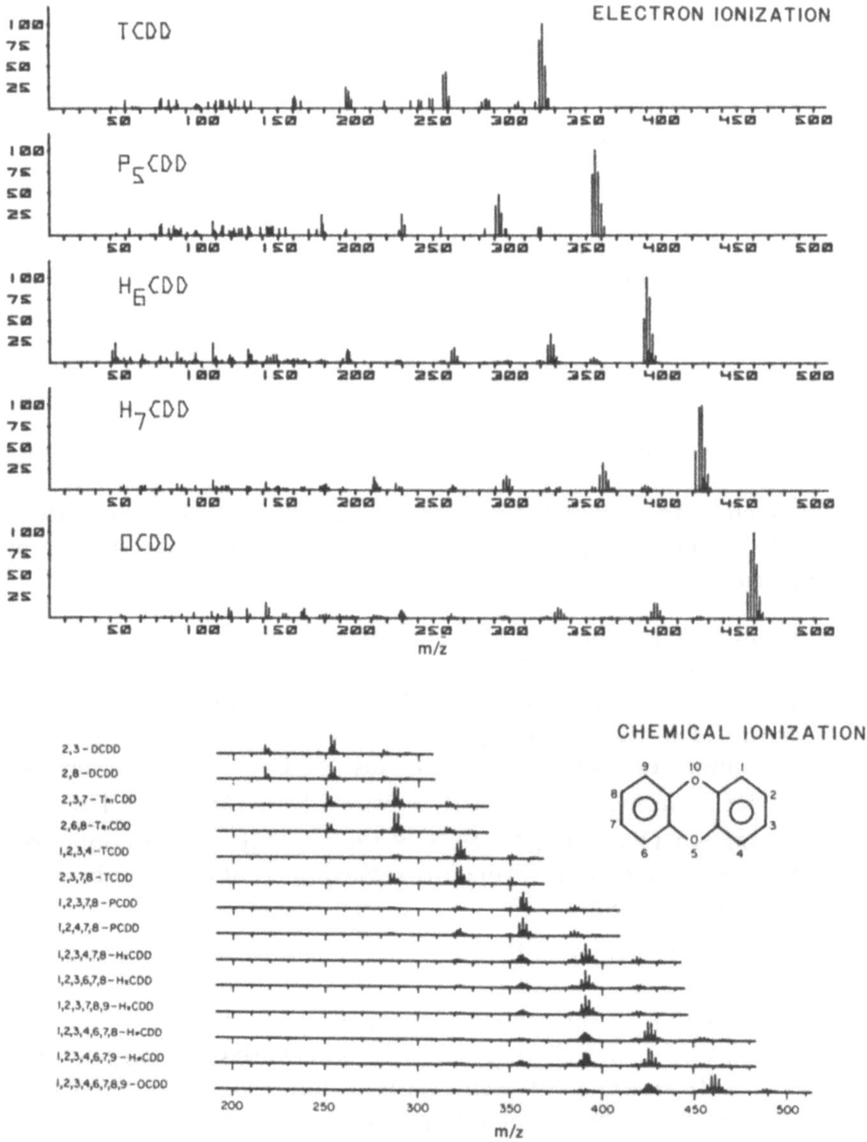


FIGURE 1. EI spectra (upper set) are compared to methane-positive CI mass spectra (lower set) of selected PCDDs. (From Ref. 2.)

dioxins.[1] In the lower mass range (*m/z* 150) the major ions are $C_4$-, $C_5$- and $C_6$-species, the latter formed by fission of the carbon–oxygen bonds of the dioxin structure followed by loss of Cl·, $Cl_2$, and so on. The resulting ions contain information on the number of chlorine atoms in each ring of the PCDD.

The characteristic spectral patterns observed in the El spectra for typical isomers of the PCDD compounds is seen in Fig. 1, along with the methane positive chemical ionization (CI) spectra. Isomers with the same number of chlorines in a specific ring give very similar mass spectra and allow classification of PCDD isomers. Although this technique cannot by itself distinguish all positional isomers, it can distinguish between mono-, di-, tri-, and tetra-substituted rings and, therefore, gives the number of chlorine atoms in each ring. The partial mass spectra of 34 PCDDs allow the number of chlorine atoms in each ring (Table 1) to be determined from the low-mass region of the mass spectra for mono through hepta isomers.

Although information is available from the low-mass region that enables the differentiation of 13 classes of PCDD isomers subdividing the 74 mono- to heptachloro isomers, careful examination of mass spectral data in this mass range is required and may be complicated by

*TABLE 1*
*Chlorine Substitution in PCDD Isomers*[a]

| Compounds | Number of chlorines in each carbon ring | | Number of isomers |
|---|---|---|---|
| | *x* | *y* | |
| mono-CDDs | 1 | 0 | 2 |
| di-CDDs | 2 | 0 | 4 |
| di-CDDs | 1 | 1 | 6 |
| tri-CDDs | 3 | 0 | 2 |
| tri-CDDs | 2 | 1 | 12 |
| tetra-CDDs | 4 | 0 | 1 |
| tetra-CDDs | 3 | 1 | 8 |
| tetra-CDDs | 2 | 2 | 13 |
| penta-CDDs | 4 | 1 | 2 |
| penta-CDDs | 3 | 2 | 12 |
| hexa-CDDs | 4 | 2 | 4 |
| hexa-CDDs | 3 | 3 | 6 |
| hepta-CDDs | 4 | 3 | 2 |
| octa-CDD | 4 | 4 | 1 |

*Structure and Chemical Abstracts* numbering system for PCDDs

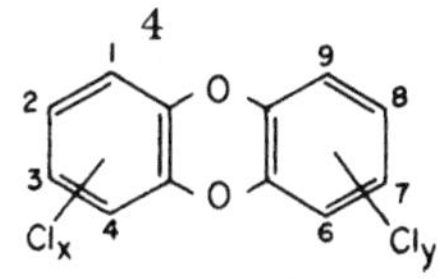


[a] Reference 1.

*TABLE 2*
*Characteristic Ions for PCDD Rings with 0–4 Chlorine Substituents*[a]

| Number of chlorines on carbon ring | *m/z* Values and composition of some characteristic ions[b] | | |
|---|---|---|---|
| 0 | 76 ($C_6H_4^+$) | | |
| | 52 ($C_4H_4^+$) | | |
| 1 | 75 ($C_6H_3^+$) | 110 ($C_6H_3Cl^+$) | |
| | 51 ($C_4H_3^+$) | | |
| 2 | 74 ($C_6H_2^+$) | 109 ($C_6H_2Cl^+$) | |
| | 50 ($C_4H_2^+$) | 97 ($C_5H_2Cl^+$) | |
| | | 85 ($C_4H_2Cl^+$) | |
| 3 | | 108 ($C_6HCl^+$) | 143 ($C_6HCl_2$) |
| | | 96 ($C_5HCl^+$) | 131 ($C_5HCl_2^+$) |
| | | 84 ($C_4HCl^+$) | |
| 4 | | 107 ($C_6Cl^+$) | 142 ($C_6Cl_2^+$) |
| | | 95 ($C_5Cl^+$) | 130 ($C_5Cl_2^+$) |
| | | | 118 ($C_4Cl_2^+$) |

[a] Reference 1.
[b] Major ions underlined.

extraneous peaks from coextractives or GC column bleed. The method will work only with pure compounds, under ideal conditions for background subtraction, or with the use of specific-ion chromatograms for characteristic fragment ions in the mass spectrum. Characteristic ions indicative of zero to four chlorines in a PCDD ring are listed in Table 2.

## 2. *CHEMICAL IONIZATION MASS SPECTRA*

### 2.1. *Positive Chemical Ionization*

Methane positive chemical ionization (MPCI) mass spectra of the PCDDs examined by Hass et al. show the $[M + 1]^+$ ion as the base peak (Fig. 1); $[M + C_2H_5]^+$ and $[M + C_3H_5]^+$ ions are also evident. Other peaks correspond to loss of $Cl^\bullet$, $H^\bullet$, and combinations of $Cl^\bullet$ and $H^\bullet$ from the $[M + 1]^+$ or $[M + C_2H_5]^+$ ions.

### 2.2. *Negative Chemical Ionization*

Electron impact and MPCI mass spectra can be as much as $10^3$-fold less sensitive than methane negative CI for the detection of PCDDs.[3] The use of oxygen, either alone, or as an adjunct to methane in NCI enhances the specificity of the method but decreases the sensitivity somewhat. The use of negative CI mass spectrometry with oxygen as the

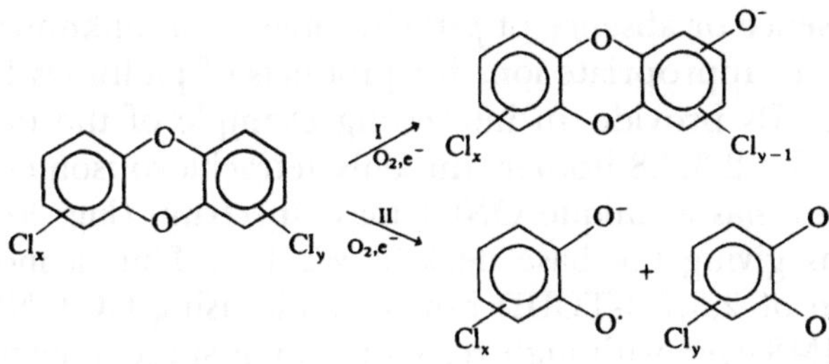


FIGURE 2. Mass spectral mechanisms for PCDDs under ONCI conditions. (From Ref. 2.)

reagent gas gives a unique spectrum for 2,3,7,8-TCDD with low-resolution mass spectrometry.[2,4]

*2.2.1. Oxygen Negative Chemical Ionization (ONCI) Mass Spectra.* In the ONCI mass spectra of 14 PCDDs, two reaction types are observed (Fig. 2).

Pathway I results in the formation of an ion at $[M - 19]^-$ corresponding to $[M - Cl + O]^-$ and is general for PCDDs with greater than two chlorine atoms. The molecular weight can thereby be established. The $[M - 19]^-$ peak is the base peak for each dioxin in this group (except OCDD) containing one or more chlorines in peri positions (positions 1, 4, 6, or 9). In the absence of *peri*-chlorines, pathway II predominates.

Pathway II leads to the formation of one or two ether cleavage ions depending upon the chlorine substitution pattern in each ring and yields information on the number of chlorine atoms in each ring. Thus, it appears possible to determine the number of chlorine atoms per ring

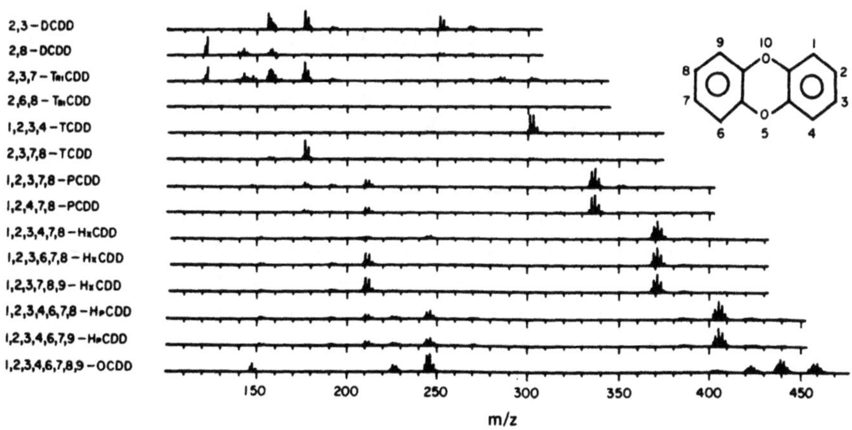


FIGURE 3. Oxygen-negative CI mass spectra of selected PCDDs. (From Ref. 2.)

and the presence or absence of *peri*-chlorines in an unknown dioxin by monitoring the appropriate ions for products of pathways I and II.

The TCDDs provide an interesting example of the utility of this procedure. The 2,3,7,8-isomer, the only tetrachloro isomer having no peri chlorines, has a unique ONCI mass spectrum (Fig. 3) with ether cleavage ions giving the base peak at *m/z* 176. Thus a more positive identification of 2,3,7,8-TCDD can be made using GC/ONCIMS than using GC/EIMS even with high-resolution mass spectrometry. With GC/ONCIMS, the GC retention data are available, and the low-resolution ONCI mass spectrum is unique. Commonly encountered interfering compounds such as chlorophenylchloroquinones would not give the ether cleavage fragment ions observed for 2,3,7,8-TCDD.

2.2.2. *Methane Negative Chemical Ionization (MNCI) Mass Spectra.* The MNCI mass spectra of the same 14 PCDDs examined in the previous section show ions predominantly consistent with loss of H, Cl, and $Cl_2$ from the negative molecular ion. In view of the differences shown in the spectra of different positional isomers, GC/MNCIMS can give reliable identification of PCDDs in most cases (Fig. 4).

The presence of ether cleavage ions and $[M - 19]^-$ ions of significant intensities for several dioxins is attributed to trace contamination of the mass spectrometer vacuum system with oxygen.

PCDFs exhibit intense negative molecular ions $[M]^-$ with limited fragmentation occurring via addition of H and loss of Cl to yield $[M - 34]^-$ ions.[5]

Cavallaro used GC/MNCIMS as a tool for the detection of PCDDs

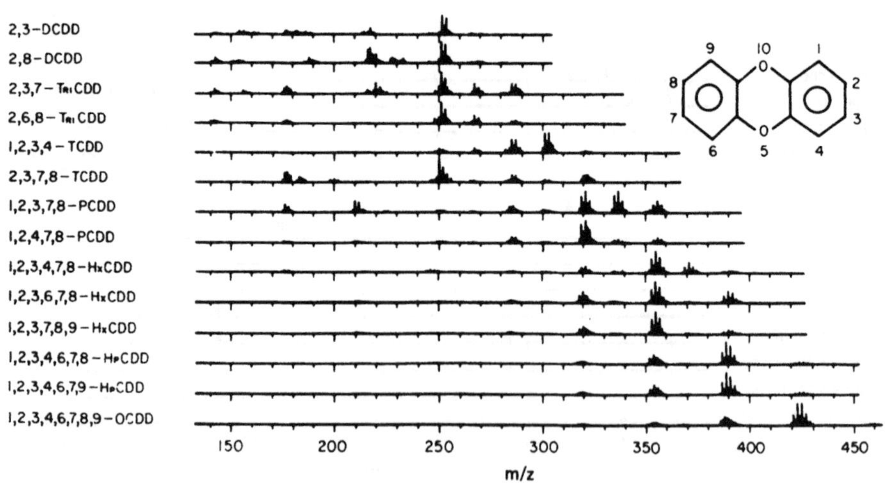


FIGURE 4. The methane-negative CI mass spectra of selected PCDDs. (From Ref. 2.)

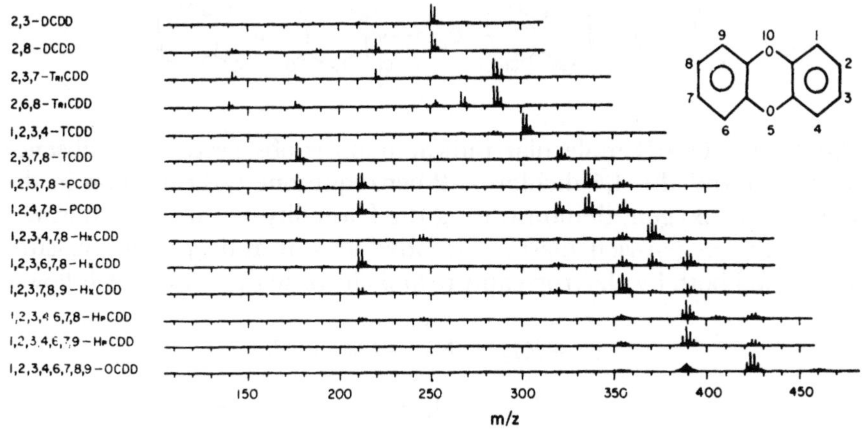


FIGURE 5. Methane–oxygen NCI mass spectra of selected PCDDs. (From Ref. 2.)

in environmental samples.[6] Stalling reported the detection of 15 PCDFs in fish tissue extracts by GC/MNCIMS and used the same technique to determine PCDFs in snapping turtle and seal fat extracts.[5]

*2.2.3. Methane–Oxygen Negative Chemical Ionization (MONCI) Mass Spectra.* MONCI spectra of PCDDs (Fig. 5) usually contain the features of both the MNCI and the ONCI spectra; thus, mass spectra with the most detailed information are produced with this mixed reagent gas. The molecular weight can readily be deduced from the $[M]^{-\cdot}$ or the $[M - H]^-$ ions, the presence or absence of peri chlorines from the intensity of the $[M - 19]^-$ peak, and the number of chlorines on each ring from the *m*/*z* values for the ether cleavage ions.

## 3. ATMOSPHERIC PRESSURE IONIZATION (API) MASS SPECTROMETRY

Atmospheric pressure ionization mass spectrometry, often effectively ONCI mass spectrometry at pressures of 1–760 torr, is a sensitive analytical technique which is highly selective in view of the thermal nature of the ion–molecule collisions involved and the long residence times of the unionized species in the ion source. Although oxygen is often the reagent gas of choice, many other reagent gases may be used.[7,8]

The reaction of 2,3,7,8-TCDD with oxygen in an electron-rich plasma was investigated by Hunt[4] and Hass[3] using low-pressure NCI conditions. The reaction likely proceeds as shown in the following equation,[9] which involves the attack of the dioxin radical anion on a neutral oxygen molecule.

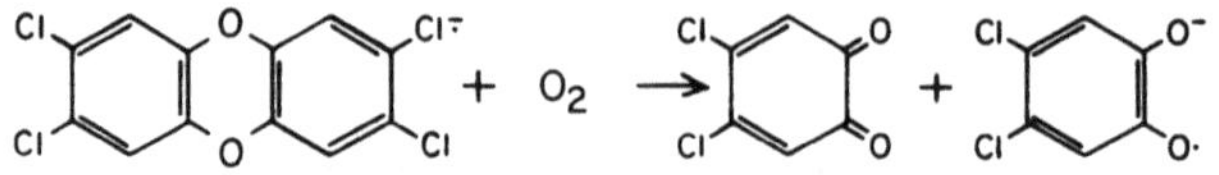

The 2,3,7,8-TCDD molecular radical anion is observed in the absence of oxygen and the 4,5-dichloro-1,2-benzoquinone radical anion in the presence of oxygen, demonstrating the feasibility of this mechanism.

In an oxygen-rich plasma, negative ions at atmospheric pressure are formed both by electron capture and clustering processes as follows:

$$e^- + O_2 + N_2 \rightarrow O_2^- + N_2$$

$$O_2^- + O_2 \rightarrow O_4^-$$

$$O_2^- + N_2 \rightarrow O_2N_2^-$$

$$O_2N_2^- + N_2 \rightarrow O_2N_4^-$$

The abundance of the negative ions is affected directly by both the ion source temperature and the concentration of oxygen in the nitrogen buffer gas. Under the conditions used by Mitchum,[9] the APNCIMS reagent ion spectrum consisted of $O_2^-$, $O_2N_2^-$, $O_4^-$, and $O_2N_4^-$ in the proportions 100 : 6 : 36 : 3.

The APNCI mass spectrum of 2,3,7,8-TCDD in an oxygen-rich plasma is dominated by the negative ion at *m/z* 176, the 4,5-dichloro-1,2-dibenzoquinone radical anion. Unlike the case at lower pressures there is no molecular ion visible. The reaction occurring under APNCI is thought to be the following:

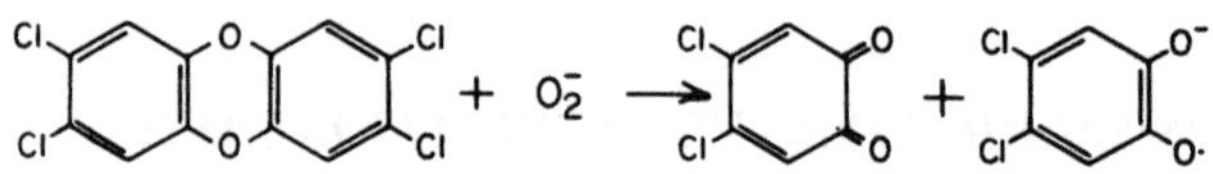

## 4. *METASTABLE IONS*

Safe and co-workers reported on the mass and ion kinetic energy (IKE) spectral data of several chlorinated dioxins. The IKE spectra of the isomeric species studied show differences, but only three TCDD isomers are studied. Chess and Gross studied the intensity of the ion due to the metastable loss of COCl from the molecular ion in the second field-free region (between the electrostatic analyzer and the magnetic sector magnet).[10] No coupling with a GC was made. One of the main disadvantages of this method is the limited scan range, making it impossible to use $^{13}C$-TCDD or $^{37}Cl$-TCDD as internal standards.

Peak switching could also not be employed because the accelerating and electrostatic analyzer voltage sources had been decoupled.

## 5. *INTERFERING COMPOUNDS IN ELECTRON IMPACT MASS SPECTRA OF PCDDs AND PCDFs*

Gas chromatography/mass spectrometry techniques allow identification of PCDDs and PCDFs by GC retention times and by the *m/z* values for molecular and daughter ions. In the case of TCDD a number of compounds eluting in the TCDD region have the correct isotopic ratios for tetrachloro compounds.

### 5.1. *Compounds Interfering with 2,3,7,8-TCDD*

Shadoff prepared a concentrated extract of 2,4,5-trichlorophenol neutral components, of which 2,3,7,8-TCDD is a well-known component.[11] By preparing this "larger than normal" sample, they were able to identify several components which, though the usual harsh cleanup procedure had been followed, have the potential to be mistaken for chlorinated dibenzo-*p*-dioxins by GC/MS.

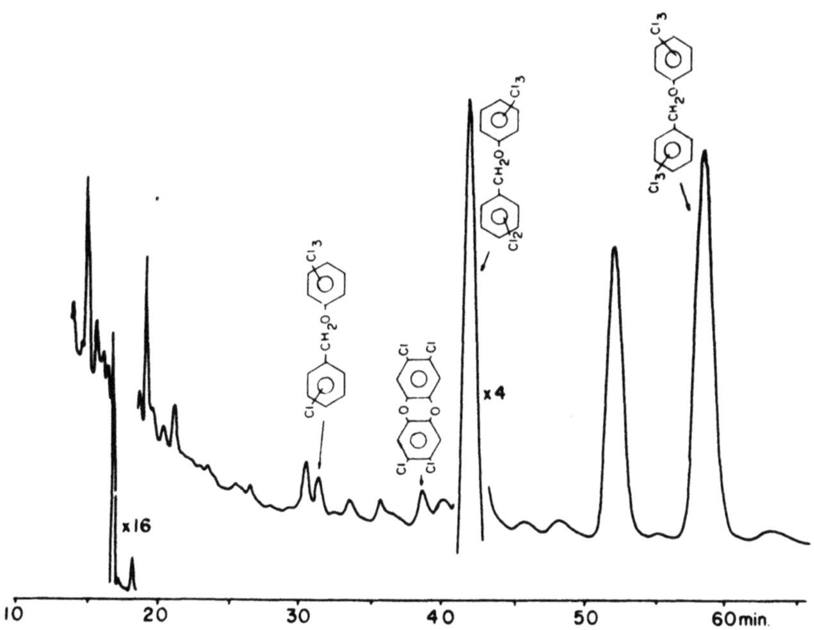


FIGURE 6. GCFID chromatogram of neutral components isolated from commercial 2,4,5-trichlorophenol. (From Ref. 11.)

The sample gave the GC-FID chromatogram shown in Fig. 6 demonstrating that chlorinated compounds can often remain after extensive cleanup and elute with retention times close to that expected for 2,3,7,8-TCDD.

Examination of the mass spectra of the compounds in this chromatogram yielded the data shown in Tables 3, 4, and 5, demonstrating that if SIM is being used to analyze for TCDDs, the chlorobenzylphenyl ethers may yield false positive results, especially if only the molecular ion cluster for TCDDs is used for identification purposes. Careful examination of the mass spectral data shows that the most abundant potentially interfering compound in this case, dichlorobenzyltrichlorophenyl ether, has peaks at *m/z* 320 and 322; the cluster for this $[M - Cl]^+$ ion (4 Cl) is, however, a 318, 320, and 322 cluster, rather than the expected 320, 322, and 324 cluster for TCDDs. Similarly, the 283, 285,

*TABLE 3*
*Mass Spectrum of Chlorobenzyl Trichlorophenyl Ether*[a]

| *m/z* | Relative abundance[b] | Possible identity |
|---|---|---|
| 320 | 1.0 | |
| 321 | 0.1 | |
| 322 | 1.8 | Molecular ion, $M^+$ |
| 323 | 0.1 | (meas. 319.9231, theor. 319.9329) |
| 324 | 0.8 | |
| 325 | 0.1 | |
| 196 | 0.8 | |
| 197 | 0.1 | $C_6H_3OCl_3^+$, trichlorophenol ion |
| 198 | 0.8 | |
| 199 | 0.1 | |
| 200 | 0.3 | |
| 167 | 2.1 | |
| 169 | 2.1 | $C_5H_2Cl_3^+$ |
| 171 | 0.8 | |
| 125 | 100.0 | |
| 126 | 7.9 | $C_7H_6Cl^+$, chlorobenzyl (tropyllium) ion |
| 127 | 39.3 | |
| 128 | 2.5 | |
| 99 | 11.5 | |
| 101 | 2.5 | $C_5H_4Cl^+$ |
| 89 | 31.8 | $C_7H_5^+$ |
| 90 | 9.9 | $C_7H_6^+$ |
| 63 | 14.4 | $C_5H_3^+$ |
| 62 | 6.8 | $C_5H_2^+$ |
| 61 | 3.1 | $C_5H^+$ |

[a] Reference 24.
[b] The relative abundances, especially for the weaker ions, may be in error by ±50% because of the presence of background which was subtracted by the data system.

*TABLE 4*
*Mass Spectrum of Dichlorobenzyl Trichlorophenyl Ether*[a]

| m/z | Relative abundance[b] | Possible identity |
|---|---|---|
| 354 | 1.2 | |
| 355 | 0.9 | |
| 356 | 1.9 | |
| 357 | 0.2 | Molecular ion, $M^+$ |
| 358 | 1.3 | (meas. 353.8806, theor. 353.8939) |
| 359 | 0.1 | |
| 360 | 0.3 | |
| 318 | 0.07 | |
| 321 | 0.08 | $(M - Cl)^+$ |
| 323 | 0.03 | |
| 318 | 0.1 | |
| 320 | 0.15 | $(M - HCl)^+$ |
| 322 | 0.05 | |
| 283 | 0.4 | |
| 285 | 0.4 | $(M - H, 2Cl)^+$ |
| 287 | 0.1 | |
| 249 | 0.3 | |
| 251 | 0.2 | $(M - 3Cl)^+$ |
| 253 | 0.05 | |
| 196 | 1.2 | |
| 198 | 1.2 | $C_6H_3OCl_3^+$, trichlorophenol ion |
| 200 | 0.3 | |
| 159 | 100 | |
| 160 | 10.6 | |
| 161 | 66.9 | $C_7H_5Cl_2^+$, dichlorobenzyl (tropyllium) ion |
| 162 | 5.3 | |
| 163 | 10.4 | |
| 164 | 0.7 | |
| 133 | 3.4 | |
| 135 | 2.3 | $C_5H_3Cl_2^+$ |
| 137 | 0.5 | |
| 124 | 5.9 | |
| 126 | 1.4 | $C_7H_5Cl^+$ |
| 89 | 13.1 | $C_7H_5^+$ |
| 63 | 9.0 | $C_5H_3^+$ |
| 62 | 6.7 | $C_5H_2^+$ |

[a] Reference 24.
[b] The relative abundances, especially for the weaker ions, may be in error by ±50% because of the presence of background which was subtracted by the data system.

TABLE 5
*Mass Spectrum of Trichlorobenzyl Trichlorophenyl Ether*[a]

| m/z | Relative abundance[b] | Possible identity |
|---|---|---|
| 388 | 0.3 | |
| 389 | 0.1 | |
| 390 | 0.5 | |
| 391 | 0.03 | Molecular ion, $M^+$ |
| 392 | 0.4 | (meas. 387.8595, theor. 387.8550) |
| 393 | | |
| 394 | 0.1 | |
| 353 | 0.05 | |
| 355 | 0.06 | $(M - Cl)^+$ |
| 357 | | |
| 352 | 0.04 | |
| 354 | 0.05 | $(M - HCl)^+$ |
| 356 | 0.04 | |
| 193 | 100 | |
| 194 | 7.9 | |
| 195 | 63.9 | $C_7H_4Cl_3^+$, trichlorobenzyl (tropyllium) ion |
| 196 | 6.2 | |
| 197 | 33.9 | |
| 198 | 2.4 | |
| 167 | 6.0 | |
| 169 | 5.8 | $C_5H_2Cl_3^+$ |
| 171 | 1.1 | |
| 157 | 8.9 | |
| 159 | 6.7 | $C_7H_3Cl_2^+$ |
| 161 | 3.8 | |
| 158 | 5.4 | |
| 160 | 4.2 | $C_7H_4Cl_2^+$ |
| 162 | 1.0 | |
| 133 | 4.1 | $C_5H_3Cl_2^+$ |
| 135 | 2.0 | |
| 123 | 8.5 | $C_7H_4Cl^+$ |
| 125 | 2.3 | $C_7H_4Cl^+$ |
| 99 | 1.8 | $C_5H_4Cl^+$ |
| 101 | 0.6 | |
| 87 | 3.0 | $C_7H_3^+$ |
| 63 | 4.4 | $C_5H_3^+$ |

[a] Reference 24.
[b] The relative abundances, especially for the weaker ions, may be in error by ±50% because of the presence of background which was subtracted by the data system.

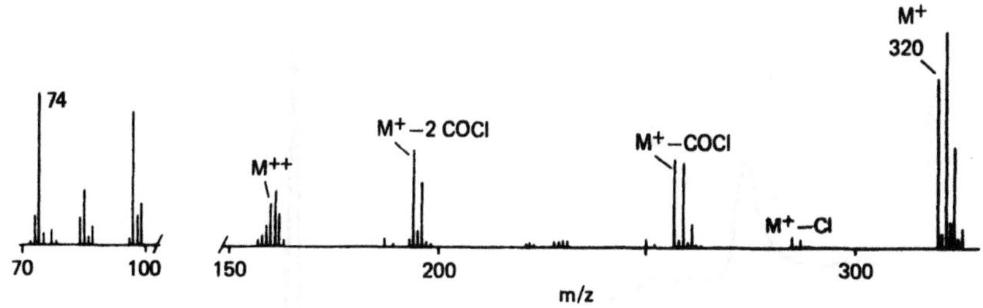


FIGURE 7. Mass spectrum of 2,3,7,8-tetrachlorodibenzo-*p*-dioxin. (From Ref. 12.)

and 287 cluster for $[M - H, 2 Cl]^+$ ion (3 Cl) in the ether is 2 mass units lower than the cluster in the mass spectrum of 2,3,7,8-TCDD (Fig. 7). In addition, the ether has peaks at 249, 251, and 253 for $[M - 3Cl]^+$; 2,3,7,8-TCDD has a weak cluster here but has a prominent 257, 259, and 261 cluster for $[M - COCl]^+$ which is absent in the ether spectrum.

The base peak for the ether is *m/z* 159; for 2,3,7,8-TCDD, it is the $[M + 2]^{+\cdot}$ peak in the molecular ion cluster. This final fact has the effect of minimizing the interference problem, since the potentially interfering peaks from the ether are ~1000 times less intense than the base peak for 2,3,7,8-TCDD. In Fig. 6 the GC-FID chromatogram shows the ether to be present to the extent of ~100 times the concentration of a TCDD; the extent to which it might play a significant role in interfering with TCDD quantitation by SIM mass spectrometry is demonstrated in Fig. 8, showing 2,3,7,8-TCDD and the pentachloro ether in a concentrated extract of 2,4,5-T.

A significant number of other compounds which may interfere with the determination of TCDD are listed in Table 6. Many of these compounds would not elute in the same fraction as a TCDD if they were present in equal quantities; however, should they be present in vast excess, a potential for interference does exist in view of their ions at *m/z* 320 and 322. Confirmation of identity of the TCDD can be effected by high-resolution mass spectrometry or the less expensive high-resolution GC low-resolution mass spectrometry.[13]

## 5.2. *Potentially Interfering Compounds for PCDDs and PCDFs*

In many cases, interfering compounds are removed in the cleanup procedure or they elute outside the proper elution range, but sometimes the observation of additional characteristic ions or the recording of com-

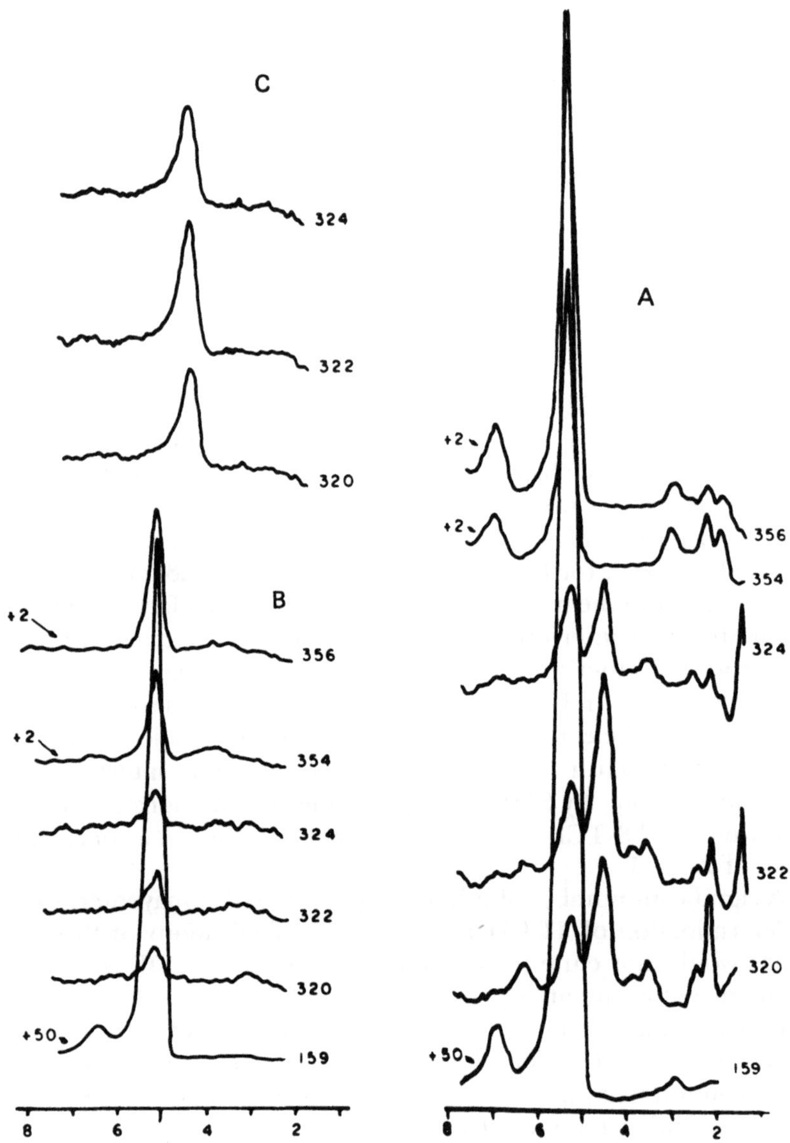


FIGURE 8. Mass chromatograms showing 2,3,7,8-TCDD and 2,6-dichlorobenzyl-2,4,5-trichlorophenyl ether: GC column temperature, 225 °C. (A) Mass chromatograms from an extract of 2,4,5-trichlorophenoxy acetic acid: propylene glyco-lisobutyl ester concentrated 10 : 1. (B) Mass chromatograms from a 3 μg/ml solution of 2,6-dichlorobenzyl-2,4,5-trichlorophenyl ether. (C) Mass chromatograms from a 0.10 μg/ml solution of 2,3,7,8-TCDD corresponding to 0.01 ppm in the sample. (From Ref. 11.)

*TABLE 6*

*Possible Interference for Tetrachlorodibenzo-p-dioxins on High-Resolution Mass Spectrometry for m/z 321.8936 and 319.8966*[a]

| Compound | Elemental composition | Ion | *m/z* | ΔM TCDD | Resolution for separation |
|---|---|---|---|---|---|
| TCDDs | $C_{12}H_4Cl_4O_2$ | $M^+$, $[M + 2]^+$ | 319.8966 | 0 | |
| | | | 321.8936 | | |
| Heptachlorobiphenyl | $C_{12}H_3\ ^{35}Cl_7$ | $[M - 2^{35}Cl]^+$ | 321.8678 | 0.0258 | 12476 |
| Nonachlorobiphenyl | $C_{12}H\ ^{35}Cl_9$ | $[M - 4^{35}Cl]^+$ | 319.8521 | 0.0445 | 7189 |
| | $C_{12}H\ ^{35}Cl_8\ ^{37}Cl$ | $[M - 3^{35}Cl^{37}Cl]^+$ | 321.8491 | 0.0445 | 7233 |
| Tetrachloromethoxy biphenyl | $C_{13}H_8\ ^{35}Cl_4O$ | $M^+$ | 319.9329 | 0.0363 | 8805 |
| | $C_{13}H_8\ ^{35}Cl_3\ ^{37}ClO$ | $M^+$ | 321.9299 | 0.0364 | 8848 |
| Tetrachlorobenzylphenyl ether | $C_{13}H_8\ ^{35}Cl_4O$ | $M^+$ | 319.9329 | 0.0363 | 8813 |
| | $C_{13}H_8\ ^{35}Cl_3\ ^{37}ClO$ | $M^+$ | 321.9300 | 0.0364 | 8843 |
| Pentachlorobenzylphenyl ether | $C_{13}H_7\ ^{35}Cl_4\ ^{37}ClO$ | $[M - H^{35}Cl]^+$ | 319.9143 | 0.01773 | 18043 |
| | $C_{13}H_7\ ^{35}Cl_3\ ^{37}Cl_2O$ | $[M - H^{35}Cl]^+$ | 321.91138 | 0.01778 | 18104 |
| DDT | $C_{14}H_9\ ^{35}Cl_3\ ^{37}Cl_2$ | $[M - H^{35}Cl]^+$ | 319.9321 | 0.03552 | 9006 |
| | $C_{14}H_9\ ^{35}Cl_2\ ^{37}Cl_3$ | $[M - H^{35}Cl]^+$ | 321.92917 | 0.03557 | 9050 |
| DDE | $C_{14}H_8\ ^{35}Cl_2\ ^{37}Cl_2$ | $M^+$ | 319.9321 | 0.03550 | 9011 |
| | $C_{14}H_8\ ^{35}Cl\ ^{37}Cl_3$ | $M^+$ | 321.92916 | 0.03556 | 9052 |
| Hydroxytetrachlorodibenzofuran | $C_{12}H_4Cl_4O_2$ | $M^+$ | 319.8966 | 0.00 | |
| | | | 321.8936 | 0.00 | |
| Tetrachlorophenylbenzoquinone | $C_{12}H_4Cl_4O_2$ | $M^+$ | 319.8966 | 0.00 | |
| | | | 321.8936 | 0.00 | |

[a] Reference 12.

*TABLE 7*
*Polychlorinated Aromatic Compounds Observed in Some Industrial and Environmental Samples Which May Interfere in the Mass Fragmentographic Analysis of PCDDs and PCDFs*[a]

| Interfering compounds | Interfering ion at $M^+$ for: PCDDs | Interfering ion at $M^+$ for: PCDFs | |
|---|---|---|---|
| PCDPEs | | $[M - Cl_2]$ | Mostly removed in cleanup, different $M^+$ |
| PCPPs (predioxins) | $[M - HCl]^+$ | | Removed in cleanup, different $M^+$, and different fragmentation |
| methylated derivatives | $[M - CH_3Cl]^+$ | | |
| PCBs | $[M - Cl_4]^+$ | | Mostly removed in cleanup, elution outside proper retention time range, different $M^+$ |
| PCNs, PCTs | | $[M + 6]^+$ | Different $M^+$ and different fragmentation |
| Polychlorinated methoxy biphenyls | $M^+$ | | Different fragmentation $[M - 15]^+$, $[M - 43]^+$, or $[M - 50]^+$ |
| Butyl ethers of chlorophenols | $M^+$ | | Elution outside proper retention time range, different fragmentation $[M - 56]^+$ |
| PCBs, DDE | $[M + 4]^+$ | | Different $M^+$ and different fragmentation, elution outside proper retention time range |

[a] Reference 14.

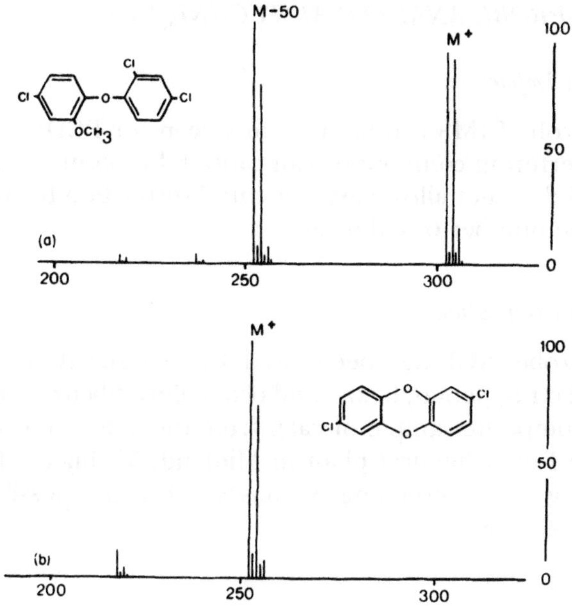


FIGURE 9. Potential for interference in analysis for a dichlorodibenzo-*p*-dioxin by a methoxy-PCPP. (a) methyl ether of Irgasan DP300. (b) 2,7-dichlorodibenzo-*p*-dioxin. (From Ref. 18.)

plete mass spectra may be required in order to confirm the presence of PCDDs or PCDFs. Table 7 lists some of the polychlorinated aromatic compounds which have been encountered in some industrial and environmental samples and that may interfere in the mass fragmentographic analysis of PCDDs and PCDFs.(14)

As indicated in Table 7, the methyl ethers of polychlorinated phenoxy phenols (PCPP, a predioxin) give fragment ions which may be mistaken for PCDD parent ions.

For example, the mass spectrum of the methyl ether of a chlorinated 2-phenoxyphenol shows a large $[M - 50]^+$ ion with one chlorine atom less than the parent ether corresponding exactly to the molecular ion of a PCDD (Fig. 9).(18)

In this case high-resolution mass spectrometry offers no advantage, since the interfering ion possesses the same elemental composition as the parent dioxin ion (i.e., $C_{12}H_6O_2Cl_2$). Cleanup methods and high-resolution GC will enable the removal of this type of interference; in addition, examination of other parts of the mass spectra of the PCDD or the PCPP ether will enable confirmation of structure.

## 6. DIRECT-PROBE ANALYTICAL TECHNIQUES

### 6.1. Electron Impact

Direct-probe EIMS can be used to screen for PCDDs and PCDFs; however, interfering compounds can cause false positive results. In addition, EIMS does not allow easy structural distinction between isomers with the same number of chlorines.

### 6.2. Chemical Ionization

Direct-probe NCI has been used by Dougherty as a screening method.[16] Tetra-, penta-, hexa-, and octa-chlorodibenzo-*p*-dioxins were found in a composite sample of carp from the Tittabawassee River adjacent to the Dow Chemical plant in Midland, Michigan. This method can only be used for screening purposes as it is not possible to differentiate between isomers.

## 7. GC/MS METHODS

### 7.1. Analysis of PCDDs and PCDFs by Packed-Column GC/MS

There are 75 PCDD isomers including 22 tetra-, 14 penta-, 10 hexa-, and 2 hepta-chloro isomers. There are also 135 PCDFs including 38 tetra-, 28 penta-, 16 hexa-, and 4 hepta-chloro isomers. Different PCDD and PCDF isomers with the same chlorine number vary significantly in their toxicological properties. Separation of homologous classes of PCDDs and of PCDFs (e.g., tetrachloro- from pentachloro- from hexachloro-isomers) is relatively straightforward and can be performed by packed-column GC (Table 8).

Nestrick used packed-column GC/MS combined with prior separation by HPLC to separate individual TCDD isomers.[17] His system begins with a reverse-phase (RP) separation and collection of groups of isomers, followed by HPLC normal-phase adsorption (silica) separation. The silica-HPLC TCDD fractions were examined by packed-column gas chromatography/low-resolution mass spectrometry (GC/LRMS).

### 7.2. Analysis of PCDDs and PCDFs by Capillary GC/MS

Gas chromatographic separation of specific tetrachloro- or hexachloro-isomers, for example, is very difficult and requires the use of capillary columns [high-resolution gas chromatography (HRGC)]. Since in some important instances—for example, the so-called "trace chemistry

*TABLE 8*

*GC Conditions for PCDDs and PCDFs for GC/MS (Packed)*[a]

| Compound | Stationary phase | Solid support | Dimensions (length × i.d.) | Temperature (°C) | Reference |
|---|---|---|---|---|---|
| 2,3,7,8-TCDD | 3% OV-225 | Chromosorb W AW-DMCS | 1.5 m × 2 mm | 230 | 19 |
| PCDD, PCDF ($Cl_{4-8}$) | 3% OV-1 | Chromosorb W AW-DMCS | 2 m × 2 mm | 140–260 (1/min) or 260 | 20, 21 |
| PCDD ($Cl_4$-2,3,7,8-) | 2% OV-17 | Diatomate CO 80–100 mesh | 1.8 m × 3 mm | 200 | 22 |
| | 1% Hi-F 8 BP | Gas Chrom Z 100–200 mesh | 1.8 m × 3 mm | 200 | 22 |
| PCDD ($Cl_4$-2,3,7,8-) | 3% OV-7 | Gas Chrom Q | 2 m × 4 mm | 275 | 23 |
| PCDD | 10% OV-101 | Gas Chrom Q 100–120 mesh | 3 m × 2 mm | 120–250 (8/min) | 24 |
| | 3% OV-3 | Gas Chrom Z 80–100 mesh | 1.8 m × 2 mm | 125–230 (10/min) or 225 | |
| PCDD ($Cl_{0-4}$) | 0.2% Carbowax 20 m | Chromosorb W 100–200 mesh | 1.8 m × 1.8 mm | 100, 120, 140 or 160 (4/min), 240 | 15 |
| PCDF ($Cl_{3-6}$) | (a) 3% OV-1 | Supelcoport 100–120 mesh | 1.8 m × 4 mm | 195 | 25 |
| | (b) 5% SP-2401 | Supelcoport 100–120 mesh | 1.8 m × 4 mm | 200 | 25 |
| | (c) 1.5% OV-17/ 1.95% QF-1 | Chromosorb W AW-DMCS 100–120 mesh | 1.8 m × 2 mm | 192 | 25 |

*(continued overleaf)*

*TABLE 8 (continued)*

| Compound | Stationary phase | Solid support | Dimensions (length × i.d.) | Temperature (°C) | Reference |
|---|---|---|---|---|---|
| PCDF ($Cl_{3-6}$) (*cont.*) | (d) 3% OV-1/5% OV-210 + 3% OV-1/ 6% OV-225 | Chromosorb W AW 60–80 mesh | 1.5 m × 4 mm<br>0.15 m × 4 mm | 213 | 25 |
| PCDF ($Cl_{3-5}$) | 3% OV-101 | Chromosorb W-HP 80–100 mesh | 1.8 m × 2 mm | 220 | 26 |
| PCDD ($Cl_{4-8}$)<br>PCDF ($Cl_{4-8}$) | 3% OV-101 | Chromosorb W-HP 80–100 mesh | 1 m × 2 mm | 200–240 (4/min) | 27 |
| PCDF ($Cl_{3-6}$) | 3% Dexil | Chromosorb W AW-DMCS 100–200 mesh | 2 m × 3 mm | 270 | 28 |
| PCDF ($Cl_{4-8}$) | 2% OV-17 | Gas Chrom Q 100–120 mesh | 2 m × 3 mm | 240–280 (3/min) | 28 |
| PCDF ($Cl_{1-5}$) | 2% OV-210 | Chromosorb W AW-DMCS 100–120 mesh | | | 28 |
| PCDD ($Cl_{4-8}$) | 0.2% Carbowax 20M | Chromosorb W 100–120 mesh | 1.8 m × 2 mm | 80 or 100 4 min (4/min), 240 | 29 |
| PCDF ($Cl_{4-8}$)<br>PCDD ($Cl_{6-8}$) | 3% OV-3 | Chromosorb W-HP | 1.8 m × 2 mm | 280 | 30 |
| | 3% OV-3 | Gas Chrom Z 100–120 mesh | 0.9 m × 2 mm | | |
| PCDD ($Cl_{4,6-8}$) | 0.60% OV-17/ 0.40% Poly S-179 | Chromosorb W-AW 80–100 mesh | 2 m × 2 mm | 230 ($Cl_4$)<br>260 ($Cl_6$)<br>275 ($Cl_{7-8}$) | 31 |

| | | | | | |
|---|---|---|---|---|---|
| PCDD ($Cl_4$) | 2% OV-17 | Chromosorb W AW-DMCS 100–120 mesh | 3.6 m × 1.8 mm | 170–250 (4/min) | 32 |
| PCDD (2,3,7,8-$Cl_4$) | 3% OV-3 | Gas Chrom Z 80–100 mesh | 1.8 m × 3 mm | 230 | 11 |
| PCDD ($Cl_{6-8}$)<br>PCDF ($Cl_{7-8}$) | 3% SP2100 | Supelcoport 80–100 mesh | 1 m × 2 mm | 270 | 33 |
| PCDF ($Cl_{4-6}$) | 1.5% Apiezon L 1.5% OV-17 or 14% SF-96<br>3% Apiezon L | Chromosorb W AW, DMCS 60–80 mesh | 5 m × 3 mm<br>2 m × 2 mm | 230 | 34 |
| PCDD ($Cl_{4-8}$) | 0.2% Carbowax 20M | Chromosorb W Acid washed | 2 m × 2 mm | 90 (4/min) 250, 15 min | 35 |
| PCDD ($Cl_{3-8}$)<br>PCDF ($Cl_{3-8}$) | 0.2% Carbowax 20M | Chromosorb W Acid washed | 2.6 m × 2 mm | 70 (4/min) 250, 15 min | 36 |
| PCDD ($Cl_{4-8}$) | 0.2% Carbowax 20M | Chromosorb W 100–120 mesh | 1.8 m × 2 mm | 80/100, 4 min (4/min), 240 | 37 |
| PCDF ($Cl_{4-7}$) | 1% OV-17 | Chromosorb W AW, DMCS 100–120 mesh | 1.5 m × 4 mm | 250 | 38 |
| PCDD ($Cl_4$) | 0.60% OV-17/ 0.40% Poly S-179 | Chromosorb W-AW Bonded to methyl silicone 80–100 mesh | 1.8 m × 2 mm | 250, 1.5 min (10/min), 240 | 10 |
| PCDD ($Cl_{4,6-8}$) | 0.60% OV-17/ 0.40% Poly S-179 | Permabond Methyl silicone, 10-cycle 80–100 mesh | 2.10 m × 2 mm | 246 ($Cl_4$) 230–300 (10/min) ($Cl_{6-8}$) | 39 |

*TABLE 9*
*Capillary GC Conditions for PCDD and PCDF Analysis*

| Compound | Stationary phase | Dimensions (length × i.d.) | Temperature (°C) | Reference |
|---|---|---|---|---|
| PCDD ($Cl_{4-8}$) | OV-61 (700*b* type) | 18 m × 0.3 mm | 180–240 (2/min) or 140 | 40,21 |
| PCDD ($Cl_{4-8}$) | OV-101 | 22 m × 0.32 mm | 225 | 41,18 |
| | OV-17 | 22 m × 0.34 mm | 225 | |
| PCDF ($Cl_{4-8}$) | Silar 10C | 22 m × 0.34 mm | 205 | |
| PCDD ($Cl_4$) | OV-17 | 25 m × 0.37 mm | 207, 220 | 42 |
| (2,3,7,8-) | OV-101 | 27 m × 0.37 mm | 207 | |
| PCDD ($Cl_{3-6}$) | OV-17 | 22 m × 0.36 mm | 207 | 43 |
| | OV-101 | 22 m × 0.36 mm | 197 | |
| PCDF ($Cl_{1-5}$) | OV-17 | 25 m × 0.36 mm | 100, 2 min (15/min), 145; (5/min), 230 | 44 |
| | OV-17 | 50 m × 0.36 mm | 100, 2 min, (15/min), 160; (2/min), 230 | 44 |
| PCDD ($Cl_{2-8}$) PCDF ($Cl_{4-8}$) | OV-17 | 25 m × 0.35 mm | 120, 2 min (15/min), 240 | 45 |
| PCDD ($Cl_{4-8}$) | OV-17 | 25 m × 0.35 mm | 200 (2/min) 240 | 44 |
| PCD ($Cl_{1-8}$) | OV-17 | 50 m × 0.36 mm | 100–160 (20/min), 2/min 250 | |
| PCDD ($Cl_{4-8}$) PCDF | OV-17 | 50 m × 0.36 mm | 200–240 (2/min) | 43 |
| PCDD ($Cl_{2-6}$) PCDF ($Cl_{2-6}$) | OV-17 | 50 m × 0.37 mm | 120, 2 min (20/min), 180; 5 min 240 or 100 (20/min) 160, 2/min, 240 | 43 |

| | | | | |
|---|---|---|---|---|
| PCDF ($Cl_{3-7}$) | OV-17 | 50 m × 0.37 mm | 120, 2 min (20/min), 180 (2 or 5/min), 250 | 45 |
| | | | | 46 |
| PCDD ($Cl_4$) | OV-101 | 20–30 m × 0.35–0.37 mm | 207–220 | 47 |
| | OV-17 | | | |
| PCDD ($Cl_4$) | Silar 10C | 35 m × 0.35 mm | 230 | 48 |
| PCDD ($Cl_4$) | SE-30 | 30 m | 80, 6 min 34/min, 245 | 49 |
| PCDF ($Cl_{4-8}$) | OV-17 | 50 m × 0.37 mm | 100 (20/min) | 46 |
| | Silar 10C | 50 m × 0.36 mm | 160 (2/min) 250 | |
| PCDD ($Cl_{4-8}$) | OV-101 | 50 m × 0.25 mm | 234 | 50 |
| PCDD ($Cl_4$) | Silar 10C | 55 m × 0.25 mm | 100, 3 min (20/min), 180 (2/min), 240 | 51 |
| | OV-17 | 50 m × 0.37 mm | As above to 250 | 51 |
| | OV-101 | 50 m × 0.40 mm | As above to 250 | 51 |
| PCDD (2,3,7,8-$Cl_4$) | SE-30 | 30 m × 0.25 mm | 80, 6 min (34/min), 265 | 52 |
| PCDD PCDF ($Cl_{7,8}$) | SE-30 | 25 m × 0.3 mm | 180, 5 min (2/min), 240 | 53 |
| PCDF ($Cl_6$) | SP-2100 | 25 m | 150 (4/min) 240 | 53 |
| PCDD ($Cl_{3-8}$) PCDF ($Cl_{3-8}$) | OV-17 | 18 m | 200 (4/min) 240 | 53 |
| PCDD ($Cl_{4-6}$) PCDF ($Cl_{3-7}$) | OV-17 | 50 m × 0.36 mm | 120, 2 min (20/min), 180 (5/min), 250 | 54 |

*(continued overleaf)*

*TABLE 9 (continued)*

| Compound | Stationary phase | Dimensions (length × i.d.) | Temperature (°C) | Reference |
|---|---|---|---|---|
| PCDD ($Cl_{4-8}$)<br>PCDF ($Cl_{4-8}$) | Silar 10C | 50 m × 0.36 mm | 100, 2 min (10/min), 140 (5/min), 240 | 47 |
| PCDF ($Cl_{3-7}$) | OV-17 | 50 m × 0.37 mm | 120, 2 min (20/min), 180 (5/min), 250 | 55 |
| PCDF ($Cl_{1-5}$) | OV-17 | 25 m × 0.36 mm | 100, 2 min (15/min), 145 (5/min), 230<br>100, 2 min (20/min), 160 (2/min), 230 | 14 |
| PCDD ($Cl_{4-8}$) | OV-101 | | 140 (10/min) | 6 |
| | OV-17/OV-1 | | 240 or 225 isothermal | |
| PCDD ($Cl_4$) | 60/40 OV-17/ Poly S-179 | 30 m × 0.5 mm | 250 | 56 |
| ($Cl_{6-8}$) | | | 280 | |

*TABLE 10*
*Elution Temperatures and Retention Times of Hexa-, Hepta-, and Octa-CDDs on an OV-61 Glass Capillary Column*[a]

| Compound | Elution temperature (°C) | Retention time[b] (min) |
|---|---|---|
| 1,2,4,6,8,9-hexa-CDD | *c* | *c* |
| 1,2,4,6,7,9-hexa-CDD | 230.8 | 27.5 |
| 1,2,3,4,6,8-hexa-CDD | 232.5 | 28.3 |
| 1,2,3,6,7,9-hexa-CDD | *c* | *c* |
| 1,2,3,6,8,9-hexa-CDD | 234.0 | 29.1 |
| 1,2,3,4,6,9-hexa-CDD | 234.7 | 29.4 |
| 1,2,3,4,7,8-hexa-CDD | 236.2 | 30.2 |
| 1,2,3,7,8,9-hexa-CDD | 236.8 | 30.5 |
| 1,2,3,6,7,8-hexa-CDD | 238.0 | 31.0 |
| 1,2,3,4,6,7-hexa-CDD | 238.4 | 31.2 |
| 1,2,3,4,6,7,9-hepta-CDD | 240 | 34.8 |
| 1,2,3,4,6,7,8-hepta-CDD | 240 | 37.1 |
| 1,2,3,4,6,7,8,9-octa-CDD | 240 | 46.0 |

[a] Reference 40.
[b] Temperature program 180–240 °C at 2°C/min.
[c] Isomers not observed in this study.

of fire" origin of PCDD during combustion processes—many isomers are formed in the same process, it is important to be able to identify the individual isomers. Glass capillary GC provides an effective means of separating many PCDD isomers and a number of stationary phases have been used (Table 9). Buser[40] has reported a glass capillary GC method for the separation of PCDD of differing chlorine number. On a semi-polar methylphenyl silicone OV-61 column programmed from 180–240 °C, tetra-CDD eluted at 200–215 °C, penta-CDD from 215–230 °C, and hexa-CDD between 230 and 240 °C, followed by hepta-CDD and octa-CDD (Table 10 and Fig. 10).

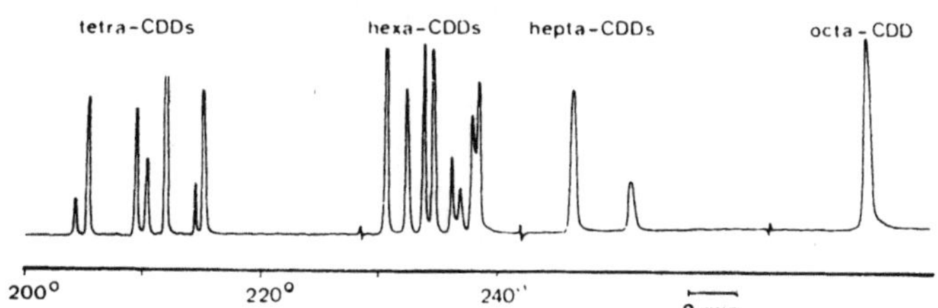


FIGURE 10. Mass fragmentograms (OV-61 glass capillary column, 200–240 °C, *m/z* 320, 388, 422, and 456) showing elution of PCDDs mixture. (From Ref. 20.)

TABLE 11
*HRGC Glass Capillary Columns Used to Separate TCDDs*[a]

| Column | Stationary phase | Type | $l^b$ (m) | i.d.[c] (mm) | $T_E$,[d] °C | $T_F$,[e] °C | Column efficiency $N^f$ | Column efficiency $K$, $T_T$ |
|---|---|---|---|---|---|---|---|---|
| I | Silar 10c | Highly polar | 55 | 0.25 | 234.6 | 240 | 192 000 | (4.7, 100) |
| II | OV-17 | Semipolar | 50 | 0.37 | 216.0 | 250 | 152 000 | (5.8, 80) |
| III | OV-101 | Nonpolar | 50 | 0.40 | 221.4 | 250 | 138 000 | (4.3, 90) |

[a] Reference 13.
[b] $l$ = length.
[c] i.d. = internal diameter of capillary column.
[d] $T_E$ = elution temperature of 2378-TCDD (temperature programmed conditions).
[e] $T_F$ = final column temperature.
[f] $N$ = theoretical plate number, $K$ = capacity ratio, $T_T$ = test temperature (°C) for naphthalene.

## 7.3. *Tetrachlorodibenzo-p-Dioxins*

Buser and Rappe[13] were able to separate all 22 TCDDs through the use of three capillary columns (50–55 m), Silar 10C, OV-17, and OV-101 (Table 11). The Silar 10C column resolved the most TCDD isomers (Fig. 11); however, the OV-17 and OV-101 columns were required to resolve all 22 isomers (Figs. 12 and 13).

The 2,3,7,8-TCDD can be readily resolved on the Silar 10C column, but on the OV-17 column it coelutes with 1,2,7,9-TCDD and is only

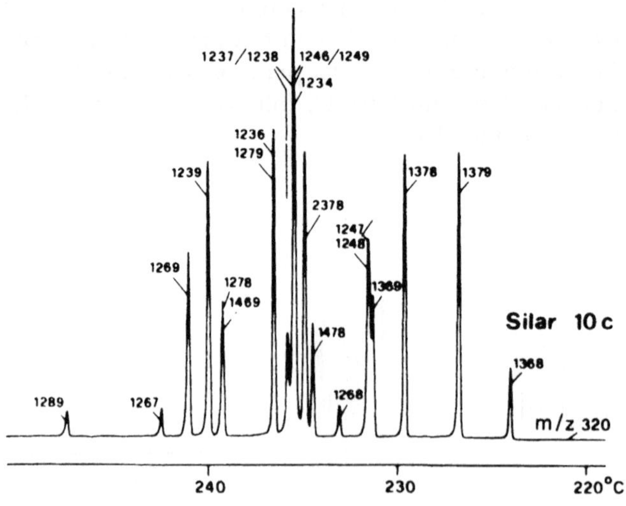


FIGURE 11. Mass fragmentograms (*m/z* 320) of a sample showing elution of all 22 TCDD isomers on a 55-m Silar 10C HRGC column. (From Ref. 13)

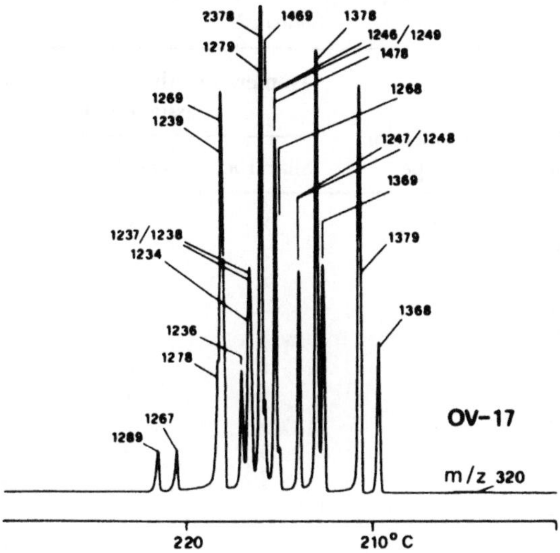


FIGURE 12. Mass fragmentograms (*m/z* 320) of a sample showing elution of all 22 TCDD isomers on a 50-m OV-17 HRGC column. (From Ref. 13.)

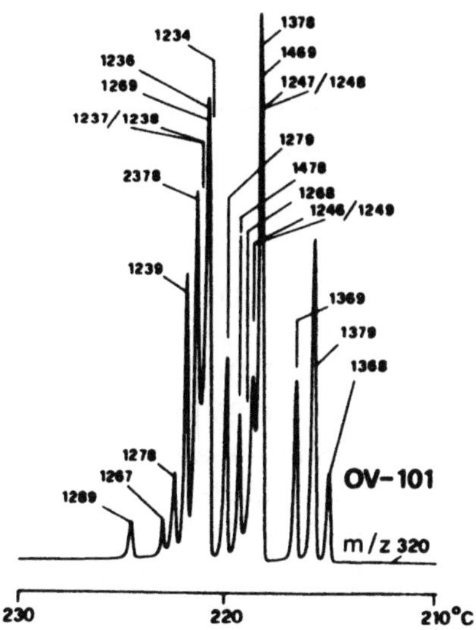


FIGURE 13. Mass fragmentograms (*m/z* 320) of a sample showing elution of all 22 TCDD isomers on a 50-m OV-101 HRGC column. (From Ref. 13.)

*TABLE 12*
*TCDD Isomer Assignments Obtainable on the Three HRGC Columns[a]*

| TCDD Isomer | Substitution type | Isomers (×) that can be uniquely assigned on Silar 10c[b] | OV-17[c] | OV-101[d] |
|---|---|---|---|---|
| 1267- | 2 : 2 | × | × | × |
| 1268- | | × | | |
| 1269- | | × | | |
| 1278- | | | | × |
| 1279- | | | | × |
| 1289- | | × | × | × |
| 1368- | | × | × | × |
| 1369- | | × | × | × |
| 1378- | | × | × | |
| 1379- | | × | × | × |
| 1469- | | | | |
| 1478- | | × | | × |
| 2378- | | × | | |
| 1236- | 3 : 1 | | × | |
| 1237- | | | | |
| 1238- | | | | |
| 1239- | | × | | × |
| 1246- | | | | |
| 1247- | | | | |
| 1248- | | | | |
| 1249- | | | | |
| 1234- | 4 : 0 | | | |

[a] Reference 13.
[b] See Fig. 11.
[c] See Fig. 12.
[d] See Fig. 13.

partially resolved from 1,4,6,9-TCDD (Fig. 12). On OV-101, 2,3,7,8-TCDD coelutes with 1,2,3,7-TCDD and 1,2,3,8-TCDD and is only partially separated from 1,2,3,9-TCDD (Fig. 13). Buser and Rappe state that a Silar 10C column 55 m × 0.25 mm i.d. is required to obtain the separation shown (Fig. 11). Shorter and wider columns did not fully separate 2,3,7,8-TCDD from 1,2,3,4-, 1,2,3,7-, 1,2,3,8-, 1,2,4,6-, and 1,2,4,9-TCDD. Table 12 lists all 22 TCDD isomers and the capillary GC columns on which most can be fully resolved by the method of Buser and Rappe. A number of TCDDs cannot be fully separated from their Smiles rearrangement isomers; that is, 1,2,4,7- and 1,2,4,8-TCDD elute together as do 1,2,4,6- and 1,2,4,9-TCDD. Differential quantitation may be used for determination of, for example, 1,4,6,9-TCDD (on Silar 10C, 1,4,6,9-TCDD plus 1,2,7,8-TCDD; on OV-101, 1,2,7,8-TCDD alone).

## 8. *QUALITATIVE AND QUANTITATIVE METHODS—GC/MS/MS*

### 8.1. *Chromatographic Cleanup and Qualitative Methods*

There is a big difference in the identification of and quantitation of dioxins between commercial products on one hand and environmental samples on the other hand.

In commercial products (e.g., 2,4,5-T) only a limited number of dioxin congeners are present, sometimes in quantities as high as ppm levels. The number of interfering compounds is limited and quite often predictable. The separation of these dioxins from the interfering substances is relatively easy. Since quantities of the dioxin contaminants are relatively high in extracts obtained from these products, it is often possible to run complete mass spectra of these dioxins by GC/MS techniques.

Environmental samples often contain low concentrations of complex mixtures of PCDDs or PCDFs in matrices containing tremendous numbers of other compounds. These compounds can interfere by generating similar mass spectral data and/or GC retention time data to those expected for specific PCDD or PCDF. It is also possible that high amounts of interfering compounds can cause solubility problems for the PCDD or PCDF present.

The separation of PCDD or PCDF from interfering compounds is a very important part of the analysis. Most cleanup procedures have been dedicated to the quantification of 2,3,7,8-TCDD. A method which gives good recoveries for tetra-, hexa-, hepta-, and octa-chlorodioxins uses the following sequence of cleanup columns after the extraction step:

(i) 4 g of 44% concentrated sulfuric acid on silica,
(ii) 2 g of 33% 1 *M* sodium hydroxide on silica, and
(iii) 1.5 g of 10% silver nitrate on silica.

This chemically modified adsorbent treatment of the extract removes the easily oxidizable species.[50]

The next step is separation of the common chemical interferences with 5 g of basic alumina. RP/HPLC fractionation of the residue is applied next to remove residual chemically similar interferences and to separate dioxins present into groups according to their degree of chlorination. A final silica-HPLC refractionation of the RP/HPLC fractions is used to provide a second high-efficiency chromatographic separation having different selectivity to remove residual interferents and to permit TCDD isomer specificity. Sometimes a specific step to remove sulfur is also used.

With samples that are very dirty the first step of the cleanup can be repeated. The GC separation can be done using packed or capillary columns.

### 8.2. *Quantification with SIM and Internal Standards*

The most elegant way to quantify the amount of chlorinated dioxins is by the addition of a known amount of $^{13}$C-labeled dioxin prior to the extraction. $^{13}$C-labeled 2,3,7,8-TCDD and $^{13}$C-labeled OCDD are commercially available (KOR Isotopes, Cambridge, Massachusetts).

The use of $^{37}$Cl-labeled 2,3,7,8-TCDD presents a number of problems.[57] The isotopic purity of the $^{37}$Cl-labeled 2,3,7,8-TCDD is often not 100%; therefore, a small response (0.75%) at *m/z* 322 is seen from the internal standard if the isotopic purity is 89%. The problem can be lessened by adding only a small amount of $^{37}$Cl-labeled 2,3,7,8-TCDD.

Another problem is due to the interference of a polychlorobiphenyl (PCB) congener at *m/z* 328. This can lead to erroneous 2,3,7,8-TCDD values. The use of $^{13}$C-labeled 2,3,7,8-TCDD avoids these problems.

For samples where the extraction efficiencies are known, use can be made of a known amount of a standard which is a compound not present in the initial sample. A mixture of known amounts of standard on PCDDs has to be run for comparison. Extraction of the sample should be as rigorous as possible.[59]

The recovery of the added $^{13}$C-labeled TCDD or $^{37}$C-labeled TCDD is not a good indication of the extraction efficiency. Sometimes only a small fraction of the original chlorinated dioxin is extracted while the added standard is extracted to an extent greater than 90%.

### 8.3. *Perchlorination*

Perchlorination, a method quite often used in the determination of PCBs, is of limited use for the determination of chlorinated dioxins for at least two reasons:

1. The great differences in toxicity of individual PCDD isomers limits the value of total dioxin data.
2. Detection limits for the individual compounds are so low at this moment that there is little need for gain in the detection limit available through perchlorination.

### 8.4. *High-Resolution versus Low-Resolution Mass Spectral Techniques*

High-resolution mass spectral (HRMS) techniques are especially useful in distinguishing PCDDs from potentially interfering compounds as discussed earlier in this chapter; however, extensive cleanup and high-resolution GC separation often accomplishes much the same thing. In the case when the interfering compounds have a different elemental composition from the dioxin molecule or fragment being looked for, but have the same numerical mass, high-resolution mass spectrometry is suitable to distinguish between the dioxin and the interfering compound (see Table 6). In the case of compounds with the same elemental composition as the dioxin in question, HRMS is of little use and HRGC is required to separate the compounds; low-resolution MS is then sufficient to characterize the separated dioxin. At very high resolution, mass spectrometry loses sensitivity, and at resolution 80,000, individual TCDDs may be resolved into a multitude of peaks because of the natural abundance of isotopes of carbon, hydrogen, and oxygen as well as chlorine.

## 9. *METABOLITES OF PCDDs AND PCDFs*

### 9.1. *Hydroxylated PCDDs as Their Methyl Ethers*

Tulp and Hutzinger synthesized a number of methyl ethers of PCDDs and investigated their mass spectra.[15] The mass fragmentation patterns of the structural isomers examined were very specific and allowed unambiguous assignment of the position of the methoxyl group in the molecule. The fragmentation patterns involved can be explained in terms of plausible mechanisms via stable-charge delocalized oxonium ions.

A major metabolic route for PCDDs in rats is conversion to hydroxylated species. Conversion of those metabolites to their methyl ethers

*TABLE 13*
*Structures, Mass Spectral Data, and GC Retention Times for PCDD Methyl Ethers*[a]

| Structure | m/z | (M)$^{+\cdot}$ | Mass spectrum[b] [M-15]$^+$ | [M-43]$^+$ | Retention time[c] |
|---|---|---|---|---|---|
| 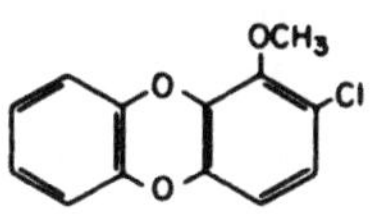 | 248 | 100 | — | 68 | 12.0(120) |
| 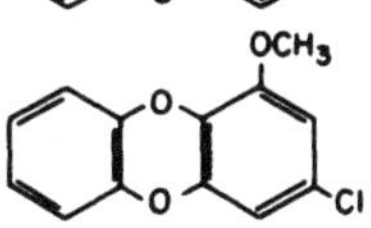 | 248 | 100 | — | 73 | 12.3(120) |
| 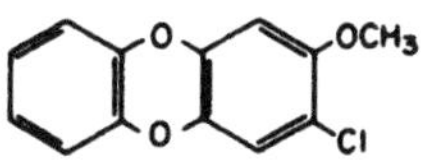 | 248 | 100 | 70 | 93 | 13.4(120) |
| 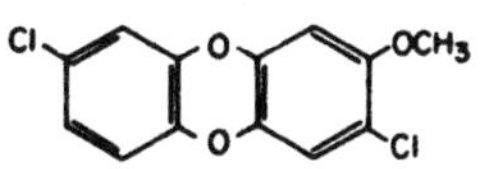 | 282 | 100 | 48 | 49 | 9.3(170) |

[a] Reference 15.
[b] Calculations based on $^{35}$Cl.
[c] Retention times in minutes. GC temperatures in parentheses.

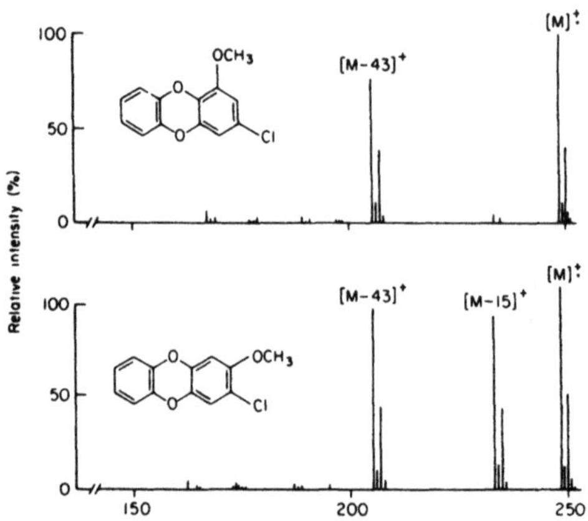


FIGURE 14. Partial mass spectra *m/z* of methoxychlorodibenzo-*p*-dioxins. (From Ref. 15.)

allows straightforward identification of the substitution pattern of hydroxyl groups in the molecule. In the dibenzo-*p*-dioxin molecule, there are two different substitution positions, the 1-position (identical to the 4-, 6-, and 9-positions) and the 2-position (identical to the 3-, 7-, and 8-positions). Structures, GC retention times, and mass spectral data for representative methoxychlorodibenzo-*p*-dioxins are given in Table 13.

Gas chromatography/mass spectrometry analyses were performed on a Hewlett-Packard 5984 system equipped with a 4 mm × 180 cm glass column packed with 0.2% Carbowax 20M on Chromosorb W. Temperatures were programmed as follows: 120 or 170 °C for 2 min and 8 °C/min to 240 °C.

The major fragment of a methoxychlorodibenzo-*p*-dioxin with the methoxy group in the 1-position is $[M - 43]^+$ (Figure 14), followed by loss of $CH_3CO$. When the methoxyl group is in the 2-position, however, $[M - 15]^+$ ions are also observed. The presence of the $[M - 15]^+$ ion in the mass spectrum of 2-methoxy isomers can be explained by the formation of a stable oxonium ion (Fig. 15). The required charge delocalization cannot occur via loss of the methyl group in 1-methoxy isomers. This difference in mass spectral fragmentation is of great use

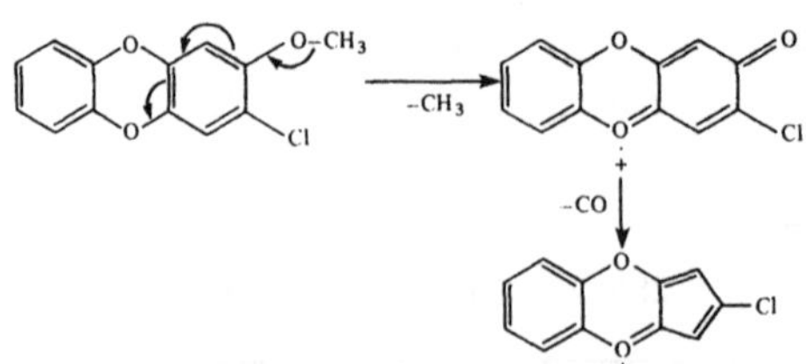


FIGURE 15. Suggested initial mass fragmentation of 2-methoxy-PCDD. (From Ref. 15.)

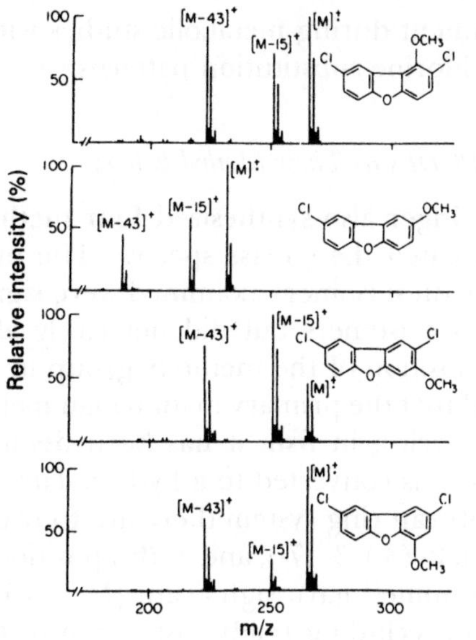


FIGURE 16. Partial mass spectra of methoxychlorodibenzofurans. (From Ref. 15.)

*TABLE 14*

*Structures, Mass Spectral Data, and GC Retention Times for PCDF Methyl Ethers*[a]

| Structure | $m/z$ | $M^{\dot{+}}$ | Mass spectrum[b] $[M\text{-}15]^+$ | $[M\text{-}43]^+$ | Retention time[c] |
|---|---|---|---|---|---|
| 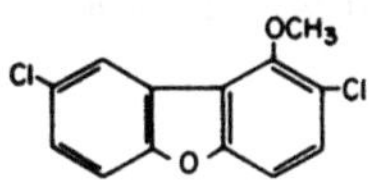 | 266 | 100 | 73 | 96 | 7.6(200 iso) |
| 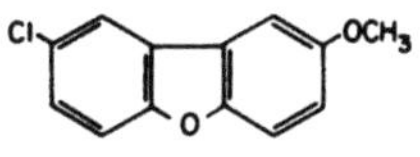 | 232 | 100 | 58 | 33 | 10.6(200 iso) |
| 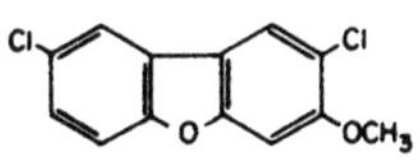 | 266 | 40 | 100 | 67 | 17.2(200 iso) |
| 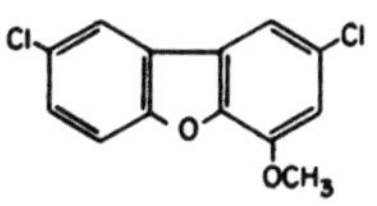 | 266 | 100 | 25 | 52 | 14.3(200 iso) |

[a] Reference 15.

[b] Calculations based on $^{35}Cl$.

[c] Retention times in minutes. GC temperatures in parentheses.

in structural assignment during metabolic studies with pure PCDD isomers with known chlorine substitution patterns.

### 9.2. *Hydroxylated PCDFs as Their Methyl Ethers*

Tulp and Hutzinger also synthesized four methyl ethers of PCDF isomers and investigated their mass spectra. The mass fragmentation patterns of the structural isomers examined were significantly different for the representative isomers but did not easily allow unambiguous assignment of the position of the methoxy group in the molecule.

It is anticipated that the primary mammalian metabolites for PCDFs are hydroxylated species; in fish, it has been demonstrated that 2,8-dichlorodibenzofuran is converted to a hydroxylated compound.

In the dibenzofuran ring system there are four unique substitution positions, the 1- (9-), 2- (8-), 3- (7-), and 4- (6-) positions. All four isomers synthesized and examined gave significant $[M - 15]^+$ and $[M - 43]^+$ ions (Fig. 16), thus precluding ready distinction of the isomers on the basis of the presence or absence of these peaks in their mass spectra. Gas chromatography retention times and mass spectral data are listed in Table 14. Gas chromatography/mass spectrometry conditions were as described for hydroxylated PCDD above. Although there are differences in relative intensities of the isomers examined, they are insufficient for structural elucidation since the relative intensities are probably also affected by the number and substitution pattern of chlorine atoms.

### 9.3. *PCDD Metabolites*

Rat metabolism studies with a series of PCDD (Fig. 17) yielded a number of hydroxylated and thiomethylated metabolites. Metabolic

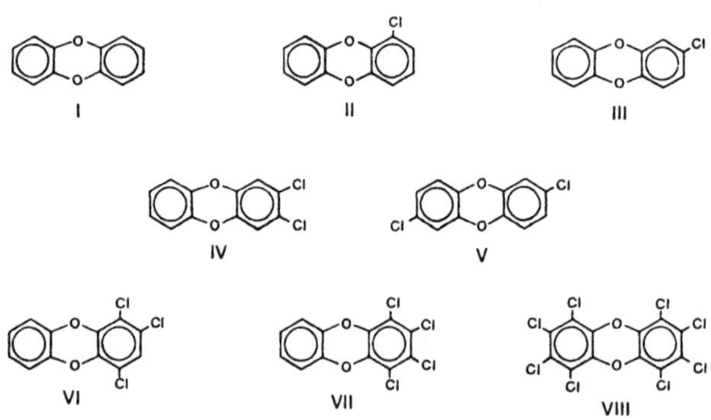


FIGURE 17. Structures of PCDDs investigated. (From Ref. 58.)

*TABLE 15*
*Mass Spectral Data of Metabolites from Compounds I–VIII*[a]

| No. | Molecular formula | Retention time[a] | *m/z* | Mass spectrum[b] M+ | M-15 | M-43 | M-46 | M-50 |
|---|---|---|---|---|---|---|---|---|
| I | $C_{12}H_8O_2$ | 13.8 (100°) | 184 | 100 | — | — | — | — |
| Ia | $C_{12}H_7O_2(OCH_3)$ | 22.2 (100°) | 214 | 100 | 44 | 66 | — | — |
| Ib | $C_{12}H_7O_2(SCH_3)$ | 22.7 (100°) | 230 | 100 | — | — | 58 | — |
| Ic | $C_{12}H_6O_2(OCH_3)_2$ | 29.7 (100°) | 244 | 100 | 41 | 33 | — | — |
| II | $C_{12}H_7O_2Cl$ | 14.0 (120°) | 218 | 100 | — | — | — | — |
| IIa | $C_{12}H_6O_2Cl(OCH_3)$ | 19.0 (120°) | 248 | 100 | 32 | 51 | — | — |
| IIb | $C_{12}H_6O_2Cl(OCH_3)$ | 21.2 (120°) | 248 | 100 | 27 | 50 | — | — |
| IIc | $C_{12}H_6O_2Cl(OCH_3)$ | 22.0 (120°) | 248 | 100 | 35 | 46 | — | — |
| IId | $C_{12}H_6O_2Cl(OCH_3)$ | 22.8 (120°) | 248 | 100 | 81 | 42 | — | — |
| IIe | $C_{12}H_6O_2Cl(SCH_3)$ | 20.0 (120°) | 264 | 100 | — | — | 80 | 40 |
| IIf | $C_{12}H_5O_2Cl(OCH_3)_2$ | 28.2 (120°) | 278 | 100 | 52 | 31 | — | — |
| IIg | $C_{12}H_5O_2Cl(OCH_3)_2$ | 29.5 (120°) | 278 | 100 | 62 | 20 | — | — |
| III | $C_{12}H_7O_2Cl$ | 13.6 (120°) | 218 | 100 | — | — | — | — |
| IIIa | $C_{12}H_6O_2Cl(OCH_3)$ | 21.4 (120°) | 248 | 100 | 34 | 52 | — | — |
| IIIb | $C_{12}H_6O_2Cl(OCH_3)$ | 21.8 (120°) | 248 | 100 | 41 | 58 | — | — |
| IIIc | $C_{12}H_6O_2Cl(OCH_3)$ | 22.8 (120°) | 248 | 100 | 76 | 31 | — | — |
| IIId | $C_{12}H_6O_2Cl(SCH_3)$ | 21.8 (120°) | 264 | 100 | — | — | 42 | 30 |
| IIIe | $C_{12}H_5O_2Cl(OCH_3)_2$ | 26.9 (120°) | 278 | 100 | 28 | 10 | — | — |
| IIIf | $C_{12}H_5O_2Cl(OCH_3)_2$ | 28.7 (120°) | 278 | 100 | 34 | 33 | — | — |
| IV | $C_{12}H_6O_2Cl_2$ | 17.9 (120°) | 252 | 100 | — | — | — | — |
| IVa | $C_{12}H_5O_2Cl_2(OCH_3)$ | 25.3 (120°) | 282 | 77 | 30 | 100 | — | — |
| IVb | $C_{12}H_4O_2Cl_2(OCH_3)_2$ | 30.5 (120°) | 312 | 100 | 32 | 10 | — | — |
| V | $C_{12}H_6O_2Cl_2$ | 18.1 (120°) | 252 | 100 | — | — | — | — |
| Va | $C_{12}H_5O_2Cl_2(OCH_3)$ | 26.0 (120°) | 282 | 100 | 47 | 55 | — | — |
| Vb | $C_{12}H_5O_2Cl_2(OCH_3)$ | 26.6 (120°) | 282 | 100 | 75 | 47 | — | — |
| Vc | $C_{12}H_5O_2Cl(OCH_3)_2$ | 28.8 (120°) | 278 | 100 | 28 | 29 | — | — |
| Vd | $C_{12}H_4O_2Cl_2(OCH_3)_2$ | 32.5 (120°) | 312 | 100 | 41 | 45 | — | — |
| VI | $C_{12}H_5O_2Cl_3$ | 16.5 (140°) | 286 | 100 | — | — | — | — |
| VIa | $C_{12}H_4O_2Cl_3(OCH_3)$ | 23.3 (140°) | 316 | 100 | 42 | 63 | — | — |
| VIb | $C_{12}H_3O_2Cl_3(OCH_3)_2$ | 27.9 (140°) | 346 | 100 | 95 | 33 | — | — |
| VII | $C_{12}H_4O_2Cl_4$ | 21.0 (140°) | 320 | 100 | — | — | — | — |
| VIIa | $C_{12}H_3O_2Cl_4(OCH_3)$ | 27.3 (140°) | 350 | 100 | 33 | 48 | — | — |
| VIIb | $C_{12}H_2O_2Cl_4(OCH_3)_2$ | 32.6 (140°) | 380 | 100 | 51 | 13 | — | — |
| VIII | $C_{12}O_2Cl_8$ | 27.0 (160°) | 456 | 100 | — | — | — | — |

[a] Reference 58.
[b] For GC conditions, see p. 288. Start temperatures are given in parentheses. Retention times are given in minutes.
[c] Calculations based solely upon $^{35}Cl$ and given as percentage of base peak (= 100%).

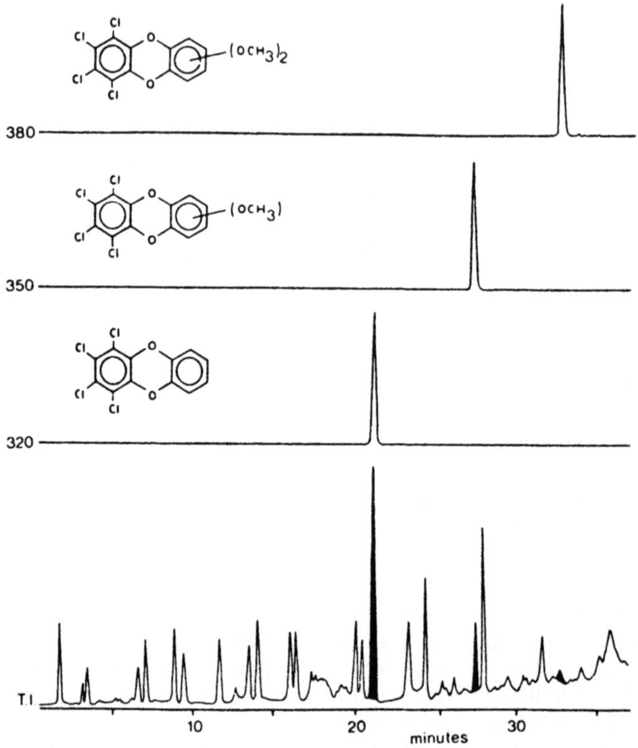


FIGURE 18. Total ion current and selected-ion chromatograms for masses 320, 350, and 380, of a fecal extract of a rat fed 1,2,3,4-TCDD. Filled peaks represent chlorine-containing compounds by a GC/MS computer technique. Masses 350 and 380 correspond to mono- and dihydroxy derivatives (as their methyl ethers) respectively. (From Ref. 58.)

products were examined using GC/MS and the data in Table 15 was generated. Gas chromatography/mass spectrometry data was screened using a computer program to selectively retrieve compounds containing chlorine; selected-ion monitoring for ions corresponding to the calculated masses of molecular ions of potential metabolites (as their methyl ethers) was also used (Fig. 18).[37]

*9.3.1. Monohydroxylated PCDDs.* All monohydroxylated PCDD metabolites (as their methyl ethers) showed $[M - 15]^+$ and $[M - 43]^+$ fragment ions, consistent with the methoxyl group being in the 2-, 3-, 7-, or 8-position in the molecule (see earlier section on PCDD methyl ethers). Compounds I, IV, and VII are symmetrical and gave only one monohydroxyl metabolite; hence, their structural assignments could be made directly from the GC/MS data. Mass spectra of their metabolites are shown in Fig. 19.

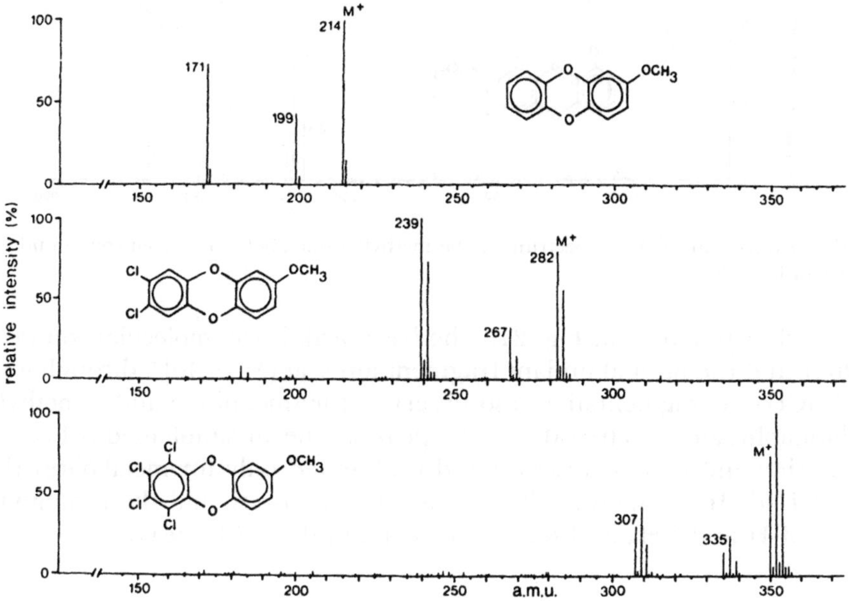


FIGURE 19. Partial mass spectra of metabolite Ia: 2-methoxydibenzo-*p*-dioxin; IVa: 2,3-dichloro-7-methoxydibenzo-*p*-dioxin; and VIIa: 1,2,3,4-TCDD. (From Ref. 58.)

*9.3.2. Dihydroxylated PCDDs.* Compounds I–VII also yielded dihydroxy metabolites which were analyzed by GC/MS as their dimethyl ethers. All dimethoxy PCDDs examined (Ic, IIf,g, IIIe,f, IVb, Vd, VIb, and VIIb) showed $[M - 15]^+$ and $[M - 43]^+$ ions formed by loss of $CH_3$ and $COCH_3$, respectively. The mass spectrum of the dihydroxy metabolite of compound IV as its dimethyl ether is shown in Fig. 20.

*9.3.3. Sulfur-Containing Metabolites of PCDDs.* Methylthiochlorodibenzo-*p*-dioxins were isolated from extracts of excreta of compounds I, II, or III.[58] The mass spectrum of the methylthio derivative of com-

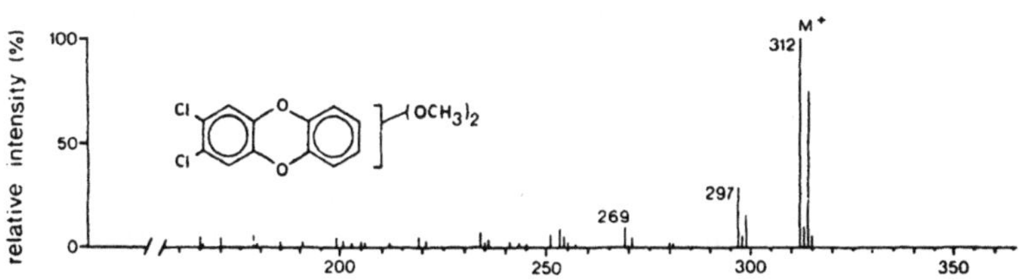


FIGURE 20. Partial mass spectrum of the dimethyl ether of the dihydroxyl metabolite of compound IV. (From Ref. 58.)

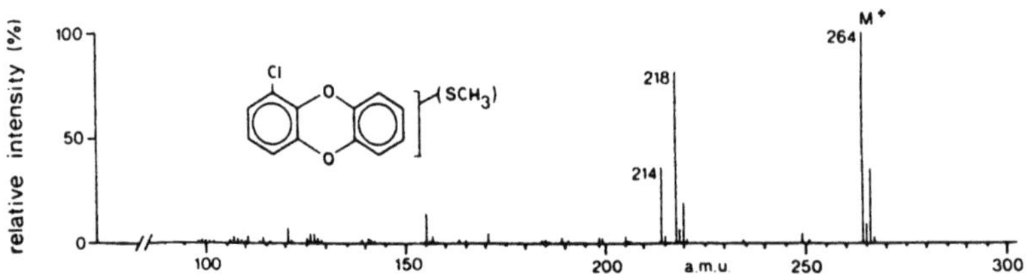


FIGURE 21. Partial mass spectrum of the methythio-metabolite IIb from compound II. (From Ref. 58.)

pound II is shown in Fig. 21. The base peak is the molecular ion (*m/z* 264), and the most abundant fragment ion is at $[M - 46]^+$ through loss of $SCH_3$, a fragmentation also observed for thioanisole and 2-methylthionaphthalene. The $[M - 50]^+$ peak is due to simultaneous loss of chlorine and a methyl radical and is observed only for metabolites IIc and IIId. In compound Ib there are no chlorines, and the analogous loss of H and $CH_3$ is observed to yield an $[M - 16]^+$ peak.

## REFERENCES

1. H. R. Buser and C. Rappe, Chemosphere **7,** 199–211 (1978).
2. J. R. Hass, M. D. Friesen, and M. K. Hoffman, *Org. Mass Spectrom.* **14,** 9–16 (1979).
3. J. R. Hass, M. D. Friesen, D. J. Harvan, and C. E. Parker, *Anal. Chem.* **50,** 1474–1479 (1978).
4. D. F. Hunt, T. M. Harvey, and J. W. Russell, *J.C.S. Chem. Comm.* 151–152 (1975).
5. D. L. Stalling, J. D. Petty, L. M. Smith, C. Rappe, and H. R. Buser, in (O. Hutzinger, R. W. Frei, E. Merian, and F. Pocchiari, eds.), *Chlorinated Dioxins and Related Compounds: Impact on the Environment,* Pergamon Press, Oxford, 1982, pp. 77–85.
6. A. Cavallaro, G. Bandi, G. Invernizzi, L. Luciani, E. Mongini, and G. Gorni, in (O. Hutzinger, R. W. Frei, E. Merian, and F. Pocchiari, eds.), *Chlorinated Dioxins and Related Compounds: Impact on the Environment,* Pergamon Press, Oxford, 1982, pp. 55–65.
7. E. C. Horning, D. I. Carroll, I. Dzidic, S.-N. Lin, R. N. Stillwell, and J.-P. Thenot, *J. Chromatogr.* **142,** 481–495 (1977).
8. H. Budzikiewicz, *Angew. Chem. Int. Ed. Engl.* **20,** 624–637 (1981).
9. R. K. Mitchum, G. F. Moler, and W. A. Korfmacher, *Anal. Chem.* **52,** 2278–2282 (1980).
10. E. K. Chess and M. L. Gross, *Anal. Chem.* **52,** 2057–2061 (1980).
11. L. A. Shadoff and R. A. Hummel, *Biomed. Mass Spectrom.* **5,** 7–13 (1978).
12. J. Ryan, Health and Welfare Canada, Ottawa, personal communication, 1980.
13. H. R. Buser and C. Rappe, *Anal. Chem.* **52,** 2257–2262 (1980).
14. H. R. Buser, Ph.D. thesis, University of Umea, 1978.
15. M. Th. M. Tulp and O. Hutzinger, *Biomed. Mass Spectrom.* **5,** 224–231 (1978).
16. R. C. Dougherty, *Biomed. Mass Spectrom.* **8,** 283–292 (1981).
17. T. J. Nestrick, L. L. Lamparski, and R. H. Stehl *Anal. Chem.* **51,** 2273–2281 (1979).
18. M. Th. M. Tulp, G. Sundström, L. B. J. M. Martron, and O. Hutzinger, *Xenobiotica* **9,** 65–77 (1979).

19. H. R. Buser and H. P. Bosshardt, *J. Chromatogr.* **90,** 71–77 (1974).
20. H. R. Buser, *J. Chromatogr.* **107,** 295–310 (1975).
21. H. R. Buser and H. P. Bosshardt, *J. Assoc. Off. Anal. Chem.* **59,** 562–569 (1976).
22. D. A. Elvidge, *Analyst* **96,** 721–727 (1971).
23. S. D. Erk, M. L. Taylor, and T. O. Tiernan, *Chemosphere* **8,** 7–14 (1979).
24. L. A. Shadoff, W. W. Blaser, C. W. Kocher, and H. G. Fravel, *Anal. Chem.* **50,** 1586–1588 (1978).
25. G. W. Bowes, M. J. Mulvihill, M. R. DeCamp, and A. S. Kende, *J. Agric. Food Chem.* **23,** 1222–1223 (1975).
26. J. A. G. Roach and I. H. Pomerantz, *Bull. Environ. Contam. Toxicol.* **12,** 338–342 (1974).
27. D. Firestone, *J. Agric. Food Chem.* **25,** 1274–1280, 7–14 (1977).
28. M. Morita, J. Nakagawa, K. Akiyama, S. Mimura, and N. Isono, *Bull. Environ. Contam. Toxicol.* **18,** 67–73 (1977).
29. K. Olie, P. L. Vermeulen, and O. Hutzinger, *Chemosphere* **6,** 455–459 (1977).
30. L. L. Lamparski, N. H. Mahle, and L. A. Shadoff, *J. Agric. Food Chem.* **26,** 1113–1116 (1978).
31. M. L. Langhorst and L. A. Shadoff, *Anal. Chem.* **52,** 2037–2044 (1980).
32. G. Sundström, S. Jensen, B. Jansson, and K. Erne, *Arch. Environ. Contam. Toxicol.* **8,** 441–448 (1979).
33. D. Firestone, M. Clower, Jr., A. P. Borsetti, R. H. Teske, and P. E. Long, *J. Agric. Food Chem.* **27,** 1171–1177 (1979).
34. H. Kuroki and Y. Masuda, *Chemosphere* **7,** 771–777 (1978).
35. G. A. Eiceman, A. C. Viau, and F. W. Karasek, *Anal. Chem.* **52,** 1492–1496 (1979).
36. G. A. Eiceman, R. E. Clement, and F. W. Karasek, *Anal. Chem.* **51,** 2343–2350 (1979).
37. K. Olie and O. Hutzinger, in (A. Frigerio, ed.), *Recent Developments in Mass Spectrometry in Biochemistry and Medicine,* Vol. 1, Plenum, New York, 1978.
38. A. Gara, C.-A. Nilsson, K. Anderson, and C. Rappe, *Chemosphere* **8,** 405–414 (1979).
39. L. L. Lamparaski and T. J. Nestrick, in (O. Hutzinger, R. W. Frei, E. Merian, and F. Pocchiari, eds.), *Chlorinated Dioxins and Related Compounds: Impact on the Environment,* Pergamon Press, Oxford, 1982, pp. 1–13.
40. H. R. Buser, *J. Chromatogr.* **114,** 95–108 (1975).
41. H. R. Buser, *Anal. Chem.* **48,** 1553–1557 (1976).
42. H. R. Buser, *Anal. Chem.* **49,** 918–922 (1977).
43. C. Rappe, A. Gara, H. R. Buser, and H. P. Bosshardt, *Chemosphere* **6,** 231–236 (1977).
44. H. R. Buser, H. P. Bosshardt, and C. Rappe, *Chemosphere* **7,** 109–119 (1978).
45. H. R. Buser, H. P. Bosshardt, C. Rappe, and R. Lindahl, *Chemosphere* **7,** 419–429 (1978).
46. H. R. Buser, H. P. Bosshardt, and C. Rappe, *Chemosphere* **7,** 165–172 (1978).
47. H. R. Buser, *Chemosphere* **8,** 415–424 (1979).
48. J. Paasivirta, J. Enqvist, S. Räisänen, and P. Passivuo, *Chemosphere* **6,** 335 (1977).
49. R. L. Harless, D. O. Oswald, and M. K. Wilkinson, 26th Annual Conference of Mass Spectrometry and Allied Topics, May 28–June 2, 1978.
50. L. L. Lamparski and T. J. Nestrick, *Anal. Chem.* **52,** 2045–2054 (1980).
51. H. R. Buser, C. Rappe, and A. Garå, *Chemosphere* **7,** 439–449 (1978).
52. R. L. Harless, E. O. Oswald, M. K. Wilkinson, A. E. Dupuy, Jr., D. D. McDaniel, and Han Tai, *Anal. Chem.* **52,** 1239–1245 (1980).
53. J. Paasivirta, M. Lahtiperä, and J. Leskijärvi, in (O. Hutzinger, R. W. Frei, E. Merian, and F. Pocchiari, eds.), *Chlorinated Dioxins and Related Compounds: Impact on the Environment,* Pergamon Press, Oxford, 1982, pp. 191–200.
54. R. Lindahl, C. Rappe, and H. R. Buser, *Chemosphere* **9,** 351–361 (1980).

55. H. R. Buser, *J. Chromatogr.* **129,** 303–307 (1976).
56. T. J. Nestrick, L. L. Lamparski, and D. I. Townsend, *Anal. Chem.* **52,** 1865–1874 (1980).
57. N. H. Mahle and R. L. Shadoff, *Biomed. Mass Spectrom.* **9,** 45–60 (1982).
58. M. Th. M. Tulp and O. Hutzinger, *Chemosphere* **7,** 761–768 (1978).
59. R. M. M. Kooke, J. W. A. Lustenhouwer, and O. Hutzinger, *Anal. Chem.* **53,** 461–463 (1981).

13

# DDT AND RELATED COMPOUNDS

## GÖRAN SUNDSTRÖM and BO JANSSON

### 1. INTRODUCTION

DDT is probably the most notorious of all compounds in the group of insecticides of the chlorinated hydrocarbon type. Due to its low price, the result of a simple synthetic route using cheap starting materials, DDT has been used on a global scale in enormous quantities since World War II. It has been estimated that more than 2 billion kilos has been used since 1940.

The first uses of DDT included attempts to eradicate mosquito-borne diseases like malaria, dengue fever, and filariasis and it has been claimed that death control from the use of DDT is perhaps the most important factor in the world population explosion.[1]

However, some of the factors that made DDT attractive as an insecticide—for example, the long duration of the treatment, the high fat solubility, and the general stability—also resulted in DDT becoming the prototype of an environmental pollutant, and both DDT and its major degradation product DDE are today found in the environment all over the world. The persistence and long-time effects, as well as the development of resistant insect strains, have today resulted in a prohibition of the use of DDT in many countries. Metcalf[1] has reviewed the history, chemistry, and biology of DDT and the development of useful analogs which still are under active investigation.

DDT is the commonly used name for 2,2-*bis*(4-chlorophenyl)-1,1,1-trichloroethane (I) and is nowadays usually more descriptively named *p,p'*-DDT (from its older name using the *p*-chlorophenyl denomination).

---

*GÖRAN SUNDSTRÖM* and *BO JANSSON* • Special Analytical Laboratory, National Swedish Environment Protection Board, Box 1302, S-171 25 Solna, Sweden.

Technical DDT is synthesized by condensation of chloral hydrate with chlorobenzene in the presence of sulfuric acid. The synthetic route gives rise to several by-products, and one of the most important, *o,p'*-DDT [2-(2-chlorophenyl)-2-(4-chlorophenyl)-1,1,1-trichloroethane, II], constitutes 10–20% of the raw product. This implies that two series of all metabolites or degradation products exist, that is, the *p,p'*- and the *o,p'*-series. In Table 1 all major degradation products and derivatives are schematically depicted.

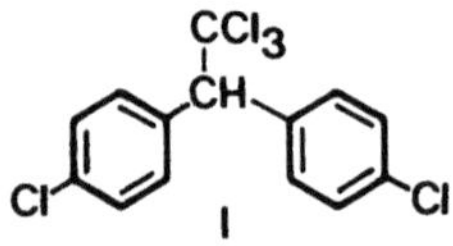
I

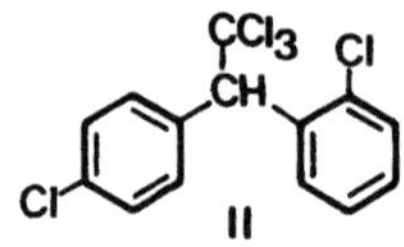
II

In a review, Fishbein[11] has treated the interconversions involved in the synthesis, metabolism, and abiotic conversions of DDT compounds, as well as the analytical aspects.

The present chapter is intended to cover data and aspects of the mass spectrometry of DDT and related compounds. Several authors have previously reviewed this matter[12–14,47] and it is our intention to only briefly discuss the data already presented and concentrate on new data and developments.

## 2. *ELECTRON IMPACT SPECTRA*

### 2.1. *DDT and Derivatives Containing Chlorophenyl Groups*

In Table 1 mass spectral data has been collected from electron impact (EI) studies utilizing detection of positive ions (EIPI) of *p,p'*-DDT, most of its common derivatives, and some related compounds. All figures of molecular weights or weights of fragments given in this chapter refer to ions containing $^{35}Cl$ only, in order to facilitate the interpretation of data.

It should be pointed out that although most data in the literature has been obtained from the *p,p'*-isomers of the compounds that will be discussed, there are usually no distinguishable differences in the EIPI spectra of the *p,p'*- and the *o,p'*-isomers.[12,2] The EIPI data in Table 1 is therefore generally valid for both isomers of each compound and special techniques have to be applied to distinguish between the isomers by mass spectrometry (see below). However, upon gas chromatographic

*TABLE 1*

*Electron Impact Positive-Ion Mass Spectral Data of p,p'-DDT, A Number of Degradation Products, and Related Compounds Containing the 4-Chlorophenyl Group (=R in the Table)*

| Name | Structure | Molecular ion | Base peak | Other major fragments |
|---|---|---|---|---|
| DDT[a] | $CCl_3$<br>\|<br>R—CH—R | 352($C_{14}H_9Cl_5$) | 235($C_{13}H_9Cl_2$) | 200($C_{13}H_9Cl$), 199($C_{13}H_8Cl$), 176($C_{14}H_8$), 165($C_{13}H_9$) |
| DDD[a] (TDE) | $CHCl_2$<br>\|<br>R—CH—R | 318($C_{14}H_{10}Cl_4$) | 235($C_{13}H_9Cl_2$) | 200($C_{13}H_9Cl$), 199($C_{13}H_8Cl$), 176($C_{14}H_8$), 165($C_{13}H_9$) |
| DDMS[b] | $CH_2Cl$<br>\|<br>R—CH—R | 284($C_{14}H_{11}Cl_3$) | 235($C_{13}H_9Cl_2$) | 165($C_{13}H_9$) |
| DDNS[c] | $CH_3$<br>\|<br>R—CH—R | 250($C_{14}H_{12}Cl_2$) | 235($C_{13}H_9Cl_2$) | 215($C_{14}H_{12}Cl$), 201($C_{13}H_{10}Cl$), 165($C_{13}H_9$) |
| DDE[a] | $CCl_2$<br>‖<br>R—C—R | 316($C_{14}H_8Cl_4$) | 246($C_{14}H_8Cl_2$) | 281($C_{14}H_8Cl_3$), 233($C_{13}H_7Cl_2$), 211($C_{14}H_8Cl$), 176($C_{14}H_8$) |
| DDMU[b] (TDEE) | CHCl<br>‖<br>R—C—R | 282($C_{14}H_9Cl_3$) | 212($C_{14}H_9Cl$) | 233($C_{13}H_7Cl_2$) |
| DDNU[b] | $CH_2$<br>‖<br>R—C—R | 248($C_{14}H_{10}Cl_2$) | 178($C_{14}H_{10}$) | 233($C_{13}H_7Cl_2$) |
| Dicofol[a] (Kelthane) | $CCl_3$<br>\|<br>R—C—R<br>\|<br>OH | 368($C_{14}H_9OCl_5$) | 139($C_7H_4OCl$) | 251($C_{13}H_9OCl_2$) |

*(continued overleaf)*

*TABLE 1 (continued)*

| Name | Structure | Molecular ion | Base peak | Other major fragments |
|---|---|---|---|---|
| DDOH[b] | $R{-}CH(CH_2OH){-}R$ | $266(C_{14}H_{12}OCl_2)$ | $235(C_{13}H_9Cl_2)$ | |
| DDCHO[d] (DDAA) | $R{-}CH(CHO){-}R$ | $264(C_{14}H_{10}OCl)$ | $235(C_{13}H_9Cl_2)$ | |
| DDA[b] | $R{-}CH(COOH){-}R$ | $280(C_{14}H_{10}O_2Cl_2)$ | $235(C_{13}H_9Cl_2)$[h] | |
| DDCN[e] | $R{-}CH(CN){-}R$ | $261(C_{14}H_9NCl_2)$ | $226(C_{14}H_9NCl)$ | $199(C_{13}H_8Cl)$, $191(C_{14}H_9N)$, $190(C_{14}H_8N)$ |
| DDM[b] | $R{-}CH_2{-}R$ | $236(C_{13}H_{10}Cl_2)$ | $201(C_{13}H_{10}Cl)$ | |
| DBH[b] | $R{-}CH(OH){-}R$ | $252(C_{13}H_{10}OCl_2)$ | $139(C_7H_4OCl)$ | $111(C_6H_4Cl)$ |
| DBP[f] | $R{-}C({=}O){-}R$ | $250(C_{13}H_8OCl_2)$ | $139(C_7H_4OCl)$ | $111(C_6H_4Cl)$ |
| Chlorobenzilate[a] | $R{-}C(COOC_2H_5)(OH){-}R$ | $324(C_{16}H_{14}O_3Cl_2)$ | $139(C_7H_4OCl)$ | $111(C_6H_4Cl)$ |

| | | | | |
|---|---|---|---|---|
| Dimite[a] | 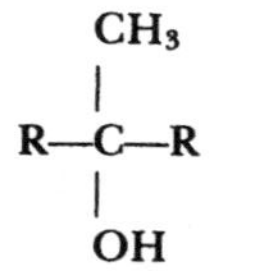 | 266($C_{14}H_{12}OCl_2$) | 251($C_{13}H_9OCl_2$) | 248($C_{14}H_{10}Cl$), 213($C_{14}H_{10}Cl$), 178($C_{14}H_{10}$), 139($C_7H_4OCl$) |
| Etoxinol[a] | 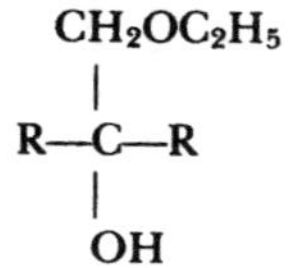 | 310($C_{16}H_{16}O_2Cl_2$) | 251($C_{13}H_9OCl$) | 139($C_7H_4OCl$) |
| DDNU-oxide[d] | 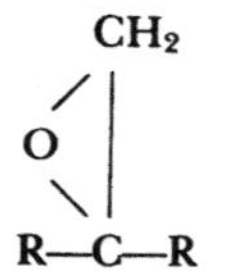 | 264($C_{14}H_{10}OCl_2$) | 199($C_{13}H_8Cl$) | |
| DDNU-diol[d] | 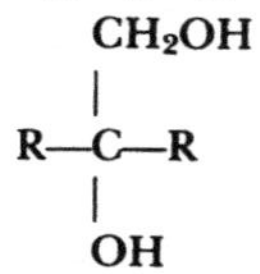 | 282($C_{14}H_{12}O_2Cl_2$) | 251($C_{13}H_9OCl_2$) | 139($C_7H_4OCl$) |
| DDA-amide[g] | 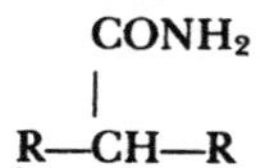 | 279($C_{14}H_{11}NOCl_2$) | 165($C_{13}H_9$) | 244($C_{14}H_{11}NOCl$), 235($C_{13}H_4Cl_2$), 199($C_{13}H_8Cl$) |

[a] Reference 2. [b] Reference 47. [c] Reference 4. [d] Reference 5. [e] References 6–8. [f] References 9 and 47. [g] Reference 10. [h] Fragment of the ethyl ester.

analysis the *o,p′*-isomers always have shorter retention times than the *p,p′* isomers.[11]

Most of the saturated compounds of the DDT-type give very weak or no molecular ions in the EIPI spectra. As is evident from Table 1 these compounds easily lose the aliphatic portion attached to the α-carbon which results in the *m/z* 235 ions. This ion generally forms the base peak in the spectrum and has been interpreted as a stabilized dichlorofluorenyl ion.[3] Analogous base peaks at *m/z* 251 are formed from the compounds of the benzhydrol type with aliphatic side chains (i.e., Dimite, Etoxinol, DDNU-diol). Another diagnostic ion in the spectra of the latter compounds is the 4-chlorophenyl carbonyl ion at *m/z* 139.

The EIPI spectra of the unsaturated compounds are characterized by base peaks formed by the loss of two Cl from the molecular ion. However, these systems also give abundant molecular ions of diagnostic value.

A common problem encountered both upon GC and MS analysis of some of the saturated systems in the DDT series is the decomposition or interconversion of compounds. This can lead to false conclusions as to the identity of components in the samples. The most familiar interconversions are those of the type DDT → DDE, and DDD → DDMU which may occur on gas chromatographic columns not properly deactivated or on other active sites in the instruments. McKinney and co-workers[47] also reported that attempts to obtain mass spectra of Dicofol via the direct probe failed and gave spectra identical to those of DBP instead.

A somewhat more surprising reaction is the formation of DDD from DDT, which has been observed to occur both in the mass spectrometer and in gas chromatographic columns. This type of conversion has been investigated and reviewed by Ott and Gunther.[15]

The use of liquid chromatography for the separation of sensitive compounds before introduction into the mass spectrometer might help to overcome the above-mentioned problems. Lovins and co-workers[16] have evaluated this possibility using *p,p′*-DDT and *p,p′*-DDD as models.

Suggs and co-workers[17] identified DDCHO as a metabolite of DDT via the mass spectrum of the 4-nitrophenylhydrazone derivative. The spectrum showed a prominent molecular ion at *m/z* 399 and a base peak at *m/z* 261.

The mass spectra of chloro-substituted benzophenones have been thoroughly investigated by George, Hassid, and Phillips.[18] The major fragments in the spectra are formed by fissions to oxonium ions. In unsymmetrically substituted isomers the base peak is formed by the oxonium ion containing the smallest number of chlorine atoms; that is, the base peak in the spectrum of 2,4-dichloro-dibenzophenone is *m/z* 105 ($C_7H_5O$) and not 173 ($C_7H_3OCl_2$).

## 2.2. *Metabolites Oxygenated in the Aromatic Rings; Amino Acid Conjugates*

In studies on the pollution of the Baltic with substances of anthropogenic origin a number of phenolic metabolites from PCB (polychlorinated biphenyls) and *p,p'*-DDE have been identified.[19–20] Further studies using rats showed that *p,p'*-DDE is metabolized to at least four phenolic compounds which were identified by comparison with synthetic material.[21]

The mass spectra of the metabolites as such, or as methyl ethers, are very similar and follow the general pattern discussed above for the unsaturated DDT systems. Only the mass spectrum of 2-methoxy-4,4'-DDE gives any further information as to substitution pattern due to the loss of 50 amu ($CH_3Cl$) from the molecular ion which probably results in a dibenzofulvene-type fragment.

Mass spectrometry has been used in several studies on the metabolism of the *o,p'*-isomers of DDD and DDT. A number of metabolites containing hydroxy and methoxy groups in the aromatic rings and with altered side chains were isolated from the excreta of rats,[23,10] chickens,[24] and humans.[25] In all these studies the identification of metabolites was mostly done after comparison with synthetic material and/or by the aid of spectroscopic methods, generally as derivatives (methyl ethers and methyl esters). A large amount of mass spectral (EIPI) data is therefore compiled in the papers mentioned. As expected the compounds mostly give mass spectra of the general type discussed above for the different types of DDT compounds. That is, the saturated systems give base peaks formed by loss of the alkyl residues and in the unsaturated systems the base peak is formed by the molecular ion. Additional information as to detailed substitution patterns (e.g., 2-, 3-, or 4-$OCH_3$) were not derived from the spectra.

By the aid of mass spectrometry the glycine, alanine, and serine conjugates of *o,p'*-DDA and/or *p,p'*-DDA has been characterized (as methyl esters) in excreta from rats,[23,10] humans,[25] syrian hamsters,[26,27] and mice.[26,27]

No molecular ion was observed in the spectrum of authentic *p,p'*-DDA-ser methyl ester but fragments at *m/z* 146 ($C_5H_8NO_4$) and 235 ($C_{13}H_9Cl_2$) together accounted for the molecular weight.[27] Another prominent fragment was formed by loss of water from the molecular ion (*m/z* 363, $C_{18}H_{15}NO_3Cl_2$). The base peak in the spectrum was *m/z* 146 using the direct-probe inlet while *m/z* 128 ($C_5H_8NO_3$) formed the base peak upon analysis via the gas chromatograph, indicating that dehydration occurs in this inlet system.

The molecular ions of the methyl esters of *p,p'*-DDA-gly (*m/z* 351) and of *p,p'*-DDA-ala (*m/z* 365) are very weak, and again, for both compounds, major fragments are found at *m/z* 235. The base peaks are formed by fragments of the amino acids at *m/z* 88 (gly) and 102 (ala).[26]

### 2.3. Sulfur-Containing Metabolites

A methyl sulfone metabolite of *p,p'*-DDE was recently detected in seals from the Baltic.[22] The identity of the metabolite as 2-(4-chloro-3-methanesulfonylphenyl)-2-(4-chlorophenyl)-1,1-dichloroethylene (III) was confirmed by synthesis which also yielded the isomeric 2-methanesulfonyl compound (IV).[28] Both compounds give rise to strong fragments at *m/z* 280 interpreted as fluorene-type ions formed by loss of $CH_3SO_2$ + Cl (114 amu). In the spectrum of the 3-$SO_2CH_3$ isomer the molecular ion (*m/z* 394) forms the base peak whereas the 2-$SO_2CH_3$ isomer exhibits the base peak at *m/z* 125 ($C_7H_6Cl$) and an intense fragment at *m/z* 359 (M − Cl). It is noteworthy that the isomerization to a sulfinate ion with subsequent fragmentations, a typical route for alkyl aryl sulfones,[29] is observed only in the case of compound IV.[28]

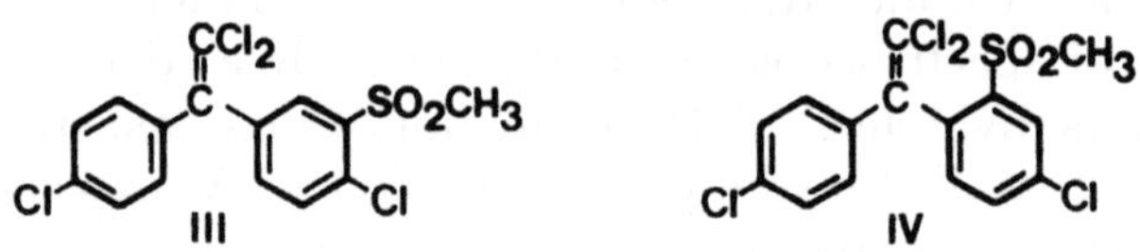


### 2.4. Photoproducts

Mass spectrometry was used by Plimmer and co-workers[30] to indicate the structure of a number of products formed upon photolysis of DDT and DDE in methanol. However, no detailed fragmentation patterns were given, only molecular ions.

The mass spectra of the photoisomers V and VI of DDE and the photodegradation product DMDE (VII) of Methoxychlor have been given by Zepp and co-workers.[31] The mass spectra of the photoproducts were indistinguishable from the isomeric parent compounds but could be separated by chromatographic methods.

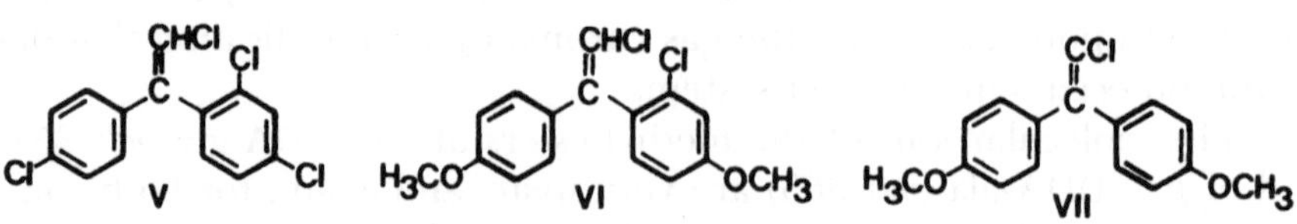

## 2.5. *Related Compounds*

In the paper by Sphon and Damico[2] not only derivatives with chlorine substituents on the aromatic rings were studied (Table 1) but also the mass spectra of Perthane (VIII) and Methoxychlor (IX). The mass spectra of these pesticides follow the general pattern of saturated DDT derivatives with weak molecular ions and the base peaks are formed by the loss of the alkyl sidechains.

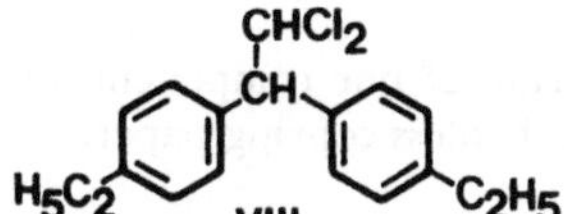

VIII

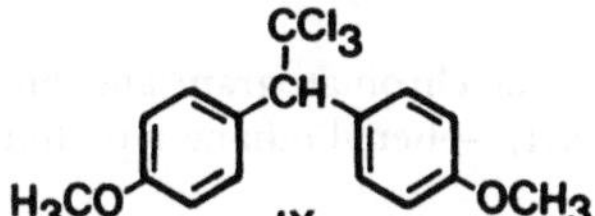

IX

Mazuch, Mitchell, and West[32] identified, by the use of mass spectrometry, tetra-*p*-anisylethylene [1,1,2,2-(4-methoxyphenyl)ethene] as a by-product in technical Methoxychlor ($M^+$, 452 amu; $M^{2+}$, 226 amu).

Sphon and Damico[2] also investigated thoroughly the mass spectra of 9-dichloromethylfluorene and 9-dichloromethylenefluorene in order to possibly establish the structure of the ions *m/z* 165 of DDT and DDD and of the ion *m/z* 246 of DDE. The involvement of fluorene-type structures in the electron impact induced breakdown of DDT compounds has also been discussed by McKinney and co-workers.[47]

A group of compounds structurally related to DDT is presently of increasing environmental interest—1,1-diaryl-substituted ethanes. Several reports have mentioned this type of compound as a possible substitute for PCB in certain applications. The mass spectrum of one diarylethane component of a solvent mixture found in Swedish copying paper is shown in Fig. 1. The chemical mixture is almost identical (as judged from GC) to a Japanese standard containing 1-phenyl-1-(3,4-dimethylphenyl)ethane (33%), 1-phenyl-1-(2,4-dimethylphenyl)ethane (60%), 1-phenyl-1-(2,5-dimethylphenyl)ethane (4%), and 1-phenyl-1-(4-ethylphenyl)ethane (3%). Both major components give identical mass spectra which are characterized by the successive loss of $CH_3$ fragments. The base peak at *m/z* 195 ($M^+ - CH_3$) has been used in mass fragmentographic analyses for the presence of phenyl-xylyl ethane compounds in the Japanese environment, with negative results.[33]

In a paper on carbon skeleton GC/MS of DDT Cooke and co-workers[3] likewise used the $M^+ - CH_3$ fragment (*m/z* 167) of 1,1-diphenylethane to monitor DDT in samples after catalytic dechlorination–hydrogenation of DDT compounds. 1,1-Diphenylethane is also formed in the dechlor-

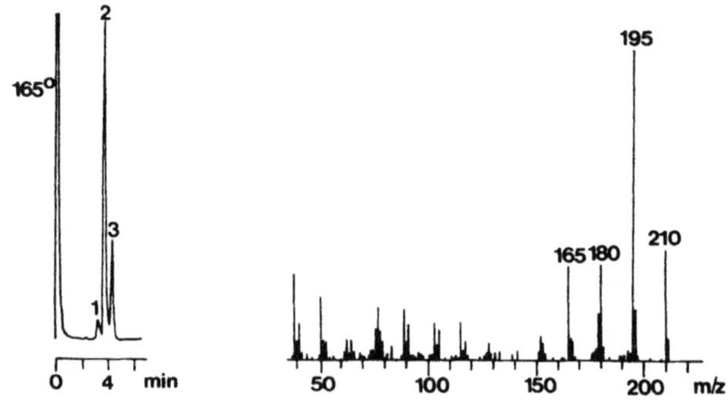


FIGURE 1. Gas chromatogram and mass spectrum of one component (3) of a solvent mixture of xylyl–phenyl ethane type found in carbonless copying paper.

ination reaction of *p,p'*-DDT with nickel boride–sodium borohydride in methanol.[4] The reaction proceeds via a number of intermediate products (DDD, DDMS, DDNS) which all were characterized by GC/MS.

Another product of the diarylethane group which have been mentioned as a possible PCB substitute is named Iralec and is manufactured by the condensation of vinyl chloride and 1,2-dichlorobenzene which results in compounds of the type X. The composition, spectral data, including the mass spectra, and photochemistry of Iralec have been described.[34] As expected, major fragments are formed by the subsequent loss of the methyl group or chlorine atoms from the parent ion.

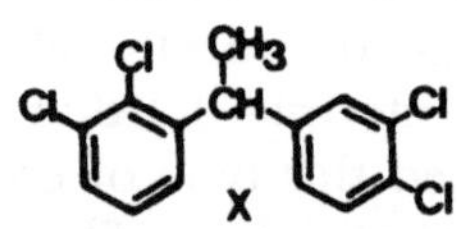


X

The mass spectra of 2,2-*bis*(4-hydroxyphenyl)-1,1-dichloroethylene at ionization energies of 70, 20, and 11 eV has been recorded and thoroughly interpreted by Malinowska and Karwowska.[35]

## 3. *ION KINETIC SPECTRA*

Ionic kinetic energy (IKE) spectroscopy can be used to distinguish the *o,p'*- and *p,p'*-isomers of DDT and DDD.[36] Thus, the ratio of certain ion decomposition peaks were shown to be significantly different for the two types of isomers; that is, the ratio between the energies of the reactions a → b and a → c were 0.31 and 0.17 for *p,p'*- and *o,p'*-DDT, respectively:

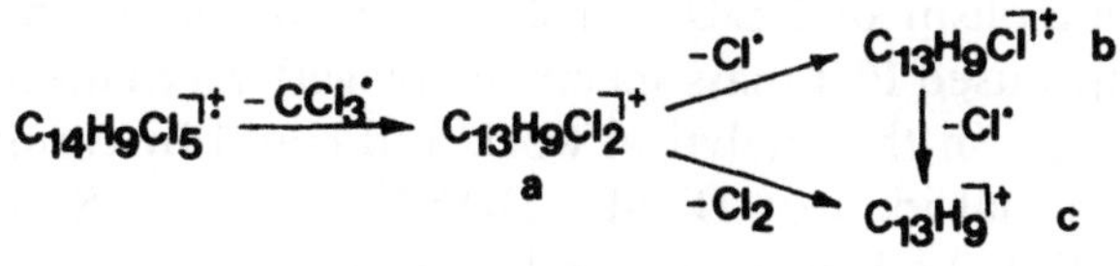


This difference was interpreted as the result of a substituent effect which at least in part is due to the different orientations of the chlorine atoms. A similar difference was found in the IKE spectra of the two DDD isomers while no significant difference was observed for the DDE isomers.[36]

## 4. CHEMICAL IONIZATION SPECTRA

An extensive study of both positive (PCI) and negative chemical ionization (NCI) mass spectra of DDT compounds have been performed by Dougherty, Roberts, and Biros.[37] They used isobutane as reagent gas and analyzed the *p,p'*- and *o,p'*-isomers of DDT, DDD, and DDE plus Perthane, Methoxychlor, DDMU, Dicofol, Dimite, and Chlorobenzilate. PCI mass spectra of the diphenylethane derivatives, except *p,p'*-DDT and Methoxychlor, gave base peaks at $(M - Cl)^+$. An interesting difference was found between *p,p'*- and *o,p'*-DDT—the former isomer gave abundant $(M - HCl)^+$ and the latter $(M - Cl)^+$ ions. No corresponding difference could be found for the two DDD isomers. On PCI all diphenylethene compounds gave intense $M^+$ ions, indicating that the ionization potential of these substances are close to that of the *tert*-butyl radical (7.42 eV). The diphenylmethanol derivatives gave base peaks at $(M - OH)^+$, but ions corresponding to loss of the alkyl chain were also present in these spectra. Several of the investigated compounds gave additive ions, such as $(M + C_3H_7)^+$ and $(M + C_4H_9)^+$. The NCI spectra were much simpler and and for most of the studied substances the $(M + Cl)^-$ ion was the only significant one. The two DDT isomers did not give identical spectra and except for the base peaks at $(M + Cl)^-$ *p,p'*-DDT gave an abundant $(M - H)^-$ ion while *o,p'*-DDT gave a similarly abundant $(M + H)^-$ ion.

Oswald, Albro, and McKinney[13] reported PCI data of some oxygen-containing DDT derivatives. Thus, DBP, DBH, and DDA ethyl ester gave base peaks at *m/z* 111 $(C_6H_4Cl)^+$ and DDOH gave a peak of 50% abundance of the base peak (*m/z* 235) at this mass number. Both the EI and PCI spectra of Dicofol were identical with those obtained from DBP (see above). The PCI spectrum of DDCHO was dominated by an intense peak at $(M + 1)^+$.

The metabolism of DDD in fish has been studied by Cairns and Parfitt.[38] They used PCI mass spectroscopy with methane as the reagent (and carrier) gas in the analytical work and could by comparisons with reference material identify DDMS, DDNU, and DDNS as metabolites. Spectra of DDD (both isomers) and DDMU are given in the original publication.

## 5. *FIELD DESORPTION (FD) SPECTRA*

Schulten and Beckey have suggested the use of FDMS for the detection and characterization of a number of environmental pollutants. Among the compounds studied were the *p,p'*-isomers of DDT, DDA, DBH, DBP, DDNU, and DDOH.[39–41] For all these compounds FD mass spectra can be obtained containing virtually only the molecular ion. At higher emitter temperatures the loss of chlorine atoms and alkyl side chains is indicated in analogy with EI mass spectrometry.

Schulten[41] also showed the possibility to simultaneously detect all the above compounds in an extract. By using suitable emitter currents (5–12 mA) composite mass spectra can be obtained containing almost exclusively the molecular ions. The amount of samples required were in the range $10^{-9}$–$10^{-6}$ g adsorbed on the FD emitter.

Suzuki, Yamato, and Koga[42] applied FDMS to extracts of soil and plants to confirm the presence of chlorinated pesticides detected by GC. Only the presence of *p,p'*-DDT and dieldrin could be confirmed by FDMS after thin-layer chromatographic cleanup of extracts.

## 6. *MASS FRAGMENTOGRAPHY*

As for a number of other important environmental pollutants, mass fragmentographic techniques have been used also for the determination of DDT compounds in environmental samples. Alford[43] analyzed *p,p'*-DDT, *o,p'*-DDD, and *p,p'*-DDE by mass fragmentography and also compared two different softwares. She found peak height to give better data than peak area for the quantitation of DDE and DDD, while the broad DDT peak was best measured by area.

Mestres, Chevallier, and Pagnon[44] used the ions *m/z* 235 and 165 for the determination of DDT and 318 and 246 for DDE, in sediments. Schaefer[45] applied mass fragmentography for the determination of DDT compounds in extracts from marine fish. *p,p'*-DDT and *p,p'*-DDD were determined using the ions *m/z* 237, 235, and 165, while the ions *m/z* 320, 318, 316 (molecular ion cluster), 248, and 246 were used for *p,p'*-DDE. Using the ions *m/z* 318, 284, 282, and 235, Bonelli, Taylor, and Morris[46] determined DDMU, DDMS, DDE, DDD, and DDT in fish extracts.

## *REFERENCES*


1. R. L. Metcalf, *J. Agric. Food Chem.* **21,** 511–519. (1973).
2. J. A. Sphon and J. N. Damico, *Org. Mass Spectrom.* **3,** 51–62. (1970).
3. M. Cooke, K. D. Khallef, G. Nickless, and D. J. Roberts, *J. Chromatogr.* **178,** 183–191. (1979).
4. W. J. Cooper and W. H. Dennis, Jr., *Chemosphere* **7,** 299–305. (1978)
5. G. Planche, A. Croisy, C. Malaveille, L. Tomatis, and H. Bartsch, *Chem.-Biol. Interactions* **25,** 157–175. (1979)
6. E. S. Albone, G. Eglington, N. C. Evans, and M. M. Rhead, *Nature* **240,** 420–421. (1972).
7. S. Jensen, R. Gothe, and M.-O. Kindstedt, *Nature* **240,** 421–422. (1972).
8. B. R. Simoneit, D. H. Smith, and G. Eglington, *Arch. Environ. Contam. Toxicol.* **3,** 385–409. (1975/76).
9. G. D. Veith, D. W. Kuehl, and J. Rosenthal, *J. Assoc. Off. Anal. Chem.* **58,** 1–5. (1975).
10. V. D. Reif and J. E. Sinsheimer, *Drug Metab. Disp.* **3,** 15–25. (1975).
11. L. Fishbein, *J. Chromatogr.* **98,** 177–251. (1974).
12. S. Safe and O. Hutzinger, *Mass Spectrometry of Pesticides and Pollutants,* CRC Press, Cleveland Ohio, 1973, pp. 113–121.
13. E. O. Oswald, P. W. Albro, and J. D. McKinney, *J. Chromatogr.* **98,** 363–448. (1974).
14. R. C. Dougherty, J. D. Roberts, H. P. Tannenbaum, and F. J. Biros, *Mass Spectrometry and NMR Spectroscopy in Pesticides Chemistry,* (R. Hague and F. J. Biros, eds.) Plenum, New York, 1974, pp. 33–47.
15. D. E. Ott and F. A. Gunther, *Res. Rev.* **10,** 70–84, (1965).
16. R. E. Lovins, S. R. Ellis, G. D. Tolbert, and C. R. McKinney, *Anal. Chem.* **45,** 1553–1556. (1973).
17. J. E. Suggs, R. E. Hawk, A. Curley, E. L. Boozer, and J. D. McKinney, *Science* **168,** 582. (1970).
18. W. O. George, D. V. Hassid, and J. Phillips, *Org. Mass Spectrom.* **5,** 605–613. (1971).
19. B. Jansson, S. Jensen, M. Olsson, L. Renberg, G. Sundstrom, and R. Vaz, *Ambio* **4,** 93–97. (1975).
20. G. Sundstrom, B. Jansson, and S. Jensen *Nature* **255,** 627–628. (1975).
21. G. Sundstrom, *J. Agric. Food Chem.* **25,** 18–21. (1977).
22. S. Jensen and B. Jansson, *Ambio* **5,** 257–260. (1976).
23. V. J. Feil, C. H. Lamoureux, E. Styrvoky, R. G. Zaylskie, E. J. Thacker, and G. M. Holman, *J. Agric. Food Chem.* **21,** 1072–1078. (1973).
24. V. J. Feil, C. H. Lamoureux, and R. G. Zaylskie, *J. Agric. Food Chem.* **23,** 382–388. (1975).
25. V. D. Reif, J. E. Sinsheimer, J. C. Ward, and D. E. Schteingart, *J. Pharm. Sci.* **63,** 1730–1736. (1974).
26. L. Wallcave, S. Bronczyk, and R. Gingell, *J. Agric. Food Chem.* **22,** 904–908. (1974).
27. R. Gingell, *Xenobiotica* **6,** 15–20. (1976).
28. A. Bergman and C. A. Wachtmeister, *Acta Chem. Scand.* **B31,** 90–91. (1977).
29. I. Pratanata, L. R. Williams, and R. N. Williams, *Org. Mass Spectrom.* **8,** 175–178. (1974).
30. J. R. Plimmer, U. I. Klingebiel, and B. E. Hummer, *Science* **167,** 67–69. (1970).

31. R. G. Zepp, N. L. Wolfe, L. V. Azarraga, R. H. Cox, and C. W. Pape, *Arch. Environ. Contam. Toxicol.* **6,** 305–314. (1977).
32. L. Mazuch, R. H. Mitchell, and P. R. West, *Chem. Ind.,* 339–340. (1975).
33. K. Sumino, *Arch. Environ. Contam. Toxicol.* **6** 365–369. (1977).
34. G. Sundstrom, K. Olie, O. Hutzinger, S. Safe, and B. Chittim, *Chemosphere* **6,** 103–108. (1977).
35. H. Malinowska and J. Karkowska, *Chemia Analityczna* **24,** 93–101. (1979).
36. S. Safe, O. Hutzinger, W. D. Jamieson, and M. Cook, *Org. Mass. Spectrom.* **7,** 217–224. (1973).
37. R. C. Dougherty, J. D. Roberts, and F. J. Biros, *Anal. Chem.* **47,** 54–59. (1975).
38. T. Cairns and C. H. Parfitt, *Bull. Environ. Contam. Toxicol.* **24,** 504–510. (1980).
39. H. D. Beckey and H. -R. Schulten, *Angew. Chemie Int. Ed.* **14,** 403–415. (1975).
40. H.-R. Schulten, in *New Methods in Environmental Chemistry and Toxicology* (F. Coulston, F. Korte, and M. Goto eds.), Int. Academic Printing, Totsuka, Tokyo, 1973, pp. 31–42.
41. H.-R. Schulten, *J. Agric. Food Chem.* **24,** 743–750. (1976).
42. M. Suzuki, Y. Yamato, and M. Koga, *Biomed. Mass Spectrom.* **5,** 518–523. (1978).
43. A. L. Alford, Environmental Protection Technology Series, EPA-660/2-74-002, U. S. Environmental Protection Agency. pp. 1–20 (1974).
44. R. Mestres, C. Chevallier, and M. Pagnon, *Trav. Soc. Pharm. Montpellier* **36,** 151–160. (1976).
45. R. G. Schaefer, *Chemiker-Zeitung* **98,** 241–247. (1974).
46. E. J. Bonelli, P. A. Taylor, and W. J. Morris, *Am. Lab.* **7(7),** 29–30, 32–35. (1975).
47. J. D. McKinney, E. O. Oswald, S. M. dePaul Palaszek, and B. J. Corbett, *Mass Spectrometry and NMR Spectroscopy in Pesticide Chemistry* (R. Hague and F. J. Biros, eds.), Plenum, New York, 1974, pp. 5–32.

14

# VOLATILE HALOCARBONS

G. A. EICEMAN

## 1. INTRODUCTION

In the last 15 years concern has grown among health authorities and environmentalists over the presence and effect of chlorinated organic compounds on the environment and human health. Since the mid-1970s a particular group of halogenated hydrocarbons, or halocarbons, has become the center of concern in two very different situations. In each instance, however, such concern was expedited through advances in the sensitivity and specificity of analytical techniques and instrumentation. These two problems were (1) partial depletion of the ozone layer through atmospheric reactions of ozone with chlorofluorocarbons and (2) generation of trihalomethanes in drinking water or municipal wastewater through the disinfection process of chlorination. Developments in sample pretreatment and techniques of mass spectrometry or gas chromatography/mass spectrometry (GC/MS) have been critical in the identification, characterization, and formulation of solutions in each environmental issue.

### 1.1. *Chlorofluorocarbons*

Chlorofluorocarbons include a series of chlorofluoro derivatives of methane and ethane which have been referred to as fluorocarbons or Freons, (Table 1) one of many trade names. Common fluorocarbons F-11 and F-12 were first prepared in 1930 for use as refrigerant material and since they exhibited excellent chemical stability and appropriate

*G. A. EICEMAN* • Department of Chemistry, New Mexico State University, Las Cruces, New Mexico 88003.

*TABLE 1*
*Fluorocarbons and Selected Constants*

| Name | Formula | m.p. (°C) | b.p. (°C) | Heat of evaporation at BP (kcal/mole) | Specific heat (kcal/mole) | Critical temp. (°C) | Critical pressure (atm) | Critical density (kg/l) |
|---|---|---|---|---|---|---|---|---|
| Freon 11 | $CCl_3F$ | −111.1 | 23.77 | 43.51 | 0.208 | 198.0 | 43.2 | 0.554 |
| Freon 12 | $CCl_2F_2$ | −155 | −29.8 | 39.9 | 0.204 | 111.5 | 40.95 | 0.555 |
| Freon 13 | $CClF_3$ | − 81.5 | | 35.65 | 0.203 | 28.8 | 39.4 | 0.581 |
| Freon 21 | $CHCl_2F$ | −135 | 8.92 | 57.85 | 0.246 | 178.5 | 51.0 | 0.522 |
| Freon 22 | $CHClF_2$ | −160 | −40.8 | 55.9 | 0.265 | 96.0 | 48.7 | 0.525 |
| Freon 23 | $CHF_3$ | −163 | −82.2 | 56 | 0.28 | 32.3 | 50.5 | — |
| Freon 113 | $C_2Cl_3F_3$ | − 35 | 47.57 | 35.04 | 0.226 | 214.1 | 33.7 | 0.576 |
| Freon 114 | $C_2Cl_2F_4$ | − 94 | 3.55 | 32.8 | 0.232 | 145.7 | 32.1 | 0.582 |
| Freon 115 | $C_2ClF_5$ | −106 | −38 | — | — | — | — | — |

physical properties they gained rapid commercial acceptance. Furthermore, fluorocarbons with different boiling points were prepared through combinations of fluorine and chlorine in different ratios for use in air conditioners and freezers. Later applications included use as propellants in aerosol dispersion of pesticides, hairsprays, deodorants, and paints.[1]

These compounds were produced through either gas or liquid-phase catalytic reactions of anhydrous HF with chlorinated methanes or ethanes:

$$\underset{}{CCl_4} \xrightarrow{HF} \underset{\text{F-11}}{CCl_3F} + \underset{\text{F-12}}{CCl_2F_2} + \underset{\text{F-13}}{CClF_3} \tag{1}$$

$$CHCl_3 \xrightarrow{HF} \underset{\text{F-21}}{CHCl_2F} + \underset{\text{F-22}}{CHClF_2} + \underset{\text{F-23}}{CHF_3} \tag{2}$$

$$C_2Cl_6 \xrightarrow{HF} \underset{\text{F-111}}{CCl_2FCCl_3} + \underset{\text{F-112}}{CCl_2FCCl_2F} + \underset{\text{F-113}}{CClF_2CCl_2F} + \underset{\text{F-114}}{CClF_2CClF_2} \tag{3}$$

Details on the production and release of F-11 and F-12 through 1975 have been summarized by McCarthy.[2] During the mid-1970s over 90% of the total production of fluorocarbons worldwide consisted of F-11, F-12, and F-22. Approximately 75% of F-11, 50% of F-12, and 25% of F-22 were used as propellants while another 35% of F-12 and 67% of F-22 were used as refrigerants.

Between 1975 and 1979 the use of fluorocarbons as propellants in nonessential items was discontinued in the United States and some other countries. These actions were based on predicted depletions of the ozone layer, which surrounds the earth about 25 km above the surface. The ozone layer absorbs excessive solar radiation in the region of the spectrum (2800–3200 Å), which is responsible for skin cancer. Although Molina and Rowland[3] first proposed the concept of ozone destruction by F-11 and F-12 in the stratosphere, numerous other researchers contributed to developing and refining the original idea. An excellent review of early research in this area was provided by Turco and Whittier.[4] The photochemistry and reactions leading to the destruction of ozone are summarized:

Formation of radicals:

$$CF_2Cl_2 + h\nu \rightarrow CF_2Cl + Cl^{\bullet} \tag{4}$$

$$CFCl_3 + h\nu \rightarrow CFCl_2 + Cl^{\bullet} \tag{5}$$

Catalytic loss of ozone:

$$Cl^{\bullet} + O_3 \rightarrow ClO^{\bullet} + O_2 \quad (6)$$

$$ClO^{\bullet} + O \rightarrow Cl^{\bullet} + O_2 \quad (7)$$

Termination of reaction:

$$Cl^{\bullet} + CH_4 \rightarrow HCl + CH_3^{\bullet} \quad (8)$$

$$Cl^{\bullet} + H_2 \rightarrow HCl + H^{\bullet} \quad (9)$$

In reactions (4) and (5) chlorine-free radicals are produced through photodissociation by 1800–2100 Å radiation. These radicals act as catalysts until they are removed from the atmosphere through wash-out of HCl, which is formed in reactions (8) or (9). While the qualitative and quantitative models proposed between 1974 and 1978 were contested, the threat to human health posed by increased exposure to harmful UV radiation from these reactions was used to justify restricting use of fluorocarbons as propellants.

### 1.2. *Trihalomethanes and Similar Compounds*

Since the early 1970s low-molecular-weight chlorinated hydrocarbons have been associated with a second serious environmental problem. The presence of trihalomethanes including chloroform, bromoform, bromodichloromethane, and other similar compounds in drinking waters has been considered a possible carcinogenic threat to humans. Many of these compounds, such as $CH_2Cl_2$, $CCl_4$, and $C_2HCl_3$, are components in industrial cleaners and have been detected in rivers and drinking waters in highly industrialized regions of the world.[5–7] While the origin of these compounds was clearly the result of industrial discharges, Rook[8] discovered a second source of trihalomethanes in drinking water. Rook concluded that chloroform and other halogenated methanes were produced during the chlorination process in water treatment plants. These results were confirmed by Bellar *et al.*[9] who found as much as 37–150 μg/l of chloroform and other trihalomethanes in finished drinking water. Results from this study are summarized in Fig. 1 and Table 2 which show that addition of chlorine to the raw river water produced chloro- and bromo-chloromethanes and that increased contact time also leads to increased concentrations of these compounds. Trihalomethanes have been formed in aqueous solution from reactions between chlorine and high-molecular-weight organic compounds which are found naturally in drinking water and are referred to broadly as fulvic acids.[10,11] These

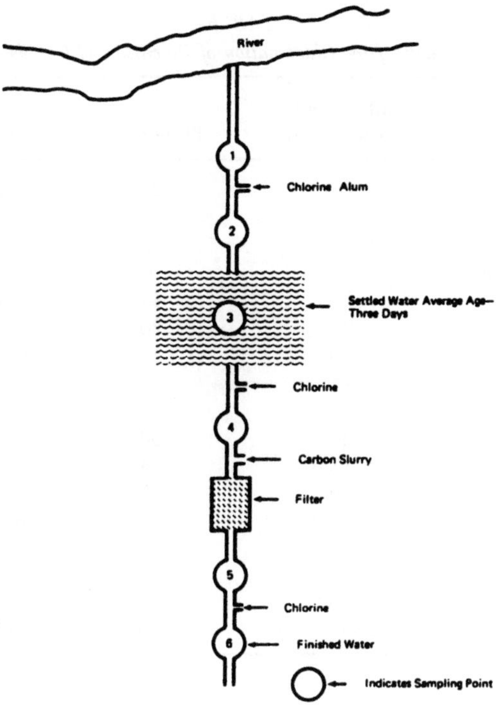


FIGURE 1. Sites of sampling in water treatment plant.

*TABLE 2*
*Trihalogenated-Methane Content of Water from Water Treatment Plant*

| Sample source | Samp. point | Free Cl (ppm) | Concentration (μg/l) | | |
|---|---|---|---|---|---|
| | | | Chloroform | Bromodichloromethane | Dibromochloromethane |
| Raw river water | 1 | 0.0 | 0.9 | *a* | *a* |
| River water treated with chlorine and alum-chlorine; contact time—80 min. | 2 | 6 | 22.1 | 6.3 | 0.7 |
| Three-day-old settled water | 3 | 2 | 60.8 | 18.0 | 1.1 |
| Water flowing from settled area to filters[b] | 4 | 2.2 | 127 | 21.9 | 2.4 |
| Filter effluent | 5 | *c* | 83.9 | 18.0 | 1.7 |
| Finished water | 6 | 1.75 | 94.0 | 20.8 | 2.0 |

[a] None detected. If present, the concentration is <0.1 μg/l.
[b] Carbon slurry added at this point.
[c] Unknown.

*TABLE 3*
*Results from Chlorination of Precursors of $CHCl_3$*

| Compound | Initial molar ratio used | Time (hr) | pH | Molar yield $CHCl_3$ (%) |
|---|---|---|---|---|
| Orcinol | 18 | 25 | 7.2 | 38 |
| | 18 | 25 | 9.2 | 70 |
| Ferulic acid | 24 | 25 | 9.9 | 58 |
| Phloroacetophenone | 9 | 4 | 7 | 52 |
| | 18 | 48 | 6,9,11 | 96–100 |
| Vanillin | 28 | 26 | 9.2 | 22 |
| | 19 | 48 | 10 | 42 |
| *o*-Vanillin | 14 | 25 | 7 | 19 |
| Syringaldehyde | 18 | 24 | 9.2 | 22 |
| | 28 | 60 | 9.2 | 17 |
| Pyrrole | 13.6 | 24 | 10 | 30 |
| Tryptophane | 20 | 7 | 7.5 | 18 |
| | 25.2 | 47 | 6–11 | 100 |
| Proline | 16.5 | 25 | 6.6–11 | 30 |
| Hydroxyproline | 28 | 170 | 5.6–10 | 110 |
| Uracil | 8.2 | 25 | 6.5–11 | 100 |
| Chlorophyll | 23 (by wt) | 100 | 9.2 | 15.5 (by wt) |

humic materials, which are too large and varied to have a single molecular formula, impart a yellow to brown hue to water and occur naturally through extraction of decaying wood, leaves, and organic matter in soil. The mechanisms of formation of trihalomethanes from humic acids have been examined by Christman *et al.*[12] and Morris and Baum.[13] Although production of chloroform through the haloform reaction with humic acids gradually became accepted as the major mechanism, Morris noted that typical haloform reactions of simple methyl ketones are too slow to fully account for the chemistry in water treatment plants. Accordingly numerous molecular structures were identified and tested as possible precursors. Table 3 contains results which show that other structural groups may also contribute to formation of chloroform. Therefore, chloroform is likely produced through reactions with more than just humic acids although humic acids are important precursor materials. Certain precursors of trihalomethanes may be added by man to raw drinking water. For example, polyelectrolyte coagulants and coagulant aids added during water treatments are also known precursors of chloroform.[14]

Long-term exposure to drinking water that contains trace concentration levels of volatile halogenated organic compounds including trihalomethanes may pose a significant threat to health since some of these compounds are known carcinogens or mutagens. However, such concepts are very difficult to answer unambiguously. Studies which address these questions have been reported by Kantor and McCabe[15] and Alavanja *et al.*[16] The November 1978 issue of *Journal of American Water Works Association* provides an in-depth survey of various aspects of production and measurement of trihalomethanes in natural waters.

## 2. *MASS SPECTROMETRY OF HALOCARBONS*

Although reports on bonding and fragmentation in halocarbons appeared in the mid-1930s, systematic characterizations of these properties were made in the 1950s for monohalomethanes,[17] trifluoromethylhalides,[18] and the series $CCl_xF_y$ where $x + y = 4$ for $x = 0$–1.[19] The most abundant positive ion for $CCl_4$ is produced through the loss of one chlorine atom using electron impact ionization.[19]

$$CCl_4 + e \rightarrow CCl_3^+ + Cl + 2e \tag{10}$$

or

$$CCl_4 + e \rightarrow CCl_3^+ + Cl^- + e \tag{11}$$

Moreover, when fluorine and chlorine occur together in $CX_4$-halogenated methanes, the most abundant ion is $(M - Cl)^+$. For example, with $CClF_3$ the abundance of $CClF_2^+$ is only 30% of the abundance of $CF_3^+$. The formation of molecular ions $CX_4^+$ for these chlorofluoromethanes is not observed for $CCl_4$, $CF_4$, or $CCl_3F$, and the base peak abundances of $M^+$ are 0.2 and 0.7%, respectively, for $CCl_2F_2$ and $CClF_3$. Negative ions were also detected in this study and two mechanisms were postulated to account for the results. The dissociative attachment mechanism (12) was found applicable with every chlorofluorocarbon. However, only $CCl_4$ showed evidence of the ion pair mechanism (13)

$$CX_4 + e \rightarrow CX_3 + X^- \tag{12}$$

$$CX_4 + e \rightarrow CX_3^+ + X^- + e \tag{13}$$

The ionization and dissociation of trifluoromethyl halides were studied using electron impact and provide additional information on mass spectra of halocarbons. Results from this work are summarized in Table 4 and also show that no molecular ion is detected for $CF_4$.

Also these data show that $(M - X)^+$ is the base peak in each trifluoromethyl halide and that loss of chlorine or bromine is favored over loss of fluorine. Comparisons of values for appearance potentials for $M^+$ and $(M - X)^+$ ions show some interesting but not surprising trends.

TABLE 4
*Fragmentation Data on Trifluoromethyl Halides*[a]

| CF$_4$ | | | CF$_3$Cl | | | CF$_3$Br | | |
|---|---|---|---|---|---|---|---|---|
| Ion | % RA | AP (eV) | Ion | % RA | AP (eV) | Ion | % RA | AP (eV) |
| $CF_4^+$ | 0.0 | | $CF_3Cl^+$ | 0.5 | 13 | $CF_3Br^+$ | 11.2 | 12.3 |
| $CF_3^+$ | 100.0 | 16.0 | $CF_3^+$ | 100.0 | 12.7 | $CF_3^+$ | 100.0 | 12.2 |
| $CF_2^+$ | 14.5 | 20.3 | $CF_2Cl^+$ | 13.3 | 15.0 | $CF_2Br^+$ | 8.9 | 15.0 |
| $CF^+$ | 3.7 | 22.6 | $CF_2^+$ | 2.4 | 20 | $CF_2^+$ | 5.8 | 19.5 |
| $C^+$ | 9.4 | 31.5 | $CF^+$ | 4.4 | 22.6 | $CF^+$ | 4.3 | 22.9 |
| $F^+$ | 5.7 | 36 | $C^+$ | 3.2 | 31 | $C^+$ | 1.8 | 33 |
| $F^-$ | 0.8 | 4.5 | $F^+$ | 1.1 | 31 | $F^+$ | 0.5 | 29 |
| $CF_3^-$ | .09 | 4.9 | $Cl^+$ | 6.5 | 21 | $Br^+$ | 5.8 | 17.6 |

[a] RA = relative abundance, AP = appearance potential.

Since molecular ions are formed through the removal of one nonbonding electron on a halogen atom, the trend in percent relative abundance of $M^+$ is $CF_3Br \gg CF_3Cl > CF_4$, which reflects the influence of two effects. These are the increase in electronegativity in Br to F and the increase in ability to stabilize a charge in F to Br. The trend in appearance potential of $(M - X)^+$, $CF_4 > CF_3Cl > CF_3Br$, reflects the increase in bond dissociation energy of C—I to C—F. The dissociation energies of C—X were measured for the methyl halides[(18)] and these values (kcal/mole) were C—F, 4.7 ± 0.2; C—Cl, 3.7 + 0.2; and C—Br, 2.9 + 0.1. Other results from this study are summarized in Table 5. In this research $M^+$ and $(M - X)^+$ were both produced in high abundance and $CH_3^+$ was produced through two mechanisms which are similar to those in

TABLE 5
*Fragmentation Data on Methyl Halides*[a]

| CH$_3$F | | | CH$_3$Cl | | | CH$_3$Br | | |
|---|---|---|---|---|---|---|---|---|
| Ion | % RA | AP (eV) | Ion | % RA | AP (eV) | Ion | % RA | AP (eV) |
| $CH_3FT^+$ | 100.0 | 12.6 | $CH_3Cl$ | 100.0 | 11.3 | $CH_3Br^+$ | 96 | 10.5 |
| $CH_3^+$ | 98 | 16.5 | $CH_3^+$ | 85 | 13.7 | $CH_3^+$ | 100.0 | 12.9 |
| $F^+$ | 1.5 | 23.0 | $Cl^+$ | 4 | 16.5 | $Br^+$ | 8 | 11.8 |
| $F^-$ | ~0.6 | 12.4 | $Cl^-$ | ~0.1 | 10 | $Br^-$ | ~1 | 0 |

[a] RA = relative abundance, AP = appearance potential.

*TABLE 6*
*Fragmentation Data on Halomethanes*

| Fragment | $CF_3Cl$ | $CF_3Br$ | $CF_3H$ | $CF_2Cl_2$ | $CF_2ClH$ | $CF_2H_2$ | $CFCl_3$ | $CFH_3$ | $CFClH_2$ | $CFCl_2H$ |
|---|---|---|---|---|---|---|---|---|---|---|
| M | | 9.0 | | 0.2 | 0.7 | 4.1 | | 39.3 | 50.1 | 4.0 |
| M − H | | | 31.4 | | 0.8 | 38.4 | | 37.0 | 1.0 | |
| M − F | 19.6 | 17.6 | 40.0 | 6.3 | 11.3 | 40.1 | 1.6 | 18.3 | 6.8 | |
| M − Cl | 52.6 | | | 64.1 | 52.6 | | 61.6 | | 19.9 | 71.5 |
| M − Br | | 46.6 | | | | | | | | |
| M − HF | | | 6.2 | | | 4.4 | | 3.0 | 1.9 | |
| $CF_3^+$ | (52.6)[a] | (46.6) | (31.4) | | | | | | | |
| $CF_2H^+$ | | | (40.0) | | (52.6) | (38.4) | | | | |
| $CF_2Cl^+$ | (19.6) | | | (64.1) | 0.8 | | | | | |
| $CF_2Br^+$ | | (17.6) | | | | | | | | |
| $CFH_2^+$ | | | | | | (40.1) | | (37.0) | (19.9) | |
| $CFClH^+$ | | | | | 11.3 | | | | 1.0 | (71.5) |
| $CFCl_2^+$ | | | | (6.3) | | | (61.6) | | | |
| $CCl_2H^+$ | | | | | | | | | | |
| $CH_2Cl^+$ | | | | | | | | | (6.8) | |
| $CCl_3^+$ | | | | | | 1.6 | | | | |
| $CH_3^+$ | | | | | | | | (18.3) | | |
| $CF_2^+$ | 8.0 | 5.8 | (6.2) | 7.9 | 3.1 | 1.4 | | | | |
| $CF^+$ | 4.8 | 4.7 | 22.0 | 6.5 | 9.2 | | 6.0 | 2.6 | | 5.6 |
| $CCl^+$ | 0.9 | | | 2.5 | | | | | | |
| $Br^+$ | | 13.8 | | | | | | | | |
| $Cl^+$ | 10.2 | | | 8.1 | 10.0 | | 10.3 | | | 5.5 |
| Max % | 1.0 | 1.0 | 1.4 | 0.4 | 5.2 | 4.4 | 1.5 | | 1.3 | 4.6 |

[a] Parentheses indicate that the percentage is also listed using another designation.

reactions (10) and (11). The $CH_3^+$ ion was formed through both mechanisms for $CH_3F$ and $CH_3Cl$, but this ion was formed only through the first (reaction 5) mechanism for $CH_3Br$. The negative ion $Br^-$ was formed only through dissociative electron capture (reaction 12), while $F^-$ and $Cl^-$ were formed through the ion pair (reaction 13) mechanism only.

In 1968 McCarthy[20] published what is perhaps the most comprehensive comparison of mass spectra fragmentation patterns for chlorofluorocarbons. Spectra from 70-eV electron impact ionization were given in tabular form for 33 fluorinated methanes, ethanes, and propanes. These results are shown in part in Table 6. In Table 6 the values for abundance of the molecular ion varied from 0 to 50.1% of total ionization and $(M - X)^+$ was always a major ion for these methane species. In every case except $CH_3F$ and $CFClH_2$ the base peak was $(M - X)^+$, and the $CX_3^+$ (X = H, F, Cl, Br) ion, which was found in greatest abundance, was formed through a loss of the halogen atom with lowest bond dissociation energy in $CX_4$. An exception of this rule was $CFH_3$. Loss of a hydrogen atom was not a favorable event particularly when chlorine was present in the compound. Furthermore, for compound $CH_{4-x}F_x$ the ion $(M - H)^+$ was lower abundance than $(M - F)^+$, except for $CFH_3$. Results from $CH_2F_2$ show these ions differ by only 1.7% total abundance.

Thus, losses of F and H are very competitive in this series. Although rules for fragmentation of the C—X bonds in $CF_3$—$CX_3$ are similar to those for fluorinated methanes, a second type of fragmentation may occur. This fragmentation is the rupture of the carbon–carbon bond with retention of charge by either $CX_3$ or $CF_3$. A third type of ion, a rearranged methyl ion, was formed through scrambling of halogen atoms. Such ions were common for these species. Therefore, prediction of molecular structure of these compounds solely on the basis of fragment ions is unsound. Fragmentation of the carbon–carbon bond was favorable in many cases and charge was retained in general by the ion with higher stabilization energy. Values for energies of dissociation for $CF_3 - X$ were measured by Noutary[21] who showed that $D(CF_3 - X)$ decreased with increasing atomic weight and increasing lengths of the C—X bond.

Mass spectral patterns of chlorine-substituted methanes and ethane were published for comparison by Bernstein *et al.*[22] Some results from this study, the earliest publication comparing mass spectra for chlorinated methanes and ethanes, are shown in Fig. 3. The base peak for $CH_2Cl_2$, $CHCl_3$, $CCl_4$, and $C_2Cl_5H$ was $(M - Cl)^+$ and virtually no molecular ion was detected in every case except $C_2Cl_5H$. The base peak for $C_2Cl_4$ was also the molecular ion, but base peaks for $C_2Cl_6$ and $C_2Cl_4Cl_2$ were ions which were produced through cleavage of the C—C bond.

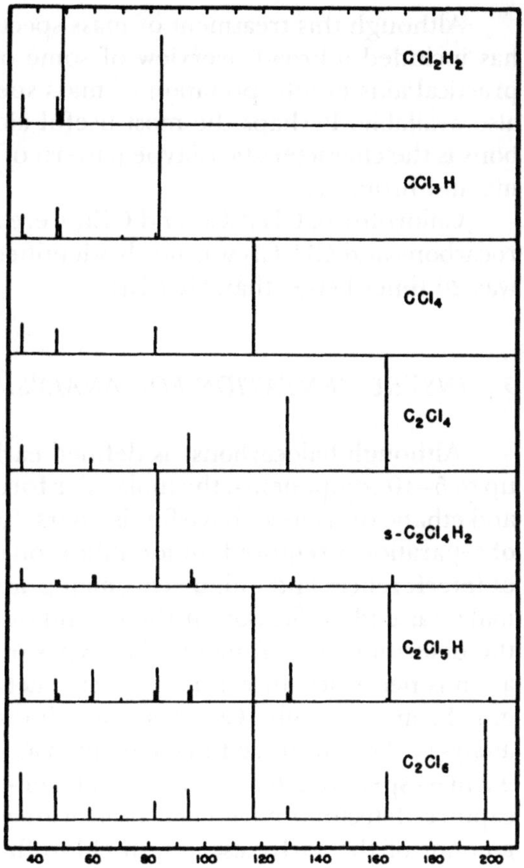


FIGURE 2. Mass spectra of chlorine-substituted methanes and ethanes.

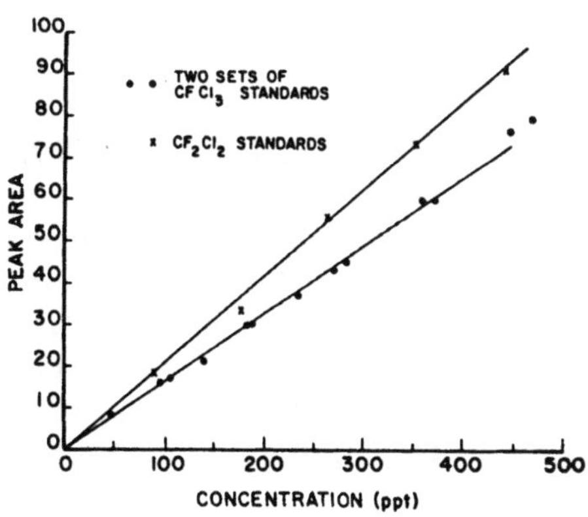


FIGURE 3. Standard plot for SIM analysis.

Although this treatment of mass spectral properties of halocarbons has included a broad overview of some of the earliest studies, certain practical aids in interpretation of mass spectra of these compounds are also available. Perhaps the most useful aid in identification of halocarbons is the characteristic isotype pattern of natural abundances of chlorine and bromine.

Chloroform, $CH_3CCl_3$ and $CCl_4$ were always detected with the fluorocarbons and $CH_2Cl_2$ was newly identified in these samples. Sensitivity was 20 times better than GCECD.

## 3. INSTRUMENTATION FOR ANALYSIS

Although halocarbons, as defined earlier, consist of a mixture with up to 5–10 components, the molecular formulas are simple and methane and ethane derivatives have few isomers. Consequently, while some form of separation is required for identification of specific species, the number of interferences and unique chromatographic properties of halocarbons make false identification of these compounds less of a problem than in the determination of many other types of compounds. Since preseparation is necessary, direct mass spectrometric analysis is uncommon, but gas chromatography (GC) has been almost universally applied in these analyses. The preferred detector for GC determination of halocarbons is a mass spectrometer. However, halogen-specific detectors (i.e., electron capture detectors) have been used widely in these analyses. In a large amount of the early research in chlorofluorocarbons, including the pioneering work of Lovelock,[28] an electron capture detector (ECD) not mass spectrometer was used with gas chromatography. The ECD is inexpensive in comparison to a mass spectrometer and thus is more available for general use. Many of the papers reported in this chapter included data from only an ECD. However, an attempt was made in particular to cite publications with results from GC/MS analyses.

## 4. CHLOROFLUOROCARBONS IN ATMOSPHERIC SAMPLES

Evidence of chlorofluorocarbons in air was first provided by Lovelock[23,24] who used direct injection of 5 ml of unconcentrated air with GC analysis in determining $CCl_3F$ in the atmosphere over the British Isles. A GC-ECD was used in these studies of atmospheric turbidity. Later, Simmons *et al.*[25] analyzed air from the Los Angeles Basin also using a GC-ECD. Identifications of four fluorocarbons were made through matching of retention times of components with standards on two different columns. Average concentrations of these compounds were: $CCl_4$, 0.2 ppb; $C_2H_3Cl_3$, 0.37 ppb; $CFCl_3$, 0.65 ppb; and $C_2Cl_4$, 1.25 ppb. Several other publications between 1972 and 1974 described fluorocarbons in the atmosphere and are summarized by Hester *et al.*[26] who also examined chemical reactivities of F-11, F-12, F-113, and F-114 under conditions simulating heavy smog. No depletions of fluorocarbons were detected under these conditions including irradiation of samples in a mixture of $SO_2$, propenes, and $NO_2$. A review of measurements of fluorocarbons in the troposphere and stratosphere through December 1974 has been given by the Interdepartment Committee for Atmospheric Sciences (1975). In this report and a second report by the Federal Task Force on Inadvertent Modification of the Stratosphere (1975) only marginal reference is made to GC/MS analysis of air for fluorocarbons.

The first reports of actual GC/MS determination of fluorocarbons in air were made by Grimsrud and Rasmussen[27,28] and Tyson.[29] Grimsrud and Rasmussen used selected-ion monitoring with GC/MS analysis for quantifying $CF_2Cl$ and $CFCl_3$ in direct injection of 20 ml of sample. An EI source was employed and the $(M - Cl)^+$ ion was monitored. The GC and MS were coupled using a heated stainless steel tube, and two 4-in. diffusion pumps removed excessive amounts of carrier gas. Samples were collected and stored in 4-l stainless steel containers which were internally electroplated and samples were stored reportedly without problems for several weeks. Results from external calibration of the GC/MS using SIM is shown in Fig. 3. These plots show the technique has good linear range and reproducibility was $\pm 5$ ppt. Repeated examinations of interferences in SIM analyses were always negative. No decreases in concentration of $CFCl_2$ or $CFCl_3$ were detected at higher altitudes and ranges were 223–239 ppt and 124–133 ppt, respectively. Furthermore, the ratios of $CF_2Cl_2$ to $CFCl_3$ were always about 1.8 in rural samples and this reflected long-range sources. Results from a second paper by Grimsrud and Rasmussen[28] are shown in Table 7.

Tyson[29] used GC/MS in the scanning mode for mass spectral identification of fluorocarbons in atmospheric samples. These samples were enriched using liquid nitrogen cooled traps containing polished copper mesh and the bulk of $CO_2$ was removed with a Mikohbite trap. The GC/MS system was equipped with a membrane separator. Tyson found high concentrations of F-11 and F-12 in densely populated areas, but at relatively short distances from these centers the compounds were found at lower but constant background levels. The presence of F-113 and $CH_3CCl_3$ was confirmed using scanning GC/MS.

Few publications on GC/MS determination of fluorocarbons were

*TABLE 7*
*Results from GC/MS Analysis of Atmospheric Samples*

| Halocarbon | Retention time (min) | Masses monitored | Concentration (ppt) |
|---|---|---|---|
| Dichlorodifluoromethane | 4.0 | 85, 87, 101 | 230 ± 10 |
| Fluorotrichloromethane | 6.0 | 101, 103 | 125 ± 8 |
| Methyl chloride | 5.6 | 47, 49, 50, 52 | 530 ± 30 |
| Dichloromethane | 3.0 | 49, 51, 84, 86 | <5 |
| Chloroform | 7.4 | 83, 85 | 20 ± 10 |
| Carbon tetrachloride[a] | 4.0 | | 120 ± 15 |
| Ethyl chloride | | 64 | <5 |
| 1,1- and 1,2-dichloroethane | | 63, 98 | <5 |
| 1,1,2-trichloroethane | | 97, 99 | <5 |
| 1,1,1-trichloroethane | 6.7 | 97, 99 | 100 ± 15 |
| 1-chloroethylene | | 62, 64 | <5 |
| as, *trans*, and 1,1-dichloroethylene | | 61, 96 | <5 |
| Trichloroethylene | | 93, 128 | <5 |
| Tetrachloroethylene | 7.7 | 164, 166 | 20 ± 10 |
| Methyl bromide | | 94 | <5 |
| Methyl iodide | | 142 | <5 |

[a] Measured by electron capture gas chromatography.

found in research journals, and as late as 1979 Cronn and Harsch[30] cited only five references. These workers expanded earlier SIM analyses in determining 11 halocarbon species. Samples of 100 ml were concentrated in a freeze-out loop packed with glass beads and introduced to the GC using a gas-switching valve. Results from this sequential SIM

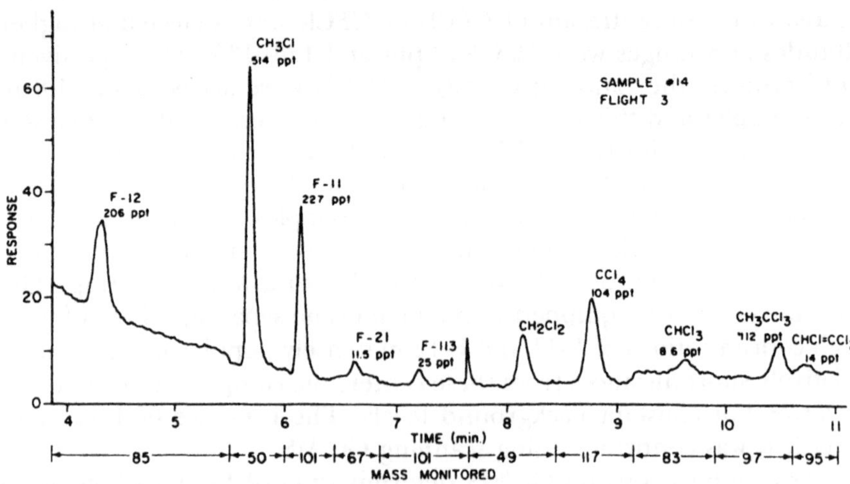


FIGURE 4. SIM plot from GC/MS analysis of stratospheric air samples.

TABLE 8
*Precision and Detection Limits of GC/MS Analysis Using SIM*

| Compound | Mixing ratio ppt[a] | % Precision (100)s/X, n = 5 | Detection limit ppt[b] |
|---|---|---|---|
| F-12 | 220 | 6 | 14 |
| F-114 | 54 | 5 | 10 |
| $CH_3Cl$ | 847 | 4 | 6 |
| F-11 | 125 | 3 | 4 |
| F-21 | 91 | 3 | 4 |
| F-113 | 50 | 10 | 12 |
| $CH_2Cl_2$ | 91 | 4 | 6 |
| $CCl_4$ | 89 | 5 | 6 |
| $CHCl_3$ | 52 | 5 | 5 |
| $CH_3CCl_3$ | 92 | 8 | 20 |
| $CHCl{=}CCl_2$ | 128 | 10 | 16 |

[a] Sample was an artificially prepared halocarbon standard.
[b] Defined as two times the noise level on a strip chart recorder.

analysis from GC/MS analysis of a stratospheric air sample are shown in Fig. 4. The total time per analysis is approximately 25 min. Detection limits and precision for this analysis are given in Table 8.

While direct analysis or cryogenic trapping of atmospheric samples has been widely applied, Elias *et al.*[31] reported sample preconcentration through the use of adsorption tubes. These tubes contained Durapak (Carbowax 400/Porasil 100–120 mesh) and were cooled to 0 °C for collecting $CBr_2F_2$ in samples as large as 1.2 l. Components were desorbed from trap to GC using a gas-switching valve and desorption temperature of 175–200 °C. Detection limits were reported as 0.5 pg.

One example of direct MS analysis of air samples has been given by Aikin and Mauer[32] who described a balloon-borne photoionization mass spectrometer. An ultraviolet source was used for photoionization of gaseous sample and cryopumps were used for creating a vacuum for the ion source and quadrupole analyzer. This system was capable of detecting $CH_3Cl$, $CH_3Br$, $CF_2Cl_2$, and $CFCl_3$.

## 5. VOLATILE HALOCARBONS IN WATER

### 5.1. Introduction

Analytical methods for the identification of volatile organic compounds including halocarbons in a water-based matrix typically have in common certain characteristic features. Generally such analyses involve

(a) extraction of the volatile compounds from solution, (b) concentration of the extract, (c) gas chromatographic separation of components in the extract, and (d) identification of various compounds using retention behavior and mass spectra. The sources for greatest modification or variation in this scheme are in the first two categories listed above. Most published methods differ only in the details of these steps, but such differences are sometimes substantial. However, in the technique of direct aqueous injection all sample pretreatment is virtually eliminated.

### 5.2. *Direct Aqueous Injection*

Since sampling handling is reduced in gas chromatographic (GC) analysis of aqueous samples using direct aqueous injection (DAI), this method may be more convenient for automated analysis or routine monitoring in comparison to extraction-based methods. Furthermore, samples could be analyzed in less time for less cost than other techniques which involve extraction and preconcentration steps. Problems with DAI may include limited sensitivity, low resolution and stability of the chro-

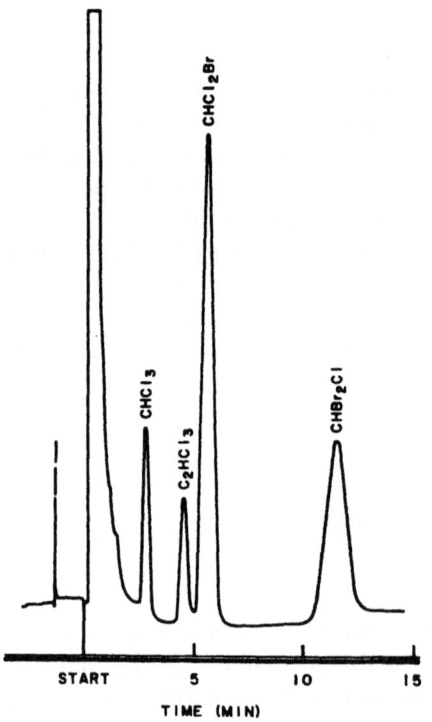


FIGURE 5. Gas chromatogram from DAI analysis.

matographic system, degradation of the MS, and interferences in mass spectra. These and other features have been investigated using GC[33] and GC/MS.[34] In GC analysis with DAI, 9 l of aqueous sample were chromatographed on a 2 m long × 2 mm i.d. glass column containing Chromosorb 101. The GC was equipped with an ECD. Isothermal conditions at 130 °C were used for the column oven and the inlet and detector temperatures were 250 °C. Results from GC analysis using DAI of a standard aqueous sample are shown in Fig. 5. Each component is present at 50 g/l in aqueous solution and the detection limits were at or below 2 μg/l for each trihalomethane. Values for sensitivity for these and other volatile halocarbons are given in Table 9 and include for comparison similar values from the technique of headspace enrichment with a modified Hall detector.

The precision of both techniques in quantitative analysis was good with relative standard deviations of 1–10%. However, the authors found the headspace technique slow and difficult to automate.[35] Since the sensitivity of DAI was lower than concentrations of compounds in many water samples, the comparatively low sensitivity was not considered a limitation.

Actual samples were analyzed using both methods and large differences in concentrations of chloroform and bromochloromethane were

*TABLE 9*
*Detection Limits from DAI and Gas Sparging Techniques*

| Compound | Directed aqueous injection | | | Gas sparging | |
|---|---|---|---|---|---|
| | Oven temp. (°C) | Retention time | Detection limit (μg/l) | Retention time | Detection limit |
| Methylene chloride | 130 | 1.7 | 500 | 6.5 | 0.1 |
| Chloroform | 130 | 3.0 | 1 | 8.1 | 0.1 |
| 1,2-dichloroethane | 130 | 4.0 | 90 | | 0.2 |
| 1,1-dichloroethane | 130 | 4.0 | 150 | | 0.2 |
| Carbon tetrachloride | 130 | 4.0 | 0.1 | | 10 |
| Trichloroethylene | 130 | 4.7 | 2 | 9.0 | 0.3 |
| Bromodichloromethane | 130 | 6.0 | 0.5 | 9.6 | 0.5 |
| Tetrachloroethylene | 130 | 13.0 | 0.5 | 11.2 | 2 |
| Chlorodibromomethane | 130 | 15.4 | 1 | 12.7 | 2 |
| Bromoform | 150 | 4.5 | 2 | 16.9 | 1.5 |
| Tetrachloroethane | 150 | 5.2 | 7 | | |
| *p*-dichlorobenzene | 150 | 9.8 | 500 | | |
| *o*-dichlorobenzene | 150 | 11.4 | 500 | | |

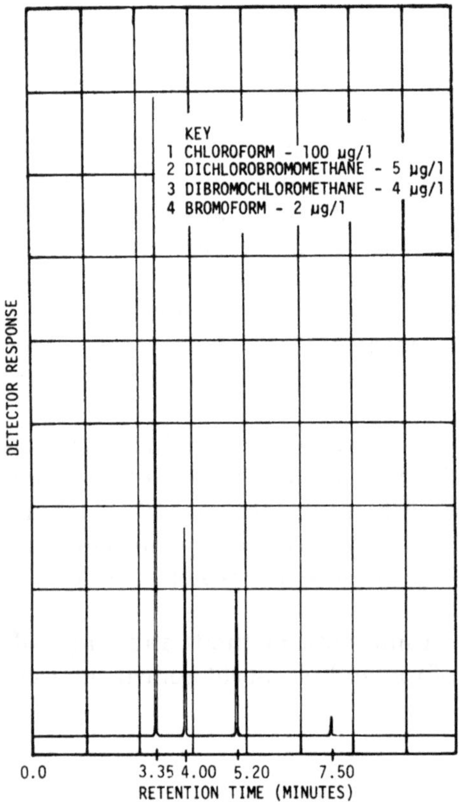


**FIGURE 6.** Gas chromatogram from analysis of extract containing halocarbons.

detected between results from each method. After matrix effects were examined using standards, the DAI technique still showed higher concentrations for trihalomethanes than the headspace enrichment method. They found that chloroform was produced during the injection process through the reaction of residual $Cl_2$ with precursors naturally present in water. The technique of DAI, therefore, is subject to severe restrictions in quantitative determination of halocarbons and may even be a source of artifacts.

While the presence of water vapor in a magnetic mass spectrometer is harmful, Harris *et al.*[34] found no detrimental effects from DAI of 1–10 l of aqueous samples on a GC quadrupole MS system with glass jet separator (Fig. 6). No effect on electron impact fragmentation patterns was detected for a wide variety of organic compounds including $HCCl_3$ and $CH_2Cl_2$. The detection limit for a 40–400 amu/5 sec scan was 1 ng/l, but 1–50 µg/l was needed to obtain a reasonably clear mass spectrum.

### 5.3. *Liquid–Liquid Extraction*

In liquid–liquid extraction of volatile organic compounds in water, an immiscible, unreactive organic solvent is mixed with aqueous sample. Organic components, including halocarbons, migrate to the organic phase and if no specific molecular interactions occur, a dynamic equilibrium or distribution may be established and described by a concentration ratio, $D_c$:

$$D_c = \frac{C^{\mathrm{org}}}{C^{\mathrm{aq}}} \tag{14}$$

where $C^{\mathrm{org}}$ is the concentration of a solute in the organic phase and $C^{\mathrm{aq}}$ is the concentration of the same component in the aqueous phase. At equilibrium the chemical potentials of the solute in each phase are equal and the thermodynamic distribution constant $K^0$ is shown in Eq. (15).

$$RT \ln K^0 = \Delta U^0 \tag{15}$$

where $\Delta U^0$ is the difference in the chemical potential of the solute in a standard state. The relationship between $K^0$, a thermodynamic concept, and $D_c$ the experimentally determined value for equilibria, is shown in Eq. (16).

$$D_c = \frac{\gamma^{\mathrm{aq}}}{\gamma^{\mathrm{org}}} K^0 \tag{16}$$

where $\gamma$ is the activity coefficient of the solute in each phase. The fraction of a component which is extracted, $\theta$, is related to $D_c$ through Eq. (17).

$$\theta = \frac{D_c V}{1 + D_c V} \tag{17}$$

where $V$ is the volume ratio $V^{\mathrm{org}}/V^{\mathrm{aq}}$ of the phases. Methods based on LLE are actually preseparation techniques since different compounds may have different values for $D_c$. A measure of the separation with two compounds $A$ and $B$ is given by the ratio $\theta_A/\theta_B$. Ideally this ratio should be very large for successful separation where $A$ is more soluble than $B$ in the organic phase.

The technique of LLE then is capable of extracting or isolating volatile organic compounds from complex mixture of salts, humic acid mixtures, and solid suspensions. Furthermore, if values for $D_c$ are large for a particular component such as chloroform, a low ratio of $V^{org}/V^{aq}$ may be used and will still be large. Thus, this is a source of not only prefractionation but also preconcentration in a simple process.

Although LLE methods are far more complex than DAI procedures, this technique of LLE is also more sensitive and flexible than DAI. Consequently, more variables must be closely controlled and optimized in quantitative analyses. Four possible approaches to LLE have been identified[(36)]: (1) manual shaking, (2) magnetic stirring, (3) mechanical shaking, and (4) continuous extraction.

Mechanical shaking and continuous extraction were chosen by Trussell *et al.*[(36)] as the preferred methods on the basis of allocation of resources of time, efficiency, and cost. Distribution ratios were determined for $CHCl_3$, $CHBrCl_2$, CHBrCl, and $CHBr_3$ using cyclohexane, hexane, *i*-octane, petroleum ether, pentane, and toluene. All of these solvents were low-boiling solvents since extraction techniques using high-boiling solvents were rejected as too time consuming. Pentane and petroleum ether showed highest values for distribution ratios and pentane was chosen for use in all later analyses. The effect of ratios of volume ($V^{org}/V^{aq}$) on extraction efficiency was also examined, and as expected, lower values for the ratio showed lower recovery efficiency. A value for the value ratio of 0.04 afforded preconcentration with average extraction efficiencies of about 80% for the trihalomethanes. Extraction times of 20 min were used since after approximately 15 min no significant improvement in recovery was noted. Values for $D_c$ were not influenced by changes in ionic strength between 0.001 and 0.03 *M* of salt, but $D_c$ for chloroform was doubled when the ionic strength was adjusted to 1–5 *M*.

Results from comparisons of precision and accuracy from analysis

*TABLE 10*
*Comparison of Results from GC Analysis Using Capillary and Packed Columns*

| Description | $CHCl_3$ (%) | $Cl_2BrCl$ (%) | $Br_2ClCH$ (%) | $CHBr_3$ (%) | Total trihalomethane average (%) |
|---|---|---|---|---|---|
| Precision | | | | | |
| Packed column | 2.0 | 3.5 | 1.5 | 4.1 | 2.8 |
| Capillary column | 3.4 | 3.1 | 2.4 | 2.1 | 2.8 |
| Accuracy | | | | | |
| Packed column | 5.8 | 4.3 | 0.3 | 2.6 | 3.3 |
| Capillary | 5.1 | 1.0 | 1.7 | 2.6 | 2.6 |

of samples on packed and capillary columns are shown in Table 10 and Fig. 6.

Values in Table 10 show that large gains in improved quantification are not achieved through the use of capillary columns. However, in analysis of more complex samples, the use of these columns may improve both identification and quantification through improved resolution of components which may coelute on packed columns. Minimum detection limits were 0.1–0.9 μg/l and time of analysis was an average 8 samples per analyst per hour.

Solvent extractions using simple manual shaking have also been developed.[37,38] Meure mixed 5.0 ml of sample with 10 ml of cyclohexane using 1 min of manual shaking. He reported reproductivity of ±2% relative standard derivation in the range 2–100 μg/l. Time of analysis was 10 min and sensitivity was undefined but better than 1 μg/l. The reported recovery for trihalomethanes was excellent: chloroform, 93%; bromodichloromethane, 92%; and chlorodibromomethane, 97%. Richard and Junk[38] used a very similar analytical approach and determined that use of larger volumes of water with a set volume of organic solvent would decrease the extraction efficiency but increase the sensitivity of analysis. Furthermore, recovery efficiencies were not influenced by the matrix effects in water from an uncontaminated well.

The use of continuous LLE has often been cited as an alternate extraction technique for halocarbons in water but few publications may be found using these techniques. Goldberg *et al.*[39] reported two continuous LLE devices for solvents with densities less than or greater than 1 g/ml. While extraction efficiencies were high for certain volatile organic compounds such as toluene, losses of up to 50% were observed when a large volume of solvent was concentrated using a Kuderna-Danish apparatus. However, this research experimentally confirmed Gilman's[40] hypothesis that as the internal pressure ($a/V^2$ in the van der Waals equation) of the solute in solvent decreases as the solubility increases.

Examples of innovative design of new methods of LLE are shown in two reports.[41,42] Although sampling and storage methods have not been discussed in this chapter, such methods are especially crucial in analysis of water samples for halocarbons. Traditional sampling and storage of samples was essentially eliminated through the use of an extraction device for *in situ* solvent extraction of river or seawater. Advantages of this system include (1) continuous sampling for more representative sample than grab sampling, (2) reduced sample manipulation, (3) no filtering of particulate matter, and (4) selectivity of sampling locations (useful for vertical profiles). A single, major limitation to this method is poor control and determination of recovery efficiencies.

### 5.4. *Headspace Analysis*

Since trihalomethanes and other volatile halocarbons have appreciable vapor pressure between 20 and 90 °C, these compounds may be determined in water samples using headspace analysis techniques. In headspace analysis an inert sealed container is partially filled with a water sample and volatile organic compounds partition between the aqueous and vapor phases. Samples of the vapor phase are then removed for analyses. Thus, in headspace analysis both extraction and preseparation steps occur in a single operation. These methods may be simple, but efficiencies can be very high. While many variations of this technique exist for aqueous samples, most methods may be conveniently classified into two separate categories: (1) Direct sampling and (2) Dynamic headspace sampling, commonly referred to as the gas-purging technique.

### 5.5. *Direct Sampling*

Halocarbons in water have been determined using direct sampling of headspace by Piet *et al.*[43] and Kaiser and Oliver.[44] Although the methods used by these two groups differed slightly, the general approach was identical. Water samples were added to a container and the air space above the liquid phase was either evacuated to 10 torr[44] or flushed with nitrogen gas.[43] The apparatus was equilibrated in a water bath at 30 °C or above temperatures. After 30 or 40 min part of the headspace was withdrawn with a syringe and analyzed using GC-ECD.

Kaiser and Oliver[44] found that results from GC quantitation of 10 μg/l of $CHCl_3$ in water had a relative standard deviation of 3% for 1-hr equilibrium periods at 70 °C. Later the effect of temperature on partition of $CHCl_3$ was determined using equilibrium times of 30 min. Actual distribution ratios were reported with temperatures: 30 °C, 0.4; 50 °C, 0.6; 70 °C, 1.0; and 90 °C, 3.7. Limit of detection using this method with 5-μl injections of headspace was 0.1 μg/l for $CHCl_3$ in solution at 70 °C. Sensitivity could reportedly be increased by simply increasing the injection volumes. This method was applied to chlorinated water samples from Lake Ontario and these compounds were determined: $CHCl_3$, 43 μg/l; $CCl_4$, 0.3 μg/l; $CHBrCl_2$, 15 μg/l; and $CHBr_2Cl$, 20 μg/l.

Piet *et al.*[43] used an equilibration temperature of only 30 °C, but used injection volumes of up to 1000 μl with a capillary column. Detection limits for 100-μl samples were $CHCl_3$, 0.1 μg/l; $CH_2BrCl$, 0.3 μg/l; $CHBrCl_2$, 0.1 μg/l; and $CHBr_2Cl$, 0.5 μg/l. The authors cited the following advantages in the use of these methods: (1) column deterioration is reduced since only volatile compounds are introduced on the column; (2) preconcentration techniques are not needed and thus no sorbent materials are required; (3) no large gas volumes are needed as in purging techniques (see next section); and (4) the methods are simple, rapid, and accurate.

### 5.6. Theoretical

When a sample of the headspace is withdrawn for analysis, the results from that analysis must be quantitatively related to the composition of the liquid phase. Such relationships are governed by Henry's law which is shown in Eq. (18).

$$P = X \cdot \text{const} \tag{18}$$

where $P$ is the partial pressure ($P = X \cdot \text{const}$) of the volatile halocarbon, $X$ is the mole fraction of the component in solution, and these are related through the Henry's law constant. This constant is simply the vapor–liquid phase distribution for dilute solutions [Equation (19)].

$$D_h = \frac{C^{\text{vap}}}{C^{\text{aq}}} \tag{19}$$

where $D_h$ is the headspace distribution ratio and $C^{\text{vap}}$ and $C^{\text{aq}}$ are values for concentrations of the solute in the vapor and aqueous phase, respectively. Attempts have been made to model values for $D_h$ as well as transfer processes for some volatile halocarbons in dilute solution.[45] Distribution ratios were defined using Equation (20).

$$D_h = \frac{16.04PM}{TS} \tag{20}$$

where $P$ is the vapor pressure of the pure solute in mm Hg, $M$ is the molecular weight of the component in grams, $T$ is the temperature in kelvin, and $S$ is the solubility of the component in water in mg/l. Values for $D_h$ were calculated using known values for $P$, $M$, and $S$ and were compared to experimentally determined values. Comparison of these values, which are given in Table 11 for a few halocarbons, shows favorable agreement between the two values.

Rates of movement of halocarbons between liquid and gas phases were also studied. Two interesting features of the model of Dilling[45] which gave good agreement with experimental data were: (1) for values of $D_h > 0.1$, large increases in $D_h$ had little effect on the overall liquid exchange constant $K_1$ and (2) the calculate half-life $t^{L}_{1/2}$ of many halo-

TABLE 11
*Values for Vapor–Liquid-Phase Distribution of Halocarbons*

| Table no. | Compound | S | P | $D_h$ Calculated | $D_h$ Found | ppm |
|---|---|---|---|---|---|---|
| 1 | $CH_3Cl$ | 6,270 | — | 0.33 | 0.30 | — |
| 2 | $CH_2Cl_2$ | 19,400 | 438 | 0.10 | 0.11 | 0.99 |
| 3 | $CHCl_3$ | 7,840 | 192 | 0.16 | 0.13 | 0.96 |
| 4 | $CCl_4$ | 800 | 113 | 1.2 | 0.87 | 0.90 |

carbons were directly related to $M_{1/2}$ and nearly independent of $D_h$ (for $D_h > 0.1$). However, exchange constants are influenced by stirring air currents and other factors.

## 5.7. *Dynamic Headspace Sampling*

The second headspace technique which has been used extensively in water analysis for volatile halocarbons is dynamic headspace sampling (DHS). In DHS a stream of inert gas is dispersed at flow rates of 10–100 ml/min through an aqueous sample. This gas is forced through the solution and is immediately displaced from the container by new volumes of gas. During this process, commonly called purging, volatile organic compounds, including halocarbons, partition into the vapor phase and are rapidly removed from the headspace for collection in cryogenic or sorbent traps. Since the purging gas is unretained in these traps, enrichment factors may be very large. Furthermore, since the concentration of compounds in the headspace is constantly being renewed, the efficiency of extractions may also be very high, according to the principles of Le Chatelier.

## 5.8. *Origins*

The origins of DHS in analysis of water samples for volatile organic compounds are not well known and some very early references may have been overlooked in this survey. However, one of the earliest reports using DHS was published by Webber and Burks.[46] These authors used $CO_2$ to purge $C_4$ hydrocarbons from boiler feedwater. Recoveries for butadiene were greater than 93% in a concentration range of 27–640 ppm.

Mass spectrometry was introduced with such analyses by Melpolder *et al.*[47] The device used for DHS in this work is shown in Fig. 7. Hydrogen gas was forced through 1.5-l samples at 200 ml/min for 30 min.

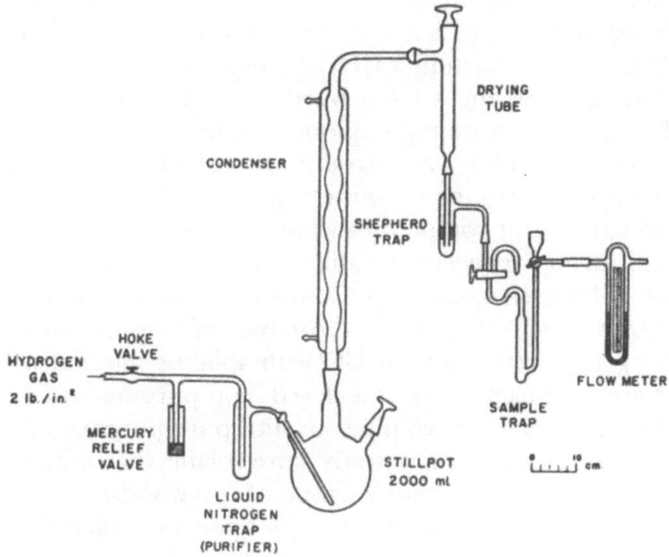


FIGURE 7. Apparatus used for DHS in early work of Melpolder.

Reflux temperatures were used for the sample and the sample trap was cooled with liquid nitrogen. The drying tube was filled with fused KOH, which removed most of the water but did not appear to absorb certain aliphatic and aromatic hydrocarbons. Several other drying materials were tested and found to absorb volatile organic compounds. Next the trap was attached to the inlet system of a mass spectrometer and heated to 200 °C. Hydrocarbons with boiling points as high as 204 °C were detected in these traps and represented concentrations in water of 10 ppb. Total time for analysis was 3.5 hr/sample. Results were reported from analysis of water from different sources including household water and the authors identified chlorinated solvents in water based on their method.

Water samples may contain complex mixtures of volatile organic compounds, and in 1973 Novak *et al.*[6] used DHS of drinking waters with GC/MS analyses. In this work compounds with boiling points as high as 140 °C were stripped from drinking waters, separated with GC, and identified using mass spectra. Conditions of purging were: temperature, 20 °C; time, 30 min; helium gas flow, 100 ml/min; and sample volume, 2 × 250 ml (500 ml total). The headspace was forced through a magnesium perchlorate drying tube and organic components were later trapped in a capillary tube which was cooled with liquid nitrogen. The design of their DHS apparatus was based on work by Swinnerton and Linnenbom.[48] Detection limits were reported as 0.1–1 ppb with an improvement in sensitivity of $10^2$–$10^3$ over solvent extraction methods. However, extraction efficiencies were only 40 ± 10%. Although numerous compounds, including $CHCl_3$, were detected in polluted drinking waters, benzene and $CHCl_3$ were also detected in pure drinking water. These results were not explored further.

Increased flexibility was introduced in DHS through the replacement of cryogenic traps for organic compounds with sorbent traps. These traps retained organic compounds while the purging gas and water were unretained. Grob[49] collected volatile organic compounds in water using DHS with a 1-mg charcoal trap. Components were removed from the charcoal trap using 10 μl of $CS_2$. Analysis of 2 μl of this extract was made using high-resolution of GC with splitless injections. Grob also demonstrated the usefulness of a closed-loop purging system and eliminated the need for large volumes of ultrapure purging gas. Unfortunately, halocarbons and compounds more volatile than toluene were not detected. Improvements in the method of using sorbent materials were made by Zlatkis *et al.*[50] who used a porous polymer called Tenax-GC.[51] Another advance was the technique of thermal desorption of volatile organic compounds from such sorbent traps through combined use of carrier gas flow and heat. Components on traps were introduced into the GC/MS in a single step through connection directly, or through switching valve of the trap and inlet port. While this approach avoided a solvent-stripping step and increased possibilities for loss or contamination, only a single analysis per trap was possible. However, traps were used to either store or transport samples conveniently.

Following the reports of Rook[8] and Bellar and Lichtenberg,[52] interest in DHS of waters for halocarbons increased dramatically since DHS was quantitatively characterized and demonstrated greater analytical usefulness. Samples with 5-ml volumes were purged for 11 min with 20 ml/min of nitrogen gas. An estimated 99% of volatile, water-insoluble (<2% soluble) compounds were purged from water, as shown in Table 12, and show that 2-butanone, a soluble compound, is not removed from solution quantitatively. The linear range was 6 μg/l to 2.5 mg/l with detection limits of 0.5 μg/l.

After 1974 the number of publications on DHS of water for halocarbons and other volatile organic compounds greatly increased. However, most of the these contributions involved refinements in the method or apparatus and further quantitative characterizations. Some of these advances will be described in brief below. Bergert *et al.*[53] used low-temperature capillary GC/MS to identify as little as 0.2 ppb using mass spectra. Janak *et al.*[54] examined the effect of water vapor on retention behavior of trace amounts of volatile organic compounds on the sorbent material Tenax-GC, which was gaining rapid acceptance as one of the

*TABLE 12*
*Quantitative Results from DHS*

| Nitrogen purge gas | | Percentage remaining in aqueous phase | | | |
|---|---|---|---|---|---|
| Flow rate (ml/min) | Volume (ml) | Methylene chloride | Chloroform | Benzene | 2-Butanone |
| 20 | 0 | 100 | 100 | 100 | 100 |
| 20 | 20 | 60 | 55 | 46 | 95 |
| 20 | 100 | 0 | 0 | 3 | 96 |
| 20 | 300 | 0 | 0 | 0 | 80 |
| 13 | 0 | 100 | 100 | 100 | 100 |
| 13 | 6.5 | 67 | 94 | 71 | 100 |
| 13 | 85 | 30 | 29 | 6 | 86 |
| 13 | 143 | 6 | 0 | 0 | 74 |
| | | | Solubility in water (%) | | |
| | | 2 | 1 | 0.08 | 35 |
| | | | Boiling point (°C) | | |
| | | 40.1 | 61.3 | 80.1 | 79.6 |

best materials for use with DHS and thermal desorption techniques.[55] Results from this work showed that the effect of water vapor on quantification with DHS was insignificant. One of a number of devices which were based on DHS in a semiautomated arrangement was described by Grob *et al.*[56] and became commercially available. These systems aided greatly in the standardization of hardware for better comparison of interlaboratory data. Hardware features of purge flow control, switching valve, sorbent trap, and desorption temperature were combined into a single unit that could be connected to most GC or GC/MS through 1.59 mm o.d. stainless steel tubing. Advances in using DHS with thermal desorption of Tenax-GC traps were made by Bertsch *et al.*[57] Critical parameters such as breakthrough capacities of adsorbent tubes, recovery efficiencies, and completeness of thermal desorption were described using capillary GC/MS with direct thermal desorption of sorbent traps. Other advances included an attempt to standardize these methods,[58] automated GC analysis,[35] and use of selective sorbent materials for amines.[59] Bellar and Licthenberg[60] extended their earlier developments through semiautomation of DHS using modifications of their basic apparatus, while Thompson *et al.*[61] made an extensive study on organic volatiles in drinking waters. The effect of sorbent trap temperature on breakthrough volumes was studied and, not surprisingly, these volumes decreased at higher temperatures. Another and final reference on using DHS for quantitative analysis with GC/MS was published by R. D. Lingg

TABLE 13
*Volatile Organic Compounds in Drinking Waters*

| Compound | Cincinnati, Ohio | Miami, Florida | Ottumwa, Iowa | Philadelphia, Pennsylvania |
|---|---|---|---|---|
| | Concentration by GC/MS (μg/l) | Concentration by GC/MS (μg/l) | Concentration by GC/MS (μg/l) | Concentration by GC/MS (μg/l) |
| Benzene | 0.3 | <0.1 | 0.1 | 0.2 |
| Chlorobenzene | 0.1 | 1 | *a* | 0.1 |
| Chloroethene (vinyl chloride) | *a* | *b* | *a* | *b* |
| 1,1-dichloroethene (vinylidene chloride) | *a* | 0.1 | *a* | 0.1 |
| *cis*-1,2-dichloroethene | <0.1 | 12 | *a* | 0.1 |
| *trans*-1,2-dichloroethene | *a* | 1 | *a* | *a* |
| Nitrotrichloromethane (chloropicrin) | 3 | 0.4 | *a* | 2 |
| Tetrachloroethene | 0.3 | <0.1 | 0.2 | 0.4 |
| Toluene | <0.1 | *a* | *a* | 0.7 |
| Trichloroethene | 0.1 | 0.3 | 0.1 | 0.5 |

[a] Not detected.
[b] Not quantitated by GC/MS.

*et al.*[62] who used mass chromatography of specific ions for quantitation. Concentrations and standard deviation with this method using standard curves are shown in Table 13 for water from five cities.

In 1979 standard methods for DHS were prepared by the U.S. Environmental Protection Agency and may be found in the Federal Register as Method 501.1 (1979).

## REFERENCES


1. M. Hudlicky, *Organic Fluorine Chemistry,* Plenum, New York, 1971, p. 69.
2. R. L. McCarthy, F. A. Bowers, and J. P. Jessen, *Atmos. Environ.* **11,** 491 (1977).
3. M. J. Molina and F. S. Rowland, *Nature (London)* **249,** 810 (1974).
4. R. P. Turco and R. C. Whittier, *Atmos. Environ.* **9,** 1045 (1975).
5. J. J. Rook, *Water Treat. Exam* **21,** 259 (1972).
6. J. Novak, J. Zlulicky, V. Kubelka, and J. Mostecky, *J. Chromatogr.* **76,** 45 (1973).
7. K. Grob and G. Grob, *J. Chromatogr.* **90,** 303 (1974).
8. J. J. Rook, *Water Treat. Exam* **23,** 234 (1974).
9. T. A. Bellar, J. J. Lichtenberg, and R. C. Kroner, *J. Am. Water Works Assoc.* **66,** 703 (1974).
10. J. J. Rook, *J. Am. Water Works Assoc.* **68,** 168 (1976).

11. A. A. Stevens, C. J. Solcum, D. R. Seager, and G. G. Robeck, in *Environmental Impact and Health Effects,* Vol. I (R. L. Jolley ed.), Ann Arbor Science, Ann Arbor, Michigan, 1978, p. 77.
12. R. F. Christman, J. D. Johnson, J. R. Hass, F. K. Pfanender, W. T. Liao, D. L. Norwood, and H. J. Alexander, in *Water Chlorination, Environmental Impact and Health Effects,* Vol. II (R. L. Jolley, H. Gorcher, and D. H. Hamilton, Jr., eds.), Ann Arbor Science, Ann Arbor, Michigan, 1978, p. 15.
13. J. C. Morris and B. Baum, in *Environmental Impact and Health Effects,* Vol. II (R. L. Jolley, H. Gorcher, and D. H. Hamilton, Jr., eds.), Ann Arbor Science, Ann Arbor, Michigan, 1978, p. 29.
14. K. L. Kaiser and J. Lawrence, *Science* **196,** 1205 (1977).
15. K. P. Kantor and L. J. McCabe, in *Environmental Impact and Health Effects,* Vol. II (R. L. Jolley, H. Gorcher, and D. H. Hamilton, Jr., eds.), Ann Arbor Science, Ann Arbor, Michigan, 1978, p. 379.
16. M. Alavanja, I. Goldstein, and M. Susses, in *Environmental Impact and Health Effects,* Vol. II (R. L. Jolley, H. Gorcher, and D. H. Hamilton, Jr., eds.), Ann Arbor Science, Ann Arbor, Michigan, 1978, p. 395.
17. V. H. Dibeler and R. M. Reese, *J. Res. NBS* **54,** 127 (1955).
18. V. H. Dibeler, R. M. Reese, and F. L. Mohler, *J. Res. NBS* **57,** 113 (1956).
19. J. D. Cragg, C. A. McDowell, and J. W. Warren, *Trans. Faraday Soc.* **48,** 1093 (1952).
20. E. R. McCarthy, *Org. Mass Spectrom.* **1,** 81 (1968).
21. C. J. Noutary, *J. Res. NBS* **72A,** 479 (1968).
22. R. B. Bernstein, G. P. Semeluk, and C. B. Arends, *Anal. Chem.* **25,** 139 (1953).
23. J. E. Lovelock, *Nature* **230,** 379 (1971).
24. J. E. Lovelock, *Atmos. Environ.* **6,** 917 (1972).
25. P. G. Simmons, S. L. Kenin, J. E. Lovelock, and F. H. Shair, *Atmos. Environ.* **8,** 209 (1974).
26. N. E. Hester, E. R. Stephens, and O. C. Taylor, *Atmos. Environ.* **9,** 603 (1975).
27. E. P. Grimsrud and R. A. Rasmussen, *Atmos. Environ.* **9,** 1010 (1975).
28. E. P. Grimsrud and R. A. Rasmussen, *Atmos. Environ.* **9,** 1014 (1975).
29. B. J. Tyson, *Anal. Lett.* **8,** 807 (1975).
30. D. R. Cronn and D. E. Harsch, *Anal. Lett.* **12**(B14), 1489 (1979).
31. L. Elias, M. McCooeye, and G. Gardner, *Geophys. Res. Lett.* **3,** 17 (1976).
32. A. C. Aikin and E. J. R. Mauer, *Rev. Sci. Inst.* **49,** 1034 (1978).
33. A. A. Nicholson, O. Meresz, and B. Lemyk, *Anal. Chem.* **49,** 814 (1977).
34. L. E. Harris, W. L. Budde, and J. W. Eichelberger, *Anal. Chem.* **46,** 1912 (1974).
35. B. J. Dowty, L. E. Green, and J. L. Laseter, *Anal. Chem.* **48,** 946 (1976).
36. A. R. Trussell, M. D. Umphres, L. Y. C. Leong, and R. R. Trussell, in *Water Chlorination, Environmental Impact and Health Effects,* Vol. II (R. L. Jolley, H. Gorcher, and D. H. Hamilton, Jr., eds.), Ann Arbor Science, Ann Arbor, Michigan, 1977, p. 543.
37. J. P. Mieure, *J. Water Tech./Qual.* **69,** 60 (1977).
38. J. J. Richard and G. A. Junk, *J. Water Tech./Qual.* **69,** 62 (1977).
39. M. C. Goldberg, L. Delong, and M. Sinclair, *Anal. Chem.* **45,** 89 (1973).
40. H. Gilman, *Organic Chemistry and Advanced Treatise,* Vol. II, Wiley, New York, 1943, p. 1755.
41. M. Ahnoff and B. Josefsson, *Anal. Chem.* **46,** 658 (1974).
42. M. Ahnoff and B. Josefsson, *Anal. Chem.* **48,** 1268 (1976).
43. G. J. Piet, P. Slingerland, and F. E. de Grunt, *Anal. Lett.* **All,** 437 (1978).
44. K. L. E. Kaiser and B. G. Oliver, *Anal. Chem.* **48,** 2207 (1976).
45. W. L. Dilling, *Env. Sci. Tech.* **11,** 405 (1977).
46. L. A. Webber and C. E. Burks, *Anal. Chem.* **24,** 1086 (1952).
47. F. W. Melpolder, C. W. Warfield, and C. E. Headington, *Anal. Chem.* **25,** 1453 (1953).

48. J. W. Swinnerton and V. J. Linnenbom, *J. Gas Chromatogr.* **5,** 570 (1967).
49. K. Grob, *J. Chromatogr.* **84,** 255 (1973).
50. A. Zlatkis, H. A. Lichtenstein, and A. Tishkee, *Chromatographia* **6,** 67 (1973).
51. R. Wan Wyk, *J. Chromatogr. Sci.* **8,** 418 (1970).
52. T. A. Bellar and J. J. Lichtenberg, *J. Amer. Water Works Assoc.* **66,** 739 (1974).
53. K. H. Bergert, V. Betz, and P. Pruggmayer, *Chromatographia* **7,** 115 (1974).
54. J. Janak, J. Ruzichova, and J. Novak, *J. Chromatogr.* **9,** 689 (1974).
55. J. M. H. Daemen, W. Dankelman, and M. E. Hendriks, *J. Chromatogr. Sci.* **13,** 79 (1975).
56. K. Grob, K. Brog, Jr., and G. Grob, *J. Chromatogr.* **106,** 299 (1975).
57. W. Bertsch, E. Anderson, and G. Holzer, *J. Chromatogr.* **112,** 701 (1975).
58. A. A. Deetman, P. Demeulemeester, M. Garcia, G. Hanck, J. I. Hollies, D. Krockenberger, D. E. Palin, H. Prigge, L. Rohrschmeider, and L. Schmidhammer, *Anal. Chim. Acta* **82,** 1 (1976).
59. C D. Chriswell and J. S. Fritz, *J. Chromatogr.* **136,** 371 (1977).
60. T. A. Bellar and J. J. Lichtenberg, ASTM Symposium, Denver, Colorado, June 19 and 20, 1978.
61. M. Thompson, M. Shoults, W. Bertsch, and G. Holzer, *J. Chromatogr.* **11,** 142 (1978).
62. R. D. Lingg, R. G. Melton, F. C. Kopfler, W. E. Coleman, and D. E. Mitchell, *J. Water Tech./Qual.* **69,** 605 (1977).
63. G. A. Burgett and L. E. Green, Hewlett-Packard Application Note #5, 1976.

15

# PHTHALATES AND RELATED PLASTICIZERS


## E. ATLAS and C. S. GIAM


### 1. INTRODUCTION

Plasticizers are chemicals which are used primarily as additives to a large variety of commercial products. Plasticizers are added to impart flexibility to a plastic resin or polymer, to facilitate processing of various resins, or to otherwise develop improved properties not present in the original material. There are numerous applications of plasticizer compounds, such as: floor coverings, wire insulation, adhesives, upholstery, automotive trim, food wraps, foamed carpet backing, rubber moldings, insect repellants, cosmetics, and floor waxes. Plasticizer additives are found in nitrocellulose, polyvinyl chloride, methacrylate, polystyrene, polyvinyl acetate, polyurethanes, and other synthetic applications.

Production of plasticizers has paralleled the increasing production and use of plastic products. United States production of all plasticizer compounds (Table 1) reached over $2 \times 10^9$ lb in 1978.[1] The most important group of plasticizers are the phthalic acid esters which account for 60% of the total production of plasticizers. Of the phthalates, di-(2-ethylhexyl)-phthalate is the most commonly used. Structures of some important plasticizers are shown in Fig. 1.

Plasticizers are introduced into the environment through several mechanisms—direct volatilization of plasticizers from resin matrices, incineration of plastic wastes, and aqueous leaching of plastics and plastic waste. The potential for environmental contamination by plasticizer compounds is significant—high production rates, mechanisms for environ-


*E. ATLAS* and *C. S. GIAM* • Department of Chemistry, Texas A & M University, College Station, Texas 77843.

*TABLE 1*
*Plasticizer Production in the United States*

| | Production ($\times 10^6$ lb) |
|---|---|
| Phthalic acid esters—total | 1259 |
| Dioctyl phthalate[a] | 409 |
| Diisodecyl phthalate | 171 |
| Diethyl phthalate | 22 |
| Dibutyl phthalate | 17 |
| *n*-hexyl-*n*-decyl phthalate | 16 |
| Dimethyl phthalate | 10 |
| Phosphoric acid esters (cyclic)—total | 99 |
| Trimellitic acid esters—total | 33 |
| Trioctyl trimellitate | 16 |
| Adipic acid esters—total | 68 |
| Di-(2-ethylhexyl) adipate | 45 |
| Epoxidized soya esters—total | 114 |
| All others | 513 |
| Total production—all plasticizers | 2086 |

[a] Mostly di-(2-ethylhexyl) phthalate

mental transport, and current use patterns all favor the introduction of plasticizers into the environment. The biological effects of plasticizers have been studied in the laboratory,[2,3] although the effects of different plasticizers in the environment have yet to be fully evaluated. However, the potential for biological effects and the widespread use of plasticizers have been impetus for further study of these compounds.

Mass spectrometry coupled with gas chromatography has been used to characterize plasticizers in environmental samples. Because of their predominance, phthalates have received the most attention.

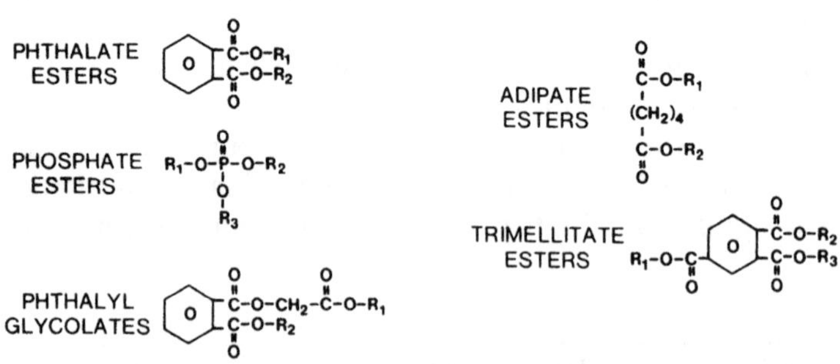


**FIGURE 1.** Structures of some common plasticizers.

*TABLE 2*
*Concentration Levels of Phthalate Esters in the Environment*

| | | Continental (urban, freshwater) | Marine (Gulf of Mexico) |
|---|---|---|---|
| Water (μg/l) | DBP | 0.1–8 | 0.003–0.3 |
| | DEHP | 0.5–600 | 0.006–0.3 |
| Sediment (μg/g) | DBP | 0–100 | 0.0001–0.015 |
| | DEHP | 1–200 | 0.003–0.014 |
| Biota (μg/kg) | DBP | 0–500 | 0.1 |
| | DEHP | 100–3200 | 1–135 |
| Air (ng/m$^3$) | DBP | 0.1–80 | 0.8–4 |
| | DEHP | 2–44 | 0.7–2 |

## 2. *OCCURRENCE OF PLASTICIZERS IN ENVIRONMENTAL SAMPLES*

To emphasize the widespread distribution of plasticizers, we present an overview of some of the reported concentrations of phthalates and other plasticizers in the environment. This presentation will also put a range on the analytical sensitivity required to detect phthalates in "unpolluted" sites. The two most commonly reported phthalates are di-*n*-butyl phthalate and di-(2-ethylhexyl)phthalate. Table 2 shows concentration ranges for these compounds in continental terrestrial samples compared to background levels found in the marine environment.(4–22) The presence of these compounds, even in open-ocean areas, is further evidence of the ubiquitous distribution of phthalates in the environment.

In addition to these two phthalates, numerous other plasticizers have been identified in environmental samples. Several comprehensive GC/MS studies of organics in drinking water and in river water near industrial areas have identified plasticizers as principal components of the synthetic organic load (Table 3). Some specific plasticizers could be traced to point sources of industrial input while other compounds, such as phthalates, appeared to have multiple or diffuse sources of input.(23–27)

The atmosphere is a potentially important mechanism for transporting plasticizers in the environment. Samples of atmospheric aerosols, particularly inside buildings, have shown significant quantities of plasticizer residue. Various citrates, phosphates, phthalates, and adipates have been found on air-borne particles.(4,5,28,29) They have also been detected in the vapor phase in air.(30)

TABLE 3
*Plasticizers Identified in Water and Atmospheric Samples from the United States*

| | |
|---|---|
| *Phthalates* | *Adipates* |
| Di-*n*-butyl phthalate | Dioctyl adipate |
| Diisobutyl phthalate | *Sebacates* |
| *n*-Isobutyl phthalate | Dibutyl sebacate |
| Dioctyl phthalate (several isomers) | Di-(2-ethylhexyl)-sebacate |
| Dimethyl phthalate | *Other* |
| Diethyl phthalate | Acetyl tributyl citrate |
| Butylbenzyl phthalate | Toluene sulfonamide |
| Dicyclohexyl phthalate | Butylphthalyl butylglycolate |
| Didecyl phthalate (several isomers) | Di-(isobutyl)-azelate |
| Numerous mixed alkyl phthalates | Tetraethylene-glycol-di-(2-ethyl hexanoate) |
| Dimethyl terphthalate | Tetraethylene-glycol-di-(2-methyl heptanoate) |
| *Phosphates* | Triethylene-glycol-di-(2-ethyl hexanoate) |
| Tributyl phosphate | |
| Triphenyl phosphate | |
| Tributoxyethyl phosphate | |
| Trioctyl phosphate | |
| 2-ethylhexyldiphenyl phosphate | |

## 3. ANALYTICAL ASPECTS

Because of the ubiquitous nature of phthalates and other plasticizers, accurate analyses, free from artifacts, are especially difficult. Laboratory equipment, supplies, and even laboratory air can contain high concentrations of phthalates which can overwhelm the actual amount in an environmental sample. Rigorous cleaning procedures and laboratory protocol are essential for accurate quantitation and source identification of plasticizers.[31,32]

TABLE 4
*Detection Limits for Phthalate Esters (DEHP)*

| Detector | Sensitivity |
|---|---|
| Liquid chromatography | |
| UV (254 nm) | 10 ng |
| UV (224 nm) | 2 ng |
| Gas chromatography | |
| FID | 1 ng |
| ECD | 20 pg |
| MS-selected ion | 10 pg |

Solvent extraction procedures for separating phthalates from sediment, biota, water, and air have been described and are similar to procedures used for pesticide or hydrocarbon extractions.[11,33–35] Subsequent separation of the phthalates from coextracted organics can be accomplished using thin-layer plates[36] or column chromatography,[37] although this step is often eliminated when using MS analysis. Final separation and detection of phthalates has been performed using liquid chromatography (LC),[37] gas chromatography (GC),[38] or combined gas chromatography/mass spectrometry (GC/MS). Chromatographic retention indices for phthalate esters and other plasticizers have been reported for aid in identification of these compounds.[39] Reported detection limits for DEHP indicate that the mass spectrometer is more sensitive and specific than other conventional detectors (Table 4).

## 4. MASS SPECTRAL PROPERTIES

### 4.1. Phthalates

The nature of mass spectra of phthalate esters has been discussed by Addison and others.[14,40–44)]

The electron impact (EI) mass spectra of most alkyl phthalates, except dimethyl phthalate, contains a prominent fragment at *m/z* 149. This fragment is used to characterize most phthalates for selected-ion mass spectrometry. The *m/z* 163 fragment is characteristic of dimethyl phthalate. The proposed pathway for the production of this ion and others characteristic of phthalates is presented in Fig. 2.

For alkyl chains of $n = 1,2,3,4$, ions at *m/z* 163, 177, 191, or 205, respectively, characterize structure III. The 167 fragment (IV) charac-

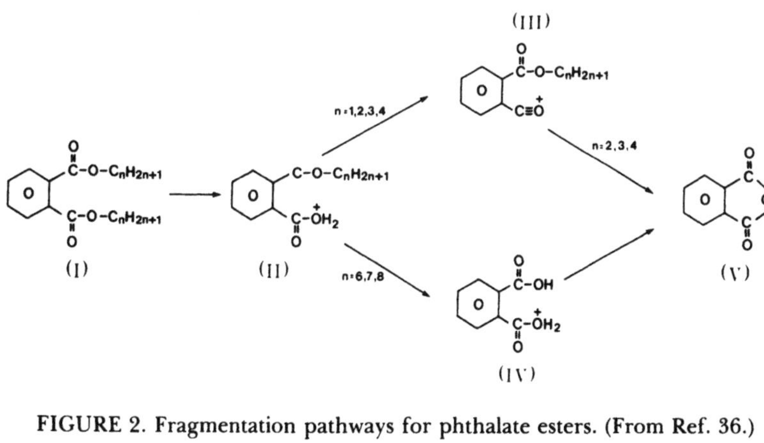


FIGURE 2. Fragmentation pathways for phthalate esters. (From Ref. 36.)

terizes larger alkyl groups. Because the concentration of phthalates may be very high relative to other trace organics in the environment, Cautreels and van Cauwenberghe suggest using secondary phthalate fragments when using the mass spectrometer to quantitate environmental phthalates since the *m/z* 149 peak may overload the mass spectrometer.[5] They suggest that diethyl phthalate can be measured using *m/z* 222 (molecular ion), dibutyl phthalates at *m/z* = 223, butylbenzyl phthalate at *m/z* = 206, and dioctyl phthalates at *m/z* = 279. Most EI spectra of phthalates show low to unobservable molecular ions. Figure 3 (a–e) shows characteristic EI spectra of common phthalate esters.

The absence of molecular ions in EI spectra of most phthalates has prompted studies of CI mass spectra of phthalate esters.[40,41] The soft ionization techniques such as CI provide relatively large protonated molecular species, $[M + H]^+$, and a simpler fragmentation pattern which

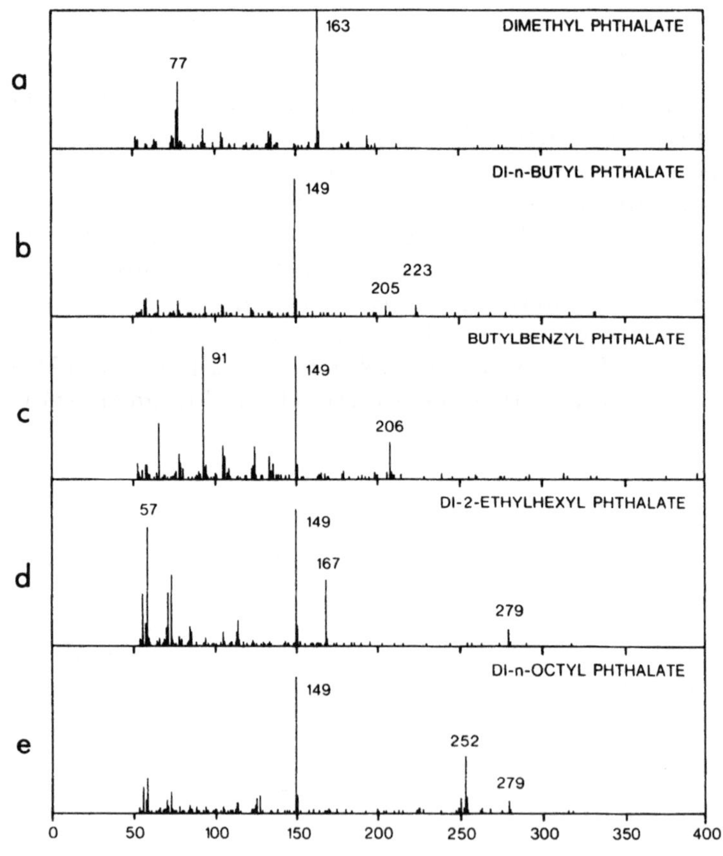


**FIGURE 3.** EI mass spectra of common phthalate esters (scan: *m/z* 50–400).

can aid in compound identification and confirmation. Figure 4 illustrates the potential for increased molecular weight information using CI techniques.

However, Addison noted a decrease in $[M + H]^+$ production at higher reaction temperatures and moderate to low reagent gas pressures used in his GC/CIMS work.(40) The decreased response is an expected effect of the plasma and is an important consideration for application of CIMS to environmental studies of phthalates. Isobutane was the preferred reagent gas producing simpler spectra and increased $[M + H]^+$ abundance (Table 5).

Other MS techniques, such as field ionization (FI) mass spectrometry, for phthalate identification have been reported, but neither CI nor FI techniques have been routinely applied to identifying phthalates in environmental samples.

### 4.2. *Phosphate Esters*

The mass spectra of phosphate ester plasticizers have not received as much attention as the phthalates, although phosphate esters can be abundant in environmental samples, particularly the indoor environment.(28) Unlike the phthalates, mass spectra of phosphate esters can be quite varied depending on the ester group. The alkyl phosphate esters (e.g., tributyl and trioctyl phosphate) have a strong fragment at *m/z* 99 which is due to the protonated species $(H_3PO_4)H^+$ (Fig. 5a). Small molecular ions are typical of the alkyl phosphates. As with the phthalates, isobutane CI can dramatically increase the $(M + H)^+$ ion, although temperature effects on the production of $(M + H)^+$ ions for phosphates have not been studied. Mixed alkyl-aryl phosphates have a more complex fragmentation pattern than the alkyl phosphate (Fig. 5b), while many

*TABLE 5*
*Relative Abundances of $[M + H]^+$ Ions of Phthalate Esters Using Chemical Ionization*

| Compound | $[M + H]^+$ *m/z* | Relative abundance (%) CI with methane | CI with isobutane |
|---|---|---|---|
| Dimethyl phthalate | 195 | 3.8 | 33.2 |
| Diethyl phthalate | 223 | 6.8 | 37.9 |
| Diisobutyl phthalate | 279 | 7.8 | 29.3 |
| Di-*n*-butyl phthalate | 279 | 8.3 | 22.2 |
| Dipentyl phthalate | 307 | 9.4 | 18.3 |
| Di-(2-ethylhexyl) phthalate | 391 | 6.5 | 12.9 |
| Dioctyl phthalate | 391 | 4.7 | 100 |

*TABLE 6*
*Typical Components of Some Commercial Aryl Phosphate Oils and Their Mass Spectra*[a]

| | Mol. wt. (% parent ion) | Other ions (and % abundance) |
|---|---|---|
| Kronitex-R | | |
| tri-*m*-cresyl phosphate | 368 (100) | 367 (49.4), 366 (34) 243 (6.3), 91 (5.3) |
| di-*m*-cresyl-*p*-cresyl phosphate | 368 (100) | 367 (54.8), 279 (6.4) 261 (8.9), 198 (9.3) |
| di-*p*-cresyl-*m*-cresyl phosphate | 368 (100) | 367 (66), 279 (9.3) 179 (17.7), 108 (18.2) |
| Santicizer-140 | | |
| triphenyl phosphate | 326 (100) | 325 (66.4), 234 (6.3) 170 (7.1), 76 (2.1) |
| *m*-cresyl diphenyl phosphate | 340 (100) | 339 (59.6), 338 (5.4) 247 (5.8), 164 (3.9) |
| phenyl dicresyl phosphate | 354 (100) | 353 (62.0), 352 (5.8) 247 (3.7), 181 (14.9) |
| Pydraul 50E | | |
| diphenyl-*p*-nonylphenyl phosphate | 452 (2.0) | 423 (10.8), 381 (27.5) 367 (100), 353 (12.8) |
| diphenyl-*p*-cumylphenyl phosphate | 444 (29.5) | 429 (100), 367 (3.9) 214 (3.2), 178 (5.4) |

[a] From Deo and Howard (1978).

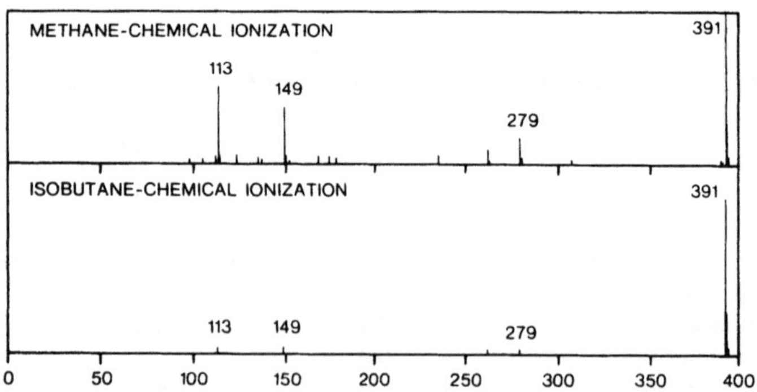


FIGURE 4. CI mass spectra of di(2-ethyl-hexyl) phthalate. (From Ref. 32.)

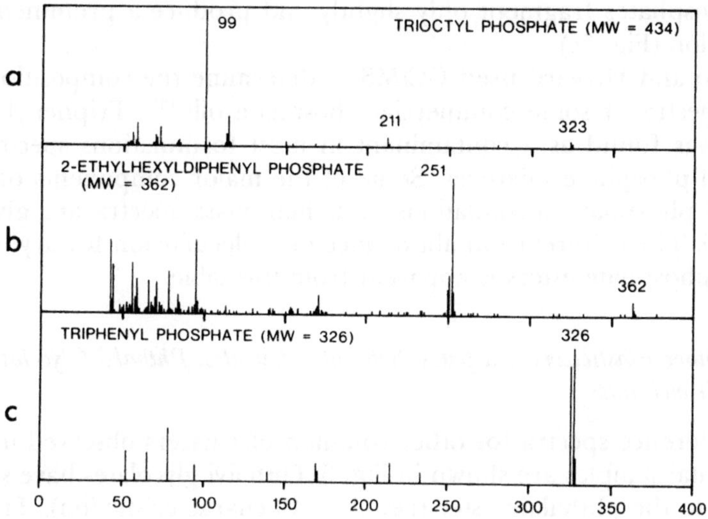


**FIGURE 5.** EI mass spectra of some common phosphate plasticizers.

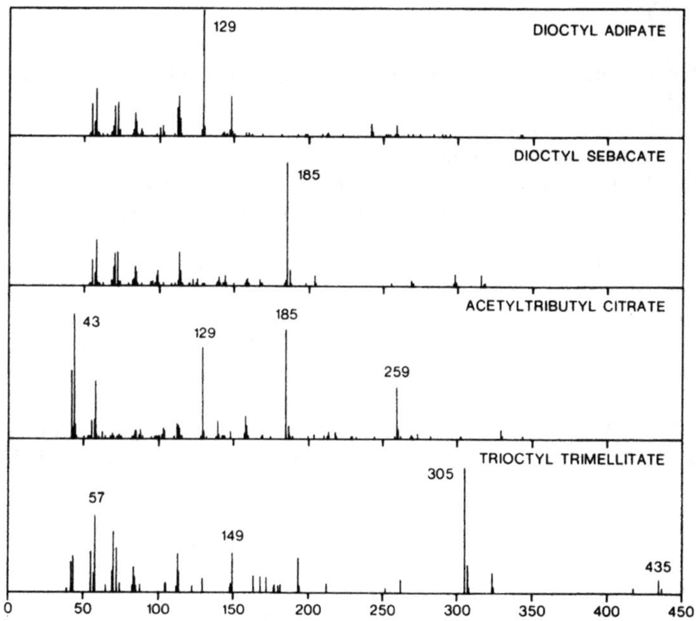


**FIGURE 6.** EI mass spectra of some widely used plasticizers.

aryl phosphates fragment only slightly and produce a prominent molecular ion (Fig. 5c).

Deo and Howard used GC/MS to determine the composition and mass spectra of some commercial phosphate oils.[45] Triphenyl phosphate was found as a contaminant in most formulations specified as tricresyl phosphate mixtures. Some of the major components of commercial phosphate formulations and their mass spectra are given in Table 6. The difference in abundance of molecular ion for aryl versus mixed phosphate esters is apparent from this table.

### 4.3. *Other Plasticizers: Adipates, Sebacates, Citrates, Phthalyl Glycolates, Trimellitates*

Reference spectra for other common plasticizers observed in environmental samples are shown in Fig. 6. Phthalyl glycolates have spectra similar to the phthalate ester (i.e., *m/z* 149 characteristic ion). Trimellitates have not been observed in environmental samples, but they may escape detection using GC because of their relatively high molecular weight and low volatility.

## *REFERENCES*

1. United States International Trade Commission, USITC Publication 921, Washington, D.C., 1979.
2. J. Autian, *Environ. Health Pers.* **3,** 3–25 (1973).
3. F. L. Mayer, Jr., and H. O. Sanders, *Environ. Health Pers.* **3,** 153–157 (1973).
4. J. L. Bove, P. Dalven, and V. P. Kukreja, *Int. J. Environ. Anal. Chem.* **5,** 187–194 (1978).
5. W. Cautreels, K. van Cauwenberghe, and L. Guzman, *Sci. Total Environ.* **8,** 79–88 (1977).
6. C. S. Giam, H. S. Chan, G. S. Neff, and E. L. Atlas, *Science* **199,** 419–421 (1978).
7. C. F. Cifrulak, *Soil Sci.* **107,** 63–69 (1979).
8. E. F. Corcoran, *Environ. Health Pers.* **3,** 13–15 (1977).
9. E. F. Corcoran and R. W. Curry, *Rev. Biol. Trop.* **26,** (Supl. 1), 125–133 (1978).
10. C. S. Giam, H. S. Chan, and G. S. Neff, in *Marine Pollutant Transfer* (H. L. Windom and R. A. Duce eds.), Heath, Lexington, MA. 1976, pp. 375–386.
11. C. S. Giam, E. L. Atlas, and H. S. Chan, Proceedings of Fourth Joint Conference on Sensing of Environmental Pollutants, New Orleans, LA, November 6–11, 1977, pp. 188–190.
12. C S. Giam, H. S. Chan, and G. S. Neff, *Mar. Poll. Bull.* **9,** 249–251 (1978).
13. C. S. Giam and E. L. Atlas, *Die Naturwiss.* **67,** 508–509 (1980).
14. R. A. Hites, *J. Chromatogr. Sci.* **11,** 570–574 (1973).
15. R. A. Hites, *Env. Health Persp.* **3,** 17–21 (1973).
16. M. Morita, H. Nakamura, and S. Mimura, *Water Res.* **8,** 781–788 (1974).
17. G. Ogner and M. Schnitzer, *Science* **170,** 317–318 (1970).
18. H. E. Schwartz, C. J. M. Anzion, H. P. M. van Vliet, J. W. Copius Peereboorns, and U. A. Th. Brinkman, *Int. J. Environ. Anal. Chem.* **6,** 133–144 (1979).

19. D. L. Stalling, J. W. Hogan, and J. L. Johnson, *Environ. Health Pers.* **3,** 17–21 (1973).
20. D. L. Stalling, J. W. Hogan, and J. L. Johnson, *Environ. Health Pers.* **3,** 159–173 (1973).
21. D. T. Williams, *J. Agr. Food Chem.* **21,** 1128–1129 (1973).
22. H. Yamasaki and K. Kuwata, *Bunseki Kagaku* **26,** 1–5 (1977).
23. W. E. Coleman, R. G. Melton, F. C. Kopfler, K. A. Barone, T. A. Aurand, and M. G. Jellison, *Environ. Sci. Tech.* **14,** 576–588 (1980).
24. R. A. Hites, G. A. Jungclaus, V. Lopez-Avila, and L. S. Sheldon, *ACS Symp. Ser.* **94,** 63–89 (1979).
25. L. H. Keith, (ed.), *Identification and Analysis of Organic Pollutants in Water,* Ann Arbor Science, Ann Arbor, Mi 1976.
26. L. S. Sheldon and R. A. Hites, *Environ. Sci. Tech.* **13,** 574–579 (1979).
27. I. W. Suffet, L. Brenner, and P. R. Cairo, *Water Res.* **14,** 853–867 (1980).
28. C. J. Weschler, *Environ. Sci. Tech.* **14,** 428–430 (1980).
29. W. Cautreels and K. van Cauwenberghe, *J. Chromatogr.* **131,** 253–264 (1977).
30. E. L. Atlas and C. S. Giam, *Science* **211,** 163–165 (1981).
31. C. S. Giam and M. K. Wong, *J. Chromatogr.* **72,** 283–292 (1972).
32. C. S. Giam and H. S. Chan, National Bureau of Standards Special Publication 422, *Accuracy in Trace Analysis: Sampling, Sample Handling, and Analysis,* pp. 701–707 (1976).
33. C. S. Giam, H. S. Chan, and G. S. Neff, *Anal. Chem.* **47,** 2225–2229 (1975).
34. C. S. Giam, H. S. Chan, and G. S. Neff, *Anal. Chem.* **47,** 2319–2320 (1975).
35. C. S. Giam, in *Strategies for Marine Pollution Monitoring* (E. D. Goldberg, ed.), Wiley-Interscience, New York, 1976, pp. 61–76.
36. M. Ehrhardt and J. Derenbach, *Marine Chem,* **8,** 339–346 (1980).
37. S. Mori, *J. Chromatogr.* **129,** 53–60 (1976).
38. C. S. Giam, E. Atlas, H. S. Chen, and G. S. Neff, *Atmos. Environ.* **14,** 65–69 (1980).
39. M. P. Friocourt, F. Berthou, D. Picart, Y. Dreano, and H. H. Floch, *J. Chromatogr.* **172,** 261–271 (1979).
40. J. B. Addison, *Analyst* **104,** 846–852 (1979).
41. H. M. Fales, G. W. A. Milne, and R. S. Nicholson, *Anal. Chem.* **43,** 1785–1789 (1971).
42. A. G. Harrison and R. K. M. R. Kallury, *Org. Mass Spectrom.* **15,** 277–283 (1980).
43. F. W. McLafferty and R. S. Gohlke, *Anal. Chem.* **31,** 2076 (1959).
44. S. Safe and O. Hutzinger, *Mass Spectrometry of Pesticides and Pollutants,* CRC Press, Cleveland, Ohio, 1973.
45. P. G. Deo and P. H. Howard, *J. Assoc. Off. Anal. Chem.* **61,** 266–271 (1978).

55. C. S. Giam ... in *Mass Spectrometry* ... Wiley-Interscience, New York, 1975, pp. 61–75.
56. M. H. [illegible] ... 6, 530–540 (1968).
57. S. Mori, *J. Chromatogr.* 129, 53–60 (1976).
58. C. S. Giam, E. Atlas, H. S. Chan, and G. S. Neff, ... 14, 65–... (1980).
59. M. [illegible] ... *J. Chromatogr.* 171, ... (1979).
60. J. [illegible] ... 166, ... (1979).
61. D. [illegible], G. W. A. Milne, and R. S. Nicholson, *Anal. Chem.* 43, 1785–1789 (1971).
62. A. G. Harrison and R. K. M. R. Kallury, *Org. Mass Spectrom.* 15, 277–283 (1980).
63. F. W. McLafferty and R. S. Gohlke, *Anal. Chem.* 31, 2076 (1959).
64. S. Safe and O. Hutzinger, *Mass Spectrometry of Pesticides and Pollutants*, CRC Press, Cleveland, Ohio, 1973.
65. P. G. Deo and P. H. Howard, *J. Assoc. Off. Anal. Chem.* 61, 860–871 (1978).

# ORGANOMETALLIC COMPOUNDS

## E. Ch. Th. GEVERS

### 1. INTRODUCTION

Organometallic compounds are defined as compounds containing at least one organic group bound to the metal atom by a metal–carbon bond. In this chapter only compounds answering this definition will be discussed.

Organometallic compounds present in the environment originate from two different sources. Under specific conditions a number of inorganic metal ions can be converted into organometallic species by biomethylation. During the past decades man-made organometallic compounds have found widespread industrial applications, particularly as biocides. Large amounts of tetraalkyllead are used as antiknock agents in gasoline fuels. These organometallic compounds, together with their degradation products, can be found in the environment.

The first organometallic compound found to be present in the environment was trimethylarsine, which appeared to be formed by microorganisms in domestic wallpapers containing Scheele's green (copper hydrogen arsenite) and Paris green (cupric acetoarsenite) dyes.[1–3]

Andreae demonstrated the occurrence of four arsenic species in seawater.[4] Two of these compounds were of an organometallic nature: dimethylarsinate and methylarsonate. From the correlation between the presence of the methylarsenic species and the depth in which they were found, it was concluded that these species were formed by photochemical methylation. Cheng and Focht incubated soil with arsenate, arsenite,

*E. Ch. Th. GEVERS* • Institute of Applied Chemistry TNO, P.O.B. 5009, 3502 JA Utrecht, The Netherlands.

methylarsonate, and dimethylarsenite. They found no methylation but a reduction to arsine.[5]

After the identification of methylmercury formed from the inorganic mercury effluent of a nearby factory as the species responsible for the Minimata disease,[6] much attention has been paid to the formation and detection of methylmercury compounds.[7–12]

Tompkins described a very sensitive method to detect antimony, germanium, and tin and was thus able to prove for the first time the biological methylation of tin in the body and in the environment. He found no proof of a biomethylation of antimony and germanium.[13]

Wong, Chau, and Luxon proved that microorganisms in lake sediments can transform certain inorganic and organometallic lead compounds into tetramethyllead.[14] A combination of gas chromatography and mass spectrometry confirmed the identity of $Me_4Pb$.

Other metals such as selenium[15] and thallium (I)[16] are reported as being methylated in the environment.

Of the man-made organometallic compounds there is a certain number that has found widespread use.

Surveys of the uses of inorganic and organometallic mercury compounds are given by D'Itry[17] and by Ulfvarson.[18] In 1968 the United States consumed more than 2.8 million kg of mercury; 360,000 kg were used in paint as bacteriocides and fungicides. Although their use has steadily decreased in recent years phenyl-, ethyl-, and methyl-mercury compounds are still used in agriculture as seed disinfectants.

Zuckerman has given a survey of the organotins annually produced in the United States since 1965.[19] In 1976 about 25 million kg of organotins were produced. Tetraorganotin compounds have found only very limited practical application, but they are used as starting materials for the synthesis of other types of organotin compounds. Triorganotin compounds display biocidal activity. Triphenyl- and tricyclo-hexyltin derivatives have found large-scale application as agricultural fungicides and miticides, respectively. Tributyltin compounds are mainly used as industrial fungicides and antifouling substances. Over two-thirds of the annual world organotin production was used as stabilizer of polyvinylchloride plastics in the form of dibutyl- and dioctyltin derivatives.

Of the more than 140 million kg of lead emitted in the United States in 1975 into the atmosphere, 127.8 million kg were emitted by gasoline combustion.[20] Tetramethyl- and tetraethyllead are the compounds used as antiknock agents in automobile fuels. Most of the tetraalkyllead is burned to inorganic lead. However, high concentrations of these compounds are found in the air in underground garages and near gasoline stations.[21]

## 2. MASS SPECTROMETRY

Organometallic compounds are generally much more toxic than their inorganic salts. Their toxicity can vary strongly according to the nature and amount of the organic groups bound to the metal atom. Therefore, it is very important to trace them as they are found in the environment.[22] Many techniques aimed at the detection of small amounts of metals—for example, atomic absorption or spark source mass spectrometry—are performed after combustion, where this direct information is lost.[7,23]

Mass spectrometry or a combination of gas chromatography and mass spectrometry is a most useful way of determining organometallics quantitatively and/or qualitatively. Despite the sensitivity and versatility of modern (computerized) mass spectrometers, relatively little use has been made of this technique for the analysis of organometallic compounds in the environment. Contamination of the mass spectrometer was not observed in the author's laboratory with a quadrupole GC/MS,

*TABLE 1*
*Natural Abundance of Carbon, Arsenic, Tin, Mercury, and Lead*

| Element | Mass | Natural abundance |
|---|---|---|
| C | 12 | 98.89 |
| | 13 | 1.11 |
| As | 75 | 100 |
| Sn | 112 | 0.96 |
| | 116 | 14.30 |
| | 117 | 7.61 |
| | 118 | 24.03 |
| | 119 | 8.58 |
| | 120 | 32.85 |
| | 122 | 4.72 |
| | 124 | 5.94 |
| Hg | 198 | 10.02 |
| | 199 | 16.84 |
| | 200 | 23.13 |
| | 201 | 13.22 |
| | 202 | 29.80 |
| | 204 | 6.85 |
| Pb | 204 | 1.48 |
| | 206 | 23.6 |
| | 207 | 22.6 |
| | 208 | 52.3 |

although a double-focusing instrument used in earlier studies showed an appreciable memory effect and required thorough baking to remove traces of the organometallics. In the MS analysis of organometallic compounds several sophisticated techniques like field desorption,[24] chemical ionization,[25–27] negative-ion [28,29] and high-resolution[30] mass spectrometry have been used.

The interpretation of the spectra of organometallic compounds is facilitated by the fact that the majority of the fragments formed bear the metal atom, which mostly has an easily recognizable isotope pattern (Table 1).[31]

## 3. ARSENIC

Cheng and Focht[5] used GC/MS in their study of the production of arsine and methylarsines in soil and culture. They emphasize the importance of a good gas chromatographic separation for a correct interpretation of the results, because the main group V $R_3M$ mass spectra display many hydride-containing ions. The mass spectra of arsine, $AsH_3$, methylarsine, $CH_3AsH_2$, dimethylarsine, $(CH_3)_2AsH$, and trimethylarsine, $(CH_3)_3As$, are presented by these authors.

## 4. MERCURY

The most abundant metal-containing ion in the mass spectrum of dimethylmercury is $HgCH_3^+$. The rearrangement ions $HHgCH_3^{+\cdot}$ (formed by an intramolecular rearrangement) and $HHg^+$ were also observed.[32] The $HHg^+$ ion has also been observed in the spectra of diethyl- and di-*n*-butylmercury. As for the main group IV metal alkyls, the elimination of R—H from $R_2M$ to form metal hydride species appears to be an energetically favored process. Hydrocarbon ions contribute to a large portion of the total ion current in all the mercury alkyls, but this may be due in part to thermal decomposition before ionization.

Bryant and Kinstle described the mass spectra of several diaryl- and dialkyl-mercury compounds as well as the fragmentation of alkylmercury chlorides.[33] Glocking extended the study of the fragmentation patterns of organomercurials with other mono- and diorgano-mercurials.[34]

Jensen and Jernelöv were the first to find that both mono- and dimethyl-mercury can be produced in bottom sediments and rotten fish.[35] Gas chromatography/selected-ion-monitoring mass spectrometry (*m/z* 215, $CH_3\,^{200}Hg^+$) was used in their study.

Several authors suggested that alkylmercury compounds decompose during the gas chromatographic analysis.[35,11] This led Baughman to an

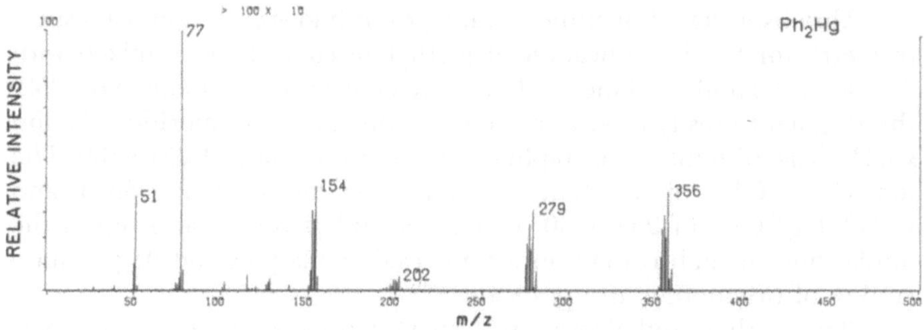


FIGURE 1. Mass spectrum of diphenylmercury.

investigation of the gas chromatographic behavior of methyl- and phenylmercury compounds by MS detection.[8] The spectra of methylmercury salts by direct-probe inlet were recorded. The mass spectra of the gas chromatographic peak of pure methylmercury chloride showed that this compound was partly converted into methylmercury iodide and methylmercury bromide in the gas chromatographic column. Phenylmercury acetate decomposed on the column into phenylmercury chloride and diphenylmercury (Figures 1 and 2). Johansson, Ryhage, and Westöo used GC/MS for the identification and quantitation of methylmercury compounds in fish extracts and compared the results with those obtained by electron capture GC and activation analysis.[9] The spectra of methylmercury chloride and methylmercury iodide are shown. Ohkoshi, Takahashi, and Sato determined methylmercury chloride in wastewater at the level of 0.1 ppb using gas chromatography/mass fragmentography at *m/z* 252 ($Me^{202}Hg^{35}Cl^+$) after a 5000-fold concentration procedure.[12] Nishi and Horimoto compared the vaporization behavior of several alkylmercury salts by GC/MS and direct-inlet MS.[11]

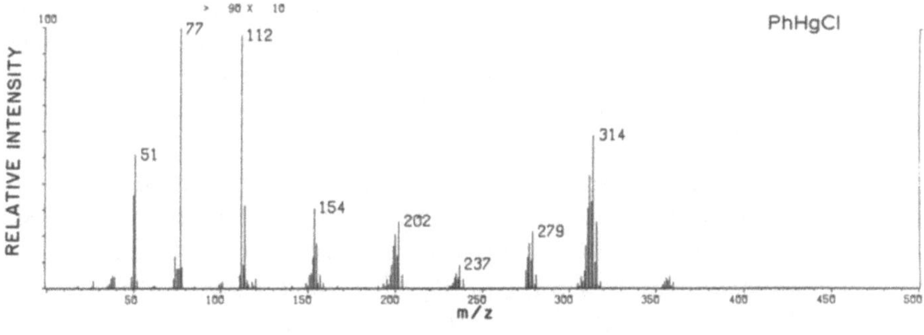


FIGURE 2. Mass spectrum of phenylmercury chloride.

Abramson has shown the advantages of high-resolution mass spectrometry for the identification of methylmercury chloride mixed with Aroclor 1260 and Lindane.[36] These two chlorides contribute to *m/z* 252, the fragment mostly used to monitor methylmercury chloride. The ion $C_{10}H_3{}^{35}Cl_3^{+\cdot}$ of heptachlorobiphenyl has an exact mass of 251.9300. The ion $C_6H_5{}^{35}Cl_5^{+\cdot}$ of Lindane has an exact mass of 251.8834 and $CH_3{}^{202}Hg^{35}Cl^+$ of 251.9630. Hutzinger and Jamieson also report the application of high-resolution MS to residue analyses and degradation studies of organomercury pesticides.[30]

Ball, Barber, and Vossen used spark source MS to determine mercury in fish.[23] Spark source MS after isotope dilution was used by Alvarez to determine mercury in orchard leaves.[37] Though this method is sensitive it lacks specificity because the method requires ashing of the sample.

Negative-ion mass spectra of trichlorovinylmercurials are described by Cohen.[28] Hutzinger, Jamieson, and Safe,[38] and Safe and Hutzinger[39] presented the spectra and discussed the fragmentation patterns of a number of phenyl- and methyl-mercury salts and of methoxyethyl- and ethoxyethyl-mercury chloride.

## 5. *TETRAALKYLTIN AND TETRAALKYLLEAD COMPOUNDS*

Some general phenomena of main group IV compounds of the type $MR_4$ (M = C, Si, Ge, Sn, Pb) have been summarized:[40]

1. Since the group IV elements are more electropositive than the R groups at dissociation, the positive charge will prefer to remain on the metal part: $MR_4^{+\cdot} \rightarrow R\cdot + R_3M^+$
2. The fragmentation of the molecular ion (odd electron) yields an even-electron, metal-containing ion by elimination of an odd-electron, radical fragment.
3. $R_3M^+$ in the mass spectra of $R_4M$ might be expected to be particularly stable because it will have an even number of electrons and the electron configuration will be that of the corresponding (stable) group III compound. The radical ions $R_2M^{+\cdot}$ are expected to be unstable.
4. Comparing $R_4Sn$ with $R_4Pb$ one can expect that the $M^{II}$ oxidation state would be more important in the Pb than in the Sn compound: the $RPb^+$ ion will be more important than $RSn^+$.
5. Molecular ions of the type $R_4M^{+\cdot}$ exhibit a low (1–10%) abundance because the odd-electron ions readily eliminate a radical to yield the $R_3M^+$ ion.

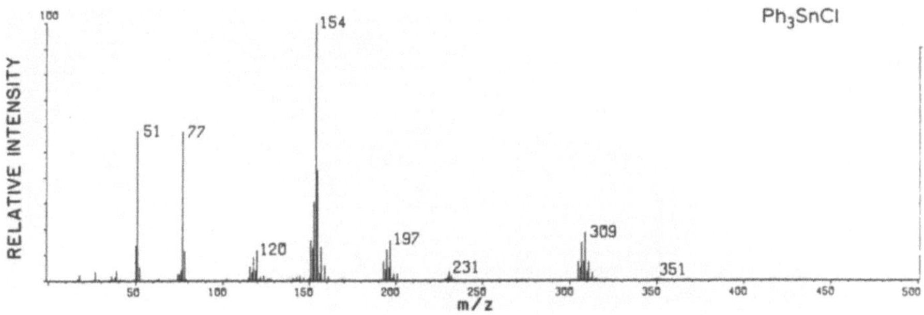


**FIGURE 3.** Mass spectrum of triphenyltin chloride.

## 6. *TIN*

The fungicide triphenyltin chloride (Figure 3) has been analyzed by Kuwahara (Figure 3) in soil using the direct-insertion selected-ion-monitoring method.(41) The fragment ion $(C_6H_5)_3Sn^+$ at *m/z* 309 was traced for the quantitation. An ionizing voltage of 30eV was used. The detection limit was 0.01 ppm. The method appears to be inapplicable to tomatoes, strawberries, and flour. The authors presume that protein renders the extraction procedure ineffective. The following fragmentation reactions of the parent molecular ion were reported by Spalding(40):

$$(C_6H_5)_3SnCl^{+\cdot} \xrightarrow{-C_6H_5\cdot} (C_6H_5)_2SnCl^+ \xrightarrow{-C_6H_5\cdot} C_6H_5SnCl^{+\cdot}$$

$$\xrightarrow{-C_{12}H_{10}} SnCl^+$$

$$\xrightarrow{-C_6H_5SnCl} C_{12}H_{10}^{+\cdot} \quad \text{(base peak)}$$

$$\xrightarrow{-SnCl} C_{18}H_{15}^{+\cdot}$$

Hutzinger(38) and Safe and Hutzinger(39) presented the spectra of triphenyltin hydroxide and tributyltin oxide (Figures 4 and 5) and discussed the fragmentation patterns of triphenyltin acetate, tributyltin acetate, tributyltin benzoate, tripropyltin acetate, and dibutyltin laurate.

Meinema studied the degradation of tributyltin oxide, which is used on a large scale as a fungicidal wood preservative.(42) A sensitive, reproducible, and interference-free method was developed for the simultaneous determination of tri-, di-, and mono-butyltin species present in aqueous systems at tin concentrations of 0.01–5 ppm. After the extraction of the butyltin species, these were converted by the Grignard method into butylmethyltin species and determined by GC/MS (Figures 6–8). Hexyldibutylmethyltin ($HexBu_2SnMe$) was used as an internal standard (Figure 9).

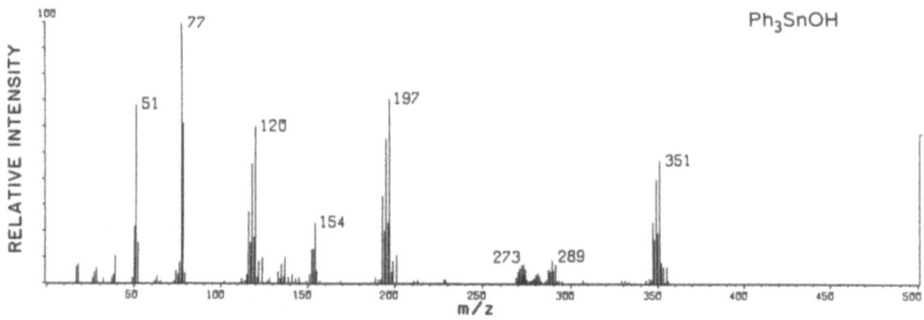


FIGURE 4. Mass spectrum of triphenyltin hydroxide.

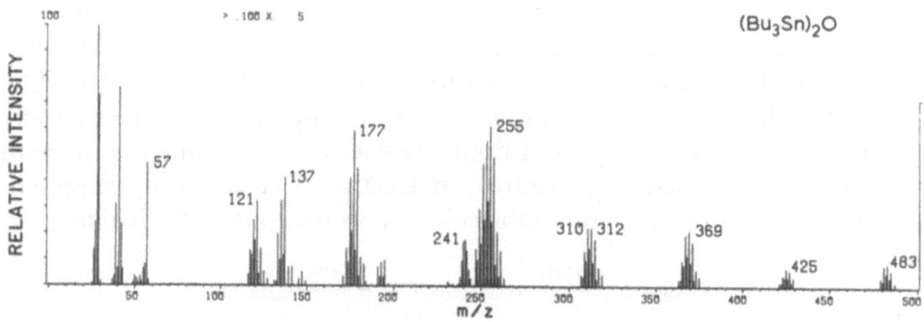


FIGURE 5. Mass spectrum of tributyltin oxide (TBTO).

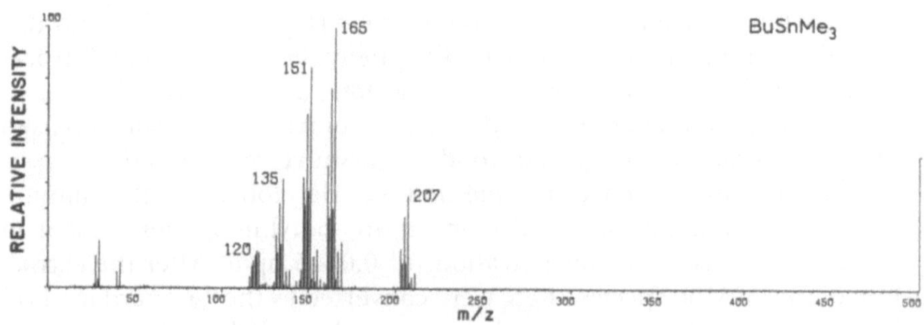


FIGURE 6. Mass spectrum of butyltrimethyltin.

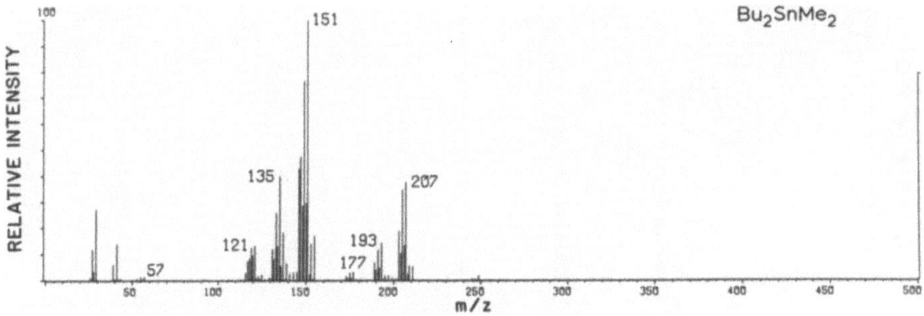


FIGURE 7. Mass spectrum of dibutyldimethyltin.

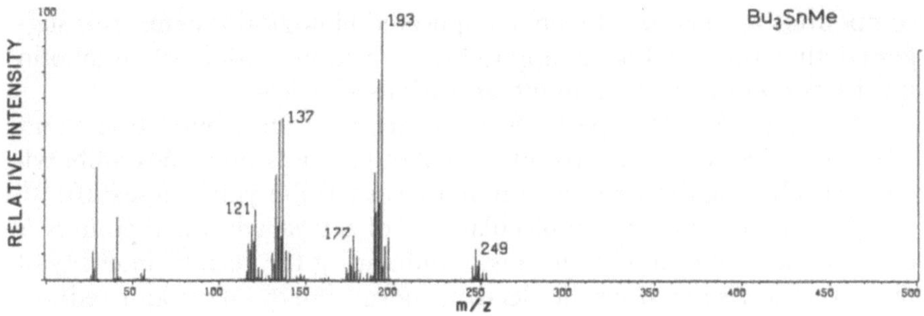


FIGURE 8. Mass spectrum of tributylmethyltin.

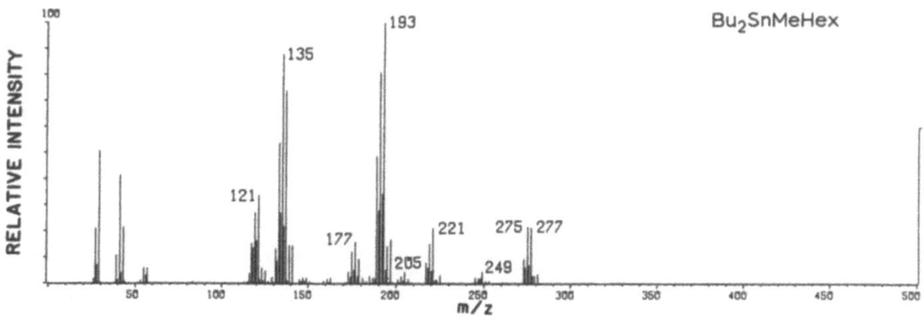


FIGURE 9. Mass spectrum of hexyldibutylmethyltin.

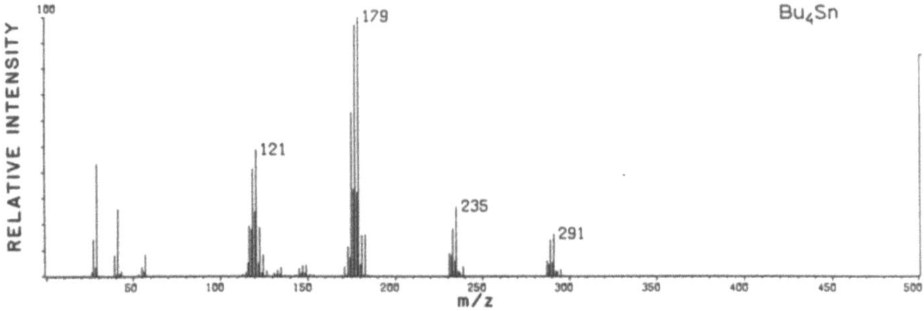


FIGURE 10. Mass spectrum of tetrabutyltin.

For the mass fragmentography, both *m/z* 135 ($Me^{120}Sn^+$) and *m/z* 193 ($BuMeSnH^+$) were monitored. The ratio of the fragments 135/193 is a characteristic constant for each compound. Following both fragments reduces sensitivity but indicates possible interferences when these tin compounds are extracted from complicated biological systems. It is suggested that this method is applicable to a large variety of organotin species present in trace amounts in various matrices.

Inorganic Sn (IV) species were determined after butylation at *m/z* 121, 177, 178, and 179. No internal standard was used, hexyldibutylmethyltin having the same retention time as tetrabutyltin (Figure 10).[42]

As stated above, the molecular ion of tetraalkyltin compounds is often weak or absent. One means of enhancing the relative intensity of the parent ion is to reduce the electron energy, but this technique reduces the absolute intensity. Another low-energy method is chemical ionization.

Fisch studied the cleavage reactions of benzyltin compounds by isobutane chemical ionization mass spectrometry.[26,27] In the case of trimethylbenzyltin an $[M + 57]^+$ ion with an abundance of 30% was found.

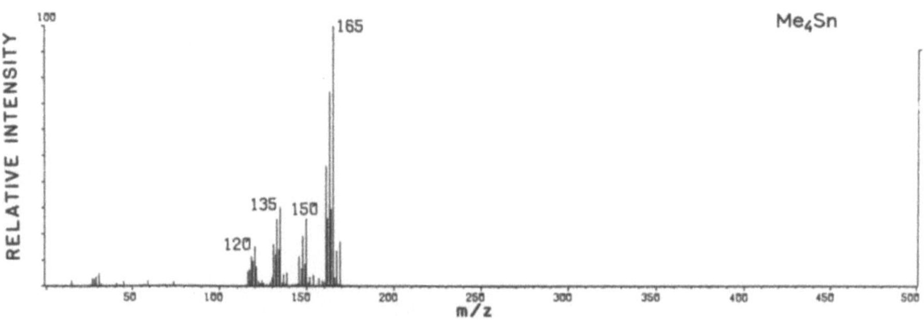


FIGURE 11. Mass spectrum of tetramethyltin.

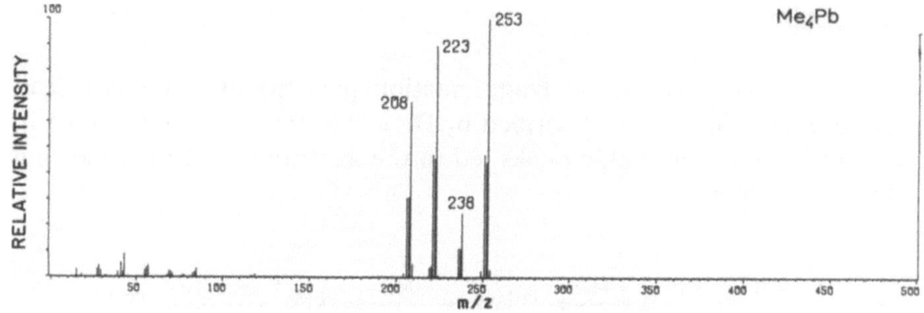


FIGURE 12. Mass spectrum of tetramethyllead.

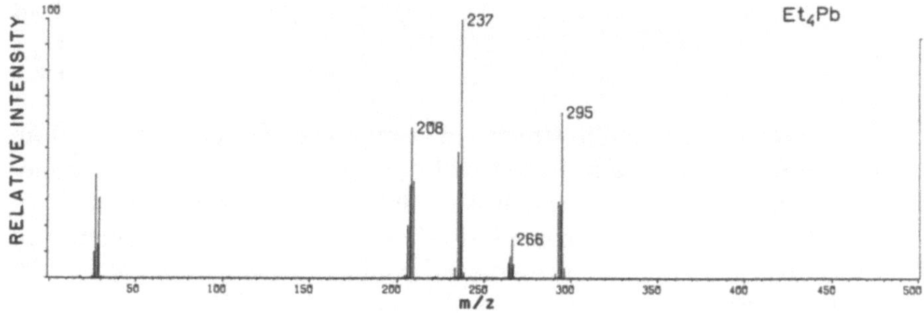


FIGURE 13. Mass spectrum of tetraethyllead.

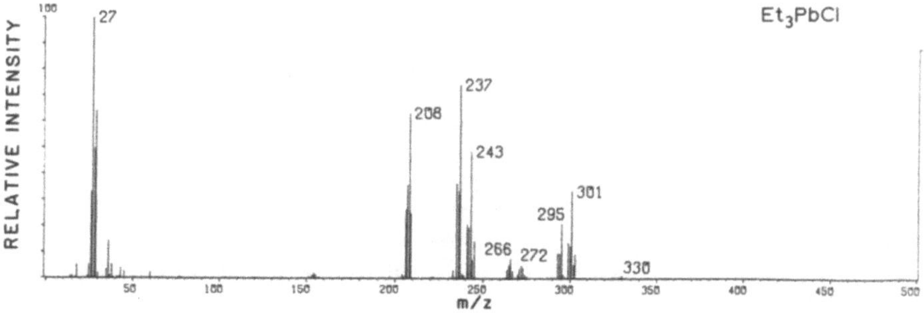


FIGURE 14. Mass spectrum of triethyllead chloride.

## 7. LEAD

The mass spectra and fragmentation patterns of tetramethyllead and tetraethyllead were described by De Ridder[43] (Figures 12 and 13). A study of the metastable peaks led to the assumption of the following dissociation patterns:

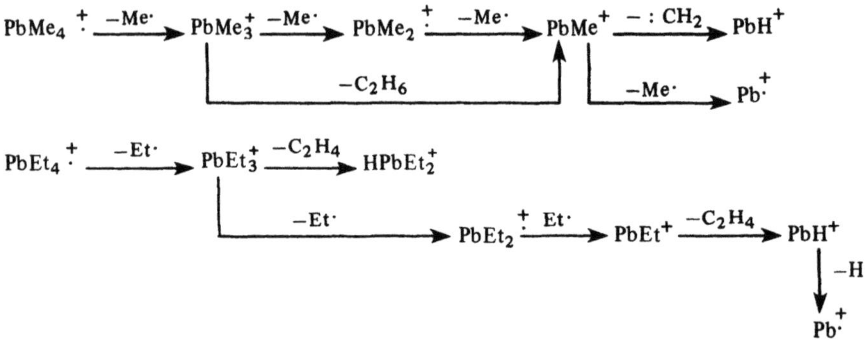


The rearrangement in which a H atom is transferred to the Pb atom can be explained by the preference of the lead ion for three- and mono-coordinationlike thallium, which is isoelectronic with the lead ion.

Wong, Chan, and Luxon used GC/MS to confirm the structure of tetramethyllead in their study on the environmental methylation of lead.[14]

Robinson, Kiesel, and Rhodes investigated the formation of triethyllead chloride ($Et_3PbCl$) from tetraethyllead ($Et_4Pb$) in seawater by a combination of gas chromatography and atomic absorption. GC/MS was used to check the GC/AA method[44](Figure 14).

## REFERENCES

1. F. Challenger, *Chem. Rev.* **36,** 315–361 (1945).
2. F. Challenger, in *Organometals and Organometalloids,* ACS Symp., Ser. 82, (F. E. Brinckman and J. M. Bellama, eds.), American Chemical Society, Washington, D.C., 1978, pp. 1–22.
3. J. M. Wood, *Science* **183,** 1049–1052 (1974).
4. M. O. Andreae, *Limnol. Oceanogr.* **24,** 440–452 (1979).
5. C. -N. Cheng and D. D. Focht, *Appl. Environ. Microbiol.* **38,** 494–498 (1979).
6. K. Irukayama, T. Kondo, F. Kai, and M. Fujiki, *Kumamoto Med. J.* **14,** 157 (1961).
7. D. F. Ball, M. Barber, and P. G. T. Vossen, *Sci. Total Environ.* **2,** 101–110 (1973).
8. G. L. Baughman, M. H. Carter, N. L. Wolf, and R. G. Zepp, *J. Chromatogr.* **76,** 471–476 (1973).
9. B. Johansson, R. Ryhage, and G. Westöo, *Acta Chem. Scand.* **24,** 2349–2354 (1970).
10. L. R. Kamps and B. McMahon, *J. Assoc. Off. Anal. Chem.* **55,** 590–595 (1972).
11. S. Nishi and Y. Horimoto *Bunseki Kagaku* **24,** 178–183 (1975).
12. S. Ohkoshi, T. Takahashi, and T. Sato, *Bunseki Kagaku* **22,** 593–595 (1973).

13. M. A. Tompkins, Thesis, University Microfilms International, Ann Arbor, 1977.
14. P. T. S. Wong, Y. K. Chau, and P. L. Luxon, *Nature* **253,** 263–264 (1975).
15. D. C. Reamer and W. H. Zoller, *Science* **208,** 500–502 (1980).
16. F. Huber and H. Kirchmann, *Inorg. Chim. Acta* **29,** L249–L250 (1978).
17. F. M. D'Itri, in *Environmental Mercury Contamination,* 2nd ed., (R. Hartung and B. D. Dinman, eds.), Ann Arbor Science, Ann Arbor, MI, 1972, pp. 5–25.
18. U. Ulfvarson, in *Fungicides,* Vol. 2, (D. C. Torgeson, ed.), Academic, New York, 1969, pp. 303–329.
19. J. J. Zuckerman, in *Organometals and Organometalloids,* ACS Symp. Ser. 82, (F. E. Brinckman and J. J. Bellama, eds.), American Chemical Society, Washington, D.C., 1978, pp. 388–424.
20. R. M. Harrison and D. P. H. Laxen, *Chem. Brit.* **16,** 316–320 (1980).
21. R. M. Harrison and D. P. H. Laxen, *Nature* **275,** 738–739 (1978).
22. A. Y. Cantillo and D. A. Segar, Metal Species Identification in the Environment, a Major Challenge for the Analyst, Int. Conf. Heavy Met. Environ., (Symp. Proc.), 1st, 183–204, 1975.
23. D. F. Ball, M. Barber, and P. G. T. Vossen, *Sci. Total Environ.* **4,** 193–200 (1975).
24. D. E. Games, A. H. Jackson, L. A. P. Kane-Maguire, and K. Taylor, *J. Organomet. Chem.* **88,** 345–349 (1975).
25. D. F. Hunt, J. W. Russell, and R. L. Torian, *J. Organomet. Chem.* **43,** 175–183 (1972).
26. R. H. Fisch, R. L. Holmstead, and H. G. Kuivila, *J. Organomet. Chem.* **137,** 175–183 (1977).
27. R. H. Fisch, R. L. Holmstead, M. Gielen, and B. de Poorter, *J. Org. Chem.* **43,** 4969–4971 (1978).
28. S. C. Cohen, *Inorg. Nucl. Chem. Lett.* **6,** 757–760 (1970).
29. J. M. Miller and G. L. Wilson, in *Advances in Inorganic Chemistry and Radiochemistry,* Vol. 18, (H. J. Emeleus and A. G. Sharpe, eds.), Academic, New York, 1976, pp. 229–285.
30. O. Hutzinger and W. D. Jamieson, in *Methods in Residue Analysis.* Proc. 2nd Int. IUPAC Congr. Pest. Chem., Vol. IV, (A. S. Tahari, ed.), Gordon and Breach, New York, 1974, pp. 7–18.
31. M. R. Litzow and T. R. Spalding, *Mass Spectrometry of Inorganic and Organometallic Compounds,* Elsevier, Amsterdam, 1973.
32. M. R. Litzow, in *Mass Spectrometry of Inorganic and Organometallic Compounds,* (M. R. Litzow and T. R. Spalding, eds.), Elsevier, Amsterdam, 1973, pp. 97–113.
33. W. F. Bryant and Th. H. Kinstle, *J. Organomet. Chem.* **24,** 573–587 (1970).
34. F. Glocking, J. G. Irwin, R. J. Morrison, and J. J. Sweeney, *Inorg. Chim. Acta* **19,** 267–270 (1976).
35. S. Jensen and A. Jernelov, *Nature* **223,** 753–754 (1969).
36. F. P. Abramson, *Anal. Chem.* **44,** 28A–33A, 35A (1972).
37. R. Alvarez, *Anal. Chim. Acta* **73,** 33–38 (1974).
38. O. Hutzinger, W. D. Jamieson, and S. Safe, *Int. J. Environ. Anal. Chem.* **1,** 85–97 (1971).
39. S. Safe and O. Hutzinger, *Mass Spectrometry of Pesticides and Pollutants,* 3rd ed. CRC Press, Cleveland, Ohio, 1976, pp. 181–188.
40. T. R. Spalding, in *Mass Spectrometry of Inorganic and Organometallic Compounds* (M. R. Litzow and T. R. Spalding, eds.), Elsevier, Amsterdam, 1973, pp. 207–319.
41. R. Kuwahara, T. Suzuki, and H. Meguro, *Agric. Biol. Chem.* **44,** 669–670 (1980).
42. H. A. Meinema, T. Burger-Wiersma, G. Versluis-de Haan, and E. Ch. Th. Gevers, *Environ. Sci. Technol.* **12,** 288–293 (1978).
43. J. J. de Ridder, Thesis, Utrecht, 1969.
44. J. W. Robinson, E. L. Kiesel, and I. A. L. Rhodes, *J. Environ. Sci. Health* **A14,** 65–85 (1979).

17

# MASS SPECTROMETRY OF CARBAMATES, THIOCARBAMATES, DITHIOCARBAMATES, UREA, THIOUREA, AND CARBOXANILIDE PESTICIDES

*FRANCIS I. ONUSKA*

## 1. INTRODUCTION

The basic skeleton of all carbamate insecticides is the hypothetical carbamic acid, the monoamide of carbon dioxide having the following structure:

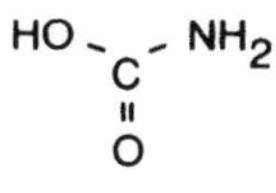


The carbamic acid can be stabilized by esterification with alkyl or aryl alcohols to form simple alkyl or aryl esters such as ethyl carbamate or phenylcarbamate commonly known as ethylurethane and phenylurethane.

If one of the protons attached to the nitrogen atom in phenylcarbamate is replaced by a methyl group, the resulting compound is toxic to insects and is known as phenyl *N*-methylcarbamate. Other aryl *N*-methylcarbamate esters with insecticidal activity can be prepared from 1-naphthol,4-hydroxybenzothiophene or 2,3-dihydro-2,2-dimethyl-7-hydroxybenzofuran. In use, mainly in Europe, are the *N,N*-dimethyl-

*FRANCIS I. ONUSKA* • National Water Research Institute, Canada Centre for Inland Waters, Burlington, Ontario L7R 4A6, Canada.

carbamates, such as 2,2-dimethylcarbamoyl-3-methyl-5-pyrazoyl *N,N*-dimethylcarbamate, known as Dimetilal, and 1-isopropyl-3-methyl-5-pyrazoyl *N,N*-dimethylcarbamate, having the trade name Pyramat.

## 2. MASS SPECTROMETRY OF CARBAMATES

Mass spectrometry (MS) has been of considerable value in the identification of methylcarbamate insecticide residues and their metabolites. In 1969 mass spectra of the Carbaryl metabolite, 5,6-dihydro-5,6-dihydroxy-1-naphthyl *N*-methylcarbamate was presented which supported the structure previously proposed.[1] Within the next few years mass spectrometric confirmation of Carbaryl residues and of several other important metabolites were reported by Liu and Bollag,[2] Durden and Bartley,[3] Bartley and co-workers,[4] and Cardona and Dorough.[5] Although electron impact (EI) mass spectra yield a weak molecular ion for these types of compounds, the fragmentation patterns often allow differentiation of very similar species. Cardona and Dorough[5] showed that 4-hydroxy-1-naphthyl *N*-methylcarbamate glucoside can be distinguished from the 5-hydroxy isomer upon careful examination of mass spectra. The use of mass spectrometry for identification and quantitation purposes of *N*-methylcarbamates has been applied by Bartley and colleagues,[4] Biros,[6] and Rankin.[7]

### 2.1. *EI Mass Spectrometry of Carbamates*

It is evident that α-cleavage to the nitrogen atom in N-substituted ethylcarbamates is possible through ethyl migration and carbon dioxide loss as follows:

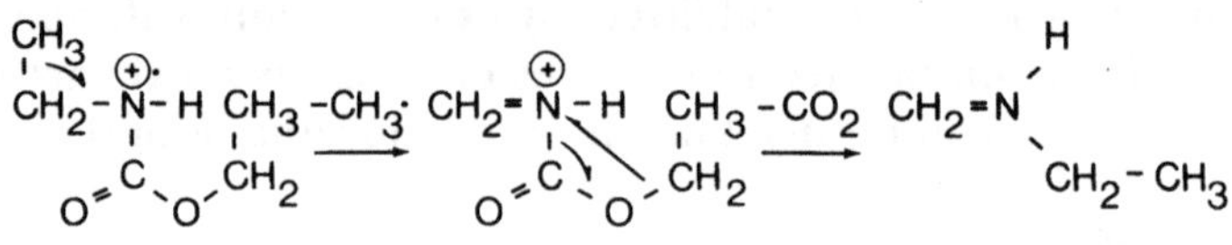


In instances where such cleavage is unfavorable, as in phenylcarbamates, the sequence of events appears to be reversed, according to Brown and Djerassi.[8]

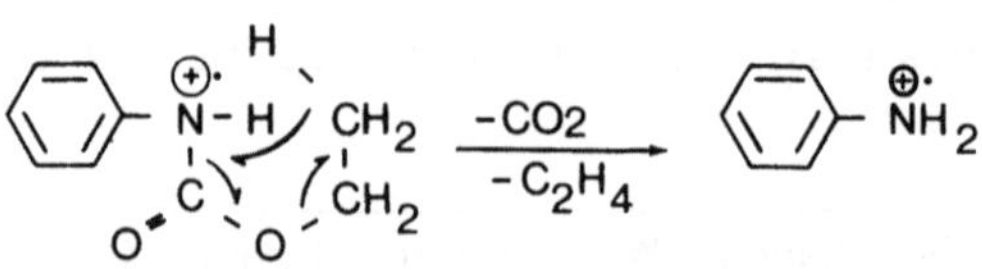

A variety of 25 *N*-substituted ethylcarbamates have been studied by Lewis.[9,10] In general, they provide distinct molecular ion peaks. The elimination of neutral ethylene or an ethyl radical from the ethoxyl moiety usually shows important fragments of variable abundance. Cleavage of the carbon–oxygen bond with charge retention by the ethyl group (*m/z* 29) gives rise to an abundant ion or base peak in ethylcarbamate mass spectra. The cleavage of an ethoxyl or carbethoxyl radical usually occurs at $[M - 45]^+$ and $[M - 73]^+$, and elimination of ethanol is also observed:

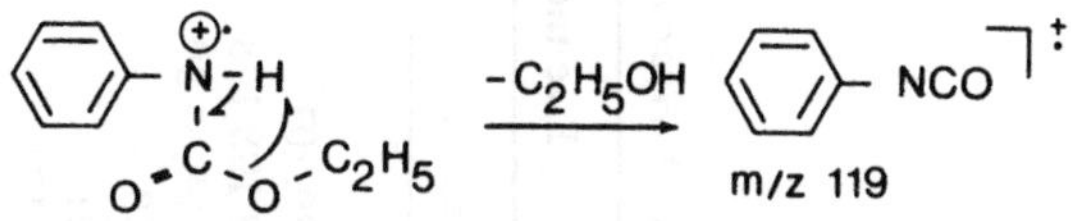


In contrast, methyl *N*-phenylcarbamate gives no skeletal rearrangement ions. It yields a spectrum very familiar to that of phenyl isocyanate according to Thomson et al.[11] Similarly, phenyl *N*-phenylcarbamate and phenyl *N*-methylcarbamate give mass spectra which essentially correspond to mixtures of the phenol and the appropriate isocyanate yet their *N,N*-diphenylcarbamates, unable to undergo phenol elimination, give the -cleavage fragments as abundant ions.

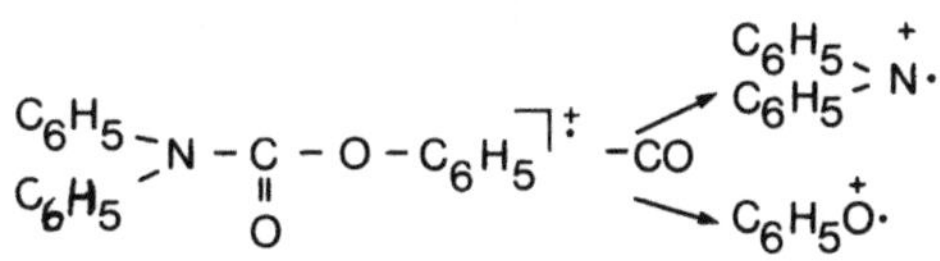


Safe and Hutzinger[12] studied some *N*-methyl carbamate pesticides containing an aryl group (Ar) and showed that the most abundant or second most abundant ion is the Ar—OH fragment. They also observed that the mass spectra of 2-(2′-propynyloxy)phenylcarbamate and 3-(2′-propynyloxy)phenylcarbamate (*m/z* 205) show weak molecular ions but an intense ion at *m/z* 148 corresponding to a loss of $CH_3NCO$ from the molecular ion. The meta-substituted isomer showed a stronger ion at *m/z* 147 than the ortho isomer and a very weak ion at *m/z* 109. This is formed by α-cleavage at the etheral oxygen which expels the propynyl moiety. The ortho isomer gives *m/z* 109 as the most abundant ion.

## 2.2. *CI Mass Spectrometry of Carbamates*

Chemical ionization mass spectra for aryl *N*-methylcarbamates show more abundant quasimolecular ions $[M + H]^+$ than the corresponding

*TABLE 1*

*Electron Impact and Chemical Ionization[a] Mass Spectra of N-Methyl, N,N-Dimethyl, and N-Hydroxy Methylcarbamates*

| No. | Chemical and trade name (carbamate) | mw | Five most abundant ions | |
|---|---|---|---|---|
| | | | Electron impact | $CH_4$ chemical ionization |
| 1 | Phenyl *N*-methyl | 151 | 94,66,39,77,151 | 95,151,94,123 |
| 2 | 1-naphthyl *N*-methyl (Carbaryl; Sevin) | 201 | 144,115,57,116,89,201 | 145,202,144,173,146 |
| 3 | 3-isopropylphenyl *N*-methyl (UC 10854) | 193 | 121,141,91,77,58,193,178 | 151,194,136,95,138,222 |
| 4 | 3,4-xylyl-6-chloro *N*-methyl | 213 | 156,121,158,141,93,65,39 | 157,214,159,158,58,216 |
| 5 | 2-isopropoxyphenyl *N*-methyl (Propoxur) | 209 | 110,152,81,58,209 | 168,111,153,110,139,210,152 |
| 6 | Carbofuran | 221 | 164,149,58,39,221 | 165,222,164,193,166,223 |
| 7 | 4-(*N,N*-dimethyl-amino)-3-methyl-phenyl *N*-methyl (Aminocarb) | 208 | 151,150,136,208,121 | 152,209,151,208,207,210 |
| 8 | 4-dimethylamino-3,5-dimethylphenyl *N*-methyl (Mexacarbate, Zectran) | 222 | 165,150,134,77,222 | 166,223,221,222,165,167 |
| 9 | 4-methylthio-3,5-dimethylphenyl *N*-methyl (Mesurol Bayer 37344) | 225 | 168,153,225,91,57 | 169,226,168,170,197,227 |
| 10 | 2-methylthio-2-methylpropanaloxime *N*-methyl (Aldicarb) | 190 | 140,86,85,58,41,165 | 89,116,86,117,90 |
| 11 | 3,4,5-trimethyl-phenyl *N*-methyl | 193 | NA | 137,136,194,138,165 |

| | | | | |
|---|---|---|---|---|
| | Aryl *N*-hydroxymethyl carbamates | | | |
| 12 | Phenyl *N*-hydroxymethyl | 167 | NA | <u>150</u>,107,95,123,168 |
| 13 | *N*-hydroxymethyl carbaryl | 217 | NA | 200,145,144,79,157,218 |
| 14 | 2,6-dimethyl-phenyl *N*-hydroxymethyl | 195 | NA | 178,123,135,122,151,196 |
| 15 | *N*-hydroxymethyl, landrin | 209 | NA | 192,149,137,165,208,210 |
| 16 | *N*-hydroxymethyl-carbofuran | 237 | NA | 220,165,177,208,193 |
| 17 | 2-hydroxyphenyl *N*-methylcarbamate | 167 | <u>110</u>,15,64,167,63 | NA |
| 18 | Isopropyl *N*-(3-chlorophenyl) (CIPC, Chlorpropham) | 213 | <u>43</u>,213,127,171,154 | NA |
| 19 | *N*-propyl, *N*-(1-phenylethyl) | 207 | <u>106</u>,164,192,207,120 | NA |
| 20 | 1-butyl *N*-(1-phenylethyl) | 221 | <u>150</u>,57,106,165,41 | NA |
| 21 | 1-ethynylcyclo-hexyl (Ethinamate) | 167 | <u>81</u>,91,106,165,41 | NA |
| 22 | 4-benzo-(β)-thienyl *N*-methyl | 207 | <u>150</u>,207,122,121,69 | NA |
| 23 | 2-methylsulphinyl-2-methylpropanal-oxime *N*-methyl | 206 | <u>86</u>,41,143,57,69 | NA |
| 24 | 3-(hydroxy-2-butyl) phenol *N*-methyl | 223 | NA | <u>206</u>,149,137,73,207 |
| 25 | 3-(hydroxy-3-butyl) phenol *N*-methyl | 223 | NA | <u>206</u>,149,122,224,107 |
| 26 | 3-(hydroxy-1-butyl) phenol *N*-methyl | 223 | NA | <u>149</u>,107,224,166,135 |
| 27 | 3-(hydroxy-4-butyl) phenol *N*-methyl | 223 | NA | <u>149</u>,224,178,167,121 |
| | Aryl and pyrazolyl *N*,*N*-dimethylcarbamates | | | |
| 28 | 1-naphthyl *N*,*N*-dimethyl | 215 | <u>72</u>,115,215,15,39 | <u>72</u>,216,57 |
| 29 | Dimethyl 5-(1-iso-propyl-3-methyl-pyrazoyl) (Isolan) | 211 | 72,45,43,41,211 | <u>72</u>,212,57 |
| 30 | 2-*N*-propyl-4-methyl pyrimidyl 6-*N*,*N*-di-Me (Pyramat) | 223 | 72,151,195,208,223 | NA |

[a] Order of ion in decreasing abundance.

parent ions in EI mass spectra (Table 1). Ions which are usually observed are:

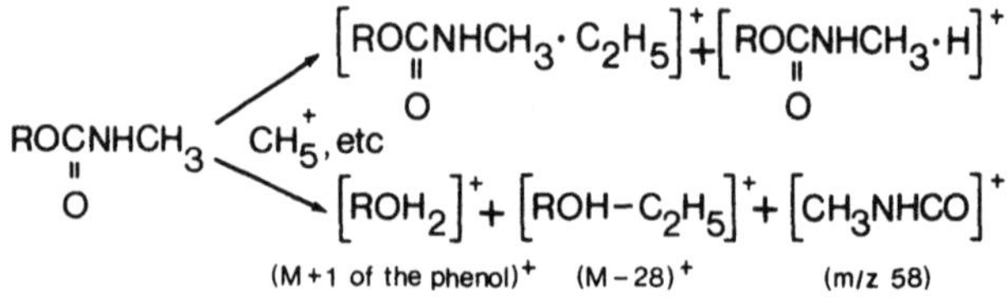


The *N,N*-dimethylamino compounds (aminocarb, mexacarbate) give similar typical CI fragment ions as aryl *N*-methylcarbamates, but their molecular ions are quite abundant and the $[M - H]^+$ ion is usually large.

Aldicarb, an oxime of a *N*-methylcarbamate, does not give an $[M + H]^+$ ion. The main CI fragments appear at *m/z* 116 corresponding to $[CH_3SC(CH_3)_2CH{=}N]^+$ and *m/z* 89 $[CH_3SC(CH_3)_2]^+$.

Aryl *N*-hydroxymethylcarbamates exhibit only small quasimolecular ions ranging from 0.1 to 6% abundance and they are usually smaller than those in EI mass spectra. The most abundant ion in their CI mass spectra are those at $[M - 17]^+$ which correspond to a loss of water from the quasimolecular $[M + 1]^+$ ion. Also, an $[M - 60]^+$ ion is evident in aryl *N*-hydroxymethylcarbamate CI mass spectra but not in EI mass spectra. This is explained by a loss of carbamoyl moiety of the molecular as shown in the following:

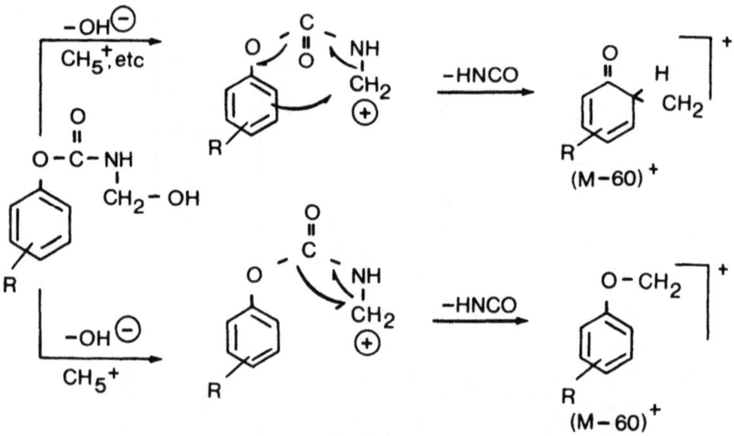


*N,N*-Dimethylcarbamates give identical CI mass spectra with the exception of the mass of their quasimolecular ion $[M + H]^+$. The *m/z* 72 ion represents the *N,N*-dimethylisocyanate ion and is the base peak in their mass spectra. A smaller *m/z* 57 ion presumably corresponds to the *N*-methylisocyanate ion.

### 2.3. Other Ionization Methods for Carbamates

Field ionization (FI) mass spectra of carbamates are simpler and have more intense molecular ions than the mass spectra obtained by means of EI ionization. Negative-ion EI mass spectra were studied by Rankin[7] and they are comparable to FI mass spectra in the number of fragment ions formed during the ionization processes. However, they produce either no, or very small, molecular ions.

Schulten[13] applied the field desorption (FD) technique to a pesticidal mixture of carbaryl and tetradifone. The molecular ion at *m/z* 201 corresponding to $C_{12}H_{11}O_2N$ (carbaryl) is the base peak in the spectrum. This is in contrast to the results obtained by Damico, Baron, and Sphon[14] by FI mass spectrometry, where a 50% relative intensity molecular ion was reported, while only 5% relative intensity molecular ion was observed in the EI mass spectrum. A prominent ion at *m/z* 143 is due to the direct bond cleavage at the naphtholic oxygen and the proton transfer in the FD mass spectrum is shown in a very abundant ion at *m/z* 144. This peak is the base peak in the EI mass spectrum of carbaryl and is approximately 80% abundant in the FI mass spectrum.

A method for the quantitative determination of carbofuran, carbofuran phenol, 3-ketocarbofuran and 3-ketocarbofuran phenol, carbaryl, Mobam (4-benzothienyl *N*-methylcarbamate), Mobam phenol (4-benzothienol), Metmercapturol (4-methylthio-3,5-xylenyl *N*-methylcarbamate), Propoxur, and Propoxur phenol in natural waters was described by Coburn and co-workers.[15] The carbamates were converted into their corresponding pentafluorobenzyl and methoxy derivatives.[16,17]

## 3. THIOCARBONATES

This group of compounds is not significant in terms of environmental pollution. The EI mass spectra of methyl-, phenyl-, diphenylthio-, dithio-, and trithio-carbonates have been extensively studied by Thomson *et al.*[11] Later, Brown and Djerassi[8] showed that loss of carbon dioxide represents the general rule in this series of compounds. They also examined the mass spectra of all possible methyl phenyl and diphenyl thiocarbonates and thiocarbamates in order to determine whether other neutral molecules containing heteroatoms could also be eliminated.

As can be seen from Fig. 1, *S*-methyl phenylthiocarbonate, in contrast with methyl phenylcarbonate, does not eliminate carbon dioxide by the loss of carbon monoxide, as shown at *m/z* 140. This ion further undergoes decomposition with the loss of sulfur at *m/z* 108, thioformaldehyde at *m/z* 94, and a methyl radical at *m/z* 125.

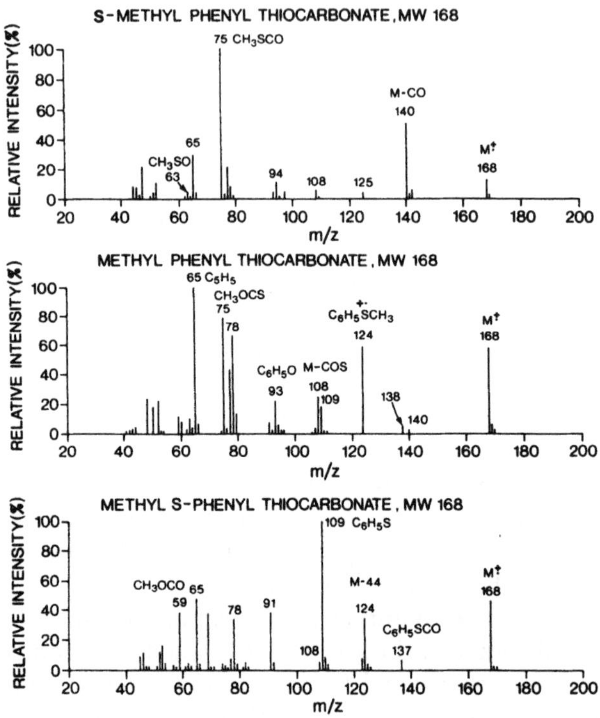


FIGURE 1. Comparison of EI mass spectra of substituted methyl phenyl thiocarbonates.

The base peak in the spectrum of methyl *S*-phenylthiocarbonate is very similar to that of the methyl phenylcarbonate, which displays a major peak at *m/z* 109 corresponding to the loss of carbon dioxide and a methyl radical from the molecular ion, except that the base peak for *S*-methyl phenylcarbonate corresponds to $C_6H_5S^+$ rather than the $C_5H_5^+$ ion.

The major rearrangement peak in the spectrum of methyl phenylthiocarbonate is found at *m/z* 124 $[M - 44]^+$ and is shown by high-resolution measurements to consist entirely of $C_7H_8S^+$. The presence of a metastable ion at *m/z* 91.5 ($124^2/168 = 91.5$) shows that carbon dioxide is released in a one-step process and thus requires an initial rearrangement of the molecular ion.

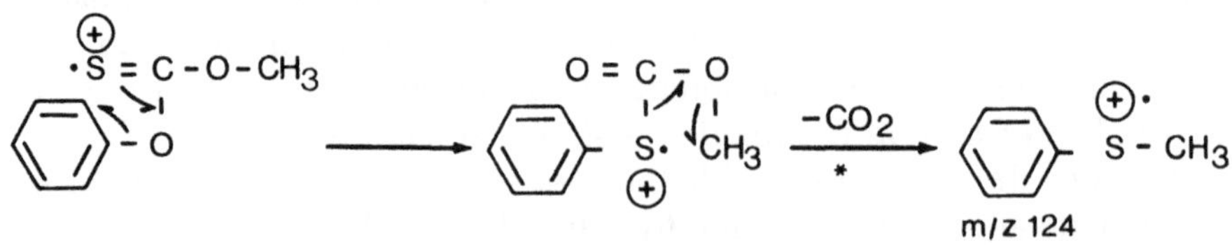

## 4. THIOCARBAMATES

The mass of spectra *N,N*-dialkyl thiocarbamates have been examined by Thomson and co-workers,(11) and sodium *N*-alkyl and *N,N*-dialkyl dithiocarbamates and related compounds by Onuska and Comba(18) (Table 2).

One of the most important groups of derivatives of dithiocarbamic acid are the salts of *N*-methyl, *N,N*-dimethyl, and ethylene-*bis*(dithiocarbamic) acids. The above-mentioned compounds are very widely used because of their relative low costs, simplicity of production, and availability of raw materials.

The *S*-methyl phenylthiocarbamate EI mass spectrum is almost identical to that of phenylisocyanate with $[M - 48]^+$ representing the elimination of methylthiol ($CH_3SH$) also being observed.

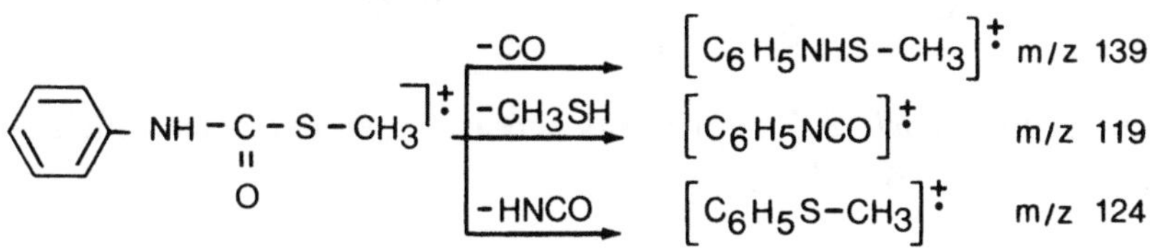


*S*-phenyl methylthiocarbamate gives an abundant $(C_6H_5S)^+$ ion (*m/z* 110). The formation of this ion from the molecular ion shows that the above ion is formed by a one-step process. This is confirmed by a metastable ion observed at *m/z* 72.5, where $110^2/167 = 72.5$. Only two other

TABLE 2
*Common and Chemical Designations of Thiocarbamates and Dithiocarbamates*

| Trade name | mw | Chemical designation |
|---|---|---|
| Eptam, EPTC | 189 | *S*-ethyl *N,N*-di-*n*-propylthiocarbamate |
| Vernam, R-1607 | 219 | *S*-*n*-propyl *N,N*-di-*n*-propylthiocarbamate |
| Tillam | 203 | *S*-*n*-propyl *N*-ethyl-*N*-butylthiocarbamate |
| Avadex, di-allate | | *S*-2,3-dichloroallyl *N,N*-diisopropylthiocarbamate |
| Vapam | 129 | Sodium *N*-methyl dithiocarbamate |
| Ferbam, Fermate | 416 | Ferric *N,N*-dimethyl dithiocarbamate |
| Marbam | | Manganese *N,N*-dimethyl dithiocarbamate |
| Ziram, Zerlate | 304 | Zinc *N,N*-dimethyl dithiocarbamate |
| Amoben | 246 | Ammonium ethylenebis dithiocarbamate |
| Nabam | 254 | Disodium ethylenebis dithiocarbamate |
| Maneb | | Manganese ethylenebis dithiocarbamate polymer |
| Zineb | | Zinc ethylenebis dithiocarbamate polymer |
| Thiram, Thiuram | 240 | Tetramethylthiuram disulfide |
| Monuram | 208 | Tetramethylthiuram monosulfide |

peaks in the mass spectrum exceed 10% relative intensity. These are the *m/z* 109 ion corresponding to $C_6H_5S^+$ and the *m/z* 66 ion corresponding to protonated cyclopentane ion.

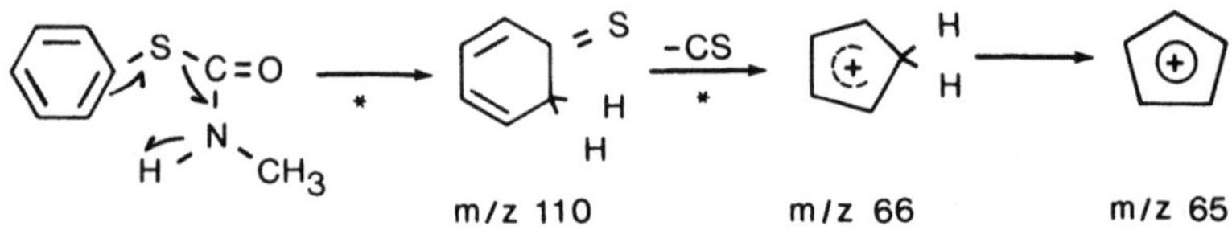


*S*-phenyl *N,N*-diphenylthiocarbamate undergoes the same fragmentation pathway as phenyl *N,N*-diphenylthiocarbamate discussed earlier. -Cleavage to the carbonyl moiety yields the major ionic species as follows:

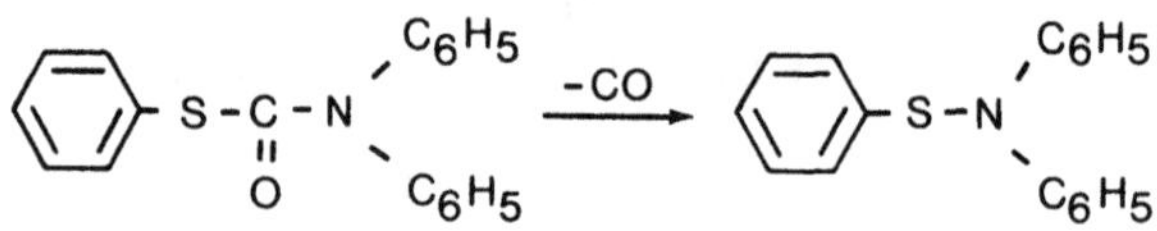


A weak ion at *m/z* 277 represents a rearrangement peak which corresponds to the $[M - CO]^+$.

## 4.1. *Dithiocarbamates*

The mass spectrometric behavior of various sodium and zinc *N*-alkyl, *N,N*-dialkyl, and *N,N*-diaryl dithiocarbamates was examined by direct inlet introduction at our laboratory. All compounds under investigation were stable enough to produce a true EI-induced mass spectra but care had to be taken to keep the temperature of the heated inlet system below 100 °C.

The mass spectra of the sodium *N*-alkyl dithiocarbamates exhibit

TABLE 3
*Metastable Ions Formed from N,N-Dialkyl Dithiocarbamates, m/z Values*

| Compound | Precursor[a] | Daughter[b] | $m^*$[c] | Precursor | Daughter | $m^*$ |
|---|---|---|---|---|---|---|
| Sodium *N,N*-dimethyl dithiocarbamate | — | — | — | — | — | — |
| Sodium *N,N*-diethyl dithiocarbamate | 116 | 88 | 66.8 | 88 | 60 | 40.9 |
| Sodium *N,N*-di-*n*-propyl dithiocarbamate | 144 | 102 | 71.3 | 102 | 60 | 35.3 |
| Sodium *N,N*-di-*n*-butyl dithiocarbamate | 172 | 116 | 78.3 | 116 | 60 | 78.3 |

[a] Initially formed product ion ($m_1$).
[b] Ion decomposition product ($m_2$).
[c] Metastable ion ($m^* = m_2^2/m_1$).

weak molecular ions. It is evident that after formation of the alkyl dithiocarbamate ion, protonation of the molecule occurs. Except for sodium *N*-methyl dithiocarbamate, the base peaks in the mass spectra of the other sodium *N*-alkyl dithiocarbamates result from the formation of the protonated alkyl isocyanate ions. *N*-methyl dithiocarbamate exhibits a base peak at *m/z* 73 arising from the loss of a neutral $H_2S$ molecule. A speculative fragmentation of the sodium *N*-alkyl dithiocarbamates is summarized as follows:

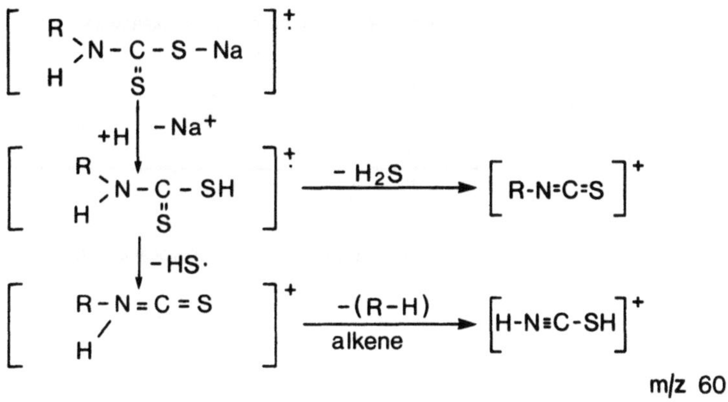


Although less prominent, the formation of the ion at *m/z* 60 is important for structural confirmation and is characteristic of the configuration of the sodium *N*-alkyl dithiocarbamates. Substantial ion abundances were also observed in the mass spectra of sodium *N*-methyl, *N*-ethyl, *N*-*n*-propyl, and *N*-*n*-butyl dithiocarbamates at *m/z* values of 15, 29, 43, and 57 for cleavage of the alkyl group from the alkyl isocyanate ion by donation of an electron pair to the nitrogen atom.

In contrast, the other sodium dialkyl dithiocarbamates gave intense ion fragments for consecutive hydrogen rearrangements, accompanied by the elimination of the corresponding alkene molecules from the alkyl isocyanatane species. These were observed at *m/z* values of 88 and 66 $[H_2NCS]^+$, 102 and 60, and 116 and 60 in the mass spectra of sodium *N,N*-diethyl, *N,N*-di-*n*-propyl, and *N,N*-di-*n*-butyl dithiocarbamates as shown in Fig. 2, respectively. Metastable ions for the above transition mechanisms are presented in Table 3.

The mass spectrum (Fig. 3) of sodium *N,N*-dimethyl dithiocarbamate exhibits successive losses of sodium ion and hydrogen sulfide from the molecular ion (Fig. 3). As outlined in the following scheme, the absence of the formation of the protonated dihydroisothiocyanate ion at *m/z* 60 confirmed the fact that *N,N*-dimethyl dithiocarbamate cannot undergo the hydrogen rearrangement.

Electron impact positive-ion mass spectra of nickel and zinc *bis*(*N,N*-

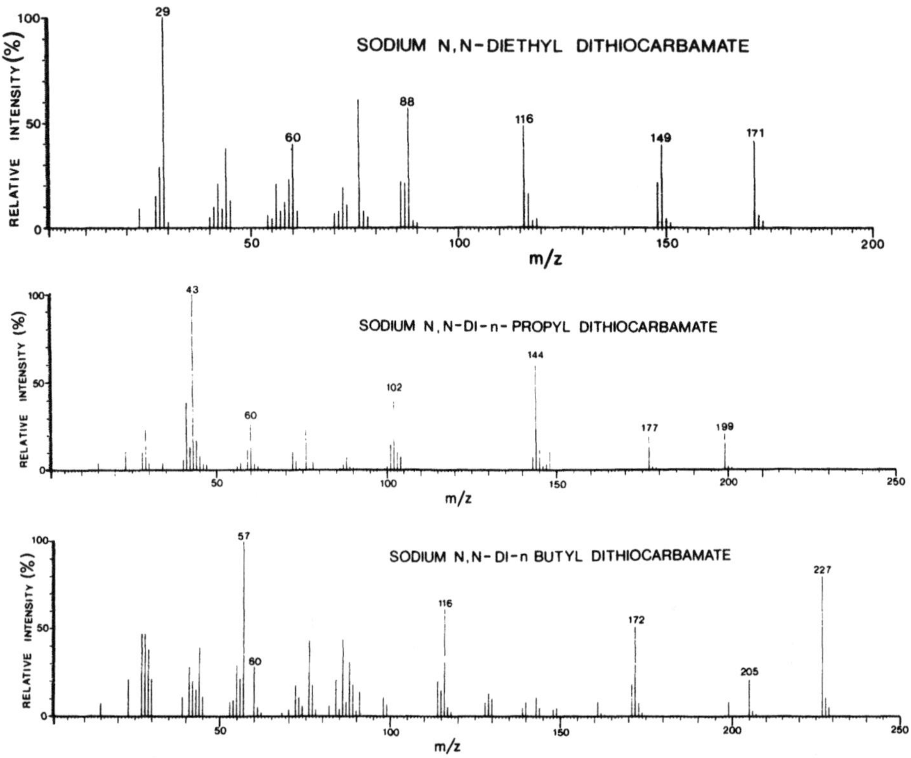


FIGURE 2. EI mass spectra of sodium *N,N*-dialkyl dithiocarbamates.

dialkyl dithiocarbamates) were extensively studied by Krupcik and colleagues.[19,20] From the mass spectra of these compounds general fragmentation pathways were derived. The divalent Zn, Ni, Co, Cu, and Cd bis(*N,N*-dialkyl dithiocarbamates) containing $C_1$ to $C_8$ for the *n*-alkyl groups and $C_3$ to $C_4$ for isoalkyl groups were analyzed using packed and

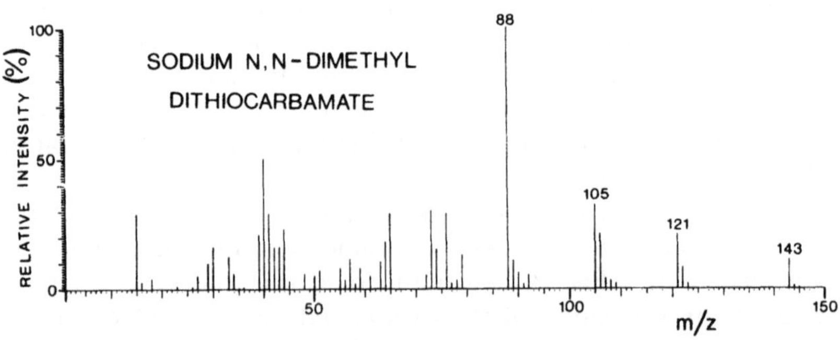


FIGURE 3. The mass spectrum of *N,N*-dimethyl dithiocarbamate.

capillary column gas chromatography. The wall-coated open tubular glass capillary column, statically coated with SE-30 and 21.5 m in length, was operated at 228 °C isothermally.

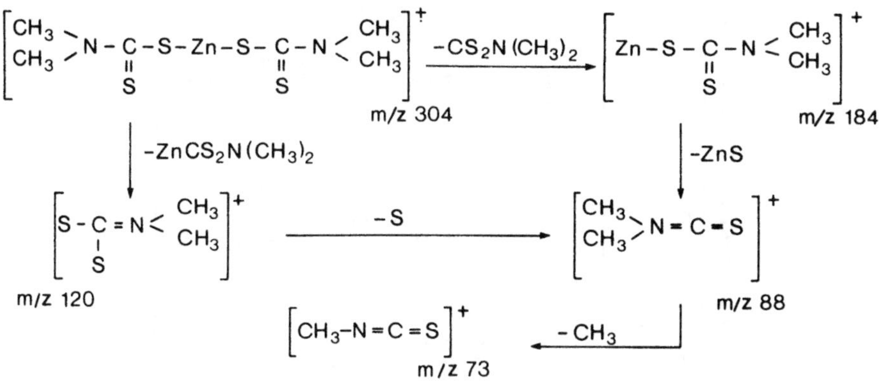


The molecular ions for nickel derivatives are more abundant than those of zinc. In general, the intensity of the ions in the high-mass region decreases with increasing size of the *n*-alkyl moieties.

Given and colleagues[21] investigated the mass spectra of a series of tris(*N,N*-dialkyl dithiocarbamate) complexes of Cr, Fe, Co, Ru, Rh, and Tl. The fragmentation patterns of these and related compounds show a loss of the dithiocarbamate ligand radical, and subsequent losses of S, $S_2$, and $R_2NCS$ species from tris and bis (dithiocarbamate) metal complex ions. The parent molecular ions are present in the mass spectra of most of these compounds, but relative abundances vary greatly.

Fraser and colleagues[22] studied the negative-ion EI mass spectra of a series of nickel bis-*N,N*-dialkyl and *N,N*-di(cycloaliphaticamino) dithiocarbamate complexes where alkyl was ethyl, *n*-propyl, and *n*-butyl, and *N,N*-dipyrrolidinyl, *N,N*-dipiperidinyl, *N,N*-dimorpholinyl, and *N*-ethyl *N*-phenyl derivatives. Intense molecular ions are observed for all compounds with most fragments originating from these ions. Metastable data indicate the $CS_2$ is eliminated from all molecular ions. Intensities of significant ions in the negative-ion mass spectra of Ni dithiocarbamate complexes are tabulated in Table 4. Positive-ion fragmentation patterns differ significantly from the negative-ion fragmentation patterns.

Onuska[23,24] studied low-resolution EI mass spectra of *S*-*n*-propyl *N*-monoalkyl dithiocarbamates and reported the separation and GC/MS of various *S*-alkyl *N,N*-dialkyl dithiocarbamates. Mass spectra for the individual *S*-*n*-propyl *N*-alkyl dithiocarbamates are shown in Fig. 4. All mass spectra yield significant molecular ions. A simplified fragmentation pathway under EI-induced fragmentation at 70 eV can be generalized.

*TABLE 4*

*Intensities of Significant Ions in the 70-eV Negative-Ion Mass Spectra of Nickel (II) Dithiocarbamate Complexes of Formula* $Ni[S_2C\text{-}NR^1R^2]_2$[a]

| $-N(R^1)(R^2)$ = | $-N(Et)(Et)$ | $-N(n\text{-}Pr)(n\text{-}Pr)$ | $-N(n\text{-}Br)(n\text{-}Bu)$ | 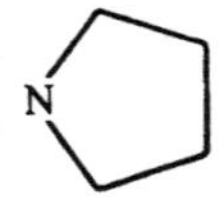 | 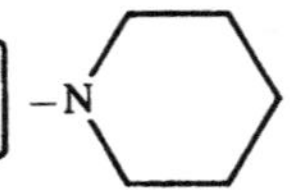 | 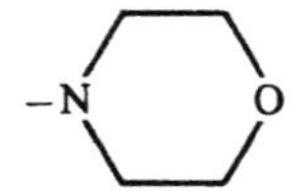 | $-N(Et)(Ph)$ |
|---|---|---|---|---|---|---|---|
| $[NiL_2]^{-\cdot}$[b] % | 100 | 100 | 100 | 100 | 100 | 100 | 49 |
| $[NiL_2\text{-}R]^-$ | 0.1 | 0.1 | 0.1 | — | — | — | — |
| $[NiL_2\text{-}S]^-$ | 0.2 | 0.7 | 0.6 | 0.1 | — | 0 | 4 |
| $[NiL_2\text{-}R^1R^2N]^-$ | 0.1 | 0.2 | 0.5 | 1.0 | 0.2 | 0.7 | 0.1 |
| $[NiL_2\text{-}CS_2]^{-\cdot}$ | 0.2 | 0.3 | 0.7 | 1.1 | 0.8 | 5 | 100 |
| $[NiL_2\text{-}CS_2\text{-}S]^{-\cdot}$ | — | — | — | — | — | — | 0.7 |
| $[NiLSH]^-$ | — | — | 0.2 | — | 0.2 | — | — |
| $[NiLS]^{-\cdot}$ | 0.1 | 0.1 | — | 0.3 | — | — | — |
| $[NiL_2\text{-}2CS_2]^{-\cdot}$ | 0.1 | 0.1 | — | — | 0.1 | 0.6 | 15 |
| $[NiLH]^-$ | 0.2 | 0.5 | 1.4 | 3.3 | 1.2 | 2.2 | 0.5 |
| $[NiL]^{-\cdot}$ | — | — | — | — | — | — | — |
| $[NiLH\text{-}R]^{-\cdot}$ | 0.1 | 0.2 | — | — | — | — | — |
| $[L]^{-\cdot}$ | 0.1 | 0.2 | 1.4 | 1.3 | 6 | 5 | 8 |

[a] All isotopes of every atom in the various clusters are considered.

[b] $Ni[S_2C\text{-}NR^1R^2]_2$ is represented by $[NiL_2]$.

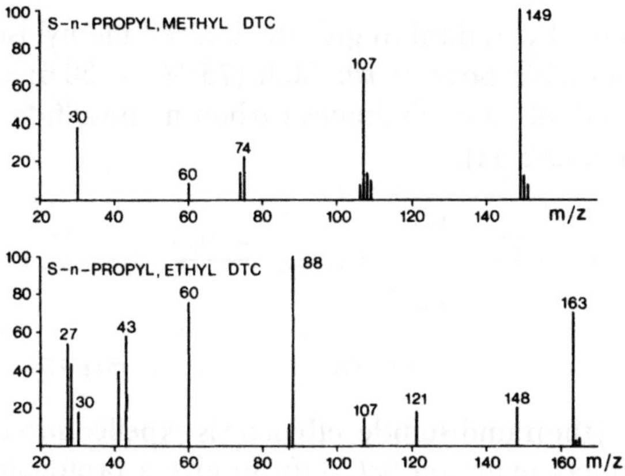


FIGURE 4. The EI mass spectra of *S*-*n*-propyl *N*-alkyl dithiocarbamates.

The quantiative determination of *S*-*n*-propyl *N*-alkyl dithiocarbamates by means of selected-ion monitoring seems to be a feasible technique for their quantitative determination in water analysis.

*S*-methyl, *S*-ethyl, and *S*-*n*-propyl *N,N*-dialkyl dithiocarbamates show EI molecular ions $M^+$. The abundance of the molecular ions of *S*-alkyl *N,N*-dialkyl dithiocarbamates depends upon the nature of the N substituent.

The shorter the chain in the *N,N*-dialkyl moiety of the dithiocarbamates, the more intense are the peaks for the molecular ions.

## 4.2. *Thiuramdisulfides*

Madsen and co-workers[25] reported mass spectra of 11 thiuramdisulfides. Generally speaking, most of the compounds investigated gave molecular ions. Only in the mass spectrum of tetrabenzyl thiuramdisulfide is a very weak peak [$M^+$ = 544] observed. The molecular ions are also relatively weak in the thiuramdisulfide with branched alkyl chains. The base peak in the majority of the thiuramdisulfide mass spectra is derived according to the following sequence:

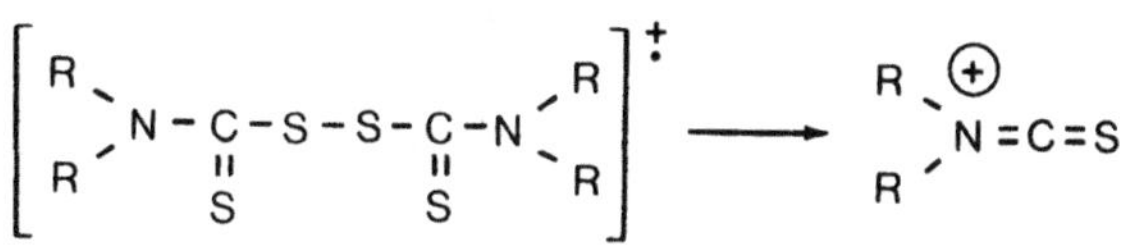

When the alkyl residues are saturated aliphatic groups larger than methyl, decomposition by elimination of alkene occurs. In tetramethyl thiuramdisulfide only two simple fragmentations are observed. The *m*/*z*

88 ion expels a methyl radical to give the ionized methyl isothiocyanate (*m/z* 73). A metastable peak at *m/z* 60.6 ($73^2/88 = 60.6$) supports this transition. The *m/z* 88 also eliminates carbon monosulfide to give a dimethyl amine ion (*m/z* 44).

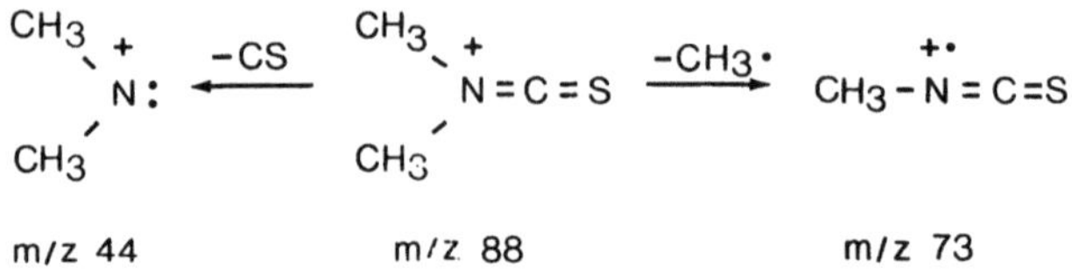


In tetraethyl thiuramdisulfide, ethylene is expelled successively from the diethyl isothiocyanate ion (*m/z* 116) to give a protonated ethyl isothiocyanate ion which is supported by a metastable peak at *m/z* 66.8 ($88^2/116 = 66.8$).

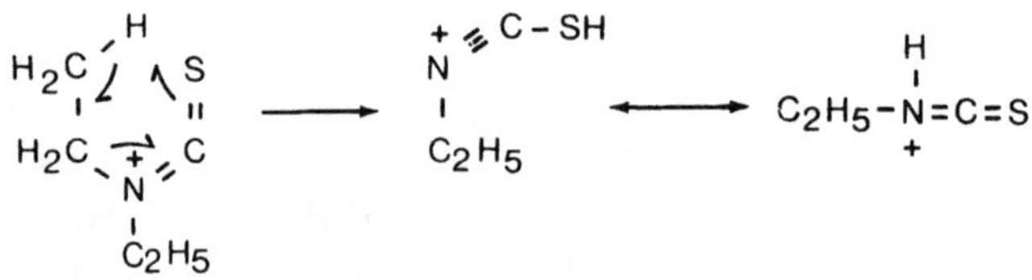


Another feature of the mass spectrum is the expulsion of carbon disulfide from diethyl thiocyanate ion (*m/z* 148) with hydrogen rearrangement to give the diethyl amine ion (*m/z* 72) supported by a metastable peak at *m/z* 35.1 ($72^2/148 = 35.1$). Ethylene can then be eliminated from the ion at *m/z* 72 as follows:

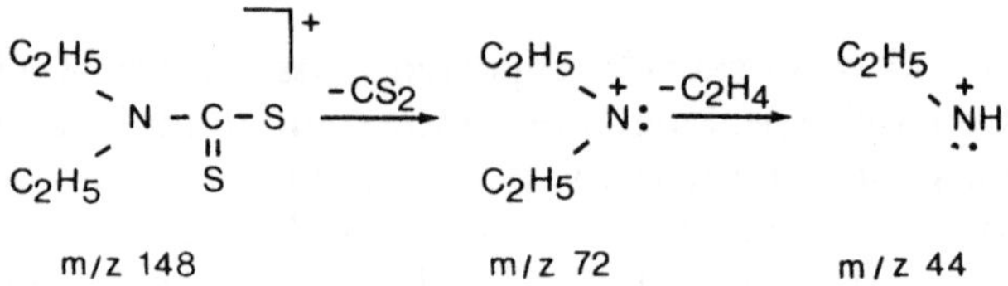


In the spectrum of the phenyl-substituted derivatives, an M − 33 ion corresponding to M − SH loss is observed. This type of elimination is not observed in other mass spectra of this series. The *N*-phenyl, *N*-methyl isothiocyanate ion (*m/z* 150) eliminates a methyl radical to give ionized phenyl isothiocyanate (*m/z* 135), which is known to decompose to $C_6H_5^+$.

The fragmentation of dimethyl diphenyl thiuramdisulfide gives rise to a phenyl migration. A possible rationalization would be charge retention in the aromatic ring and formulation of *m/z* 182, which could then undergo four-member ring decomposition to form an ion *m/z* 109, and expel neutral methyl isothiocyanate.

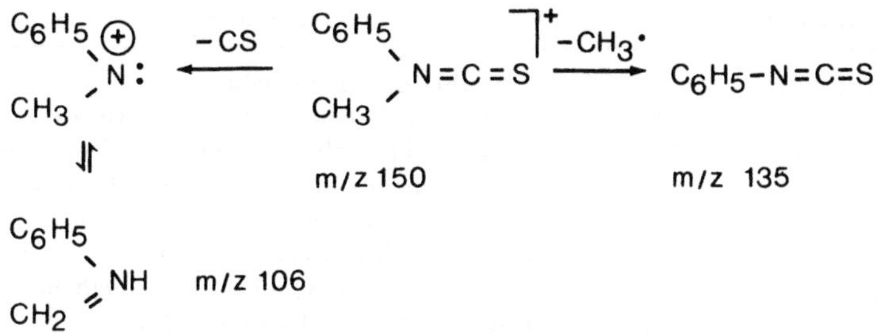


In the mass spectrum of tetrabenzyl thiuramdisulfide the stable benzylic ion dominates, but the ions expected on the basis of the foregoing discussion are also found. A peak at *m/z* 123 corresponds to $C_7H_7S^+$ which would require the migration of a benzyl group to sulfur subsequent to electron impact. It appears that the presence of aromatic substituents among thiuramdisulfides may give rise to deep-seated skeletal rearrangements upon electron impact.

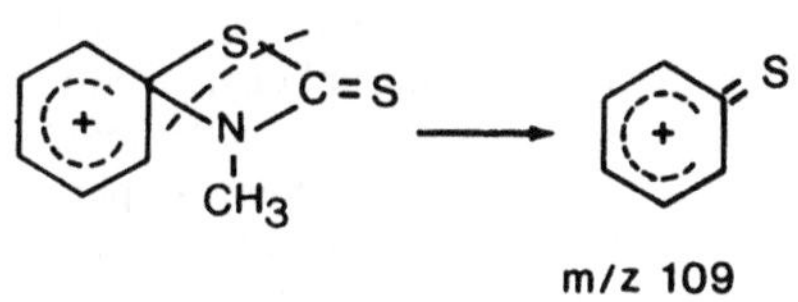


## 5. UREA AND THIOUREA HERBICIDES

Since their discovery in the fifties, the substituted ureas have represented the prominent and diversified groups of herbicides. A list of common names of pesticides is included in Table 5.

The first important phenyl ureas were chlorine-substituted dimethyl derivatives with a high order of inherent phytotoxicity. In more recent compounds ring substituents and alkyl substituents have become increasingly complex. Since they have low water solubility and thus unfavorable partition coefficients between organic phases and water, their movement in plant and soil systems is limited. Losses caused by rapid leaching through soil layers or by evaporation and air transport have not yet been demonstrated for urea herbicides.

The fragmentation pattern of substituted *N*-alkyl ureas has been reported by Baldwin and colleagues.[26] He also reported ionization potentials of *N*-alkyl ureas and thioureas, and found that methylated ureas and thioureas form $(C_2H_6N)^+$ and $(CH_4N)^+$ ions by direct cleavage

*TABLE 5*
*Common and Chemical Names of Substituted Urea Herbicides*

| Common name | mw | Chemical name |
|---|---|---|
| Benzthiazuron | | 3-(2-benzothiazolyl)-1-methylurea |
| Buturon | 236 | 3-(4-chlorophenyl)-1-isobutinyl-1-methylurea |
| Chlorbromuron | 293 | 3-(4-bromo-3-chlorophenyl)-1-methoxy-1-methylurea |
| Chloroxuron | 290 | 3-4-(4-chlorophenoxy)phenyl-1,1-dimethylurea |
| Chlortoluron | | 3-(3-chloro-4-methylphenyl)-1,1-dimethylurea |
| Cycluron | | 3-cyclooctyl-1,1-dimethylurea |
| Difenoxuron | | 3-4-(4-methoxyphenoxy)phenyl-1,1-dimethylurea |
| Diuron | 232 | 3-(3,4-dichlorophenyl)-1,1-dimethylurea |
| Fenobenzuron | | 3-benzoyl-3(3,4-dichlorophenyl)-1,1-dimethylurea |
| Fenuron | 164 | 3,3-dimethyl-1-phenylurea |
| Fenuron TCA | | 3,3-dimethyl-1-phenyluronium trichloroacetate |
| Fluometuron | 232 | 3-(3-trifluoromethylphenyl)-1,1-dimethylurea |
| Linuron | 249 | 3-(3,4-dichlorophenyl-1-methoxy-1-methylurea |
| Methabenzthiazuron | | 3-benzothiazol-2-yl-1,3-dimethylurea |
| Metobromuron | 256 | 3-(4-bromophenyl-1-methoxy-1-methylurea |
| Metoxuron | 228 | 3-(3-chloro-4-methoxyphenyl)-1,1-dimethylurea |
| Monolinuron | 214 | 3-(4-chlorophenyl)-1-methoxy-1-methylurea |
| Monuron | 198 | 3-(4-chlorophenyl)-1,1-dimethylurea |
| Monouron TCA | 362 | 3-(4-chlorophenyl)-1,1-dimethyluronium trichloroacetate |
| Neburon | 250 | 3-(3,4-dichlorophenyl)-1-butyl-1-methylurea |
| Noruron (Norea) | 222 | 3-(hexahydro-4,7-methanoindan-5-yl)-1,1-dimethylurea |
| Siduron | 232 | 3-(2-methylcyclohexyl)-1-phenylurea |
| Trimeturon | | 3-(4-chlorophenyl)-2,1,1-trimethylisourea |

under electron impact. For urea, *N*-methyl urea, and *N*-dimethyl urea, ionization potentials (IPs) have values of 12.9, 13.25, and 13.7 eV. Baldwin and co-workers[27] showed that the methylation process of urea causes a change in IPs, but in thioureas this phenomenon produces considerable less change in the IP of individual methyl-substituted thioureas.

Basically, the mass spectra of urea and its methyl derivatives can be accounted for by radical abstraction ($R_2N^{\cdot}$) and the formation of abundant $[R_2N{=}C{=}O]^+$ and $R_2NH^+$ even-electron ions. The replacement of a methyl group with a hydrogen atom leads to the following process.

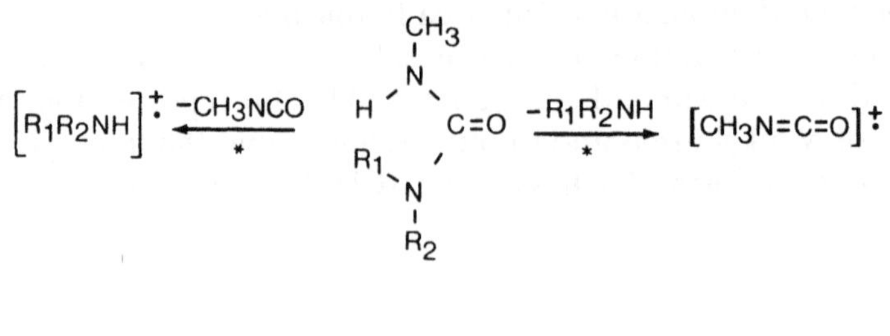

Longer chain aliphatic substituents in alkyl ureas will produce more complex mass spectra, since reactions involving the alkyl substituents can occur.

Matsumoto and co-workers[28] reported analysis of 13 alkylureas by GC/MS. Alkylureas were treated with trifluoroacetic acid. The trifluoroacetylated alkylureas were separated on 7% diethyleneglycol succinate liquid phase modified with 1% $H_3PO_4$.

### *5.1. N-Phenylureas*

Similarly, Baldwin and colleagues[27] studied the mass spectra of the phenylurea series. For *N*-phenylureas which contain a hydrogen atom on the symmetrical nitrogen, the base peak corresponds to the aromatic amine, and often the phenylisocyanate ion (*m/z* 119) is present. This type of fragmentation is evident in the mass spectrum of Siduron. The base peak corresponds to the anilino ion (*m/z* 93) and a weak phenylisocyanate ion occurs at *m/z* 119.

Benson and Damico[28] reported that *N*-aryl *N'*-methylureas fragment to yield a peak corresponding to the hydrogen rearrangement forming the aniline ionic species due to the proton transfer from the nitrogen moiety:

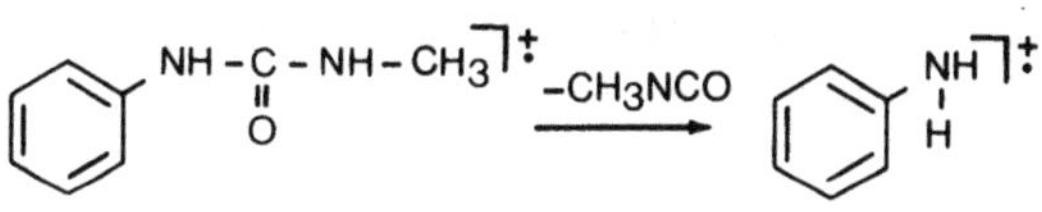


Even though the other proton transfer process is theoretically possible to form a phenyl isocyanate ion, it does not represent a significant process due to the stability of the aniline ion being more energetically stable than the methylamine ion.

The *N*-aryl *N'*-dimethylureas give substituted *N*-methyl isocyanate and *N*-methylamine ions as shown above. In the mass spectra of *N*-chlorophenyl *N'*-dimethylurea and *N*-3-trifluoromethyl phenyl *N,N'*-dimethylurea, the base peak is the dimethyl isocyanate ion (*m/z* 72) and only a very weak peak is seen at *m/z* 44 corresponding to the dimethylamine ion. The replacement of an N-substituted methyl group with an N methoxy group, however, significantly alters the fragmentation pathway.

Brophy and colleagues[29] compared EI and CI mass spectra of a series of 18 *N,N'*-diaryl ureas (Table 6).

The main fragments in the EI mass spectra of *N,N'*-diaryl ureas are

*TABLE 6*
*Five Most Abundant Ions in the EI Mass Spectra of* N,N'-*Diaryl Ureas*

| | | | | | |
|---|---|---|---|---|---|
| *N,N'*-diphenyl urea | 93(100) | 119(40) | 91(25) | 66(20) | 212(17) |
| *N*-Me-*N,N'*-di-Ph urea | 107(100) | 77(45) | 119(40) | 91(33) | 64(20) |
| *N*-(2-MeOPh)-*N'*-Ph-urea | 123(100) | 108(83) | 149(80) | 93(80) | 119(48) |
| *N*-(3-MeOPh)-*N'*-Ph-urea | 93(100) | 149(60) | 123(60) | 119(60) | 242(25) |
| *N*-(4-MeOPh)-*N'*-Ph-urea | 123(100) | 108(05) | 93(80) | 119(45) | 242(60) |
| *N*-Me-*N*-(2-MeOPh)-*N'*-Ph-urea | 137(100) | 122(95) | 119(85) | 91(60) | 256(25) |
| *N*-Me-*N*-(3-MeOPh)-*N'*-Ph-urea | 137(100) | 256(50) | 135(40) | 119(33) | 108(25) |
| *N*-Me-*N*-(4-MeOPh)-*N'*-Ph-urea | 122(100) | 137(78) | 256(60) | 119(20) | 91(13) |
| *N*-Me-*N'*-(2-MeOPh)-*N*-Ph-urea | 106(100) | 149(35) | 119(32) | 256(18) | 76(18) |
| *N*-Me-*N'*-(3-MeOPh)-*N*-Ph-urea | 107(100) | 256(47) | 149(13) | 134(24) | 77(20) |
| *N*-Me-*N'*-(4-MeOPh)-*N*-Ph-urea | 106(100) | 149(50) | 134(46) | 256(30) | 79(16) |
| *N*-(4-F-Ph)-*N'*-Ph-urea | 111(100) | 93(85) | 119(45) | 137(40) | 230(15) |
| *N*-(4-Cl-Ph)-*N'*-Ph-urea | 117(100) | 93(63) | 153(50) | 109(45) | 246(17) |
| *N*-(4-Br-Ph)-*N'*-Ph-urea | 171(100) | 93(75) | 119(50) | 197(45) | 290(20) |
| *N*-(4-Ac-Ph)-*N'*-Ph-urea | 193(100) | 119(70) | 64(70) | 156(25) | 254(2) |
| *N*-(4-MePh)-*N'*-Ph-urea | 107(100) | 93(85) | 226(40) | 77(30) | 119(22) |
| *N*-(4-Nitro-Ph)-*N'*-Ph-urea | 93(100) | 119(95) | 64(35) | 138(20) | 257(3) |
| *N*(4-Amino-Ph)-*N'*-Ph-urea | 108(100) | 93(80) | 119(55) | 134(50) | 227(40) |
| *N*-(4-BiPh)-*N'*-Ph-urea | 195(100) | 169(95) | 93(41) | 119(35) | 288(5) |

formed from rearrangement reactions. The fragments correspond to aromatic amine and aryl isocyanate ions. Where the urea is unsymmetrically substituted, two sets of amine and isocyanate ions are formed from the molecular ion and they are confirmed by metastable ions, as shown in the following:

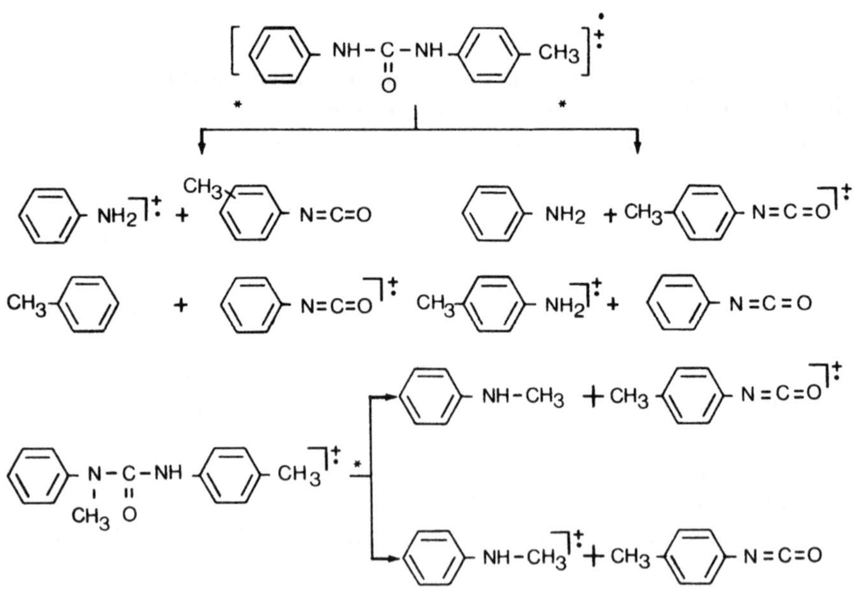

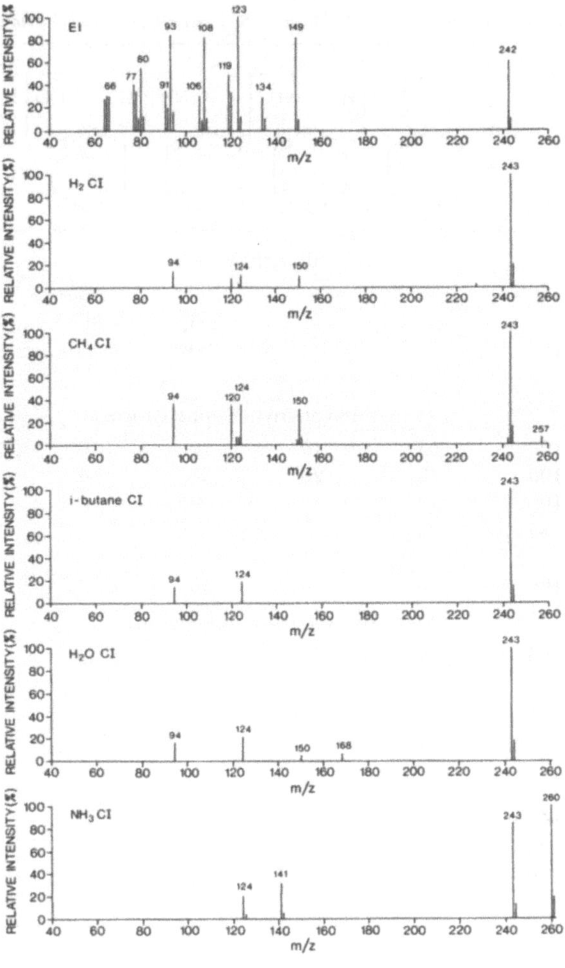


FIGURE 5. The EI and CI mass spectra of *N*-4-methoxyphenyl-*N′*-phenylurea.

The relative abundance of these ions is dependent on the substituents in the aryl ring. The EI mass spectrum of *N*-4-methoxyphenyl-*N′*-phenylurea is not very complex. The homolytic cleavage is negligible in the fragmentation of molecular ions (see Fig. 5).

Chemical ionization mass spectra of these compounds show that a significant amount of fragmentation occurs when hydrogen is the reagent gas, and both possible protonated isocyanates and protonated amines are formed. Methane CI produces similar fragmentation with the $[MH]^+$ ion once again being the most abundant peak. The commonly observed addition ions in methane CI mass spectra, $[M + 29]^+$ and $[M + 41]^+$,

*TABLE 7*
*Prinicpal Ions in the CI Spectra of N-Substituted N-Phenyl Ureas with Various Reagent Gases*

| Reagent gas | $[MH]^+$ | 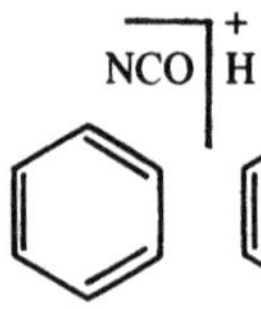 | 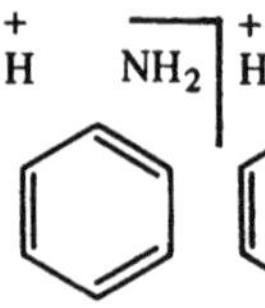 | 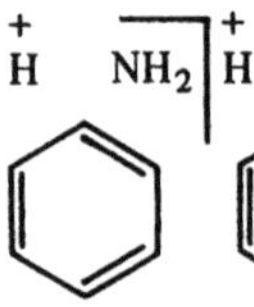 | 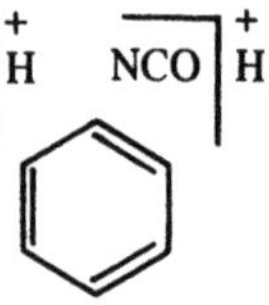 | Other |
|---|---|---|---|---|---|---|
| | | | *N,N'*-diphenyl urea | | | |
| $H_2$ | 60% | | | 100% | 15% | |
| $CH_4$ | 30 | | | 100 | 55 | |
| $C_4H_{10}$ | 100 | | | 20 | | |
| $NH_3$ | 30 | | | | | $[M + NH_4]^+$ 100 |
| | | | *N*-(3-methoxyphenyl)-*N'*-phenyl urea | | | |
| $H_2$ | 100 | 13 | 10 | 16 | 10 | |
| $CH_4$ | 100 | 30 | 28 | 30 | 30 | |
| $C_4H_{10}$ | 100 | 22 | | 22 | | |
| $NH_3$ | 85 | 20 | | | | 260(100) |
| | | | | | | 141(30) |
| $H_2O$ | 100 | 20 | 5 | 15 | | 168(8) |
| | | | *N*-(4-nitrophenyl)-*N'*-phenyl urea | | | |
| $H_2$ | 55 | 50 | 25 | 55 | 45 | 228(100) |
| | | | | | | 137(15) |
| | | | | | | 109(20) |
| $C_4H_{10}$ | 16 | 10 | | 90 | | 228(100) |
| | | | | | | 156(60) |
| | | | | | | 123(25) |
| | | | | | | 109(20) |
| $NH_3$ | 20 | 15 | | | | 275(100) |
| $D_2/D_2O$ | 100 | 65 | | 79 | 25 | 223(60) |
| | | | | | | 114(15) |

are present. Reagent gases with higher proton affinities produce less fragmentation. With isobutane the two possible protonated amine ions are formed, but not the ions corresponding to the protonated isocyanatane. Water as a reagent gas also yields the protonated amines but there is also a small peak at *m/z* 168 which is presumably due to protonated carbamic acid formed by addition of hydroxonium ion $[H_3O]^+$ to the isocyanate which loses water from the methoxyphenyl isocyanate ion. With ammonia, $[M + NH_4]^+$ ions are formed which lose $NH_3$ to yield $[MH]^+$ ions, and these fragments to a minor extent to give protonated 4-methoxyaniline.

The amount of fragmentation of CIMS is much more dependent on temperature and pressure than with EIMS.

TABLE 8
*Relative Sensitivity of EI and CI for N-Methyl-N,N'-Diphenylurea with Various Reagent Gases*

| Mode | Reagent gas | Integral collector reading | Relative sensitivity |
|---|---|---|---|
| EI | *a* | 16250 | |
| | | 16450 | 18.7 |
| | | 12700 | |
| EI | *b* | 662 | |
| | | 955 | 1 |
| EI | Hydrogen | 41100 | |
| | | 42200 | 47.4 |
| | | 31700 | |
| CI | Methane | 57560 | |
| | | 55700 | 70.0 |
| CI | Isobutane | 123300 | |
| | | 93300 | 134.0 |
| CI | Ammonia | 88800 | |
| | | 94500 | 113.4 |

[a] MS-12 EI mass spectrometer.
[b] MS-9 CI mass spectrometer in EI mode.

Comparisons of the CI mass spectra of the *o*-methoxy and *p*-methoxy compounds indicate that the ortho effect may be associated with the formation of $[MH]^+$ ions where a proton is bound between the ortho nitrogen and methoxyl oxygen atom as shown:

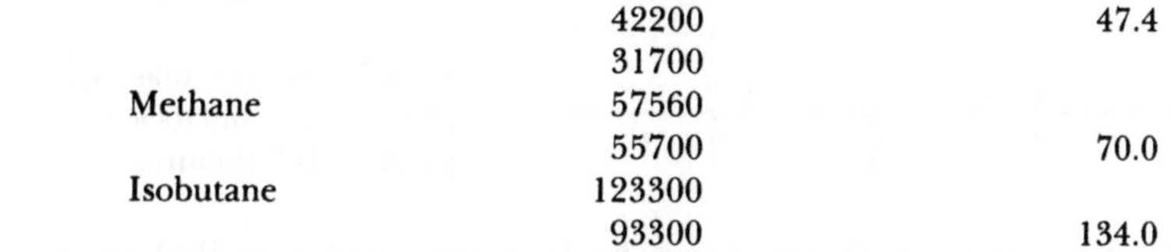


Principal peaks in the CI mass spectra of some compounds and various reagent gases are shown in Table 7.

The relative sensitivity of EIMS and CIMS using hydrogen, methane, isobutane, and ammonia have been studied by monitoring the quasimolecular ion $[MH]^+$ with time, employing *N*-methyl *N,N'*-diphenyl urea as a test substance. The results show the maximum sensitivity of the CI isobutane and $NH_3$ CI mass spectra (Table 8).

### 5.2. *Thioureas*

Grehn[30,31] has studied mass spectra of ortho-substituted 1-phenyl-2-thioureas. In contrast to previous studies by Shapiro,[32] Baldwin and Loudon,[33,34] and Baldwin and colleagues,[35] Grehn established that in the mass spectra of arylureas, the hydrogen in sulfhydryl radical ab-

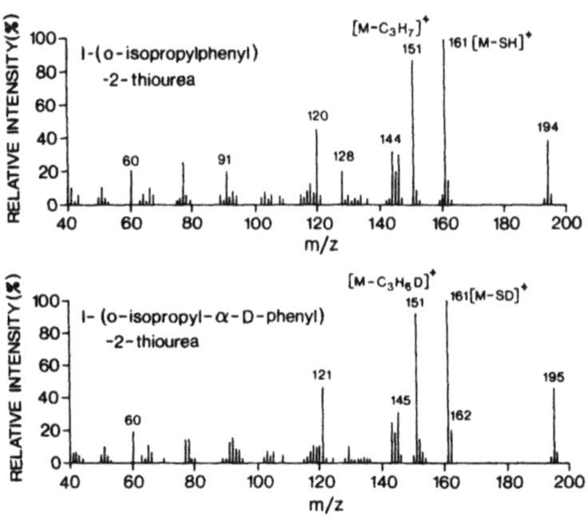


FIGURE 6. The mass spectra of D-labeled and unlabeled 1-(*o*-isopropylphenyl)-2-thiourea.

straction originates from the α-hydrogens in the *o*-alkyl group. The intensity of the fragment ion formed by loss of the *o*-alkyl group increases with the stability of the corresponding alkyl radical. The loss of SH is facilitated by the increasing lability of the α-hydrogen on from $CH_3$— to $(CH_3)_2CH$—. Moreover, the *t*-butyl compound, lacking α-hydrogens, has a significantly lower ratio of $[M - SH]^+/[M]^+$ than the isopropyl derivative.

Deuterium labeling of the -position of the isopropyl derivative shows clearly that the hydrogen lost in the HD radical comes mainly from this position. Another striking feature of the mass spectra of ortho-substituted alkylphenylthioureas is the facile loss of the alkyl substituent. The mass spectra of unlabeled and unlabeled 1-(*o*-isopropylphenyl)-2-thiourea are illustrated in Fig. 6.

Schulten[36] has investigated herbicides, insecticides, and fungicides of the phenylurea-, carbamate-, and thiocarbamate-type compounds by field desorption MS. The spectra show high-molecular-ion intensities and a characteristic fragmentation pattern. The optimal methodological parameters for trace analyses of these compounds in low- and high-resolution field desorption MS employing electrical rather than photoplate scanning require 1–3-ng samples. Using indirect heating (e.g., a laser), 900 pg of Metoxuron can be ionized, producing a suitable mass spectrum with a 10 : 1 signal-to-noise ratio. The corresponding sensitivity for phenylurea derivatives is better than $5 \times 10^{-11}$ C/g.

If the FD mass spectra are obtained at the best anode temperature, only molecular ions as base peaks are observed.

Proton transfer reactions are characteristic of the FD mass spectra of polar compounds. The double-charged ions $[2M + H]^+$ are observed in their mass spectra. It is important, especially for interpretation of the FD mass spectra, to establish whether the $M^+$ ion or $[M + H]^+$ ion is observed. In contrast, for aryl-substituted phenylureas, formation of $[2M]^+$ is observed and it should be considered if a questionable molecular

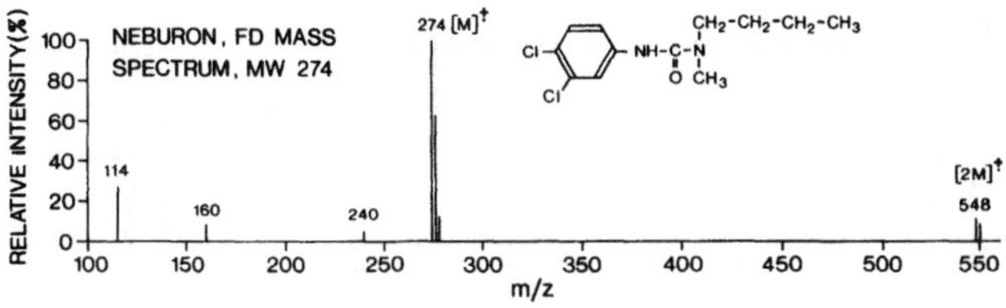


FIGURE 7. The FD mass spectrum of Neburon.

ion $[M]^+$ or protonated molecular ion $[M + H]^+$ is to be positively identified.

The FD mass spectrum of Neburon shown in Fig. 7 indicates two complementary fragments at *m/z* 114 and 160. Similarly, the FD mass spectrum of Metobromuron is given in Fig. 8 and simple fragmentation is observed. Some low-intensity ions are visible, and they must be carefully examined for correct interpretation of the mass spectrum. An example is shown in this mass spectrum, where two fragment ions at *m/z* 88 and 170 are depicted. Also, the intramolecular hydrogen rearrangement ions are shown at *m/z* 61 and 197. The first ion corresponds to the secondary methoxymethyl amine moiety and the second ion at *m/z* 129, add to the unambiguous identification of the molecular ion.

Schulten[36,37] showed that FD mass spectra of various biocides can be obtained without their separation. In Fig. 9 a test mixture consisting of 10 ng of each of three phenylurea biocides (Chlortoluton, Metoxuron, and Buturon) gives mass spectra consisting of their molecular ions when an electrical recording of the mass spectra is performed in the mass range from *m/z* 200 to 250. Schulten[37] applied the combination of FDMS and liquid chromatography for the determination and identification of urea and carbamate herbicides in Rhine river water. For identification of the biocides, they reported low- and high-resolution FDMS. Reliable identification of the pesticides is achieved with sample amounts corresponding to 10–20 ppt concentrations in the river water samples.

The metabolic studies of some phenylurea compounds were performed by Tweedy and co-workers[38] who described the comparative metabolism of metobromuron by four soil microorganisms *Talaromyces*

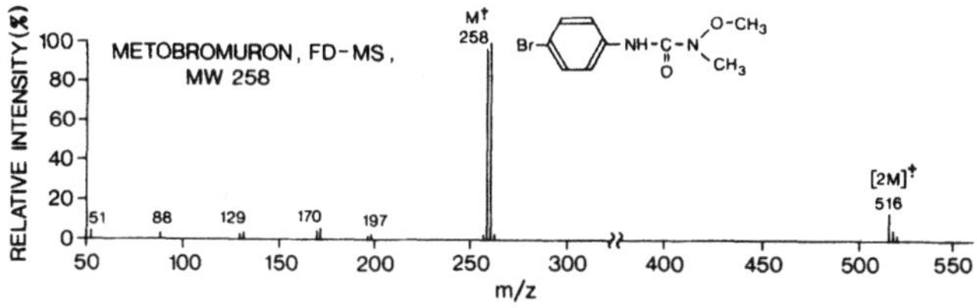


FIGURE 8. The FD mass spectrum of metobromuron.

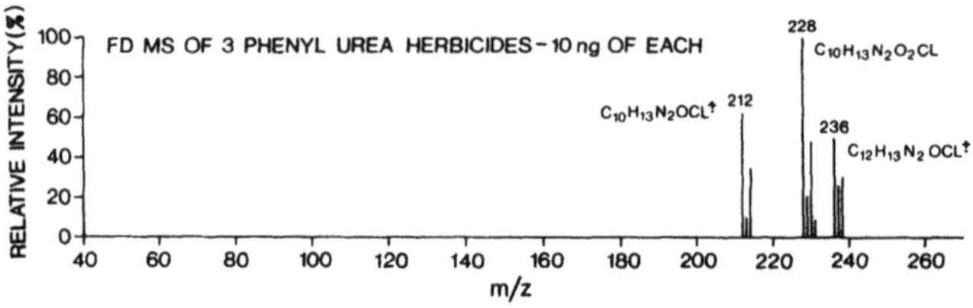


FIGURE 9. FD mass spectra of three phenylurea herbicides (Chlortoluton, Metoxuron, and Buturon).

*wortmanii, Fusarium oxysporum, Chlorella vulgaris,* and a species of *Bacillus.* The metabolites produced by *T. wortmanii* were 1-(*p*-bromophenyl)-3-methoxyurea, 1-(*p*-bromophenyl-3-methylurea, *p*-bromophenylurea, and *p*-bromoacetanilide with identification made by thin-layer chromatography and mass spectrometry.

The metabolism of ring-$^{14}$C-, carbonyl-$^{14}$C-, and methyl-$^{14}$C-labeled diuron by corn seedlings was described by Onley and colleagues.[(39)] The presence of 3,4-dichloronitrobenzene, 3-(3,4-dichlorophenyl)-1-methylurea, 3-(3,4-dichlorophenyl) urea, and 3,4-dichloroaniline were identified as metabolites in corn seedlings which were exposed to diuron.

The metabolic fate of Siduron in dogs and rats fed at a dietary level of 2500 ppm for 14 months was studied by Belasco and Reiser.[(40)] The metabolites of Siduron were found in the dog's urine. The conjugates of 1-(*p*-hydroxyphenyl)-3-(2-methylcyclohexyl) urea; 1-(4-hydroxy-2-methylcyclohexyl)-3-(*p*-hydroxyphenyl) urea; and 1-(4-hydroxy-2-methylcyclohexyl)-3-phenylurea were identified by mass spectrometry. Their structures are shown below:

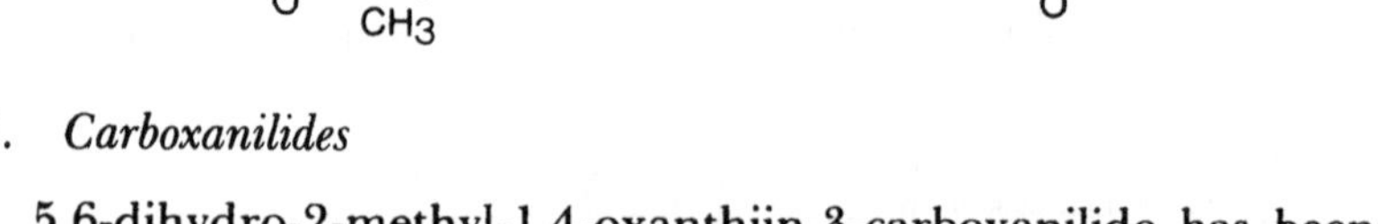

## 5.3. *Carboxanilides*

5,6-dihydro-2-methyl-1,4-oxanthiin-3-carboxanilide has been subjected to extensive study since its preparation in Canada and found to be an active fungicide for treatment of smuts and bunts in cotton and peanuts and against *Rhizoctonia* in vegetables and in seed treatment of cereals. Onuska, Harrison, and Comba[(41)] and Harrison and Onuska[(42)] studied the mass spectra of various 1,4-oxathiin-3-carboxanilides and their metabolites by means of EI and CI mass spectrometry.

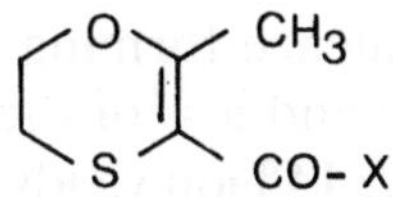


| | | |
|---|---|---|
| X = $NHC_6H_5$ | X = $NHC_6H_5$ | X = $NHC_6H_5$ |
| X = OH | X = OH | X = OH |
| X = $OCH_2C{\equiv}CH$ | X = $OCH_2C{\equiv}CH$ | X = $OCH_2C{\equiv}CH$ |
| X = $NH_2$ | X = $NHC_6H_4CH_3(p)$ | X = $NH_2$ |
| X = $NHCH_3$ | X = $N(CH_3)C_6H_5$ | X = $NHCH_3$ |
| X = $NHC_6H_4CH_3(p)$ | | X = $NHC_6H_4CH_3(p)$ |
| X = $N(CH_3)C_6H_5$ | | X = $N(CH_3)C_6H_5$ |

Jardine and Fenselau[43] suggested that the inverse correlation with proton affinity may only be valid where other fragmentation reactions are negligible, where the departing X group is small, and where the initial site of protonation is located on X.

$$[RX \cdot H]^+ \rightarrow [R]^+ + HX$$

*5.3.1. Sulfides.* The EI and CI mass spectra of sulfide are shown in Fig. 10. An intense molecular ion is observed in the EI spectrum with

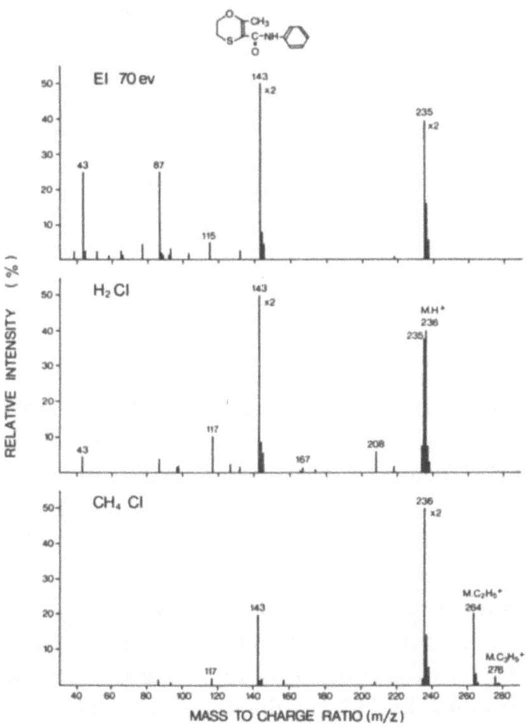


FIGURE 10. EI and CI mass spectra of 5,6-dihydro-2-methyl-1,4-oxathiin-3-carboxanilide.

the major fragmentation sequence forming the base peak, *m/z* 143, by rupture of the amide linkage and loss of the anilino radical.

Loss of $C_2H_4$ from the *m/z* 143 ion yields the main component (90%) at *m/z* 115 due to $[C_4H_3O_2S]^+$ while loss of CO yields the minor component (10%), $[C_5H_7OS]^+$. Metastable refocusing studies at 5000 resolution showed a metastable ion for the loss of $C_2H_4$, but not for the loss of CO, from the *m/z* 143 ion. The subsequent fragmentation of the *m/z* 115 ion is:

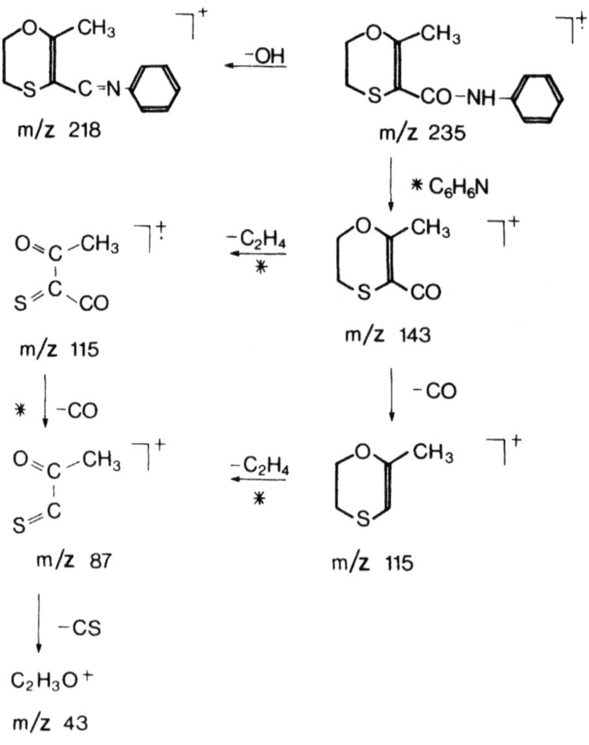


A low-intensity peak is observed at *m/z* 218 $[M - OH]^+$, while significant ion currents are observed at *m/z* 132 and 103. No metastable peaks indicating the origin of these latter peaks are observed. The loss of $C_2H_4$ from $[M]^+$, a retro Diels–Alder reaction, occurs only to a minor extent, leading to the ion *m/z* 207 with an intensity 0.2% of the base peak.

In the methane CI spectrum of the sulfide, the protonated molecule forms the base peak, the major fragmentation reaction being loss of aniline from $[MH]^+$ to form *m/z* 143, as in the EI spectrum. The more exothermic protonation by $H_3^+$ results in a reduced $[MH]^+$ ion intensity and formation of *m/z* 143 as the base peak in the $H_2$ CI spectrum, with further fragmentation of *m/z* 143 as observed in the EI spectrum. Loss of $H_2O$ (*m/z* 218) and loss of $C_2H_4$ (*m/z* 208) from $[MH]^+$ are observed in both CI spectra. The ion at *m/z* 117 in both CI spectra corresponding to loss of $C_6H_5NCO$ from $[MH]^+$ and is accompanied in the methane CI spectrum by peaks at *m/z* 145 and 157 corresponding to elimination of the same neutral from $[M + C_2H_5]^+$ and $[M + C_3H_5]^+$. The $[M]^+$ intensity is quite high in the $H_2$ CI spectrum indicating significant charge transfer from $H_3^+$, consistent with an ionization potential for the sulfide of 9.3 eV, the recombination energy of $H_3^+$.

As was observed, the major fragmentation reaction involves loss of neutral HX from the protonated molecule, the product ion comprising the base peak in all spectra. Relatively abundant molecular ions $[M]^+$ formed by charge transfer from $[H_3]^+$ are observed as well as lower abundance $[M - H]^+$ ions, presumably originating by hydride ion abstraction. Minor fragmentation pathways of the protonated molecule lead to $[MH - 28]^+$ ions, presumably involving loss of $C_2H_4$ and to *m/z* 117, which can be formulated according to the following reactions:

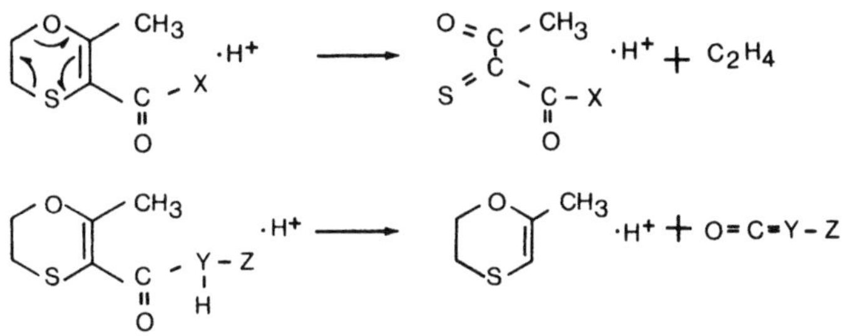


It is interesting to note that this reaction is not observed for $X = N(CH_3)C_6H_5$ which contains no transferable H on the nitrogen adjacent to the carbonyl function, but rather a significant ion current is observed at *m/z* 134, presumably corresponding to the $[C_6H_5N(CH_3)CO]^+$ fragment ion.

In the methane CI mass spectra of the sulfides, the $[RX \cdot H]^+$ and $[R]^+$ ions dominate the spectra. The extent of reaction (1) decreases dramatically with the proton affinity of HX, with the result that $[R]^+$ accounts for 67% of the total ionization when X = OH but only 7 percent when X is a methyl-substituted anilino group.

In the isobutane CI mass spectra of the sulfides only X = OH and $X = OH_2C$ CH showed significant ion signals for $[R]^+$. In view of the methane CI spectra, this result is not surprising. For all the sulfides, the $[RX \cdot H]^+$ ion is the base peak and apart from a small $[C_3H_3]^+$ cluster peak was the only peak observed in the spectra of the remaining compounds.

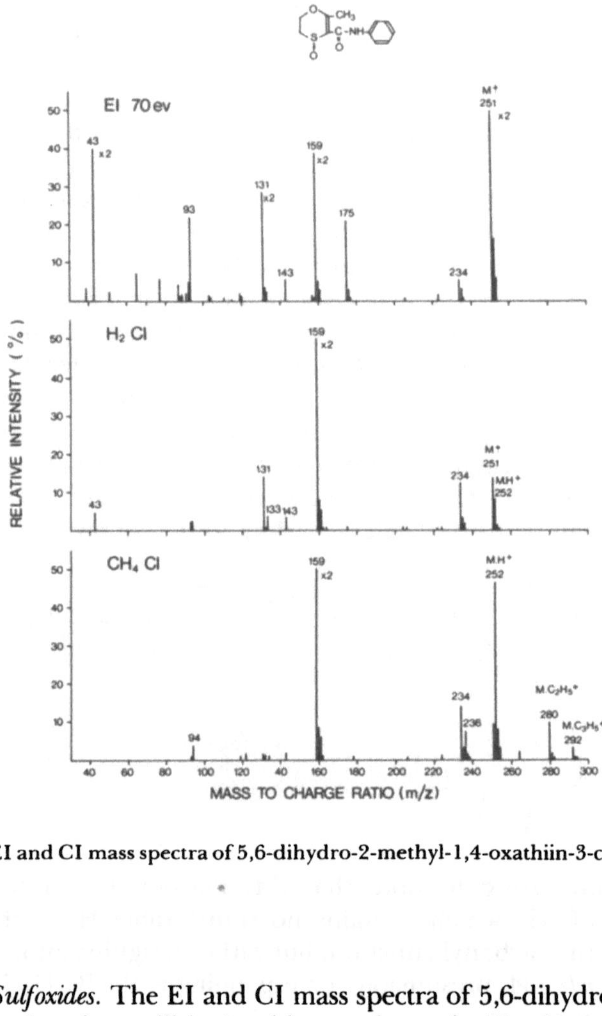


**FIGURE 11.** EI and CI mass spectra of 5,6-dihydro-2-methyl-1,4-oxathiin-3-carboxanilide-4-oxide.

*5.3.2. Sulfoxides.* The EI and CI mass spectra of 5,6-dihydro-2-methyl-1,4-oxathiin-3-carboxanilide-4-oxide are shown in Fig. 11.

The EI-induced fragmentation sequence is outlined below. The major fragmentation sequence is identical to that of the sulfide and involves the loss of the anilino radical to give *m/z* 159, which readily loses $C_2H_4$ yielding the abundant *m/z* 131 ion. It should be noted that *m/z* 131 is a singlet and there is no evidence for loss of CO from the $[M - C_6H_5NH]^+$ ion, in contrast to the sulfide case where loss of both $C_2H_4$ and CO appear to occur. Loss of OH from the molecular ion is observed, as is loss of an oxygen atom. The spectrum of the sulfoxide does not change appreciably when the source temperature varied from 100 to 169 °C, indicating that thermal decomposition of the same was negligible. An interesting fragmentation sequence confirmed by both mass measurements and metastable transitions is the RDA reaction involving loss of $C_2H_4$ leading to *m/z* 223 ion, which shows a facile loss of SO yielding the abundant *m/z* 175 fragment.

The methane CI mass spectrum of sulfoxides shows an abundant $[MH]^+$ ion accompanied by the $[M + C_2H_5]^+$ and $[M + C_3H_5]^+$ cluster ions. As for the sulfide, the major fragmentation mode for $[MH]^+$ is loss of aniline, yielding *m/z* 159 as the base peak. The same fragment is abundant in the EI mass spectrum. Losses of $H_2O$ (*m/z* 224) and $H_2O + C_2H_4$ (*m/z* 206) from $[MH]^+$ also are observed. An interesting and unusual fragmentation is the loss of an oxygen atom from both $[MH]^+$ and $[M + C_2H_5]^+$ yielding *m/z* 236 and 264, respectively. The protonated aniline ion, *m/z* 94, is more abundant than in the sulfide spectrum and is accompanied by an ion at *m/z* 122 $[C_6H_5NH_2 \cdot C_2H_5]^+$ and *m/z* 134 $[C_6H_5NH_2 \cdot C_3H_5]^+$. The *m/z* 133 ion presumably arises by loss of $C_6H_5NCO$ from $[MH]^+$.

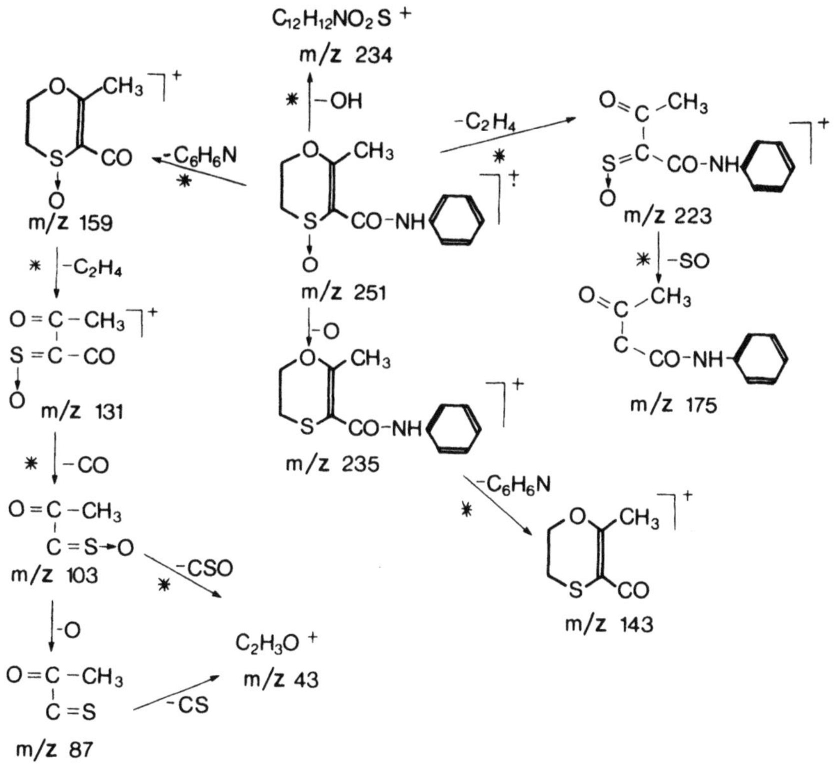


The $H_2$ CI mass spectrum of the sulfoxide is similar to the methane

CI mass spectrum but shows more extensive fragmentation and hence a lower $[MH]^+$ ion, as well as the absence of the ionic species resulting from the clustering reactions and fragmentation of these clusters. In addition, an abundant $[M]^+$ ion is observed resulting from the charge transfer reaction of $H_3^+$ with the sulfoxide.

In contrast to the sulfides discussed above, in the limited series of sulfoxides studied the $[R^+/RX \cdot H]^+$ ratio shows no consistent trend with changes in the proton affinity of the departing HX molecule. This is true for both the $H_2$ CI results and the methane CI results. There are other significant ion signals in many of the CI mass spectra ranging from abundant *m/z* 131 peaks $[R - 28]^+$ in the $H_2$ CI mass spectra of X = OH and X = $-OCH_2C{\equiv}CH$, to abundant $[MH - O]^+$ peaks in the methane CI mass spectra of X = $-NCH_6H_4CH_3$ and X = $N(CH_3)C_6H_5$. Pre-

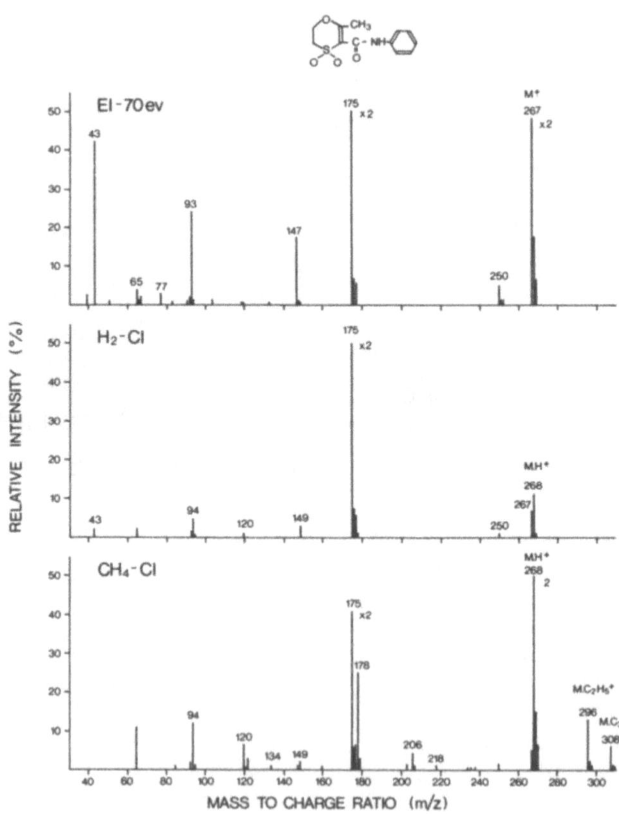


FIGURE 12. EI and CI mass spectra of 5,6-dihydro-2-methyl-1,4-oxathiin-3-carboxanilide-4,4-dioxide.

sumably, because of these competing and/or consecutive reactions the $[R^+/RX \cdot H]^+$ ratio is no longer a good measure of the relative importance of reaction (1) as the functional group X is varied. This result tends to support Jardine and Fenselau's[(43)] suggestion that the correlation will only be observed where other complicating fragmentation reactions are of minor importance. It also is possible that the sulfoxide spectra are more susceptible to variation with source and solid-probe temperature since sulfoxides are thermally unstable.

In the isobutane CI mass spectra $[RX \cdot H]^+$ accounted for 95% of the total ionization in all cases.

*5.3.3. Sulfones.* The EI and CI mass spectra obtained for 5,6-dihydro-2-methyl-1,4-oxathiin-3-carboxanilide-4,4-dioxide are shown in Fig. 12.

The EI fragmentation scheme is derived from the mass measurements and observed metastable transitions. The major fragmentation route being cleavage of the amide bond with loss of the aniline radical, the resultant *m/z* 175 ion fragments further by loss of $C_2H_4$. The aniline molecular ion of *m/z* 93 is of increased importance in the EI mass spectrum of sulfone than that of sulfoxide or sulfide.

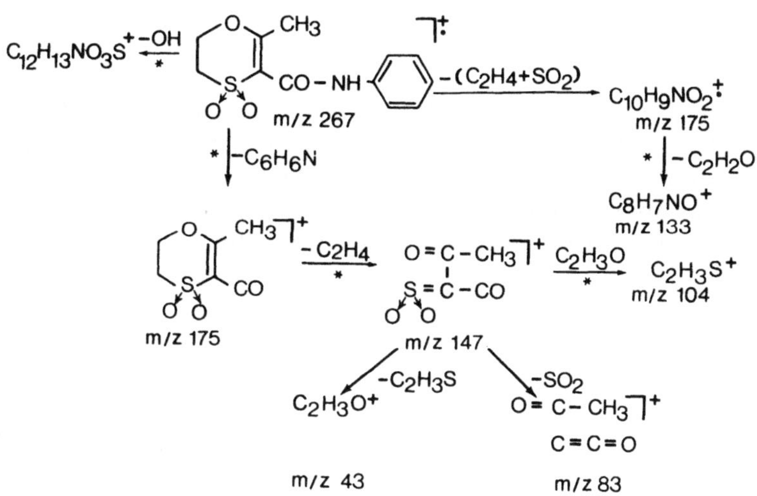


The methane CI spectrum of the sulfone shows $[MH]^+$ as the base peak and the expected $[M + C_2H_5]^+$ cluster ions. As for the sulfide and sulfoxide, the major fragmentation of $[MH]^+$ involves loss of aniline to give the same fragment ion as observed in the EI spectrum. The protonated aniline ion (*m/z* 94) is more pronounced than in the spectra of sulfide and sulfoxide, and is accompanied by $[C_6H_5NH_2^+\ C_2H_5]^+$ (*m/z* 122) and $[C_6H_5NH_2^+\ C_3H_5]^+$ (*m/z* 134). A significant fragment ion is that

TABLE 9
*Methane CI Mass Spectra of Sulfones (3)*[a]

| X | $[RX \cdot H]^+$ | $[R]^{+b}$ | $[R]^+/[RX \cdot H]^+$ | PA(HX) | Other ions (abundance)[c] |
|---|---|---|---|---|---|
| OH | 1.9 | 42.8 | 22.5 | 169 | 159(2.2), 149(11.8), 143(3.6), 103(8.7), 85(12.6) |
| $OCH_2C{\equiv}CH$ | 17.4 | 80.5 | $4.6_3$ | ~185 | |
| $NH_2$ | 24.7 | 63.2 | $2.5_6$ | 201 | $[M]^+$(1.9) |
| $NHC_6H_5$ | 38.5 | 31.0 | $0.8_1$ | 210 | $[M]^+$(4.0), 178(9.6), 94(4.7) |
| $NHCH_3$ | 62.0 | 20.3 | $0.3_3$ | 212 | $[M]^+$(2.6), $[M - H]^+$(1.1), 116(1.2) |
| $N(CH_3)C_6H_5$ | 67.6 | 16.0 | $0.2_4$ | 216 | |
| $NHC_6H_4CH_3$ | 55.4 | 20.3 | $0.3_7$ | 216 | $[M]^+$(5.4), $[M - H]^+$(3.1) |

[a] Intensities as percent of total additive ionization.
[b] $m/z$ 175.
[c] $[M \cdot C_2H5]^+$ (~9%) and $[M \cdot C_2H_5]^+$ (~2%) not listed.

observed at $m/z$ 178, presumably corresponding to loss of $SO_2$ + $C_2H_2$ from $[MH]^+$, with ions at $m/z$ 206 and 218 corresponding to the same loss from the $[M + C_2H_5]^+$ and $[M + C_3H_5]^+$ ions. The $H_2$ CI spectrum of sulfone shows a much reduced $[MH]^+$. Again a moderately intense $[M]^+$ ion signal resulting from charge transfer from $H_3^+$ is observed. The $m/z$ 149 ion corresponds to the $[MH - C_6H_5NCO]+$ ion, a species which is also observed in the CI spectra of 1 and 2.

The main features of the $H_2$ and methane CI mass spectra of the sulfones are summarized in Table 9. By far the most important fragmentation reaction in both the $H_2$ and methane CI systems is reaction (1). As shown in the table, in both systems there is a consistent trend of decreasing fragmentation by reaction (1) as measured by the $[R]^+/[RX \cdot H]^+$ ratio, with increasing proton affinity of the departing neutral HX molecule. The isobutane CI mass spectra of the sulfones all show $[RX \cdot H]^+$ as the base peak with only relatively minor fragmentation by reaction (1).

### 5.4. *Trifluralin*

Trifluralin (α,α,α-trifluoro-2,6-dinitro-*N*,*N*-dipropyl-*p*-toluidine) is the selective dinitroaniline herbicide having the following structure:

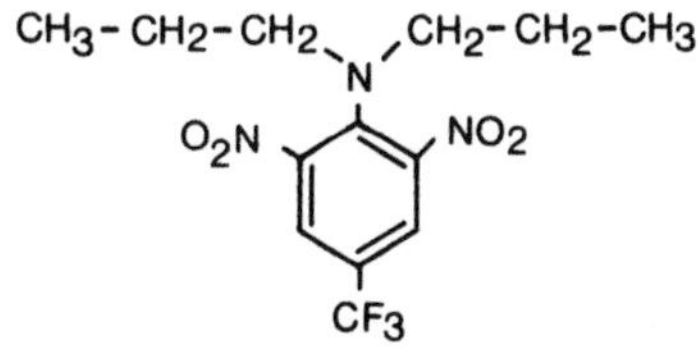


The fate of trifluralin and other related herbicides in soil, plants, animals, water, and air has been investigated, and the mass spectrometry of several degradation products described by Golab *et al.*[44,45] In soil 28 products derived from trifluralin have been detected. The identities of these compounds have been determined either by synthesizing possible candidates and matching these chromatographically and by spectral methods with soil products or by direct spectral structure determination followed by comparison with synthetic compounds.

Of the foregoing compounds some were isolated from soil in sufficient quantity for identification by spectroscopic techniques. The most abundant degradation product was α,α,α-trifluoro-2,6-dinitro-*N*,*N*-dipropyl-*p*-toluidine, which resulted from decomposition of 2% of the applied trifluralin. Products identified by chromatographic comparison are α,α,α-trifluoro-2,6-dinitro-*N*,*N*-dipropyl-*p*-toluidine, α,α,α-trifluoro-2,6-dinitro-*N*-propyl-*p*-toluidine, 7-nitro-1-propyl-5-(trifluoromethyl)benzimidazole, 4-(dipropylamino)-3,5-dinitrobenzoic acid, *N*-propyl-2,2′-azoxybis(α,α,α-trifluoro-6-nitrotoluidine), and some of their respective mass spectra are shown in Fig. 13.

Low- and high-resolution measurements on the degradation products of trifluralin labeled with $^{14}C$ in the aromatic ring and the trifluoromethyl group indicated that substituted benzenes, benzimidazoles, and azo and azoxy compounds were formed beside the *N*,*N*-dialkylamines. They show major peaks due to amine α-cleavage and due to amino α-cleavage followed by carbon–nitrogen bond fission with rearrangement of a hydrogen atom.

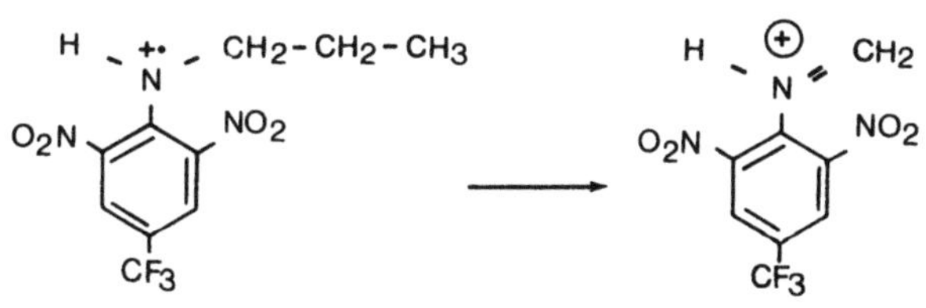


Compounds having nitro groups yield mass spectra containing peaks due to intraionic oxidation by the nitro groups. In some cases, propionyl ions formed by this process are major ions in the mass spectra. Azo and azoxybenzenes resulting from condensation of two trifluraline molecules through nitro-nitrogens fragment mainly via the nitrogen–nitrogen bond cleavage.

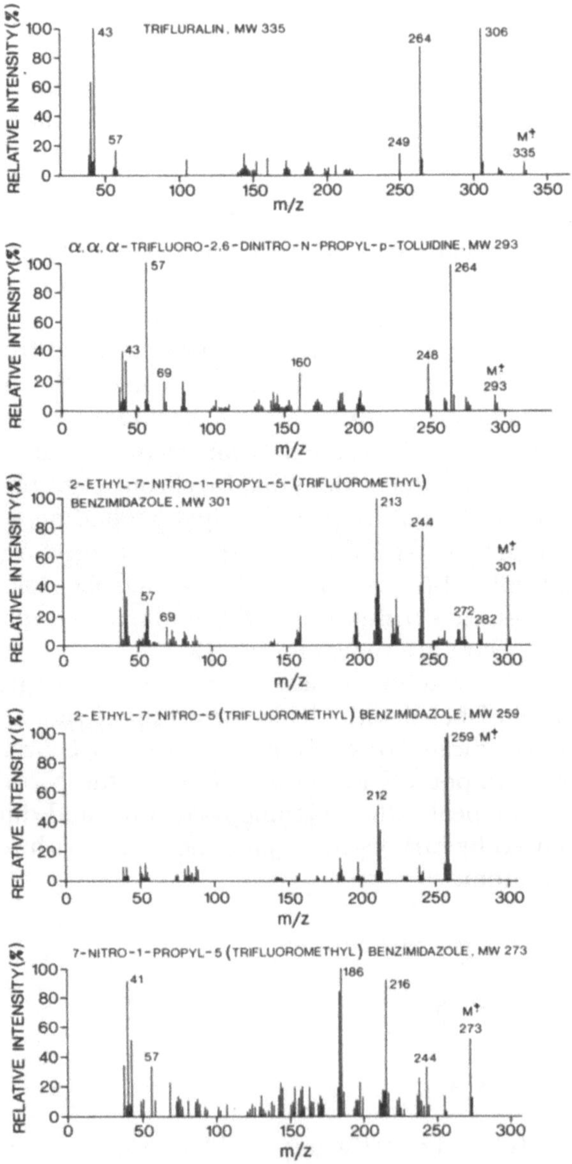


FIGURE 13. EI mass spectra of Trifluralin and its degradation products.

## REFERENCES


1. R. L. Baron, J. A. Sphon, J. J. Chen, E. Lustig, J. D. Doherty, E. A. Hansen, and S. H. Kolbye, *J. Agric. Food Chem.* **17,** 883–887 (1969).
2. S. Y. Liu and J. M. Bollag, *J. Agric. Food Chem.* **19,** 487–490 (1971).
3. J. A. Durden, Jr., and W. J. Bartley, *J. Agric. Food Chem.* **19,** 441–444 (1971).
4. W. J. Bartley, N. R. Andrawes, E. L. Chancey, W. P. Bagley, and H. W. Spurr, *J. Agric. Food Chem.* **18,** 446–453 (1970).
5. R. A. Cardona and H. W. Dorough, *J. Agric. Food Chem.* **21,** 1065–1071 (1973).
6. F. J. Biros, *Adv. Chem. Ser.* **104,** 132–144 (1971).
7. P. C. Rankin, *J. Assoc. Off. Anal. Chem.* **54,** 1340–1348 (1971).
8. P. Brown and C. Djerassi, *Angew. Chem. Internat. Edit.* **6**(6), 477–580 (1967).
9. C. P. Lewis, *Anal. Chem.* **36,** 176–180 (1964).
10. C. P. Lewis, *Anal. Chem.* **36,** 1582–1588 (1964).
11. J. B. Thomson, P. Brown, and C. Djerassi, *J. Am. Chem. Soc.* **88,** 4049 (1966).
12. S. Safe, and O. Hutzinger, *Mass Spectrometry of Pesticides and Pollutants,* CRC Press, Cleveland, Ohio, 1973.
13. H. R. Schulten, *J. Agric. Food Chem.* **24,** 743–750 (1976).
14. J. N. Damico, R. P. Baron, and J. A. Sphon, *Int. J. Mass Spectrom. Ion Phys.* **2,** 161–175 (1969).
15. J. A. Coburn, B. D. Ripley, and A. S. Y. Chau, *J. Assoc. Off. Anal. Chem.* **59**(1), 188–196 (1976).
16. R. Greenhalgh, W. D. Marshall, and R. R. King, *J. Agric. Food Chem.* **24,** 266–270 (1976).
17. J. Lawrence, *J. Agric. Food Chem.* **24,** 1236–1238 (1976).
18. F. I. Onuska and M. E. Comba, *J. Assoc. Off. Anal. Chem.* **59**(3), 622–632 (1976).
19. J. Krupcik, J. Garaj, S. Holotik, D. Oktavec, and M. Kosik, *J. Chromatogr.* **112,** 189–196 (1975).
20. J. Krupcik, P. A. Leclercq, J. Garaj, and J. Masaryk, *J. Chromatogr.*, **171,** 285–304 (1979).
21. K. W. Given, B. M. Mattson, G. L. Miessler, and L. H. Pignolet, *J. Inorg. Nucl. Chem.* **39,** 1309–1316 (1977).
22. I. W. Fraser, J. L. Garnett, I. K. Gregor, and K. J. Jessop, *Org. Mass Spectrom.* **10,** 69–73 (1975).
23. F. I. Onuska and W. R. Boos, *Anal. Chem.* **45,** 967–971 (1973).
24. F. I. Onuska and M. E. Comba, *J. Chromatogr.* **99,** 291–299 (1974).
25. J. O. Madsen, S. O. Lawesson, A. M. Duffield, and C. Djerassi, *J. Org. Chem.* **32,** 2054–2058 (1967).
26. M. A. Baldwin, A. Kirkien-Konasiewicz, A. G. Loudon, A. Maccoll, and D. Smith, *Chem. Commun.*, 574–578 (1966).
27. M. A. Baldwin and A. G. Loudon, *Org. Mass Spectrom.* **2,** 549–552 (1969).
28. H. Matsumoto, S. Fukushima, and N. Kunita, *Osaka-furitsu Koshu Eisei Kenkyusho Kenkyu Hokoku Shokuhin Eisei Hen* **9,** 25–34, (1979); *Mass Spec.* **3,** 92:35227f (1980).
29. W. R. Benson and S. N. Damico, *J. Assoc. Off. Anal. Chem.* **51,** 347–365 (1968).
30. J. J. Brophy, D. Nelson, J. S. Shannon, and S. Middleton, *Org. Mass Spectrom.* **14**(7), 379–386 (1979).
31. L. Grehn, *Org. Mass Spectrom.* **14**(7), 379–386 (1977).
32. L. Grehn, *Org. Mass Spectrom.* **12**(5), 267–268 (1977).
33. R. H. Shapiro, *J. Org. Chem.* **33,** 243–250 (1968).
34. M. A. Baldwin, A. G. Loudon, K. S. Webb, and P. C. Cardell, *Org. Mass Spectrom.* **12**(5), 280–282 (1977).

35. M. A. Baldwin, A. M. Kirkien, A. G. Loudon, and A. Maccoll, *Org. Mass Spectrom.* **4,** 81–88, (1970).
36. H. R. Schulten, *Fresenius Z. Anal. Chem.* **293,** 273–281 (1978).
37. H. R. Schulten and J. Stoeber, *Fresenius Z. Anal. Chem.* **293**(5), 370–376 (1978).
38. B. G. Tweedy, C. Leoppky, and J. A. Ross, *J. Agr. Food Chem.* **18,** 851–853 (1970).
39. J. H. Onley, G. Yip, and M. H. Aldridge, *J. Agric. Food Chem.* **16,** 426–433 (1968).
40. I. J. Belasco and R. W. Reiser, *J. Agric. Food Chem.* **17,** 1000–1003 (1969).
41. F. I. Onuska, A. G. Harrison, and M. E. Comba, *Biomed. Mass Spectrom.* **3,** 248–254 (1976).
42. A. G. Harrison and F. I. Onuska, *Org. Mass Spectrom.* **13**(1) 35–38 (1978).
43. I. Jardine and C. Fenselau, *J. Am. Chem. Soc.* **98,** 5086–5089 (1976).
44. J. Golab and J. L. Occolonitz, *Biomed. Mass Spectrom.* **6**(1), 1–9.

# MASS SPECTROMETRY OF CHLORINATED POLYCYCLIC HYDROCARBON PESTICIDES

*FRANCIS I. ONUSKA*

## 1. *INTRODUCTION*

Organochlorine pesticides and their biodegradation and photodegradation products as environmental and food contaminants have been extensively reviewed.[1–3] They were observed in water,[2] food,[3] sediment,[4] and fatty tissues.[5]

Gas chromatography and thin-layer chromatography procedures generally employed for the detection of chlorinated polycyclic hydrocarbons, however, do not provide an unambiguous identification.[6]

The characteristic chlorine isotopic patterns and retro Diels-Alder (RDA) rearrangement ions associated with all chlorinated polycyclic hydrocarbon pesticides enable the recognition of potentially hazardous pesticides and their metabolites through the use of gas chromatography/mass spectrometry (GS/MS). When coupled with a data-handling system, this approach provides a high rate of information output and a rapid and sensitive analytical procedure. Structural elucidation of metabolites and biodegradation products by auxiliary techniques such as nuclear magnetic resonance and infrared spectroscopy further extends these capabilities.

## 2. *DIMETHANONAPHTHALENES*

It is relevant to the energetics of ion formation and to the relationship between fragmentation and structure to discuss preferred frag-

*FRANCIS I. ONUSKA* • National Water Research Institute, Canada Centre for Inland Waters, Burlington, Ontario L7R 4A6, Canada.

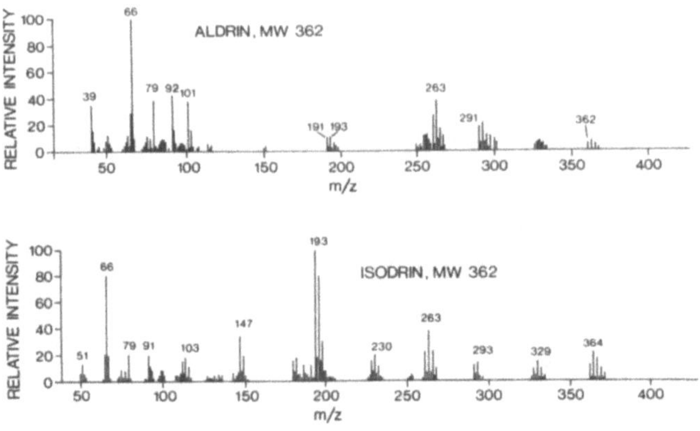


FIGURE 1. EI mass spectra of aldrin and isodrin.

mentation of dimethanonaphthalenes. As can be predicted according to Stevenson's rule, the positive charge remains on the more substituted fragment—that is, the one with the lower ionization potential.[7]

The EI mass spectra of parent cycloaromatic chlorohydrocarbon pesticides have been investigated in detail.[8–13] Using aldrin as the basic system for study, three sets of RDA are observed.

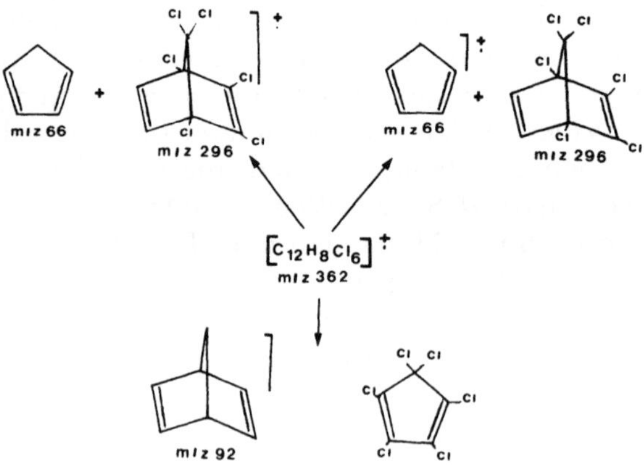


Depending on which bonds are broken during the ring fusion, cyclopentadiene (*m/z* 66) or hexachloro cyclopentadiene ions can be formed. They are further accompanied by one or the other bornadiene moiety, hexachloronorbornadiene (*m/z* 296) and bornadiene itself (*m/z* 92).[14] A very similar mass spectrum, differing only in the relative abundance of the *m/z* 193 ion, is exhibited by the isomers aldrin and isodrin (Fig. 1).

The fragmentation pattern in the mass spectra of two similar epoxides, dieldrin and endrin, has been discussed in detail.[5] For dieldrin the most abundant ion is seen at *m/z* 79, which corresponds to an elimination of CHO moiety from the norbornadiene oxide ion (*m/z* 108). The mass spectra are presented in Fig. 2.

Mass spectrometry has been used in metabolic and photodegradation studies of chlorinated dimethanonaphthalene and methanoindene products (Table 1).

Biros and his colleagues have reported the methane chemical ionization (CI) mass spectra of 10 cycloaromatic chlorohydrocarbon insecticides and their metabolites.[15] Not surprisingly, the $[MH]^+$ ion intensities are small, the base peak normally being $[MH - HCl]^+$, except for those compounds containing a hydroxyl group, where $[MH - H_2O]^+$ constitutes the base peak. Further loss of HCl from $[MH - HCl]^+$ was observed in several cases as was a RDA rearrangement. It was suggested that this latter fragmentation might be useful in distinguishing between isomers, namely, aldrin versus isodrin, *cis*-chlordane versus *trans*-chlordane. The main use of this feature of mass spectra has been confirmed in a study of oxygen-containing derivatives by McKinney and co-workers.[16]

The EI and CI mass spectra show that there is no general rule regarding formation of a quasimolecular ion in the CI mass spectra of polycyclodiene compounds and their metabolites. Both $[M - H]^+$ and to a lesser extent $[MH]^+$ are noted in the CI mass spectra. $[MH]^+$ formation seems to be largely dependent on the presence of specific oxygen-containing species such as in the case of epoxides. Further, the isotopic distribution pattern for these systems containing six chlorine atoms is highly visible in the quasimolecular ion region. The lower energy requirements of CIMS favor less extensive fragmentation and a predom-

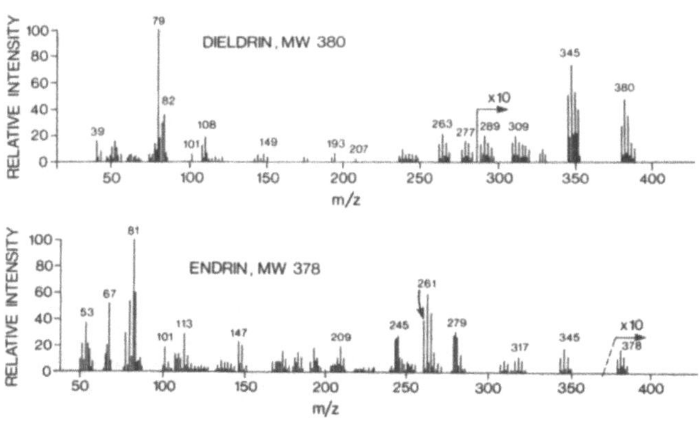


FIGURE 2. EI mass spectra of dieldrin and endrin.

*TABLE 1*
*EI Mass Spectra Data for Dimethanonaphthalene and Methanoindene Pesticides*

| Compound | mw | Most abundant ions |
|---|---|---|
| Aldrin | 362 | 66, 79, 92, 101, 263 |
| Aldrin photoproduct | 364 | 329, 152, 293, 257 |
| Isodrin | 362 | 193, 66, 147, 263, 364 |
| Dieldrin | 378 | 79, 82, 108, 263, 345 |
| Endrin | 378 | 81, 263, 261, 279, 245 |
| Endrin photoproduct | 350 | 67, 101, 139, 183, 243, 279, 315 |
| Heptachlor | 370 | 100, 39, 65, 102, 272, 237, 335, 337 |
| Heptachlor epoxide | 386 | 81, 353, 237, 272, 217 |
| Heptachlor-exo-epoxide (2-chloro) | 386 | 81, 351, 353, 153, 115 |
| Heptachlor-exo-epoxide (1-chloro) | 386 | 115, 81, 53, 237, 109 |
| Chlordane | 406 | 100, 135, 65, 109, 272, 373 |
| Oxy-chlorane | 420 | 115, 51, 105, 261, 149, 185, 237, 387 |
| 1-Hydroxychlordene | 352 | 81, 82, 319, 217, 253 |
| 1-OH Chlordene photoproduct | 316 | 217, 182, 253, 281 |
| *trans*-Chlordene photoproduct | 336 | 196, 303, 204, 100 |
| *trans*-Nonachlor | 440 | 409, 100, 109, 263, 239, 97 |
| *trans*-Nonachlor photoproduct 1 | 370 | 335, 230, 266, 204, 301 |
| *trans*-Nonachlor photoproduct 2 | 370 | 238, 134, 266, 337, 299 |
| 1-OH Chlordene photoproduct | 316 | 219, 183, 255, 283 |
| α-Epoxichlordene | 352 | 36, 183, 217, 63 |
| α-Chlordene (dideCl) | 268 | 43, 197, 162, 51, 199, 233 |
| Chlordene, exo-epoxide | 352 | 81, 117, 53, 119, 89, 319, 263, 237 |
| Chlordene, endo-epoxide | 352 | 81, 149, 53, 109, 219, 255, 183 |
| Heptachlor, exo-epoxide (3Cl) | 386 | 81, 183, 109, 135, 217, 253, 289, 353 |
| 1-OH Chlordene-exo-epoxide | 386 | 97, 235, 263, 81 |
| Chlordene, α-ketone | 352 | 183, 75, 109 |
| Chlordene, α-epoxide | 352 | 183, 181, 145, 73 |
| Chlordene, α-dechloro | 302 | 196, 241, 277, 16, 97, 65 |
| Chlordene, β-dechloro | 302 | 196, 231, 269, 75, 145, 97 |
| Photo-α-chlordene | 336 | 230, 169, 267, 303, 36 |
| Chlordene, dichlor-exo-epoxide | 420 | 272, 237, 387, 115 |
| Chlordene, 1-keto-exo-epoxide | 366 | 333, 205, 68, 97, 207, 262, 237 |
| Endrin aldehyde | 378 | 343, 281, 243, 219, 173, 149, 67 |
| Endrin alcohol | 378 | 343, 209, 193, 149, 139, 129, 111 |
| Endrin ketone | 378 | 343, 315, 281, 251, 209, 67 |
| Photoheptachlor | 370 | 335, 299, 270, 100 |
| Photoaldrin | 362 | 186, 221, 79, 151, 291, 257, 327 |
| Mirex | 540 | 270, 235, 402, 357 |
| 8-Monohydromirex | 506 | 272, 238, 203, 474, 370 |
| 10-Monohydromirex | 506 | 238, 272, 203, 368, 478 |
| 2,8-Dehydromirex | 472 | 238, 203, 375, 334, 441 |
| Kepone | 486 | 270, 235, 216, 316 |
| Kepone hydrate | 504 | 270, 235, 353, 433, 451, 486 |
| Kepone, hemiketal | 518 | 270, 235, 447, 353, 483 |
| Kepone, monohydro | 452 | 270, 235, 201, 182, 249, 317, 417 |
| Kepone, dihydro | 418 | 236, 201, 182, 215, 383 |

inance of the characteristic initial fragmentation processes. The mass spectra exhibit ions corresponding to a set of RDA processes as well as ions resulting from successive losses of Cl, HCl, or both, and ions produced by combinations of a RDA process with preliminary or subsequent loss of Cl or HCl. The CI mass spectrum of aldrin shows a quasimolecular ion $[M - H]^+$ and has a base peak of *m/z* 326 corresponding to the loss of HCl. There is less extensive fragmentation exhibited in the CI mass spectrum as compared to the EI mass spectrum.

### 2.1. *Dieldrin*

Dieldrin yields an ion at *m/z* 79 as a base peak in its EI mass spectra. This also occurs in several other systems related to dieldrin. Apparently, the favored RDA arrangement is the one which produces the norbornadiene oxide ion, which undergoes further fragmentation with loss of CHO to produce a cyclohexadiene ion $(C_6H_7)^+$.

As shown by McKinney and co-workers, aldrin ketone (oxodihydro aldrin) yields an EI mass spectrum containing *m/z* 79 as the base peak, but in its methane CI mass spectrum, the quasimolecular ion represents the base peak.(16) The CI mass spectrum also contains the characteristic recombination fragments, in addition to a fragment corresponding to the combined losses of HCl and $CH_2CO$.

Schulten and Beckey demonstrated that by using the FD technique, one can differentiate between aldrin-4,5-*cis*- and aldrin-4,5-*trans*-diol.(17) The *trans*-isomer displays a peak at *m/z* 378 representing the $[M - H_2O]^+$ ion, which is more than 20 times more abundant than that of the *cis*-isomer. They also clearly identified dieldrin by FD mass spectrometry. All mass spectra of standards show prominent $[M]^+$, $[M(^{35}Cl_{n-1}{}^{37}Cl)]^{+\cdot}$; and $[M(^{35}Cl_{n-2}{}^{37}Cl_2)]^+$ ions. Suzuki and colleagues confirmed these results in analyses of water and soil samples.(18,19) The same research group also claims that the FD mass spectrum of aldrin cannot be obtained because of its high volatility.(20)

Photodieldrin yields a base peak in its EI mass spectrum at *m/z* 81, representing a protonated cyclopentadienone moiety. The CI mass spectrum of photodieldrin does not yield an appreciably more intense quasimolecular ion but contains a base peak in the $[M - Cl]^+$ region. A characteristic fragment is also found at *m/z* 163 which resembles a two-chlorine pattern ion representing a protonated dichlorocyclohexadienone moiety.

Dieldrin also contains the *m/z* 81 ion in its methane CI and *i*-butane CI mass spectra suggesting that the quasimolecular ion $[M - H]^+$ formed may resemble a photodieldrin-type structure. Henderson and Crosby (1967) speculated that one of the dichloroethylene chlorine atoms is lost

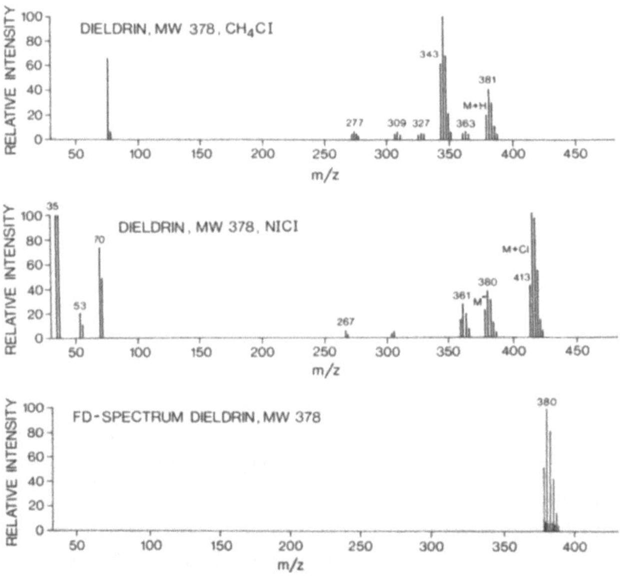


FIGURE 3. CI and FD mass spectra of dieldrin.

first, since photolysis of aldrin and dieldrin have yielded such dechlorinated products.[21] Figure 3 shows CI and FD mass spectra of dieldrin as a representative example of the compounds discussed. The FD mass spectrum shows no fragments, except for $M^+$, in contrast to the significant fragmentation of the EI-induced mass spectrum.

## 2.2. *Endrin*

Endrin, a stereoisomer of dieldrin, gives the most complex EI mass spectrum of the compounds discussed. Qualitatively, very different mass spectra can be obtained depending on the mode of introduction of the sample. The complexity is related to the fact that it rearranges readily under a variety of conditions to form at least two other products, -ketoendrin, having an EI base peak at *m/z* 67 and a CI base peak at *m/z* 343, and endrin aldehyde, EI base peak at *m/z* 67. Onuska and Comba have presented spectra of endrin ketone, endrin aldehyde, and endrin alcohol under EI conditions.[10]

Figure 4 illustrates the advantages of the CI mode of operation for producing mass spectra of endrin in which rearrangement processes are minimized while, at the same time, the most abundant ion is the quasimolecular ion $[MH]^+$. Schulten and Beckey recorded the FD mass spec-

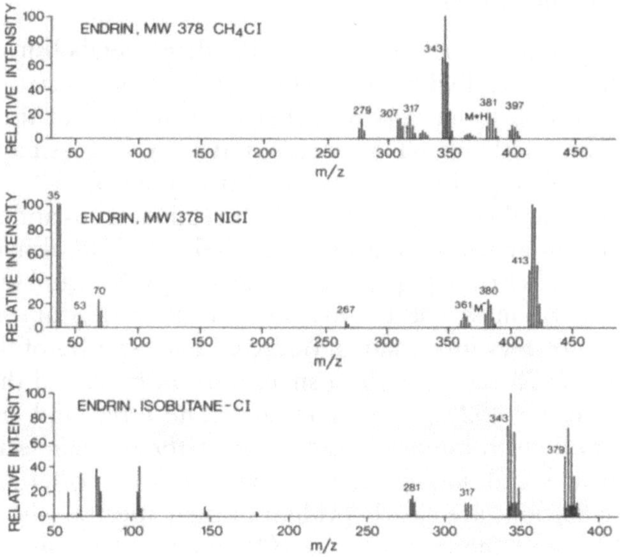


FIGURE 4. CI and FD mass spectra of endrin.

trum of endrin (Fig. 5), which contains no fragmentation in the mass range between *m/z* 300 and the molecular ion pattern.[17] The spectrum is, however, temperature dependent, which indicates a thermally induced fragmentation especially when the emitter temperature is increased over 200 °C. A characteristic fragment ion of endrin in the EI mass spectrum can be seen at *m/z* 261, corresponding to $C_7H_2Cl_5$, and in a similar way a characteristic fragment of δ-ketoendrin is shown at *m/z* 315, representing loss of CO and Cl.

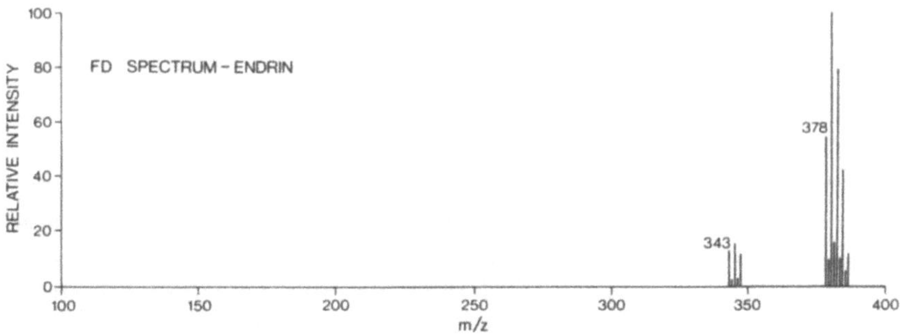


FIGURE 5. The FD mass spectrum of endrin.

### 2.3 Dieldrin and Aldrin

The major fecal metabolite from dieldrin metabolism in rats has been identified as 12-hydroxydieldrin. The EI mass spectrum shows no molecular ion due to an apparent thermal instability of this molecule. However, the methane CI spectrum indicates a quasimolecular ion $[M + H]^+$ implying a hydroxylated dieldrin molecule.[22]

McKinney and colleagues showed separation of *cis*- and *trans*-aldrin alcohols and *cis*- and *trans*-aldrindiols as their silyl and disilyl ethers, on two different gas chromatographic columns (3% OV-1; and QF-1 + OV-225).[16] The EI mass spectra of the silyl ether derivatives of aldrin alcohols are not very informative. In the CI mass spectra of the alcohols, the peak at *m/z* 73 accounts for a small portion only, and the quasimolecular ion at *m/z* 452 is very abundant. The *trans*- and *cis*-aldrindiol disilyl ethers furnish intense ions at or near the quasimolecular ion region. The $[M - Cl]^+$ ion becomes the base peak for both diols and both quasimolecular ions are clearly evident in their spectra, where the *trans*-isomer is 2% and the *cis*-isomer is 10% abundant relative to the base peak.

Oxidation of *cis*-aldrindiol produced two related and apparently isomeric products. The reaction mixture was analyzed by GC/MS using the CI mode of operation, and the mass spectra revealed a quasimolecular ion $[M - H]^+$ at *m/z* 393 (22% abundance) and a base peak of *m/z* 95 corresponding to the loss of CHO, CHOH, and CHO-CHOH-Cl. The gas chromatogram showed apparently isomeric products not very well resolved. The hydroxy-aldrin ketone structure was assigned to the first peak and the other component in the chromatogram exhibited a quasimolecular ion $[MH]^+$ at *m/z* 395 and a base peak at *m/z* 79 corresponding to a $C_6H_7$ moiety. This compound was identified as the corresponding ene-diol having the following structure:

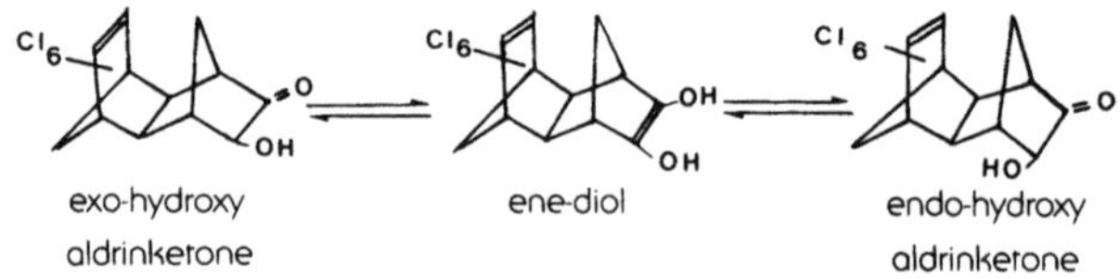


Korte and Arent identified a metabolite of dieldrin in rats as aldrin-4,5-*trans*-diol.[23] Using FD mass spectrometry, it was possible to distinguish between *cis*- and *trans*-isomeric structures. The *trans* diol shows a peak at *m/z* 378 corresponding to $[M - H_2O]^+$, which is 20 times more abundant than the same ion of the *cis*-isomer. An isodrin metabolite isolated from white cabbage leaves was identified as hexachloro-5-norbornene-2,3-dicarboxylic acid. There is no $M^+$ ion in the EI mass spectrum, while it is the base peak in the FD spectrum, having the composition $C_9H_4O_4Cl_6$. Also, high-resolution FD mass spectra of isodrin-4,5-*cis*-diol shows the molecular ion at *m/z* 396 corresponding to $C_{12}H_{10}O_2Cl_6$.

### 2.4. *Isodrin, Heptachlor, Heptachlor Epoxide, Nonachlor, and Chlordanes*

The application of negative-ion chemical ionization (NICI) mass spectrometry to the analysis of pesticide residues in crude extracts of environmental samples has not been extensively used. The main reason for this is the fact that the molecular sensitivity of different classes of pesticides to NICI mass spectrometry can vary greatly. Many of the chlorinated polycyclodiene insecticides do not respond to isobutane and methane CI, presumably because their proton affinities are less than that of the reactant ion.[15,35] The positive- and negative-ion methane CI mass spectra of various chlorinated polycyclic hydrocarbon pesticides are presented in Figs. 6–13.

Dougherty and colleagues described NICI processes observed for aldrin using methane as a reactant gas as shown in the following scheme[24]:

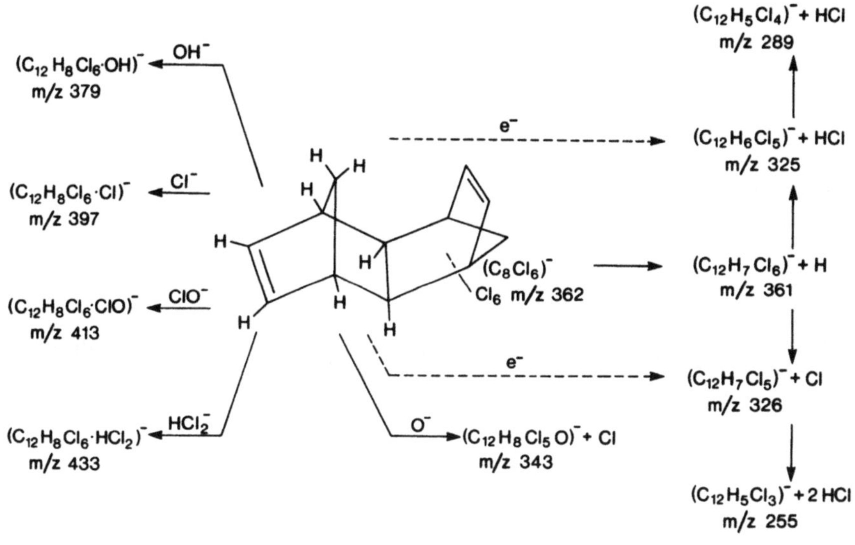


Ion molecular attachment reactions are shown to the left, while resonance capture, dissociative resonance capture, and ion displacement reactions are shown to the right.

### 2.5. *Mirex and Kepone*

Alley and co-workers[25] and Kaiser[26] reported that a substantial part of Lake Ontario has been contaminated with mirex (dodecachloro octahydro-1,3,4-metheno-2*H*-cyclobuta-[*c,d*]pentalene). Also, some spe-

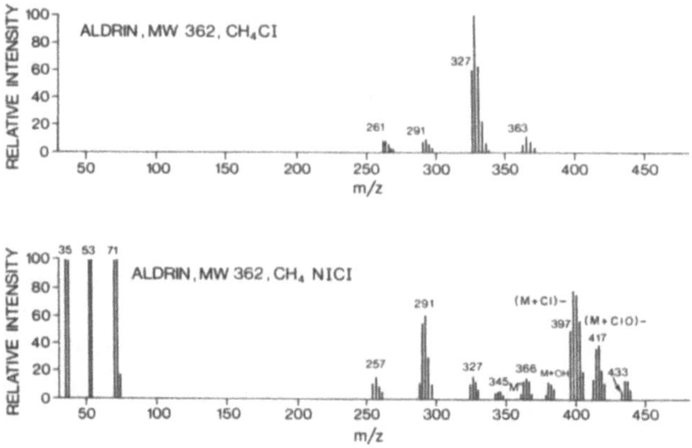


FIGURE 6. The positive- and negative-ion methane CI mass spectra of aldrin.

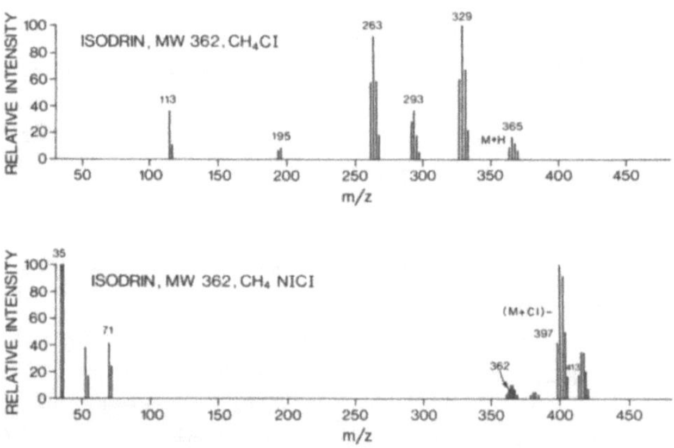


FIGURE 7. The positive- and negative-ion methane CI mass spectra of isodrin.

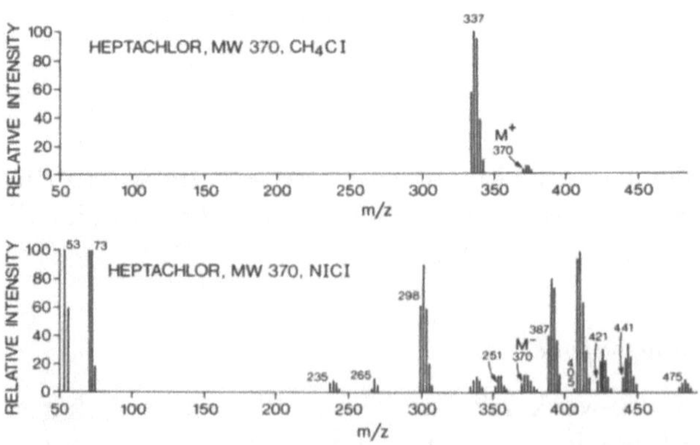


FIGURE 8. The positive- and negative-ion methane CI mass spectra of heptachlor.

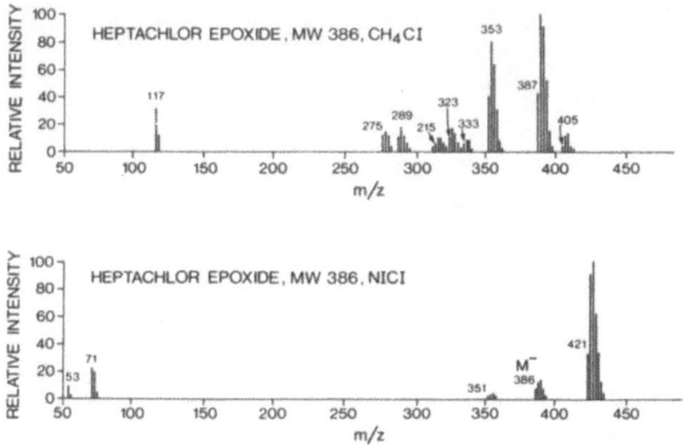


FIGURE 9. The positive- and negative-ion methane CI mass spectra of heptachlor epoxide.

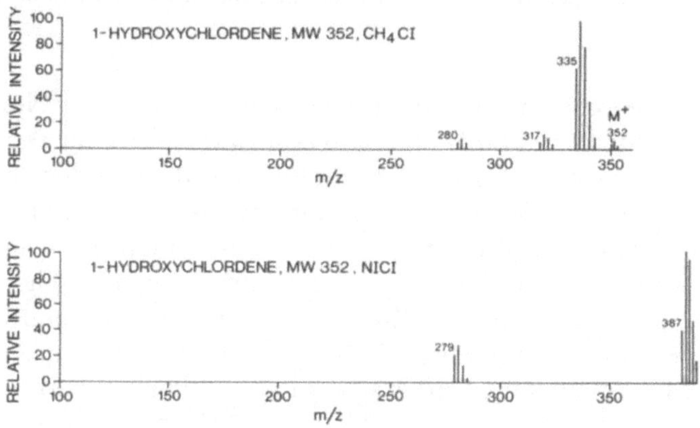


FIGURE 10. The positive- and negative-ion methane CI mass spectra of 1-hydroxychlordene.

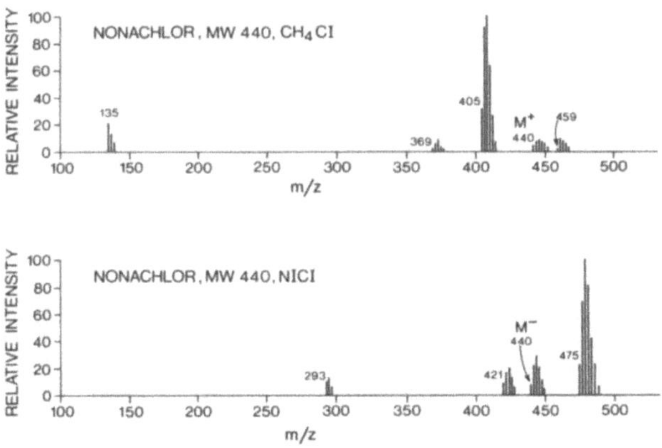


FIGURE 11. The positive- and negative-ion methane CI mass spectra of nonachlor.

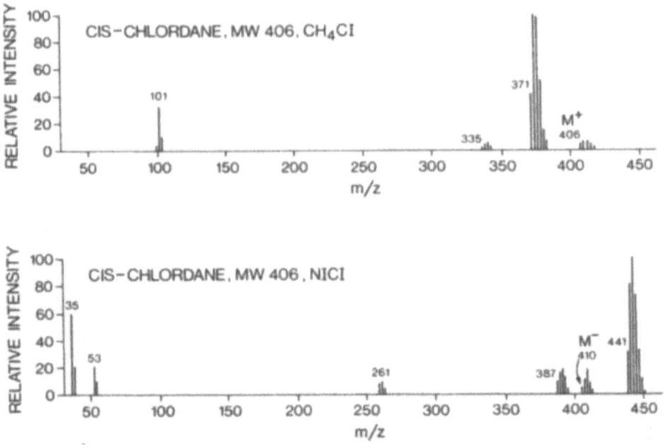


FIGURE 12. The positive- and negative-ion methane CI mass spectra of *cis*-chlordane.

cific wildlife exposures to mirex have been confirmed by Norstom, Hallett, and Onuska[27] and Hallett and co-workers.[28] They showed that 8-monohydromirex is formed by photolytic reductive dechlorination of mirex. There is evidence that 8-monohydromirex, 2,8-dihydromirex, and 10-monohydromirex are present in Lake Ontario fish and sediment samples according to Onuska, Comba, and Coburn[29] (Fig. 14).

Oswald, Albro, and McKinney used methane CI and *i*-butane CI mass spectrometry for mirex.[9] No $M^+$ or $MH^+$ ions are observed in either EI or methane CI mass spectra and only a weak $M^+$ ion in the *i*-butane CI mass spectrum is visible. An OV-17 packed column and pro-

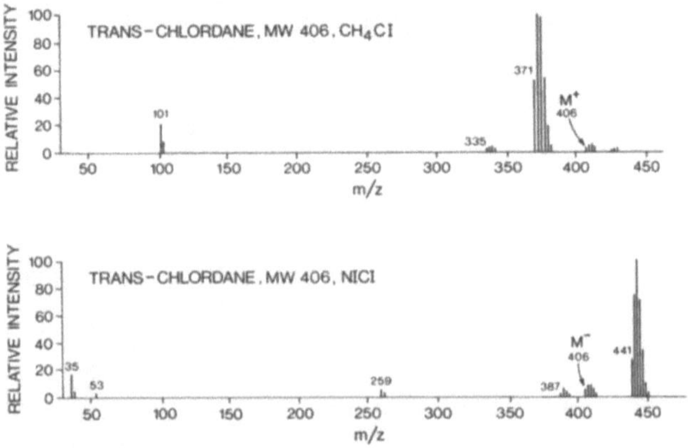


FIGURE 13. The positive- and negative-ion methane CI mass spectra of *trans*-chlordane.

grammed temperature from 220 to 260 °C at 10 °C/min was employed in GC/MS. The $[M - Cl]^+$ ion corresponding to $C_{10}Cl_{11}$ is reported to be weak in both mass spectra with $C_5Cl_6^+$ as the base peak in the *i*-butane CI spectrum and $C_5Cl_5^+$.

Alley and Layton have reported on mirex and kepone photoproducts.[25] The structure of the monohydromirex $C_{10}Cl_{11}H$ indicated that the fragmentation pathways of the photoproduct were the same as those for mirex. The CI mass spectrum of mirex has ion clusters corresponding to $C_{10}Cl_{12}^+$, $C_{10}Cl_{11}^+$, and $C_5Cl_5^+$ fragments. The basic configuration of the monohydromirex molecule is shown below.

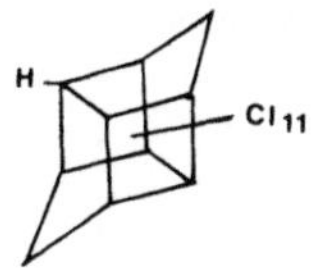


8-monohydromirex

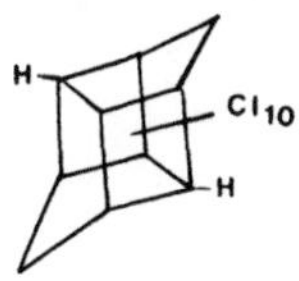


2,8-dihydromirex

Hydrogen substitution is possible at position 8, 9, and 10. For the dihydromirex photoproduct, a major peak corresponding to $C_5Cl_5H^+$ ion but none of $C_5Cl_4H^+$ or $C_5Cl_6^+$ is observed, thus leaving the possible identities as 2,8-dihydro, 3,8-dihydro, 6,9-dihydro, 4,9-dihydro, 5,10-

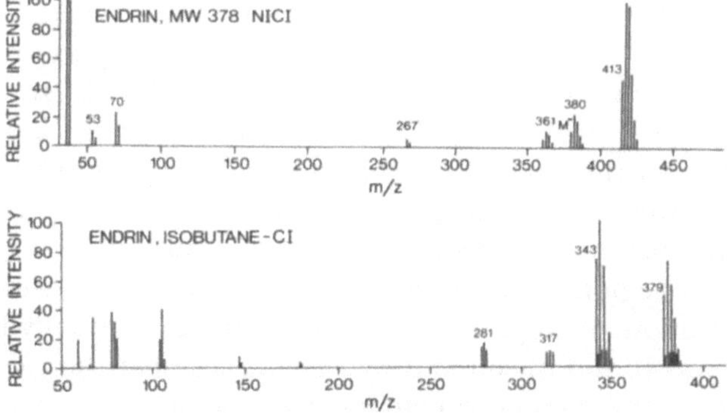


FIGURE 14. The EI mass spectra of 8-monohydro-, 10-monohydro-, and 2,8-dihydromirex.

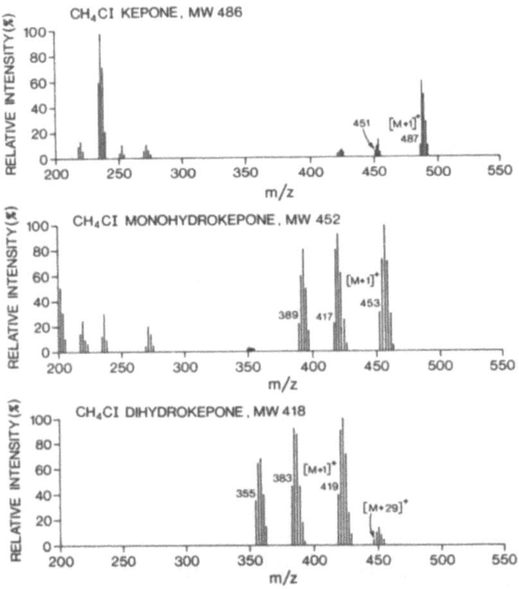


FIGURE 15. The methane CI mass spectra of kepone and reduction products.

dihydro, and 6,10-dihydromirex isomers. Similarly, kepone (1,1a,3,3a,4,5,5,5a,5b,6-decachloro octahydro-1,3,4-metheno-2*H*-cyclobuta[*c,d*]pentalen-2-one) has a fragmentation pattern analogous to that of mirex according to Dilling and Dilling.(30) The mass spectrum of kepone shows a molecular ion at *m/z* 486 corresponding to $(C_{10}Cl_{10}O)^+$, however, it is a very weak ion (less than 1%). When the pentacyclic skeleton of kepone is cleaved, $C_5Cl_6^+$ and $C_5Cl_4O^+$ fragment ions are observed at *m/z* 270 and 216, respectively. The mass spectrum of monohydrokepone ($C_{10}HCl_9O$) has major peaks for $C_{10}Cl_8HO^+$ ion clusters (*m/z* 417), $C_5Cl_6^+$ (*m/z* 236) and a small ion cluster corresponding to formula $C_{10}Cl_8H_2^+$ (*m/z* 418). The methane CI mass spectra of kepone, monohydrokepone, and dihydrokepone shown in Fig. 15 are each characterized by $[M + H]^+$, $[M - Cl]^+$, and $[M - COCl]^+$ ions. The selected-ion-monitoring technique utilizing GC/CIMS for kepone was employed by Harless and colleagues,(31) who monitored the four largest characteristic ions at *m/z* 488.7, 490.7, 492.7, and 494.7 in the quasi-molecular ion region. The minimum detectable level obtained from a standard solution of kepone is 700 pg. Further, monohydrokepone, $C_{10}Cl_9HO$, was found in soil and water samples near the vicinity of the manufacturing site. Gas chromatography/mass spectrometry analysis is performed on a 183 cm long, 2 mm i.d. stainless steel GC column packed with 3% Dexsil 300 on 80–100 mesh Chromosorb W HP. Using helium

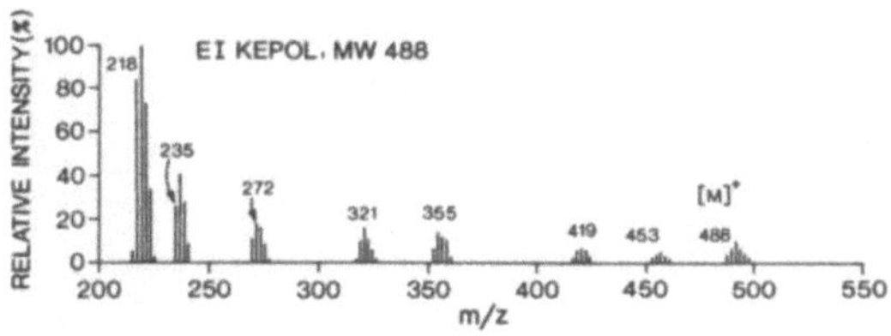


FIGURE 16. The EI mass spectrum of kepol.

as a carrier gas at 15 ml/min and a column temperature of 210 °C, an artifact containing a cluster corresponding to nine chlorine atoms (*m/z* 453) is observed. Methane CIGC/MS results show that this compound elutes from the GC column at approximately 2.5 times the retention time of kepone. The mass shift of two mass units from kepone suggests reduction of the carbonyl to a secondary alcohol called Kepol. It was proven that solutions of kepone in methanol can yield Kepol as a reduction product (Fig. 16). The FD mass spectra obtained with 2 μg/μl solutions of kepone produced an intense molecular ion cluster at *m/z* 504 representing kepone hydrate when the sample is applied from benzene solution, and a less abundant ion cluster at *m/z* 518 representing kepone hemiketal when applied from methanol.

### 2.6. *Toxaphene*

The identification, chemical characterization, and quantitative analysis of toxaphene is a challenging task to pesticide analysts due to the complexity of the mixture. Toxaphene is a complex mixture of at least 177 chlorinated terpenes produced by the chlorination of technical camphene. It may be characterized as having an average empirical formula of $C_{10}H_{10}Cl_8$ and corresponding molecular weight of 410.

Toxaphene is the most widely used chlorinated insecticide in the United States. In 1976 toxaphene was identified in lake trout *(Salvelinus namaycush)*. Because of the extreme complexity of toxaphene stereoisomers, only four major toxic components were reported by Casida and co-workers[32] and Anagnostopoulos, Parlar, and Korte.[33]

Stalling initially examined toxaphene using EI mass spectrometry in tandem with gas chromatography.[34] Toxaphene peaks were eluted from a 274 cm × 2 mm i.d. column. The column was packed with 3% OV-7 on Chromosorb W HP and the temperature was programmed from 150 to 200 °C at 4 °C/min. Eluted peaks were scanned over a mass range of 18–550, and a total of 151 scans were run. Electron impact mass spectra from several major toxaphene components were examined in order to obtain characteristic ion fragments. The mass chromatograms corresponding to *m/z* 291 and 293 ions resembled the peak envelope

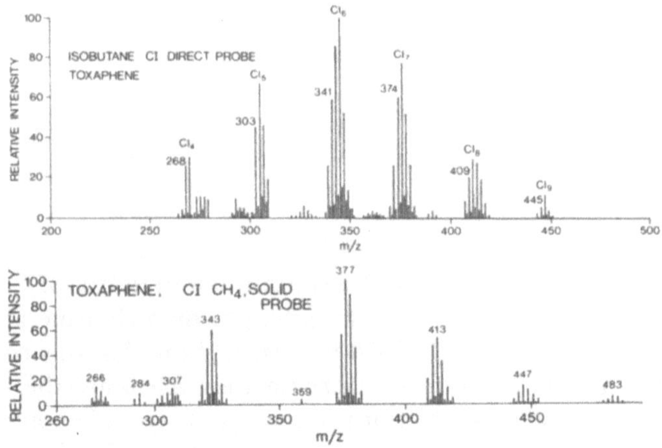


FIGURE 17. The methane and *i*-butane CI mass spectra of toxaphene.

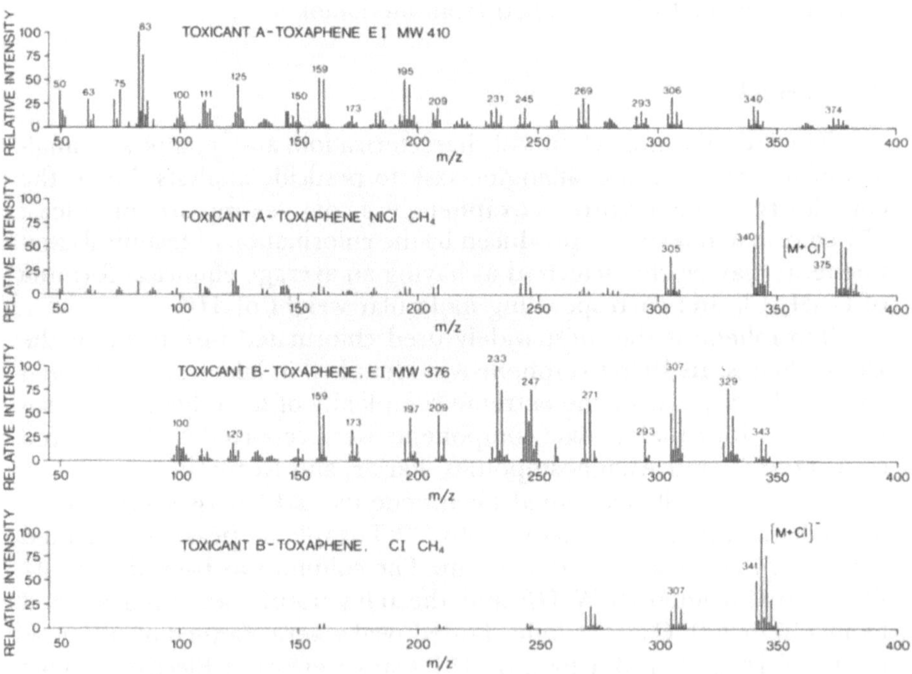


FIGURE 18. The EI and CI mass spectra of Toxicant A and Toxicant B.

recorded from the total ion monitor. The mass chromatogram GC trace corresponding to *m/z* 117 ($CCl_3^+$) generally eluted earlier.

Mass spectrometric characterization of purified toxaphene constituents was reported by Casida.[32] Chemical ionization mass spectra of two purified constituents were identified as empirical formulas of $C_{10}H_{10}Cl_8$ and $C_{10}H_{11}Cl_7$. Each CI mass spectrum indicated loss of a chlorine atom from the molecule to form $[M - Cl]^+$ and $[M - Cl_2]^+$ ions. Examination of toxaphene by negative-ion $CH_4$ and *i*-butane CI direct probe resulted in a mass spectra characterized by multiple quasimolecular ion clusters. These clusters indicate an isomeric series of $C_{10}H_5Cl_9$, with variable numbers of hydrogen atoms for each degree of chlorine substitution as shown in Fig. 17. Stalling estimated the concentration of the various isomeric series.[34] Two series of nonachloro constituents were apparent in the spectrum and they corresponded to $C_{10}H_7Cl_9$ and $C_{10}H_9Cl_9$ moieties. The two most important constituents identified in a toxaphene mixture are known as Toxicant A, which consists of two compounds identified as 2,2,5-endo-6-exo-, 8,8,9,10-octachlorobornane and 2,2,5-endo-6-exo, 8,9,10-heptachlorobornane. Their EI and CI mass spectra are shown in Fig. 18 with the EI and methane CI mass spectra of Toxicant B, another important constituent of toxaphene.

## REFERENCES


1. A. L. Burlingame, T. A. Baillie, P. J. Derrick, and O. S. Chizhov, *Anal. Chem.* **52**(5), 214R–258R (1980).
2. A. Alford, *Biomed. Mass Spectrom.* **4**(1), 1–22 (1977).
3. R. Self, *Biomed. Mass Spectrom.* **6**(9), 361–373 (1979).
4. S. Safe, in *Mass Spectrometry*, Vol. 5 (R. A. W. Johnstone, Sr. Reporter), The Chemical Society, London, 1979, pp. 234–249.
5. S. Safe and O. Hutzinger, *Mass Spectrometry of Pesticides and Pollutants*, CRC Press, Cleveland, Ohio, 1973.
6. L. Fishbein, *Chromatography of Environmental Hazards*, Vol. 3, Elsevier, Amsterdam, 1975.
7. D. P. Stevenson, *Disc. Faraday Soc.* **10,** 35–43 (1951).
8. J. N. Damico, R. P. Baron, and J. M. Ruth, *Org. Mass Spectrom.* **1,** 331–341 (1968).
9. E. O. Oswald, P. W. Albro, and J. D. McKinney, *J. Chromatogr.* **98,** 363–448 (1974).
10. F. I. Onuska and M. E. Comba, *Biomed. Mass Spectrom.* **2,** 169–175 (1975).
11. F. I. Onuska and M. E. Comba, *Biomed. Mass Spectrom.* **2,** 176–182 (1975).
12. F. I. Onuska and M. E. Comba, *J. Assoc. Off. Anal. Chem.* **59**(3), 622–632 (1976).
13. A. Demayo and M. E. Comba, *Bull. Environ. Contam. Technol.* **8**(4), 212–216 (1972).
14. P. Story, *J. Org. Chem.* **26**(2), 287 (1961).
15. F. J. Biros, R. C. Dougherty, and J. Dalton, *Org. Mass Spectrom.* **6,** 1161–1170 (1972).
16. J. D. McKinney, E. O. Oswald, S. M. de Paul Palaszek, and B. J. Corbett, in *Mass Spectrometry and NMR Spectrometry in Pesticide Chemistry* (R. Haque and F. J. Biros, eds.), Plenum, New York, 1974, pp. 5–32.
17. R. H. Schulten and H. D. Beckey, *J. Agric. Food Chem.* **21,** 372–375 (1973).

18. M. Suzuki, Y. Yamamoto, and T. Akiyama, *Water Res.* **8,** 643 (1974).
19. M. Suzuki, Y. Yamamoto, and T. Watanabe, *Pestic. Monit. J.* **11,** 88 (1977).
20. M. Suzuki, Y. Yamamoto, and M. Koga, *Biomed. Mass Spectrom.* **5**(9), 518–523 (1978).
21. G. L. Henderson and D. G. Crosby, *J. Agric. Food Chem.* **15,** 888–893 (1967).
22. J. D. McKinney, H. B. Matthews, and L. Fishbein, *J. Agric. Food Chem.* **20**(3), 597–602 (1972).
23. F. Korte and H. Arent, *Life Sci.* **4,** 2017 (1965).
24. R. C. Dougherty, J. D. Roberts, H. P. Tannenbaum, and F. J. Biros, in *Mass Spectrometry and NMR Spectrometry in Pesticide Chemistry* (R. Haque and F. J. Biros, eds.), Plenum, New York, 1974, pp. 33–47.
25. E. G. Alley and B. R. Layton, in *Mass Spectrometry and NMR Spectrometry in Pesticide Chemistry* (R. Haque and F. J. Biros, eds.), Plenum, 1974, pp. 81–90.
26. K. L. E. Kaiser, *Science* **185,** 523–525 (1974).
27. D. H. Hallett, R. J. Norstrom, F. I. Onuska, M. E. Comba, *Agric. Food Chem.* **24,** 1189–1193 (1976).
28. D. J. Hallett, K. S. Khera, D. R. Stoltz, I. Chu, D. C. Villeneuve, and G. Trivett, *J. Agric. Food Chem.* **26,** 388–391 (1978).
29. F. I. Onuska, M. E. Comba, and J. A. Coburn, *Anal. Chem.* **52**(14), 2272–2275 (1980).
30. W. L. Dilling and M. L. Dilling, *Tetrahedron* **23,** 1225 (1967).
31. R. L. Harless, D. E. Harris, G. W. Sovocool, R. D. Zehr, N. K. Wilson, and E. O. Oswald, *Biomed. Mass Spectrom.* **5**(3), 232–237 (1978).
32. J. E. Casida, R. L. Holmstead, S. Khalifa, J. R. Knox, E. Ohsawa, K. J. Palmer, and R. Y. Wong, *Science* **183,** 520 (1974).
33. M. L. Anagnostopoulos, H. Parlar, and F. Korte, *Chemosphere* **2,** 65 (1974).
34. D. L. Stalling, Proceedings of the 165th ACS Meeting, Div. of Pesticide Chemistry, Dallas, Texas, April, 1973, Abstr. No. 77, 1973.
35. R. C. Dougherty, J. Dalton, and F. J. Biros, *Org. Mass Spectrom.* **6,** 1171–1182 (1972).

19

# TRIAZINE HERBICIDES

D. C. G. MUIR

## 1. INTRODUCTION

Gas chromatography/mass spectrometry (GC/MS) has been used frequently for the determination of triazine herbicides and their N-dealkylated degradation products since these compounds are volatile, thermally stable, and easily separated on a variety of GC columns. Safe and Hutzinger[1] have reviewed mass spectral studies of triazine herbicides published prior to 1973. This chapter will focus on more recent applications of mass spectrometry to the analysis of triazine and triazinone herbicides (Table 1). The mass spectra of substituted *s*-triazines, a diverse group of industrial compounds[2] will not be discussed here. Porter and Baldas[3] have discussed the mass spectra of 1,3,5-*s*-triazines.

## 2. ATRAZINE AND SIMAZINE

Pape and Zabik[4] identified photodegradation products of chloro-, methoxy-, and methylthio-*s*-triazines in alcohol and water systems by use of GC/MS. The compounds were separated on a column of 1% SE-30 on Gas Chromosorb Q (column temperature, 125–175 °C) and spectra obtained on an LKB 9000 MS. Karlhuber and co-workers[5] reported the use of GC/mass fragmentography in the determination of triazine herbicides. Atrazine was chromatographed on a column (1.0 m × 3 mm i.d.) packed with a mixture of 2% NPGS and 2% free-fatty-acid phase (1 : 1) on Chromosorb G and detected by monitoring ions at *m/z* 200 and 215. Mestres and co-workers[6] reported a similar technique in which chloro-*s*-triazines were separated on a column of 1–3% OV-101 on Chro-

*D. C. G. MUIR* • Department of Fisheries and Oceans, Freshwater Institute, 501 University Crescent, Winnipeg Manitoba R3T 2N6, Canada.

TABLE 1
*Common Names and Chemical Names of Triazine Herbicides and Degradation Products*

| | |
|---|---|
| Atrazine | 2-chloro-4-(ethylamino)-6-(isopropylamino)-*s*-triazine |
| Atratone | 2-(ethylamino)-4-(isopropylamino)-6-(methoxy)-*s*-triazine |
| Chlorazine | 2-chloro-4,6-bis(diethylamino)-*s*-triazine |
| Cyanazine | 2-chloro-4-(ethylamino)-6-(1-cyano-1-methylethylamino)-*s*-triazine |
| Cyanazine IV | 2-hydroxy-4-(1-carboxy-1-methylethylamino)-6-(ethylamino)-*s*-triazine |
| Cyanatryn | 2-(1-cyano-1-methylethylamino)-4-(ethylamino)-6-(methylthio)-*s*-triazine |
| Cypratone | 2-(isopropylamino)-4-(cyclopropylamino)-6-(methoxy)-*s*-triazine |
| Cyprazine | 2-chloro-4-(cyclopropylamino)-6-(isopropylamino)-*s*-triazine |
| DA | 6-(1,1-dimethylethyl)-3-(methylthio)-1,2,4-triazin-5(4H)-one |
| DADK | 6-(1,1-dimethylethyl)-1,2,4-triazin-3,5(2*H*,4*H*)-dione |
| DK | 4-amino-6-(1,1-dimethylethyl)-1,2,4-triazin-3,5-(2*H*,4*H*)-dione |
| Hexazinone | 3-cyclohexyl-6-(dimethylamino)-1-methyl-1,3,5-triazine-2,4(1*H*,4*H*)-dione |
| Hydroxyatrazine | 2-hydroxy-4-(ethylamino)-6-(isopropylamino)-*s*-triazine |
| Ipazine | 2-chloro-4-(isopropylamino)-6-(diethylamino)-*s*-triazine |
| Metribuzin | 4-amino-3-(methylthio)-6-*tert*-butyl-1,2,4-triazin-5-one |
| Procyazine | 2-chloro-4-(cyclopropylamino)-6-(1-cyano-1-methylethylamino)-*s*-triazine |
| Prometone | 2,4-bis(isopropylamino)-6-(methoxy)-*s*-triazine |
| Simazine | 2-chloro-4,6-bis(ethylamino)-*s*-triazine |
| Simetone | 4,4-bis(ethylamino)-6-(methoxy)-*s*-triazine |
| Terbutryn | 2-(*tert*-butylamino)-4-(ethylamino)-6-(methylthio)-*s*-triazine |
| Trietazine | 2-chloro-4-(ethylamino)-6-(diethylamino)-*s*-triazine |

mosorb W (80–100 mesh) and detected by multiple-ion monitoring (*m/z* 200 and 215 for atrazine and *m/z* 210 and 186 for simazine). Keith and co-workers[7] identified atrazine, simazine, and N-deethylated atrazine in water samples from New Orleans city reservoirs by GC/MS. Large volumes of water (up to 25.5 $m^3$) were extracted by passing through a carbon adsorption column; sorbed compounds were recovered by extraction of the activated carbon with chloroform. Sample extracts were chromatographed on a column (3.0 m × 2 mm i.d. glass) of 3% SP-2100 on Supelcon AW/DMCS (80–100 mesh) programmed from 40 to 280 °C (6 °C/min) or on a 30 m × 0.4 mm i.d. glass capillary column (containing SP-2100) and analyzed with a Finnigan 1015 spectrometer/data system. Atrazine, simazine, and N-deethylated atrazine were identified in the presence of a variety of volatile pesticides and industrial organics by computer matching of spectra and by relative retention time. Aaronson and co-workers[8] identified atrazine in pond water, from an area treated with cyanazine by use of GC/MS. Atrazine was chromatographed on a column 1.8 m × 4 mm i.d. packed with 4% SE-30/6% OV-210 on

Gas Chromosorb Q and identified by library search with a Finnigan 4023 data system.

Khan and Marriage[9] confirmed the presence of hydroxyatrazine and N-dealkylated hydroxytriazines in soil extracts by identification of their methylated derivatives using GC/MS. Atratone and N-deethylated atratone (formed by diazomethane methylation of their hydroxy analogs) were separated on a column (1.5 m × 4 mm i.d. glass) packed with 3% Carbowax 20M on Chromosorb W-HP (80–100 mesh) at 220 °C. Atratone was identified by a molecular ion at *m/z* 211 and loss of neutral methyl, ethylene, and isopropyl fragments from the alkylamino substituents on the triazine ring. Khan and Foster[10] separated atrazine and several methylated (diazomethane) hydroxytriazines on a column (1.5 m × 4 mm i.d.) of 0.3% EGA on Chromosorb W-HP (80–100 mesh) for identification of atrazine metabolites in chicken tissues by GC/MS. Lusby and Kearney[11] obtained the mass spectra of several trimethylsilyl (TMS) derivatives of degradation products of simazine and atrazine by GC/MS. The compounds were derivatized with *bis*(trimethylsilyl)trifluoroacetamide; following derivatization they were taken up in hexane and chromatographed on a column (1.8 m × 2 mm i.d.) packed with 3% OV-17 operated from 155 to 235 °C (4 °C/min). Mass spectra were obtained with Dupont 491 B GC/MS (70 eV) with a source temperature 220 °C. They proposed the fragmentation scheme outlined in Fig. 1 for *N*-ethylammeline (TMS derivatized), based on a detailed examination of the mass spectrum of this compound. The molecular ion (*m/z* 299) can lose a methyl radical and ethylene to give an ion at *m/z* 256. Loss, via McLafferty-type rearrangement, of the ethylamino group will also give an ion of mass 256. Loss of TMS-HNCN from the ion at *m/z* 284 resulted in an intense ion at *m/z* 170 due to a four-membered ring structure. Elimination of CO from the ion at *m/z* 199 could give rise to a five-member ring structure (*m/z* 171).

Stoks and Schwartz[12] analyzed TMS derivatives of the triazine derivatives, cyanuric acid, ammelide, ammeline, and melamine by GC/MS. The compounds were chromatographed on a column (1.5 m × 3 mm i.d. glass) containing 3% OV-17 on Chromosorb G-AW (100–120 mesh) at 140–190 °C (4 °C/min). The mass spectra of the TMS derivatives showed intense molecular ions at *m/z* 345, 344, 343, and 342 corresponding to *O,O,O-tris*(TMS)cyanuric acid, *N-O,O-tris*(TMS)ammelide, *N,N,O-tris*(TMS)ammeline and *N,N,N-tris*(TMS)melamine.

Bakke and co-workers[13] identified several hydroxy and N-dealkylated hydroxy metabolites of atrazine and hydroxyatrazine in rat urine by mass spectrometry of their TMS derivatives. The volatile TMS derivatives were chromatographed on a column (1.8 m × 2 mm i.d. glass) packed with 3% SE-30 on Chromosorb W (60/80 mesh) and collected

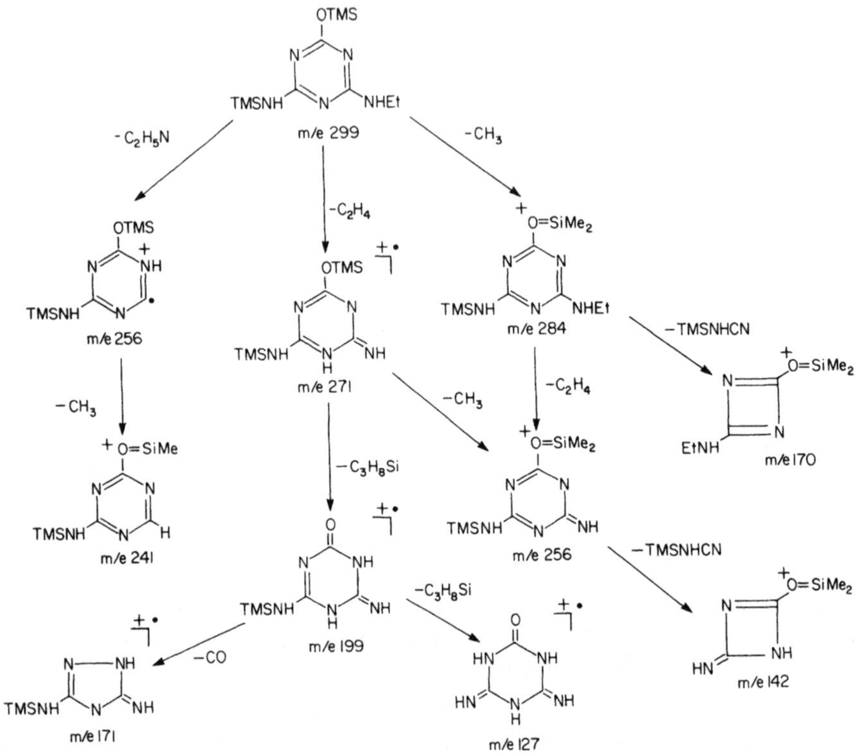


FIGURE 1. Proposed fragmentation scheme for the TMS derivative of *N*-ethylammeline. (From Ref. 11.)

individually from an effluent splitter in glass tubes for mass spectral analysis.

Studies on the metabolism of atrazine by sorghum shoots, utilized mass spectrometry for identification purposes.[14,15] Glutathione, lanthionine, and cystine conjugates of atrazine metabolites were prepared for mass spectrometry by purification on thin-layer chromatography (TLC) plates followed by alkylation or acetylation to yield volatile derivatives which were introduced into the mass spectrometer with a solid sample probe (heated from 25 to 250 °C). A key intermediate in the degradation pathway of atrazine in sorghum was tentatively identified as *S*-(4-ethylamino-6 -isopropylamino-*s*-triazinyl-2) cysteine. Preparation of larger amounts of the product by the action of -glutamyltranspeptidase on -glutamyl-*S*-(4-ethylamino-6-isopropylamino-*s*-triazinyl) cysteine followed by purification on cation exchange resin columns resulted in a cystine rather than the expected cysteine derivative because of dimerization of the product on the resin column. The mass spectrum of the

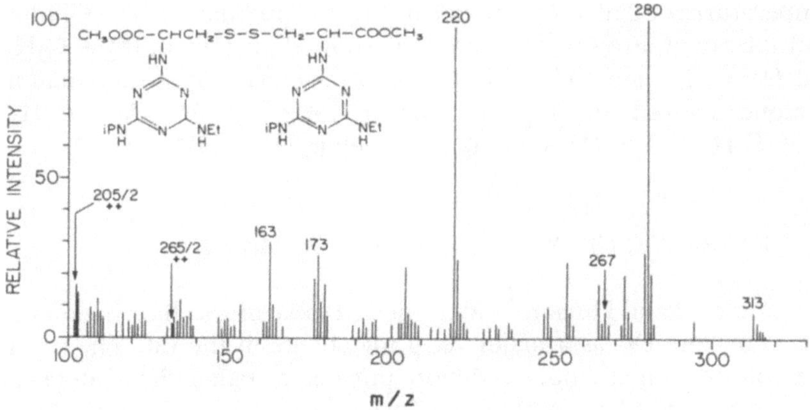


FIGURE 2. Mass spectrum of the dimethyl ester of a disulfide dimer of an atrazine conjugate. (From Ref. 15.)

dimethyl ester of the cystine derivative (Fig. 2) showed an ion at *m/z* 313 formed by cleavage of the disulfide bond. Metastable daughters of the *m/z* 313 ion were detected at 280 and 267 indicating a common origin for these ions. The intense peaks at *m/z* 280 and 220 correspond to cleavage to the disulfide bond and loss of $HCOOCH_3$ from the *m/z* 280 ion, respectively. The much less intense ion at *m/z* 267 was thought to be formed by cleavage to the disulfide bond. Loss of neutral fragments $CH_3^{\bullet}$, $C_2H_4^{\bullet}$, and $C_3H_6NH^{\bullet}$ from the alkylamino- side chains of the 280 and 220 ions gave rise to peaks at 205, 192, 178, and 163, respectively. Doubly charged ions at 102.5 and 132.5 correspond to the loss of the methyl radical from doubly charged ions of mass 220 and 280.

## 3. METHOXY-*s*-TRIAZINES

Bakke and co-workers[16] identified several N-dealkylated hydroxy metabolites of 2-methoxy-4-ethylamino-6-*sec*-butylamino-*s*-triazine in cow and goat tissues by mass spectral analysis of TMS derivatives. Urine and fecal metabolites were isolated by ion exchange or paper chromatography, derivatized with Regisil, and further purified by gas chromatography before being introduced into the mass spectrometer. Bakke and Price[17] used similar purification and derivatization techniques for identification of TMS derivatives of metabolites of prometone in rat urine.

LeClercq and Pacakova[18] reported the chemical ionization (CI) mass spectra of several bis(alkylamino)-*s*-triazines including atratone and simetone. The spectra were obtained on a Finnigan 4000 MS with a direct-sample introduction probe heated from 40 to 120 °C and source temperatures of 250 °C (methane) and 150 °C (isobutane). The CI (methane) spectra of atratone and simetone showed prominent $(M + C_2H_5)^+$ and $(M + H)^+$ ions. Using isobutane as a reagent gas, simetome and atratone showed small peaks due to $(M + C_4H_9)^+$, $(M + C_3H_7)^+$, $(M + C_3H_5)^+$, and $(M + C_3H_3)^+$, as well as $(M + H)^+$.

## 4. CYANAZINE, CYANATRYN, AND PROCYAZINE

Crayford and Hutson[19] identified a cyanazine dechlorination product, cyanazine IV, as a major fecal metabolite in the rat. The product was isolated by methanol extraction, purified by paper chromatography, and methylated (diazomethane) before direct introduction into the mass spectrometer. The mass spectrum of the methylated derivative of cyanazine IV (Fig. 3) showed a weak molecular ion at *m/z* 269, an intense peak at *m/z* 210 ($M^{\dot{+}} - COOCH_3$) and an abundant ion at *m/z* 169 due to a loss of a neutral $(CH_3)_2CHCOOCH_3$ group. Loss of the ethylamino group from the ion at *m/z* 169 with the transfer of a hydrogen to the ring gave rise to the ion at *m/z* 126. Elimination of methyl radicals from the prominent ions in the spectrum and from the molecular ion explained the presence of ions at *m/z* 254, 195, and 154.

Determination of cyanazine by GC/MS using selective-ion detection has been reported for confirmation of residues of the herbicide.[20] Cyanazine was chromatographed on a column (1.8 m × 4 mm i.d. glass) packed with a mixture of 3% OV-210 and 2% BDS (both on Supelcoport,

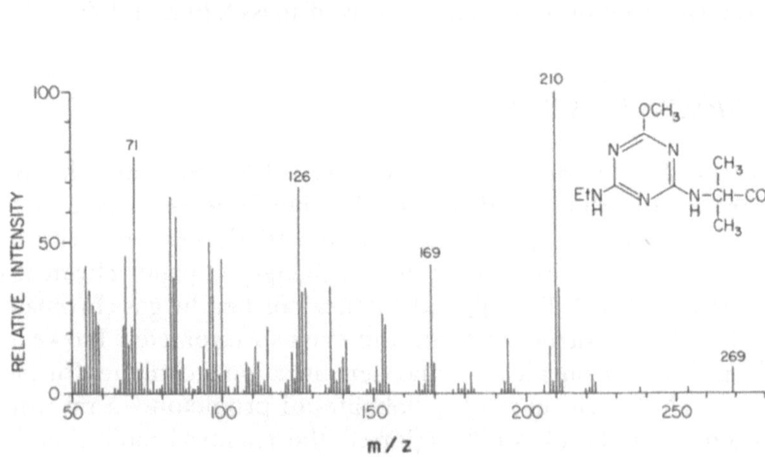


FIGURE 3. Mass spectrum of the dimethyl derivative of cyanazine IV. (From Ref. 22.)

80–100 mesh) which was operated at 225 °C and mass peaks at *m/z* 212 and 214 were monitored.

Bedford and co-workers[21] isolated a degradation product of cyanatryn having identical properties to the S-oxide derivative, from a rat liver microsome preparation. The compound was purified by TLC and derivatized with 3,4-dichlorothiophenol. GC/MS identification was carried out on a column (0.25 m × 2 mm i.d. glass) containing 4% OV-17 on Gas Chromosorb Q. The chlorinated derivative had molecular ions at *m/z* 382/384 and a major fragment at *m/z* 211. Erickson and co-workers[23] analyzed pig urine metabolites of procyazine by GC/MS. Volatile metabolites of procyazine were separated on a column (1.2 m × 4 mm i.d. glass) packed with 3% OV-101 and operated at 170–210 °C but could not be positively identified from the low-resolution mass spectra of each peak.

## 5. *CYPRAZINE*

Larsen and Bakke[24] isolated cyprazine metabolites from rat urine and identified them by a combination of gas chromatographic purification and mass spectrometry of their TMS derivatives. The silanized compounds were separated on a column (1.8 m × 2 mm i.d.) of 3% SE-30 on Chromosorb W with an effluent splitter (10% to detector), collected in glass tubes, and assayed by MS using a solid-sample inlet system. Two compounds (Fig. 4) had intense base peaks at *m/z* ($M^+$ − $CH_2OTMS$) which indicated the presence of primary alcohol groups on the alkylamino side chains. The positioning of the OH group on the cyclopropylamino substituent was thought unlikely because it would not yield a ($M^{+\cdot}$ − 103) fragment ion as a base peak without extensive rearrangement.

Muir and Baker[25] confirmed the presence of hydroxycyprazine in soil by GC/MS analyses of methylated soil extracts and comparison with

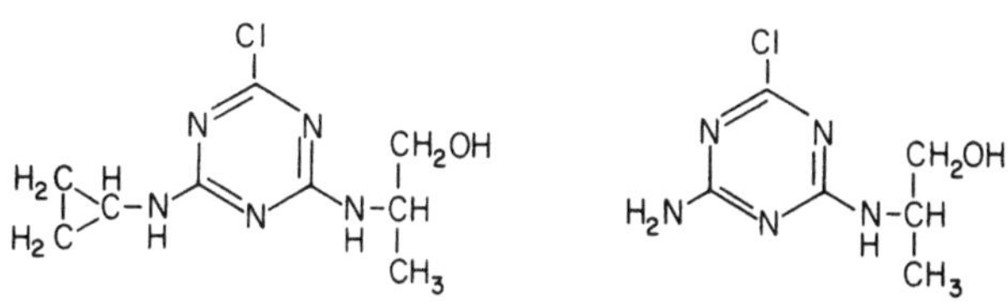

FIGURE 4. Structures of two cyprazine degradation products from rat urine. (From Ref. 24.)

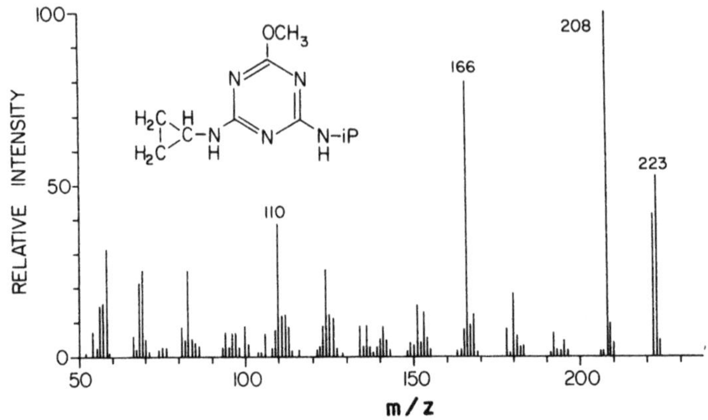


FIGURE 5. Mass spectrum of diazomethane-methylated hydroxy-cyprazine (cypratone). (From Ref. 22.)

the spectrum of authentic cypratone. The methylated soil extracts were chromatographed on a column (1.8 m × 4 mm i.d. glass) packed with 3% Carbowax 20M on Chromosorb W-HP (80–100 mesh) at 220 °C. Cypratone (Fig. 5) exhibited a molecular ion at *m/z* 223 and abundant ions at *m/z* 222, 208 (base peak), and 166 due to loss of H•, $CH_3^{\bullet}$, and the isopropylamino substituent (via McLafferty rearrangement in which the alkylamino group was replaced with a hydrogen), respectively.

## 6. *TERBUTRYN*

Larsen and co-workers[26] identified several terbutryn metabolites in rat and goat excreta by gas chromatography and mass spectrometry of their TMS derivatives. Extracts were derivatized, chromatographed on a column of 3% SE-30 on Chromosorb W (80–100 mesh; 1.8 m × 2 mm i.d.), and individual derivatives were collected from the GC effluent, in glass tubes, for mass spectral analysis. Comparison of the fragmentation of one of the oxidation products (Fig. 6) with other terbutryn degradation products revealed that the substituent on the sulfur influ-

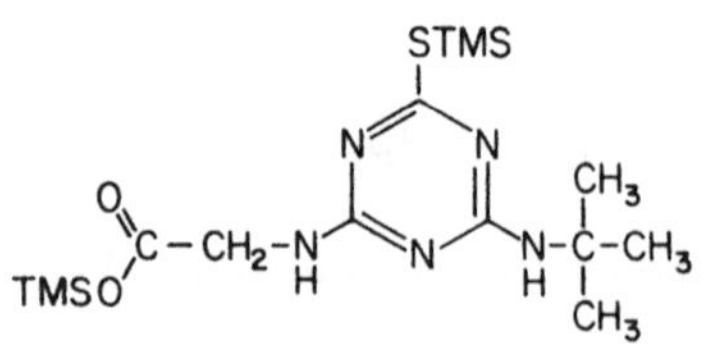


FIGURE 6. Structure of terbutryn degradation product in rat and goat urine. (From Ref. 26.)

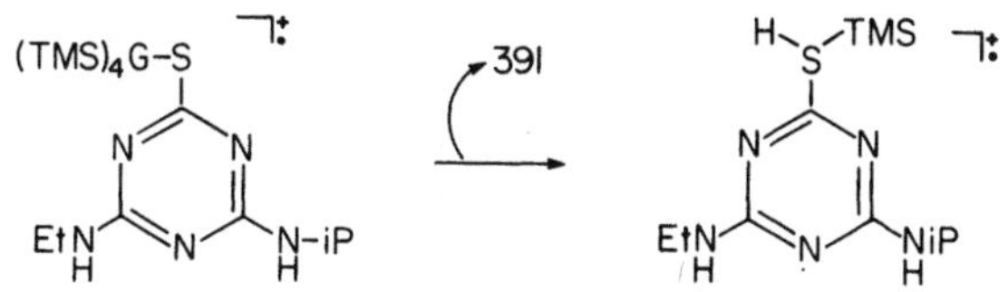


FIGURE 7. Fragmentation of *S*-glucuronide derivatives of terbutryn. (From Ref. 27.)

enced the type of fragment ions produced by the oxidized ethylamino side chains. The loss of $\cdot CH_2COOR$ (R = $CH_3$ or TMS) was not observed when the same compound having a thiomethyl group (instead of the S-TMS group shown in Fig. 6) was examined. Loss of the $CH_2COOTMS$ radical gave rise to a base peak at *m/z* 270. The *N*-triazinyl glycine structure was confirmed by the elimination of a *tert*-butyl radical [$CH_2 = C(CH_3)_2$] from the molecular ion to give an ion at *m/z* 345. Larsen and Bakke[27] identified glucuronide conjugates of terbutryn metabolites from goat and rat urine by interpretation of the mass spectra of the TMS derivatives. Compounds containing *S*-glucuronide substitution (Fig. 7) showed intense peaks at ($M^{+\cdot}$ − 391) due to loss of a $(TMS)_4$ glucuronic acid. An *N*-glucuronide structure was considered unlikely by the authors because of the instability of the N-glucosidic linkage in the acid conditions used to isolate the conjugates.

## 7. *DIALKYLAMINO-SUBSTITUTED CHLORO-s-TRIAZINES*

LeClercq and Pacakova[18] reported the mass spectra of three chloro-*s*-triazines, ipazine, trietazine, and chlorazine having *sec*-alkylamino substituents on the triazine ring. The compounds showed unique base peaks at ($M^{+\cdot}$ − $C_2H_5$), as well as the loss of $CH_3^{\cdot}$ and $C_2H_4$ which are always observed with mono-alkylamino-substituted *s*-triazines. Loss of $C_2H_6$ from the ($M^{+\cdot}$ − $C_2H_5$) ion was also observed. The CI spectra (methane) of ipazine and trietazine had prominent ($MH^+$ − HCl) ions while chlorazine showed a base peak at ($M^+$ + H). With isobutane as reagent gas the three triazines had prominent ($M^+$ + H) ions; all other fragments were less than 10% of the base peak.

## 8. *HEXAZINONE*

Reiser and co-workers[28] identified degradation products of the *s*-triazinone herbicide, hexazinone, in fish and rat tissues by MS. Tissue

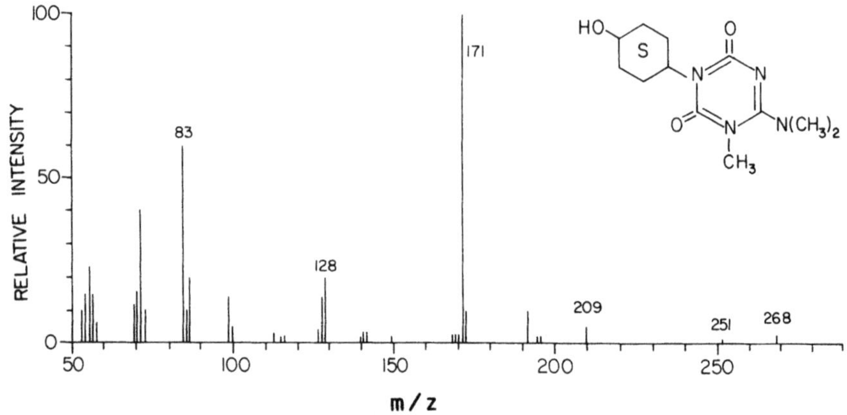


FIGURE 8. Mass spectrum of hexazinone metabolite (obtained on a Finnigan 1015 direct probe at 70 eV).

extracts were purified by TLC and the $^{14}C$-labeled metabolites were recovered from the plate, dissolved in methanol/chloroform, and analyzed by MS using direct-sample probe introduction. The mass spectrum of metabolite A, the major hexazinone transformation product in fish and in rat urine[(29)] (Fig. 8), was characterized by loss of a hydroxycyclohexyl moiety and the transfer of two hydrogens to the carbonyl oxygens through six-membered cyclic transition states to give a base peak at *m/z* 171. Loss of $CH_3^{\cdot}$ and $N(CH_3)_2^{\cdot}$ fragments from the molecular ion yielded a peak at *m/z* 209. A relatively abundant peak at *m/z* 128 was due to the loss of the dimethylamino fragment from the base peak with subsequent rearrangement of the hydrogen atom. The metabolite was identified as the 4-hydroxy isomer by comparison with the mass spectrum of the synthetic 4-hydroxy derivative.

## 9. *METRIBUZIN*

Pape and Zabik[(30)] analyzed metribuzin and its photoproducts in water, methanol, carbon tetrachloride, and benzene by use of GC/MS. The compounds were separated on a column (1.8 m × 2 mm i.d.) of 1% SE-30 on Gas Chromosorb Q (60–80 mesh) operated at 150–200 °C. The mass spectrum of DA, the major photoproduct of metribuzin, had a molecular ion at *m/z* 199, 15 mass units less than metribuzin, suggesting the loss of NH by photolysis. Abundant ions were observed at $M^{+\cdot} - SCH_3$ and $M^{+\cdot} - CH_3$. An ion at *m/z* 144 ($M^{+\cdot} - 55$) appeared to be due to loss of the *t*-butyl moiety with rearrangement of two hydrogens onto the triazine ring. DADK, the major metabolite of metribuzin in plants and soils[31] was also tentatively identified as a photoproduct by Pape and Zabik.[30] This compound (Fig. 9A) has a molecular ion at *m/z* 169 and a base peak at *m/z* 83 due to the $(CH_3)_3CCN$ fragment. Peaks at *m/z* 154 and 113 are due to the loss of $CH_3^{\cdot}$ and $CH_2$—$C(CH_3)_2$ radicals, respectively. The presence of a $NH_2$ group on the triazinone structure influences the fragmentation pattern of these compounds since DK (Fig. 9B) shows a base peak at *m/z* 57 due to the *t*-butyl radical. An intense ion at *m/z* 114 appears to be due to the loss of the *t*-butyl group from the *m/z* 169 ion with rearrangement of two hydrogens. Similar $M^{\ddagger} - 70$ ions were observed for metribuzin,[30] but not for DADK.

Prestel and co-workers[32] identified several degradation products of metribuzin in plant and soil extracts by use of GC/MS. The degradation products were isolated by TLC and were gas chromatographed on a column of 5% Carbowax 20M on Gas Chromosorb Q (130–180 °C) interfaced to an LKB 9000 mass spectrometer.

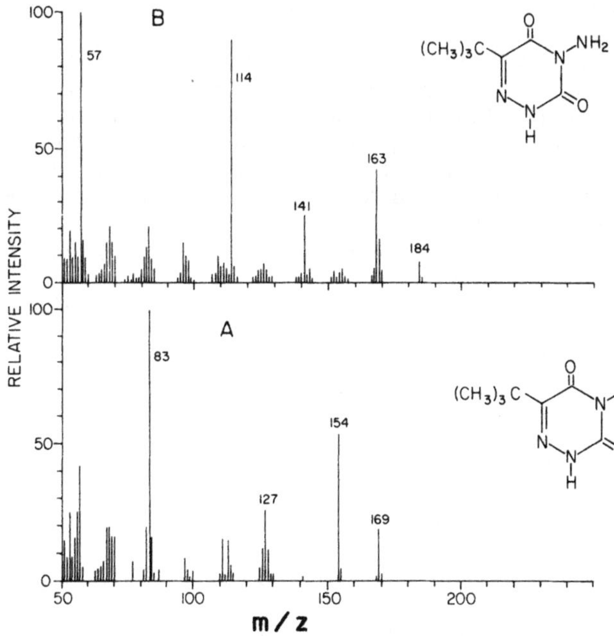


FIGURE 9. Mass spectrum of metribuzin degradation products (A) DADK and (B) DK (obtained on a Finnigan 1015 with direct probe at 70 eV).

Engelhardt and Wallnofer[33] identified desamino-metamitron in microbial cultures incubated with metamitron, an analog of metribuzin. Desamino-metamitron was isolated by TLC and introduced directly into a mass spectrometer following extraction from the TLC plate. The compound had a strong molecular ion at *m/z* 187 and a base peak at *m/z* 104 due to the loss of $C_6H_5NH$ from the molecular ion.

## REFERENCES


1. S. Safe and O. Hutzinger, *Mass Spectrometry of Pesticides and Pollutants,* CRC Press, Cleveland, 1973, pp. 169–175.
2. L. Fishbein, *Chromatogr. Rev.* **12,** 167–238 (1970).
3. Q. N. Porter and J. Baldas, *Mass Spectrometry of Heterocyclic Compounds,* Wiley-Interscience, New York, 1971.
4. B. E. Pape and M. J. Zabik, *J. Agric. Food Chem.* **18,** 202–207 (1970).
5. B. A. Karlhuber, W. D. Hormann, and K. A. Ramsteiner, *Anal. Chem.* **47,** 2450–2452 (1975).
6. R. Mestres, Ch. Chevallier, Cl. Espinoza, and R. Cornet, *Ann. Fals. Exp. Chim.* **70,** 177–188 (1977).
7. L. H. Keith, A. W. Garrison, F. R. Allen, M. H. Carter, T. L. Floyd, J. D. Pope, and A. D. Thruston, Jr., in *Identification and Analysis of Organic Pollutants in Water* (L. H. Keith, ed.), Ann Arbor Science, Ann Arbor, MI, 1976, pp. 329–373.
8. M. J. Aaronson, K. W. Kirby, and J. D. Tessari, *Bull. Environ. Contam. Toxicol.* **25,** 492–497 (1980).
9. S. U. Khan and P. B. Marriage, *J. Agric. Food Chem.* **25,** 1408–1413 (1977).
10. S. U. Khan and T. S. Foster, *J. Agric. Food Chem.* **24,** 768–771 (1976).
11. W. R. Lusby and P. C. Kearney, *J. Agric. Food Chem.* **26,** 635–638 (1978).
12. P. G. Stoks and A. W. Schwartz, *J. Chromatogr.* **168,** 455–462 (1979).
13. J. E. Bakke, J. D. Larson, and C. E. Price, *J. Agric. Food Chem.* **20,** 602–607 (1972).
14. G. L. Lamoureux, L. E. Stafford, R. E. Shimabukuro, and R. G. Zaylskie, *J. Agric. Food Chem.* **21,** 1020–1030 (1973).
15. R. H. Shimabukuro, W. C. Walsh, G. L. Lamoureux, and L. E. Stafford, *J. Agric. Food Chem.* **21,** 1031–1036 (1973).
16. J. E. Bakke, J. D. Robbins, and V. J. Feil, *J. Agric. Food Chem.* **19,** 462–466 (1971).
17. J. E. Bakke and C. E. Price, *J. Agric Ed. Chem.* **21,** 640–644 (1973).
18. P. A. LeClerq and V. Pacakova, *J. Chromatogr.* **178,** 193–207 (1979).
19. J. V. Crayford and D. H. Hutson, *Pestic. Biochem. Physiol.* **2,** 295–307 (1972).
20. Shell Development Corp., "Analytical Methods for Pesticides and Plant Growth Regulators," (G. Zweig and J. Sherma, eds.), 1978, pp. 275–292.
21. C. T. Bedford, M. J. Crawford, and D. H. Hutson, *Chemosphere* **5,** 311–316 (1975).
22. S. L. M. Skinner and R. Greenhalgh, *Mass Spectra of Insecticides, Herbicides and Fungicides and Their Metabolites,* Agriculture Canada, Chemistry and Biology Research Institute, Ottawa, 1977.
23. M. D. Erickson, C. W. Frank, and D. P. Morgan, *J. Agric. Food Chem.* **27,** 743–745 (1979).
24. G. L. Larsen and J. E. Bakke, *J. Agric. Food Chem.* **23,** 388–392 (1975).
25. D. C. G. Muir and B. E. Baker, *Weed Res.* **18,** 111–120 (1978).
26. G. L. Larsen, J. E. Bakke, and V. J. Feil, *Biomed. Mass Spectrom.* **5,** 382–390 (1978).
27. G. L. Larsen and J. E. Bakke, *Biomed. Mass Spectrom.* **5,** 391–394 (1978).

28. R. W. Reiser, I. J. Belasco, and R. C. Rhodes, Presented at American Society of Mass Spectrometry, Seattle, June, 1979.
29. R. C. Rhodes and R. A. Jewell, *J. Agric. Food Chem.* **28,** 303–306 (1980).
30. B. E. Pape and M. J. Zabik, *J. Agric. Food Chem.* **20,** 72–75 (1972).
31. J. S. Thornton and C. W. Stanley, *J. Agric. Food Chem.* **25,** 380–386 (1977).
32. D. Prestel, I. Weisgerber, W. Klein, and F. Korte, *Chemosphere* **5,** 137–144 (1976).
33. G. Englehardt and P. R. Wallnofer, *Chemosphere* **7,** 463–466 (1978).

# 20

# OTHER HERBICIDES
## DIPHENYL ETHERS, DINITROANILINES, DINITROPHENOLS, SUBSTITUTED BENZOIC ACIDS, PHOSPHONATES, URACILS, AND DIPYRIDYLIUMS

D. C. G. MUIR and G. R. B. WEBSTER

### 1. DIPHENYL ETHER HERBICIDES

#### 1.1. Introduction

Most diphenyl ether herbicides (Table 1) contain a halide or nitro-substituted ring (A), with the substituent in the ortho position, and a parasubstituted (B) ring (substituents such as $NO_2$, $CH_3$, alkyl acids, alkyl amides).(1) Mass spectral studies of these herbicides have focused mainly on the identification of metabolites of these compounds that have been isolated from plant and animal tissues. Diphenyl ether herbicides and many of their degradation products are volatile and thermally stable and can be analyzed by gas chromatography/mass spectrometry (GC/MS). However, for identification of metabolites most workers have used solid-probe sample introduction.

The mass spectra of diphenyl ethers have been discussed by Budzikiewicz and co-workers.(2) Diphenyl ethers exhibit a characteristic loss

*D. C. G. MUIR* • Department of Fisheries and Oceans, Freshwater Institute, 501 University Crescent, Winnipeg, Manitoba R3T 2N6, Canada. *G. R. B. WEBSTER* • Pesticide Research Laboratory, Department of Soil Science, University of Manitoba, Winnipeg, Manitoba R3T 2N2, Canada.

of 28 mass units due to the elimination of CO and rearrangement to yield a benzotropylium ion. More complicated diaryl ethers also exhibit M − H, M − CO, and M − HCO ions as well as the formation of aryl ions with the more highly substituted aryl ion predominating.[2] The presence of chlorine on the diphenyl ether rings results in a greatly altered fragmentation pattern in comparison with the unsubstituted molecule.[3]

## 1.2. *Fluorodifen*

Shimabukuro and co-workers[4] identified an *S*-glutathione metabolite of fluorodifen in peanut plants by mass spectrometry using a solid-sample probe (source, 210 °C, 70 eV). The mass spectrum of the hydrolysis product of this conjugate is shown in Fig. 1. Upon hydrolysis, the conjugate yielded a disulfide with a parent ion at *m/z* 444. The ratio of intensities at *m/z* 444/446 confirmed the presence of sulfur and the *M* − 19 peak suggested the loss of F, confirming a trifluoromethyl-substituted compound. A base peak at *m/z* 222 was due to the 2-nitro-4-trifluoromethylthiophenoxy ion. The relatively weak ion at *m/z* 334 was reported to be due to a 4,4′-di(trifluoromethyl)-2-nitro-biphenyl ion formed by rearrangement of the disulfide.

Lamoureux and Davison[5] reported the mass spectrum of the methyl ester of a mercapturic acid metabolite of fluorodifen isolated from rat urine.

The product was extracted from concentrated urine samples by cation exchange resin chromatography (AG50W × 2, Aminex), eluted with a pyridine-acetate buffer gradient and purified by thin-layer chro-

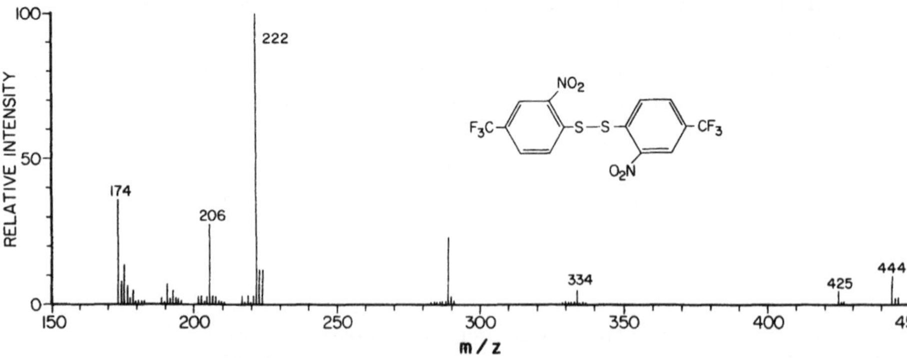


FIGURE 1. Mass spectrum (EI) of bis(2-nitro-4-trifluoromethylphenyl)disulfide derived from a glutathione conjugate of fluorodifen. (From Ref. 4.)

matography (TLC). Mass spectral analyses were carried out by solid-sample probe.

This compound underwent similar fragmentation to the disulfide product and was confirmed to be a mercapturic acid derivative by the presence of intense ions at *m/z* 88 and *m/z* 118 associated with the *N*-acetyl cysteine side chain of the fluorodifen metabolite.

Ruzo and co-workers[6] identified photoproducts of fluorodifen and several other related diphenyl ether herbicides in aqueous suspensions by GC/MS. The major photoproducts of fluorodifen were *p*-nitrophenol, which was identified by prominent ions at *m/z* 139 (molecular ion), (M − H), (M − OH), and (M − $NO_2$) and 4(trifluoromethyl)-2-aminophenol (molecular ion at *m/z* 177). The compounds were separated on a stainless steel column (0.7 m × 4 mm i.d.) containing a mixture of 15% OF-1 and 10% DC200 on Gas Chromosorb Q (80–100 mesh) which was operated at 70–230 °C (7–5 °C/min).

### 1.3. Diclofop-Methyl

Shimabukuro and co-workers[5] confirmed the identity of two metabolites of diclofop-methyl isolated from wild oat and wheat plants by mass spectrometry. The plants were extracted with aqueous methanol and the extract was partitioned with ether. The ether extract was purified by chromatography on Sephadex and XAD columns and fractions of the eluents were ethylated. The mass spectra were obtained using a solid-sample probe. Derivatization (ethanolic HCl) of one of the metabolites yielded the ethyl analog of diclofop-methyl (Fig. 2A) with a molecular ion at *m/z* 354, indicating that the metabolite was a carboxylic acid (diclofop). A base peak at *m/z* 120 was due to the phenoxyethyl ion ($C_6H_5OCHCH_3^+$). Intense ions at *m/z* 253 and 281 were due to the loss of $CH_3CHCOOCH_3$ and $COOCH_3$, respectively, from the molecular ion. The mass spectrum of the second diclofop-methyl metabolite (Fig. 2B) had an intense molecular ion (base peak) at *m/z* 356. Ions at *m/z* 297 and 269 were 16 mass units greater than the corresponding ions in the diclofop-methyl spectrum (Fig. 2B), suggesting the presence of oxygen on the diphenyl ether group. An ion at *m/z* 177 corresponding to $C_6H_3Cl_2O_2^+$ confirmed the presence of a hydroxy group on the dichlorophenoxy ring. The compound was identified as a hydroxy-ring conjugate because upon derivatization (diazomethane) only the carboxylic acid group was methylated, whereas after hydrolysis, methylation gave the dimethyl derivative. The exact location of the hydroxy group on the ring was not determined.

Gorbach and co-workers[7] reported the identification of carboxylic acid and hydroxy metabolites of diclofop-methyl using chemical ioni-

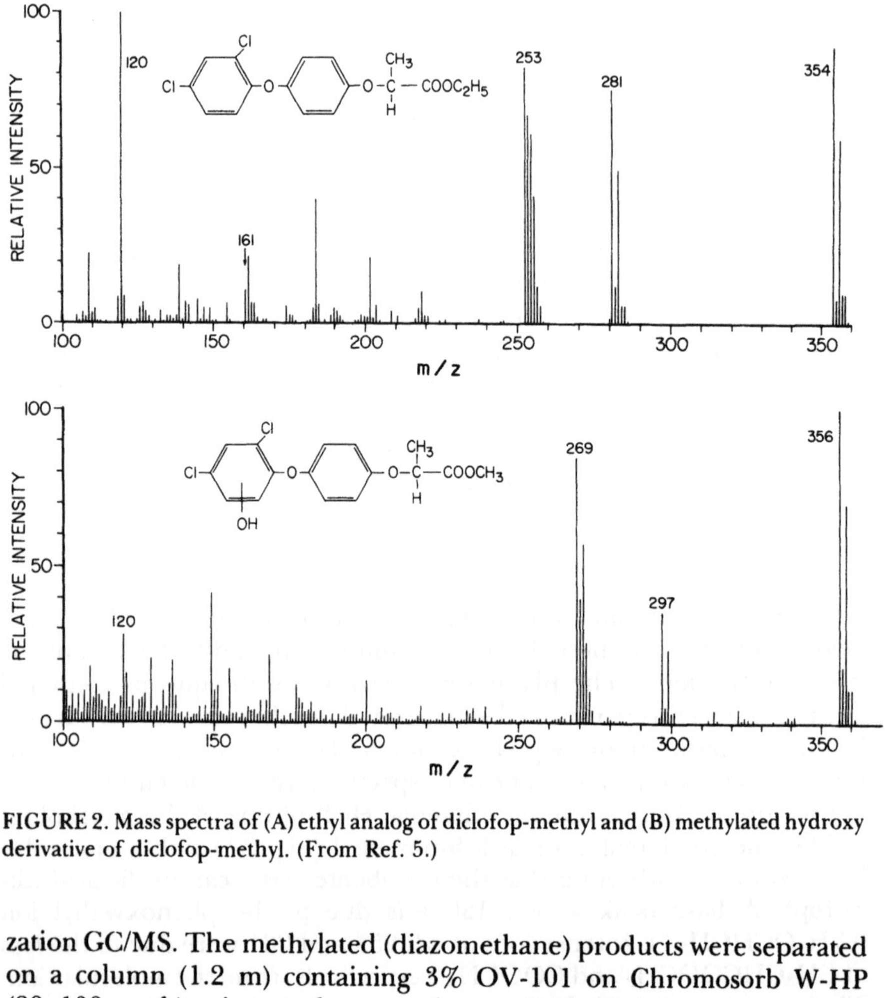


FIGURE 2. Mass spectra of (A) ethyl analog of diclofop-methyl and (B) methylated hydroxy derivative of diclofop-methyl. (From Ref. 5.)

zation GC/MS. The methylated (diazomethane) products were separated on a column (1.2 m) containing 3% OV-101 on Chromosorb W-HP (80–100 mesh) using methane carrier gas (15 ml/min) at an oven temperature of 150–300 °C. The dimethyl derivative of hydroxy-diclofop-methyl had an intense ion at *m/z* 371 (M + 1) and a base peak at *m/z* 311 due to the loss of $COOCH_3$ from the M + 1 ion. The authors concluded, from gas chromatographic data, that the compound was substituted at the 5-position on the 2,4-dichlorophenoxy ring.

### 1.4. *Oxyfluorfen, Nitrofen, and CNP*

Alder and co-workers[8] identified several degradation products of oxyfluorfen in rat tissue extracts by use of GC/MS. Actual gas chromatographic conditions were not reported. Oxyfluorfen had (Fig. 3) a base peak at *m/z* 252 due to the loss of Cl, $CH_2CH_2$, and $NO_2$ from the

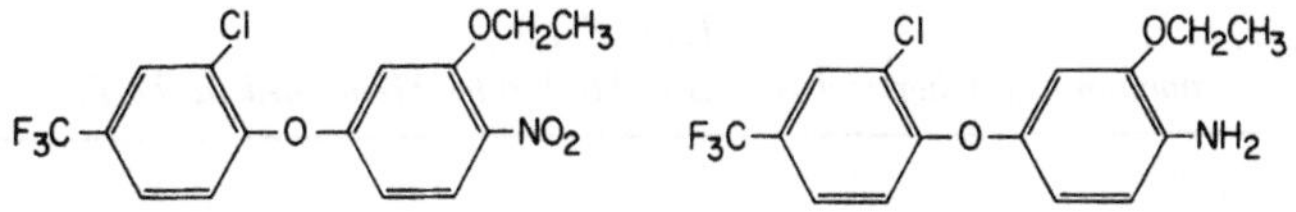


FIGURE 3. Structures of oxyfluorfen and amino oxyfluorfen.

molecular ion. An intense ion at *m/z* 69 was observed due to the $CF_3$ group. The 4-amino metabolite (Fig. 3) had a distinctly different mass spectrum possibly due to the influence of the $NH_2$ group. An intense ion observed at *m/z* 302 may be due to the loss of $CH_2CH_2$ and a hydrogen from the molecular ion. An ion at *m/z* 274 was observed which may result from the expulsion of CO from the *m/z* 302 ion. An ion resulting from diphenyl ether cleavage was also observed at *m/z* 123 ($ClC_6H_5^+$).

Nakagawa and Crosby[9] reported the identification by GC/MS of a large number of photoproducts of nitrofen in aqueous suspensions exposed to sunlight or simulated sunlight. A gas chromatographic column (1.0 m × 2 mm i.d. glass) containing 6% SE-30 on Chromosorb G (HMDS treated; 60–80 mesh) was used for separation of the photoproducts. The mass spectrum of nitrofen, unlike oxyfluorfen, was characterized by loss of NO and the presence of an abundant $C_6H_4Cl_2O^+$ ion (*m/z* 162). Kuwahara and coworkers[10] monitored nitrofen and CNP in soil extracts by mass spectrometry using a single-ion monitoring technique. Hexane extracts of soil were introduced into the mass spectrometer via a solid-sample probe and ions at *m/z* 283 and 317 were monitored for nitrofen and CNP, respectively.

Yamato and co-workers[11] reported the use of FDMS for the determination of CNP in environmental samples. Water extracts were applied to an FD emitter (prepared by high-temperature activation of a 10-micron tungsten wire under benzonitrile atmosphere) and the spectra were compared to those obtained with analytical standards of the herbicide. The FD spectrum of CNP had a base peak at the molecular ion (*m/z* 317) and intense peaks at M + 2, M + 4, and M + 6 due to the presence of three chlorines on the molecule.

## 2. *DINITROANILINE AND DINITROPHENOL HERBICIDES*

### 2.1. *Introduction*

Dinitroaniline herbicides are 2,6-dinitro-substituted anilines with alkyl substituents on the aniline nitrogen and various alkyl or alkyl halide substituents at the 4-position of the ring (Table 1). The mass spectrometry of various aromatic dinitro compounds has been discussed by Bud-

*TABLE 1*
*Common and Chemical Names of Herbicides Mentioned in the Text*

| Common name/class | Chemical name |
|---|---|
| *Diphenyl ethers* | |
| Bifenox | Methyl-5-(2,4-dichlorophenoxy)-2-nitrobenzoate |
| CNP | 2,4,6-trichlorophenoxy-4-nitrobenzene |
| Diclofop-methyl | Methyl-2-(4-(2′,4′-dichlorophenoxy)phenoxy)propanoate |
| Fluorodifen | 2-nitro-1-(4-nitrophenoxy)-4-(trifluoromethyl)benzene |
| Nitrofen | 2,4-dichloro-1-(4-nitrophenoxy)benzene |
| Oxyfluorfen | 2-chloro-1-(3-ethoxy-4-nitrophenoxy)-4-(trifluoromethyl)benzene |
| *Dinitroanilines and dinitrophenols* | |
| Benefin | *N*-butyl-*N*-ethyl-2,6-dinitro-4-(trifluoromethyl)benzenamine |
| Butralin | *N*-sec-butyl-*N*-*tert*-butyl-2,6-dinitro-4-(trifluoromethyl)-benzenamine |
| Chlornidine | *N*,*N*-(2-chloroethyl)-2,6-dinitro-4-toluidine |
| Dinitramine | $N^1$,$N^1$-diethyl-2,6-dinitro-4-(trifluoromethyl)-,3-benzenediamine |
| Fluchloralin | *N*-(2-chloroethyl)-2,6-dinitro-*N*-propyl-4-(trifluoromethyl)-benzenamine |
| Profluralin | *N*-(cyclopropylmethyl)-2,6-dinitro-*N*-propyl-4-(trifluoromethyl)-benzenamine |
| Trifluralin | 2,6-dinitro-*N*,*N*-dipropyl-4-(trifluoromethyl)benzenamine |
| Dinoseb | 2-(1-methylpropyl)-4,6-dinitrophenol |
| DNOC | 2-methyl-4,6-dinitrophenol |
| *Benzoic acids and benzonitriles* | |
| Chloramben | 3-amino-2,5-dichlorobenzoic acid |
| Dicamba | 3-6-dichloro-2-methoxybenzoic acid |
| Bromoxynil | 3,5-dibromo-4-hydroxybenzonitrile |
| Glyphosate | *N*-(phosphonomethyl)glycine |
| *Uracils* | |
| Bromacil | 5-bromo-6-methyl-3-(1-methylpropyl)-2,4(1*H*,3*H*)-pyrimidinedione |
| Terbacil | 5-chloro-3-(1,1-dimethylethyl)-6-methyl-2,4(1*H*,3*H*)-pyrimidinedione |
| *Pyridyliums* | |
| Cyperquat | 1-methyl-4-phenylpyridinium chloride |
| Diquat | 6,7-dihydropyridol(1,2-*a*:2′,1-*c*)pyrazinediium cation |
| Paraquat | 1,1′-dimethyl-4,4′-bipyridinium ion |

zikiewicz and co-workers[2] and by Plimmer and Klingebeil.[12] Golab and Occolowitz[13] presented the mass spectra of 28 compounds which had been identified as degradation products of trifluralin in soil and discussed the fragmentation of these products. The present discussion will focus on the application of GC/MS for the identification of trifluralin and other dinitroanilines and their degradation products.

Dinitrophenol herbicides share some common structural features with dinitroaniline herbicides. In fact, a dinitrophenol metabolite of trifluralin has been isolated from soil.(13) The metabolism of dinitrophenols such as DNOC and dinoseb has been studied extensively,(14) but there are few published reports on the use of mass spectrometry for the identification of these herbicides or their degradation products.

### 2.2. *Dinitroanilines*

Golab and co-workers(15) and Golab and Occolowitz(13) employed GC/MS and direct-probe sample introduction for the identification of trifluralin and a large number of soil degradation products by mass spectrometry. Soil samples from field plots treated with $^{14}$C-trifluralin were extracted with methanol and the extracts were chromatographed on thin-layer plates. Radioactive products were further purified by column chromatography, TLC, and high-pressure liquid chromatography. The mass spectra of trifluralin and other volatile degradation products were obtained by GC/MS using columns (1.8 m × 3 mm i.d.) packed with 3 and 5% Carbowax 20M or 1.5 and 5% XE 60 both on Chromosorb W-HP (100–200 mesh) operated at ambient to 240 °C.

The major ions in the mass spectrum of trifluralin (Fig. 4) were observed at *m/z* 306 due to (M − $C_2H_5$) and at *m/z* 43 (base peak) due to the *n*-propyl ion. Loss of *n*-propyl from the *m/z* 306 ion with hydrogen rearrangement yielded the ion at *m/z* 264. Similar fragmentation was observed for other *N-N*-dipropyl amines that were degradation products of trifluralin.(13) The ion at *m/z* 57 was found by high-resolution MS to have the composition $C_3H_5O$. This ion was produced in dinitroanilines having at least one *N*-propyl group and an ortho-substituted nitro group and was believed to arise by intraionic oxidation of the aliphatic group.

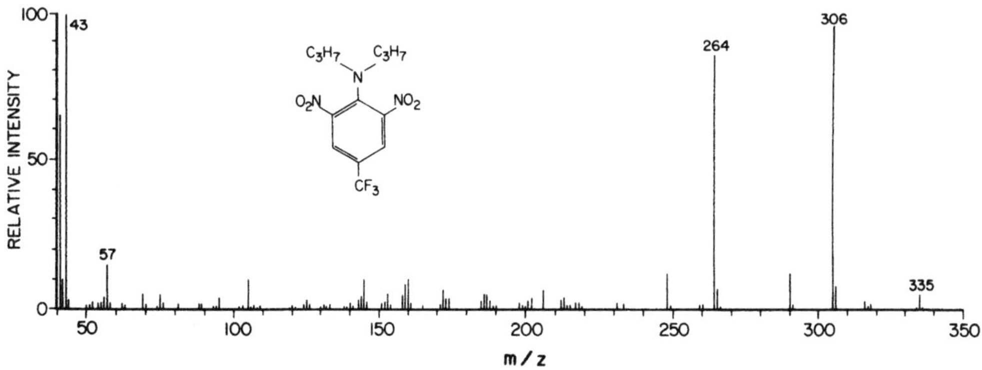


FIGURE 4. Mass spectrum (EI) of trifluralin. (From Ref. 13.)

Additional confirmation of this mechanism was obtained by mass spectral determinations using the deuterated analog of N-depropylated trifluralin. The presence of the base peak at *m/z* 43 was thought to arise from a $C_3H_7^+$ ion generated by a charge site reaction[16] in which the electron-withdrawing nitro groups destabilize the molecular ion and the condition is relieved by a charge site reaction yielding an aniline radical.

Plimmer and Klingebeil[12] obtained low-resolution mass spectra of dinitroaniline derivatives including trifluralin and dinitramine by GC/MS. The compounds were separated on an open tubular capillary column (15 m × 0.51 mm i.d.) coated with SE-30 on Chromosorb W. Downer and co-workers[17] reported the analysis of benefin and trifluralin in soil extracts by quantitative GC/MS. The two herbicides were chromatographed on a column (1.5 m × 4 mm i.d. glass) containing 1.2% SE-30 on Gas Chromosorb Q (100–120 mesh) operated at 190 °C. The compounds had identical retention times under these conditions but could be identified individually by monitoring characteristic fragments at *m/z* 292 (benefin) and *m/z* 306 (trifluralin). Benefin and trifluralin were determined at 0.01 to 0.08 g/g levels in soils by use of this multiple-ion detection procedure.

Kearney and co-workers[18] identified soil degradation products of trifluralin, profluralin, dinitramine, butralin, fluchloralin, and chlornidine by use of GC/MS and comparison of mass spectra with those of authentic standards. The compounds were chromatographed on a column (1.85 m × 6 mm i.d. glass) containing 5% SE-30 on Chromosorb W, operated at 180 °C, and full spectra of each product were obtained. Kearney and co-workers[19] obtained high- and low-resolution mass spectra of butralin metabolites in soil and in culture solutions of a soil fungus. The compounds were isolated by column chromatography and TLC before direct-probe mass spectrometry or GC/MS. The major metabolite of butralin had a molecular formula of $C_{14}H_{21}N_3O_5$ indicating the addition of one oxygen to the molecule and a base peak at *m/z* 266 ($C_{12}H_{16}N_3O_4$) due to loss of a $C_2H_5O$ moiety from the *N*-alkyl chain of the molecular ion. The compound was subsequently identified as a secondary alcohol metabolite by nuclear magnetic resonance studies.

Lusby and co-workers[20] identified metabolites of 2,6-dinitro-4-(trifluoromethyl) benzenamine, a major breakdown product of several dinitroaniline herbicides, in bacterial cultures by use of high- and low-resolution mass spectrometry. The culture solutions were extracted with ethyl acetate and the extract was chromatographed on a dry column of silica to isolate the metabolites. Samples were analyzed by direct-probe introduction into the mass spectrometer.

Heck and co-workers[21] determined deuterated trifluralin in rat tissue extracts by field ionization mass spectrometry. Trifluralin was extracted by homogenizing rat tissue with hexane–ether (9 : 1) and separated from lipid coextractives by high-pressure gel permeation chromatography. Mass spectra were obtained with a multipoint field ionization source operated at an ionizing potential of 1.5 kV, using a solid-sample inlet probe for sample introduction.

### 2.3. *Dinitrophenols*

Nojima and co-workers[22] have reported the detection by GC/MS of the methylated derivative of DNOC in rainwater, following ultraviolet irradiation. The compounds were separated on a 2.0-m column containing 10% SE-30 on Uniport B and specific-ion monitoring was used. The partial mass spectrum (major ions) of 2-methyl-4,6-dinitroanisole was characterized by an intense molecular ion, which is typical of aromatic ethers,[2] and a base peak of M − NO. An intense ion at *m/z* 165 was due to the loss of $NO_2$ and hydrogen from the molecular ion; this fragmentation pattern was also observed with trifluoro-methyl-2,6-dinitro-*p*-cresol, a soil metabolite of trifluralin.[13]

Skinner and Greenhalgh[23] determined the mass spectra of dinoseb and several other dinitrophenol herbicides by use of direct-probe sample introduction. The mass spectrum of dinoseb (Fig. 5) has an intense ion at *m/z* 211 due to the loss of $CH_2CH_3$ from the alkyl side chain. Loss of $H_2$, a characteristic of ortho-substituted alkyl phenols,[16] and NO from the *m/z* 211 ion results in an intense fragment at *m/z* 163. Abundant ions at *m/z* 147 and 117 appear to be the result of elimination of $NO_2$ from the *m/z* 211 and 163 ions, respectively. Loss of CO from the *m/z* 117 ion

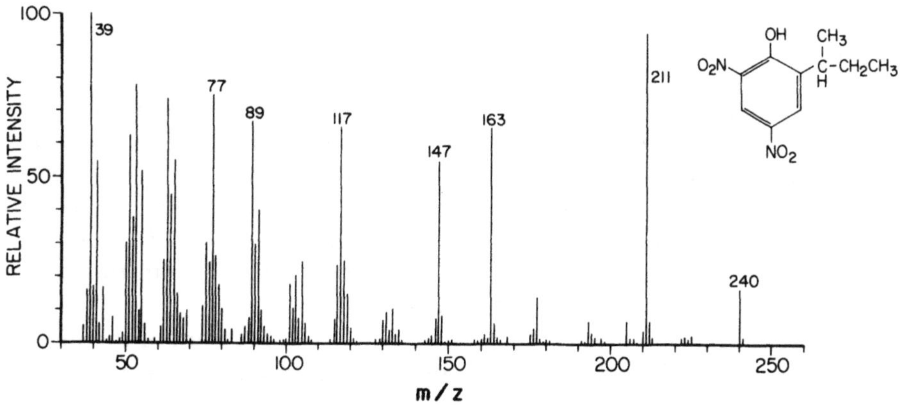


FIGURE 5. Mass spectrum (EI) of dinoseb. (From Ref. 23.)

may give rise to the ion at *m/z* 89, while the series of intense ions at *m/z* 39, 51, 53, 63, and 77 is due to fragmentation of the aryl group. Other dinitrophenol herbicides such as dinitrocresol (DNOC) and dinosam are structurally similar to dinoseb and would be expected to have similar mass spectral and gas chromatographic characteristics.[14,24]

Schulten[25] reported the field desorption mass spectrum of DNOC. The spectrum was produced on a modified CEC-21 110B instrument using a photographic detection system. The field desorption emitter was prepared by high-temperature activation of a 10-m tungsten wire. The herbicide had a base peak at the molecular ion (*m/z* 198). No fragmentation of DNOC was observed by this technique because desorption occurred before enough thermal energy could be transferred to induce fragmentation.[25]

## 3. SUBSTITUTED BENZOIC ACID AND BENZONITRILE HERBICIDES

### 3.1. Introduction

The substituted benzoic acid herbicides chloramben and dicamba are widely used for control for broad-leaved and grassy weeds in soybean and cereal crops, respectively.[26] Bromoxynil, a bromine-substituted benzonitrile, is used for broadleaf weed control in cereals. Mass spectrometry has not been used extensively for residue analysis or metabolism studies with these compounds with the exception of studies of chloramben degradation in plants.[27] Skinner and Greenhalgh[23] have reported the mass spectra of several benzoic acids and benzonitrile herbicides, using solid-sample probe determinations of analytical standards.

### 3.2. Chloramben

Skinner and Greenhalgh[23] reported the mass spectrum of chloramben (Fig. 6). The mass spectrum has an intense molecular ion at *m/z* 205 and prominent M + 2 and M + 4 peaks indicative of a compound containing two chlorines. Prominent ions at *m/z* 188 (M − OH) and *m/z* 160 (M − $CO_2$, − H) are also observed. Loss of HC1 from the *m/z* 160 ions accounts for the ion at *m/z* 90. Loss of HCN from the *m/z* 90 ion may give rise to a $C_5H_3^+$ fragment (*m/z* 63).

Frear and co-workers[27] obtained the mass spectrum of the methyl ester of chloramben by GC/MS. The compound was chromatographed on a column (1.8 m × 2 mm i.d. glass) which contained 1.5% OV-17 on Gas Chromosorb Q (operated at 175–225 °C at 5 °C/min). The ester had a molecular ion at *m/z* 219 and a base peak at *m/z* 188 (M − $OCH_3$). Additional ion fragments at *m/z* 160 and *m/z* 124 were identical to those

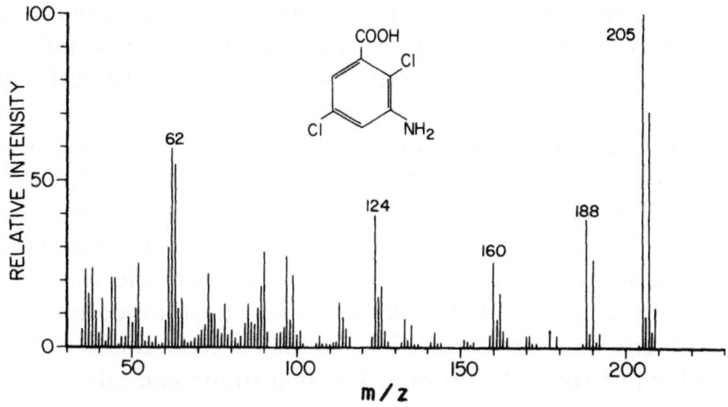


FIGURE 6. Mass spectrum of chloramben. (From Ref. 23.)

in chloramben itself. Chloramben-amide had a molecular ion and a base peak at *m/z* 204. Other ions at *m/z* 188 (M − $NH_2$) and *m/z* 160 (M − $CONH_2$, − Cl) were the same as those observed for chloramben. The authors also isolated a glucose ester of chloramben from plants. The mass spectrum of the peracetylated derivative of this compound had a small molecular ion corresponding to the pentaacetate derivative (*m/z* 577) and characteristic chloramben ions at *m/z* 160.

### *3.3. Bromoxynil*

Skinner and Greenhalgh[23] reported the mass spectrum of bromoxynil (Fig. 7). Bromoxynil has a base peak at the M + 2 molecular ion (*m/z* 277). The intense M + 2 and M + 4 peaks are characteristic of

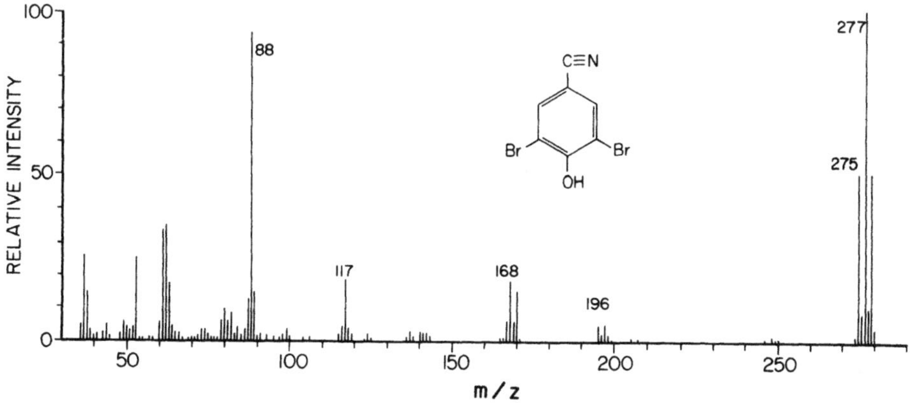


FIGURE 7. Mass spectrum of bromoxynil. (From Ref. 23.)

a compound containing two bromines. Loss of HCN results in a minor peak at *m/z* 247, characteristic of a cyano-containing compound.[16] Ions at *m/z* 196 and *m/z* 168 are due to the loss of a bromine radical from the molecular ion and the ion at *m/z* 247, respectively. Loss of two bromines from the molecular ion results in the ion at *m/z* 117. The intense ion at *m/z* 88 may be due to a $C_6H_5N^+$ ion resulting from the elimination of CO from *m/z* 117, a characteristic of phenols. Tong and co-workers[28] reported the determination of bromoxynil in soil extracts (0.03–510 g/kg) by spark source mass spectrometry. The herbicide was extracted from soil with acetone in a Soxhlet extractor, partitioned into hexane, and the solvent was concentrated for mass spectral analysis. The concentrated hexane solution was mixed with graphite and taken to dryness on a gold wire which was sparked against a second gold electrode.

### 3.4. *Dicamba*

Oehler and Ivie[29] identified several degradation products of dicamba in the urine and feces of a lactating cow by use of GC/MS. Dicamba and its major metabolite, 2,6-dichloro-2-hydroxybenzoic acid were chromatographed on a column (1.8 m × 2 mm i.d.) of 3% SE-30 on Varaport 30 (100–200 mesh) at a temperature of 170 °C. A second metabolite, 2,5-dichlorophenol, was chromatographed on the same column and confirmed by comparison of mass spectra with that of an authentic standard.

## 4. *PHOSPHONATE HERBICIDES*

### 4.1. *Introduction*

The most widely used phosphonate herbicide is glyphosate, a broad-spectrum herbicide. Determination of glyphosate residues in soil and water is difficult because it is nonvolatile and strongly sorbed to soil particles. Mass spectrometry of the alkylated *N*-trifluoroacetylated derivative offers one means of analysis of the herbicide. The mass spectra of volatile derivatives of glyphosate and its major degradation product, aminomethylphosphonic acid, have been described by Rueppel and co-workers[30,31]

### 4.2. *Glyphosate*

Rueppel and co-workers[30] reacted glyphosate sequentially with diazo-*n*-butane and trifluoroacetic anhydride to produce the tri-*n*-butyl-*N*-TFA derivative of the herbicide. Aminomethyl phosphonic acid gave rise to a di-*n*-butyl-*N*-TFA derivative under the same reaction conditions. The derivatives were chromatographed on a column (1.8 m × 2 mm i.d.) containing 1.5% OV-17 which was temperature programmed from 120 °C (10 °C/min). The column was connected to the mass spectrometer via a Watson–Biemann two-stage separator held at 300 °C. The mass spectrum of tri-*n*-butyl-*N*-TFA glyphosate (Fig. 8) was characterized by cleavage of the *n*-butyl groups from the molecular ion (*m/z* 433) with rearrangement of hydrogens to give ions at *m/z* 378 (M − $C_4H_7$), *m/z* 322 (M − 2 ($C_4H_7$)) and *m/z* 266 (*m/z* 322 − $C_4H_8$). A base peak at *m/z* 247 resulted from the loss of butanol and two butenes from the molecular ion. The fragmentation pathways were confirmed by high-resolution mass spectrometry and by preparation of $^{13}C$ and $^{14}C$ derivatives as well as *N*-heptafluoro derivatives of glyphosate. Di-*n*-butyl-*N*-TFA-aminomethyl phosphonate showed similar fragmentation to that observed for the glyphosate derivative but the molecular ion was not observed.[30]

The trimethyl *N*-TFA glyphosate derivative, which is prepared for analytical determinations of the herbicide in soil extracts,[32] has similar mass spectral fragmentation to the tri-*n*-butyl ester.[33] The trimethyl ester has a molecular ion at *m/z* 307 and prominent ions at *m/z* 275 (M − $CH_3OH$), *m/z* 248 (M − $CH_3OCO$), and *m/z* 210 (M − $CF_3CO$). A base peak at *m/z* 124 (on a scan of *m/z* 100–350) may be due to a $HNCH_2PO(OCH_3)_2^+$ ion which loses $CH_3$ to give an ion at *m/z* 109.

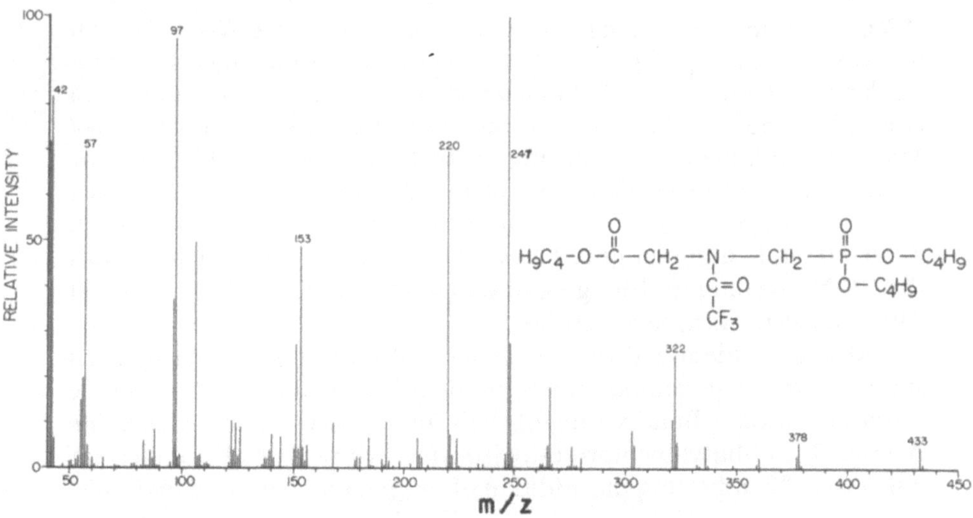


FIGURE 8. Mass spectrum of tri-*n*-butyl-*N*-TFA phosphonomethyl glycine. (From Ref. 31.)

## 5. URACIL HERBICIDES

### 5.1. Introduction

The substituted uracils, terbacil and bromacil, are widely used for control of annual and perennial weeds in sugarcane and fruit crops. The mass spectral characteristics of substituted uracils have been discussed in detail by Reiser[34] and reviewed by Safe and Hutzinger.[35] In more recent work GC/MS has been used for determination of residues of these herbicides in soil, plants, and animals.

### 5.2. Terbacil and Bromacil

Tong and co-workers[28] used spark source mass spectrometry for the determination of bromacil in soil extracts by detection of bromine atoms on the molecule. Bromacil was extracted from soil by Soxhlet extraction with acetone and partitioned into hexane. The hexane solution was concentrated, mixed with graphite, and evaporated to dryness on a gold wire as described for the analysis of bromoxynil. Bromacil was determined in fortified soil samples at concentrations ranging from 0.05 to 670 g/kg (bromine).

Khan[36] reported the determination of terbacil in soil by gas chromatography with electron capture detection and by GC/MS. Terbacil was extracted from soil by shaking with 1.5% NaOH, partitioned into ethyl acetate, and separated from coextractive material by column chromatography on Florisil. Extracts were analyzed by GC/MS using a column (1.5 m × 4 mm i.d.) containing 3% OV-17 on Chromosorb W-HP (80–100 mesh) which was operated at 215 °C. Terbacil was identified by a molecular ion at *m/z* 216, and characteristic ions at M + 2 (due to one Cl atom), M − 55, and M − 56. Suzuki and co-workers[37] compared GC/MS using multiple-ion monitoring, with electron capture detection for determination of bromacil in soil, plant, and animal samples. Bromacil was chromatographed on a column of 5% XE-60 on Gas Chromosorb Z (60–80 mesh) and monitored at *m/z* 260 and 205. Detection limits with the GC/MS technique (1.0 μg/kg in soil) were superior to those obtained with the electron capture detector.

Rhodes[38] identified two metabolites of terbacil in alfalfa by mass spectrometry. The compounds were identified by direct-probe sample introduction on a Bendix time-of-flight mass spectrometer. Metabolite A (Fig. 9) (3-*t*-butyl-5-chloro-6-hydroxymethyluracil) had a molecular ion at *m/z* 232 indicating the addition of oxygen to the terbacil molecule. Major peaks at *m/z* 177 and 176 (M − 55; M − 56) were due to the loss of the butyl group with a double or single H rearrangement as observed for terbacil.

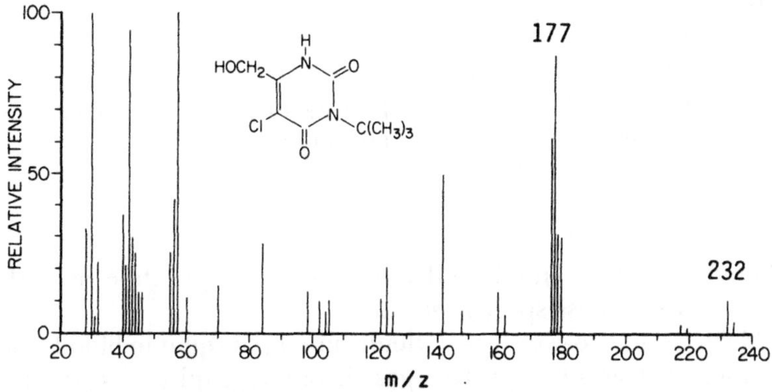


FIGURE 9. Mass spectrum of terbacil metabolite A. (From Ref. 38.)

## 6. PYRIDYLIUM HERBICIDES

### 6.1. Introduction

The dipyridylium herbicides, diquat and paraquat, are widely used as preharvest crop desiccants and for postemergent nonselective weed control in a variety of crops.[39] Both herbicides have also been used extensively for aquatic weed control. Diquat and paraquat as well as the other pyridylium herbicide cyperquat are nonvolatile cationic salts unsuitable for direct gas chromatographic separation. Interest in the fate of these herbicides in soil, water, and in cases of human poisoning has led to the development of methods for preparation of volatile derivatives and to the use of GC/MS techniques for residue analysis.

### 6.2. Diquat, Paraquat, and Cyperquat

Khan[40] described an analytical procedure for the determination of diquat and paraquat in soil extracts by gas chromatography with alkali-flame or mass spectrometer detection. Diquat and paraquat were hydrogenated ($PtO_2$ catalyst) to perhydro-dipyrido (1,2a : 2,1′c) pyrazine (cis and trans isomers) and 1,1′-dimethyl-4,4′-bipiperidine, respectively. The derivatives were separated on a column (1.5 m × 4 mm i.d. glass) packed with 3% SE-30 or with 3% Carbowax 20M + 1% KOH, both on Chromosorb W-HP, at 140–160 °C. Mass spectra of the pyrazine and bipiperidine derivatives showed parent ions at *m/z* 194 and 196, respectively. Hydrogenation of cyperquat[41] yielded two compounds; GC/MS analysis indicated that these compounds had molecular ions at *m/z* 181 and 175, respectively. The two compounds were identified as 1-methyl-

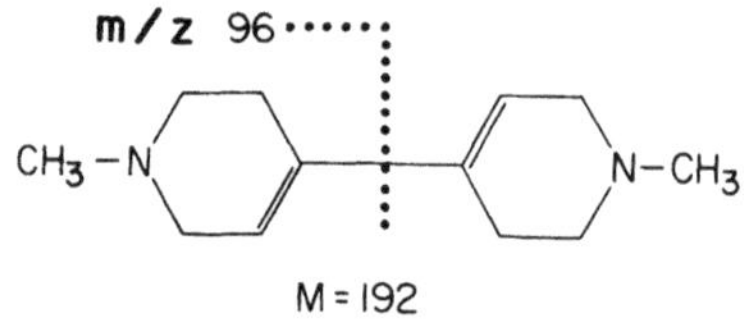


FIGURE 10. Structure of paraquat reduction product. (From Ref. 43.)

4-phenylpiperidine and 1-methyl-4-cyclohexylpiperidine by NMR and high-resolution mass spectral studies.

Martens and co-workers(42) determined paraquat in blood urine and lung tissue extracts by pyrolysis gas chromatography/mass fragmentography. Paraquat was thermally dealkylated to 4,4′-bipyridine in the injection port of the gas chromatograph at 400 °C and the pyrolysis product was determined quantitatively by selective-ion monitoring (*m/z* 156). Draffan and co-workers(43) determined paraquat in human plasma by use of GC/MS. Paraquat was reduced with sodium borohydride which yielded an ether-soluble product; the mass spectrum of this product had a molecular ion at *m/z* 192 and a base peak at *m/z* 96 indicating a *bis*(dehydropiperidine) structure (Fig. 10).

The reduction product was chromatographed on a column (1.5 m × 2 mm i.d. glass) containing 3% OV-1 on Gas Chromosorb Q (80–100 mesh) at an oven temperature of 180 °C and detected by monitoring the molecular ion (*m/z* 192) and the M − 2 ion. The use of GC/MS enabled detection of paraquat at trace (5–10 ng/ml) levels in plasma. Wall and Brine(44) used similar procedures for the determination of paraquat in contaminated marihuana samples by GC/MS.

Hughes and co-workers(45) compared field desorption, field ionization, and electron impact mass spectrometry for the determination of paraquat and diquat. They concluded that field desorption proved superior to other procedures for mass spectral analysis of these herbicides by use of direct-probe sample introduction. McEwen and co-workers(46) reported the electron impact mass spectrum of paraquat as part of a study on the use of mass spectrometry for the determination of structure of organic salts. Mass spectra of paraquat (dichloride salt) were found to contain ions at *m/z* 186, 171, and 165 formed by thermally promoted electron transfer and demethylation of the herbicide. Ryan and co-workers(47) used secondary ion mass spectrometry for the characterization of paraquat, diquat, and other organic salts. The herbicides were burnished onto a silver or platinum foil, exposed to a beam of argon ions, and the mass spectra obtained using a quadrupole mass analyzer.

## REFERENCES

1. S. Matsunaka, *Herbicides,* Vol. 2 (P. C. Kearney and D. D. Kaufman, eds.), Marcel Dekker, New York, 1976, pp. 709–740.
2. H. Budzikiewicz, C. Djerassi, and D. H. Williams, *Mass Spectrometry of Organic Compounds,* Holden-Day, San Francisco, 1967.
3. A. Curley, R. W. Jennings, V. W. Burse, and E. C. Villanueva, *Mass Spectrometry and NMR Spectrometry in Pesticide Chemistry,* (R. Haque and F. J. Biros, eds.), Plenum, New York, 1974, pp. 71–80.
4. R. H. Shimabukuro, G. L. Lamoureux, H. R. Swanson, W. C. Walsh, L. E. Stafford, and D. S. Frear, *Pestic. Biochem. Physiol.* **3,** 483–494 (1973).
5. R. H. Shimabukuro, W. C. Walsh, and R. A. Hoerauf, *J. Agric. Food Chem.* **27,** 615–623 (1979).
6. L. O. Ruzo, J. K. Lee, and M. H. Zabik, *J. Agric. Food Chem.* **28,** 1289–1292 (1980).
7. S. G. Gorbach, K. Kuenzler, and J. Asshauer, *J. Agric. Food Chem.* **25,** 507–511 (1977).
8. I. L. Alder, B. M. Jones, and J. P. Wargo, Jr., *J. Agric. Food Chem.* **25,** 1339–1341 (1977).
9. M. Nakagawa and D. G. Crosby, *J. Agric. Food Chem.* **22,** 849–853 (1974).
10. R. Kuwahara, T. Suzuki, and H. Meguro, *Agric. Biol. Chem.* **44,** 1195–1196 (1980).
11. Y. Yamato, M. Suzuki, and X. Watanabe, *Biomed. Mass Spectrom.* **6,** 205–207 (1979).
12. J. R. V. Plimmer and U. K. Klingebeil, in *Mass Spectrometry and NMR Spectroscopy in Pesticide Chemistry* (R. Haque and F. J. Biros, eds.), Plenum, New York, 1974, pp. 99–111.
13. T. Golab and J. L. Occolowitz, *Biomed. Mass Spectrom.* **6,** 1–9 (1979).
14. D. D. Kaufman, *Herbicides,* Vol. 2 (P. C. Kearney and D. D. Kaufman, eds.), Marcel Dekker, New York, 1976, pp. 665–708.
15. T. Golab, W. A. Althaus, and H. L. Wooten, *J. Agric. Food Chem.* **27,** 163–179. (1979).
16. F. W. McLafferty, *Interpretation of Mass Spectra,* Benjamin, Reading, MA, 1973.
17. B. B. Downer, M. Hall, and D. N. B. Mallen, *J. Agric. Food Chem.* **24,** 1223–1225 (1976).
18. P. C. Kearney, J. R. Plimmer, W. B. Wheeler, and A. Kontson, *Pestic. Biochem. Physiol.* **6,** 229–238 (1976).
19. P. C. Kearney, J. R. Plimmer, V. P. Williams, U. I. Klingebeil, A. R. Isensee, T. L. Laanio, G. E. Stolzenberg, and R. G. Zaylskie, *J. Agric. Food Chem.* **22,** 856–859 (1974)
20. W. R. Lusby, J. E. Oliver, and P. C. Kearney, *J. Agric. Food Chem.* **28,** 641–644 (1980).
21. H. d'A. Heck, R. L. Dyer, A. C. Scott, and M. Anbar, *J. Agric. Food Chem.* **25,** 901–908 (1977).
22. K. Nojima, K. Fukaya, S. Fukui, S. Kanno, S. Nishiyama, and Y. Wada, *Chemosphere* **7,** 25–30 (1976).
23. S. L. M. Skinner and R. Greenhalgh, *Mass Spectra of Insecticides, Herbicides and Fungicides and Their Metabolites,* Agriculture Canada, Chemistry and Biology Research Institute, Ottawa, 1977.
24. R. C. Gardner and R. L. McKellan, *J. Agric. Food Chem.* **28,** 258–261 (1980).
25. H. R. Schulten, *J. Agric. Food Chem.* **24,** 743–749 (1976).
26. D. S. Frear, in *Herbicides,* Vol. 2 (P. C. Kearney and D. D. Kaufman, eds.), Marcel Dekker, New York, 1976, pp. 541–664.
27. D. S. Frear, H. R. Swanson, E. R. Mansager, and R. G. Wien, *J. Agric. Food Chem.* **26,** 1347–1351 (1978).
28. S. C. Tong, W. H. Gutenmann, L. E. St. John, Jr., and D. J. Lisk, *Anal. Chem.* **44,** 1069–1072 (1972).
29. D. D. Oehler and G. W. Ivie, *J. Agric. Food Chem.* **28,** 685–689 (1980).

30. M. L. Rueppel, L. A. Suba, and J. T. Marvel, *Biomed. Mass Spectrom.* **3,** 28–31 (1976).
31. M. L. Rueppel, B. B. Brightwell, J. Schaefer, and G. T. Marvel, *J. Agric. Food Chem.* **25,** 517–528 (1977).
32. Monsanto Company, Method C., Monsanto Agricultural Products Co., St. Louis, MO, 1975.
33. M. Povoledo and D. A. Murray, Freshwater Institute, Winnipeg, Manitoba, Canada, 1978.
34. R. W. Reiser, *Org. Mass Spectrom.* **2,** 467–479 (1969).
35. S. Safe and O. Hutzinger, (1974). *Mass Spectrometry of Pesticides and Pollutants,* CRC Press, Cleveland, OH, 1974, pp. 177–179.
36. S. U. Khan *Bull. Environ. Contam. Toxicol.* **18,** 83–88 (1977).
37. T. Suzuki, K. Umedzu, Y. Itagaki, and K. Tuzimura, *Agric. Biol. Chem.* **41,** 775–779 (1977).
38. R. C. Rhodes, *J. Agric. Food Chem.* **25,** 1066–1068 (1977).
39. A. Calderbank and P. Slade, *Herbicides,* Vol. 2. (P. C. Kearney and D. D. Kaufman, eds.), Marcel Dekker, New York, 1976, pp. 501–504.
40. S. U. Khan, *J. Agric. Food Chem.* **22,** 863–867 (1974).
41. S. U. Khan and K. S. Lee, *J. Agric. Food Chem.* **24,** 684–686 (1976)
42. M. A. Martens, C. H. van Peteghem, and A. Heyndrickx, *Meded. Fac. Landouwwet., Rijksuniv. Gent.* **40,** 1149–1160 (1975).
43. G. H. Draffan, R. A. Clare, D. L. Davies, G. Hawksworth, S. Murray, and D. S. Davies, *J. Chromatog* **139,** 311–320 (1977).
44. M. E. Wall and D. R. Brine (G. G. Nahas and W. D. M. Paton, eds.), Pergamon, New York, 1979, pp. 15–43.
45. J. F. J. Hughes, N. Evans, D. E. Games, M. J. E. Hewlins, A. H. Jackson, J. R. Jackson, N. A. Khan, S. A. Matlin, M. Rossiter, R. G. Saxton, H. A. Swaine, and K. T. Taylor *Adv. Mass Spectrom. Biochem.* **1,** 357–368 (1976).
46. C. N. McEwen, S. F. Layton, and S. K. Taylor, *Anal. Chem.* **49,** 922–926 (1977).
47. T. M. Ryan, R. J. Day, and R. G. Cooks, *Anal. Chem.* **52,** 2054–2057 (1980).

21

# ORGANOPHOSPHORUS COMPOUNDS

J. M. DESMARCHELIER and M. J. LACEY

## 1. MASS SPECTROSCOPY AS A TOOL IN ANALYTICAL CHEMISTRY

Mass spectroscopy (MS) is examined in this review as a part of systems of extraction, purification, identification, and quantitation, that is, analysis of synthetic organophosphorus esters and their alteration products (OPs). This placement of MS in the broader context of analytical chemistry is necessary because no amount of sophistication in MS can compensate for errors in sampling, extraction, and workup.

Because of the wide scope of the subject of MS in analysis of OPs, references that do not significantly contribute to general principles have not been cited and particular points are supported by only one reference, which has often been chosen arbitrarily from a large number of relevant references. Books and conference papers are cited only when they illustrate principles that are not found in journals. A glossary of the notation used in the text is included at the end of this review.

The analysis of OPs has been extensively reviewed(1,2) and is regularly reviewed in journals.(3–5) Official procedures for analysis of individual pesticides have been summarized,(6) and high-performance liquid chromatography (HPLC),(7) gas chromatography (GC),(8) and thin-layer chromatography (TLC)(9) have been reviewed. Chemical derivatization procedures(10) have also been reviewed.

The most common method of analysis of OPs involves extraction, cleanup by liquid–liquid partitioning, and column chromatography, then determination by GC with the aid of a phosphorus-specific detector.

---

J. M. DESMARCHELIER and M. J. LACEY • CSIRO Division of Entomology, P.O. Box 1700, Canberra, A. C. T. 2601, Australia.

Although this procedure has proved adequate for most intact OPs, it is often unsatisfactory for polar OPs, is time consuming, requires large amounts of solvents, and does not provide unambiguous proof of identity. In the case of suspected violation, some confirmatory method is required.

Gel permeation chromatography is increasingly replacing column chromatography for cleanup, at least in unofficial procedures. Apart from the practical advantages of gel permeation chromatography in reducing solvent requirement, there is less chance of OPs being retained by gel permeation than there is with column chromatography. Although OPs can be separated by HPLC, the absence in HPLC of the equivalent of the flame photometric detector used in GC limits the usefulness of HPLC for multiresidue (MR) analysis. Increasing use is being made of thin-layer chromatography (TLC) in the detection of OPs as a result of improvements in equipment and spray reagents.

Mass spectroscopy has a potential for wider application in the analysis of OPs principally because of its sensitivity and specificity, but also because it can be integrated with each of the methods of purification and identification that have been discussed. These include direct coupling with packed-column gas chromatography (GC)[(11)] or capillary GC;[(12,13)] off-line and on-line coupling of high-performance liquid chromatography (HPLC) to MS[(14–16)] with off-line coupling illustrated for OPs by Lin and others;[(17)] on-line reversed-phase HPLC/MS[(18)] purification by TLC followed by probe insertion [(19,20)] or GC[(21)] and off-line cleanup by ion exchange,[(22)] gel permeation,[(12,13)] liquid–liquid partitioning,[(11)] or column chromatography[(23)] followed by MS.

Possible ionization modes in MS that can be used for qualitative and quantitative detection of OPs include electron impact (EI) ionization, positive-ion chemical ionization (PCI), negative-ion chemical ionization (NCI), field desorption (FD), and field ionization (FI). We include the formation of EI-type mass spectra by charge exchange reactions in PCI and electron capture processes in NCI. For involatile or thermally unstable OPs, a number of strategies are available for obtaining their mass spectra, including in-beam techniques[(24)] and desorption ionization methods.[(25)] The contribution of MS as part of a system of analysis is discussed by Ohkawa and Miyamoto[(26)] and Shafer, Lucas, and Jakobsen,[(27)] and comparisons are made by the latter authors between GC/MS, HPLC/FTIR (Fourier transform infrared spectroscopy) and GC/FTIR. The combination of radiogas chromatography with MS[(28)] has considerable potential for metabolic studies. Coupling of MS with specific GC or HPLC detectors clearly increases the data input, while reactant ion monitoring in GC/MS using PCI or NCI[(29)] provides a more versatile approach to the selective detection of chemical classes in the qualitative analysis of complex mixtures.

Systematic problems (i.e., other than those due to carelessness) in analysis of OPs occur in MR determination of unknown compounds, especially polar compounds. Budde and Eichelberger[30] suggest that, while other techniques serve for identification of target compounds, for MR determinations GC/mass spectroscopy "may be the only viable choice." To these comments must be added the observations that target compounds cannot always be separated by GC[11–13,31,32] and that MS serves for the identification of natural compounds, including undesirable ones, as well as pollutants.[33]

## 2. GENERAL PRINCIPLES OF THE MASS SPECTRA OF OPs

Most insecticidal OPs, especially the older type, contained a good (in the chemical sense) leaving group which is required for irreversible inhibition of acetylcholinesterase. Because such groups do not favorably support a positive charge, and because they favor radical formation, only a small proportion of the ion current in EI mass spectra is carried by the $M^{+\cdot}$ ion. The majority of the ion current, particularly for higher *m/z* values, is carried by ions that contain phosphorus and not by, for example, aromatic ions with substituents having positive values for Hammett's $\sigma$ constant. A principal advantage of PCI, NCI, FI, and FD is the greater proportion of ion current, relative to EI, that is carried by the high-mass ions in the spectra.

In the extreme, complete concentration of the ion current in a single ion such as $M^{+\cdot}$ would optimize sensitivity in the determination of known compounds, but provide no structural information on unknowns. There must be a compromise between the sensitivity achieved from concentrating all information in one peak and the diagnostic information obtained from several peaks. That is, EI, PCI, NCI, FI, and FD are inherently complementary to each other, and not antagonistic. The optimal compromise varies with circumstances but the presence of different methods of ionization is far more useful than is the presence of different detectors with GC alone. Furthermore, apart from one or two examples,[34] little use has been made with OPs of instrumental procedures that decipher ion fragmentations occurring in the flight path of a mass spectrometer rather than in the source itself. These techniques of tandem mass spectrometry or MS/MS[35] have considerable potential for determination of OPs at much reduced chemical noise, and ionization modes that enhance the abundances of single high-mass ions in the source offer distinct advantages for quantitation.

The comments made on the effect of the leaving group on the EI mass spectra of intact OPs do not necessarily apply to alteration products such as amino fenitrothion,[36] where the aryl substituent has a -ve $\sigma$ constant, or dimethylphosphoric acid. Although these alteration products are discussed later, it is worth noting here that fragmentations of such alteration products may be different from those of intact esters. Also, if a derivatization procedure is used, one can select a procedure suitable for the ionization mode in MS.[37,38]

## 3. EI MASS SPECTRA OF OPs

The EI mass spectra of OPs can be rationalized in terms of the stabilities of carbonium ions and leaving groups, even though such assignments have not been supported theoretically. A few "rules," given below, are useful guides for interpretation of spectra of intact OP esters with good leaving groups. Common ions in the spectra of these and other compounds are given in Table 1. Further details are given in reviews:[20–26] for cyclophosphamide,[11,39–42] for heterocyclic phosphine oxides,[43] for flame retardants and plasticizers,[44] and for cyclic phosphorus esters.[45] Additional references are given in Table 1.

The mass spectra of OP fungicides,[46,47] herbicides,[48] hypophosphates,[49] phosphites,[50] and phosphonates[51] are not extensively discussed in this present review.

The rules listed in Secs. 3.1 to 3.7 refer to compounds of the type A·B· PY-XL where XL is the best leaving group (e.g., *O*-aryl, *S*-alkyl, or aryl); A and B are either alkoxy, alkyl, or thioalkoxy; Y is either *O* or *S*; and X is either *O, S, NH,* or *N*-alkyl.

### 3.1. *Proportion of Ion Current Containing Phosphorus Ions*

Other things being equal, the percentage of P-containing ions in spectra of A·B· PY-XL (cf. above) is as follows: (a) increases with increasing values of the Hammett $\sigma$ constant of aryl substituents in XL; (b) decreases with increasing ability of A or B to support positive charges. In practice, this means decreasing for S-alkyl rather than O-alkyl; propyl or propoxy rather than methyl or methoxy; or aryl rather than alkyl. For example, whereas the mass spectrum of chlorpyrifosmethyl (Fig. 1) contains largely phosphorus ions (Table 1), the mass spectrum of chlorpyrifosethyl (Fig. 2) contains an abundant ion at *m/z* 197 due to the 3,5,6-trichloropyrid-2-ol moiety. Fragmentations from the loss of olefins (e.g., fragment ions of *m/z* 286, 256, compare Sec. 3.4) are another distinguishing feature of the mass spectrum of chlorpyrifosethyl.

*TABLE 1*

*Common Phosphorus Ions in the EI Mass Spectra of OPs*[a]

| m/z | Other ions at m/z | Formal structure | Rationalization | References |
|---|---|---|---|---|
| 185 | 153, 125, 97, 63 | $(EtO)_2PS - SL$ | $(EtO)_2PS - S$ | 11, 39, 52 |
| 185 | 169, 157, 141 | $Ph \cdot PS(OEt)(OL)$ | $Ph \cdot PS - OEt$ | 11, 40 |
| 169 | 185, 157, 141 | | $Ph \cdot PO \cdot OEt$ | |
| 157 | 185, 169, 141 | | $Ph \cdot PS \cdot OH$ | 39, 40, 52 |
| 157 | 125, 93, not 185 | $(MeO)_2PS - SL$ | $(MeO)_2PS_2$ | 11, 39, 52 |
| 153 | 125, 97, 65 | $(EtO)_2PS \cdot XL$ | $(EtO)_2PS$ | 11, 39, 40 |
| 141 | 185, 169, 157 | | $Ph \cdot PO \cdot OH$ | |
| 137 | 109, 93, 81 | $(EtO)_2PO \cdot XL$ | $(EtO)_2PO$ | 11, 39, 40 |
| 125 | 153, 97, 65 | | $EtO \cdot PS \cdot OH$ | |
| 125 | 109, 93, not 97, 153 | $(MeO)_2 \cdot PS \cdot XL$ | $(MeO)_2 \cdot PS$ | 11, 39, 40, 52 |
| 115 | | $(RO)(R'O) \cdot PX \cdot YL$ | $(HO)_3PSH$ | 40 |
| 109 | 137, 81, not 125, 79 | | $EtO \cdot PO \cdot OH$ | |
| 109 | 79, not 81, 125 | $(MeO)_2PO \cdot XL$ | $(MeO)_2PO$ | (39, 40, 52 |
| 109 | 125, 93, not 97, 153 | | | |
| 99 | 81 | $(RO)_3 \cdot PO$ | $(HO)_4P$ | 39, 40, 53 |
| 97 | 153, 125, 65, not 83 | | $(HO)_2PS$ | |
| 93 | 125 | | $(MeO)_2P$ | |
| 93 | 153, 125, 97 | | | |
| 83 | | $(RO)_2PO \cdot R$ | $(HO)_3P \cdot H$ | 11, 55, 56 |
| 81 | 137, 109 | | $(HO)_2PO$ | |
| 79 | 109, or 125, 93 | | $MeO \cdot P \cdot OH$ | 51, 57 |
| 65,63 | | various | $P(OH)_2,PS$ | 39, 52, 55 |

[a] X = O, S; R = alkyl; L = leaving group

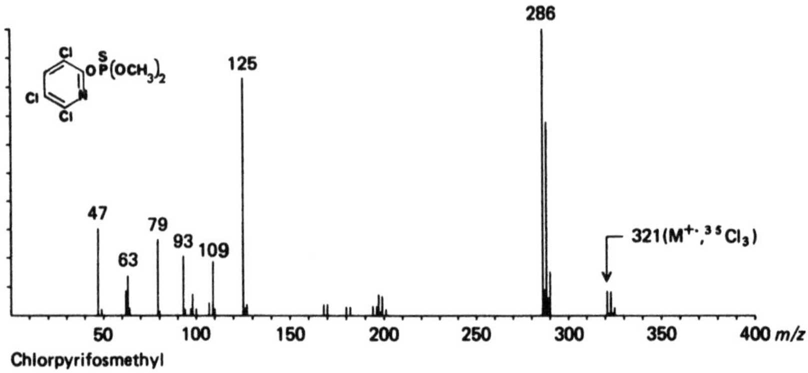


FIGURE 1. EI mass spectrum (70 eV, source temperature 160 °C, direct insertion) of chloropyrifosmethyl. Ions below *m/z* 40 have been omitted fr^m Figs. 1 to 6. All spectra obtained on VG Micromass 70-70 mass spectrometer.

## 3.2. *Molecular Ion Abundance*

The percentage ion current carried by $M^{+\cdot}$ increases with its ability to sustain a positive charge and decreases with favorable energetics of fragmentation processes. If, in Sec. 3.1, "the percentage of ion current carried by $M^{+\cdot}$" is substituted for "the percentage of P-containing ions," statements 3.1(a) and 3.1(b) hold. The ionization energy is lowered for S rather than O in X or Y.

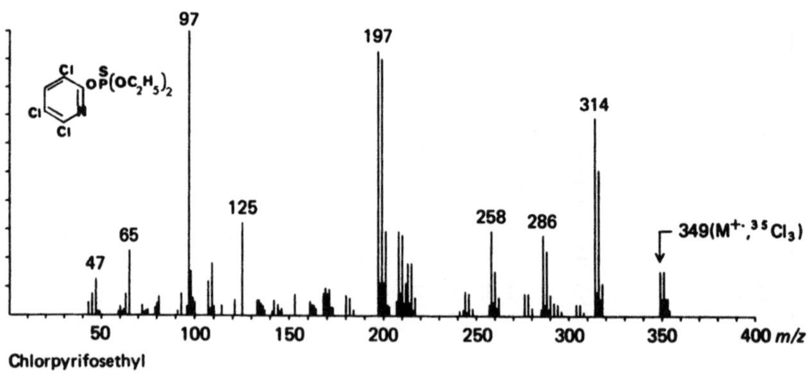


FIGURE 2. EI mass spectrum (70 eV, source temperature 140 °C, direct insertion) of chlorpyrifosethyl.

### 3.3. *Losses of Radicals (R˙) From A or B*

Loss of R˙ is more important where R is (a) aryl rather than alkyl; (b) *S*-alkyl or *N*-alkyl rather than *O*-alkyl; and (c) methyl rather than ethyl or propyl.

### 3.4. *Loss of R—H*

Loss of R—H is more important where R is (a) ethyl or propyl rather than methyl or aryl; and (b) *S*-methyl rather than *O*-methyl.

### 3.5. *Loss of Mass 32 (S)*

Fragment ions, but not $M^{+\cdot}$, can lose mass 32 (S) if the fragment ion can be formalized as $A^{\cdot}B^{\cdot}PS^+$. Loss of S seems to occur from even-electron ions but not from radical ions.

### 3.6. *Maximization of PXH Groups*

Loss of alkyl (other than methyl) groups from P-alkyl, P-*O*-alkyl or P-*S*-alkyl is accompanied by single hydrogen transfer (i.e., loss of olefin) or by double hydrogen transfer (i.e., loss of olefinic radical), such that the resulting ion contains the maximum number of PXH groups [e.g., $^+P(OH)_4$]. All abundant ions listed in Table 1 conform to this rule; for example, $(EtO)_2PS$-*O*-aryl forms $(HO)_2PS^+$ whereas $(MeO)_2PS$-*O*-aryl forms $(MeO)_2PS^+$. Likewise, the base peak in the mass spectrum of chlorpyrifosethyl (Fig. 2) is at *m/z* 97 $[(HO)_2PS^+]$, whereas the base peak

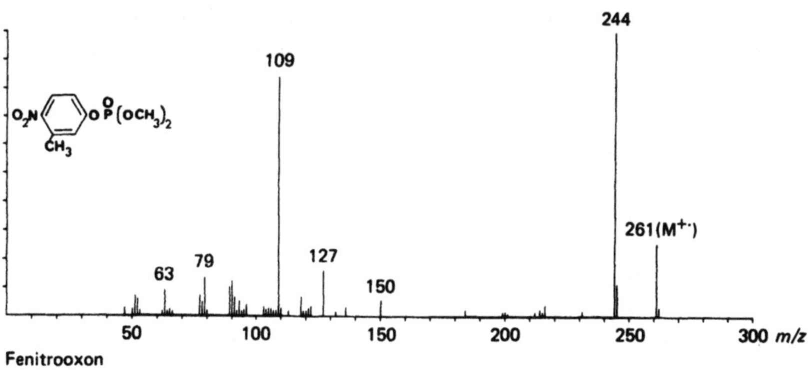


FIGURE 3. EI mass spectrum (70 eV, source temperature 130 °C, direct insertion) of fenitrothion.

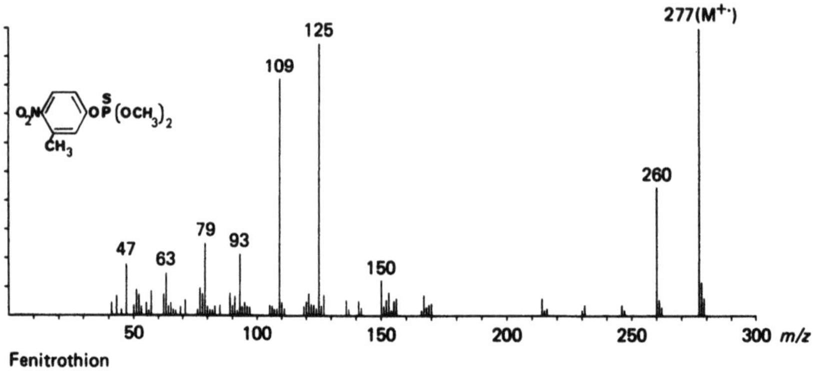


FIGURE 4. EI mass spectrum (70 eV, source temperature, 170 °C, direct insertion) of fenitrooxon.

in the mass spectrum of chlorpyrifosmethyl (Fig. 1) is at *m/z* 286 [$(M - Cl)^+$] and its second most abundant peak is at *m/z* 125 [$(MeO)_2PS^+$]. This latter ion is abundant in the mass spectrum of fenitrothion (Fig. 3), and the corresponding ion at *m/z* 109 [$(MeO)_2PO^+$] is abundant in the mass spectrum of fenitrooxon (Fig. 4).

### *3.7. P(S)-OR versus P(O)-SR*

Compounds of type P(S)-OR fragment by competitive loss of RO$^\bullet$ and RS$^\bullet$ (cf. *m/z* 125, 109, Figs. 1 and 3) whereas the thiolo isomers lose RS$^\bullet$ only.[40,58,59] Loss of (RS − H) is more important for the thiolo isomers, but abundances of $[M - (RS\text{-}H)]^{+\cdot}$ are often low.

Of the rules presented, those in Secs. 3.1 and 3.6 are especially important in screening for unknown compounds as they rationalize the common ions listed in Table 1. Some of the common ions feature also in the spectra of compounds with poor leaving groups, such as *O,O*-diethyl-*O*-4-methylphenyl phosphorothioate and its thiolo isomer (Figures 48, 49[40]), though the proportion of the ion current carried by such ions is much less than is the case for ethylparathion or its S-ethyl isomer (Figures 50, 51[40]).

The molecular ion abundance, the most important specific parameter, is usually sufficiently great in the EI mass spectra of OP esters to establish identity. It was, however, not detected in the spectrum of dioxathion.[39] $M^{+\cdot}$ was also of low abundance in the spectra of leptophos oxon and leptophos[40] though the $(M - Cl)^+$ ions were abundant in each spectrum.

The question of thiono-thiolo isomerization is of importance because small amounts of thiolo isomers can increase the mammalian toxicity of OP esters of low toxicity, such as malathion. These isomers can be distinguished by the rule in Sec. 3.7 and, for S-methyl isomers, Sec. 3.4(b). In addition the thiolo isomers have much lower $R_f$ values on TLC than thiono isomers, are more readily identified than thiono isomers by enzyme inhibition on TLC prior to oxidation, are eluted on silica, alumina, and florisil columns much later than the thiono isomers, and usually have different retention times on GC to thiono isomers.

The chemical interpretation of EI mass spectra can also be applied to phosphoramidates. Ammonia and amines are better nucleophiles toward phosphorus than alcohols or thiols[41] and the P–N bond is present in many ions in the spectra of amino, *N*-methylamino, *N,N*-dimethylamino, *N*-acetylamido, and acetamido phosphoramidates.[40,60–64] The common ion in these spectra at *m/z* 63 or 64 is often not abundant, though fortunately it is for the insecticide Monitor (Fig. 9[40]) where the base peak ($CH_3O\text{-}PO\text{-}NH^+{}_2$) is formed by loss of $\cdot SCH_3$. In general, the mass spectra of phosphoramidates can be rationalized by the rule in Sec. 3.6 and retention of the P–N link(s). The ion *m/z* 80 [$(HO)_2PNH_2{}^+$] is tentatively suggested for monitoring for the ROPX(XR)-NHR group (R-alkyl or H) and the ion *m/z* 78 ($MeOPNH_2{}^+$) for the MeOPONHR group.

In general, phosphoramidates have not been extensively investigated in MR studies, though one would expect, on chemical grounds, that they would result from the reaction of OP esters with ammonia (e.g., in sediments, soils). Phosphoramidates are not recovered by cleanup procedures suitable for nonpolar OP esters and are often intractable on GC although they can be derivatized.[65,66] MS could play an important role in such studies, but more work is required to establish the generality of the suggested common ions.

## 4. PROCEDURES FOR QUANTITATIVE MASS SPECTROSCOPY

Strategies for quantitative MS have been discussed in a recent monograph by Millard.[46] As well as use of MS to confirm identity in MR analyses[67] and use of MS plus chemical derivatization to confirm structures,[68] both Japanese[11] and German[12,13] workers have applied GC/MS to MR analysis of OP esters. Both EI[11] and PCI[12,13] have been used. Stan[12,13] used a gel permeation cleanup, similar to procedures used for the cleanup of many types of pollutants.[69–72] Johansson[67] used simplified column chromatography and Kobayashi and others[11] used only a water–hexane extraction for cleanup. Despite this minimal cleanup, sensitivity with both EI and PCI was sufficient to detect nanogram quantities in crops. The limit of detection[32,73] can be reduced by selected-ion monitoring and a feature of its use is the simplification of the cleanup procedure compared to, for example, acetonitrile–hexane partitioning and florisil chromatography. Lee and Westcott[74] used selected-ion monitoring and an internal standard to analyze dimethoate. Kobayashi and others[11] also report that GC/MS was able to distinguish pesticides that were not well separated by GC. The importance of the simplified cleanup with GC/MS is the reduction of errors associated with inadequate recovery of OPs.[30]

## 5. DETECTION OF METABOLITES IN MR ANALYSIS OF PHOSPHORUS ACIDS AND HYDROGEN PHOSPHITES

Among the common alteration products of OP esters are the acids formed by hydrolysis and oxidative hydrolysis (e.g., diethylphosphoric and diethyl thiophosphoric acid from parathion), and O-dealkylated, especially desmethyl, acids. Hydrogen phosphites and phosphorothioites are also formed from soil transformation of dialkyl-thiophosphoric and dithiophosphoric acids.[75]

GC/MS has not been used for the analysis of these acidic metabolites in MR determinations, but all the necessary groundwork has been done. The acids can be chromatographed directly[75] or after derivatization to alkyl[41] or aryl derivatives.[76,77] The EI mass spectra of two methyl derivatives of common metabolites are shown in Figs. 5 and 6. Inorganic phosphate can be separated by ion exchange chromatography[77] or derivatized for GC/MS either off-line[78] or on-line.[79]

In addition, thio- or dithio-phosphoric acids including desmethyl

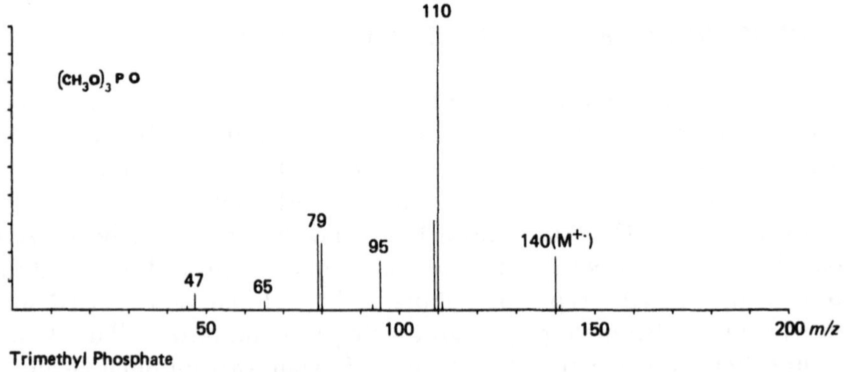


FIGURE 5. EI mass spectrum (70 eV, source temperature 150 °C, heated glass reservoir inlet) of trimethyl phosphate.

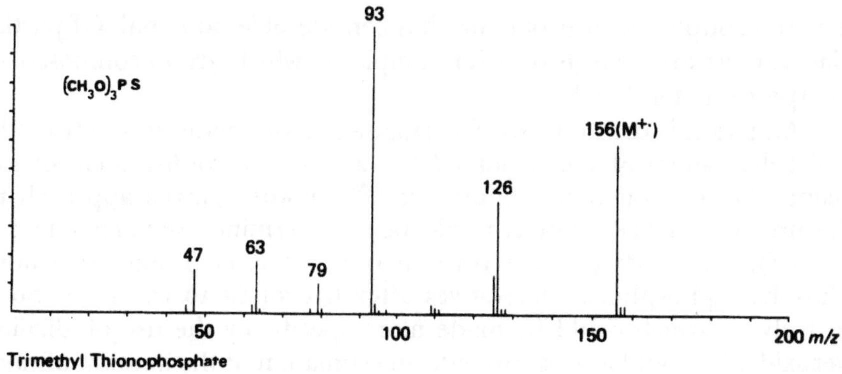


FIGURE 6. EI mass spectrum (70 eV, source temperature 120 °C, heated glass reservoir inlet) of trimethyl thionophosphate.

pesticides react rapidly with alkyl halides to form exclusively thiolo derivatives(40) and the alkyl esters(40,41,75) and can be detected by common and specific ions. Because no $M^{+\cdot}$ ion was detected in the EI mass spectra of $(PrO)_2PXH$ (X = O,S) where M − XH had the highest observed *m/z* value(75) and because false $M^{+\cdot}$ peaks were observed in the mass spectra of some,(80) but not other,(40,55,56,81) acids, derivatization may be necessary or an alternative mode of ionization used. Reaction with alkyl halides followed by methylation or benzylation with an electron-withdrawing substituent(76,82) [cf. Sec. 3.1(b)] may prove suitable. Finally, because the acids partition into aqueous alkali from hexane, it is possible to separate acids from esters. This could be important in view of the environmental methylation of phosphorus acids(75,83) which could be masked by a derivatization of acids to methyl esters.

Of particular concern in MR analysis of environmental samples is the loss of polar alteration products during cleanup or GC. This problem is partly solved by the simpler cleanup used but the GC problem remains.(31) The problem is largely confined to MR analysis because known compounds can be purified by TLC and analyzed in the mass spectrometer by direct insertion. Coupling of HPLC to MS may be the answer to this problem,(14) and there is a need for more data on the MS of metabolites of OP esters.

In metabolism studies (where speed of analysis is not important) frequent use has been made of purification by TLC and probe insertion of the TLC extract. The advantages of TLC over other chromatographic procedures(84) are considerable. A variety of reagents(2) is available as a help for detection of OPs. In view of the significant advances in densitometry(85,86) and TLC(9,84) a screening program based on densitometry, coupled with probe MS, has considerable potential. Of particular interest are charge transfer complexes which are decomposed in the spectrometer.[87–90]

Attention is drawn to specific fragmentation modes in the trimethylsilyl derivatives of glucuronides[91,92] and to the confirmation of exchangeable hydrogen by deuteration.[93] Another partial approach to the problem of MR analysis would be to determine nonpolar OPs (by GC/MS), free acids (after extraction into water, etc.), and total non-phosphate phosphorus derivatives after hydrolysis to the acids. Such hydrolyses, which could be made more specific by the use of alkaline peroxide,[41] would at least provide an estimation of the amount of non-analyzed OPs and would identify, by difference, the relevant phosphorus acid.

Another approach, which perhaps may provide the solution to rapid MR screening, is to use the mass spectrometer for direct mixture analysis after minimal cleanup.[35] The comments made on this subject (MS/MS) in our review on pyrethroids can be equally applied to OPs.

## 6. *POSITIVE-ION CHEMICAL IONIZATION (PCI)*

PCI has been used extensively in metabolism studies[94] and Holmstead and Casida[95] reported the PCI (methane) mass spectra of 29 OPs, obtained via insertion probe. Abundant $(M + H)^+$ ions were formed for OPs with aryl leaving groups (e.g., ronnel), or phosphates (e.g., dichlorvos) but not for OPs with stabilized sulfhydryl leaving groups (e.g., malaoxon, azinophos methyl, phorate) or EPN. Although common fragment ions as reported for EI spectra (Table 1) were also common in PCI (with the exception that *m/z* 126 was moved to 127) such common ions were of low or negligible abundance in the spectra of ethyl and methyl parathion and in the spectra of OPs with leaving groups able to stabilize a positive charge (e.g., phosmet, malathion). On the other hand, ions of formal structure $(RO)_2PX\text{-}S\text{-}CH_2^+$ were abundant in the spectra of compounds of type $(RO)_2PX\text{-}S\text{-}CH_2R$, except where the majority of the ion current was formed by ions containing R (e.g., azinophos methyl).

Stan[12] reported the PCI mass spectra of 23 OPs using methane, isobutane, and methanol as reagent gases. The use of isobutane increased the relative abundance of the $(M + H)^+$ ion over that obtained by methane to 100% for all compounds tested except azinophos methyl (42%) and ethyl (32%). These included malathion and carbophenothion, whose $(M + H)^+$ ions were hardly detectable (0 and 2.4%, respectively) with methane.

Sass and Fisher[96] also found that use of isobutane resulted in a more abundant $(M + H)^+$ ion (e.g., 100% for dipinacoyl methylphosphonate) than did methane (approximately 1%) but reported that the absolute sample ion current with isobutane was only one-hundredth of that obtained with methane, and only one-tenth of that obtained with ethylene. Sass and Fisher preferred ethylene and isobutane to methane because the former gases gave a more abundant $(M + H)^+$ ion. Ethylene was preferred to isobutane because it was more sensitive and provided more fragmentation.

The conclusion of the article by Sass and Fisher provides a fair summary of the present use of PCI in analysis of OPs. That is, each reagent gas, and EI, has its own advantages and disadvantages. In analysis of known compounds, the use of isobutane or ethylene enables determination via the unambiguous $(M + H)^+$ ion, but in analysis of unknown compounds EI offers the advantage of a more extensive fragmentation pattern for structural characterization.[11,73]

PCI (and also EI and NCI) can be used with special extended probe tips[24,97,98] as a method for detecting chemicals that are too involatile for the normal direct-insertion probe or for GC/MS. This technique is possible without alteration to the usual mode of operation of the spectrometer. The dependence of such spectra on both time and temperature should be noted.[99]

Despite the advantages of PCI, its temperature dependence[96,100] requires that care be taken in obtaining reproducible PCI spectra. The former authors point out, in a discussion of biologically important compounds including Delnav and phosphamidon, that EI, PCI, FI, and FD have their respective advantages and that one is not obligated to a specific technique.

## 7. *NEGATIVE IONIZATION (NI) AND NEGATIVE-ION CHEMICAL IONIZATION (NCI)*

Although Rankin[101] reported that the NI spectra at low source pressure of four OPs of type $(RO)_2PS\text{-}XL$ were simple and governed by carbanion chemistry, relatively little attention has been paid to the technique because the spectra usually depend on electron voltages and source design, and because the absolute sensitivity is much lower than it is with EI.

At typical pressures in a CI source, there are large numbers of low-energy electrons, and associative and dissociative electron capture processes can be much more efficient than at low source pressures. Total ion currents for molecules containing electronegative groups can be two or three orders of magnitude higher in electron capture NCI than in

EI or PCI[102,103] and this can facilitate detection and quantitation of compounds present in the subpicogram range.

The bulk of the ion current in the electron capture NCI spectrum of malathion[31] was carried by the ion $(MeO)_2PS_2^-$ and the $(M - 15)^-$ ion. Common P-containing daughter ions were not abundant in the EI[39] or PCI[95] spectra of malathion. The (M − 15) ion was also abundant in the negative atmospheric pressure ionization mass spectrum of methyl parathion.[104]

The versatility of reactant ions in PCI also applies in NCI, and the use of different gases or gas mixtures can lead to a variety of reactant negative ions, such as $OH^-$ and $Cl^-$, for reaching different analytical objectives.[31,105] For instance, $OH^-$ reacts mainly by proton abstraction to give $(M - H)^-$ ions with a wide range of organic compounds,[106] while $Cl^-$ is much more selective in its reactivity.[31] The information obtained on molecular weight and structure is complementary to that obtained by PCI and EI.

The simultaneous recording of PCI and electron capture NCI spectra, as in the pulsed method of Hunt and others[103] would, subject to further data, result in an $(M + H)^+$ ion from PCI and an $(RO)_2PXY^-$ ion from NCI. Not all OPs are sensitive to electron capture GC and hence may be insensitive to electron capture NCI. However, because the perfluorobenzyl derivatives of phenols are sensitive to electron capture,[103] the confirmational procedure of Coburn and Chan[107] via this derivative could be extended to electron capture GC and NCI. The sensitivity of NCI to such derivatives is superior to that of PCI.[103] Similar comments probably apply to derivatization of phosphorus-containing acids, where there is a choice of triazene agents for derivatization.[76,77]

## 8. DESORPTION AND IONIZATION TECHNIQUES

The great advantages of field desorption (FD) and other desorption ionization methods[25] over some conventional techniques of ionization are their ability to ionize involatile compounds,[108] including, for example, phosphonium halides,[109] and their ability to detect polar materials in unpurified biological extracts (e.g., cyclophosphamide).[110] These advantages are offset in MR determinations by the inability to couple directly with HPLC though a thermospray interface is possible.[15] Field desorption and especially field ionization typically have a lower absolute sensitivity than EI, PCI, or NCI.[111] Because there may be little fragmentation in the FI mass spectra of OPs, especially (from the limited data available) in the FI mass spectra of phosphorothiolates,[57] FI would not necessarily detect common ions of unknowns. FD mass spectroscopy would be one of the most direct methods of detecting impurities in formulations, where lack of fragmentation would be an advantage and sensitivity is not critical; for instance, Schulten[112] detected diazoxon in a commercial formulation of diazinon. The importance of contaminants in formulations is emphasized in the reviews by Alford.[113–115]

Desorption ionization modes coupled with tandem mass spectrometry[35] have considerable potential for the analysis of metabolites and could eliminate the need for hydrolysis of conjugates to provide material amenable to GC/MS. Such uses are relatively common in studies on pharmaceuticals.[116] The strategy has the potential to enable identification of OPs after minimal cleanup. There are, however, comparatively little data on the desorption ionization spectra of OPs compared with data on, for example, EI mass spectra of OPs or FD mass spectra of carbomates.[117,118] It can be reasonably expected that the potential of desorption ionization methods will be increasingly appreciated as more data become available.

## *9. PHOSPHORYLATION IN THE ANALYSIS OF AMINES, PHENOLS, ALCOHOLS, AND THIOLS*

Phosphorylation has been used as means of derivatizing alcohols and phenols for GC[119] or for GC/MS.[120] Phosphorylation enables the use of the robust and specific flame photometric detector. Phosphorylation of amines has been shown to be, and phosphorylation of alcohols and phenols would probably be, catalyzed by dialkylaminopyridines.[121] The preferred derivative for determination of amines, alcohols, and phenols would probably be the dimethylthionophosphoryl ester, which would be more stable to hydrolysis and more quickly eluted than the dimethylphosphoryl derivative, and would undergo less fragmentation in EI than other derivatives such as the diethyl- or the diethylthionophosphoryl derivative (compare Figs. 1 and 2, as derivatives of 3,5,6-trichloro-2-pyridol and Figs. 5 and 6, as derivatives of methanol).

## *10. MASS SPECTROSCOPY IN ANALYSIS OF OPs—AN OVERVIEW*

Analysis of environmental samples is undertaken for many reasons, which include screening to ensure absence of levels in excess of maximum residue limits, measurement of average daily intakes, and metabolic studies. It is not an optimal use of resources (both man and machine) to obtain the greatest precision and sensitivity in all of these analyses because such precision withdraws resources that might be better employed in, for example, sampling. Such considerations were discussed in two symposia at the 1978 IUPAC Conference on Pesticide Science.[122]

In screening studies on intact pesticides the use of GC coupled to a phosphorus-specific detector is usually sufficient to detect possible violative levels. The ideal method of confirmation of such violative levels is MS, and machine time can be minimized by off-line methods of purification such as TLC or HPLC. The suggestion of Budde and Eichelberger[30] that MS be used routinely as a check on the accuracy of GC assignments is another method of efficient use of MS.

In certain studies, such as in forensic science or on average daily intakes, high levels of precision, sensitivity, and accuracy are required and one needs to determine unsuspected as well as suspected chemicals. In such studies several chromatographic (e.g., retention times, $R_f$ values) and mass spectral parameters should be used for unambiguous identification.

Dougherty and Hett[31] claim that in such situations any cleanup other than gel permeation or liquid–liquid partitioning can lead to destruction or retention of unknowns. We believe that this assessment is correct but point out that liquid–liquid partitioning does not always give quantitative separation of polar OPs. As discussed in our review on pyrethroids, the use of MS/MS may ultimately alleviate the problems of chromatography or cleanup, or at least provide a check on cleanup procedures. Because MS does not distinguish enantiomers,[26] some chemical procedure of derivatization or a chiral column[123] will be needed to supplement even MS/MS.

The simplicity of the gel[12,13] permeation or the liquid–liquid partitioning[11] cleanup may make GC/MS the cheapest[30] method of analysis of intact OPs and the applicability of gel permeation[4] is under constant review. Where MS is to be used, the whole procedure of cleanup, derivatization, and resolution should be selected with MS in mind. For example, because of the greater resolution achieved by selected-ion monitoring, complete separation by GC is not required in GC/MS. It has only been comparatively recently[40] that the emphasis in MS studies on OPs has centered on the use of MS in analytical chemistry and the amount of background data that has been accumulated for GC[6] has not been accumulated for GC/MS. It can be reasonably expected that such data will be accumulated over the next few years and that MS will increasingly be regarded as an essential part of the analytical chemistry of OPs.

## *ACKNOWLEDGMENT*

One of us (J.M.D.) thanks the Australian Wheat Board for financial assistance.

## NOTATION

| | |
|---|---|
| A,B | Any group attached by a single bond to a phosphorus ion |
| EI | Electron impact ionization |
| GC | Gas chromatography or gas–liquid chromatography or gas–capillary chromatography |
| HPLC | High-performance liquid chromatography |
| L | cf. XL |
| $M^{\ddagger}$ | The (positive) molecular ion |
| MR | Multiresidue |
| MS | Mass spectroscopy |
| NCI | Negative-ion chemical ionization |
| OPs | Synthetic organophosphorus esters, related compounds such as phosphoramidates and their alteration products that contain phosphorus |
| PCI | Positive-ion chemical ionization |
| R | Any group, for example, alkyl, aryloxy |
| TLC | Thin-layer chromatography or high-performance thin-layer chromatography |
| X | O, S, NH, or NR attached by a single bond to phosphorus |
| XL | The best leaving group attached to phosphorus |
| Y | O or S attached to phosphorus by a double bond |

## REFERENCES


1. H. P. Burchfield and E. E. Storrs, *J. Chromatogr. Sci.* **13**, 202–211 (1975).
2. J. H. A. Ruzicka, *Pestic. Sci.* **4**, 417–423 (1973).
3. P. E. Corneliussen, *J. Assoc. Off. Anal. Chem.* **62**, 381–382 (1979).
4. K. A. McNully, *J. Assoc. Off. Anal. Chem.* **63**, 283–285 (1980).
5. W. Thornburg, *Anal. Chem.* **51**, 196R–210R (1979).
6. B. McMahon and J. A. Burke, *J. Assoc. Off. Anal. Chem.* **61**, 640–652 (1978).
7. J. F. Lawrence and D. Turton, *J. Chromatogr. Sci.* **159**, 207–226 (1978).
8. S. P. Cram and T. H. Risby, *Anal. Chem.* **50**, 213R–243R (1978).
9. G. Zweig and S. Sherma, *Anal. Chem.* **50**, 50R–65R (1978).
10. W. P. Cochrane, *J. Chromatogr. Sci.* **17**, 124–137 (1979).
11. H. Kobayashi, M. Kojima, O. Matano, and S. Goto, *J. Pestic. Sci.* **4**, 463–472 (1979).
12. H.-J. Stan, *Fresenius Z. Anal. Chem.* **287**, 104–111 (1977).
13. H.-J. Stan, *Z. Lebensm. Unters. Forsch.* **164**, 153–159 (1977).
14. P. J. Arpino and G. Guiochon, *Anal. Chem.* **51**, 682A–701A (1979).
15. C. R. Blakley and M. L. Vestal, *Anal. Chem.* **55**, 750–754 (1983).
16. E. R. Schmid, I. Fogy, F. Heresch, E. Kenndler, and J. F. K. Huber, *Mikrochimica Acta* **2**, 207–217 (1979).
17. S.-N. Lin, C.-Y. Chen, S. D. Murphy, and R. M. Caprioli, *J. Agric. Food Chem.* **18**, 85–88 (1980).
18. B. L. Karger, D. P. Kirby, P. Vouros, R. L. Foltz, and B. Hidy, *Anal. Chem.* **51**, 2324–2328 (1979).
19. D. Ambrosi, P. C. Kearney, and J. A. Macchia, *J. Agric. Food Chem.* **25**, 342–347 (1977).
20. T. B. Waggoner, *J. Agric. Food Chem.* **20**, 157–160 (1972).
21. D. L. Bull and G. W. Ivie, *J. Agric. Food Chem.* **24**, 143–146 (1976).
22. J. P. Mieure and M. W. Dietrich, *J. Chromatogr. Sci.* **11**, 559–570 (1973).
23. B. G. Luke and C. J. Dahl, *J. Assoc. Off. Anal, Chem.* **59**, 1081–1087 (1976).

24. G. D. Daves, Jr., *Acc. Chem. Res.* **12,** 359–365 (1979).
25. K. L. Busch and R. G. Cooks, *Science* **218,** 247–254 (1982).
26. H. Ohkawa and J. Miyamoto, *J. Pestic. Sci.* **4,** 71–81 (1979).
27. K. H. Shafer, S. V. Lucas, and R. J. Jakobsen, *J. Chromatogr. Sci.* **17,** 464–470 (1979).
28. I. M. Campbell, *Anal. Chem.* **51,** 1012A–1021A (1979).
29. F. Hatch and B. Munson, *Anal. Chem.* **49,** 731–733 (1977).
30. W. L. Budde and J. W. Eichelberger, *Anal. Chem.* **51,** 567R–574R (1979).
31. R. C. Dougherty and E. A. Hett, *Environ. Sci. Res.* **12,** 339–350 (1978).
32. G. Vander Velde and J. F. Ryan, *J. Chromatogr. Sci.* **13,** 322–329 (1975).
33. B. Binnemann and D. Jahr, I. *Fresenius Z. Anal. Chem.* **297,** 341–356; II, *ibid.* **298,** 337–348 (1979).
34. A. Vincze, P. Bel, and L. Gefen, 28th Annual Conference, American Society for Mass Spectrometry, New York, Paper WAMOA8, 1980.
35. F. W. McLafferty, (ed.), *Tandem Mass Spectrometry,* Wiley, New York, 1983.
36. K. Mihara, Y. Misaki, and J. Miyamoto, *J. Pestic. Sci.* **4,** 175–185 (1979).
37. A. J. Francis, E. D. Morgan, and C. F. Poole, *Org. Mass Spectrom.* **13,** 671–674 (1976).
38. C. F. Poole and A. Zlatkis, *J. Chromatogr Sci.* **17,** 115–123 (1979).
39. J. N. Damico *J. Assoc. Off. Anal. Chem.* **49,** 1027–1040 (1966).
40. J. M. Desmarchelier, D. A. Wustner, and T. R. Fukuto, *Resid. Rev.* **63,** 77–185 (1976).
41. M. Eto, *Organophosphorus Pesticides: Organic and Biological Chemistry,* CRC Press, Cleveland, 1974, pp. 113–121.
42. C. Fenselau, *J. Assoc. Off. Anal. Chem.* **59,** 1028–1036 (1976).
43. G. L. Kenyon, D. H. Eargle, Jr., and C. W. Koch, *J. Org. Chem.* **41,** 2417–2421 (1976).
44. P. G. Deo and P. H. Howard, *J. Assoc. Off. Anal. Chem.* **61,** 266–271 (1978).
45. N. Fukuhara and M. Eto, *J. Pestic. Sci.* **5,** 63–67 (1980).
46. B. J. Millard, *Quantitative Mass Spectrometry,* Heyden, London, 1978.
47. Y. Uesugi and H. D. Sisler, *Pestic. Biochem. Physiol.* **9,** 247–254 (1978).
48. H. Ohkawa, N. Mikami, and J. Miyamoto, *Agric. Biol. Chem.* **40,** 2125–2127 (1976).
49. W. J. Stec, B. Zielinska, and J. R. van Wazer, *Org. Mass Spectrom.* **10,** 485–493 (1975).
50. H. R. Harless, *Anal. Chem.* **33,** 1387–1389 (1961).
51. J. L. Occolowitz and J. M. Swan, *Aust. J. Chem.* **19,** 1187–1195 (1966).
52. J. Jörg, R. Houriet, and G. Spiteller, *Monatsh. für Chem.* **97,** 1064–1087 (1966).
53. A. Quayle, in *Advances in Mass Spectrometry,* (J. D. Waldron, ed.), Pergamon, New York, 1959, pp. 365–380.
54. D. A. Bafus, E. J. Gallegos, and R. W. Kiser, *J. Phys. Chem.,* 2614–2619 (1966).
55. P. Haake and P. S. Ossip, *Tetrahedron* **24,** 565–573 (1968).
56. P. Haake, M. J. Frearson, and C. E. Diebert, *J. Org. Chem.* **33,** 788–794 (1968).
57. J. N. Damico, R. P. Barron, and J. A. Sphon, *Int. J. Mass Spectrom. Ion Phys.* **2,** 161–169 (1969).
58. R. G. Cooks and A. F. Gerrard, *J. Chem. Soc. (B),* 1327–1333 (1968).
59. E. Santoro, *Org. Mass Spectrom.* **7,** 589–599 (1975).
60. M. Eto, S. Okake, Y. Ozoe, and K. Maekawa, *Pestic. Biochem. Physiol.* **7,** 367–377 (1977).
61. D. J. Harvey and M. G. Horning, *Org. Mass Spectrom.* **9,** 111–124 (1974).
62. D. J. Harvey and M. G. Horning, *Org. Mass Spectrom.* **9,** 955–969 (1974).
63. P. Jacobsen, S. Treppendahl, and J. Wieczorkowski, *Org. Mass. Spectrom.* **6,** 1303–1312 (1972).
64. W. J. McMurray, S. R. Lipsky, C. Zioudrou, and G. L. Schmir, *Org. Mass Spectrom.* **3,** 1031–1034 (1970).
65. M. Look, *J. Agric. Food Chem.* **29,** 888–889 (1980).
66. M. L. Rueppel, L. A. Suba, and J. T. Marvel, *Biomed. Mass Spectrom.* **3,** 28–31 (1976).
67. C. E. Johansson, *Pestic. Sci.* **9,** 313–322 (1978).

68. J. Singh and W. P. Cochrane, *J. Assoc. Off. Anal. Chem.* **62,** 751–756 (1979).
69. J. A. Ault, C. M. Schofield, L. D. Johnson, and R. H. Waltz, *J. Agric. Food. Chem.* **27,** 825–828 (1979).
70. L. D. Johnson, R. H. Waltz, J. P. Ussary, and F. E. Kaiser, *J. Assoc. Off. Anal. Chem.* **59,** 174–186 (1976).
71. J. Pflugmacher and W. Ebing, *J. Chromatogr.* **160,** 213–220 (1978).
72. W. Specht and M. Tillkes, *Fresenius Z. Anal. Chem.* **301,** 300–307 (1980).
73. G. Y. LeBel, D. T. Williams, G. Griffith, and F. M. Benoit, *J. Assoc. Off. Anal. Chem.* **62,** 241–249 (1979).
74. Y. W. Lee and N. D. Westcott, *J. Assoc. Off. Anal. Chem.* **62,** 782–785 (1979).
75. T. Masuda and S. Endo, *J. Pestic. Sci.* **4,** 505–510 (1979).
76. C. G. Daughton, A. M. Cook, and M. Alexander, *Anal. Chem.* **51,** 1949–1953 (1979).
77. D. Y. Takade, J. M. Reynolds, and J. H. Nelson, *J. Agric. Food Chem.* **27,** 746–753 (1979).
78. N. Zinbo and W. R. Sherman, *Tetrahedron Lett.* **33,** 2811–2813 (1969).
79. P. M. Wiese and R. H. Hanson, *Anal. Chem.* **44,** 2393–2394 (1972).
80. R. H. Wiley, *Org. Mass. Spectrom.* **5,** 675–685 (1971).
81. H. Budzikiewiez and Z. Pelah, *Monatsh. für Chem.* **96,** 1739–1745 (1965).
82. C. G. Daughton, A. M. Cook, and M. Alexander, *J. Agric. Food Chem.* **27,** 1375–1382 (1979).
83. J. B. McBain, L. J. Hoffman, and J. J. Menn, *Pestic. Biochem. Physiol.* **1,** 356–365 (1971).
84. E. Stahl, *J. Chromatogr.* **165,** 59–73 (1979).
85. U. Hezel, *Angew. Chem. Intern. Ed.* **12,** 298–306 (1973).
86. A. Zlatkis and R. E. Kaiser, *HPTLC, High Performance Thin Layer Chromatography,* Elsevier, Amsterdam, 1977.
87. R. W. Frei and P. E. Belliveau, *Chromatographia* **5,** 296–300 (1972).
88. O. Hutzinger, *Anal. Chem.* **41,** 1662–1665 (1969).
89. O. Hutzinger and W. D. Jamieson, *J. Assoc. Off. Anal. Chem.* **54,** 1100–1109 (1971).
90. J. D. Macneil, R. W. Frei, and O. Hutzinger, *Int. J. Environ. Anal. Chem.* **1,** 205–220 (1972).
91. J. E. Bakke, C. E. Price, and R. G. Zaylskie, *Biomed. Mass Spectrom.* **3,** 329–334 (1976).
92. S. Billets, P. S. Lietman, and C. Fenselau, *J. Med. Chem.* **16,** 30–33 (1973).
93. J. Zulalian and R. C. Blinn, *J. Agric. Food Chem.* **25,** 1033–1039 (1977).
94. R. G. Spear, W. J. Popendorf, J. T. Leffingweel, and D. Jenkins, *J. Agric. Food Chem.* **23,** 808–815 (1975).
95. R. L. Holmstead and J. E. Casida, *J. Assoc. Off. Anal. Chem.* **57,** 1050–1055 (1974).
96. S. Sass and T. L. Fisher, *Org. Mass Spectrom.* **14,** 257–264 (1979).
97. D. I. Carroll, I. Dzidic, M. C. Horning, F. E. Montgomery, J. G. Nowlin, R. N. Stillwell, J. P. Thenot, and E. C. Horning, *Anal. Chem.* **51,** 1858–1860 (1979).
98. R. J. Cotter, *Anal. Chem.* **51,** 317–318 (1979).
99. G. Hansen and B. Munson, *Anal. Chem.* **50,** 1130–1134 (1978).
100. H. M. Fales, G. W. A. Milne, H. V. Winkler, H. D. Beckey, J. N. Damico, and R. Barron, *Anal. Chem.* **47,** 207–219 (1975).
101. P. C. Rankin, *J. Assoc. Off. Anal. Chem.* **54,** 1340–1348 (1971).
102. D. F. Hunt, and F. W. Crow, *Anal. Chem.* **50,** 1781–1784 (1978).
103. D. F. Hunt, G. C. Stafford, Jr., F. W. Crow, and J. W. Russell, *Anal. Chem.* **48,** 2098–2105 (1976).
104. E. C. Horning, D. I. Carroll, I. Dzidic, S.-N Lin, R. B. Stillwell, and J.-P Thenot, *J. Chromatogr.* **142,** 481–495 (1977).
105. F. H. Field, 28th Annual Conference, American Society for Mass Spectrometry, New York, Paper PL2, 1980.

106. A. L. Smit and F. H. Field, *J. Amer. Chem. Soc.* **99,** 6471–6483 (1977).
107. J. A. Coburn and A. S. Y. Chan, *J. Assoc. Off. Anal. Chem.* **57,** 1272–1278 (1974).
108. H. D. Beckey and H.-R. Schulten, *Angew. Chem. Int. Ed.,* 403–415.
109. G. W. Wood, J. M. McIntosh, and P.-Y Lau, *J. Org. Chem.* **40,** 636–639 (1975).
110. W. D. Lehmann and H.-R Schulten, *Fresenius Z. Anal. Chem.* **290,** 121–122 (1978).
111. D. F. Hunt, J. Shabanowitz, F. K. Botz, and D. A. Brent, *Anal. Chem.* **49,** 1160–1164 (1977).
112. H.-R. Schulten, *J. Agric. Food Chem.* **24,** 743–750 (1976).
113. A. Alford, *Biomed. Mass Spectrom.* **2,** 229–253 (1975).
114. A. Alford, *Biomed. Mass Spectrom.* **4,** 6–20 (1977).
115. A. Alford, *Biomed. Mass Spectrom.* **5,** 259–283 (1978).
116. D. A. Brent, *J. Assoc. Off. Anal. Chem.* **59,** 1006–1015 (1976).
117. H.-R. Schulten, *Fresenius Z. Anal. Chem.* **293,** 273–281 (1978).
118. H.-R. Schulten and I. Stöber, *Fresenius Z. Anal. Chem.* **293,** 370–376 (1978).
119. P. G. Deo and P. H. Howard, *J. Assoc. Off. Anal. Chem.* **61,** 210–212 (1978).
120. E. E. Knaus, R. T. Coutts, and C. W. Kazakoff, *J. Chromatogr. Sci.* **14,** 525–530 (1976).
121. G. Höfle, W. Steglich, and H. Vorbrüggen, *Angew. Chem. Int. Ed.* **17,** 569–583 (1978).
122. H. Frehse and H. Geissbühler, *Pesticides Residues, a Contribution to their Interpretation, Relevance and Legislation,* Pergamon, Oxford, 1979.
123. N. Oi, M. Horiba, and H. Kitahara, *Agric. Biol. Chem* **43,** 2403–2404 (1979).

# FUNGICIDES

## B. D. RIPLEY

### 1. INTRODUCTION

Fungicides represent only 5 to 10% of synthetic organic pesticide sales and because of their high $LD_{50}$ values are often considered environmentally safe. However, the biological and toxicological properties of a number of fungicides—their ubiquitous nature, persistence, and concentration in the food chain—may constitute a significant source of environmental hazard to man.[1] Moreover, many fungicides contain byproducts or contaminants as a result of the synthetic process or may degrade to products that are inimical. Such products include ethylenethiourea from ethylene-bis-dithiocarbamates; chlorinated dibenzo-*p*-dioxins and dibenzofurans in chlorinated phenols; hexachlorobenzene (HCB), a former fungicide and contaminant in many chlorinated fungicide formulations; and nitrosamines in formulations or due to *in vivo* metabolism.

Chemicals may be harmful in several ways, and today both new and older fungicides are being examined by industry and government to ensure that they do not have adverse effects not only in the short term but also due to repeated use, manufacture over a period of time, and by contamination of food, water, or the environment.[2] The development of a new agricultural chemical is a long and costly process and involves synthesis, testing for efficacy on plants, animals and the environment, identification and evaluation of metabolic and decomposition products, as well as analytical and residue studies.[3]

Agricultural fungicides are widely used on fruits and vegetables which are usually grown in concentrated, intensively cultivated areas. In Ontario in 1978, for example, approximately $1.88 \times 10^5$ kg of fungicides

*B. D. RIPLEY* • Provincial Pesticide Residue Testing Laboratory, Ontario Ministry of Agriculture and Food, University of Guelph, Guelph, Ontario N1G 2W1, Canada.

were applied to fruits and $1.84 \times 10^5$ kg were applied to vegetables; this represented 51 and 44% of all pesticides applied to these crops, respectively.[4] Despite this widespread use of fungicides, terminal residues on foods are low and environmental residues are infrequently found. Unfortunately, few fungicides are analyzed by the commonly employed pesticide screening procedures and hence a paucity of data is available on monitoring studies. In a monitoring study of 11 agricultural watersheds in southern Ontario, the analytical procedures could detect 79 pesticides which represented 74% of the insecticides, 47% of the herbicides, and 10% of the fungicides used in the areas studied.[4a] However, on a volume of use basis this represented 93% of the insecticides, 81% of the herbicides, but only 0.2% of the fungicides.

The environmental applications of mass spectrometry have been recently reviewed and although these three reviews contain over 500 references, few pertain to fungicides and only a little discussion on their contaminants is included.[5–7] Mass spectrometry has become an important analytical tool being used both as a sensitive, specific detector and as a spectroscopic method for the elucidation or confirmation of structure. The advantages and major use of mass spectrometry to pesticide studies apply equally to the fungicide class, and these focus on the detection, confirmation, and quantitation of the parent compound and its transformation products. Fungicides belong to many different chemical classes and additional MS data is presented in a previous review.[8]

## 2. *ANALYTICAL CONSIDERATIONS*

Gas–liquid chromatography is the method of choice for the analysis of pesticide residues. Hence, GC/MS techniques have been widely used for the quantitative analysis or confirmation of a specific pesticide residue. Packed or capillary columns are employed to provide the desired separation of the compound from other residues or coextractives. In this manner, the mass spectrometer may be considered as one of the most specific and sensitive detectors for gas chromatography. Use of selected-ion monitoring (SIM) adds to the sensitivity of the approach and also allows generation for data of a specific compound in the presence of coextractives.

Unfortunately, fungicides in general are more labile than other classes of pesticides such as the organochlorines or the chlorinated aromatic hydrocarbon pollutants. Many fungicides are thermally unstable or nonvolatile and hence are not amenable to direct gas chromatographic analysis. Chemical derivatization affords an alternative approach which involves the preparation of thermally stable and volatile compounds for

GC. However, many analytical methodologies for fungicides still rely on colorimetry, TLC, or other techniques. Reviews on the analysis of fungicides discuss some of these limitations.[9–14] The question remains, therefore: how can a fungicide residue be analyzed or confirmed by MS when it cannot be introduced by GC?

In many metabolic studies TLC is employed to isolate a specific compound and the spot is scraped and introduced into the MS via a direct probe. Careful control of probe temperature may also be employed to preferentially obtain the mass spectrum of a specific compound in a crude extract. Soil extracts containing triphenyltin chloride, for example, were subjected to MS using 30-eV ionization with monitoring of *m/z* 309 while the temperature of the direct-inlet system was programmed at about 30 °C/min from ambient. Unfortunately, proteins interfered and this method was unsuitable for biological tissue.[15] Recently Cotter [16] reviewed the desorption of nonvolatile compounds from an extended probe located in the MS source area. These techniques are not conveniently applied to routine pesticide analyses and may suffer if other pesticides or metabolites are present, although they may be useful for screening sample composites or a few selected samples when MS confirmation is necessary. High-pressure liquid chromatography (HPLC) is becoming widely accepted as a useful means of separating labile compounds such as many fungicides. Under its ambient temperature conditions, the compounds tend to elute intact. Development of LC/MS interfaces and commercially available instruments may provide a satisfactory answer for the MS analysis of many fungicides. Although this technique is still in its infancy, it has been shown to be applicable to labile carbamate pesticides.[17,18]

Compilations of mass spectra are usually based on direct-probe, electron impact (EI) studies and often these spectra do not correspond directly to the species obtained by GC/MS. Hence, the use of data systems and library searches of standard spectra may result in poor matches or misinterpretation of an unknown spectrum. Benson and Damico used time-of-flight mass spectrometry with a vacuum leak direct-inlet system and a vaporizing temperature of 100–180 °C to study the dithiocarbamate and ethylene-bis-dithiocarbamate (EBDC) fungicides.[19] The dithiocarbamates (ferbam, ziram, thiram) have a fragmentation pattern similar to the aryl *N,N*-dimethylcarbamates[20] and show a characteristic cleavage with loss of a $(CH_3)_2NCS_2$ species (*m/z* 120) and fragment ions at *m/z* 88 and 44. All three compounds exhibited a molecular ion. However, that of ferbam was weaker probably due to thermal instability; ziram also had isotope peaks due to Zn. Onuska and Comba observed a similar fragmentation for ziram.[21] The EBDCs (zineb and maneb) did not produce a molecular ion due to thermal decomposition and hence

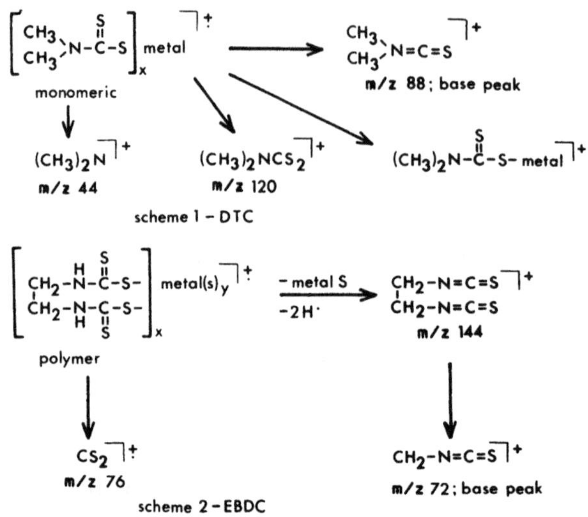


FIGURE 1. Fragmentation pattern for dithiocarbamate (DTC; scheme 1) and ethylene-bis-dithiocarbamate (EBDC; scheme 2) fungicides.

their mass spectra are similar (Fig. 1). The EBDCs are not amenable to direct gas chromatography[22] as they pyrolyze to ethylenethiourea (ETU) and minor amounts of ethylenethiuram monosulfide (ETM) and other components.[23]

Mass spectrometry, therefore, is useful for confirming the intact elution of a compound from the gas chromatograph or to identify the eluting species so that subsequent MS analysis can be employed to correctly confirm residues. Fungicides such as DPX-3217 [1-(2-cyano-2-

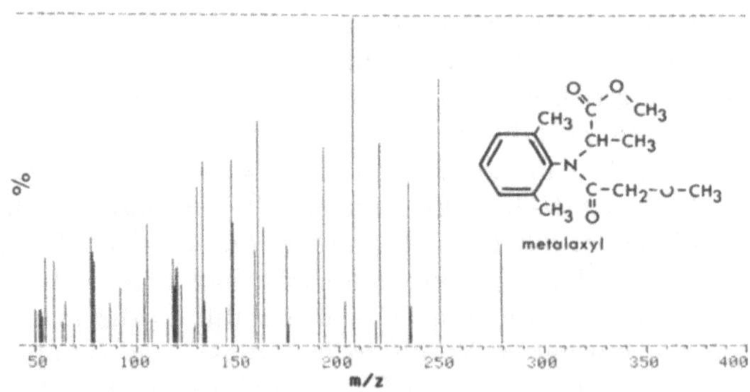


FIGURE 2. Mass spectrum of metalaxyl obtained on a Hewlett-Packard 5992 GC/MS.

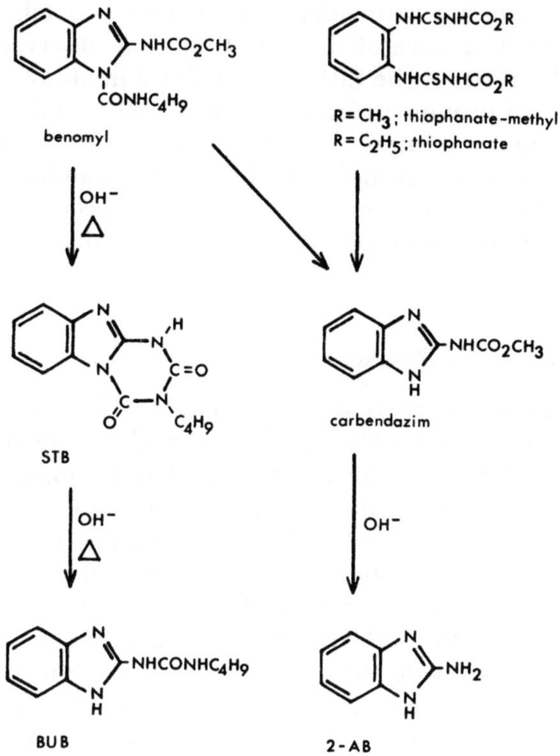


FIGURE 3. Some benzimidazole fungicides and their reactions.

methoxyiminoacetyl)-3-ethylurea],[24] chlorothalonil,[25] and triarimol,[26] have been shown to elute intact from the gas chromatograph. The mass spectrum of the acylalanine fungicide metalaxyl obtained by GC/MS is shown in Fig. 2. Captan is routinely analyzed by gas chromatography, but there is some evidence that it may partially degrade on the GC column to tetrahydrophthalimide.[27] Other studies have shown that the benzimidazole fungicides (Fig. 3), in particular benomyl[28] and ETM,[29] are thermally labile and may break down completely during gas chromatographic analysis.

## 2.1. *Electron Ionization*

The ionization process or source conditions can also significantly affect the mass spectrum of fungicides. Mass spectrometry of most fungicides has been conducted using 70-eV ionization. However, there are indications that some fungicides may be thermally decomposed. Chiba and Doornbos showed that benomyl was thermally unstable and at 70 eV and 180 °C only a carbendazim spectrum was obtained, whereas at 20 eV and 80 °C a trace of the parent ion was observed.[30] A similar effect was observed for the EBDC metabolite 2-imidazolin-2-yl sulfenate which exhibited prominent ions at *m/z* 100 (100%) and 102 (15%) at 75 eV, whereas in a low-energy (15-eV) spectrum, *m/z* 118 ($M^+$) and 117 (M − 1) were also observed.[31] Many fungicides exhibit weak molecular ions during EI studies even at low ionization energies and hence other ionization processes may be more valuable in identifying these compounds.

## 2.2. *Other Ionization Processes*

Field desorption mass spectrometry (FDMS) has been applied to compounds of low volatility and has been shown to strongly reduce thermal degradation of the compound. The EI mass spectra of the phthalimide fungicides (captan, folpet, and captafol) exhibit weak molecular ions, the characteristic chlorine isotopic clusters, loss of chlorine from the molecular ion, and a base peak at *m/z* 79 except folpet, which gave a base peak of *m/z* 76.[32] Using FDMS[33] it was found that the base peak of captan was the molecular ion at *m/z* 298.934. Varying the anode temperature of the field desorption emitter can lead to a number of fragmentation processes (Fig. 4). Due to prominent ions in the higher mass range, impurities can be identified and hence FDMS is ideally suited to the analysis of mixtures. FDMS can also be applied to the analysis of salts, and with dodine (I) the *N*-dodecylguanidinium ion at *m/z* 228.244 is displayed as the base peak while *N*-dodecylguanidine is also observed (*m/z* 227.236) at comparable intensity and yields a doubly charged ion at *m/z* 113.618. Dodine is nonvolatile and hence must be derivatized for gas chromatographic analysis. Newsome[34] used hexafluoroacetylacetone to prepare 2-dodecylamino-4,6-bis(trifluoromethyl)pyrimidine (II) which had a molecular ion (60%) at *m/z* 399 and a base peak at *m/z* 244.

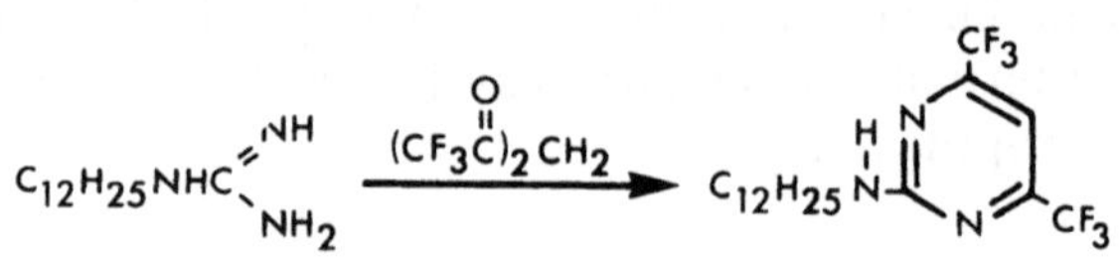


I  II

Negative-ion mass spectrometry has been applied to pentachlorophenol and HCB but its general use for fungicide analysis has been limited.[35,36] Chemical ionization mass spectrometry (CIMS) may be more valuable due to formation of the quasimolecular ion. CIMS has been

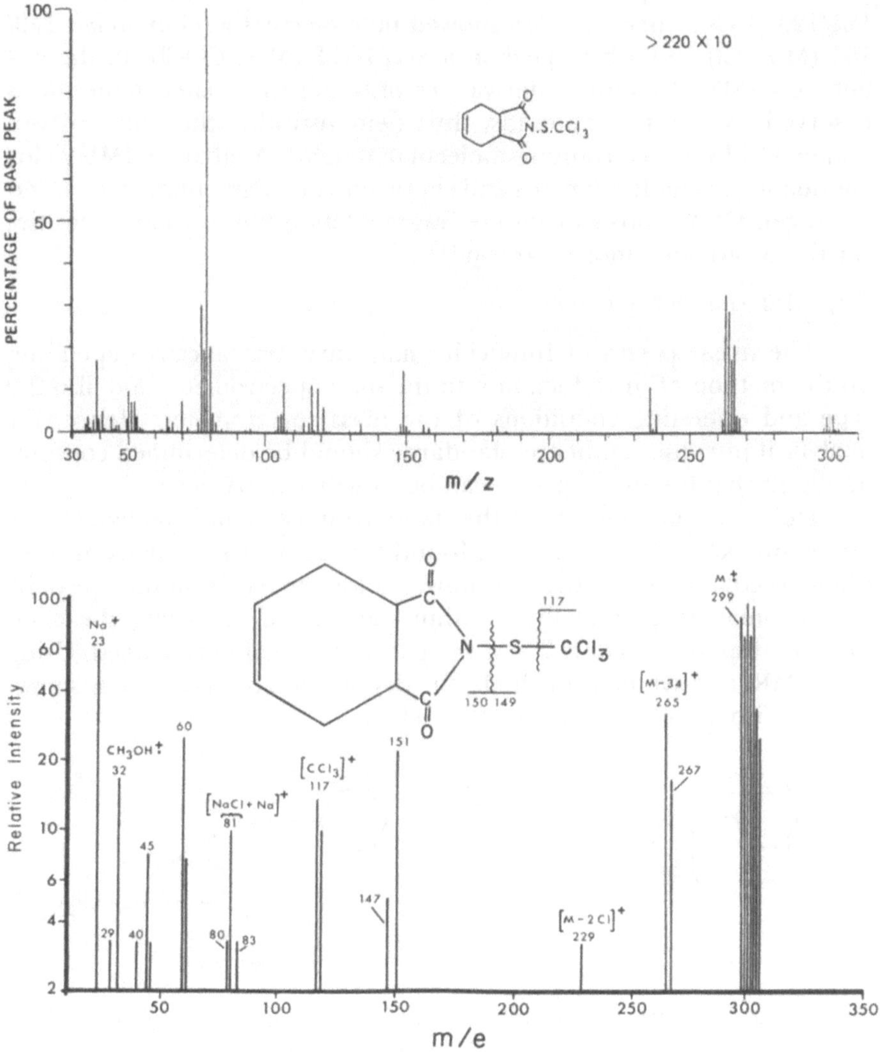


FIGURE 4. Mass spectra of captan. Top, EI, quadrupole MS, and direct probe. Bottom, FD mass spectrum of 91% active ingredient in methanol, best anode temperature, 8–12 mA.

shown to be applicable to the organophosphorus and thermally labile *N*-methylcarbamate pesticides.[37,38] In synthesizing some metabolites of the fungicide triforine it was found that one intermediate, dimeric glyoxylyl chloride, had the following structure as suggested by mass spectrometry and $^{1}H$ NMR (H.CO.CO.O.CHCl.CO.Cl).[39] Isobutane CIMS indicated $[M + H]^+$ at *m/z* 185/187/189 (2 Cl) and a base peak at *m/z* 121/123 (1 Cl), whereas EIMS showed no molecular ion but an *m/z* 149/151 (M − Cl) and a base peak at *m/z* 121/123 (M − COCl). In the isobutane CIMS of oxathiin derivatives only the protonated molecule is observed for most compounds, thus demonstrating the utility of isobutane CIMS in determining molecular weights. Methane CIMS yields abundant quasimolecular ions and characteristic fragmentation ions, and hydrogen CIMS shows extensive fragmentation but low ion intensities for the protonated molecular ion.[40]

### 2.3. *Authenticated Standards*

The mass spectra of fungicides may vary dramatically depending on the method of introduction into the mass spectrometer and also the type and operating conditions of the mass spectrometer. Hence, for analytical purposes authentic standards should be determined concomitantly so that the mass spectra can be exactly compared.

However, the integrity of the standard must be maintained. Cooke and co-workers,[41] for example, found that standard solutions of iprodione in ethanol (or methanol) showed a longer retention time peak by gas chromatographic analysis within minutes of dissolving the compound; after 10 days only 7% of the parent compound remained. Using MS, NMR, and synthetic methods, the degradation compound was shown to be an isomeric rearrangement product.

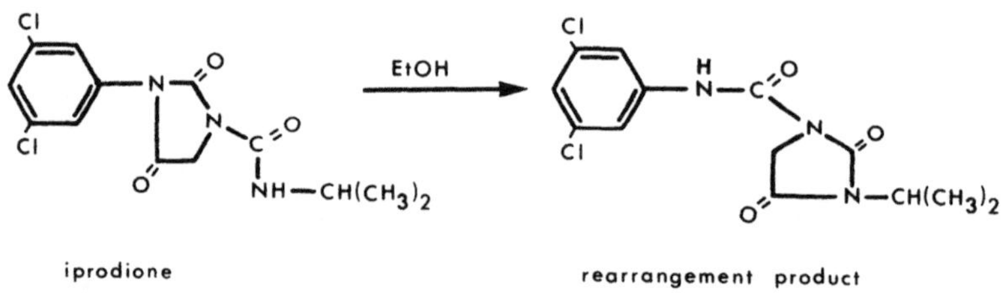


Using direct-insertion MS, the mass spectrum of iprodione showed a weak $M^+$ ion (1%) at *m/z* 329/331/333 (2 Cl) and a base peak at *m/z* 43. Other peaks occurred at *m/z* 314/316/318 (2 Cl, 15%), 244/246/248 (2 Cl, 7%), 187/189/191 (2 Cl, 20%), 124/126 (1 Cl, 11%), 84 (10%), 69 (40%), and 56 (97%). In contrast, the rearrangement product under identical conditions had the same molecular ion cluster (2 Cl, 8%) but a base peak at *m/z* 127 (no Cl), and other ions at *m/z* 187/189/191 (2 Cl, 50%), 142 (70%), 124/126 (1 Cl, 31%), 99/101 (1 Cl, 65%), 69 (45%), 56 (75%), and 43 (97%). The iprodione isomer was much less fungitoxic compared to the parent compound, thus explaining the observed difference in mode of action of iprodione and the related dicarboximide fungicides vinclozolin and procymidone. The other *N*-(3,5-dichlorophenyl)dicarboximide fungicides show similar fragmentation patterns.

Mass spectrometry should also be used with other spectroscopic techniques to confirm the structure of an unknown since any one tool may not provide all the necessary data to completely elucidate the structure. Once the structure and properties of an unknown, a derivative, or a synthetic standard have been determined, less rigorous analysis can be applied to the compound. Therefore, using an authenticated standard, TLC or GC analysis can be routinely used to monitor or determine these components. Subsequent mass spectral analysis can also be better applied knowing the spectrum or ions expected.

Several fungicides undergo chemical or biological transformations which account for their fungitoxic activity.[42,43] Furthermore, many of these compounds are not amenable to direct analytical techniques and hence they have to be chemically modified or derivatized prior to detection. While some of these labile compounds demonstrate a mass spectrum indicative of the parent material using direct-probe introduction, environmental or field-incurred residues are usually not confirmed by direct-probe MS since extensive cleanup and isolation of the residue would be required. Routinely, many of these fungicides are analyzed by GC or GC/MS despite the fact that the GC eluting species may not be the injected compound.

## 3. *ANALYTICAL APPLICATIONS*

Many fungicides and their metabolites are thermally labile or nonvolatile and cannot be determined by direct GC. Chemical derivatization is a useful technique for forming thermally stable, volatile, and/or sensitive compounds for GC analysis.[22] Mass spectrometry or GC/MS is invaluable for confirming the identity of these derivatives for subsequent GC analysis or confirmatory purposes. Derivatization also affords a way to increase the molecular weight of the compound or to produce indicative ions for SIM or mass fragmentography (MF). Derivatization reactions usually focus on active hydrogen atoms in amine, amide, phenol, or carboxylic acid substituents in parent compounds and more often in their more polar metabolites. A wide variety of reagents are available for such purposes[44,45] and some information is available on the MS of these derivatives.[8,46]

Most derivatization reactions proceed under simple, facile conditions and result in the expected product. Sometimes the result is unexpected and MS is imperative to identify the product. Thiourea (III), for example, often produces a tailing peak on direct GC and hence derivatization is indicated. Attempts to prepare the methyl, acetyl, trimethylsilyl, trifluoroacetyl, and sulfonyl derivatives resulted in multiple peaks.[47] Benzoylation produced a unique, sharp, single peak indicative of the monobenzoyl thiourea.

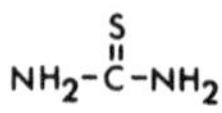

III

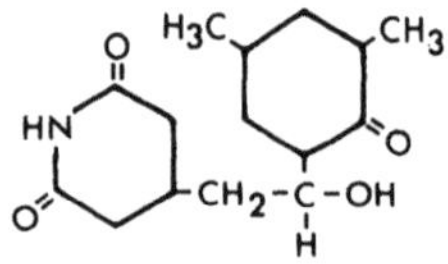

IV

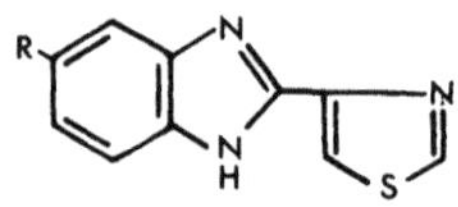

V: R=H
Va: R=OH

Brown showed that the antibiotic cycloheximide (IV) produced a mono and bis derivative on direct silylation.[48] This bis derivative could not be formed quantitatively and was thus considered very labile. Addition of isopropyl alcohol to the derivatization reagent resulted in quantitative formation of the mono-TMS derivative. Gas chromatography/mass spectrometry and derivatization of model compounds were employed to determine that the two TMS groups on the bis derivative were located on the hydroxyl oxygen and the imide nitrogen, and after addition of the alcohol the imide TMS was removed. The mass spectra indicated that pentafluorobenzylation of two metabolites of captan in milk and meat resulted in formation of the *N*-PFB derivative of tetrahydrophthalimide and a *N,O*-di-PFB derivative of tetrahydrophthalamic acid.[49]

The mass spectrum of thiabendazole (V) exhibits a molecular ion at *m/z* 201 and fragment ions at *m/z* 174, 129, 117, and 90 corresponding to sequential loss of HCN, $SCH^{\bullet}$, $C^{\bullet}$, and HCN.[50] N-methylation using dimethylformamide dimethylacetal in $CH_3CN$ medium results in formation of *N*-methylthiabendazole that has an intense molecular ion at *m/z* 215 and weak ions at *m/z* 188, 155, 143, 131, 104, and 90 corresponding to sequential loss of HCN, $SH^{\bullet}$, $C^{\bullet}$, $C^{\bullet}$, HCN, and $CH_2^{\bullet}$ fragments. Rouchaud and Decallonne described a gas chromatographic method to determine residual carbendazim [or as the hydrolysis product of benomyl (Fig. 3)] as the trifluoroacetate derivative.[51] Gas chromatography/mass spectrometry indicated that carbendazim-TFA was monotrifluoroacetylated at the 1-position of the benzimidazole ring. Pyysalo derivatized carbendazim (from benomyl) using acetic anhydride and confirmed by GC/MS that the esterification yielded mainly 2-AB acetate (*m/z* 175, 133, 105, 78) and lesser amounts of carbendazim acetate (*m/z* 233, 218, 191, 175, 160, 133, 105, 90).[52] Quantitation was based on the sum of the two derivatives. Tjan and Jansen prepared the pentafluorobenzyl derivatives of thiabendazole and carbendazim and confirmed the structure by GC/MS.[53] Thiabendazole formed a mono-PFB derivative with a molecular ion at *m/z* 381 and fragment ions at *m/z* 362 (M − F), 214 (M − $C_6F_5$), and 200 (M − $CH_2C_6F_5$) whereas carbendazim formed a di-PFB derivative with a molecular ion at *m/z* 551 and ions at 532 (M − F), 492 (M − $COOCH_3$), 370 (M − $CH_2C_6F_5$), 339 (M − $CH_2C_6F_5$, − $OCH_3$), and 292 (M − $CH_2C_6F_5$, − $OCH_3$, − F, − CO). A *p*-nitrobenzyl derivative for the HPLC confirmation of thiabendazole was characterized by an intense molecular ion at *m/z* 336 and weaker peaks at *m/z* 290 (M − $NO_2$), 214 (M − $C_6H_4NO_2$), 200 (M − $CH_2C_6H_4NO_2$), 174 (M − $CH_2C_6H_4NO_2$, − NCH), and 149 (M − H, − $C_6H_4NC_4H_2NS$).[54] Nose *et al.* formed the monopentafluorobenzoyl derivative of thiabendazole.[55] Gas chromatography/mass spectrometry showed the expected molecular ion at *m/z* 395 and loss of HCN (*m/z* 368), as well as the characteristic pentafluorobenzamide fragments at *m/z* 195 (base peak, $COC_6F_5$) and 167 ($C_6F_5$) and the thiabendazole fragment (*m/z* 200).[56]

The EBDCs have recently come under critical review because of ethylenethiourea (ETU, VI) which has been shown to be carcinogenic and is a product of manufacture, processing, and metabolism of the EBDC fungicides. Mass spectrometry has been employed to identify or confirm EBDCs or ETU and their breakdown products or derivatives after metabolism in rats[31] or plants[57,58] and during thermal breakdown[59] or oxidation inactivation.[60,61] The various mass spectral studies on EBDC breakdown products indicate that many of the metabolites have intense molecular ions or characteristic fragmentation patterns that may be used for confirmatory purposes. Many of these compounds are thermally decomposed and must be introduced to the mass spectrometer ion source intact. Characteristic ions are *m/z* 102 for ETU, 176 for ETM, 70 for 2-imidazoline, and 86 for ethyleneurea. ETU is often analyzed by GC as a chemical derivative.[62–68] Mass spectrometry is used to elucidate the structure of the derivative and confirm residues. In the reaction scheme below, either the first derivative, *S*-(*m*-trifluoromethylbenzyl)ETU (VII) with an abundant molecular ion at *m/z* 260 and fragment ions at *m/z* 159 (100%), 109, 72, and 70, or the second derivative after *N*-trifluoroacetylation (VIII) which has a molecular ion at *m/z* 356 and fragment ions at *m/z* 287 (M − $CF_3$), 159 (base peak; $CH_2C_6H_4CF_3$), 109, and 72 can be used for analysis or confirmation.[63] Variable low-residue analytical data and backgrounds from different substrates make mass spectrometry extremely important for confirmation of ETU residues particularly at the tolerance limit.[69,70]

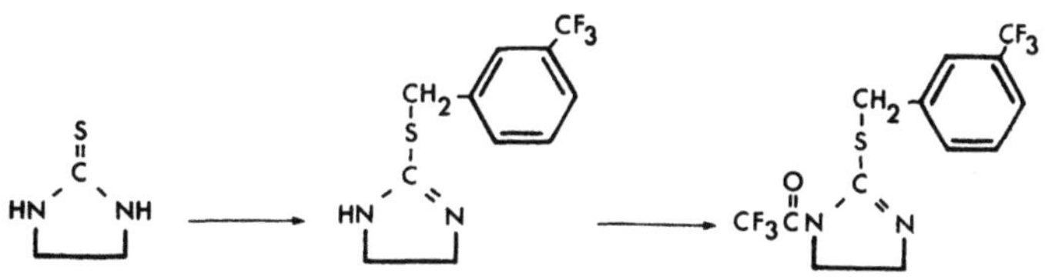


VI VII VIII

Chlorinated benzenes and phenols are widely used industrial chemicals, pesticides, and pesticide precursors which have been reported in all components of the global ecosystem. Use of mass spectrometry for chlorinated pesticides is ideally suited since many of the compounds yield intense molecular ions or fragment ions that contain characteristic chlorine isotopic clusters. Furthermore, many of these compounds can be gas chromatographed intact and since plants do not contain naturally occurring organochlorine components, GC/MS may be used for analysis and confirmation of trace amounts. The mass spectra of these compounds have been reported in detail elsewhere and is only briefly mentioned below.

The chlorinated aromatic fungicides exhibit abundant molecular ions and many contain fragment ions for $C_6Cl_4$ (*m/z* 212), $C_6Cl_3$ (177), $C_6Cl_2$ (142), and $C_6Cl$ (107) species.[71] The $C_6Cl^+$ ion and its precursors must be highly unsaturated species whose structures are most likely noncyclic and acetylenic in nature. It has been proposed that ions derived from certain aromatic and acetylenic compounds may be acyclic.[8] The ubiquitous chlorinated benzenes (in particular HCB (IX) exhibit this fragmentation pattern) feature the stepwise expulsion of chlorine atoms, whereas the polychloronitro compounds such as quintozene (X) initially lose their nitro groups, followed by the stepwise loss of chlorines (Fig. 5). The electron-withdrawing cyano groups in chlorothalonil (XI) result in loss of four chlorines in succession to produce an *m/z* 124 fragment. Pentachlorophenol (XII) fragmentation is influenced by the electron-

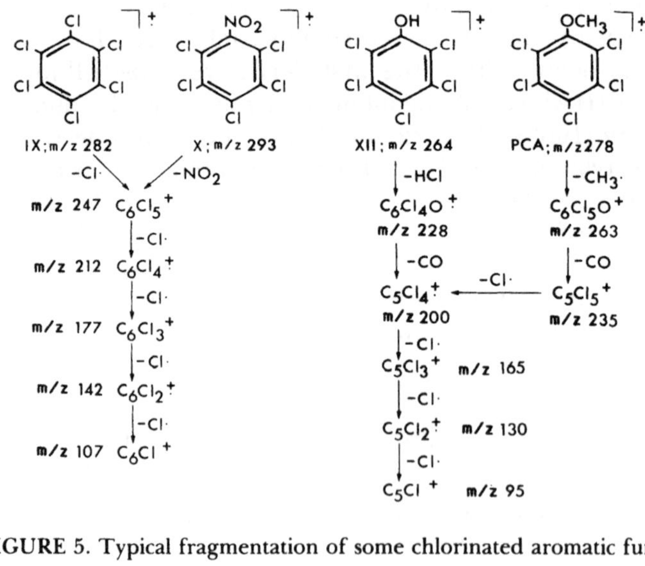


FIGURE 5. Typical fragmentation of some chlorinated aromatic fungicides.

donating hydroxyl group. The molecular ion (base peak) loses HCl to form the ketonic species which then loses CO to produce a $C_5Cl_4$ ion that fragments through successive loss of $Cl^{\bullet}$ analogous to the $C_6Cl_4$ species (Fig. 5).

Phenols can be chromatographed intact using acidic GC columns and GC/MS can be used to identify the free phenols in a wide variety of substrates.[72,73] Better volatility and peak shape is often obtained after derivatization to the methyl ether using diazomethane or dimethyl sulfate. Chlorinated anisoles have major mass spectral peaks for the molecular ion, M − 15 ($-CH_3$), and M − 43 ($-CH_3$, −CO), with each peak exhibiting the characteristic chlorine cluster.[74] Chlorophenolic metabolites as their methyl ethers were confirmed in urine extracts from a pentachlorophenol feeding study with rats and from a human population by GC/MS.[75] Chemical ionization using methane reagent gas produced fairly strong M + 1 quasimolecular ion isotope clusters beginning at *m/z* 245 for tetrachlorophenol isomers, *m/z* 279 for PCP, *m/z* 295 for pentachlorothiophenol, and *m/z* 275 for tetrachloropyrocatechol and tetrachlorohydroquinone. A trichlorodihydrobenzene isomer with $[M + 1]^+$ at *m/z* 240 was tentatively identified in the feeding study samples. In contrast to the simple fragmentation of chlorothalonil (XI), the methyl ether of its 4-hydroxy metabolite (XIII) is strongly influenced by the methoxy group. No M − 15 is observed. However, the *m/z* 217 ion (M − $CH_3$, − CO) is intense and it subsequently fragments with loss of $Cl_2$ and $Cl^{\bullet}$, while there is a minor pathway involving M − 28 (90%) followed by loss of one chlorine.

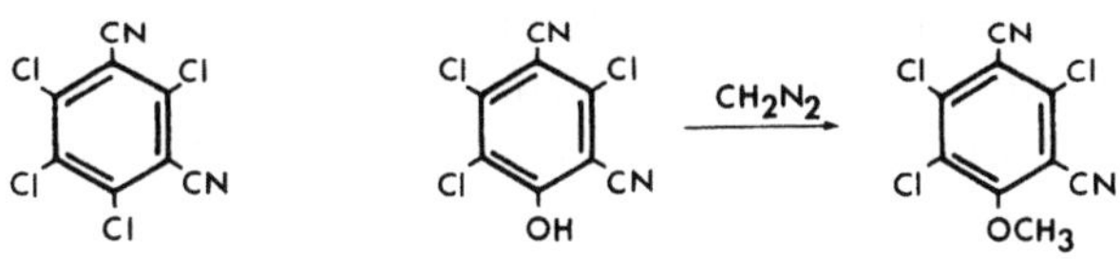


XI XIII

Other derivatives of phenolic compounds are equally feasible for analytical or confirmatory purposes.[22] Chau and Coburn prepared the acetate derivatives of PCP and other phenols using acetic anhydride.[76] PCP-acetate had a molecular ion at *m/z* 306 and after loss of the acetate moiety with rearrangement to PCP, the fragmentation was as expected (Fig. 5).

Microbial methylation converts tetra- and pentachlorophenols into their corresponding polychloroanisoles which have been implicated as causing musty taints in the meat of chickens reared on PCP-treated wood shavings. GC/MS was used to identify microbial decomposition products of sodium pentachlorophenolate.[77] The principal metabolite identified was pentachlorophenol acetate (up to 6.2%), while minor amounts (1%) of tetrachlorocatechol diacetate and penta and tetrachloroanisoles were also found. Tetrachlorophenol isomers, tetrachlororesorcinol, and tetrachlorohydroquinone were identified as their acetyl derivatives. PCP, pentachloroanisole (PCA), HCB, and some dioxins were found in the blood and milk of cows fed technical pentachlorophenol.[78] The negative ion mass spectrum of PCA contained an abundant molecular ion as well as an abundant M − 19 fragment ion due to addition of oxygen coupled with the loss of chlorine from the molecular ion.

Single-ion, multiple-ion, or mass fragmentography is often used with GC to specifically detect and quantitate trace amounts of a component in a mixture of structurally similar compounds. Buser and coworkers used this technique to determine polychlorinated dibenzo-*p*-dioxins and dibenzofurans in commercial pentachlorophenol.[79,80] Buser and Rappe used high-resolution GC to separate the 22 isomers of tetrachlorodibenzo-*p*-dioxin and employed SIM (*m*/*z* 320) to monitor these compounds in several 2,4,5-trichlorophenol and/or hexachlorophene contaminated environmental samples.[81] Similarly, quintozene-treated soils were examined for possible polychloroazobenzenes using the molecular ions at *m*/*z* 388, 422, 458, 492, and 526 for the hexa, hepta, octa, nona, and decachloro isomers, respectively.[82] Hsia and Burant recorded the EIMS of tetrachloroazobenzene and its azoxy analog both which had a base peak of *m*/*z* 145 corresponding to the dichlorophenyl cation, whereas the hydrazo analog had a base peak at *m*/*z* 160 that was assigned as the dichloroanilino radical cation.[83]

Mass spectrometry is often employed as a confirmatory assay to validate positive initial analytical data since the presence of the appropriate mass spectrum at the expected retention time of a compound is convincing evidence of the presence of the compound in a sample. VandenHeuvel *et al.*[84] used such an approach to confirm the anthelmintic and antifungal agent thiabendazole (TBZ; V) and its 5-hydroxy metabolite (HTBZ; Va) after on-column methylation using trimethylanilinium hydroxide in methanol.[84] For quantitative purposes a full scan may not be useful. However, selected-ion monitoring of characteristic ions with their appropriate relative intensities and GC retention time was found to conclusively confirm the presence of these two compounds in animal tissues. The 20-eV spectrum of the *N*-methyl derivative of TBZ is dominated by a characteristic cluster of ions at *m*/*z* 214 (M − 1, 32%), 215 (M, 100%) and 216 (M + 1, 14%), whereas the *N,O*-dimethylated product of HTBZ has intense ions at *m*/*z* 245 (M, 100%) and 230 (M − 15, 42%). These ions and intensities were used for qualitation while *m*/*z* 215 and 245 were used for quantitation.

## 4. CONFIRMATION OF STRUCTURE

The structure of an unknown metabolite may be elucidated using mass spectrometry. Often the change in molecular weight and fragmentation pattern may be sufficient to indicate that the parent material underwent a process such as hydrolysis, oxidation, hydroxylation, or conjugation. Preparation of a synthetic reference material and comparison of the mass spectrum could confirm the structure. On the other hand, many metabolites could be isomeric forms and not distinguishable by mass spectroscopy alone. Hence MS serves as one technique to supplement other spectroscopic evidence on the structure.

Ethylenethiuram monosulfide (ETM; XIV) is an alteration product of the EBDCs and has been shown to be highly fungitoxic. Several structures have been assigned to this compound based on IR and UV data and elemental analysis.[85,85a] Pluijgers *et al.* reexamined the structure, and based on a molecular weight of 176 and other spectroscopic evidence reassigned the structure as XIV.[86] Benson *et al.* confirmed this structure using mass, IR, Raman, and NMR spectroscopy.[87] Structure XIV (*N,N'*-dimethylene-5-imino-1,2,-dithia-4-azolidine-3-thione or 5,6-dihydro-3*H*-imidazo-[2,1-*c*]-1,2,4-dithiazole-3-thione) is consistent with the data. The molecular ion is *m/z* 176 and the ion at *m/z* 178 is due to the sulfur isotope. Although the mass fragmentation pattern (Fig. 6) could be due to another bicyclic structure the S–S bond is indicated in the Raman data, and this structure was later confirmed by chemical methods. This mass spectral datum was obtained by direct probe; Newsome reported that ETM does not chromatograph intact, and using GC/MS and cochromatography showed that the decomposition product was ethylenebis(isothiocyanate).[29]

With compounds that produce possible isomeric products on breakdown or after chemical reactions, mass spectrometry is useful in determining a molecular weight and a general structural formula. Such information could be gleaned for the photochemical product of chlorothalonil.[88] However, other chemical or spectroscopic techniques

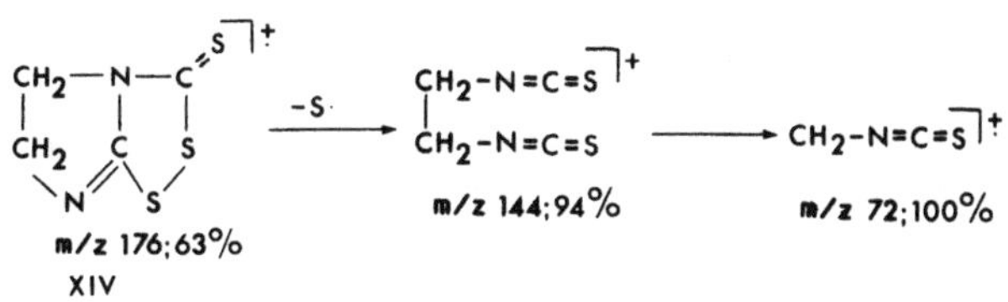


FIGURE 6. Major fragmentation of ETM (XIV).

such as $^{13}C$ NMR were required to elucidate the structure of the product[89] and confirm that substitution of the chlorine at C − 4 (ortho and para to CN) occurred. FT p.m.r. aided in the identification of the photoproducts of quinomethionate.[90]

## 5. BREAKDOWN AND METABOLIC STUDIES

Apart from the numerous biochemical transformations, the principal factors affecting the degradation of a pesticide in the environment are the influence of water, light, and oxygen.[91] Hence, many studies make use of mass spectrometry to monitor or identify the breakdown products of fungicides during hydrolysis or UV irradiation. Such studies have recently been conducted on captan,[92] triforine,[39] piperazine,[93] dichlofluanid,[94] chlorothalonil and its 4-hydroxy metabolite,[88,95,96] and quinomethionate.[90]

Darda *et al.* isolated and identified the major decomposition products of triforine after hydrolysis and photolysis.[39] Triforine hydrolyzed rapidly over the range pH 4.7–9.2 to chloride ion and other acidic hydrolysis products (Fig. 7). After 13 weeks, compounds XVI, XVIII, and XXI were found and these were identified by 70-eV mass spectrometry and TLC with synthetic reference materials. The mass spectrum of XIX, which was not synthesized, showed a molecular ion at *m/z* 148 and fragment ions at *m/z* 114 and 85. Compound XV, the major early hydrolysis product could not be isolated and generally decomposed to XVI. The mass spectrum of XV was essentially identical to XVI with an observed highest mass ion at *m/z* 315/317/319 (3 Cl) which indicated that XV decomposed in the mass spectrometer. A similarly formed compound was examined by high-resolution and field desorption MS and showed a molecular ion at *m/z* 360 (3 Cl). Intermediates XVI and XVIII each attained a maximum level of 20–30% while XIX was the major product containing the piperazine ring. When the acidic hydrolysis solution was concentrated or made strongly alkaline, piperazine was quantitatively formed. Triforine was degraded slowly by sunlight but using UV light, XVII, XXI, *N*-(2,2-dichlorovinyl)formamide (XXII), and 2,2,2-trichloroethane-1,1-diol were isolated and identified. Compound XXII had a molecular ion at *m/z* 139/141/143 (2 Cl) and a base peak at *m/z* 76. After 1 hr in aqueous solution, XXII was found but it was photolabile and quickly decomposed. XXII was shown by TLC to a fungitoxic photoproduct of triforine.

Darda studied the absorption, metabolism, and excretion of triforine in rats.[97] The major urinary metabolite was isolated by extraction and cleaned up by preparative TLC prior to 70-eV mass spectrometry using

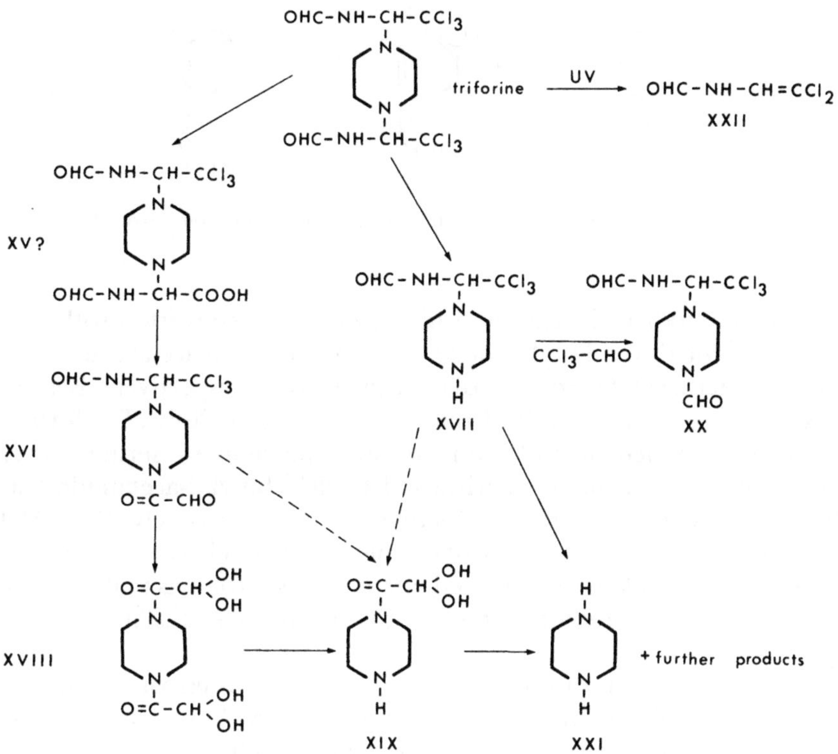


FIGURE 7. Degradation of triforine in water. (From Ref. 39.)

a direct-insertion probe. Despite this cleanup, urea (*m/z* 60) and xanthine (*m/z* 152) were always detected in the background of the mass spectra. The metabolite was identical to synthetic *N*-[2,2,2-trichloro-1-(piperazin-1-yl)ethyl]formamide (XVII, Fig. 7) and was characterized by a molecular ion at *m/z* 259/261/263/265 (3 Cl), a base peak at *m/z* 142 (M − $CCl_3$), and intense ions at *m/z* 215/217/219/221 (M − NHCHO), 188/189 (M − HCl, −Cl), and 85 ($C_4H_9N_2$). Although large amounts of impurities were present in acetone extracts of feces, typical triforine ions containing chlorine at *m/z* 270/272/274/276 and 315/317/319/321 were identified which indicated elimination of some unchanged triforine.

Rouchaud *et al.* examined the metabolism in barley and photodecomposition in water of piperazine, the major breakdown product of triforine.[98,98a] The breakdown products and their chemical derivatives were synthesized and characterized by their significant and diagnostic mass spectral ions. TLC residues were successively treated with *n*-butanol for esterification of acids and then acetic anhydride to acetylate amino and alcoholic functional groups. In this manner, residues of oxalic acid,

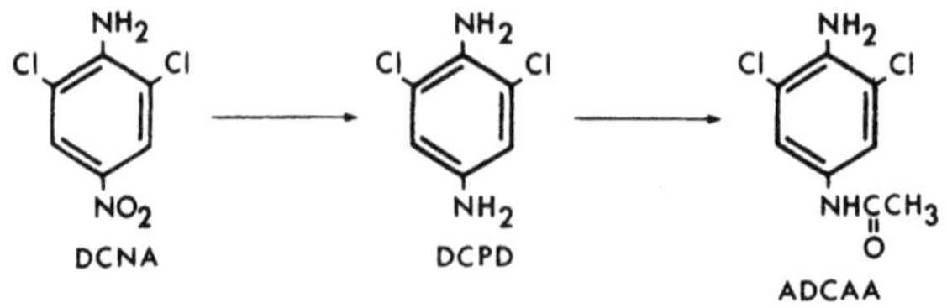


FIGURE 8. Metabolism of DCNA by microorganisms and in soil. (From Ref. 101.)

glycine, iminodiacetic acid, and piperazine were converted with better than 90% yield to dibutyl oxalate, butyl *N*-acetamidoacetate, dibutyl *N*-acetyliminodiacetate, and 1,4-diacetylpiperazine, respectively. Based on a knowledge of the standard materials, the metabolism of triforine in barley was studied by TLC and two other metabolites, serine and ethanolamine, were similarly derivatized to yield butyl 2-acetamido-3-acetoxypropionate and 2-acetamidoethyl acetate, respectively.[99,100] Metabolic results indicated that triforine is metabolized in barley to piperazine, which subsequently is metabolized into nontoxic products that are water-soluble polar compounds, and enter the metabolic pool. Piperazine was not found.

The metabolism of the fungicide 2,6-dichloro-4-nitroaniline (DCNA) by microorganisms and in soil was studied by Van Alfen and Kosuge.[101,102] The identity of the major metabolites (Fig. 8) was ascertained by mass spectrometry and other spectroscopic techniques to be 4-amino-3,5-dichloroacetanilide (ADCAA) and its precursor 2,6-dichloro-*p*-phenylenediamine (DCPD). ADCAA showed a molecular ion at *m/z* 218/220/222 (2 Cl) and major fragments at *m/z* 176 ($NH_2PhCl_2NH_2^+$, due to rearrangement), 175 ($NH_2PhCl_2N^+H$), and 43 ($CH_3C^+{=}O$). The only ion greater than 10% intensity in the spectrum of DCPD was the molecular ion at *m/z* 176/178/180 (2 Cl). In soil, a pink fluorescent compound was also detected and was apparently formed by oxidative dimerization of DCPD since the accurate mass of the parent peak (345.9347) suggested the molecular formula of $C_{12}H_6Cl_4N_4$ (345.9348).

In animals benomyl is eliminated via the urine and feces within 72 hr.[103] The major metabolite was identified as 5-hydroxy-2-benzimidazolecarbamate (5-HBC) which is present in urine as glucuronide and/or sulfate conjugates. Other identified metabolites were carbendazim and 4-hydroxy-2-benzimidazolecarbamate (4-HBC). Mass spectrometry was used to confirm the structure of 5-HBC and 4-HBC which show a similar breakdown pattern (Fig. 9). The *m/z* 148 fragment is considerably stronger in the spectrum of 5-HBC due to stabilization by a *p*-quinone resonance form whereas the *o*-quinone resonance form in 4-HBC is less stable. Fragmentation of other benzimidazoles is similar.[8]

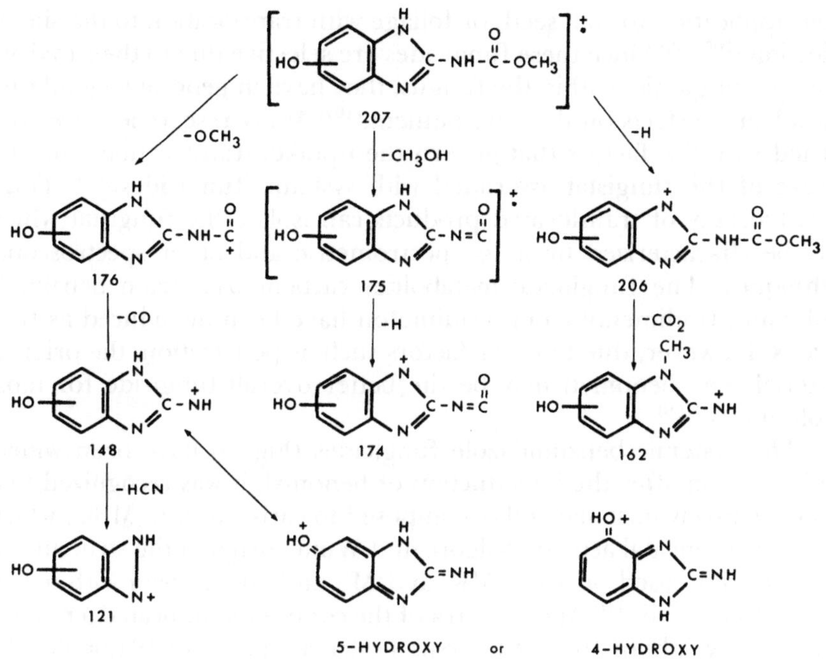


FIGURE 9. Mass fragments for 5-HBC and 4-HBC. (From Ref. 103.)

Hawkins *et al.* examined the metabolic fate of $^{14}$C-tridemorph (XXIII) in rats.[104] Although tridemorph was incompletely absorbed, a negligible amount of unchanged material was excreted and several more polar metabolites were observed in urine, bile, and feces. Poorly defined mass spectra of metabolites with no prominent molecular ion were obtained, but an intense ion at *m/z* 128 was present. This ion was also present as the base peak in tridemorph and is due to the 2,6-dimethylmorpholinylmethylene fragment.

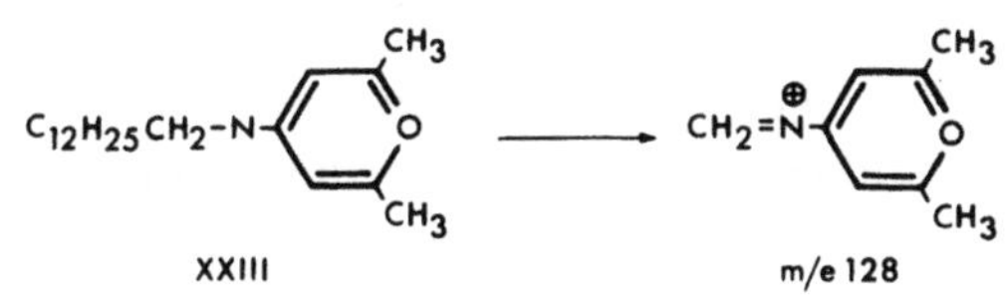


## 6. SYSTEMIC FUNGICIDE STUDIES

Systemic fungicides, which comprise about one-third of the total fungicides in use, are becoming increasingly important due to their selective toxicity against disease, low amounts required for control, and their application to soil, seed, or foliage with translocation to the site of infection.[105,106] Since these fungicides are selective due to their toxicity to only a single site within the fungus, they have in general engendered few adverse effects on the environment.[107] Many researchers are concerned with the factors that govern the uptake, translocation, and the nature of the fungistat associated with systemic fungicides.[108] Thus, TLC bioassay of translocated products can isolate the fungistat which may be characterized by mass spectrometric and other spectroscopic techniques. The fungitoxic metabolites carbendazim from benzimidazoles and tradimenol from triadimefon have been introduced as fungicides. However, due to other factors such as penetration, the original material (i.e., benomyl) may be the better overall fungicide for most applications.[109]

The systemic benzimidazole fungicides (Fig. 3) have been widely studied. Soon after the introduction of benomyl, it was recognized that it was relatively unstable and decomposed to carbendazim (MBC) which also had fungicidal activity. Kilgore and White assigned the structure of carbendazim based on IR, NMR, and MS and comparison with a synthesized standard.[110] Mass spectra of the compound indicated a relative molecular weight of 191.0681 versus a calculated mass of 191.0693. The chemical fate of benomyl under thermal and alkaline conditions was studied[28] and two hydrolysis products, STB and BUB, were isolated and synthesized. Spectroscopic methods indicated their structure (Fig. 3). STB had a molecular weight of 258.28 which was 32 mass units lower than the parent compound due to loss of $CH_3OH$. STB decomposed to BUB which was characterized by a molecular ion at *m/z* 232 and major peaks at *m/z* 159 ($M - H, -HNC_4H_9$) and 131 ($M - H, -CONHC_4H_9$). Chiba examined the instability of benomyl under various conditions and found it to be unstable in various solvents; the resultant precipitate was confirmed by MS to be carbendazim.[30]

The metabolic breakdown of benomyl in plants generally follows

$$\text{Benomyl} \rightarrow \text{carbendazim} \rightarrow \text{2-AB} \rightarrow \text{benzimidazole}$$

and 2-aminobenzonitrile, *o*-phenylenediamine, aniline, and glucosidic conjugates of carbendazim and 2-AB were also reported in melon plants. Rouchaud *et al.* followed this study by examining the fate of benomyl in carrots, strawberries, and apples using GC, GC/MS, and trifluoroacetylation.[93,93a] Treated and control samples, with and without TFA derivatization, were compared by GC to find possible metabolites. Beside the above metabolites, two new metabolites were found in strawberry fruit and calyces, apple pulp and peel, and carrot leaves following the benomyl treatment. Based on the mass spectral data (Fig. 10), one of the unknowns was suggested to be 3-amino-2-(3-methoxycarbonylureido)phenol.

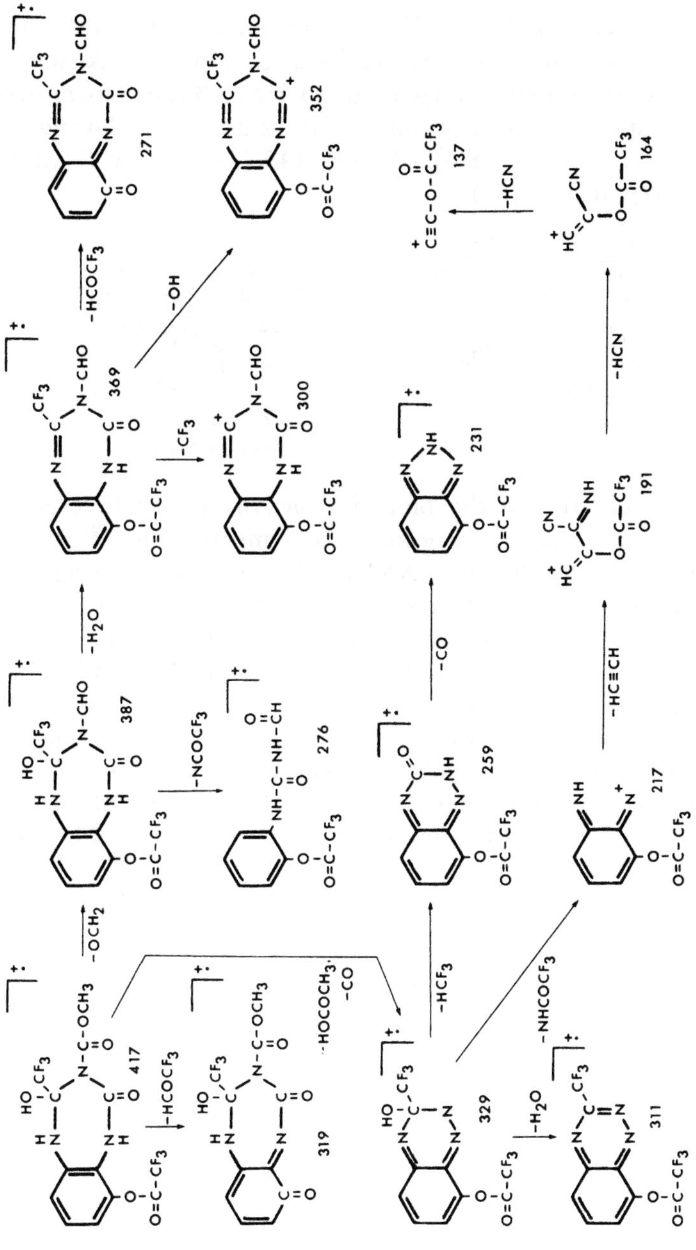


FIGURE 10. Possible fragmentation pattern of 3-amino-2-(3-methoxycarbonylureido)phenol as its trifluoroacetyl derivative. (From Ref. 93.)

Using a complement of techniques (bioassay, GC, GC/MS), Cooke *et al.* showed that procymidone (XXIV) is translocated unaltered from strawberry roots to flowers in sufficient amounts to protect the latter from *Botrytis cinerea* infection.[111] GC/MS analysis of extracts of flower and leaf indicated that XXIV eluted intact and other peaks were due to naturally occurring plant components such as 2,3-dihydrogeraniol. The 70-eV mass spectrum of procymidone indicated the following ions and intensities: *m/z* 283 ($M^+$, 8%), 187 (3), 145 (2), 124/126 (1 Cl, 6/2), 96 (100), 67 (62), and 53 (35).

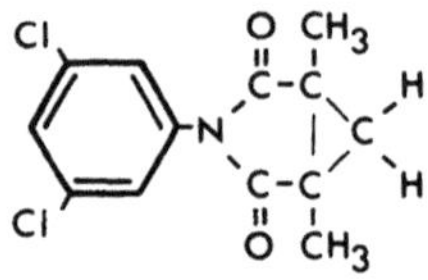

XXIV

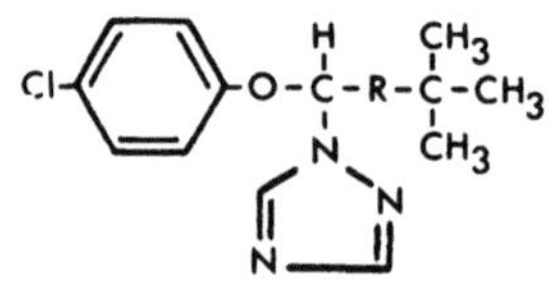

XXV: R = CO
XXVI: R = CHOH

Clark *et al.* studied the biological activity of triadimefon and its breakdown products after photolysis and metabolism.[112] A fragmentation pattern for triadimefon (XXV) is shown in Fig. 11. Fragmentation occurs due to α-cleavage to the carbonyl group and results in a base peak at *m/z* 57. The enzymatic metabolic product was shown to be triadimenol (XXVI), the diastereoisomeric alcohol of XXV, which was systemic and equally as fungitoxic as XXV. Triadimenol had a base peak at *m/z* 112 and significant ions at *m/z* 57, 128/130, 181/183, 208/210, and at *m/z* 238/240 as a result of α-cleavage at the alcohol function. High-resolution mass spectral analysis of *m/z* 226/228 from XXVI indicated the triazole had been eliminated from the molecular ion. Metabolism of

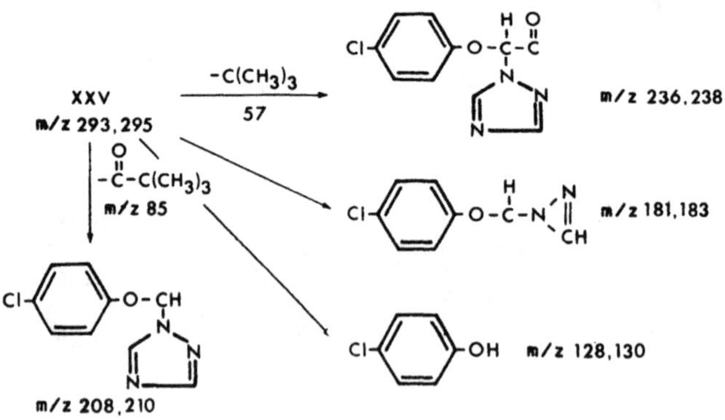


FIGURE 11. Major mass fragments for triadimefon (XXV).

XXV by *Aspergillus niger* resulted in a compound with a mass spectrum similar to XXV but with a strong ion at *m/z* 182 and differing in 14 amu in the molecular ion and in fragments containing the *t*-butyl group. This suggested formation of the isopropyl analog due to demethylation as a result of triadimefon participation in fungal biosynthesis of ergosterol. Photolysis of XXV resulted in formation of the nonfungitoxic 1,2,4-triazole, 4-chlorophenol, and 4-chlorophenyl methyl carbonate.

The 1,4-oxathiin derivatives, carboxin (XXVII) and oxycarboxin (XXVIII), are important systemic fungicides for controlling bunt and smut in cereals and plant rusts, respectively. Carboxin has been shown to readily oxidize to the sulfoxide (XXIX) and sulfone (XXVIII) in plants, soils, and water[113,113a] by cultures of *Rhizopus japonicus*[114] and through photodegradation.[115] Onuska *et al.* examined the CI and high-resolution EI mass spectra of these compounds and confirmed that they fragment as anilides.[116] Compounds XXVII–XXIX have abundant molecular ions in the EI mode and the major fragmentation mode involves loss of the anilino radical ($C_6H_6N^{\bullet}$, *m/z* 92) followed by loss of $C_2H_4$, whereas in the CI mode, abundant ions are observed for $[MH - C_6H_5NH_2]^+$ and $[C_6H_5NH_3]^+$ species.

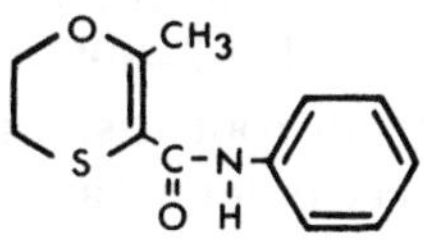

XXVII

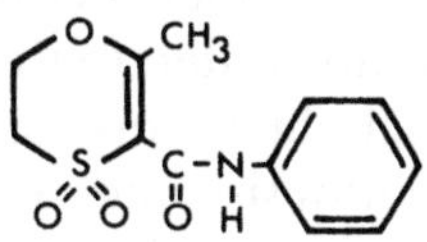

XXVIII

Greenhalgh *et al.* studied the trifluoroacetylation of pesticides and their metabolites that contained a sulfoxide moiety.[117] Unlike other compounds that formed a trifluoroacetoxymethyl sulfide analog, the sulfoxide of XXVII (XXIX) underwent both an additive and transannular Pummerer reaction depending on the reaction conditions.[118] At room temperature and 15 min, XXIX formed the Pummerer intermediate (XXX) due to substitution across the double bond (Fig. 12); XXX

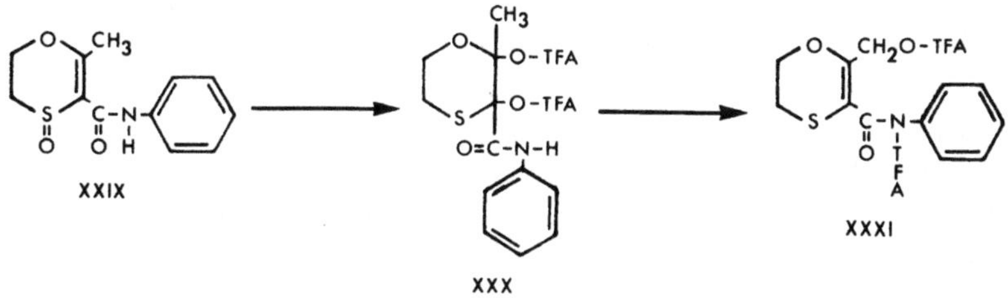


FIGURE 12. Trifluoroacetylation of the sulfoxide of carboxin. (From Ref. 117.)

slowly rearranged to a second di-TFA derivative (XXXI). Reaction at 100 °C for 45 min resulted in formation of XXXI, but various side reactions also occurred.

Using mass spectroscopic and other spectroscopic techniques, the degradation of carboxin and oxycarboxin by *Pseudomonas aeruginosa* isolated from soil was studied.(119) The bacterium hydrolyzed oxycarboxin via 2-(vinylsulfonyl) acetanilide liberating 2-(2-hydroxyethylsulfonyl) acetic acid and aminophenol. Carboxin was similarly hydrolyzed following oxidation to XXIX and then to XXVIII. Briggs *et al.* studied the metabolism of carboxin in growing barley and identified its major soluble metabolite as the *p*-hydroxy phenyl derivative.(120) This metabolite which was monohydroxylated in the aniline ring at the para position had a distinct TLC $R_f$ value, GC retention time as the methyl ether, and mass spectrum which showed main peaks at *m/z* 237, 158, 143, and 108. Some other unstable phenolic compounds were observed and were suggested as dihydroxylated derivatives based on their molecular ion at *m/z* 267.

The metabolism and translocation of pentachloronitrobenzene (PCNB, quintozene, X) and mass spectral characterization of seven glutathione-related conjugates was conducted by Lamoureux and Rusness.(121) Using $^{14}C$-labeled PCNB, only 4–7% of the root-absorbed radioactivity was translocated to the shoots while 57–89% remained in the roots. Polar metabolites were isolated from root tissue and were characterized by synthesis and mass spectrometry as *S*-(pentachlorophenyl)glutathione (XXXII), *S*-(*ar*-tetrachloronitrophenyl)glutathione, *S*-(pentachlorophenyl)cysteine, *S,S'*-(*ar*-tetrachlorophenylene) dicysteine, *S*-(pentachlorophenyl)-*N*-malonylcysteine, *S*-(*ar*-tetrachloronitrophenyl)-*N*-malonylcysteine, and two isomeric forms of *S,S'*-(*ar*-tetrachlorophenylene)diglutathione. Use of different reagents for esterification and acylation of these products aided in their identification and gas chromatography. The mass spectra of the different derivatives, for example, verified that XXXII contained two free carboxyl groups, one free amino group, and a pentachlorothiophenyl nucleus. Molecular ion and fragment ion clusters suggested the number of chlorines on the aromatic nucleus. High-resolution MS of some fragment ions for the amino acid side chains suggested the molecular composition, while key ions for the peptide fragmentation such as *m/z* 371, 359, 239, 200, and 158 for the glutathione side chain or *m/z* 202, 189, 170, 101, and 88 for malonylcysteine were indicative. Figure 13 shows the mass spectrum and ion identity for the acetylated ethyl ester derivative of XXXII. The initial metabolism of quintozene appears to involve three competing reactions: (1) aryl nitro reduction, (2) nucleophilic displacement of the nitro group, and (3) nucleophilic displacement of a chloro group. The metabolism in shoots appears to follow the same pattern as observed in the roots.

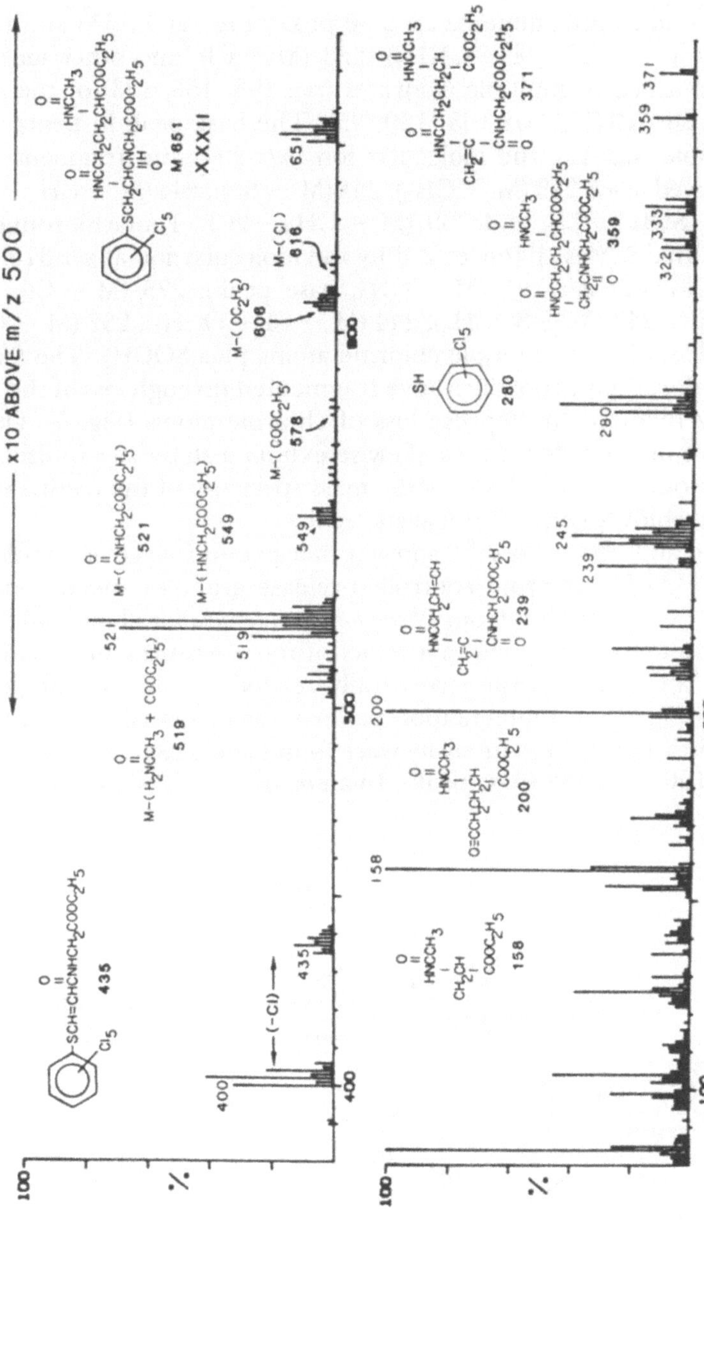


FIGURE 13. Mass spectrum of the acetylated ethyl ester derivative of *S*-(pentachlorophenyl)glutathione. (From Ref. 121.)

The 59.2% of chloroform-soluble radioactivity from quintozene in peanut roots was comprised of 28.7% quintozene, 22.5% pentachloroaniline, 3.1% pentachlorothioanisole, 2.6% pentachlorothiophenol, 0.5% pentachlorophenyl methyl sulfoxide, and lesser amounts of other products.(122) Pentachloroaniline had a base peak at *m/z* 263 ($M^+$) and fragment ions at *m/z* 235 ($M - CNH_2$), 228 (M − Cl), and other ions resulting from loss of multiple chlorines (*m/z* 193, 158, 123) or multiple chlorines plus $CNH_2$ (*m/z* 165, 130, 95). The base peak in pentachlorothioanisole was also the molecular ion (*m/z* 294) and fragment ions were observed at *m/z* 279 ($M - CH_3$), 261 (M − SH), 244 ($M - CH_3, -Cl$), 212 ($M - SCH_3, -Cl$), and 209 ($M - CH_3, -2Cl$). Pentachlorophenyl methyl sulfoxide was characterized by the molecular ion at *m/z* 310 and fragment ions at *m/z* 295 ($M - CH_3$, base peak), 275 (M − Cl), 263 ($M - SCH_3$), 247 ($M - SOCH_3$), 244 ($M - Cl, -OCH_3$), 232 ($M - CCl, -OCH_3$), and loss of multiple chlorine atoms plus $SOCH_3$. The *S*-acetylpentachlorothiophenol derivative fragmented through loss of the acetyl moiety followed by stepwise loss of chlorine atoms (Fig. 5). Other metabolites or their derivatives likewise exhibit a stepwise expulsion of chlorine atoms. Figure 14 shows the mass spectrum of the methyl ester of *S*-(pentachlorophenyl)-2-thioacetic acid.

Cooke and co-workers(123) showed that potato foliage accumulated 0.1–0.5 μg/g ofurace from controlled-release granules and the leaves were afforded protection from *Phytophthora infestans* for about 100 days. GC/MS was used to confirm that intact ofurace was present in leaf extracts. By analogy with the structurally related acylalanine fungicide, metalaxyl (Fig. 2), a fragmentation pattern was proposed. For ofurace, characteristic ions at high intensity were found at *m/z* 281/283 (1 Cl, $M^+$), 186/188, 160, 144, 132 (base peak). In an examination of six acylalanine

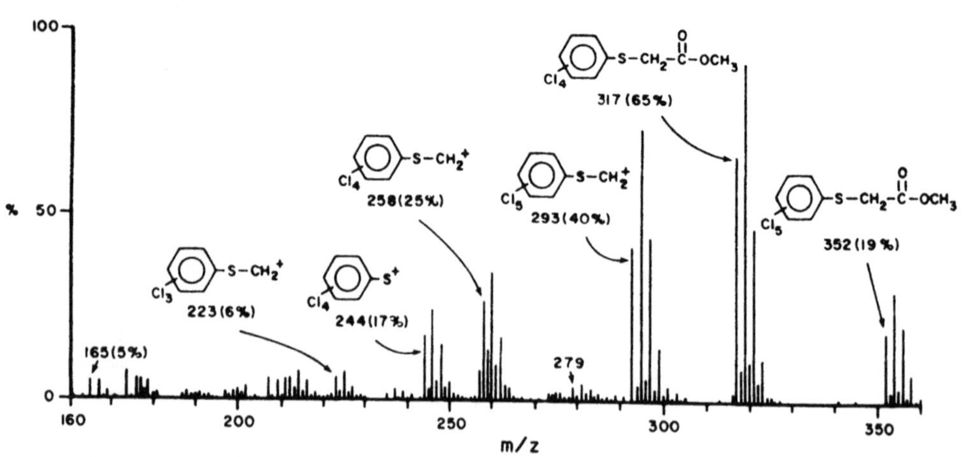


FIGURE 14. GC mass spectrum of the methyl ester of *S*-(pentachlorophenyl)-2-thioacetic acid. (From Ref. 121.)

fungicides and their synthetic intermediates and derivatives, Ripley[124] confirmed all the compound fragments similarly, producing a large number of characteristic and intense ions (Fig. 2).

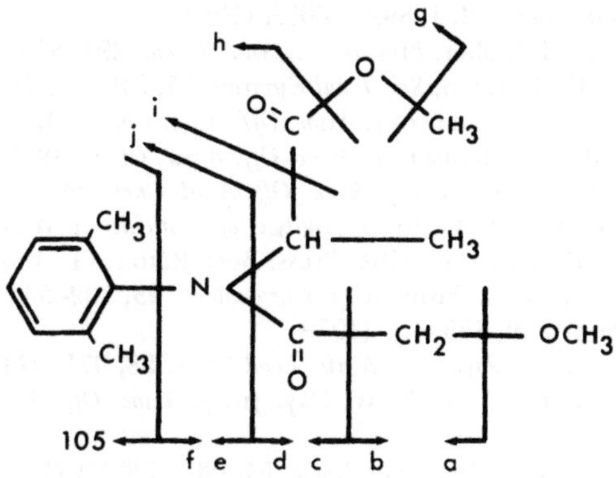


metalaxyl

The acylalanine fungicides show moderately intense molecular ions and the base peak often arises from fragmentation between b and c, (i.e., *m/z* 45 or M − $CH_2OCH_3$ or M − $CH_2R$). Common ions for the c–j N-substituted *N*-2,6-dimethylaniline moiety are *m/z* 160, 132, 117, 105, 91, and 77.

## REFERENCES


1. L. Fishbein, in *Antifungal Compounds,* Vol. 2 (M. R. Siegel and H. D. Sisler, eds.), Marcel Dekker, New York, 1977, pp. 537–598.
2. W. C. von Meyer and R. J. Lacoste, in *Antifungal Compounds,* Vol. 1 (M. R. Siegel and H. D. Sisler, eds.), Marcel Dekker, New York, 1977, pp. 507–529.
3. C. J. Delp, in *Antifungal Compounds,* Vol. 1 (M. R. Siegel and H. D. Sisler, eds.), Marcel Dekker, New York, 1977, pp. 69–88.
4. N. F. Roller, *Survey of Pesticide Use in Ontario,* Economics Branch, Ontario Ministry of Agriculture and Food, Toronto, 1978.

4a. R. Frank, H. E. Braun, G. Sirons, M. V. H. Holdrinet, B. D. Ripley, D. Onn, and R. Coote, *Stream Flow Quality—Pesticides in Eleven Agricultural Watersheds in Southern Ontario, Canada, 1974–77,* International Joint Commission Windsor, Ontario, 1978.

5. A. Alford, *Biomed. Mass Spectrom.* **5,** 259–286 (1978).
6. A. L. Burlingame, T. A. Baillie, P. J. Derrick, and O. S. Chizhov, *Anal. Chem.* **52,** 214R–258R (1980).
7. S. Safe, in *Mass Spectrometry,* Specialist Periodical Report, Vol. 5 (R. A. Johnstone, Sr. Reporter), The Chemical Society, London, 1979, pp. 234–249.
8. S. Safe and O. Hutzinger, *Mass Spectrometry of Pesticides and Pollutants,* CRC Press, Cleveland, 1973.
9. D. J. Austin, K. A. Lord, and I. H. Williams, *Pestic. Sci.* **7,** 211–222 (1976).
10. P. B. Baker and R. A. Hoodless, *Pestic. Sci.* **5,** 465–472 (1974).
11. H. P. Burchfield and E. E. Storrs, in *Antifungal Compounds,* Vol. 1 (M. R. Siegel and H. D. Sisler, eds.), Marcel Dekker, New York, 1977, pp. 463–505.
12. J. Sherma, *J. Chromatogr.* **113,** 97–137 (1975).

13. D. A. M. Watkins, *Pestic. Sci.* **7,** 184–192 (1976).
14. G. Zweig and J. Sherma (eds.), in *Analytical Methods for Pesticides and Plant Growth Regulators* (G. Zweig, ed.), Vol. VIII, Academic, New York, 1976, pp. 249–331.
15. R. Kuwahara, T. Suzuki, and H. Meguro, *Agric. Biol. Chem.* **44,** 669–670 (1980).
16. R. J. Cotter, *Anal. Chem.* **52,** 1589A–1606A (1980).
17. H.-R. Schulten and I. Stöber, *Fresenius Z. Anal. Chem.* **293,** 370–376 (1978).
18. I. Stöber and H.-R. Schulten, *Sci. Total Environ.* **16,** 249–262 (1980).
19. W. R. Benson and J. N. Damico, *J. Assoc. Off. Anal. Chem.* **51,** 347–365 (1968).
20. J. N. Damico and W. R. Benson, *J. Assoc. Off. Agric. Chem.* **48,** 344–354 (1965).
21. F. I. Onuska and M. E. Comba, *J. Assoc. Off. Anal. Chem.* **59,** 622–632 (1976).
22. B. D. Ripley and A. S. Y. Chau, in *Analysis of Pesticides in Water,* Vol. 3, (A. S. Y. Chau and B. K. Afghan, eds.), CRC Press, Boca Raton, FL, 1982, pp. 1–182.
23. W. L. Zielinski, Jr., and L. Fishbein, *J. Chromatogr.* **23,** 302–304 (1965).
24. R. F. Holt, *Pestic Sci.* **10,** 455–459 (1979).
25. J. Northover and B. D. Ripley, *J. Agric. Food Chem.* **28,** 971–974 (1980).
26. R. Frank, O. D. Decker, and E. W. Day, Jr., *J. Assoc. Off. Anal. Chem.* **56,** 11–13 (1973).
27. A. A. Carlstrom, *J. Assoc. Off. Anal. Chem.* **54,** 688–696 (1971).
28. E. R. White, E. A. Bose, J. M. Ogawa, B. T. Manji, and W. W. Kilgore, *J. Agric. Food Chem.* **21,** 616–618 (1973).
29. W. H. Newsome *J. Agric. Food Chem.* **23,** 348–350 (1975).
30. M. Chiba and F. Doornbos, *Bull. Environ. Contam. Toxicol.* **11,** 273–274 (1974).
31. K. Savolainen and H. Pyysalo, *J. Agric. Food Chem.* **27,** 1177–1181 (1979).
32. S. I. M. Skinner and R. Greenhalgh, *Mass Spectra of Insecticides, Herbicides and Fungicides and Metabolites,* Chemistry and Biology Research Institute, Research Branch, Agriculture Canada, Ottawa, 1977.
33. H.-R. Schulten, *J. Agric. Food Chem.* **24,** 743–750 (1976).
34. W. H. Newsome, *J. Agric. Food Chem.* **24,** 997–999 (1976).
35. R. C. Dougherty and K. Piotrowska, *J. Assoc. Off. Anal. Chem.* **59,** 1023–1027 (1976).
36. P. C. Rankin, *J. Assoc. Off. Anal. Chem.* **54,** 1340–1348 (1971).
37. R. L. Holmstead and J. E. Casida, *J. Assoc. Off. Anal. Chem.* **57,** 1050–1055 (1974).
38. R. L. Holmstead and J. E. Casida, *J. Assoc. Off. Anal. Chem.* **58,** 541–547 (1975).
39. S. Darda, R. L. Darskus, D. Eichler, W. Ost, and M. Wotschokowsky, *Pestic. Sci.* **8,** 183–192 (1977).
40. A. G. Harrison and F. I. Onuska, *Org. Mass Spectrom.* **13,** 35–38 (1978).
41. B. K. Cooke, R. S. T. Loeffler, and A. C. Pappas, *Pestic. Sci.* **10,** 393–398 (1979).
42. A. Kaars Sijpesteijn, H. M. Dekhuijzen, and J. W. Vonk, in *Antifungal Compounds,* Vol. 2 (M. R. Siegel and H. D. Sisler, eds.), Marcel Dekker, New York, 1977, pp. 91–147.
43. G. D. Paulson, in *Antifungal Compounds,* Vol. 2 (M. R. Siegel and H. D. Sisler, eds.), Marcel Dekker, New York, 1977, pp. 149–208.
44. K. Blau and G. S. King (eds.), *Handbook of Derivatives for Chromatography,* Heyden, London, 1977.
45. D. R. Knapp, *Handbook of Analytical Derivatization Reactions,* Wiley, New York, 1979.
46. C. F. Poole and A. Zlatkis, *J. Chromatogr. Sci.* **17,** 115–123 (1979).
47. M. Toyoda, S. Ogawa, Y. Ito, and M. Iwaida, *J. Assoc. Off. Anal. Chem.* **62,** 1146–1149 (1979).
48. L. W. Brown, *J. Agric. Food Chem.* **21,** 83–86 (1973).
49. J. H. Onley, *J. Assoc. Off. Anal. Chem.* **60,** 679–681 (1977).
50. A. Tanaka and Y. Fujimoto, *J. Chromatogr.* **117,** 149–160 (1976).
51. J. P. Rouchaud and J. R. Decallonne, *J. Agric. Food Chem.* **22,** 259–260 (1974).
52. H. Pyysalo, *J. Agric. Food Chem.* **25,** 995–997 (1977).

53. G. H. Tjan and J. T. A. Jansen, *J. Assoc. Off. Anal. Chem.* **62,** 769–773 (1979).
54. F. Tafuri, C. Marucchini, M. Patumi, and M. Businelli, *J. Agric. Food Chem.* **28,** 1150–1153 (1980).
55. N. Nose, S. Kobayashi, A. Tanaka, A. Hirose, and A. Watanabe, *J. Chromatogr.* **130,** 410–413 (1977).
56. B. D. Ripley, B. J. French, and L. V. Edgington, *J. Assoc. Off. Anal. Chem.* **65,** 1066–1072 (1982).
57. R. E. Hoagland and D. S. Frear, *J. Agric. Food Chem.* **24,** 129–133 (1976).
58. R. C. Rhodes, *J. Agric. Food Chem.* **25,** 528–533 (1977).
59. W. D. Marshall, *J. Agric. Food Chem.* **25,** 357–361 (1977).
60. W. D. Marshall and J. Singh, *J. Agric. Food Chem.* **25,** 1316–1320 (1977).
61. W. D. Marshall, *J. Agric. Food Chem.* **26,** 110–115 (1978).
62. L. D. Haines and I. L. Alder, *J. Assoc. Off. Anal. Chem.* **56,** 333–337 (1973).
63. R. R. King, *J. Agric. Food Chem.* **25,** 73–75 (1977).
64. R. G. Nash, *J. Assoc. Off. Anal. Chem.* **57,** 1015–1021 (1974).
65. R. G. Nash, *J. Assoc. Off. Anal. Chem.* **58,** 566–571 (1975).
66. W. H. Newsome, *J. Agric. Food Chem.* **20,** 967–969 (1972).
67. J. H. Onley and G. Yip, *J. Assoc. Off. Anal. Chem.* **54,** 165–169 (1971).
68. J. H. Onley, L. Giuffrida, N. F. Ives, R. R. Watts, and R. W. Storherr, *J. Assoc. Off. Anal. Chem.* **60,** 1105–1110 (1977).
69. H. L. Pease and R. F. Holt, *J. Agric. Food Chem.* **25,** 561–567 (1977).
70. B. D. Ripley and D. F. Cox, *J. Agric. Food Chem.* **26,** 1137–1143 (1978).
71. O. Hutzinger, W. D. Jamieson, and S. Safe, *J. Assoc. Off. Anal. Chem.* **54,** 178–186 (1971).
72. W. F. Barthel, A. Curley, C. L. Thrasher, V. A. Sedlak, and R. Armstrong, *J. Assoc. Off. Anal. Chem.* **52,** 294–298 (1969).
73. A. Stark, *J. Agric. Food Chem.* **17,** 871–873 (1969).
74. D. T. Williams, *J. Assoc. Off. Anal. Chem.* **56,** 200–203 (1973).
75. T. R. Edgerton, R. F. Moseman, R. E. Linder, and L. H. Wright, *J. Chromatogr.* **170,** 331–342 (1979).
76. A. S. Y. Chau and J. A. Coburn, *J. Assoc. Off. Anal. Chem.* **57,** 389–393 (1974).
77. B. Rott, S. Nitz, and F. Korte, *J. Agric. Food Chem.* **27,** 306–310 (1979).
78. D. Firestone, M. Clower, Jr., A. P. Borsetti, R. H. Teske, and P. E. Long, *J. Agric. Food Chem.* **27,** 1171–1177 (1979).
79. H.-R. Buser, *J. Chromatogr.* **107,** 295–310 (1975).
80. H.-R. Buser and H.-P. Bosshardt, *J. Assoc. Off. Anal. Chem.* **59,** 562–569 (1976).
81. H.-R. Buser and C. Rappe, *Anal. Chem.* **52,** 2257–2262 (1980).
82. H.-R. Buser and H.-P. Bosshardt, *Pestic. Sci.* **6,** 35–41 (1975).
83. M. T. S. Hsia and C. F. Burant, *J. Assoc. Off. Anal. Chem.* **62,** 746–750 (1979).
84. W. J. A. VandenHeuvel, J. S. Wood, M. DiGiovanni, and R. W. Walker, *J. Agric. Food Chem.* **25,** 386–389 (1977).
85. G. D. Thorn and R. A. Ludwig, *Can. J. Chem.* **32,** 872–879 (1954).
85a. G. D. Thorn, *Can. J. Chem.* **38,** 2349–2358 (1960).
86. C. W. Pluijgers, J. W. Vonk, and G. D. Thorn, *Tetrahedron Lett.* **18,** 1317–1318 (1971).
87. W. R. Benson, R. D. Ross, J.-Y. T. Chen, R. P. Barron, and D. Mastbrook, *J. Assoc. Off. Anal. Chem.* **55,** 44–46 (1972).
88. R. W. Binkley, G. L. Kirstner, V. C. Opaskar, and P. Olynyk, *Chemosphere,* **6,** 163–166 (1977).
89. R. S. T. Loeffler, *Pestic. Sci.* **9,** 310–312 (1978).
90. T. Clark, and R. S. T. Loeffler, *Pestic Sci.* **11,** 451–457 (1980).
91. D. Woodcock, in *Antifungal Compounds,* Vol. 2 (M. R. Siegel and H. D. Sisler, eds.), Marcel Dekker, New York, 1977, pp. 209–249.

92. N. L. Wolfe, R. G. Zepp, J. C. Doster, and R. C. Hollis, *J. Agric. Food Chem.* **24,** 1041–1045 (1976).
93. J. P. Rouchaud, G. J. Lhoest, M. J. Mercier, and J. A. Meyer, *Pestic Sci.* **8,** 23–30 (1977).
94. T. Clark and D. A. M. Watkins, *Pestic Sci.* **9,** 225–228 (1978).
95. R. W. Binkley, C.-Y. J. Tu, and J. C. Johnston, *Chemosphere* **6,** 793–796 (1977).
96. M. B. Szalkowski and D. E. Stallard, *J. Agric. Food Chem.* **25,** 208–210 (1977).
97. S. Darda, *Pestic. Sci.* **8,** 193–202 (1977).
98. J. Rouchaud, J. R. Decallonne, and J. A. Meyer, *Pestic. Sci.* **9,** 139–145 (1978).
98a. J. Rouchaud, C. Moons, J. R. Decallonne, and J. A. Meyer, *Pestic. Sci.* **9,** 305–309 (1978).
99. J. Rouchaud, C. Moons, and J. A. Meyer, *Pestic Sci.* **10,** 438–446 (1979).
100. J. Rouchaud, C. Moons, and J. A. Meyer, *Pestic. Sci.* **10,** 509–518 (1979).
101. N. K. van Alfen and T. Kosuge, *J. Agric. Food Chem.* **22,** 221–224 (1974).
102. N. K. van Alfen and T. Kosuge, *J. Agric. Food Chem.* **24,** 584–588 (1976).
103. J. A. Gardiner, J. J. Kirkland, H. L. Klopping, and H. Sherman, *J. Agric. Food Chem.* **22,** 419–427 (1974).
104. D. R. Hawkins, W. H. Down, L. F. Chasseaud, and J. D. Lewis, *Pestic. Sci.* **5,** 535–542 (1974).
105. K. J. Bent, *Endeavour* (New Series) **3,** 7–14 (1979).
106. L. V. Edgington, R. A. Martin, G. C. Bruin, and I. M. Parsons, *Plant Disease* **64,** 19–23 (1980).
107. R. J. W. Bryde and D. V. Richmond, *Pestic Sci.* **7,** 372–378 (1976).
108. L. V. Edgington and C. A. Peterson, in *Antifungal Compounds,* Vol. 2, (M. R. Siegel and H. D. Sisler, eds.), Marcel Dekker, New York, 1977, pp. 51–89.
109. F. J. Baude, J. A. Gardiner, and J. C. Y. Han, *J. Agric. Food Chem.* **21,** 1084–1090 (1973).
110. W. W. Kilgore and E. R. White, *Bull. Environ. Contam. Toxicol.* **5,** 67–69 (1970).
111. B. K. Cooke, A. C. Pappas, V. W. L. Jordan, and N. M. Western, *Pestic. Sci.* **10,** 467–472 (1979).
112. T. Clark, D. R. Clifford, A. H. B. Deas, P. Gendle, and D. A. M. Watkins, *Pestic. Sci.* **9,** 497–506 (1978).
113. W.-T. Chin, G. M. Stone, and A. E. Smith, *J. Agric. Food Chem.* **18,** 709–712 (1970).
113a. W.-T. Chin, G. M. Stone, and A. E. Smith, *J. Agric. Food Chem.* **18,** 731–732 (1970).
114. P. R. Wallnöfer, M. Königer, S. Safe, and O. Hutzinger, *J. Agric. Food Chem.* **20,** 20–22 (1972).
115. A. W. Wolkoff, F. I. Onuska, M. E. Comba, and R. H. Larose, *Anal. Chem.* **47,** 754–756 (1975).
116. F. I. Onuska, M. E. Comba, and A. G. Harrison, *Biomed. Mass Spectrom.* **3,** 248–254 (1976).
117. R. Greenhalgh, R. R. King, and W. D. Marshall, *J. Agric. Food Chem.* **26,** 475–480 (1978).
118. R. R. King, R. Greenhalgh, and W. D. Marshall, *J. Org. Chem.* **43,** 1262–1263 (1978).
119. R. H. Balasubramanya, R. B. Patil, M. V. Bhat, and G. Nagendrappa, *J. Environ. Sci. Health* **B15,** 485–505 (1980).
120. D. E. Briggs, R. H. Waring, and A. M. Hackett, *Pestic Sci.* **5,** 599–607 (1974).
121. G. L. Lamoureux and D. G. Rusness, *J. Agric. Food Chem.* **28,** 1057–1070 (1980).
122. D. G. Rusness and G. L. Lamoureux, *J. Agric. Food Chem.* **28,** 1070–1077 (1980).
123. L. R. Cooke, D. R. Clifford, A. H. B. Deas, and M. E. Holgate, *Pestic. Sci.* **13,** 686–692 (1982).
124. B. D. Ripley, Translocation and fate of acylalanine fungicides in potatoes (in preparation).

# PYRETHROIDS

## J. M. DESMARCHELIER and M. J. LACEY

### 1. ANALYSIS OF PYRETHROIDS

Although the word *pyrethroid* is used to describe insecticidal analogs of the cyclopropane carboxylate esters of natural pyrethrins, almost the only common chemical feature of pyrethroids is that they are insecticidal carboxylic acid esters.[1–3] This point is also illustrated by the structures of the pyrethroids discussed in the text and summarized in Table 1.

Methods have been developed for the analysis of trace amounts of all pyrethroids, but they are not included in routine determinations of environmental contaminants in multiresidue (MR) programs. There are various reasons for this omission, including their relatively small, although increasing, usage, their great selectivity which results in low levels of application or application of moderate levels of compounds of low mammalian toxicity,[3] and their extensive metabolism in higher organisms.[4] A further reason for the lack of interest in MR analysis of pyrethroids is the photolability of the "old" pyrethroids, that is, esters of pyrethric and chrysanthemic acid (cf. Table 1). However, the "newer" pyrethroids such as permethrin, decamethrin, and fenvalerate are much more photostable[3,5] and even the derivatives of chrysanthemic acid are persistent in certain circumstances, such as in the mothproofing of wool[6] or on stored grain.[7,8]

A further reason for the noninclusion of pyrethroids in MR screening has been the absence of a method suitable for all pyrethroids in the presence of other contaminants. For example, decomposition can occur on some gas chromatography (GC) columns.[9] Unlike natural[10,11] and several synthetic[12,13] pyrethroids, resmethrin is not sensitive to electron

*J. M. DESMARCHELIER* and *M. J. LACEY* • CSIRO Division of Entomology, P.O. Box 1700, Canberra, A.C.T. 2601, Australia.

TABLE 1
*Pyrethroids of Structure $ACO_2B$*[a]

| Name | Moiety A | | Moiety B | |
|---|---|---|---|---|
| Chrysanthemic acid | 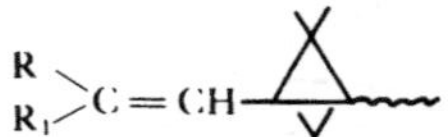 | $R,R_1 = CH_3$ | H | |
| Allethrin | 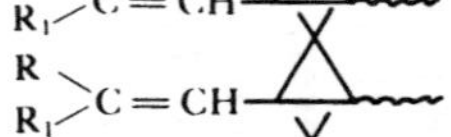 | $R,R_1 = CH_3$ | 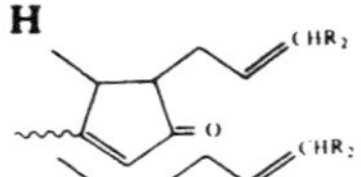 | $R_2 = H$ |
| Cinerin[b] | 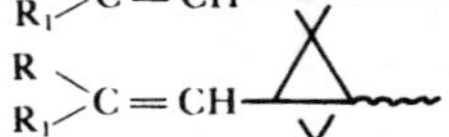 | [b] | 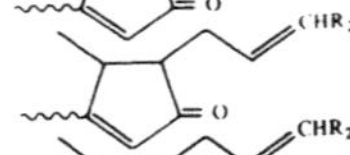 | $R_2 = CH_3$ |
| Jasmolin[b] | 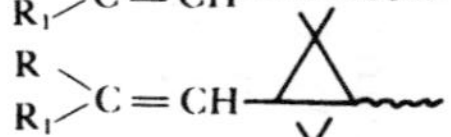 | [b] | 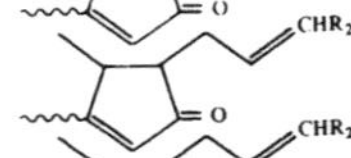 | $R_2 = C_2H_5$ |
| Pyrethrin[b] | 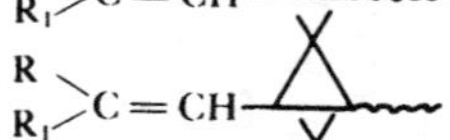 | [b] | 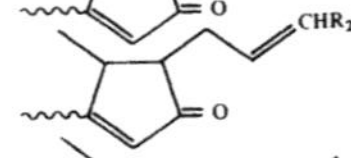 | $R_2 = C_2H_3$ |
| Isopyrethrin[b] | 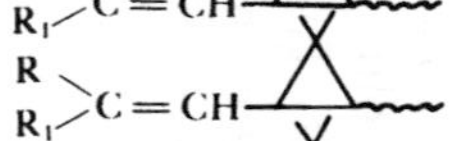 | [b] | 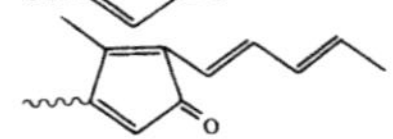 | |
| Resmethrin | 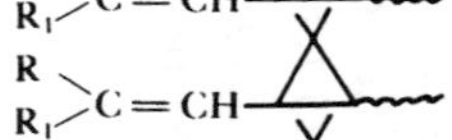 | $R,R_1 = CH_3$ | 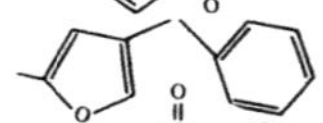 | |
| Tetramethrin | 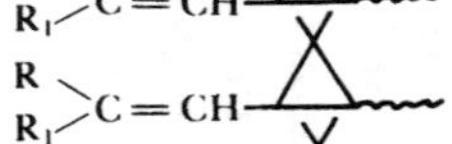 | $R,R_1 = CH_3$ | 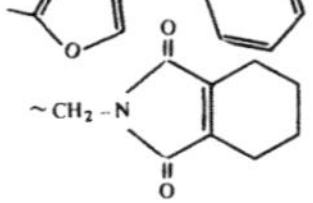 | |
| Pyrethric acid | 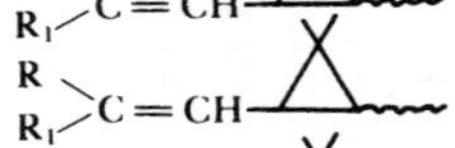 | $R = CH_3$, $R_1 = CO_2CH_3$ | H | |
| Phenothrin | 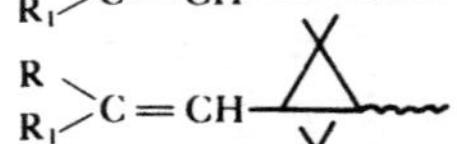 | $R,R_1 = CH_3$ | 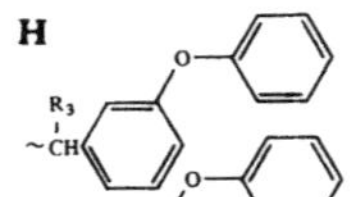 | $R_3 = H$ |
| Permethrin | 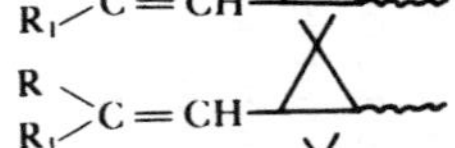 | $R,R_1 = Cl$ | 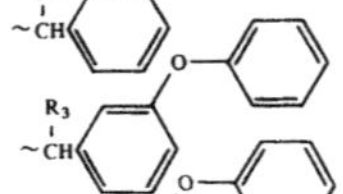 | $R_3 = H$ |
| Cypermethrin | 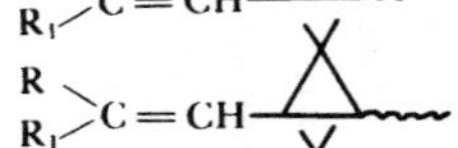 | $R,R_1 = Cl$ | 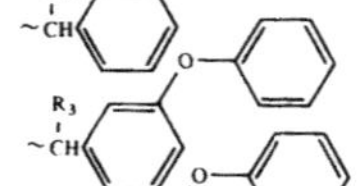 | $R_3 = CN$ |
| Decamethrin | 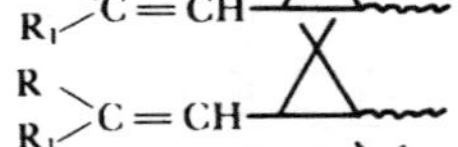 | $R,R_1 = Br$ | 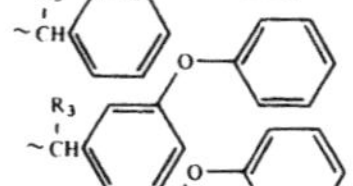 | $R_3 = CN$ |
| Fenvalerate | 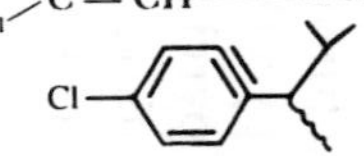 | | 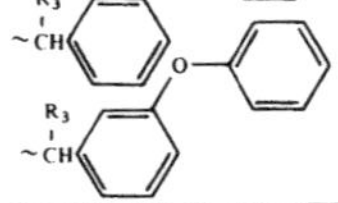 | $R_3 = CN$ |

[a] Stereochemistry has not been indicated. – methyl group; ~ point of attachment.

capture. The flame ionization detector can be used to detect pyrethroids,[9] but is not very sensitive to bioresmethrin and lacks selectivity.[14,15] High-performance liquid chromatography (HPLC) has been used to detect several pyrethroids[16–19] but, as the maximum absorbance at approximately 225 nm is not specific, ambiguities could occur in MR determinations.

Because of the greater use of pyrethroids, the need to include them in MR studies, or at least on average daily intake, is increasing. The potential of mass spectroscopy (MS) for such studies is explored in this review, where the emphasis is on MS as an analytical tool.

## 2. *GENERAL PRINCIPLES FOR APPLICATIONS OF MASS SPECTROSCOPY*

Mass spectroscopy (MS) can be coupled directly with GC[20] and HPLC to provide sensitive and unambiguous identification of specific compounds in environmental analysis. Procedures for quantitation by MS of trace levels of compounds in complex mixtures have been summarized in an excellent monograph by Millard.[21] Selected-ion monitoring lowers the limit of detection by real-time selection of diagnostic ions, and the chemical specificity of the mass spectrometer can also be enhanced by removing interfering ions at higher resolving powers, despite the resulting decreased absolute sensitivity.[22] Cautionary comments on the difference between ion current detection and unambiguous identification are apposite and opportune.[23]

MS is a highly versatile analytical technique, and alternative modes of ionization that could be appropriate for the analysis of pyrethroids and their metabolic or alteration products include electron ionization (EI), positive-ion chemical ionization including charge exchange (PCI), negative-ion chemical ionization including electron capture (NCI), field desorption (FD), and fast atom bombardment (FAB). Some of these modes are exemplified for pyrethroids in this review.

Adsorption chromatographic techniques coupled with MS for MR screening studies of complex mixtures are time consuming and may be compromised by retention or destruction of unknowns.[24,25] A number of methods are presently available for ionizing involatile components in the mass spectrometer[26,27] and could be applied to obtain characteristic mass spectra of less tractable pyrethroids or alteration products. Furthermore, rapid MR screening is possible using a tandem mass spectrometer as the sole source of chemical separation[28] and this is discussed later in the review.

## 3. EI MASS SPECTRA OF PYRETHROIDS

The general features of the EI mass spectra (direct insertion) of the separated natural pyrethroids (pyrethrin I and II, cinerin I and II, and jasmolin I and II, cf. Table 1) have been discussed.[29,30] The $M^{+\cdot}$ ions are of weak relative abundance and the primary fragmentation modes are mainly associated with fission at the internal ester linkage. Indeed, the fragmentation patterns are essentially a superimposition of those observed for the component acid and alcohol moieties.[29,30]

A formal representation of the primary fragmentation modes of the natural pyrethroids is shown in Scheme I. The mass spectra of pyrethrin I, cinerin I, and jasmolin I are dominated by a common base peak due to the chrysanthemyl ion of *m/z* 123. The analogous common

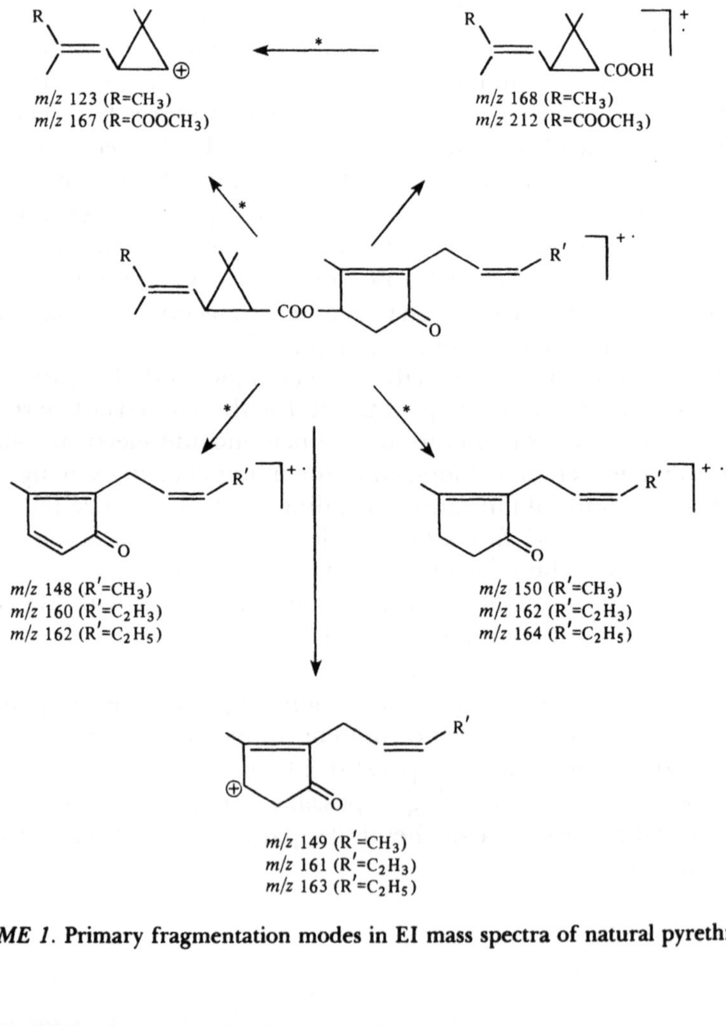


*SCHEME 1*. **Primary fragmentation modes in EI mass spectra of natural pyrethroids**

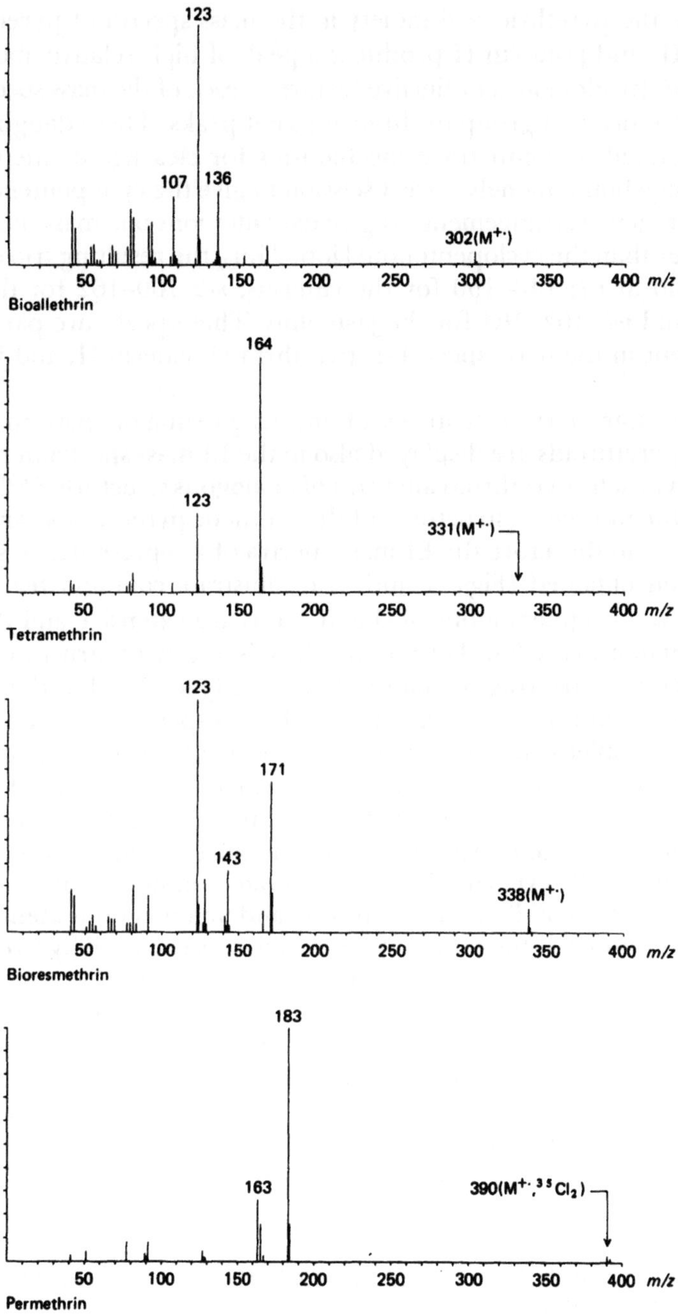


FIGURE 1. EI mass spectra (70 eV) of some synthetic pyrethroids. VG Micromass 70-70 mass spectrometer, direct insertion, source temperatures (140–180 °C. Ions below *m/z* 40 have been omitted from the spectra.

ion from the pyrethric acid moiety in the mass spectra of pyrethrin II, cinerin II, and jasmolin II produces a peak of high relative abundance at *m/z* 167. In addition, a distinctive feature of each of the mass spectra[29,30] is the presence of a group of three adjacent peaks. These daughter ions are attributed to competitive mechanisms for cleavage of the *O*-cyclopentenonyl bond, namely, direct scission to give the cyclopentenonyl ion and hydrogen rearrangements to give daughter ions one mass unit higher and lower than the cyclopentenonyl ion. Thus, the resulting trio of peaks are found at *m/z* 148–150 for the cinerins, *m/z* 160–162 for the pyrethrins, and *m/z* 162–164 for the jasmolins. These peaks are particularly prominent in the mass spectra of pyrethrin II, cinerin II, and jasmolin II.

The characteristic features of the fragmentation pattern for the natural pyrethroids are displayed also in the EI mass spectrum reported for the synthetic pyrethroid allethrin of analogous structure.[29,31] Nevertheless, the molecular structures of the synthetic pyrethroids vary widely (Table 1) and therefore the EI mass spectra of a representative selection have been obtained (Figs. 1 and 2) to illustrate common features. In general, the fragmentations of the $M^{+\cdot}$ ions are extensive and their relative abundances are low. For the pyrethroids of general structure $ACO_2B$ (Table 1), primary fragmentation is essentially localized at the internal ester group, and the base peak in the EI mass spectrum is due either to $A^+$ ions (bioallethrin, bioresmethrin, cypermethrin, fenvalerate, decamethrin) or to $B^+$ ions (tetramethrin, permethrin). Common fragment ions in the mass spectra can be attributed to the presence of a common acid moiety A (bioallethrin, tetramethrin, bioresmethrin) or a common alcohol moiety B (cypermethrin, fenvalerate, decamethrin).

The use of GC/MS in the separation and unambiguous identification of small amounts of the natural pyrethroids was reported by Crombie[32] and the distinctive features of the previously mentioned trio of peaks in the EI mass spectra were again emphasized. The retention time in GC/MS provides an added dimension of information to establish identity when selected ion monitoring is used for quantitative estimations. By monitoring the base peak (*m/z* 123) in GC/MS of the synthetic pyrethroid phenothrin, the limit of detection (S/N ~ 3) achieved by the present authors was 50 pg (unpublished work, VG Micromass 70-70 mass spectrometer).

## 4. *POSITIVE-ION CHEMICAL IONIZATION MASS SPECTRA OF PYRETHROIDS*

The reported absence of some $M^{+\cdot}$ ions of natural pyrethroids in GC/MS using EI[32] was circumvented by Holmstead and Soderlund[33]

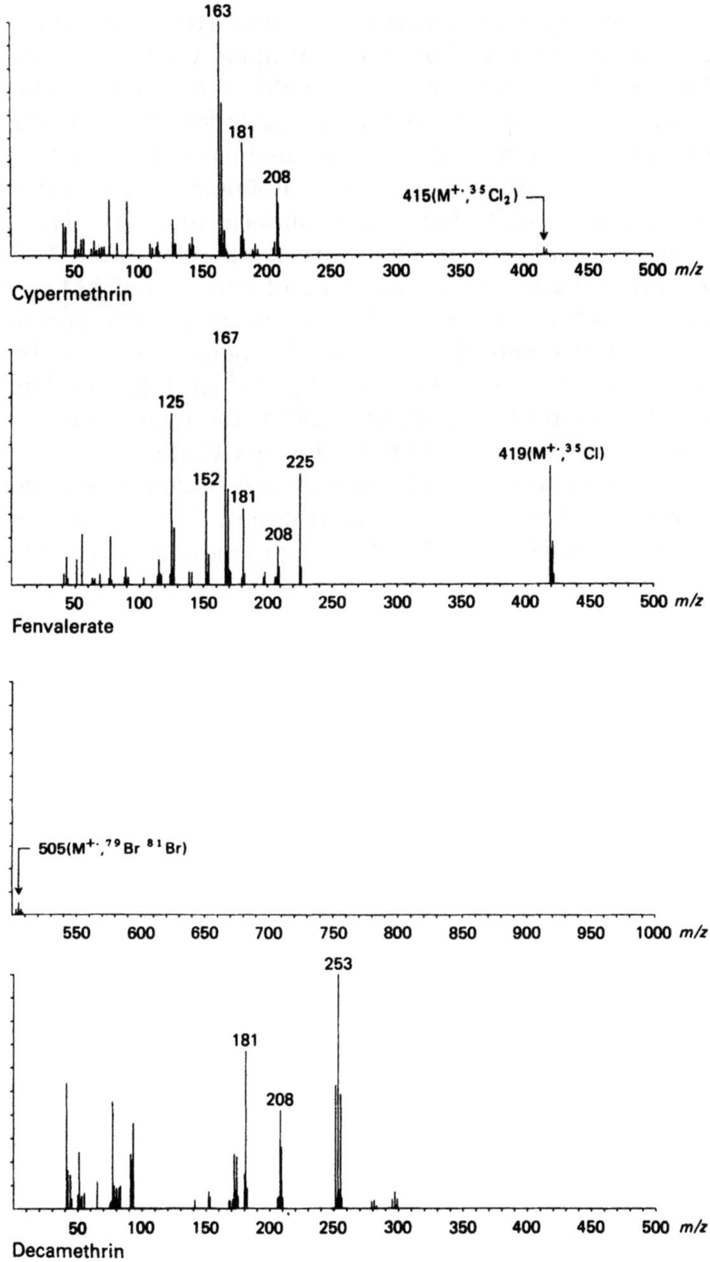


FIGURE 2. EI mass spectra (70 eV) of some synthetic pyrethroids. (See footnote to Fig. 1.)

who used positive-ion chemical ionization (PCI) in the GC/MS analysis of the natural pyrethroids to increase the proportion of total ion current carried by the higher mass ions and to simplify the fragmentation patterns. The number of fragmentation modes in the PCI (methane) and PCI (isobutane) mass spectra are considerably fewer than in the corresponding EI mass spectra, and the protonated molecular ions [$(M + H)^+$] usually provide the base peak in the PCI (isobutane) mass spectra. Thus, using selected-ion monitoring of the characteristic $(M + H)^+$ ions in PCI (isobutane) mass spectra of the six component esters, Holmstead and Soderlund[33] achieved a lower limit of detection of 114 ng of total pyrethrum extract injected onto the column for GC/MS. Furthermore, they also achieved a ready detection of 11.4 ng of total extract by single-ion monitoring of the $(M + H)^+$ ion of pyrethrin I. It is evident therefore that PCI can provide additional aid in the unambiguous identification and trace detection of pyrethroids by GC/MS.

A formal representation of the principal fragmentation modes reported for the PCI (isobutane) mass spectra of the natural pyrethroids is shown in Scheme II. Several of the daughter ions appear to arise from

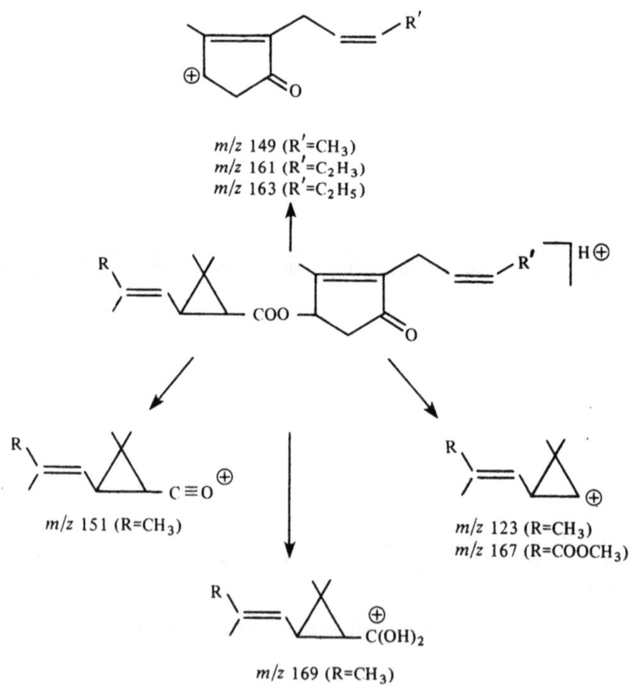


*SCHEME 2.* Principal fragmentation modes in the PCI (isobutane) mass spectra of natural pyrethroids.

initial proton transfer to the internal ester group and, as would be expected, even-electron daughter ions are generally formed in the fragmentations.

The thermal isomerization of a pyrethrin mixture was explored by Holmstead and Soderlund[33] using GC/MS with PCI (isobutane) to examine the products. $(M + H)^+$ ions in the PCI (isobutane) mass spectra of iso-pyrethrin I and iso-pyrethrin II exhibited more extensive fragmentation than those of their progenitors. The increased likelihood that PCI can provide molecular mass information has also encouraged the use of this technique in studies of photochemical transformations of synthetic pyrethroids[34,38] and of their metabolism.[39]

The PCI mass spectra of a number of synthetic pyrethroids have been obtained by the present authors (unpublished work) to illustrate general features. The PCI (isobutane) mass spectra of tetramethrin and bioresmethrin (Fig. 3) demonstrate the considerably increased relative

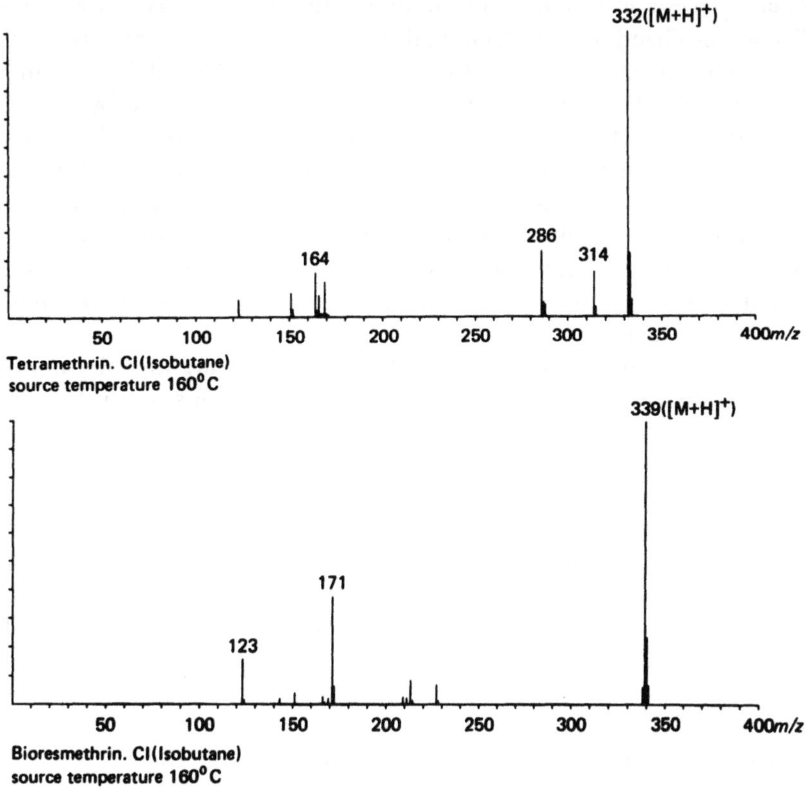


FIGURE 3. PCI (isobutane) mass spectra of tetramethrin and bioresmethrin. Source pressure 60 Pa, VG Micromass 70-70 mass spectrometer. Reactant ions omitted from spectra in Figs. 3–6. Source pressures finely adjusted to maximize abundances of higher mass ions.

abundances of the $(M + H)^+$ ions compared to the $M^{+\cdot}$ ions in the EI mass spectra (Fig. 1). Characteristic even-electron fragment ions are also formed, which can be used in addition to the $(M + H)^+$ ions for identification and selected-ion monitoring.

The PCI (isobutane) mass spectra of other synthetic pyrethroids (fenvalerate, decamethrin, bioallethrin, cypermethrin, and permethrin) have also been obtained (present authors, unpublished work), but these show more extensive fragmentations of the respective $(M + H)^+$ ions. Nevertheless the fragment ion currents are generally concentrated among fewer ion types than in their EI mass spectra (Figs. 1 and 2) and, apart from fenvalerate, the relative abundances of $(M + H)^+$ ions are generally higher than those of $M^{+\cdot}$ ions in the EI mass spectra.

It is possible to gain a greater concentration of total sample ion current in high-mass ions in PCI by choosing a reactant ion such as $NH_4^+$ whose conjugate base has a closer gas phase proton affinity than isobutane to that of the pyrethroid. Under the conditions of PCI (ammonia), parent ions from the analyte such as $(M + H)^+$ ions and/or collision-stabilized ion–molecule adducts $[(M + NH_4)^+]$ may be formed with much less excess energy than the respective $(M + H)^+$ ions in PCI (isobutane) or PCI (methane), and fragmentation is thereby curtailed. For instance, PCI (ammonia) has been used recently by Wright and others[18] to confirm molecule masses of conjugates in the metabolism of cypermethrin. PCI (ammonia) is a highly selective mode of soft ionization and other components of an environmental sample having significantly lower gas phase proton affinities or basicities than the pyrethroids may be transparent to the ionization. Chemical noise in the analysis is thereby reduced.

The PCI (ammonia) mass spectra of fenvalerate and decamethrin (Fig. 4) show the total ion current in each case almost entirely concentrated in the $(M + NH_4)^+$ ions. The minor fragmentation contrasts strikingly with that in their respective EI mass spectra (Fig. 2). Characteristic even-electron fragment ions (cf. Figs. 1 and 2) are more evident in the PCI (ammonia) mass spectra of bioallethrin and cypermethrin (Fig. 5). While the $(M + NH_4)^+$ ion still provides the base peak in the PCI (ammonia) mass spectrum of cypermethrin, the ion current in the high-mass region of the spectrum of bioallethrin is shared between the $(M + H)^+$ and $(M + NH_4)^+$ ions, possibly reflecting the higher gas phase proton affinity of bioallethrin than that of cypermethrin.

Unlike PCI with more acidic reactant ions such as $CH_5^+$, the source temperature can be a crucial factor in controlling the fragmentation patterns in PCI (ammonia). This is exemplified in Fig. 6 for the PCI (ammonia) mass spectra of permethrin at different source temperatures. At 180 °C, the $(M + NH_4)^+$ ion abundance is relatively small, even though it is more significant than the $M^{+\cdot}$ ion abundance in the EI mass

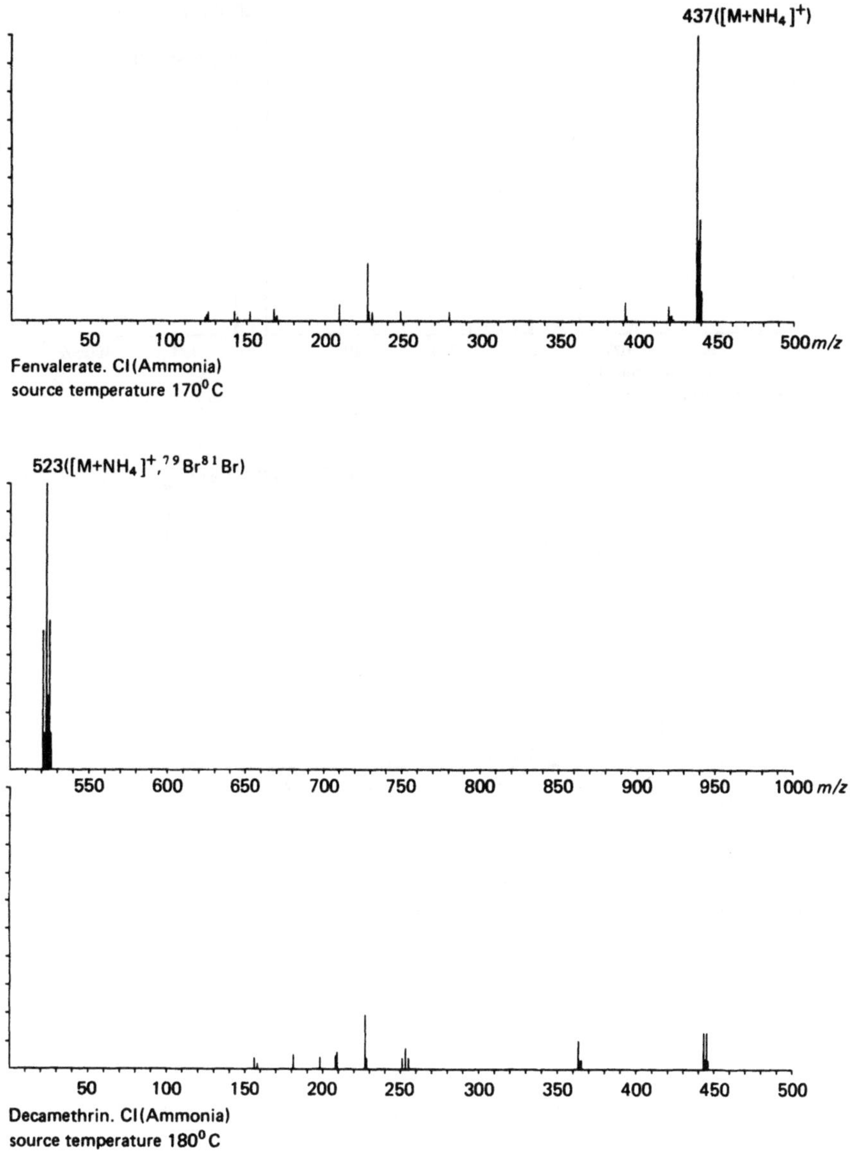


FIGURE 4. PCI (ammonia) mass spectra of fenvalerate and decamethrin. Source pressure 40 Pa, VG Micromass 70-70 mass spectrometer.

spectrum (Fig. 1), and characteristic even-electron fragment ions dominate the spectrum. It is possible that the novel ion of *m/z* 200 is $BNH_3^+$ (compare Table 1), formed from the $(M + NH_4)^+$ ion by a nucleophilic displacement with $NH_3$.[(40)] Reduction of the source temperature to 80 °C produces a dramatic change in the fragmentation pattern

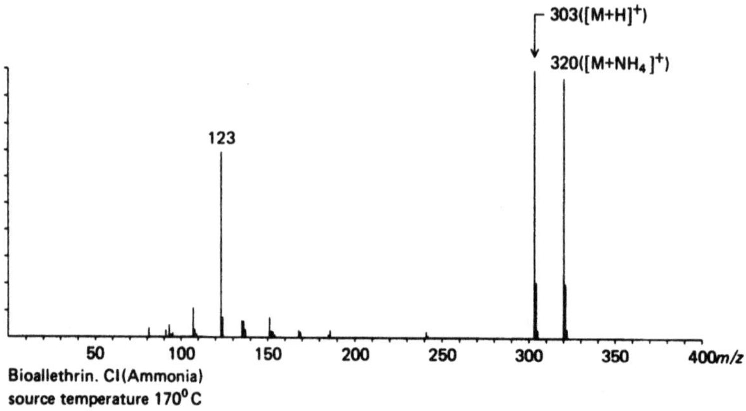


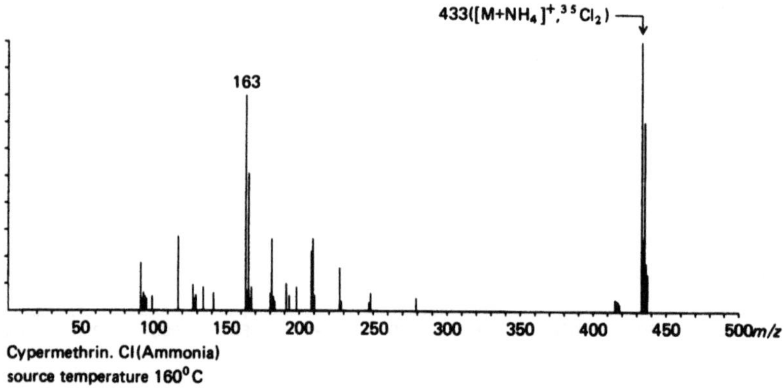


FIGURE 5. PCI (ammonia) mass spectra of bioallethrin and cypermethrin. Source pressure 40 Pa, VG Micromass 70-70 mass spectrometer.

and $(M + NH_4)^+$ ions are now the only significant ions in the PCI (ammonia) mass spectrum (Fig. 6).

PCI (ammonia) presents clear advantages in detection of pyrethroids by increasing the relative abundances of higher mass ions and by reducing the chemical noise. It has the disadvantage that the absolute sample ion currents may be lower than those with more energetic plasmas such as PCI (methane) because sensitivity in PCI appears to be related to reaction exothermicity.[41] An alternative approach that could be applied to pyrethroids to increase the absolute sensitivity and still maintain the selectivity is NCI.

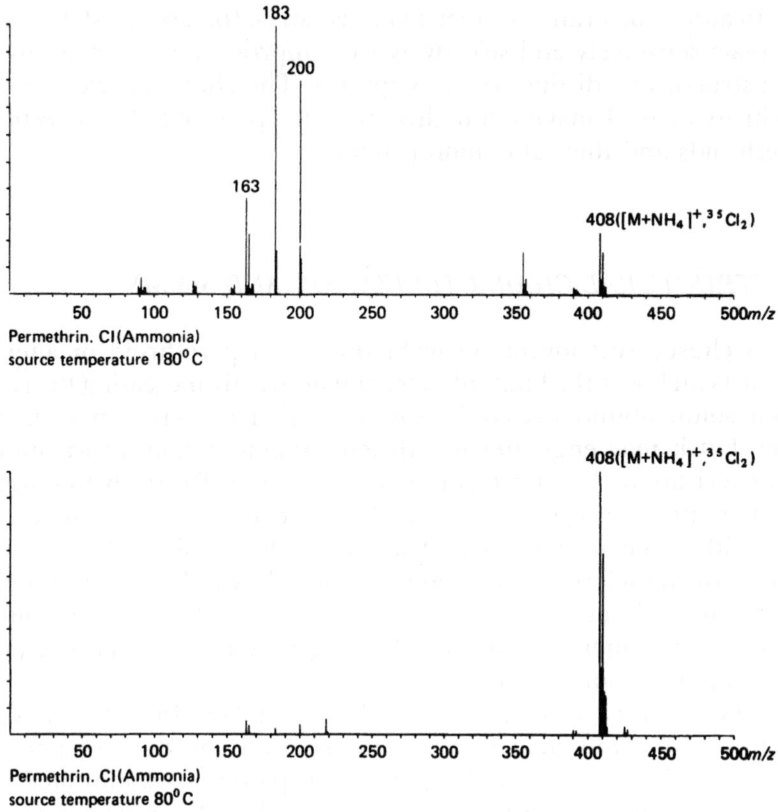


FIGURE 6. PCI (ammonia) mass spectra of permethrin at different source temperatures. Source pressure 40 Pa, VG Micromass 70-70 mass spectrometer.

## 5. NEGATIVE-ION CHEMICAL IONIZATION (NCI)

A chemical ionization source operating at moderately high pressure contains large numbers of quasithermal electrons, which may react associatively or dissociatively with analyte molecules of high electron affinities to form $M^{-\cdot}$ ions and/or fragment negative ions. In general, samples that give an excellent response with an electron capture detector in GC (as do several pyrethroids) will also give an intense electron capture NCI mass spectrum.[42] In favorable cases for electron capture, the absolute ion yield from the sample may be two or three orders of magnitude greater than in EI or PCI.[43]

In addition, a range of reactant ions can be formed in NCI[41] which can react sensitively and selectively with appropriate molecules to produce structurally distinctive mass spectra. The chemical selectivity and sensitivity of NCI make it a highly attractive possibility for detection of pyrethroids and their alteration products.

## 6. STRUCTURAL CHARACTERIZATION AND MS/MS

A chosen soft ionization technique for a pyrethroid or alteration product can lower the limits of detection by greatly increasing the relative and absolute abundances of ions in the higher mass region of the spectrum, but it may engender insufficient fragmentation information for characterization. This problem can be overcome effectively by enforcing fragmentations of the parent ions downstream of an ion source rather than within the source itself. Analysis of the resulting fragmentation pattern of the selected parent ion is achieved by independent operation of tandem analyzers of a mass spectrometer, and selected-ion monitoring using such fragment ions is much less aggravated by chemical noise[24,44] than fragmentations in the source.

The recently developed methodology of MS/MS[28,44,45] makes efficient use of a tandem mass spectrometer as the sole device for chemical separation, by deciphering fragmentation pathways as indicated above for components of an ion mixture generated in the source. Separation is virtually instantaneous since it is based on *m/z* rather than retention time, and the MS/MS methodology offers comparable or better resolution of components to that available by GC, with less restrictive requirements of volatility.[46,47] Technical refinements are being made to tandem mass spectrometers to improve their versatility and efficacy in MS/MS[28,48] and to minimize cross-talk between the spectra of components, including operation at high mass resolution.[49,50] MS/MS is compatible with most methods of soft ionization, including FD, and promises a bright future for rapid and effective MR screening with minimal pretreatment.

## REFERENCES

1. J. E. Casida, *Pyrethrum—The Natural Insecticide,* Academic, New York, 1973.
2. M. Elliott, *ACS Symposium Series* **42,** 1–28 (1977).
3. M. Elliott and N. F. Janes, in *Advances in Pesticide Science,* Part 2, Pergamon, Oxford, 1979, pp. 166–173.

4. J. E. Casida, L. C. Gaughan, and L. O. Ruzo, in *Advances in Pesticide Science,* Part 2, Pergamon, Oxford, 1979, pp. 182–189.
5. M. Elliott, A. W. Farnham, N. F. Janes, P. H. Needham, D. A. Pulman, and S. H. Stevenson, *Nature* **246,** 169–170 (1973).
6. R. E. Bry and R. A. Simonaitis, *Textile Chem. Color.* **7,** 28–29 (1975).
7. J. M. Desmarchelier, *J. Pestic Sci.* **5,** 901–908 (1980).
8. J. M. Desmarchelier, M. Bengston, and R. Sticka, *Pyr. Post.* (in press).
9. D. M. Soderlund and J. E. Casida, *Pestic. Biochem. Physiol.* **7,** 391–401 (1977).
10. S. W. Head, *Pyr. Post.* **8,** 3–7 (1966).
11. P. Tetenyi, E. Hethelyi, T. Okuda, and I. Szilagyi, *Pyr. Post.* **11,** 29–48 (1971).
12. R. A. Chapman and C. R. Harris, *J. Chromatogr.* **166,** 513–518 (1979).
13. R. A. Chapman and H. S. Simmons, *J. Assoc. Off. Anal. Chem.* **60,** 977–979 (1977).
14. J. M. Desmarchelier, *J. Stored Prod. Res.* **12,** 245–252 (1976).
15. R. A. Simonaitis and R. S. Cail, *J. Assoc. Off. Anal. Chem.* **58,** 1032–1036 (1975).
16. D. Mourot, B. Delépbine, J. Boisseau, and G. Gayot, *J. Chromatogr.* **168,** 277–279 (1979).
17. D. Mourot, B. Delépbine, J. Boisseau, and G. Gayot, *J. Chromatogr.* **173,** 412–414 (1979).
18. A. N. Wright, T. R. Roberts, A. J. Dulton, and M. U. Doig, *Pestic. Biochem. Physiol.* **13,** 71–80 (1980).
19. J. M. Zehner and R. A. Simonaitis, *J. Chromatogr. Sci.* **14,** 326–327 (1976).
20. C. J. W. Brooks and B. S. Middleditch, in *Mass Spectrometry,* Vol. 5, (R. A. W. Johnstone, Sr. Reporter), The Chemical Society, London, Chapter 6, pp. 142–185, 1979.
21. B. J. Millard, *Quantitative Mass Spectrometry,* Heyden, London, 1978.
22. W. F. Haddon, in *High Performance Mass Spectrometry Chemical Applications,* ACS Symposium Ser. 80, (M. L. Gross, ed.), ACS, Washington, D.C., 1978, pp. 97–119.
23. A. L. Burlingame, T. A. Baillie, P. J. Derrick, and O. S. Chizhov, *Anal. Chem.* **52,** 214R–258R (1980).
24. R. G. Cooks, *Amer. Lab.* **10,** 111–123 (1978).
25. R. C. Dougherty and E. A. Hett, *Environ. Sci. Res.* **12,** 339–350 (1978).
26. G. D. Daves, Jr., *Acc. Chem. Res.* **12,** 359–365 (1979).
27. K. L. Busch and R. G. Cooks, *Science* **218,** 247–254 (1982).
28. F. W. McLafferty, P. J. Todd, D. C. McGilvery, and M. A. Baldwin, *J. Amer. Chem. Soc.* **102,** 3360–3363 (1980).
29. T. A. King and H. M. Paisley, *J. Chem. Soc. C,* 870–874 (1969).
30. G. Pattenden, L. Crombie, and P. Hemesley, *Org. Mass Spectrom.* **7,** 719–735 (1973).
31. M. Elliott and J. E. Casida *J. Agric. Food Chem.* **20,** 295–299 (1972).
32. L. Crombie, G. Pattenden, and D. J. Simmonds, *Pestic. Sci.* **7,** 225–230 (1976).
33. R. L. Holmstead and D. M. Soderlund, *J. Assoc. Off. Anal. Chem.* **60,** 685–689 (1977).
34. R. L. Holmstead, J. E. Casida, L. O. Ruzo, and D. G. Fullmer, *J. Agric. Food Chem.* **26,** 590–595 (1978).
35. R. L. Holmstead and D. G. Fullmer, *J. Agric. Food Chem.* **25,** 56–58 (1977).
36. K. Ohsawa and J. E. Casida, *J. Agric. Food Chem.* **27,** 1112–1120 (1979).
37. L. O. Ruzo, R. L. Holmstead, and J. C. Casida, *J. Agric. Food Chem.* **25,** 1385–1394 (1977).
38. K. Ueda, L. C. Gaughan, and J. C. Casida, *J. Agric. Food Chem.* **22,** 212–220 (1974).
39. T. Shono, L. Ohsawa, and J. E. Casida, *J. Agric. Food Chem.* **27,** 316–325 (1979).
40. Y. Y. Lin and L. L. Smith, *Org. Mass Spectrom.* **5,** 604–611 (1978).
41. F. H. Field, 28th Annual Conference American Society for Mass Spectrometry, New York, Paper PL 2, 1980.
42. R. C. Dougherty and K. Piotrowska, *J. Assoc. Off. Anal. Chem.* **59,** 1023–1027 (1976).
43. D. F. Hunt and F. W. Crow, *Anal. Chem.* **50,** 1781–1784 (1978).

44. R. W. Kondrat and R. G. Cooks, *Anal. Chem.* **50,** 81A–92A (1978).
45. R. A. Yost and C. G. Enke, *Anal. Chem.* **51,** 1251A–1264A (1979).
46. G. L. Glish, V. M. Shaddock, K. Harmon, and R. G. Cooks, *Anal. Chem.* **52,** 165–167 (1980).
47. D. F. Hunt, J. Shabanowitz, and A. B. Giordani, *Anal. Chem.* **52,** 386–390 (1980).
48. T. H. Maugh II, *Science* **209,** 675–677 (1980).
49. F. W. McLafferty, *Acc. Chem Res.* **13,** 33–39 (1980).
50. D. H. Russell, E. H. McBay, and T. R. Mueller, *Int. Lab.,* April, 49–61 (1980).

# MASS SPECTROMETRY OF SOME FEMALE LEPIDOPTERAN SEX PHEROMONES AND RELATED COMPOUNDS

D. L. STRUBLE

## 1. INTRODUCTION

The first female lepidopteran sex pheromone identified was the main component of the silkworm moth (*Bombyx mori*), (E)-10,(Z)-12-hexadecadienol.[1] Since then, progress has been excellent on the identification, syntheses, and applications of female sex pheromones of other lepidopterans and this progress has been adequately reviewed.[2,3] It is interesting to note that a second minor pheromone component of the silkworm moth, (E)-10,(Z)-12-hexadecadienal, was reported 19 years after the identification of the major component.[4] It is now generally accepted that the sex pheromones of lepidopterans usually consist of a blend of two or more biologically active components.[5–7]

Pheromones are of great interest because they provide a species-specific technique for studying the behavior of moths, for monitoring population densities of pest species, and for providing more efficient use of pesticides through timing of pesticide applications in integrated pest management, and they are showing promise for control of pests by a mating disruption technique and by mass trapping.

In general, the sex pheromones of female lepidopterans are combinations of geometrical and positional isomers of unsaturated long-chain (e.g.,$C_{10}$–$C_{18}$) aliphatic primary alcohols, acetates, and aldehydes. These components are usually present in nanogram amounts per female,

*D. L. STRUBLE* • Agriculture Canada Research Station, Lethbridge, Alberta T1J 4B1, Canada.

and their structural identification has been accomplished from complementary information afforded by several techniques such as behavioral bioassay; electrophysiological data recorded from olfactory receptors of the insect; thin-layer, liquid, and gas–liquid chromatography; and infrared, ultraviolet, nuclear magnetic resonance, and mass spectrometry.[7] The ultimate test is the efficiency of the synthetic pheromone, compared to unmated females for the attraction of males, under natural conditions.

This article is a summary of some developments and applications of mass spectrometry that pertain to the identification of pheromones. Most examples are taken from the sex pheromones of female lepidopterans, but the techniques discussed here also apply to pheromones of other insect species.

## 2. GAS–LIQUID CHROMATOGRAPHY/MASS SPECTROMETRY (GLC/MS) INSTRUMENTATION

Combined GLC/MS has been one of the most powerful analytical tools available to organic chemists and it has been used extensively in the identification of pheromones. GLC/MS instruments were used predominantly with packed GLC columns during the 1960s and early 1970s, and the coupling of high-resolution capillary GLC columns with rapid-scanning mass spectrometers has recently been accomplished.

Packed GLC columns required carrier gas flow rates of 5–20 ml/min, and it was necessary to concentrate the GLC effluent by removal of the carrier gas to maintain a reasonable pressure in the ion source of the mass spectrometer. There were various transfer systems for enriching the GLC effluent, but these often led to loss of resolution of the components in the effluent and a loss of sensitivity by the mass spectrometer. Due to the limited resolving power of the packed GLC columns and the problems associated with transfer systems, it was often not possible to obtain mass spectra of the nanogram quantities of mixtures of geometrical and positional isomers that are common in pheromones directly by GLC/MS. In most instances, the isomeric pheromone components had to be partially purified before the GLC/MS analyses.[8]

The direct coupling of high-resolution capillary GLC columns to rapid-scanning (e.g., 40–400 amu/sec) mass spectrometers, equipped with data systems to store and manipulate the vast quantity of data generated in a single GLC/MS analysis of a complex mixture, such as a crude pheromone extract, has contributed tremendously to pheromone research. Mass chromatograms based upon selected ions and the total ion current can be reconstructed by the data system for an entire GLC/MS analysis of a mixture of compounds or natural products. Differential (background subtracted) mass spectra of the compounds of interest can be obtained and compared to those of authentic samples that may be stored in a data system library.

The application of this type of GLC/MS instrumentation for the identification of some insect sex pheromone components has been described.[9–15] The mass spectra of the pheromone components provide important structural information such as their molecular weight and at least an indication of their functionality. The isomeric assignments of unsaturated pheromonelike compounds can be partially based upon the mass spectral data and the retention time of the compounds on high-resolution capillary columns in combined GLC/MS.[15,16] Some confirmatory data regarding the geometrical and positional isomeric assignments from other methods such as those outlined in this article will always be advantageous.

## *3. MASS SPECTROMETRY OF SATURATED AND UNSATURATED LONG-CHAIN ALIPHATIC ALCOHOLS, ACETATES, AND ALDEHYDES*

In general, the published data on electron ionization (EI), chemical ionization (CI), or negative chemical ionization (NCI) mass spectrometry of unsaturated aliphatic compounds does not provide sufficient information for the unambiguous diagnosis of the positional and geometrical isomers. The identification of positional isomers has been partially accomplished by mass spectrometry of derivatives of the double bonds, but the assignment of the geometrical isomers has been based on data from other techniques. The diagnostic fragmentation ions of saturated and unsaturated alcohols, acetates, and aldehydes by EI, CI, and NCI are considered here.

### *3.1. Mass Spectrometry of Alcohols, Acetates, and Aldehydes*

The fragmentation mechanisms involved in EI mass spectrometry of some saturated and unsaturated aliphatic alcohols, acetates, and aldehydes have been reviewed.[17] The main diagnostic ions of these functional groups are summarized here.

*3.1.1. Alcohols.* The molecular ion $M^+$ is always weak or undetectable in saturated and monounsaturated alcohols, and it is usually weak but detectable in diunsaturated alcohols. A major ion is always $(M - 18)^+$ due to the loss of water, and frequently there is a detectable ion at $(m - 18 - 28)^+$ due to the loss of water and ethylene. In long-chain primary *n*-alkanols, the loss of a second ethylene molecule is common to give a major ion at $(M - 74)^+$. The other main fragmentations arise from oxygen-containing and hydrocarbon ions. The hydrocarbon ions usually predominate and they often obscure the typical alcohol fragmentations. The mass spectra of these compounds resemble those of the corresponding alkanes with peak groups corresponding to the series $C_nH_{2n-1}$, $C_nH_{2n}$, and $C_nH_{2n+1}$ at 14 mass units ($CH_2$), which decrease in intensity with increasing fragment weight. The fragmentations of alkanes have been studied in detail recently by Lavanchy.(18)

Typical examples of pheromone components that had these diagnostic ions were (Z)-7-dodecenol,(19) (Z)-11-hexadecenol,(12) (Z) -5,(E)-7-dodecadienol,(20) and (E)-8,(E)-10-dodecadienol. (21) A weak molecular ion $M^+$ and major ions $(M - 18)^+$ and $(M - 46)^+$, due to the loss of water and loss of water plus ethylene, were reported in the spectrum of (E)-11-tetradecenol.(22)

*3.1.2. Acetates.* The molecular ion $M^+$ is usually very weak or undetectable in saturated and monounsaturated acetates, but it is usually detectable in compounds having two or more double bonds. Three diagnostic ions are $(M - 60)^+$, $(M - 60 - 28)^+$, and $61^+$. The $(M - 60)^+$ is due to the loss of acetic acid and the $61^+$ ion is protonated acetic acid. The loss of acetic acid involves a McLafferty rearrangement. Protonation of the acetic acid fragment to form the acetyloxonium ion, $CH_3CO_2H_2^+$, occurs by a nonspecific abstraction of a proton from the alkyl chain of the alcohol moiety.(23) The $(M - 60 - 28)^+$ is presumably due to the loss of acetic acid and ethylene. The remaining fragmentations resemble those of a hydrocarbon chain and there are insufficient characteristic ions for diagnosis of the positional or geometrical isomers in unsaturated acetates.

The presence of an $(M - 18)^+$ due to the loss of water was reported in ethyl acetate(24) and the proposed mechanism was confirmed by Folk.(25) The $(M - 18)^+$ ion has not been reported in long-chain unsaturated acetates.

Typical examples of some pheromone components that exhibited these characteristic ions were (Z)-5- and (Z)-7-decenyl acetates;(26) dodecanyl, (E)-5-, (Z)-7-, and (E)-9-dodecenyl acetates;(27) (Z)-5-, (Z)-7-, (Z)-9-, (E)-11-, and (Z)-11-tetradecenyl acetates;(28,29) (Z)-9- and (Z)-11-hexadecenyl acetates;(14,30) (E)-7- and (Z)-9-dodecadienyl acetate;(21) (E)-9,11- and (Z)-9,11-dodecadienyl acetates;(31) (Z)-9, (E)-11-, and (Z)-9,(E)-12-tetradecadienyl acetates;(32) (E)-6,(Z)-11-hexadecadienyl acetate;(33) and (E)-4,(Z)-7-tridecadienyl acetate and (E)-4,(Z)-7,(Z)-10-tridecatrienyl acetate.(34)

*3.1.3. Aldehydes.* The molecular ion $M^+$ is usually very weak or undetectable in saturated and monounsaturated aldehydes but, like the corresponding alcohols and acetates, it is detectable in compounds having two carbon–carbon double bonds.

Diagnostic ions are $(M - 18)^+$, $(M - 28)^+$, $(M - 28 - 18)^+$, and $(M - 44)^+$ due to the loss of water, ethylene, ethylene and water, and $C_2H_4O$. Major ions are also common at $29^+$, $43^+$, $44^+$, and $57^+$ due to $CHO^+$, $C_2H_3O^+$, $C_2H_4O^+$, and $C_3H_5O^+$. The mechanisms involved with these fragmentations have been discussed.[35]

The loss of water from the molecular ion $(M - 18)^+$ is extensive only in *n*-alkanals with six or more carbon atoms and it involves the loss of hydrogen atoms from carbon atoms 4, 5, and 6. Three mechanisms have been suggested for the loss of water.[36] The $(M - 28)^+$ ion, due to the loss of ethylene from the molecular ion, occurs only in *n*-alkanals of five or more carbon atoms and it involves the loss of carbon atoms 2 and 3. In higher *n*-alkanals (e.g., *n*-nonanal) the ion $(M - 28 - 18)^+$ is observed and this gives rise to the olefinic series $C_nH_{2n-2}$.[37] The $(M - 44)^+$ is a predominant ion in all *n*-alkanals except propanal and it is formed by the McLafferty rearrangement, β-cleavage with rearrangement. The charge may occur on either fragment, that is, $(M - 44)^+$ or $44^+$. The $C_2H_4O^+$ ion also fragments to give the $C_2H_3O^+$ ion. The prominent $57^+$ may be due to $C_3H_5O^+$ or $C_4H_9^+$. These lower weight fragmentations are often due to the hydrocarbon ions that may obscure the more diagnostic oxygencontaining ions.

Typical examples of some pheromone components that exhibited these characteristic ions were tetradecanal, (Z)-9-tetradecenal and hexadecanal, and (Z)-7-, (Z)-9-, and (Z)-11-hexadecenal.[12,13] Weak molecular ions, $M^+$, were detected in these monounsaturated aldehydes, but not in the saturated aldehydes. Weatherston and Maclean reported that (E)-11-tetradecenal exhibited a detectable $M^+$ ion and prominent ions at $(M - 18)^+$, $(M - 28)^+$, and $(M - 44)^+$ as expected.[22] The lower fragments were not reported for all spectra.

### 3.2. *Chemical ionization of Saturated and Unsaturated Long-Chain Alcohols, Acetates, and Aldehydes*

Chemical ionization (CI) mass spectrometry involves ion–molecule reactions between the molecules of substance under investigation and ionized reactant gas. CI mass spectra contain a relatively small number of ions and an important feature is the enhancement of the quasimolecular ions of most compounds compared to the EI molecular ions, which are often weak or nondetectable. This provides a good means of determining the molecular weight of the compounds under investigation. The published CI mass spectral data of unsaturated long-chain aliphatic alcohols, acetates, and aldehydes do not provide sufficient fragment ions for the diagnosis of the double-bond position or geometry. This must be achieved by mass spectrometry of suitable derivatives or by other means.

Different spectra are produced by the use of different ionizing reactant gases. Methane and isobutane have been the most commonly used reactants for pheromonelike compounds.

The methane ion plasma consists of $CH_5^+$, $C_2H_5^+$, and $C_3H_5^+$ ions, which react with the molecules under investigation by transfer of protons, hydride ions, or alkyl carbonium ions. The mass spectra of compounds will often contain $(M + 29)^+$ and $(M + 41)^+$ ions due to reactions with $C_2H_5^+$ and $C_3H_5^+$.

Isobutane reactant gas generates a plasma of mainly $C_4H_9^+$ with some $C_3H_3^+$ ions. These act as Bronsted acids and are much milder protonating agents than the ions in the methane plasma. Common ions in the CI isobutane spectra are $(M + 39)^+$ and $(M + 57)^+$ due to reactions with $C_3H_3^+$ and $C_4H_9^+$.

The temperature of the mass spectrometer ion source has a marked effect on the amount of fragmentation that occurs with isobutane as the reactant gas; more fragmentation occurs at higher source temperatures. The fragmentation with methane is less dependent upon temperature of the ion source; however, there is more fragmentation than with isobutane. This has been attributed to the fact that the $CH_5^+$ ion in the methane plasma is a much stronger Bronsted acid than the $C_4H_9^+$ ion in the isobutane plasma. Much more dissociation is effected by the chemical energy of the $CH_5^+$ ion and additional thermal energy from the ion source has only a minor influence on the methane CI spectra.

The characteristic fragmentation ions produced by methane and isobutane CI of saturated and unsaturated long-chain aliphatic alcohols, acetates, and aldehydes are summarized here.

*3.2.1. Alcohols—CI (Methane).* Characteristic ions of primary alcohols are $(M - 1)^+$, $[(M + 1) - 18]^+$, $(M + 29)^+$, and $(M + 41)^+$. The $[(M + 1) - 18]^+$ is due to the loss of water and is the most abundant ion in the high-mass region. The $(M - 1)^+$ ion is formed by hydride ion abstraction at carbon atom 1 in primary alcohols and it was not observed in tertiary alcohols.[38] On the other hand, the $(M + 1)^+$ ion is observed in secondary alcohols but not in primary alcohols. The rest of the hydrocarbon chain gives two series of ions $C_nH_{2n-1}$ and $C_nH_{2n+1}$, and a relatively small amount of oxygen-containing fragments.

The mass spectrum of dodecanol was reported to have a weak $[(M + 1) - 18]^+$ ion and lower mass fragments typical of alkanes.[39] The ions $(M + 1)^+$, $(M - 1)^+$, and $[(M - 1) - 18]^+$ were reported for (Z)-11-hexadecenol by mass fragmentography.[40] These ions and a major $[(M + 1) - 18]^+$ ion were confirmed in the spectrum of this alcohol.[41]

*3.2.2. Alcohols—CI (Isobutane).* The main ions at a source temperature of 180 °C were $(M - 1)^+$, $(M + 1)^+$, $[(M + 1) - 18]^+$, $(M + 39)^+$, and

$(M + 57)^+$, and their relative abundance depends upon the structure of the alcohol. The $(M + 1)^+$ was the major ion in the lower alcohols (propanol), but it decreased as the molecular weight of the alcohols increased. It was not detected in heptanol and higher alcohols. The $(M + 39)^+$ was more intense in saturated secondary alcohols than in primary, and the $(M + 57)^+$ was larger in primary alcohols. The relative intensities of these ions provide some structural evidence of the alcohols. The $(M - 1)^+$, $[(M + 1) - 18]^+$, $(M + 39)^+$, and $(M + 57)^+$ are the expected diagnostic fragmentation ions for unsaturated alcohols, but mass spectra of these have not been reported.

*3.2.3. Acetates—CI (Methane or Isobutane).* The characteristic ions of saturated and unsaturated long-chain aliphatic acetates with either reactant gas are $(M + 1)^+$ and $[(M + 1) - 60]^+$ due to the loss of acetic acid and $61^+$ due to protonated acetic acid. Weak ions of $(M + 29)^+$ and $(M + 41)^+$ are common with methane as the reactant gas and $(M + 39)^+$ and $(M + 57)^+$ are often detected with isobutane. The loss of acetic acid from the $(M + 1)^+$ ion involves hydrogen migration from the alcohol to the carboxyl portion of the compound. Benoit and Harrison reported that hydrogen migration originated from carbon atoms 1, 2, and 3 of propyl acetate in 27, 23, and 50%, respectively.[(42)] They suggested that this migration was independent of the ester and the mode of ionization. A $(2M + 1)^+$ ion due to the protonated dimer of the molecular ion has been reported for benzyl and *t*-amyl acetates at low (42 °C) ion source temperatures. The $(2M + 1)^+$ ion has not been reported in pheromonelike acetates, possibly because of the higher ion source temperature used in most CI mass spectrometry.

Some pheromone components that had these characteristic fragmentations were decanyl; dodecanyl; (E)-5-, (Z)-5-, (Z)-7-, (E)-9-, and (Z)-9-dodecenyl acetates; [(39,43,44)] (Z)-11-tetradecenyl acetate;[(9)] (Z)-11-hexadecenyl acetate;[(14)] and (Z)-7,(E)-11-hexadecadienyl acetate.[(7)]

*3.2.4. Aldehydes—CI (Methane or Isobutane).* The characteristic ions of saturated and unsaturated long-chain aliphatic aldehydes with either reactant gas are $(M + 1)^+$ and $[(M + 1) - 18]^+$, with possibly a weak intensity ion due to $(M - 1)^+$. Ions occur at $(M + 29)^+$ and $(M + 41)^+$ with methane reactant gas and at $(M + 39)^+$ and $(M + 57)^+$ with isobutane.

The CI (methane) mass spectra of pheromone components (Z)-9-tetradecenal, (Z)-11-hexadecenal, and (Z)-11,(Z)-13-hexadecadienal were reported to have characteristic ions of $M^+$, $(M + 1)^+$, $[(M + 1) - 18]^+$, $(M + 29)^+$, and $(M + 41)^+$. [(45)] The spectrum of (Z)-9-tetradecenal exhibited a weak $(M - 1)^+$ ion.

Fales *et al.* studied the fragmentations leading to the loss of water in heptanal and a series of deuterated heptanals with methane and isobutane reactant gases.[46] They concluded that the loss of water from the quasimolecular ion $(M + 1)^+$ followed the same pattern as in the EI, and hydrogen was lost from all positions of the chain except carbon atoms 1 and 2. About 28% of the water eliminated contained a proton from the reactant gas. Audier reported the study of two different mechanisms for the loss of water from protonated long-chain aldehydes by $CD_4$ and $D_2O$ chemical ionization.[47] Their results are in agreement with those of Fales.

A detailed study of hydrogen scrambling in the fragmentation of acetaldehyde by EI and CI using hydrogen and deuterium as reagent gases has been reported.[48]

### 3.3. *CI (Nitric Oxide)*

Nitric oxide, or mixtures of nitric oxide and nitrogen, has been used as a reactant gas which produces nitrosonium ions $(NO)^+$. This functions as an electron acceptor, hydride abstracting agent, and an electrophile towards organic compounds in the ion source.

CI (nitric oxide) of saturated and unsaturated long-chain alcohols, acetates, and aldehydes has not been reported. It would probably not provide information for the diagnosis of the double-bond positions or geometry in unsaturated compounds, but it should provide some useful fragmentations for diagnosis of various functional groups.

It is possible to differentiate primary, secondary, and tertiary alcohols by CI (nitric oxide) mass spectrometry. The major ions for alcohols are as follows: primary—$(M - 1)^+$, $(M - 17)^+$, $[(M - 2) + 30]^+$, and $(M - 3)^+$; secondary—$(M - 1)^+$, $(M - 17)^+$, $[(M - 2) + 30]^+$; and tertiary—$(M - 17)^+$. Also, $(M - 1)^+$ and $(M + 30)^+$ are common ions for aldehydes, while ethers give only $(M - 1)^+$ ions. The $(M + 30)^+$ is the major ion produced with ketones, esters, and carboxylic acids, while the latter have an abundant $(M - 17)^+$ ion.

The CI (nitric oxide) of alkenes was reported to give $M^+$, $(M + 30)^+$, and $(M - 1)^+$ as the major ions, but there were insufficient fragmentations to assist in the diagnosis of the double-bond position or geometry.[49] On the other hand, alkynes gave both hydrocarbon and NO-containing fragmentation ions.[50] This was reported to be more reliable for determining the position of the triple bond than CI (methane). CI (nitrous oxide) may be helpful for characterizing the acetylenic intermediates in the syntheses of many olefinic pheromonelike compounds.[51]

### 3.4. *Negative Chemical Ionization (NCI) of Saturated and Unsaturated Alcohols, Acetates, and Aldehydes*

Preliminary results with decanyl to hexadecanyl acetates and the corresponding monounsaturated compounds, with NCI using nitrous oxide, methane, isobutane, and mixtures of nitrous oxide and methane or isobutane as reactant gases gave $(M - 1)^-$ ions as the only major ions.[52] NCI appears to be useful for increasing the sensitivity toward these types of compounds and for confirming their molecular weight. This does not provide any information on the location or geometry of the double bond in unsaturated compounds.

## 4. *MASS SPECTROMETRY OF DERIVATIVES OF UNSATURATED COMPOUNDS*

Several derivatives of unsaturated pheromonelike compounds have been used to determine the number and position of the double bonds. It is essential that any of the derivatives used for this purpose can be prepared on a microgram or preferably on a few nanograms of pheromone component.

### 4.1. *Hydrogenation or Deuteration*

The direct hydrogenation of unsaturated compounds in the inlet of a GLC has been reported.[53] This has been adapted to combined GLC/MS where the number of double bonds in an unsaturated compound can be determined by comparing the mass spectra of the reduced and unsaturated compounds.[45]

The position of double bonds has been determined from diagnostic fragmentation ions in deuterated branched alkanes[54] and in linear *vic*-dideuterioalkanes.[55] Thermal fragmentation followed by mass spectrometry of deuterated unsaturated long-chain fatty acid esters has been used to locate the double-bond positions.[56] These authors pointed out that, in EI mass spectrometry, molecular rearrangements occur with hydrogen exchange and migration. Thus, even when double bonds were saturated with deuterium, the deuterium atoms migrated and it was not possible to determine the original double-bond position. Thermal fragmentation did not cause serious rearrangements of the deuterium atoms. By studying the mass spectra of the thermal fragmentations, they were able to determine the double-bond positions in deuterated mono- and poly-unsaturated fatty acids. This technique has not been reported with pheromonelike compounds. It would probably be necessary to study the pyrolysis/mass spectrometry of a series of model compounds before attempting to use this technique for identifying positional isomers.

### 4.2. *Ozonolysis*

Microscale ozonolysis of purified unsaturated pheromonelike components and comparison of the products by their retention time on GLC with those of authentic samples have been used extensively for the diagnosis of the double-bond positions. In some instances the ozonolysis products of the pheromone components have been compared to authentic samples by mass spectrometry. The main EI fragmentations of 4-acetoxybutanal to 11-acetoxyundecanal were reported to be $(M - 43)^+$, $(M - 60)^+$, $(M - 43 - 60)^+$, $61^+$, and base peak $43^+$.[57] The first three ions of this series were presumably due to the loss of $CH_3CO$, $CH_3CO_2H$, and of both groups. The $61^+$ was probably due to protonated acetic acid. The $(M - 43 - 60)^+$ ion was only detected in 6-acetoxyhexanal to 11-acetoxyundecanal, while the $(M - 60)^+$ ion was only detected in 4-acetoxybutanal to 6-acetoxylhexanal. These major ions were also reported in 9-acetoxynonanal and 11-acetoxyundecanal by other workers.[30,58] Ando also reported the presence of hydrocarbon fragment ions that corresponded to $C_nH_{2n}$ and a series $C_nH_{2n-1}$, which were typical for the alkane part of the molecule.

Wakamura reported these major ions in the spectra of 5-acetoxypentanal and 7-acetoxyheptanal.[26] A $(M - 60 - 28)^+$ ion was also reported in the latter spectrum, which would be expected since it was a common ion in the corresponding acetates.

The EI mass spectrum of 1,11-undecanedial from the ozonolysis of (E)- or (Z)-11-hexadecenal had major ions at $(M - 18)^+$, $55^+$, and $41^+$.[59]

The CI (methane) spectra of 1,9-nonanedial and 1,11-undecanedial were reported to have similar fragmentations. The spectrum of 1,9-nonanedial exhibited major ions at $(M + 1)^+$, $(M + 29)^+$, $(M + 41)^+$, $[(M + 1) - 18]^+$ and $[(M + 1) - 36]^+$.[60] These were the expected ions for an aldehyde, except $[(M + 1) - 36]^+$, which was presumably due to the loss of two molecules of water from the $(M + 1)^+$ ion of the dialdehyde. A similar loss of two molecules of water was not reported in the EI spectrum of 1,11-undecanedial.[59]

### 4.3. *Epoxidation*

Many insect pheromones are mixtures of compounds with different carbon chain lengths and positional and geometrical isomers. Preparation of derivatives (e.g., ozonolysis) that lead to cleavage of the components before subsequent analysis necessitates the purification of each component before derivatization. This can be extremely difficult when nanogram amounts of isomeric compounds are involved. It is often desirable to make derivatives that do not involve cleavage of the original compounds, so that the derivatives can be resolved directly (e.g., by GLC) before analysis by mass spectrometry. Epoxides are suitable derivatives for this purpose.

*4.3.1. EI Mass Spectrometry of Epoxides.* The EI fragmentations of *vic*-dialkyl and trialkyl epoxides were reported by Bierl[61] and aliphatic epoxides have been summarized by Budzikiewicz.[62] The epoxide position was determined in vic-dialkyl epoxide by the maxima in plots of the percent abundance vs $m/z^+$ for various series of ions. The first series of ions was due to the cleavage of the carbon chain at the α-position to the epoxide group, which gave the series $C_nH_{2n-1}$, ($m/z^+$ 113, 127, 141, . . . ). The second series was due to transannular cleavage of the epoxide ring plus the loss of water, and was of the general formula $C_nH_{2n-2}$ ($m/z^+$ 110, 124, 138, . . . ). In trialkyl compounds the epoxide location was determined from similar plots of ions corresponding to α-cleavage plus hydrogen transfer—for example, $C_nH_{2n}O$, ($m/z^+$ 72, 86, 100, . . . )—and transannular cleavage with hydrogen transfer to give the series $C_nH_{2n+1}O$, ($m/z^+$ 115, 129, 143, . . . ). The mass spectra of epoxides of geometrical isomers were virtually identical, so that EI mass spectrometry was only useful for determining the double-bond position.

Bierl–Leonhardt reported a method of locating the double-bond position in monounsaturated long-chain aldehydes, acetates, and alcohols by EI mass spectral analysis of their epoxide derivations.[63] Although this method was based upon a limited number of compounds, the mass spectra of the epoxide derivatives had diagnostic ions that corresponded to cleavages at the α-position to the epoxide groups. The relative abundances of these fragment ions ranged from 1 to 10% in epoxyaldehydes, 0.2 to 4% in epoxyacetates, and 1 to 11% in an epoxyalcohol. In the epoxyaldehydes and epoxyacetates, the low $m/z^+$ fragment ions formed by α-cleavages were more abundant than the corresponding high $m/z^+$ fragment ions. On the other hand, in the spectrum of *cis*-11,12-epoxyhexadecanol, the higher $m/z^+$ fragment ion, the epoxyalcohol, was the most abundant ion of the α-cleavage products. The mass spectra of other epoxyalcohols were not reported.

In all of these derivatives the alkyl epoxide ions were of diagnostic value when their $m/z^+$ was sufficiently large that they could be recognized among the other hydrocarbon fragmentation ions. The spectra of epoxyaldehydes did not have $M^+$ and $(M - 18)^+$ ions, and the $M^+$ and $(M - 60)^+$ ions were not reported for the epoxyacetates.

The epoxide derivatives of a mixture of (Z)-7-, (Z)-9-, and (Z)-11-

hexadecenals were sufficiently resolved by high-resolution GLC to obtain identifiable mass spectra of each compound.[63] The relative ease of resolving positional and geometrical isomeric epoxides by GLC is an important aspect of these derivatives.[64]

The epoxyacetates were further characterized by oxidative cleavage of the epoxide group with iodic acid ($HIO_3$) to give an alkyl aldehyde and an acetoxyaldehyde. The mass spectra of the acetoxyaldehydes did not have detectable $M^+$ and $(M - 60)^+$ ions, but they had diagnostic ions at $(M - 43)^+$, $(M - 78)^+$, $(M - 88)^+$, and $(M - 103)^+$. These were attributed to the loss of $CH_3CO$, acetic acid plus water, acetic acid plus carbon monoxide, and acetic acid plus $C_2H_3O$, respectively. Except for the $(M - 88)^+$ ion, these were the major diagnostic ions reported in the EI spectra of acetoxyaldehydes produced by ozonolysis of monounsaturated acetates.

The positional and geometrical isomer assignments of the pheromone components (Z)-5-tetradecenyl acetate, (Z)-7-, (Z)-9-, and (Z)-11-hexadecenals, and (Z)-11-hexadecenol were mainly based upon the GLC retention times and the EI spectra of these components and their epoxide derivatives.[12,13] The epoxide derivative of (Z)-5-tetradecenyl acetate had a diagnostic ion at $155^+$ which corresponded to the α-cleavage product with the loss of acetic acid [i.e., $(157 - 60)^+$]. The three epoxyaldehydes had diagnostic ions at $141^+$, $169^+$, and $197^+$, which corresponded to the α-cleavage products with loss of the alkyl group. The epoxyalcohol from (Z)-11-hexadecenol gave a similar diagnostic ion at $199^+$. The major ions for the other α-cleavage products were not reported for these epoxyaldehydes.

*4.3.2. CI (Methane and Isobutane) Mass Spectrometry of Epoxides.* CI (methane and isobutane) mass spectrometry of dialkyl epoxides, epoxyalkanyl, and epoxyalkenyl acetates were reported to have major ions that correspond to $(M + 1)^+$, $[(M + 1) - 60]^+$, and $[(M + 1) - 60 - 18]^+$.[44] The mass spectra with methane as reactant gas had less abundant ions and more fragmentations than with isobutane, as would be expected. Diagnostic ions for locating the epoxide group in epoxyalkanyl acetates were attributed to transannular cleavage of the epoxide ring. The fragment containing the hydroxyl and acetate groups fragmented further by loss of acetic acid. A mechanism was proposed for these fragmentations.

The epoxyalkenyl acetates gave similar fragmentations when the double bond was separated from the epoxy group by more than one methylene group and when it is not between the two oxygen-containing groups. The double bond did affect the fragmentation when it was between the oxygen-containing groups and when it was one methylene group from the epoxide ring. This technique provides valuable structural information about diunsaturated pheromonelike compounds, but it would be advisable to study the fragmentations of some model compounds before attempting to assign the geometrical isomerism by this method.

The epoxyalkyl acetates have different fragmentation pathways under EI and CI (methane or isobutane) mass spectrometric conditions. These methods of ionization should provide useful complementary spectral data for determining the position of the epoxide group in these derivatives.

### 4.4. *Silylation*

Trimethylsilyl ether derivatives have been used extensively to increase the volatility of long-chain alcohols for their analysis by GLC and GLC/MS. CI (methane) major fragmentations of saturated and some mono- and di-unsaturated $C_{14}$ alcohols were reported to be $(M + 1)^+$, $(M - 1)^+$, $[(M + 1) - 16]^+$, and $(M + 29)^+$.[(65)] This provided useful confirmatory structural information, but it did not assist in determining the location and geometry of the double bonds in unsaturated compounds.

### 4.5. *Methoxymercuration–Demercuration*

Methoxymercuration–demercuration has been reported to be a useful technique for location of double-bond positions in esters of long-chain fatty acids and insect-derived alkenes.[(66)] In EI mass spectrometry fragmentation occurred at the carbon atoms to the methoxy ether group giving major ions that were diagnostic for the location of the original double bond. The fragmentations in partially or fully methoxylated alkadienes and diunsaturated fatty acid esters were more complex, but the major fragmentations were still α to the methoxy ether group. It was usually possible to assign the double-bond positions.

Plattner reported that the partial oxymercuration of fatty acid methyl esters containing one to four double bonds by reaction with mercuric acetate followed by hydrogenation directly in the GLC gave a mixture of the monomethoxy alkanoates.[(67)] The EI mass spectrum of this mixture was simpler than the spectrum of completely methoxylated polyenoates and more definitive for locating all of the olefinic positions. To my knowledge, these derivatives have not been studied by CI mass spectrometry nor have they been applied to unsaturated pheromonelike compounds.

### 4.6. Trimethylsilyloxy and O-Isopropylidene Derivatives

Trimethylsilyloxy derivatives of monounsaturated fatty acid esters have been used to locate the double-bond position by EI and CI (isobutane) mass spectrometry.[68] These derivatives fragment at the carbon bond between the two trimethylsilyloxy groups and the resulting major ions mark the position of the original double bond. These derivatives have also been used to determine the double-bond positions in monounsaturated hydroxy methyl esters of fatty acids.[69] The mass spectra of the geometrical isomers were not sufficiently different to permit the assignment of their geometry.

O-Isopropylidene derivatives of unsaturated fatty acid esters have also been used to locate the position of the double bond by EI mass spectrometry.[70] Neither of these derivatives have been reported for pheromonelike compounds.

## REFERENCES


1. A. Butenandt, R. Beckmann, D. Stamm, and E. Hecker, *Z. Naturforsch* **14b,** 283–284 (1959).
2. M. C. Birch (ed.), *Pheromones,* American Elsevier, New York, 1974.
3. H. H. Shorey and J. J. McKelvey, Jr. (eds.), *Chemical Control of Insect Behavior,* Wiley, New York, 1977.
4. G. Kasang, K. E. Kaissling, O. Vostrowsky, and H. J. Bestmann, *Angew. Chem. Int. Ed.* **17,** 60 (1978)
5. E. Priesner, *Ann. Zool. Ecol. Anim.* **11,** 533–546 (1979).
6. Y. Tamaki, in *Chemical Ecology: Odour Communication in Animals* (F. J. Ritter, ed.), Elsevier/North-Holland Biomedical, 1979, pp. 169–180.
7. J. H. Tumlinson and R. R. Heath, *J. Chem. Ecol.* **2,** 87–99 (1976).
8. M. Beroza, B. A. Bierl, and H. R. Moffitt, *Science* **183,** 89–90 (1974).
9. H. Arn, E. Priesner, H. Bogenschütz, H. R. Buser, D. L. Struble, S. Rauscher, and S. Voerman, *Z. Naturforsch.* **34c,** 1281–1284 (1979).
10. H. J. Bestmann, O. Vostrowsky, K. H. Koschatzky, H. Platz, T. Brosche, I. Kantardjiew, M. Rheinwald, and W. Knauf, *Angew. Chem.* **90,** 815–816 (1978).
11. H. J. Bestmann, O. Vostrowsky, K. H. Koschatzky, H. Platz, and A. Szymanska, *Tetrahedron Lett.* **19,** 605–608 (1978).
12. J. A. Klun, B. A. Bierl-Leonhardt, J. R. Plimmer, A. N. Sparks, M. Primiani, O. L. Chapman, G. Lepone, and G. H. Lee, *J. Chem. Ecol.* **6,** 177–183 (1980).
13. J. A. Klun, J. R. Plimmer, B. A. Bierl-Leonhardt, A. N. Sparks, M. Primiani, O. L. Chapman, G. H. Lee, and G. Lepone, *J. Chem. Ecol.* **6,** 165–175 (1980).
14. D. L. Struble, H. Arn, H. R. Buser, E. Städler, and J. Freuler, *Z. Naturforsch,* **35c,** 45–48 (1980).
15. D. L. Struble, H. R. Buser, H. Arn, and G. E. Swailes, *J. Chem. Ecol.* **6,** 573–584 (1980).
16. R. R. Heath, G. E. Burnsed, J. H. Tumlinson, and R. E. Doolittle, *J. Chromatogr.* **189,** 199–208 (1980).
17. F. W. McLafferty, *Interpretation of Mass Spectra,* 2nd ed., Benjamin, London, 1973.
18. A. Lavanchy, R. Houriet, and T. Gaumänn, *Org. Mass Spectrom.* **14,** 79–85 (1979).

19. J. H. Tumlinson, E. R. Mitchell, S. M. Browner, M. S. Mayer, N. Green, R. Hines, and D. A. Lindquist, *Environ. Entomol.* **1,** 354–357 (1972).
20. M. H. Vu, T. Ando, N. Takahashi, S. Tatsuki, A. Yamane, T. Ikeda, and S. Yamazaki, *Agric. Biol. Chem.* **44,** 231–233 (1980).
21. H. R. Buser and H. Arn, *J. Chromatogr.* **106,** 83–95 (1975).
22. J. Weatherston and W. Maclean, *Can. Entomol.* **106,** 281–284 (1974).
23. W. Benz and K. Biemann, *J. Am. Chem. Soc.* **86,** 2375–2378 (1964).
24. A. N. H. Yeo, *Chem. Comm.*, 1154–1155 (1970).
25. T. L. Folk, L. G. Wideman, and K. S. Cottman, *Org. Mass Spectrom.* **14,** 286–288 (1979).
26. S. Wakamura, *Appl. Ent. Zool.* **13,** 290–295 (1978).
27. R. G. Smith, G. E. Daterman, G. Doyle Daves, Jr., K. D. McMurtrey, and W. L. Roelofs, *J. Insect Physiol.* **20,** 661–668 (1974).
28. H. J. Bestmann, O. Vostrowsky, H. Platz, T. Brosche, K. H. Kochatzky, and W. Knauf, *Tetrahedron Lett.* **20,** 497–500 (1979).
29. H. Sugie, K. Yaginuma, and Y. Tamaki, *Appl. Entomol. Zool.* **12,** 69–74 (1977).
30. T. Ando, K. Kishino, S. Tatsuki, and N. Takahashi, *Agric. Biol. Chem.* **44,** 765–775 (1980).
31. B. F. Nesbitt, P. S. Beevor, R. A. Cole, R. Lester, and R. G. Poppi, *J. Insect Physiol.* **21,** 1091–1096 (1975).
32. Y. Tamaki and T. Yushima, *J. Insect Physiol.* **20,** 1005–1014 (1974).
33. J. Kochansky, J. Tette, E. F. Taschenberg, R. T. Cardé, K. E. Kaissling, and W. L. Roelofs, *J. Insect Physiol.* **21,** 1977–1983 (1975).
34. C. J. Persoons, S. Voerman, P. E. J. Verwiel, F. J. Ritter, W. J. Nooyen, and A. K. Minks, *Entomol. Exp. Appl.* **20,** 289–300 (1976).
35. J. G. Pritchard, *Org. Mass Spectrom.* **8,** 103–108 (1974).
36. S. Meyerson, C. Fenselau, J. L. Young, W. R. Landis, E. Selke, and L. C. Leitch, *Org. Mass Spectrom.* **3,** 689–707 (1970).
37. A. G. Harrison, *Org. Mass Spectrom.* **3,** 549–555 (1970).
38. M. S. B. Munson and F. H. Field, *J. Am. Chem. Soc.* **88,** 2621–2630 (1966).
39. J. P. Kochansky, W. L. Roelofs, and P. Sivapalan, *J. Chem. Ecol.* **4,** 623–631 (1978).
40. E. W. Underhill, W. F. Steck, and M. D. Chisholm, *Environ. Entomol.* **5,** 307–310 (1976).
41. H. R. Buser and D. L. Struble, unpublished data.
42. F. M. Benoit and A. G. Harrison, *Org. Mass Spectrom.* **13,** 128–131 (1978).
43. W. F. Steck, E. W. Underhill, M. D. Chisholm, and H. S. Gerber, *Environ. Entomol.* **8,** 373–375 (1979).
44. J. H. Tumlinson, R. R. Heath, and R. E. Doolittle, *Anal. Chem.* **46,** 1309–1312 (1974).
45. J. A. Coffelt, K. W. Vick, P. E. Sonnet, and R. E. Doolittle, *J. Chem. Ecol.* **5,** 955–966 (1979).
46. H. M. Fales, C. Fenselau, and J. H. Duncan, *Org. Mass Spectrom.* **11,** 669–674 (1976).
47. H. E. Audier, A. Milliet, and C. Perret, *Org. Mass Spectrom.* **14,** 132–134 (1979).
48. B. G. Keyes and A. G. Harrison, *Org. Mass Spectrom.* **9,** 221–246 (1974).
49. D. F. Hunt and T. M. Harvey, *Anal. Chem.* **47,** 2136–2141 (1975).
50. E. Busker and H. Budzikiewicz, *Org. Mass Spectrom,* **14,** 222–226 (1979).
51. C. A. Henrick, *Tetrahedron* **33,** 1845–1889 (1977).
52. D. L. Struble, unpublished results.
53. M. Beroza and R. Sarmiento, *Anal. Chem.* **38,** 1042–1047 (1966).
54. D. S. Winberg and M. W. Scoggins, *Org. Mass Spectrom.* **2,** 552–565 (1969).
55. D. G. Earnshaw, F. G. Doolittle, and A. W. Decora, *Org. Mass Spectrom.* **5,** 801–815 (1971).
56. K. K. Sun, H. W. Hayes, and R. T. Holman, *Org. Mass Spectrom.* **3,** 1035–1042 (1970).
57. R. Yamaoka, H. Fukami, and S. Ishii, *Agr. Biol. Chem.* **40,** 1971–1977 (1976).

58. H. Noguchi, Y. Tamaki, and T. Yushima, *Appl. Entomol. Zool.* **14,** 225–228 (1979).
59. Y. Tamaki, K. Kawasaki, H. Yamada, T. Koshihara, N. Ôsaki, T. Ando, S. Yoshida, and H. Kakinohana, *Appl. Entomol. Zool.* **12,** 208–210 (1977).
60. J. H. Tumlinson, D. E. Hendricks, E. R. Mitchell, R. E. Doolittle, and M. M. Brennan, *J. Chem. Ecol.* **1,** 203–214 (1975).
61. B. A. Bierl and M. Beroza, *J. Amer. Oil Chem. Soc.* **51,** 466–469 (1974).
62. H. Budzikiewicz, C. Djerassi, and D. H. Williams, *Mass Spectrometry of Organic Compounds,* Holden-Day, San Francisco, 1967.
63. B. A. Bierl-Leonhardt, E. D. DeVilbiss, and J. R. Plimmer, *J. Chromatogr. Sci.* **18,** 364–367 (1980).
64. B. A. Bierl, M. Beroza, and C. W. Collier, *Science* **170,** 87–89 (1970).
65. E. W. Underhill, A. P. Arthur, M. D. Chisolm, and W. F. Steck, *Environ. Entomol.* **8,** 740–743 (1979).
66. G. J. Blomquist, R. W. Howard, C. A. McDaniel, S. Remaley, L. A. Dwyer, and D. R. Nelson, *J. Chem. Ecol.* **6,** 257–269 (1980).
67. R. D. Plattner, G. F. Spencer, and R. Kleiman, *Lipids* **11,** 222–227 (1976).
68. E. Araki, T. Ariga, and T. Murata, *Biomed. Mass Spectrom.* **3,** 261–264 (1976).
69. G. Eglinton and D. H. Hunneman, *Phytochemistry* **7,** 313–322 (1968).
70. J. A. McCloskey and M. J. McClelland, *J. Amer. Chem. Soc.* **87,** 5090–5093 (1965).

AUTHOR'S NOTE: This chapter does not include literature beyond October 1980.

25

# MASS SPECTROMETRY OF N-NITROSAMINES

## L. M. LIBBEY and R. A. SCANLAN

### 1. INTRODUCTION

Magee and Barnes[1] reported that dimethylnitrosamine (NDMA) caused liver tumors when fed to rats. Ender and colleagues[2] reported that toxic herring meal contained NDMA and they speculated that the nitrosamine arose from the reaction of sodium nitrite and the methyl amines present in fish meal. *N*-Nitrosamines (NAs) are now known to be a widely acting group of animal carcinogens[3] and are found, usually at trace levels, in many diverse places in the environment, such as: nitrite-cured meats, polluted air, tobacco smoke, cutting oils, hydraulic fluids, cosmetics, herbicides, pharmaceuticals, beer, Scotch whiskey, and rubber products. This is not to say that nitrosamines will always be found in the sources listed, and it should be noted that due to research and subsequent corrective action, the levels of volatile nitrosamines have been reduced in certain products.

One aspect of current concern is *in vivo* nitrosation of amines in the gastrointestinal tract. An example would be the reaction of gastric nitrite with nitrosatable amines that were ingested as drugs or food.[3,4]

Collin[5] published the first mass spectral study of nitrosamines; he examined five dialkyl *N*-nitroso and two *C*-nitroso compounds. Recently, Rainey and co-workers[6] published condensed mass spectra for 146 nitrosamines which they had synthesized. Complete mass spectra for these compounds had been published earlier.[7,8] Having this large database to draw upon, Rainey and colleagues[6] were able to make a detailed study of nitrosamine fragmentation and elucidated several helpful gen-

*L. M. LIBBEY* and *R. A. SCANLAN* • Department of Food Science and Technology, Oregon State University, Corvallis, Oregon 97331.

eralizations. Earlier, Pensabene and co-workers[9] obtained the mass spectra of 25 *N*-nitrosamines; the complete spectra were published as a separately available supplement to their paper. Schroll and colleagues[10] studied the mass spectra of 13 *C*-nitroso and 10 *N*-nitroso compounds; exact mass measurements and metastable transitions were used to support their fragmentation mechanisms. ApSimon and Cooney[11] obtained the mass spectra of seven cyclic *N*-nitrosamines; they also used high-resolution and metastable measurements to support the suggested fragmentation schemes. Saxby[12] obtained the mass spectra of 24 dialkyl *N*-nitrosamines; deuterium labeling, as well as high-resolution and metastable measurements were used to document proposed fragmentation mechanisms.

As a technique, mass spectrometry has many of the advantages sought for in a method to analyze trace levels of *N*-nitrosamines in environmental samples. In view of the uncertainty of some of the early nitrosamine analyses, it might be well to set down some of the advantages and shortcomings of mass spectrometry in nitrosamine analysis.

*1. Specificity.* Mass spectrometry is basically very specific, as it is quite rare to find two different compounds with the same mass spectrum. High-resolution mass spectrometry is the most specific form. However, it is sometimes difficult to obtain high resolution and high sensitivity at the same time; often a compromise is made whereby only one characteristic ion of the nitrosamine in question is monitored at a resolution of 5000–7000. Although fragment ions such as *m/z* 30 are sometimes monitored, this ion is usually not sufficiently characteristic and/or intense to be generally useful.[13] Monitoring the molecular, or some other characteristic ion is usually preferred, and for NDMA the accurate mass of $M^+$ at *m/z* 74.0480 is often monitored. If precautions are taken to exclude an interferent arising from silicone, $Si^{29}(CH_3)_3$ of mass 74.0469, unequivocal results can be obtained at moderate resolution (approximately 7000).

2. *Sensitivity.* Volatile *N*-nitrosamines can be confirmed in complex media, food for example, down to the low μg/kg level using full-scan low-resolution mass spectrometry.[14] High-resolution (approximately 10,000) single-ion monitoring has a detection limit for NDMA of approximately 0.3 pg.[15]

## 2. *DIALKYL N-NITROSAMINES*

The mass spectra of four dialkyl *N*-nitrosamines of environmental interest are reproduced in Fig. 1.

As is fairly common in homologous series, the first member of the

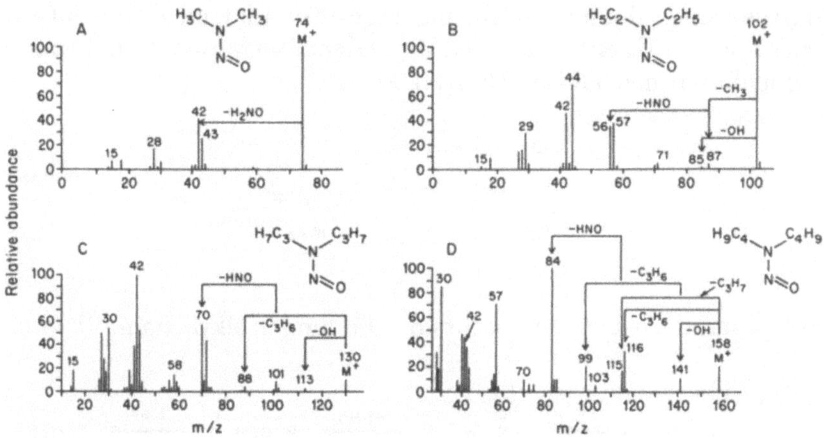

FIGURE 1. Mass spectra of dimethyl (A) and diethylnitrosamine (B) (Ref. 9, with permission); di-*n*-propylnitrosamine (C) (Ref. 10, with permission); and di-*n* -butylnitrosamine (D) (Ref. 12, with permission).

series is somewhat atypical; for example dimethylnitrosamine cannot undergo the α-cleavage common in most dialkyl *N*-nitrosamines. The molecular ion in the commonly found dialkyl *N*-nitrosamines is usually prominent, and chemical ionization is not usually needed. In fact, the molecular ion is of sufficient intensity and specificity to be useful for high-resolution monitoring. Two ions commonly found in the spectra of dialkyl *N*-nitrosamines are *m*/*z* 30 (NO) and *m*/*z* 42 ($H_2C{=}N^+{=}CH_2$); these ions are fairly characteristic of many nitrosamines; computer-aided specific-ion searches of acquired unit resolution GC/MS data is a powerful technique for locating a nitrosamine in a complex total ionization chromatogram. A convenient and rapid scheme to locate a specific nitrosamine would be to obtain overlays of specific-ion searches, for example $M^+$, *m*/*z* 30 and 42. Coincidence of peak maxima for the specific-ion searches at the appropriate retention time indicates the probable nitrosamine location; this would then require confirmation by obtaining a satisfactory full-scan spectrum for the peak in question. Low-resolution mass fragmentography (monitoring one or more ions at unit resolution) is a common data acquisition technique in GC/MS, and can be quite useful in screening, but the inherent lack in specificity limits its utility. The major advantage of mass fragmentography is a marked increase in sensitivity as the spectrometer spends all of its scanning time on only those few ions of interest; the disadvantage is lowered specificity, as obviously the analyst cannot observe full spectra or examine ions which were not collected.

Saxby[12] made a careful study of the fragmentation of 24 dialkyl *N*-nitrosamines; deuterium labeling, high-resolution measurements, and metastable transitions were used as evidence to support the proposed fragmentation mechanisms. For NDMA:

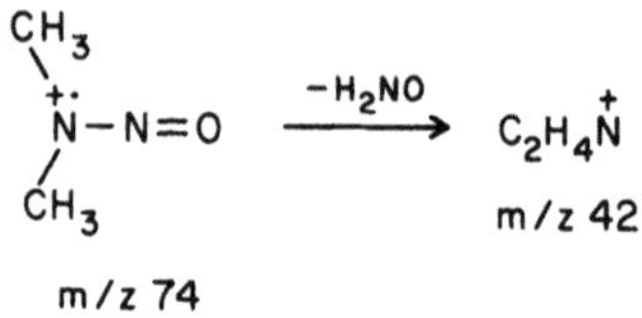


More generally, α-cleavage occurs on the longer alkyl chain.[12]

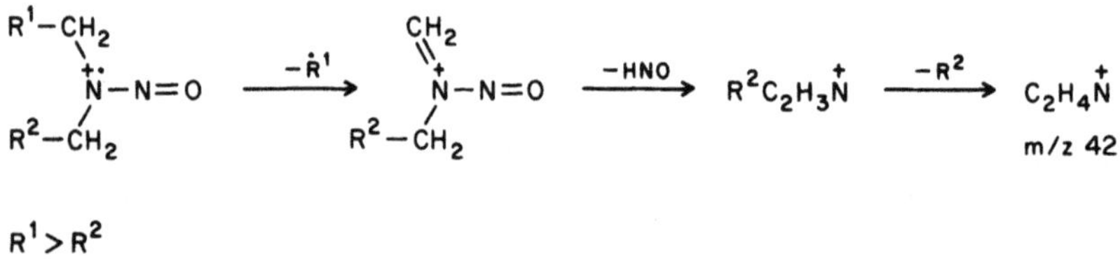


Two other generalizations on the fragmentation of dialkyl *N*-nitrosamines should be mentioned; the relative size of the molecular ion decreases with increasing length of the alkyl chains, and an ion, due to the loss of 17 (OH) from the molecular ion, becomes more intense as the alkyl chain is increased (Fig. 1). It appears that the loss of OH proceeds through a cyclic transition in which a hydrogen is transferred from a β-carbon to the oxygen.[6]

## 3. HETEROCYCLIC N-NITROSAMINES

*N*-Nitrosopyrrolidine (NPYR) is frequently found at trace (ppb) levels in cooked bacon.[16] The legal mandate requiring mass spectral confirmation for the presence of NPYR in bacon requires reliable standard spectra.[17] There are marked discrepancies between many of the published mass spectra for NPYR.[7,17,18] Apparently NPYR is thermally labile and the published spectra often exhibit varying degrees of thermal decomposition.

In Fig. 2, A exhibits little thermal decomposition, while B is strongly suspected of thermal breakdown. The ion at *m*/*z* 28 definitely arises from NPYR and is not artifactual (air leak, etc.). Minimal thermal decomposition is evidenced by a comparison of the relative ion intensities: *m*/*z* 68 and 69 and *m*/*z* 100 and 30.[17] Accurate mass measurement indicated that *m*/*z* 68 arose from M − $H_2NO$.[11]

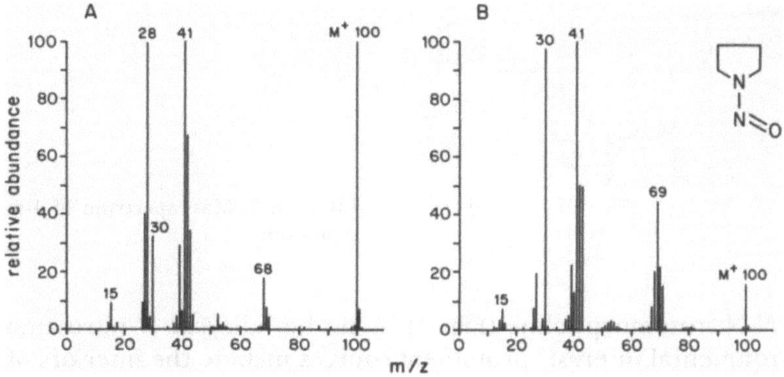


FIGURE 2. Mass spectra of *N*-nitrosopyrrolidine. A, standard spectrum of NPYR (Ref. 17, with permission); B, standard spectrum of NPYR suspected of thermal degradation (Ref. 18, with permission).

## 4. *ALICYCLIC N-NITROSAMINES*

1-Nitrosopiperidine became of interest when it was found in certain sausages that had been manufactured with a spice premix.[19] Two components of the premix, piperine (found in the black pepper) and sodium nitrite, reacted to form 1-nitrosopiperidine.

The mass spectrum under EI for 1-nitrosopiperidine is given in Fig. 3; the fragmentation is quite typical of an alicyclic *N*-nitrosamine.[6] A prominent molecular ion is noted at *m/z* 114. The loss of 30 (NO) leads to *m/z* 84, while M − NOH yields the ion at *m/z* 83. Rainey and coworkers have rationalized the fragmentation of 1-nitrosopiperidine as follows.[6]

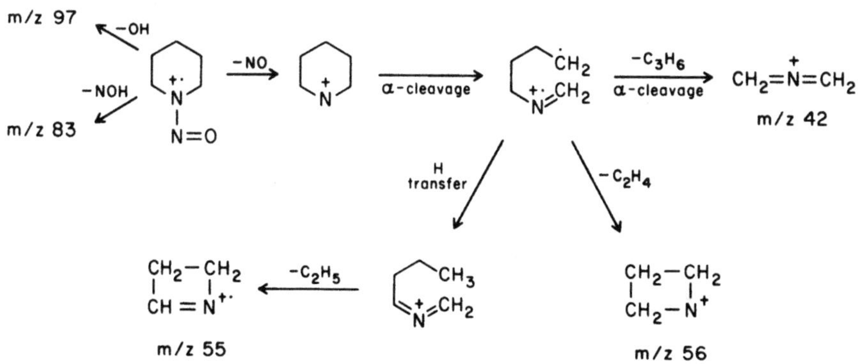

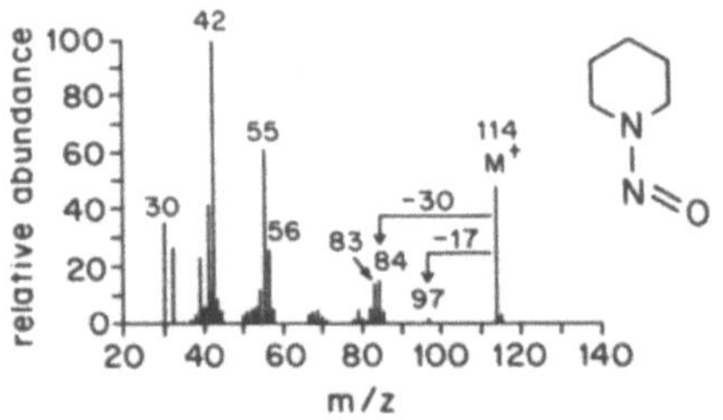


FIGURE 3. Mass spectrum of 1-nitrosopiperidine.

*N*-Nitrosomorpholine (Fig. 4) is another alicyclic *N*-nitrosamine of environmental interest; prominent sources include the interiors of new cars,[20] and diesel crankcase oil.[21] Schroll and his colleagues[10] included *N*-nitrosomorpholine in the mass spectral study of *N*-nitrosamines; precise mass measurements and metastable transitions were used to support the suggested fragmentation scheme.

## 5. *AROMATIC N-NITROSAMINES*

*N*-Nitrosamines with an aromatic moiety are also of great environmental interest. Aromatic amines are common industrial chemicals and are used in such products as dyes, pigments, pharmaceuticals, agricultural chemicals, and so on. *N*-Nitrosamines can be formed via side reactions during manufacture, or nitrosatable amines in the product can later react, for example with nitrite during storage or after usage.

The mass spectrum of *N*-nitrosomethylphenylamine has been obtained.[6,10] The mass spectrum used with permission from the latter reference (Fig. 5) exhibits ions typical of a phenyl compound (*m/z* 39,

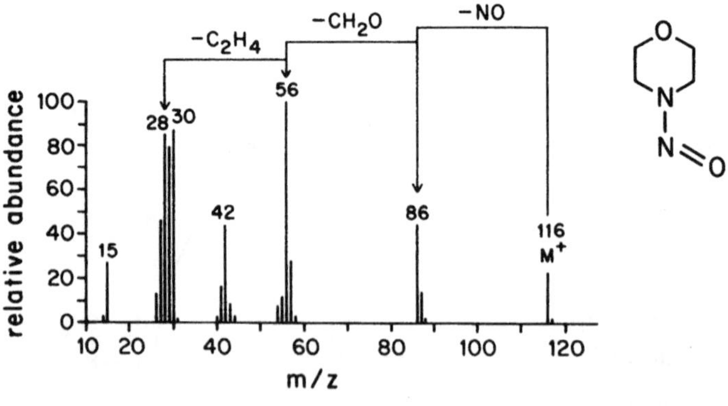


FIGURE 4. Mass spectrum of *N*-nitrosomorpholine (Ref. 10, with permission).

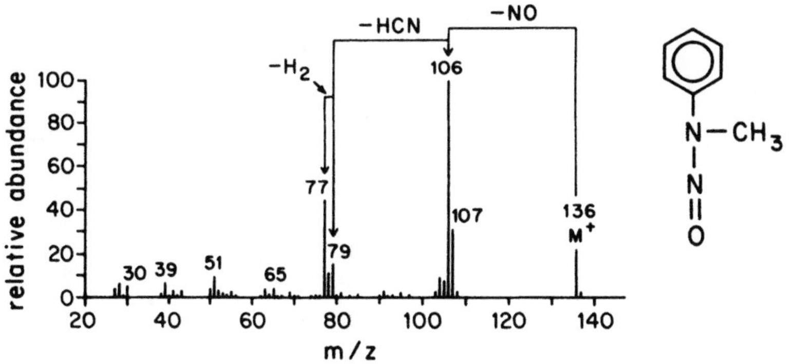


FIGURE 5. Mass spectrum of *N*-nitrosomethylphenylamine.

51, 77) and also exhibits a facile loss of *m/z* 30 (NO) from the molecular ion to yield the base peak at *m/z* 106.

Schroll and colleagues[10] have pointed out the facile thermal decomposition of aromatic *N*-nitrosamines during mass spectrometry. This decomposition is evidenced in Fig. 5 by hydrogen abstraction and loss of NO to the corresponding amine (methylphenylamine) having a molecular ion at *m/z* 107.

## 6. NONVOLATILE N-NITROSAMINES

The relatively volatile *N*-nitrosamines have so far received most of the attention of environmental scientists. This is due largely to the comparative ease with which volatile nitrosamines can be isolated (i.e., by distillation) and the relative ease by which these compounds can be separated by gas chromatography. There is little doubt that the more polar or nonvolatile *N*-nitrosamines deserve and will receive more analytical attention as state-of-the-art technologies, such as HPLC and TEA (thermal energy analyzer)/HPLC, are applied to their analysis.[22] Tandem HPLC/MS also appears very promising.

*N*-Nitrosopyrrolidin-3-ol (NHPYR), a polar *N*-nitrosamine, almost certainly has hydroxyproline as its precursor, and has been isolated from cured meats[23] and from fried bacon.[24,25] Lee and colleagues[26] had previously obtained mass spectral evidence that clearly showed heated 4-hydroxy-*N*-nitrosoproline gave rise to *N*-nitrosopyrrolidin-3-ol. Although the polar NHPYR can be gas chromatographed with its hydroxy group free (underivatized), trace analysis of environmental samples for NHPYR usually require derivatization of the hydroxy function.[24,25]

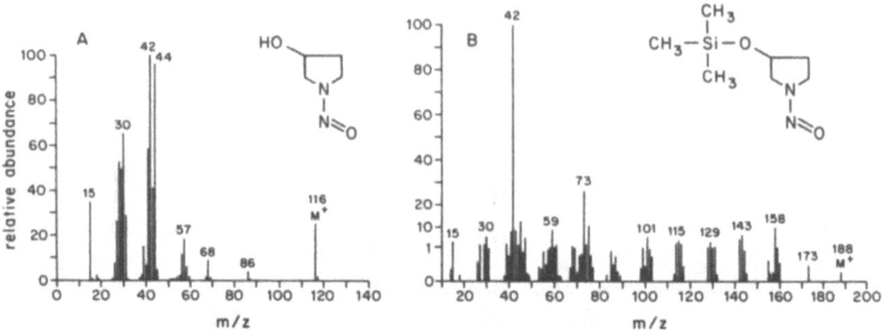


FIGURE 6. A, NHPYR; B, silyl derivative of NHPYR.

Mass spectra of NHPYR and of its trimethylsilyl derivative are given in Fig. 6: A, NHPYR;[26] B, silyl derivative of NHPYR.[25]

Although the underivatized NHPYR gives a relatively simple mass spectrum containing several intense ions of diagnostic value, due to chromatographic tailing it is very difficult to obtain confirmatory spectra for NHPYR by GC/MS on the "dirty" samples common in environmental analysis. Even though the spectrum contains many ions of low abundance, and is diagnostically more difficult, the TMS derivative of NHPYR gives a much sharper GC peak and the detection limit in a dirty sample is much lower. Sen and colleagues[24] have successfully determined NHPYR in food samples by making the methyl ether derivative.

Nitrosoiminodiethanol (NDELA), a polar *N*-nitrosamine, is of current interest as the amine precursors, *bis*(2-hydroxyethyl)amine and *tris*-(2-hydroxyethyl)amine are common ingredients in many toiletries and

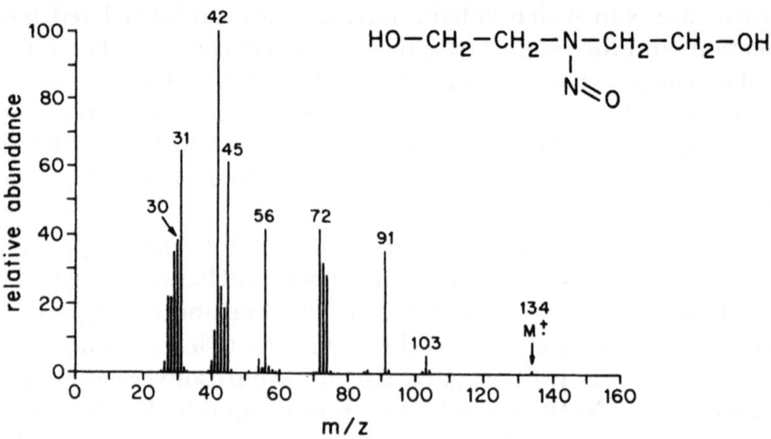


FIGURE 7. Mass spectrum of *N*-nitrosoiminodiethanol (NDELA).

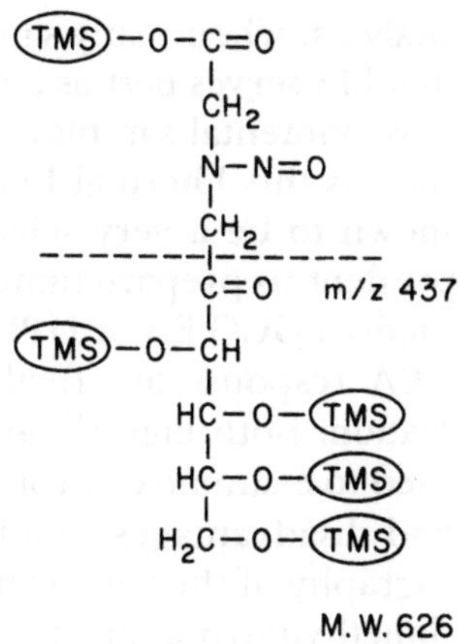


FIGURE 8. Structure of the persilylated *N*-nitroso Amadori compound from glucose plus glycine; TMS indicates -$Si(CH_3)_3$.

cosmetics, and some consumer products have been shown to contain NDELA.[27] Some of the cutting oils used by machinists have been found to contain up to 3% NDELA.[28] The mass spectrum of NDELA has been obtained by Rainey and colleagues[8] and also by Fan and co-workers;[27] the spectrum (modified, with permission) from the latter reference is given in Fig. 7.

*N-Nitroso sugar-amino acids:* Amadori compounds are formed in nonenzymatic (Maillard) browning reactions between aldoses and amino acids. The N-nitrosation of several sugar amino acid compounds, including Amadori compounds, has been studied by Heyns and colleagues.[29] Mass spectra were obtained after persilylation of the *N*-nitroso sugar amino acid compounds. The structure of the silyl derivative of the *N*-nitroso Amadori compound from glucose plus glycine is given in Fig. 8.

The EI spectra gave diagnostically useful ions, particularly at high mass: M − 15 ($CH_3$) and M − 30 (NO); for the compound in Fig. 8, which has a mw of 626, these ions were, respectively, *m/z* 611 and 596. In addition, it was suggested that the ion at *m/z* 437 could be used to differentiate the *N*-nitroso glucose amino acid compounds from those containing fructose.[29]

## 7. *MASS SPECTROMETRY FOR THE CONFIRMATION OF N-NITROSAMINES*

This section is concerned with MS as a technique to confirm the presence of NAs in environmental samples that have been screened and found tentatively positive by another technique. Particularly in routine analyses, when many samples must be examined, we feel quite strongly that MS serves best as a confirmatory technique, and prior screening of environmental samples for nitrosamines is best done on an instrument such as the Thermal Energy Analyzer (TEA). Since the TEA has been shown to be a very selective and highly sensitive detector,[30] it seems prudent to prepare samples with only minimal cleanup and first screen them by GC/TEA or HPLC/TEA. Only those samples that show a positive TEA response are further cleaned up and submitted for MS confirmation. Both high-[31] and low-resolution[14] MS have been successfully used to examine environmental samples for *N*-nitrosamines; advantages and disadvantages could be listed for each approach. For gas chromatography of the volatile nitrosamines, a group that roughly encompasses dimethylnitrosamine to *N*-nitrosomorpholine, both packed and capillary columns have been used with the MS; packed columns can handle larger samples, but with only moderate chromatographic resolution; conversely, capillary columns will tolerate only relatively small samples, but with greatly increased resolution. For our GC/MS confirmatory work with nitrosamines, we have found that long (150–180 m) wide-bore (0.50–0.75 mm i.d.) stainless steel or glass capillaries are a good compromise. Direct syringe or trap injections are well tolerated without an injection splitter and sufficient chromatographic resolution is available to permit confirmation of nitrosamines in most TEA-positive samples. In confirming samples for volatile NAs by GC/MS at the ppb level, we have found adequate cleanup of the samples and good chromatography to be critical; spurious ions of any mass and from any source restrict the sensitivity of the MS, while sharp well-resolved chromatographic peaks are necessary for clean unambiguous mass spectra. To avoid compromising the use of MS in confirmation, we consider NAs as confirmed by low-resolution MS when all of the following identification criteria have been satisfied; namely, the sample must be TEA positive, the compound must have the correct retention time on two GC columns, and the complete full-scan low-resolution mass spectrum must closely match that of the authentic NA.[14]

An analytical method that embodies most of the above considerations has recently been published,[14] where the unequivocal confirmation of low μg/kg levels of volatile NAs in several foods was detailed. Basically, the method consists of grinding the frozen sample, adding ammonium sulfamate, adjusting the pH of the sample to 2.5 or lower, adding mineral oil, and vacuum distilling the mixture. The sample volatiles containing any volatile NAs were condensed in liquid nitrogen traps. The thawed distillate was extracted with dichloromethane (DCM), and the DCM extract containing the NAs was washed successively with aqueous sodium bisulfite, 3 *N* HCl and 1.5 *N* NaOH. After backwashing the bisulfite and

HCl washes, the combined DCM extracts were dried over anhydrous $Na_2SO_4$. Prior to GC/TEA analysis, the DCM extract was concentrated, first in a Kuderna-Danish apparatus, and finally the volume was reduced to 1 ml by fitting the concentrator tube with a micro-Snyder column and using a $N_2$ stream. Screening of the concenterate for NAs was accomplished on the GC/TEA. Betweeen 4 and 8 ml of concentrate was injected onto a 9m × 2.0 mm i.d. stainless steel column packed with 11% Carbowax 20M on 80–120 mesh Chromosorb. This column was used at either 100 or 160 °C (depending on the retention time of the NA in question); the GC effluent was directed via a glass-lined stainless steel tube into the model 502L TEA. Tentative identification and quantification was accomplished by injecting known amounts of standard NA solutions. After repeated injections of unknown and standard, a comparison of retention times was used to assign probable identities to the unknown NAs, and concentrations of NAs in the unknown were calculated. If confirmation is required for the TEA-positive samples, further cleanup of the concentrate was required. This was accomplished on an activity II alumina column, using DCM to elute the NAs. Prior to GC/MS, the cleaned-up fraction was injected onto the packed Carbowax column of the GC/TEA, and the TEA-positive fraction was trapped. As soon as the TEA detector indicated the start of the NA peak, a Carle valve was used to divert the gas stream containing the NA into an external trap consisting of a loop of nickel tubing cooled in a bath containing dry ice-methoxyethanol. Only the fractions containing NAs were trapped, and the trapping process was repeated several times to insure adequate sample (~ 50 ng) for unequivocal mass spectral confirmation. The contents of the nickel trap, containing the pooled NA, were kept chilled and taken to the GC/MS. The cold trap was inserted into a valved bypass, and the contents of the trap were flashed onto the wide-bore capillary column of the GC/MS. We have used this trapping–reinjection technique for many years, and it is described in more detail elsewhere.[14,32]

The utility of computers and data systems as an aid in the acquisition and interpretation of GC/MS nitrosamine data can hardly be overestimated. Prior to data acquisition the desired parameters can be specified so that the system will automatically collect the data. In trace organic analysis we have found a program called IFSS (integration as a function of signal strength) very useful in maintaining a relatively constant signal-to-noise ratio while avoiding signal saturation on large GC peaks.[33] The computer-aided analysis of acquired GC/MS nitrosamine data can be greatly facilitated by asking the computer for specific ion searches for ions characteristic of the nitrosamine in question. For example if NDMA is sought in a complex total ionization chromatogram, one might overlay the search for *m/z* 74 on the total ionization chart; if NDMA is present, *m/z* 74 should show a peak at the retention time for NDMA. Confirmatory specific-ion searches might be made for *m/z* 30 and/or *m/z* 42. A variety of output programs are commercially available to permit rapid visualization on a video terminal; often the cleanest confirmatory spectrum is determined semiempirically by trial and error at the video terminal prior to the plotting of permanent hard-copy.

## REFERENCES

1. P. N. Magee and J. M. Barnes, *Br. J. Cancer* **10,** 114 (1956).
2. F. Ender, G. Harve, A. Helgebostad, N. Koppang, R. Madsen, and L. Ceh, *Naturwissenschaften* **51,** 637 (1964).
3. W. Lijinsky, in *N-Nitrosamines* (J.-P. Anselme, ed.), American Chemical Society, Washington, D.C., 1979, pp. 165–173.
4. T. Wang, T. Kakizoe, P. Dion, R. Furrer, A. J. Varghese, and W. R. Bruce, *Nature* **276,** 280–281 (1978).
5. J. Collin, *Bull. Soc. Roy. Sci. Liège* **23,** 201–220 (1954).
6. W. T. Rainey, W. H. Christie, and W. Lijinsky, *Biomed. Mass Spectrom.* **5,** 395–408 (1978).
7. W. Lijinsky, W. H. Christie, and W. T. Rainey, Jr., Mass Spectra of *N*-nitroso Compounds, Oak Ridge Nat. Lab., Tech. Report No. TM-4359, 1973, pp. 1–106.
8. W. T. Rainey, W. H. Christie, C. A. Pritchard, and W. Lijinsky, Mass Spectra of *N*-Nitroso Compounds, Oak Ridge National Laboratory, Tech. Report No. TM-5500, 1976, pp. 1–63.
9. J. W. Pensabene, W. Fiddler, C. J. Dooley, R. C. Doerr, and A. E. Wasserman, *J. Agric. Food Chem.* **20,** 274 (1972).
10. G. Schroll, R. G. Cooks, P. Klemmensen, and S.-O. Lawesson, *Arkiv för Kemi* **28,** 413–422 (1968).
11. J. W. ApSimon and J. D. Cooney, *Can. J. Chem.* **49,** 1367–1371 (1971).
12. M. J. Saxby, *J. Assoc. Off. Anal. Chem.* **55,** 9–12 (1972).
13. R. Self, *Biomed. Mass Spectrom.* **6,** 361–373 (1979).
14. J. H. Hotchkiss, L. M. Libbey, and R. A. Scanlan, *J. Assoc. Off. Anal. Chem.* **63,** 74–79 (1980).
15. K. S. Webb, T. A. Gough, A. Carrick, and D. Hazelby, *Anal. Chem.* **51,** 989–992 (1979).
16. T. Fazio, R. H. White, L. R. Dusold, and J. W. Howard, *J. Assoc. Off. Anal. Chem.* **56,** 919–921 (1973).
17. L. M. Libbey and R. A. Scanlan, *J. Assoc. Off. Anal. Chem.* **64,** 1255–1256 (1981).
18. S. R. Heller and G. W. A. Milne, *EPA/NIH Mass Spectral Data Base,* Department of Commerce, U.S. Government, Washington, D.C., 1978.
19. N. P. Sen, W. F. Miles, B. Donaldson, T. Panalaks, and J. R. Iyengar, *Nature* **245,** 104 (1973).
20. D. P. Rounbehler, J. Reisch, and D. H. Fine, *J. Food Cosmet. Toxicol.* **18,** 147–151 (1980).
21. E. U. Goff, J. R. Coombs, D. H. Fine, and T. M. Baines, *Anal. Chem.* **52,** 1833–1836 (1980).
22. T. Y. Fan, I. S. Krull, R. D. Ross, M. H. Wolf, and D. H. Fine, in *Environmental Aspects of N-nitroso Compounds,* (E. A. Walker, M. Castegnaro, L. Griciute, R. E. Lyle, and W. Davis, eds.), International Agency for Research on Cancer, Lyon, 1978, pp. 3–17.
23. C. Janzowski, G. Eisenbrand, and R. Preussmann, *J. Chromatogr.* **150,** 216–220 (1978).

24. N. P. Sen, W. F. Miles, S. Seaman, and J. F. Lawrence, *J. Chromatogr.* **128,** 169–173 (1976).
25. J. S. Lee, L. M. Libbey, R. A. Scanlan, and D. D. Bills, *Bull. Environ. Contam. Toxicol.* **19,** 511–517 (1978).
26. J. S. Lee, D. D. Bills, R. A. Scanlan, and L. M. Libbey, *J. Agric. Food Chem.* **25,** 422–423 (1977).
27. T. Y. Fan, U. Goff, L. Song, D. H. Fine, G. P. Arsenault, and K. Biemann, *J. Food Cosmet. Toxicol.* **15,** 423 (1977).
28. T. Y. Fan, J. Morrison, D. P. Roundehler, R. Ross, D. H. Fine, W. Miles, and N. P. Sen, *Science* **196,** 170–171 (1977).
29. K. Heyns, S. Röper, H. Röper, and B. Meyer, *Angew. Chem. Int. Ed. Engl.* **18,** 878–880 (1979).
30. D. H. Fine, F. Rufeh, D. Lieb, and D. P. Rounbehler, *Anal. Chem.* **47,** 1188–1191 (1975).
31. T. A. Gough and K. Webb, in *Environmental Carcinogens: Selected Methods of Analysis.* Vol. 1. (H. Egan, R. Preussmann, M. Castegnaro, E. A. Walker, and A. E. Wasserman, eds.), International Agency for Research on Cancer, Lyon, 1978, pp. 141–150.
32. R. A. Scanlan, R. G. Arnold, and R. C. Lindsay, *J. Gas Chromatogr.* **6,** 372–373 (1968).
33. W. L. Budde and J. W. Eichelberger, *Organics Analysis Using Gas Chromatography Mass Spectrometry,* Ann Arbor Science, Ann Arbor, MI, 1979, pp. 35–37.

26

# OTHER INDUSTRIAL POLLUTANTS AND NATURAL TOXINS

S. SAFE and J. SPARLING

## 1. INDUSTRIAL POLLUTANTS

*Tris*(1,3-dichloro-2-propyl)phosphate(I) and *tris*(2,3-dibromopropyl)-phospate(II) are industrial flame retardants which have been used to impregnate fabrics. The carcinogenicity of II resulted in its removal from commercial products. A key scientific study, which indicated human exposure to II, was the negative-ion atmospheric pressure ionization MS detection of 2,3-dibromopropanol, a hydrolysis product of II, in the urine of children wearing pajamas treated with this flame retardant.(1) The analytical procedure involved selective-ion monitoring of the *m/z* 81 and 79 bromine isotopes. The rat also metabolized II to give a diester urinary metabolite which was identified by the mass spectra of its corresponding methyl derivative. The EIMS of this compound did not give a molecular ion but a relatively intense $[M - Br]^+$ ion at *m/z* 429 (3 Br) and several diagnostic fragment ions were observed. The chlorinated flame retardant I was also metabolized in the rat to a bis phosphate which was identified by the MS of the trimethylsilyl derivative.(2) Recently, negative chemical ionization mass spectral screening of human seminal fluid has identified I by its *m/z* 463 ion (i.e., $M^+ + Cl$) which exhibited an ion cluster pattern characteristic of a $Cl_7$ species.(3)

Alkylated aromatics (biphenyl, bibenzyl, and naphthalene) are being used as PCB replacement compounds, and selective-ion monitoring has been used to quantitate diisopropylnaphthalene in sea mud and fish.(4)

---

S. SAFE and J. SPARLING • The Guelph–Waterloo Centre for Graduate Work in Chemistry, Department of Chemistry, University of Guelph, Guelph, Ontario N1G 2W1, Canada.

Ethylene glycol derivatives and related polyoxyethylene polymers $R_1$ $(OCH_2CH_2)$ $R_2$ are widely used commercial products with a total production of about two billion pounds. In a study of the chemical pollutants present in the Delaware River, Sheldon and Hites have identified several low-molecular-weight ethylene glycol-derived industrial compounds and confirmed their structures by mass spectrometry (MS).[5] The EIMS of 2-(2-[2-(*p*-1′,1′,3′,3′-tetramethylbutylphenoxy)-ethoxy]ethoxy)ethanol and the 1-chloro analog exhibit intense high mass ions at m/z 267 and 285 respectively which correspond to the

$$\begin{array}{c} \phantom{M^{+}-CH_2-}CH_3\phantom{-CH_3} \\ \phantom{M^{+}-CH_2-}|\phantom{-CH_3} \\ M^{+}-CH_2-C-CH_3 \\ \phantom{M^{+}-CH_2-}|\phantom{-CH_3} \\ \phantom{M^{+}-CH_2-}CH_3\phantom{-CH_3} \end{array}$$

ionic species. Successive cleavage at the ether linkage all result in the generation of fragment ions which aid in structure confirmation (Fig. 1).[5,6]

Polyoxyethylene glycol alkyl and alkylphenyl ethers are higher molecular weight ethylene glycol derivatives which are used as nonionic surfactants. Conventional EIMS analysis is not suitable for identification of these compounds, however field description (FD) MS has proven to be successful for confirmatory analysis.[7] For a series of polymers with the general formula $C_nH_{2q+1}C_6H_4O(C_2H_4O)_mH$ molecular ions could be detected for polymers with molecular weights as high as 700–800. Moreover, the technique was applied for the identification of these polymers in polluted streams.

Aromatic amines are widely used industrial compounds which would not be expected to have significant environmental lifetimes due to diverse chemical and biodegradation pathways. Analysis of single aromatic amines or defined mixtures is conveniently carried out by derivatization (e.g., to acyl derivatives) followed by gas chromatography. However, structure confirmation of more complex matrices is greatly facilitated by GC/MS analysis. Westoo and Noren have identified 2′,3,4,4′,5-pentachloro- and 2′,3,4,4′,5,6-hexachloro-2-aminodiphenyl ethers as residues in fish by their mass spectra, which exhibit characteristic molecular ions and chlorine isotopic distribution patterns.[8] GC/MS analysis of basic organics from Buffalo River sediments confirmed the presence of 11 aromatic amines including *N,N*-dimethylaniline; 4,4′-bis(dimethylamino)biphenyl; 4-(dimethylamino)benzaldehyde; 4-(dimethylamino)benzophenone; aniline; 4-(dimethylamino)diphenylmethane; bis(4-dimethylaminophenyl)-methane; bis(4-dimethylaminophenyl)phenylmethane; tris(4-dimethylaminophenyl)methane; 1-naphthylamine; and *N*-ethyl-*N*-phenylbenzylamine.[9] Typically the EIMS for most of these compounds exhibited intense molecular ions, $[M - N(CH_3)_2]$ ions and M-phenyl ions as well

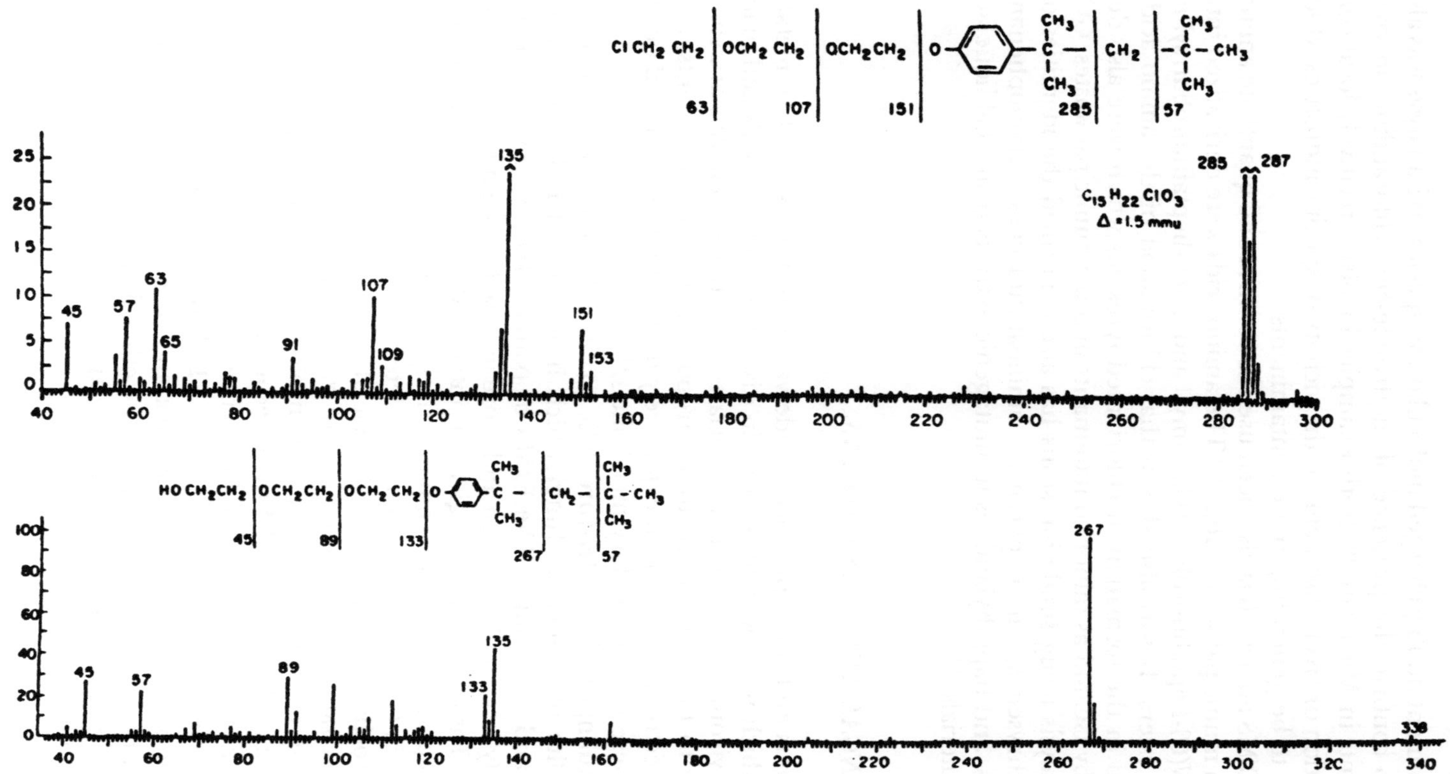


FIGURE 1. Mass spectra of ethylene glycol derived compounds present as pollutants in Delaware River.[6]

as prominent doubly charged molecular ion species. MS analysis was also used to confirm the presence of methylene-*bis*(2-chloraniline) in river sediments in Michigan.[10] Both examples involved relatively local contamination of river biosystems in the vicinity of specific industries which appear to be generating these contaminants.

GC/MS analysis has also been used to confirm and quantitate amino acids in municipal wastewater.[11] The amino acids were derivatized into their *N(O)*-heptafluorobutyrylisoamyl and *N(O)*-heptafluorobutrsyl-*n*-propyl esters. It was also shown that chlorination of the amino acids resulted in the formation of chlorinated tyrosines which were also detected by GC/MS as minor contaminants in some municipal wastes. GC/MS analysis using single ion scans has also confirmed the identities of aminochrysenes, aminopyrenes, aminoanthracenes, aminophenanthrenes, and naphthylamines in mutagenic extracts from coal liquefaction materials.[12]

## 2. FUNGAL AND PLANT TOXINS

Plants and fungi produce a wide variety of toxic secondary metabolites which have been implicated in diseases of both animals and man. The environmental problems associated with these chemicals are comparable to those attributed to anthropogenic chemical pollutants, and mass spectrometry has been extensively used both to confirm and quantitate these toxic secondary metabolites.[13–15]

Patulin[4-hydroxy-4*H*-furo 3,2*c* pyran-2(6*H*)-one] is a mycotoxin metabolite of numerous fungi which have been identified in rotting apples and apple juice. The EIMS of patulin(III) ($M^+$ 154) exhibits a relatively weak molecular ion and more intense fragment ions at *m/z* 110 and 55. Patulin acetate(IV) gives an intense M − HOAc ion at *m/z* 136 and this peak can be used to quantitatively estimate gram levels of patulin acetate via GC/MS total ion current detection.[16,17]

Mirocha[18,19] made extensive use of mass spectrometry both for the characterization and quantitation of zearalenone[2,4-dihydroxy-6-(10-hydroxy-6-oxo-*trans*-1-undecenyl)benzoic acid u-lactone] and related metabolites. This group of fungal secondary metabolites are produced by several *Fusarium* strains and have been implicated as causal agents in some domestic animal diseases. The EIMS fragmentation patterns for this group of compounds is complex and structure dependent; zearalenone undergoes ring cleavage at the ether linkage followed by β-ring fission to give an intense *m/z* 188 ion which contains the overall aromatic functionality.

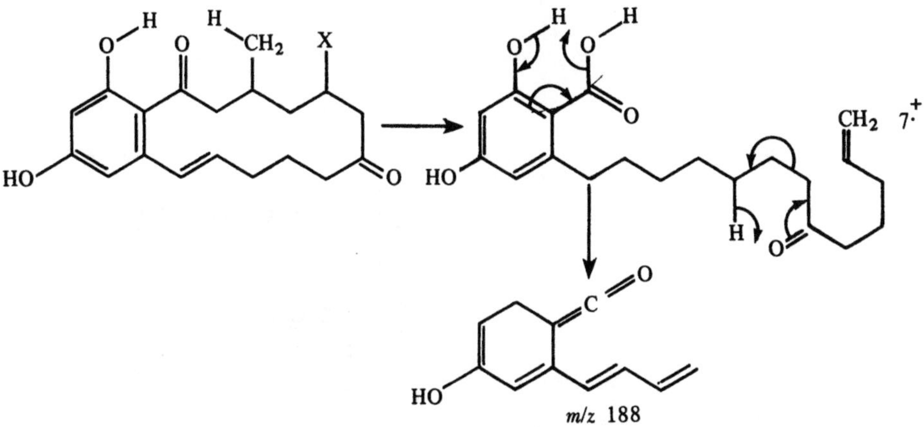


Although this fragmentation pattern dominates the EIMS of zearalenone and the major metabolites, this technique is primarily used for structural elucidation studies; quantitation is primarily via GC analysis although FDMS (which gives the molecular ion as the major peak) may be useful for future quantitative analysis.[14]

Mass spectrometry has also been used to characterize and confirm the identification of the potent hepatocarcinogen aflatoxin $B_1$ and related metabolites which are produced by *Aspergillus flavus*.[20] The EI-induced mass spectra of most aflatoxins gives intense molecular ions and the application of FDMS for the analysis of crude mixtures for aflatoxins may have considerable analytical potential in the future.[14] Currently GC and LC techniques are used for routine sample analysis.

Sterigmatocystin(V) is another highly toxic and carcinogenic secondary metabolite which has been isolated from several *Aspergillus* isolates. Nonderivatized sterigmatocystin can be analyzed by TLC and GLC and a combination GC/MS specific-ion-monitoring method has also been described.

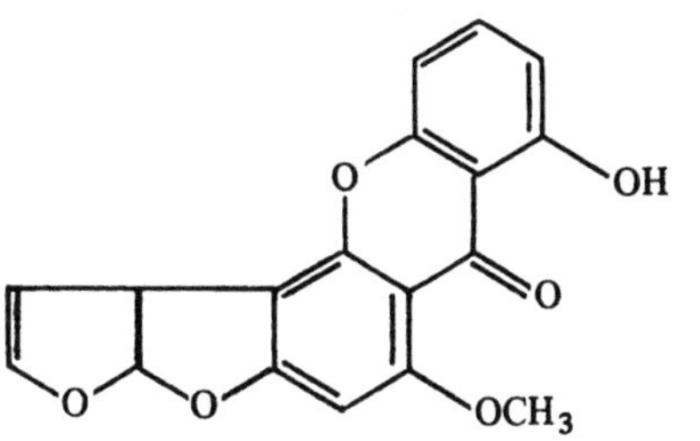


V

The EI-induced fragmentation of this toxin features three major high-

*TABLE 1*

*Chemical Derivatives of Various Tricothecenes* ($OAc{=}CH_3COO{-}$)[a]

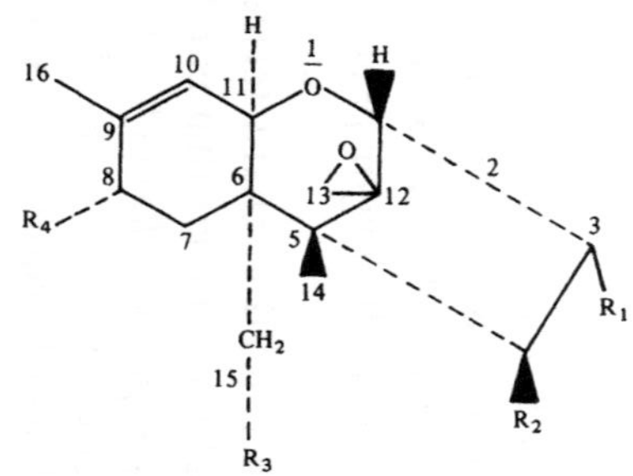


| | $R_1$ | $R_2$ | $R_3$ | $R_4$ | Formula | $M^+$ |
|---|---|---|---|---|---|---|
| 1. 4-Acetoxyscirpendiol | OH | OAc | OH | H | $C_{17}H_{24}O_6$ | 324 |
| 2. Monoacetylneosolaniol | OH | OAc | OAc | OAc | $C_{21}H_{28}O_9$ | 424 |
| 3. Monoacetoxyscirpenol | OH | OH | OAc | H | $C_{17}H_{24}O_6$ | 324 |
| 4. Diacetoxyscirpenol | OH | OAc | OAc | H | $C_{19}H_{26}O_7$ | 366 |
| 5. Scirpentriol | OH | OH | OH | H | $C_{15}H_{22}O_5$ | 282 |
| 6. Neosolaniol | OH | OAc | OAc | OH | $C_{19}H_{26}O_8$ | 382 |
| 7. T-2 tetraol | OH | OH | OH | OH | $C_{15}H_{22}O_6$ | 298 |
| 8. T-2 toxin | OH | OAc | OAc | $OCOCH_2CH(CH_3)_2$ | $C_{24}H_{34}O_9$ | 466 |
| 9. Diacetylneosolaniol | OAc | OAc | OAc | OAc | $C_{23}H_{30}O_{10}$ | 466 |

[a] With additions, from *J. Agric. Food Chem.* **26,** 650, 1978, with permission.

mass ions at *m/z* 324, 306, and 295 which presumably are derived from the losses of $H_2O$ and HCO from the molecular ion species (*m/z* 324). Using these three ions as monitors, GC/MS analysis can be used to detect 1 ppb of sterigmatocystin in spiked whole wheat and rice samples.[21]

Toxicoses from moldy corn and cereal grains have also been attributed to a group of tricothecenes which are secondary metabolites produced by *Fusarium roseum* and related fungi.[22–24] A summary of various tricothecenes is given in Table 1 and their structure elucidation was, in part, dependent on their EI-induced fragmentations. Their relative ion intensities and decomposition pathways are structure dependent and feature losses of $CH_3COOH$ from the $M^{+\cdot}$ or lower mass fragment ions.

Mass spectrometry has also played a key role in the structure determination of toxic cyclodepsipeptides from *Tricothecium roseum*[25] and *Alternaria mali*.[26] The latter study describes in detail the fragmentation of a number of related cyclodepsipeptides and the structures are determined by a combination of end-group and side-chain derivatizations, specific NH–CO cleavages, and diagnostic side-chain fragmentation reactions.

## REFERENCES


1. A. Blum, M. D. Gold, B. N. Ames, C. Kenyon, F. R. Jones, E. A. Hett, R. C. Dougherty, E. C. Horning, I. Dzidic, D. I. Carroll, R. N. Stillwell, and J.-P. Thenot, *Science* **201,** 1020–1023 (1978).
2. R. K. Lynn, K. Wong, R. G. Dickenson, N. Gerber, and J. M. Kennish, *Res. Commun. Chem. Pathol. Pharmacol.* **28,** 351–360 (1980).
3. T. Hudec, J. Thean, D. Kuehl, and R. C. Dougherty, *Science* **211,** 951–952 (1981).
4. K. Sumino, *Arch. Environ. Contam. Toxicol.* **6,** 365–369 (1977).
5. S. Sheldon and R. A. Hites, *Sci. Total Environ.* **11,** 279–286 (1979).
6. S. Sheldon and R. A. Hites, *Environ. Sci. Technol.* **12,** 1188–1194 (1978).
7. A. Otsuki and H. Shirarshi, *Anal. Chem.* **51,** 2329–2332 (1979).
8. G. Westoo and K. Noren, *Ambio* **6,** 232–234 (1977).
9. C. R. Nelson and R. A. Hites, *Environ. Sci. Technol.* **14,** 1147–1149 (1980).
10. G. E. Parris, G. W. Diachenko, R. C. Entz, J. A. Poppiti, P. Lombardo, T. K. Rohrer, and J. L. Hesse, *Bull. Environ. Contam. Toxicol.* **24,** 497–503 (1980).
11. J. L. Burleson, G. R. Peyton, and W. H. Glaze, *Environ. Sci. Technol.* **14,** 1354–1359 (1980).
12. B. W. Wilson, R. Pelrog, and J. T. Cresto, *Mut. Res.* **79,** 193–202 (1980).
13. D. G. I. Kingston, *J. Assoc. Off. Anal. Chem.* **59,** 1016–1022 (1976).
14. J. A. Sphon, P. A. Dreifuss, and H. R. Schulten, *J. Assoc. Off. Anal. Chem.* **60,** 73–82 (1977).
15. R. Self, *Biomed. Mass Spectrom.* **6,** 361–373 (1979).
16. J. W. Ralls and R. M. Lane, *J. Food Sci.* **42,** 1117–1119 (1977).
17. K. R. Price, *Biomed. Mass Spectrom.* **6,** 573–574 (1979).
18. C. J. Mirocha, B. Schauerhamer, and S. V. Pathre, *J. Assoc. Off. Anal. Chem.* **57,** 1104–1110 (1974).

19. J. A. Steele, C. J. Mirocha, and S. V. Pathre, *J. Agric. Food Chem.* **24,** 89–97 (1976).
20. W. F. Haddon, M. S. Masri, V. G. Randall, R. H. Elsken, and B. J. Meneghelli, *J. Assoc. Off. Anal. Chem.* **60,** 107–113 (1977).
21. A. S. Salhab, G. F. Russell, J. R. Coughlin, and D. P. H. Hsieh, *J. Assoc. Off. Anal. Chem.* **59,** 1037–1044 (1976).
22. S. V. Pathre, S. J. Mirocha, C. M. Christensen, and J. Behrens, *J. Agric. Food Chem.* **24,** 97–103 (1976).
23. H. M. Stahr. P. F. Ross, W. Hyde, and W. Obioha, *Appl. Spectr.* **32,** 167–171 (1978).
24. K. Ishii, S. V. Pathre, and C. J. Mirocha, *J. Agric. Food Chem.* **26,** 649–653 (1978).
25. G. W. Engstrom, J. V. DeLance, J. L. Richard, and A. L. Baetz, *J. Agric. Food Chem.* **23,** 244–253 (1975).
26. T. Ueno, T. Nakashima, M. Uemoto, H. Fukami, S.-N. Lee, and N. Izumiya, *Biomed. Mass Spectrom.* **4,** 134–142 (1977).

# APPENDIX

| Compound | Ionization technique | Page |
|---|---|---|
| *N*-acetyl-Met-Gly-Met-Met methyl ester | CI | 86 |
| Acetyltributyl citrate | EI | 349 |
| Aldrin | EI | 406 |
| | CI, NCI | 414 |
| Anthracene | EI | 196 |
| 9,10-Anthracenedione | EI | 201 |
| 9-Anthrone and isomers | EI | 178 |
| Aroclor 1242 | EI | 213 |
| Aroclor 1254 | EI | 213, 222, 224 |
| Aroclor 1254 and four pesticides | EI | 222, 233 |
| Aroclor 1260 | PPNICI | 218 |
| Atrazine conjugate, dimethyl ester of a disulfide dimer of | EI | 427 |
| 7H-Benzo[de]anthracene-7-one | EI | 202 |
| Benzo[c]cinnoline | EI | 202 |
| Benzo[a]fluorene | EI | 198 |
| Benzo[a]pyrene | EI | 196 |
| Benzo[e]pyrene | EI | 196 |
| Bioallethrin | EI,CI | 509, 516 |
| Bioresmethrin | EI,CI | 509, 513 |
| Bis(2-nitro-4-trifluoromethylphenyl) disulfide | EI | 438 |
| Bis(trimethylsilyl)trifluoroacetamide (BSTFA) and trifluoroacetic anhydride (TFA)-treated Tenax adsorbent, extract of | EI | 171 |
| Bromoxynil | EI | 447 |
| Buturon, chlortoluton, and metoxuron (mixture) | FD | 392 |
| Butylbenzyl phthalate | EI | 346 |
| Butyltrimethyltin | EI | 360 |
| Captan | EI,FD | 481 |
| Chloramben | EI | 447 |
| 3-Chloroaniline-lignin copolymer | EI | 98 |

| Compound | Ionization technique | Page |
|---|---|---|
| 1,4-Chloronitrobenzene | CI,NCI | 32 |
| Chlorpyrifosethyl | EI | 460 |
| Chlorpyrifosmethyl | EI | 460 |
| Chlortoluton, metoxuron, and buturon (mixture) | FD | 392 |
| Cis-chlordane | CI,NCI | 416 |
| Cyanazine IV, dimethyl derivative of | EI | 428 |
| Cypermethrin | EI | 511 |
| | CI | 516 |
| DADK (metribuzin degradation product) | EI | 433 |
| *pp*′-DDT | FD | 67 |
| Decamethrin | EI,CI | 511, 515 |
| Delnav | EI,CI,FI,FD | 68 |
| Dibutyldimethyltin | EI | 360 |
| Di-*n*-butylnitrosamine | CI | 539 |
| Di-*n*-butyl phthalate | EI | 346 |
| 2′,5′-Dichloro-4-biphenylol | EI | 243 |
| 2′,5′-Dichloro-4-biphenylol, ethyl ether of | EI | 243 |
| 2,7-Dichlorodibenzo-*p*-dioxin (DCDD) | EI | 273 |
| 2,3-Dichlorodibenzo-*p*-dioxin, dihydroxyl metabolite of, dimethyl ether of | EI | 293 |
| 3,3′-Dichloro-4′-methoxy-4-biphenylol | EI | 243 |
| 3,3′-Dichloro-4′-methoxy-4-biphenylol, ethyl ether of | EI | 243 |
| 2,3-Dichloro-7-methoxydibenzo-p-dioxin | EI | 293 |
| Diclofop-methyl, ethyl anlog of | EI | 440 |
| Diclofop-methyl, methylated hydroxy derivative of | EI | 440 |
| Dieldrin | EI,CI,NCI,FD | 407, 410 |
| Di-2-ethylhexyl phthalate | EI,CI | 346, 348 |
| Diethylnitrosamine | EI | 539 |
| Dihydrokepone | CI | 418 |
| 5,6-Dihydro-2-methyl-1,4-oxathiin-3-carboxanilide | EI,CI | 393 |
| 5,6-Dihydro-2-methyl-1,4-oxathiin-3-carboxanilide-4,4-dioxide | EI,CI | 398 |
| 5,6-Dihydro-2-methyl-1,4-oxathiin-3-carboxanilide-4-oxide | EI,CI | 396 |
| Dimethylnitrosamine | EI | 539 |
| Dimethyl phthalate | EI | 346 |
| Dinoseb | EI | 445 |
| Dioctyl adipate | EI | 349 |
| Di-*n*-octyl phthalate | EI | 346 |

| Compound | Ionization technique | Page |
|---|---|---|
| Dioctyl sebacate | EI | 349 |
| Diphenyl mercury | EI | 357 |
| Di-*n*-propylnitrosamine | EI | 539 |
| DK (metribuzin degradation product) | EI | 433 |
| Endrin | FI,FD | 65 |
| | EI | 407 |
| | FD | 411 |
| | CI | 411, 417 |
| | NCI | 411, 417 |
| Ethylene glycol derived compounds (pollutants in Delaware River) | EI | 553 |
| 2-Ethylhexyldiphenyl phosphate | EI | 349 |
| 2-Ethyl-7-nitro-1-propyl-5-(trifluoromethyl) benzimidazole | EI | 402 |
| 2-Ethyl-7-nitro-5-(trifluoromethyl) benzimidazole | EI | 402 |
| Exhaust (vehicle) gases extract | EI | 186,187 |
| Fenitrooxon | EI | 462 |
| Fenitrothion | EI | 461 |
| Fenvalerate | CI | 515 |
| Fish extract, Lake Ontario trout | NCI | 89 |
| Fish extract, Tittabawassee River carp | NCI | 89 |
| Fluoranthene | EI | 196 |
| Fluorene | EI | 198 |
| 9H-Fluoren-9-one | EI | 202 |
| Gastric fluid extract | CI | 87 |
| Heptachlor | CI,NCI | 414 |
| Heptachlor epoxide | CI,NCI | 415 |
| Heptachlorodibenzo-*p*-dioxins ($H_7CDD$) | EI,CI | 258, 261–263, 281 |
| Hexachlorodibenzo-*p*dioxins ($H_6CDD$) | EI,CI | 258, 261–263, 281 |
| Hexazinone metabolite | EI | 432 |
| Hexyldibutylmethyltin | EI | 361 |
| Human adipose tissue, chlorinated compounds | NCI | 90 |
| 1-Hydroxychlordene | CI,NCI | 415 |
| Hydroxycyprazine, diazomethane-methylated (cypratone) | EI | 430 |
| 4-D-Hydroxysphinganine-bis-methylneboronate | CI | 85 |
| 3-Hydroxy-2,2′,5,5′-tetrachlorobiphenyl, TMS derivative of | EI | 245 |

| Compound | Ionization technique | Page |
|---|---|---|
| Irgasen DP300, methyl ether of | EI | 273 |
| Isodrin | EI | 406 |
| | CI,NCI | 414 |
| (1-*o*-Isopropylphenyl)-2-thiourea | EI | 390 |
| 1-(*o*-Isopropyl-α-D-phenyl)-2-thiourea | EI | 390 |
| Kepol | EI | 419 |
| Kepone | CI | 418 |
| Metalaxyl | EI | 478 |
| Methane | CI | 78 |
| Methoxychlorodibenzo-*p*-dioxins | EI | 288 |
| Methoxychlorodibenzofurans | EI | 289 |
| 2-Methoxydibenzo-*p*-dioxin | EI | 293 |
| *N*-4-Methoxyphenyl-*N'*-phenylurea | EI,CI | 387 |
| 2-Methylanthracene | EI | 199 |
| *N*-Methylated *N*-acetyl-Met-Gly-Met-Met methyl ester | NCI | 87 |
| 1-Methylfluorene | EI | 199 |
| Methyl phenyl thiocarbonate | EI | 374 |
| *S*-Methyl phenyl thiocarbonate | EI | 374 |
| Methyl *S*-phenyl thiocarbonate | EI | 374 |
| Metobromuron | FD | 391 |
| Metoxuron, chlortolutron, and buturon (mixture) | FD | 392 |
| 1-Monochlorodibenzo-*p*-dioxin, methylthio-metabolite of | EI | 294 |
| Monohydrokepone | CI | 418 |
| Naphthalene | EI | 196 |
| Neburon | FD | 391 |
| 9-Nitroanthracene | EI | 200 |
| 2-Nitrofluorene | EI | 200 |
| 2-Nitronaphthalene | EI | 200 |
| 7-Nitro-1-propyl-5-(trifluoromethyl) benzimidazole | EI | 402 |
| 1-Nitropyrene | EI | 200 |
| *N*-Nitrosoiminodiethanol (NDELA) | EI | 544 |
| *N*-Nitrosomethylphenylamine | EI | 543 |
| *N*-Nitrosomorpholine | EI | 542 |
| 1-Nitrosopiperidine | EI | 542 |
| *N*-Nitrosopyrrolidine (NPYR) | EI | 541 |
| *N*-Nitrosopyrrolidin-3-ol (NHPYR) | EI | 544 |
| | EI | 544 |
| Nonachlor | CI,NCI | 415 |
| Octachlorodibenzo-*p*-dioxin (OCDD) | EI,CI | 258, 261–263, 287 |

| Compound | Ionization technique | Page |
|---|---|---|
| PCB isomers | EI | 222 |
| Pentachlorodibenzo-*p*-dioxin ($P_5CDD$) | EI,CI | 210, 258, 261–263 |
| 2,2′,4,5,5′-Pentachlorobiphenyl | EI | 210 |
| *S*-(Pentachlorophenyl)glutathione, acetylated ethyl ester derivative of | EI | 499 |
| *S*-(Pentachlorophenyl)-2-thioacetic acid methyl ester of | EI | 500 |
| Permethrin | EI,CI | 509, 517 |
| 9-Phenanthrenecarboxaldehyde | EI | 201 |
| Phenvalerate | EI | 511 |
| Phenylmercury chloride | EI | 357 |
| Picene | EI | 196 |
| Pollutants, EPA priority | EI | 129 |
| *S*-*n*-Propyl-*N*-ethyl dithiocarbamate | EI | 381 |
| *S*-*n*-Propyl-*N*-methyl dithiocarbamate | EI | 381 |
| Pyrene | EI | 196 |
| Rhine water extract | FD | 72, 73 |
| Satratoxin H | FD | 61 |
| Sediment extract | EI | 224 |
| Sodium *N*,*N*-di-*n*-butyl dithiocarbamate | EI | 378 |
| Sodium *N*,*N*-diethyl dithiocarbamate | EI | 378 |
| Sodium *N*,*N*-dimethyl dithiocarbamate | EI | 378 |
| Sodium *N*,*N*-di-*n*-propyl dithiocarbamate | EI | 378 |
| Terbacil metabolite A | EI | 451 |
| Tetrabutyltin | EI | 361 |
| 2,2′,5,5′-Tetrachlorobiphenyl | EI | 210 |
| Tetrachlorodibenzo-*p*-dioxins (TCDD) | EI,CI | 258,261–263, 269, 270, 281–284, 292–293 |
| 1,2,3,4-Tetrachlorodibenzo-*p*-dioxin (TCDD) | EI | 292, 293 |
| 2,3,7,8-Tetrachlorodibenzo-*p*-dioxin (TCDD) | EI | 269, 270 |
| Tetraethyllead | EI | 363 |
| Tetramethrin | EI | 509 |
| | CI | 513 |
| Tetramethyllead | EI | 363 |
| Tetramethyltin | EI | 362 |
| Toxaphene | CI | 420 |
| Toxicant A-toxaphene | EI,CI | 420 |
| Toxicant B-toxaphene | EI,CI | 420 |
| Trans-chlordane | CI,NCI | 416 |

| Compound | Ionization technique | Page |
|---|---|---|
| Tributylmethyltin | EI | 361 |
| Tri-*n*-butyl-*N*-TFA phosphonomethyl glycine | EI | 449 |
| Tributyltin oxide (TBTO) | EI | 360 |
| 2,2′,3-Trichlorobiphenyl | EI | 213 |
| 2,4′,5-Trichlorobiphenyl | EI | 214 |
| Tridecanoic acid, methyl ester | CI,EI | 30 |
| Triethyllead chloride | EI | 363 |
| Trifluoroacetic anhydride (TFA) and bis(trimethylsilyl)trifluoroacetamide (BSTFA) treated Tenax adsorbent, extract of | EI | 171 |
| α,α,α-Trifluoro-2,6-dinitro-*N*-propyl-*p*-toluidine | EI | 402 |
| Trifluralin | EI | 402,443 |
| Trimethyl phosphate | EI | 464 |
| Trimethyl thionophosphate | EI | 465 |
| Trioctyl phosphate | EI | 349 |
| Trioctyl trimellitate | EI | 349 |
| Triphenyl phosphate | EI | 349 |
| Triphenyltin chloride | EI | 359 |
| Triphenyltin hydroxide | EI | 359 |
| 9H-Xanthen-9-one | EI | 202 |

[a]KEY: EI, electron impact; CI, chemical ionization (positive); NCI, negative-ion chemical ionization; FD, field desorption; FI, field ionization; PPNICI, pulsed positive/negative-ion chemical ionization; CIMS, chemical ionization mass spectrometry.

# INDEX

MIX
Papier aus verantwortungsvollen Quellen
Paper from responsible sources
FSC® C105338

If you have any concerns about our products,
you can contact us on
**ProductSafety@springernature.com**

In case Publisher is established outside the EU,
the EU authorized representative is:
**Springer Nature Customer Service Center GmbH**
**Europaplatz 3, 69115 Heidelberg, Germany**

Printed by Libri Plureos GmbH
in Hamburg, Germany